# 21st Century Canon of
# Solar Eclipses

## DELUXE
## BLACK & WHITE
## EDITION

### Fred Espenak

Edition 1.0
August 2020

21st Century Canon of Solar Eclipses – Deluxe Black & White Edition

Astropixels Publishing
P.O. Box 16197
Portal, AZ 85632

www.astropixels.com/pubs


This book may be ordered at: www.astropixels.com/pubs/21CCSE.html


More about solar eclipses of the 21st Century can be found at:

www.eclipsewise.com/solar/SEcatalog/21CCSEcat.html



Printed in the United States of America



Astropixels  Publication Number: AP023


ISBN 978-1-941983-22-5


First Edition (Version 1.0a)

Front Cover: A time sequence shows the partial phases and totality during the total solar eclipse of 2001 Jun 21 from Chisamba, Zambia. Photo copyright ©2001 by Fred Espenak.

Back Cover: Portrait of Fred Espenak, Copyright © 2016 by Patricia Totten Espenak.

# Table of Contents

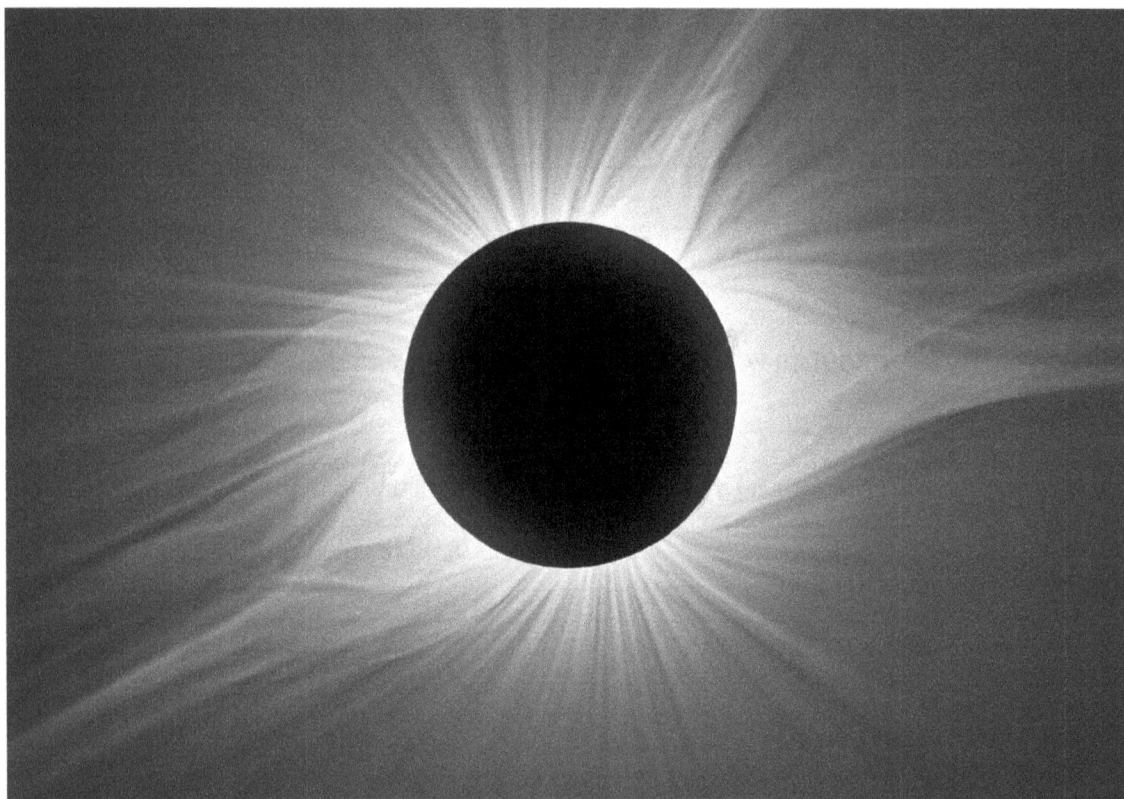

*The Sun's corona is only visible to the naked eye for a few minutes during a total eclipse of the Sun.*
*This is an HDR composite image of the total solar eclipse of August 21, 2017 from Casper, Wyoming.*
*© 2017 F. Espenak, www.MrEclipse.com*

# Section 1: Solar Eclipse Fundamentals

## 1.1 Introduction

The Moon's orbit is tilted about 5.1° to Earth's orbit around the Sun. The points where the two orbits appear to cross are called the nodes. When the New Moon occurs near one of these nodes, the Moon's shadow can sometimes fall on some portion of Earth and a solar eclipse takes place.

The Moon's shadow is composed of three cone-shaped components. The outer or penumbral shadow is a zone where the Sun's rays are partially blocked. Nested within the penumbra is the umbral shadow — a region where direct rays from the Sun are completely blocked. The conical umbra tapers to a point beyond which extends an expanding cone called the antumbra. From within this third shadow, the Moon appears smaller than the Sun and is seen in silhouette against the solar disk.

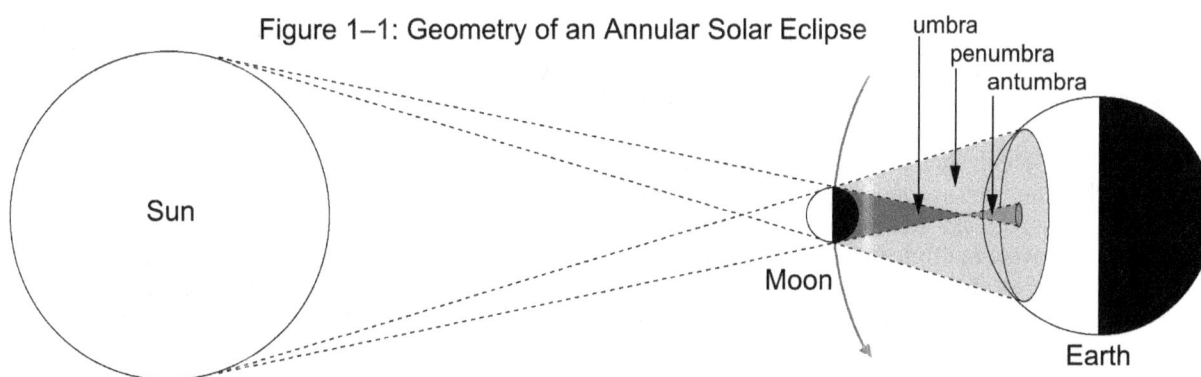

Figure 1–1: Geometry of an Annular Solar Eclipse

*Figure 1–1 illustrates the geometry of an annular solar eclipse. A partial eclipse is visible from within the large penumbral shadow, while the annular eclipse is confined to the much smaller antumbral shadow.*

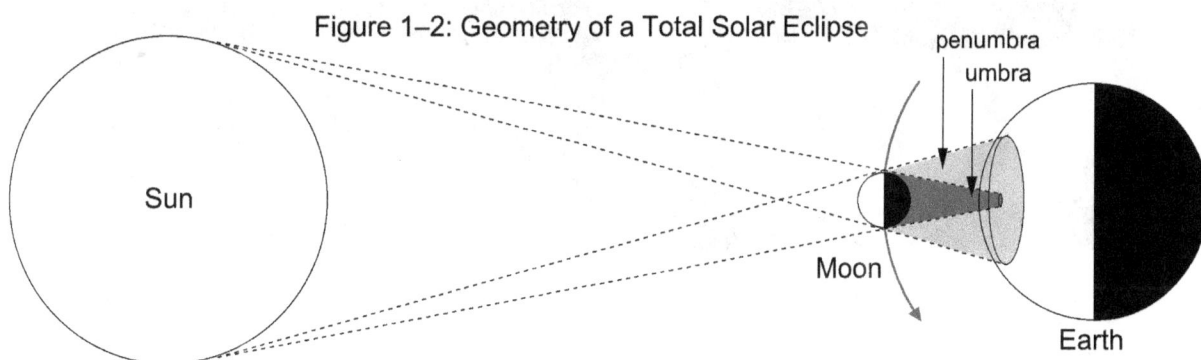

Figure 1–2: Geometry of a Total Solar Eclipse

*Figure 1–2 illustrates the geometry of a total solar eclipse. A partial eclipse is visible from within the large penumbral shadow, while the total eclipse is only seen from the much smaller umbral shadow.*

## 1.2 Classification of Solar Eclipses

There are four basic types of solar eclipses:

1. **Partial Solar Eclipse** — The Moon's penumbral shadow traverses Earth while the umbral and antumbral shadows miss Earth. A portion of the Sun's disk is obscured from within the penumbra.
2. **Annular Solar Eclipse** — The Moon's penumbral and antumbral shadows traverse Earth. The Moon's umbral shadow completely misses Earth. The Moon's disk appears smaller than the Sun so a bright

     ring surrounds the Moon when viewed from within the antumbral shadow. A partial eclipse is seen within the penumbral shadow (Figure 1–1).

3. **Total Solar Eclipse** — The Moon's penumbral and umbral shadows traverse Earth. The Moon's antumbral shadow extends beyond Earth's surface. The Moon's disk appears larger than the Sun and completely covers the solar disk when viewed from within the umbral shadow. A partial eclipse is seen within the penumbral shadow (Figure 1–2).

4. **Hybrid Solar Eclipse** — The Moon's penumbral, umbral and antumbral shadows all traverse parts of Earth. The curvature of Earth's surface brings some regions into the umbra and others into the antumbra. The eclipse is total within the umbra and annular within the antumbra. A partial eclipse is seen within the penumbral shadow. Hybrid eclipses are also known as annular-total eclipses.

Annular, total and hybrid eclipses are sometimes referred to as *central* eclipses[1]. However, on rare occasions it is possible to have an annular or total eclipse that is non-central (Sect. 3.1).

## 1.3 Visual Appearance of Partial Solar Eclipses

When the Moon's penumbral shadow strikes Earth, a partial eclipse of the Sun is visible from that region. The Moon's apparent motion with respect to the Sun is relatively slow — the partial phases can last up to three hours or more. During this time, the Moon's dark limb slowly creeps across the Sun's disk.

Partial eclipses are dangerous to look at with the unprotected eye because the Sun is still extremely bright. Special techniques are needed to safely view the eclipse (Sect. 1.6). Even when a partial eclipse reaches its maximum phase, the sky and landscape remain bright. But careful inspection of dappled sunlight beneath a leafy tree will reveal multiple images of the eclipse. The gaps between the tree leaves act like pinhole cameras and project images of the eclipse onto the ground below. If the eclipse occurs on a mostly cloudy day, brief views of the partial phases may be possible as the Sun passes though the more translucent clouds.

The Moon's penumbral shadow is 6700 to 7300 kilometers in diameter and can cover a significant fraction of the daytime hemisphere of Earth. Consequently, partial eclipses may be visible from large geographic areas as the penumbra sweeps across Earth's surface.

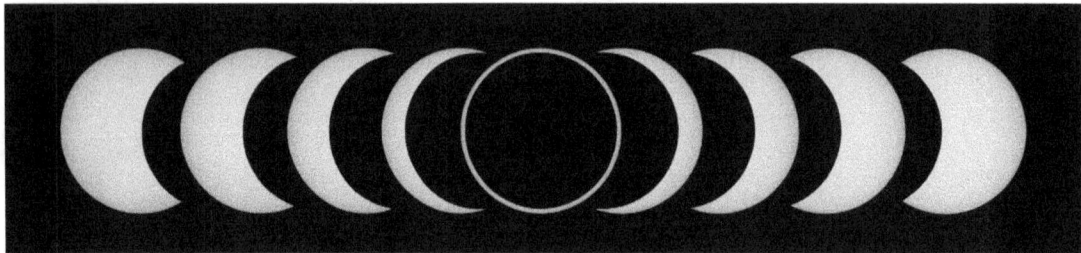

*Photo 1–1 shows various phases of the Annular solar eclipse of 2005 Oct 03. ©2005 F. Espenak*

## 1.4 Visual Appearance of Annular Solar Eclipses

During an annular eclipse, the Moon's penumbral and antumbral shadows sweep across Earth. Compared to the penumbra, the antumbra is much smaller and has a maximum diameter of 374 kilometers. Because of this, the antumbra covers a tiny fraction of Earth's surface.

A partial eclipse is visible within the penumbral shadow, but only observers located in the much narrower track of the antumbra will see an annular eclipse. For this reason, the antumbra's trajectory across Earth is called the path of annularity.

---

[1] A central eclipse is one in which the central axis of the Moon's umbral/antumbral shadow intersects with Earth's surface.

All annular eclipses begin with a series of partial phases lasting about an hour. At the peak of the eclipse, the Moon's disk can be seen in complete silhouette against the Sun. The remaining solar photosphere appears as a ring of intensely bright light surrounding the Moon. The annular phase can last a maximum of 12 ½ minutes, but is more typically 3 to 6 minutes in length. After annularity, another series of partial phases occurs as the Moon gradually uncovers the Sun.

Special precautions must be used to watch the eclipse (Sect. 1.6). Even during the annular phase, the Sun is dangerously bright and cannot be viewed without a solar filter. The landscape and sky remain bright throughout the eclipse, giving little indication of the celestial event in progress.

*Photo 1–2 shows various phases of the total solar eclipse of 1999 Aug 11. ©1999 F. Espenak*

## 1.5 Visual Appearance of Total Solar Eclipse

During a total eclipse, the Moon's penumbral and umbral shadows fall upon Earth. The umbra has a maximum diameter of 273 kilometers. The narrow track traced out as the umbra sweeps across Earth's surface is called the path of totality. The Sun is completely obscured by the Moon from within this zone. The total phase can last up to 7 ½ minutes, but is more typically 2 to 3 minutes in length.

Total eclipses all begin and end with a series of partial phases lasting about an hour. But this is where the resemblance between partial and annular eclipses ends — the total phase is the most spectacular astronomical event visible to the naked eye. During totality, the Sun's outer atmosphere — the solar corona — appears as a gossamer halo surrounding the Moon, and bright stars and planets are visible.

The eclipse begins to take on a unique character about five minutes before the total phase commences. Sunlight takes on a foreboding quality and casts abnormally sharp shadows. The approaching lunar umbra darkens the western sky and the air temperature is noticeably cooler. A minute before totality, ghostly shadow bands[2] ripple across the ground.

The ambient light grows feeble even though the crescent Sun is still too bright to see. In the final seconds, the Sun's corona emerges from the glare as the solar crescent shrinks to a brilliant jewel. This celestial diamond ring[3] lingers for a moment before the sunlight is extinguished and totality begins.

The Sun's glorious corona is now displayed to full advantage in the darkened sky. Standing within the Moon's umbra affords the rare and unprecedented opportunity to gaze directly at the glowing million-degree plasma surrounding our star. Twisted, tortured, and constrained by the Sun's enormous magnetic fields, the solar corona is revealed to the naked eye only during the brief seconds when the Moon completely blocks the brilliant disk of the Sun. An eerie twilight bathes the landscape and the colors of dusk surround the horizon.

Minutes race by like seconds. Suddenly, a sparkling bead of sunlight reappears along one edge of the Moon and quickly grows to blindingly bright proportions. Daylight returns as the corona fades and the total phase ends. Another hour of partial phases remains before the eclipse ends, it is anticlimactic after totality.

---

[2] Shadow bands are caused by the thin slit-like solar crescent illuminating Earth's atmosphere moments before and after a total solar eclipse. They appear as thin parallel undulating lines of alternating light and dark that race across the ground. This motion is caused by atmospheric winds.

[3] This phenomenon is commonly known as the *diamond ring effect.*

While filters are required for viewing the partial phases, they must be removed for totality. The total phase is the only time it is completely safe to look directly at the Sun without protection. In fact, the total phase is not even visible through solar filters because the Sun's corona is a million times fainter than the photosphere[4].

### Figure 1–3: Path of Totality

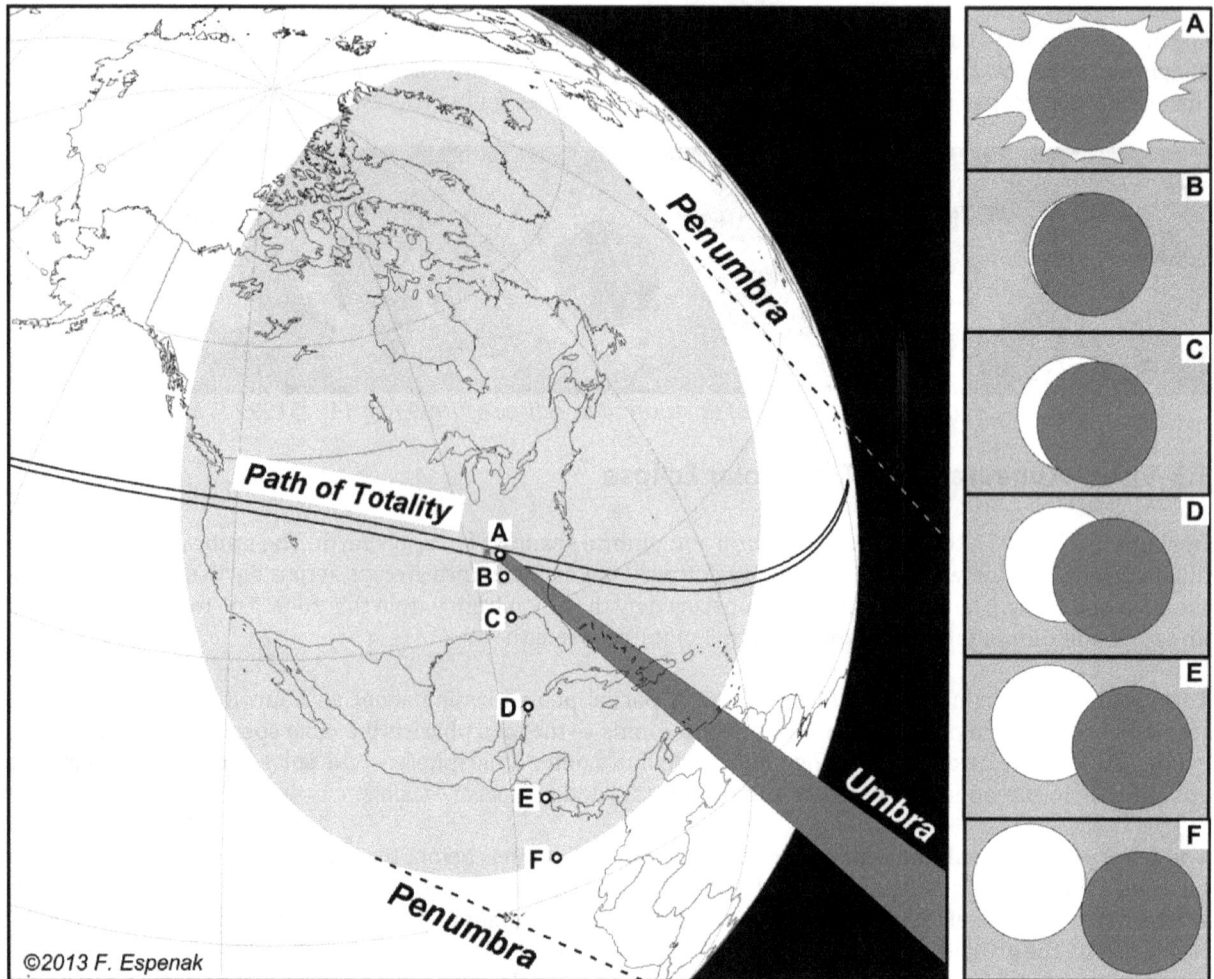

*Figure 1–3 illustrates the path of totality for the total solar eclipse of 2017 Aug 21. The appearance of maximum eclipse from six different geographic locations on the map (labeled A through F) is shown in the side bar. The total eclipse is visible from position A because it lies in the path of totality.*

## 1.6 Safely Observing Solar Eclipses

Partial eclipses, annular eclipses, and the partial phases of total eclipses are never safe to watch without special precautions. Even when 99% of the Sun's surface is obscured during the partial phases, the remaining photospheric crescent is intensely bright and cannot be viewed safely without eye protection (Chou, 1981; Marsh, 1982). Do not attempt to observe the partial or annular phases of any solar eclipse with the naked eye. Failure to use appropriate filtration may result in permanent eye damage or blindness. The only time it is safe to view the Sun directly with the naked eye is during the brief period of the total eclipse phase when the Sun's disk is completely covered by the Moon.

---

[4] The photosphere is the visible surface of the Sun's disk.

The same equipment and techniques used to observe the Sun outside of eclipse can be used for viewing partial phases and annular eclipses (Littmann, Espenak & Willcox, 2008, Reynolds & Sweetsir, 1995; Pasachoff & Covington, 1993). The most inexpensive of these methods is by projection — a pinhole or small opening is used to project the Sun's image onto a screen placed two or more feet beyond the opening. Projected images of the Sun can even be seen on the ground by creating small openings between interlaced fingers, or in the dappled sunlight beneath a leafy tree.

Observing the Sun directly is possible when using filters specifically designed for this purpose. Such filters usually have a thin layer of metal deposited on their surfaces that attenuates both visible and infrared energy. Another widely available filter is a number 14 welder's glass, obtained from welding supply outlets. Black polymer has recently become a popular, inexpensive alternative. This material can be cut with scissors and adapted to any kind of viewing device. No filter is safe to use with an optical device (i.e., telescope, binoculars, etc.) unless it has been specifically designed for that purpose. Sources for filters can be found at:

*eclipsewise.com/extra/equipment.html#Solar_Filters*

Local science museums, planetariums and amateur astronomy clubs are good sources for additional information. Remember, only the total phase of an eclipse can be safely viewed without a filter.

## 1.7 Central Line and Duration of Totality

The axis of the Moon's shadow determines the central line of the path of totality (and annularity). For the purposes of this discussion, it is assumed that the central line lies in the middle of the eclipse path.[5]

The duration of totality is longest on the central line, so great effort is often made to get as close to it as possible. However, the duration actually drops off quite slowly with distance from the central line. For example, a location 20% from the center to the edge of the path still has a totality lasting 98% of the central line duration. Even if one travels half way to the path limit, the duration is still 87% of the central line value.

Although figure 1–4 plots data for the 2017 total solar eclipse, the same relationship holds for all total eclipse paths. The solid curve is calculated assuming a smooth lunar limb. In reality, the Moon's limb profile has mountains that extend the duration and valleys that shorten it. These topographic features typically change the duration by a second or two for most of the path — their effects become significantly larger near the umbral path limits. The dotted line in figure 1–4 has been calculated using the Moon's true profile.

The following expression is useful for calculating the duration of totality for any location at a perpendicular distance of $\delta$ kilometers from the central line. It assumes a smooth profile for the Moon's limb.

$$d = D \times [\, 1 - (\, \delta/Z \,)^2 \,]^{\frac{1}{2}} \text{ seconds} \qquad\qquad (1\text{–}2)$$

Where:   d = duration of totality at point of interest (seconds)
         D = duration of totality on the central line (seconds)
         $\delta$ = perpendicular distance from the central line to position of interest (kilometers)
         Z = perpendicular distance from the central line to the path edge (kilometers)

The Moon's limb profile varies with the lunar libration, which changes from eclipse to eclipse. Consequently, the effects of the limb profile are different at each eclipse.

---

[5] The position of the central line is actually offset from the center. The difference is greatest in cases where the Sun's altitude is low. Curvature of Earth's surface is responsible for this shift.

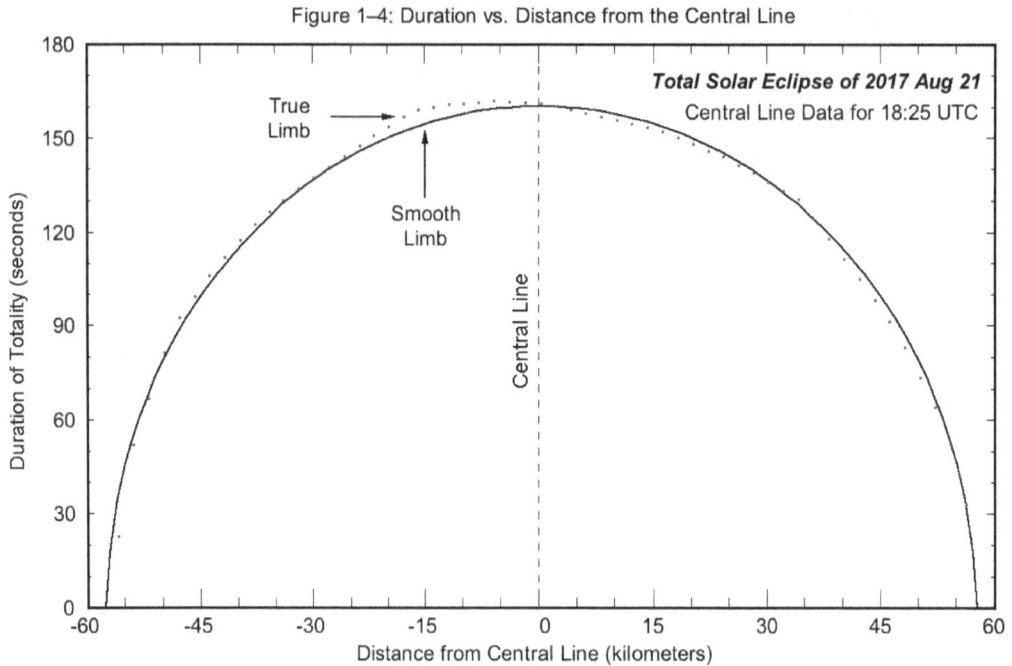

Figure 1–4: Duration vs. Distance from the Central Line

*Figure 1–4 shows how the duration of totality changes with distance from the central line. The solid curve represents the duration using the Moon's mean limb. At a distance ~87% from the central line to the path limit, the duration drops to 50%. It is only when the distance from the central line to the path limit reaches ~90% that the duration takes a precipitous drop. But even at a distance of 96%, the duration is still 28% of the central line value. The dotted curve uses the Moon's true limb to calculate the duration using Kaguya/Herald limb data.*

Photo 1–5 *Pat and Fred Espenak gaze at the Sun's corona during the total solar eclipse of 2008 August 01 from Jinta, China. ©2008 F. Espenak*

# Section 2: Solar Eclipse Predictions

## 2.1 Solar Eclipse Contacts

During the course of a solar eclipse, the instants when the Moon's disk becomes tangent to the Sun's disk are known as eclipse contacts. They mark various stages or phases of a solar eclipse.

Partial solar eclipses have two primary contacts.

**First Contact (C1)** — Instant of first exterior tangency of the Moon with the Sun (Partial Eclipse Begins)
**Fourth Contact (C4)** — Instant of last exterior tangency of the Moon with the Sun (Partial Eclipse Ends)

Central solar eclipses (total, annular or hybrid) have four primary contacts. Contacts C2 and C3 mark the instants when the Moon's disk is first and last internally tangent to the Sun. These are the times when the annular or total phase of the eclipse begins and ends, respectively.

**First Contact (C1)** — Instant of first exterior tangency of the Moon with the Sun (Partial Eclipse Begins)
**Second Contact (C2)** — Instant of first interior tangency of the Moon with the Sun
         (Annular or Total Eclipse Begins)
**Third Contact (C3)** — Instant of last interior tangency of the Moon with the Sun
         (Annular or Total Eclipse Ends)
**Fourth Contact (C4)** — Instant of last exterior tangency of the Moon with the Sun (Partial Eclipse Ends)

## Figure 2–1: Solar Eclipse Contacts

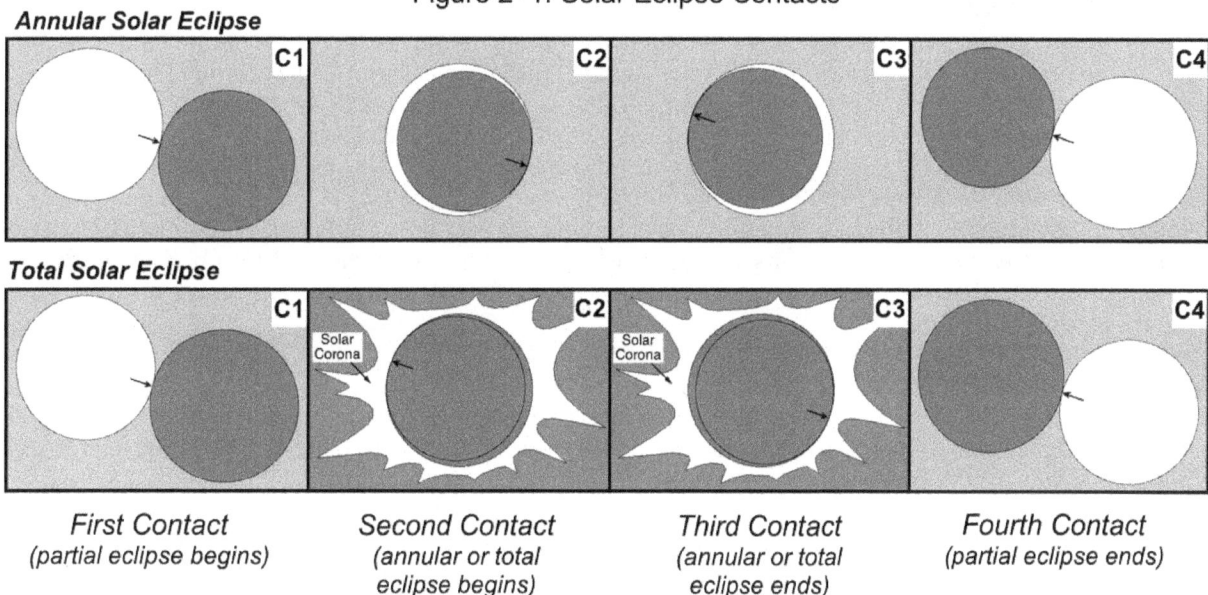

*Figure 2–1 illustrates the four contacts for annular and total solar eclipses. The arrows indicate the contact point of the Sun's limb in each diagram.*

## 2.2 Mean Lunar Radius

The International Astronomical Union (IAU) has adopted a value of k=0.2725076 for the mean lunar radius. This value is currently used by the Nautical Almanac Office for all solar eclipse predictions (Fiala and Lukac, 1983) and is believed to be the best mean radius, averaging mountain peaks and low valleys along the Moon's rugged limb. However, this introduces a problem in predicting the character and duration of central eclipses, particularly total eclipses. A total eclipse can be defined as an eclipse in which the Sun's disk is completely occulted by the Moon. This cannot occur so long as any photospheric rays are visible through deep valleys along the Moon's limb (Meeus, Grosjean and Vanderleen, 1966).

This work uses the IAU's accepted value of k (k=0.2725076) for all penumbral (exterior) contacts. In order to avoid eclipse type misidentification and to predict central durations, which are closer to the actual durations observed at total eclipses, we depart from convention by adopting the smaller value for k (k=0.272281) for all central (interior) contacts. This is consistent with predictions published in *Thousand Year Canon of Solar Eclipses* (Espenak 2014). Consequently, the smaller k produces shorter central durations and narrower paths for total eclipses when compared with calculations using the IAU value for k. Similarly, the smaller k predicts longer central durations and wider paths for annular eclipses.

## 2.3 Solar and Lunar Coordinates

The coordinates of the Sun and the Moon used in the eclipse predictions presented here have been calculated with the JPL DE405 (Jet Propulsion Laboratory Developmental Ephemeris 405). The DE405 is based upon the International Celestial Reference Frame (ICRF), the adopted reference frame of the IAU. The DE405 includes both nutation or libration and has an absolute accuracy of several kilometers for planetary positions,. In most cases this corresponds to a small fraction of an arc-second.

The Moon's center of figure does not coincide with its center of mass. To compensate, an empirical correction is sometimes added to the Moon's center of mass position. Unfortunately, the large variation in lunar libration from one eclipse to the next minimizes the effectiveness of this empirical correction. Because of this, no correction has been made to the Moon's center of mass position in the *21st Century Canon*.

## 2.4 Measurement of Time

The most natural form of time measurement is the solar day (usually measured from solar noon to solar noon). The length of the solar day varies during the year because of the eccentricity of Earth's orbit around the Sun. Mean solar time resolves this problem by using an average to define the mean solar day.

In 1884, Greenwich Mean Time (GMT) — the mean solar time on the Greenwich Meridian (0° longitude) — was adopted as the standard reference time for clocks around the world. A fundamental basis of GMT is the assumption that Earth's rotation on its axis is constant. It wasn't until the mid-twentieth century that astronomers realized the rotation period is gradually increasing. Earth is slowing down because of tidal friction with the Moon.

For purposes of orbital calculations, time using Earth's rotation was abandoned for a more uniform time scale based on Earth's orbit about the Sun. In 1952, Ephemeris Time was introduced to address the problem. The ephemeris second was defined as a fraction of the tropical year[6] for 1900 Jan 01 as calculated from Newcomb's *Tables of the Sun* (1895). Ephemeris Time was used for Solar System ephemeris calculations until 1979.

Terrestrial Dynamical Time (TD) is the modern replacement for Ephemeris Time and is used in theories of planetary motion in the Solar System. TD is based on International Atomic Time (TAI), which is a high-precision standard using several hundred atomic clocks worldwide. To ensure continuity with Ephemeris Time, TD was defined to match ET for the date 1977 Jan 01. In 1991, the IAU refined the definition of TD to make it more precise. It was also renamed Terrestrial Time (TT) although the author prefers to use the older name Terrestrial Dynamical Time.

Civilian time used throughout the world is still based on mean solar time, although indirectly. While Greenwich Mean Time was determined though observations of the Sun, its modern day replacement, Universal Time (actually UT1) is based on Earth's rotation using observations of distant quasars. UT1 is a nonuniform time

---

[6] The tropical year is the length of time that the Sun takes to return to the same position in the cycle of seasons, as seen from Earth (e.g., the time from vernal equinox to vernal equinox).

because Earth is gradually slowing down at an irregular rate. At present (2016), the accumulated error in the rotation of Earth in the course of one year is ~0.5 seconds.

Coordinated Universal Time (UTC) is derived from International Atomic Time (TAI). The length of the UTC second is defined in terms of an atomic transition of cesium and is accurate to approximately one nanosecond (billionth of a second) per day. UTC was defined to closely parallel UT1. However, the two time systems are intrinsically incompatible since UTC is uniform while UT1 is based on Earth's rotation, which is gradually slowing. In order to keep the two times within 0.9 seconds of each other, a leap second is added to UTC as needed (currently once every few years).

Today, UTC is the time standard used to define time zones around the world. It is the time reference for GPS satellites and aviation, and is used to synchronize the clocks of computers across the Internet.

## 2.5 ΔT (Delta T)

The orbital positions of the Sun and the Moon, required by eclipse predictions, are calculated using Terrestrial Dynamical Time (TD) because it is a uniform time scale. However, world time zones and daily life are based on Universal Time[7] (UT1). In order to convert eclipse predictions from TD to UT1, the difference between these two time scales must be known. The parameter ΔT (Delta T) is the arithmetic difference, in seconds, between the two as:

$$\Delta T = TD - UT1 \qquad\qquad (2-1)$$

Past values of ΔT can be deduced from historical records. In spite of their relatively low precision, these data represent the only evidence for the value of ΔT prior to 1600. In the centuries following the introduction of the telescope (circa 1609), thousands of high quality observations have been made of lunar occultations of stars, affording valuable data with increased accuracy in the determination of ΔT.

In modern times, the determination of ΔT is made using atomic clocks and radio observations of quasars. From 1955 to 2010, the average 1-year change in ΔT ranges from 0.18 seconds to 1.06 seconds. Future changes in ΔT are unknown since theoretical models of the physical causes are imprecise. Extrapolations from the table weighted by the long period trend from tidal braking of the Moon offer estimates of +71 seconds in 2024, +85 seconds in 2050, and +127 seconds in 2100.

Polynomial expressions for ΔT based on this data can be found at: *eclipsewise.com/help/deltatpoly2014.html*

## 2.6 Date Format

There are a number of ways to write the calendar date through variations in the order of day, month, and year. The International Organization for Standardization's (ISO) 8601 advises a numeric date representation, which organizes the elements from the largest to the smallest. The exact format is YYYY-MM-DD where YYYY is the calendar year, MM is the month of the year between 01 (January) and 12 (December), and DD is the day of the month between 01 and 31. For example, the 27th day of April in the year 1943 would then be expressed as 1943-04-27. The ISO convention is adopted here, but the month number has been replaced with the three-letter English abbreviation of the month name for additional clarity. From the previous example, the date then is expressed as 1943 Apr 27.

---

[7] World time zones are actually based on Coordinated Universal Time (UTC). It is an atomic time synchronized and adjusted to stay within a second of astronomically determined Universal Time (UT1) through the addition of an occasional "leap second" to compensate for the gradual slowing of Earth's rotation.

# Section 3: Solar Eclipse Statistics

## 3.1 Statistical Distribution of Eclipse Types

Eclipses of the Sun can only occur during the New Moon phase. It is then possible for the Moon's penumbral, umbral and/or antumbral shadows to sweep across Earth's surface thereby producing an eclipse. There are four types of solar eclipses:

1. **Partial** — Moon's penumbral shadow traverses Earth (umbral and antumbral shadows miss Earth)
2. **Annular** — Moon's penumbral and antumbral shadows traverse Earth (Moon is too far from Earth to completely cover the Sun)
3. **Total** — Moon's penumbral and umbral shadows traverse Earth (Moon is close enough to Earth to completely cover the Sun)
4. **Hybrid** — Moon's penumbral, umbral and antumbral shadows traverse Earth (eclipse is annular or total along different sections of its path). Hybrid eclipses are also known as annular-total eclipses.

During the 100-year period from 2001 to 2100, Earth experiences 224 eclipses of the Sun. The statistical distribution of the four eclipse types over this interval is shown in Table 3–1.

### Table 3–1: Distribution of Basic Eclipse Types

| Eclipse Type | Abbreviation | Number | Percent |
|---|---|---|---|
| All Eclipses | — | 224 | 100.0% |
| Partial | P | 77 | 34.4% |
| Annular | A | 72 | 32.1% |
| Total | T | 68 | 30.4% |
| Hybrid | H | 7 | 3.1% |

All partial eclipses are events in which some portion of the Moon's penumbral shadow passes across Earth's surface. In comparison all annular, total and hybrid eclipses can be characterized as events in which some portion of the Moon's umbral and/or antumbral shadow crosses Earth.

In the case of umbral or antumbral eclipses (annular, total or hybrid), they can be further categorized as:

1. **Central (two limits)** — The central axis of the Moon's umbral or antumbral shadow traverses Earth thereby producing a central line in the eclipse track. The umbra or antumbra falls entirely upon Earth producing a ground track with both a northern and southern limit.
2. **Central (one limit)** — The central axis of the Moon's umbral or antumbral shadow traverses Earth. However, a portion of the umbra or antumbra misses Earth throughout the eclipse thereby producing a ground track with just one limit.
3. **Non-Central** — The central axis of the Moon's umbral or antumbral shadow misses Earth. However, one edge of the shadow grazes Earth producing a ground track with one limit and no central line.

Using these categories, the distribution of the 72 annular eclipses appears in Table 3–2.

### Table 3–2: Statistics of Annular Eclipses

| Annular Eclipses | Number | Percent |
|---|---|---|
| All Annular Eclipses | 72 | 100.0% |
| Central (two limits) | 68 | 94.4% |
| Central (one limit) | 2 | 2.8% |
| Non-Central (one limit) | 2 | 2.8% |

Central annular eclipses with one limit include: 2003 May 31, and 2044 Feb 28. Non-central annular eclipses include: 2014 Apr 29, and 2043 Oct 03.

Similarly, the distribution of the 68 total eclipses is shown in Table 3–3.

**Table 3–3: Statistics of Total Eclipses**

| Total Eclipses | Number | Percent |
|---|---|---|
| All Total Eclipses | 68 | 100.0% |
| Central (two limits) | 67 | 98.5% |
| Central (one limit) | 0 | 0.0% |
| Non-Central (one limit) | 1 | 1.5% |

There are no central total eclipses with a single limit during the 21st Century. However, there is one non-central total eclipse (one limit) on 2043 Apr 09.

All 7 hybrid eclipses are central with two limits. Hybrid eclipses with a single limit (both central and non-central) are exceedingly rare. An estimate of the mean frequency of non-central hybrid eclipses is one out of every 600 million eclipses or once every 250 million years (Meeus, 2002).

## 3.2 Eclipse Frequency and the Calendar Year

There are 2 to 5 solar eclipses in every calendar year. Table 3–4 shows the distribution in the number of eclipses per year for the 100 years covered in the *21st Century Canon*.

**Table 3–4: Number of Eclipses per Year**

| Number of Eclipses per Year | Number of Years | Percent |
|---|---|---|
| 2 | 82 | 73.2% |
| 3 | 12 | 16.1% |
| 4 | 6 | 10.7% |
| 5 | 0 | 0.0% |

When two eclipses occur in one calendar year, they can be any combination of P, A, T or H (partial, annular, total or hybrid) with the one exception — they can not both be T.

When three eclipses occur in one calendar year, there are 14 possible combinations of P, A, T or H. In most cases all three eclipses are partial.

Years in which four eclipses take place are: 2011, 2029, 2047, 2065, 2076, and 2094.

The maximum number of five solar eclipses in one calendar year is quite rare. No year during the 21st Century has five solar eclipses. The last time this occurred was in 1935 and the next instance is in 2206.

## 3.3 Extremes in Greatest Central Duration and Eclipse Magnitude

The longest and shortest central eclipses of the century as well as largest and smallest partial eclipses appear in Table 3–5.

**Table 3–05: Extremes in Central Duration and Eclipse Magnitude**

| Extrema Type | Date (Dynamical Time) | Central Line Duration | Eclipse Magnitude |
|---|---|---|---|
| Longest Annular | 2010 Jan 15 | 11m08s | — |
| Shortest Annular | 2085 Dec 16 | 00m19s | — |
| Longest Total | 2009 Jul 22 | 06m39s | — |
| Shortest Total | 2068 May 31 | 01m06s | — |
| Longest Hybrid | 2013 Nov 03 | 01m40s | — |
| Shortest Hybrid | 2067 Dec 06 | 00m08s | — |
| Largest Partial | 2051 Apr 11 | — | 0.9849 |
| Smallest Partial | 2098 Oct 24 | — | 0.0057 |

*Central line duration at the instant of greatest eclipse

## 3.4 Eclipse Seasons

Because of its ~5.1° inclination the Moon's orbit crosses the ecliptic at two points or nodes. If New Moon takes place within approximately 17° of a node[8], a solar eclipse will be visible from some location on Earth.

The Sun makes one complete circuit of the ecliptic in 365.24 days, so its average angular velocity is 0.99° per day. At this rate, it takes 34.5 days for the Sun to cross the 34° wide eclipse zone centered on each node. Because the Moon's orbit with respect to the Sun has a mean duration of 29.53 days, there will always be one and possibly two solar eclipses during each 34.5-day interval when the Sun passes through the nodal eclipse zones. These time periods are called eclipse seasons.

The mid-point of each eclipse season is separated by 173.3 days because this is the mean time for the Sun to travel from one node to the next. The period is a little less that half a calendar year because the lunar nodes slowly regress westward by 19.3° per year.

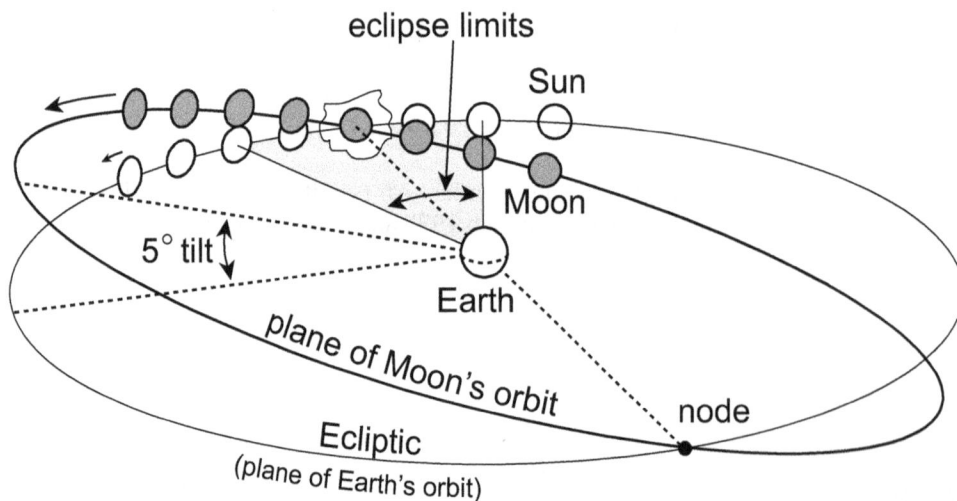

Figure 3–1: Eclipse Seasons and Orbital Nodes

---

[8] The exact angular distance from the node depends on the distances of the Sun and the Moon from Earth, which determine their angular diameters.

## 3.5 Quincena

The mean time interval between New Moon and Full Moon is 14.77 days. This is less than half the duration of a 34.5-day eclipse season. As a consequence, the same Sun–node alignment geometry responsible for producing a solar eclipse always results in a complementary lunar eclipse within a fortnight. The lunar eclipse may precede or succeed the solar eclipse. In either case, the pair of eclipses is referred to here as a quincena[9]. The QLE (Quincena Lunar Eclipse parameter) identifies the type of the lunar eclipse and whether it precedes or succeeds a particular solar eclipse. There are three basic types of lunar eclipses:

1. penumbral lunar eclipse (n) — Moon passes partly or completely within Earth's penumbral shadow
2. partial lunar eclipse (p) — Moon passes partly within Earth's umbral shadow
3. total lunar eclipse (t) — Moon passes completely within Earth's umbral shadow

The QLE is a two character string consisting of one or more of the above lunar eclipse types. The first character in the QLE identifies the type of lunar eclipse preceding a solar eclipse. The second character identifies the type of lunar eclipse succeeding a solar eclipse. In most instances, one of the two characters is "–" indicating no lunar eclipse occurs. For example, a QLE of "–p" means that no lunar eclipse precedes a solar eclipse, but a partial lunar eclipse follows the solar eclipse 15 days later.

On rare occasions, a double quincena occurs in which a solar eclipse is both preceded and succeeded by a lunar eclipse.

Photo 4–2 Fine details in the solar corona are captured during the total solar eclipse of 2006 March 29 from Jalu LIBYA. © 2006 F. Espenak, www.MrEclipse.com

---

[9] Quincena is a Spanish word meaning *a period of fifteen days*. This also happens to be the time interval, rounded to the nearest day, between New Moon and Full Moon, or Full Moon and New Moon. So *quincena* is a convenient and appropriate term for describing a pair of eclipses (one solar and one lunar) separated by this period.

# Section 4: Explanation of Solar Eclipse Catalog in Appendix A

## 4.1 Introduction

Earth experiences 224 eclipses of the Sun during the 21st Century. The catalog in *Appendix A* summarizes the principal characteristics of each eclipse and complements the maps in *Appendices B, C* and *D*.

Each line in the catalog corresponds to a single eclipse and provides concise parameters to characterize the eclipse. The calendar date and time of the instant of greatest eclipse are given, along with the value of Delta T. The lunation number and Saros series are listed along with the eclipse type. Gamma is the distance of the shadow axis from Earth's center at greatest eclipse, while the eclipse magnitude is the fraction of the Sun's diameter obscured at that instant. The geographic latitude and longitude of the umbral or antumbral axis are given for greatest eclipse, along with the Sun's altitude and azimuth, the width of the path, and the central line duration of totality or annularity. For both partial and non-central total or annular eclipses, the latitude and longitude correspond to the point closest to the shadow cone axis at greatest eclipse. The Sun's altitude is always 0° at this location. Detailed descriptions of each field in the catalog appear in the following sections.

## 4.2 Cat Num (Catalog Number)

The catalog number is the sequential number assigned to each eclipse from 1 to 224.

## 4.3 Calendar Date

Gregorian calendar date at greatest eclipse is given in the ISO order of YEAR-MONTH-DAY.

## 4.4 TD of Greatest Eclipse (Terrestrial Dynamical Time of Greatest Eclipse)

The instant of greatest eclipse occurs when the distance between the axis of the Moon's shadow cone and the center of Earth reaches a minimum. For partial eclipses, this instant differs slightly from the instant of greatest magnitude due to Earth's flattening. For total eclipses, this instant differs slightly from the instant of greatest duration, although the differences are relatively small.

Greatest eclipse is given in Terrestrial Dynamical Time or TD (Sect. 2.4), which is a time system based on International Atomic Time. As such, TD is the modern equivalent to its predecessor Ephemeris Time and is used in theories of planetary motion in the Solar System. To determine the geographic visibility of an eclipse, TD is converted to Universal Time (UT1) using the parameter Delta T (Sect. 2.5).

## 4.5 ΔT (Delta T)

ΔT (Delta T) is the arithmetic difference, in seconds, between Terrestrial Dynamical Time (TD) and Universal Time (UT1). For more information on ΔT, see Section 2.5.

## 4.6 Luna Num (Lunation Number)

The lunation number is the number of synodic months or lunations since New Moon on 2000 Jan 06. It can be converted to the Brown Lunation Number[10] by adding 953.

---

[10] The *Brown Lunation Number* defines lunation 1 as beginning at the first New Moon of 1923, the year when Ernest W. Brown's lunar theory was introduced in the major national astronomical almanacs.

## 4.7 Saros Num (Saros Series Number)

Each eclipse belongs to a Saros series using a numbering system first introduced by van den Bergh (1955). The eclipses with an odd Saros number take place at the ascending node of the Moon's orbit; those with an even Saros number take place at the descending node. This relationship is reversed for *lunar* eclipses.

The Saros is a period of 223 synodic months (~ 18 years, 11 days, and 8 hours). Eclipses separated by this period belong to the same Saros series and share similar geometry and characteristics.

## 4.8 Ecl Type (Solar Eclipse Type)

The first value in this 2-character parameter gives the eclipse type. The four basic types of solar eclipses are:

1. **Partial Solar Eclipse (P)** — The Moon's penumbral shadow traverses Earth; the Moon's umbral and antumbral shadows completely miss Earth
2. **Annular Solar Eclipse (A)** — The Moon's penumbral and antumbral shadows traverse Earth; the Moon is too far from Earth to completely cover the Sun
3. **Total Solar Eclipse (T)** — The Moon's penumbral and umbral shadow traverse Earth; the Moon is close enough to Earth to completely cover the Sun
4. **Hybrid Solar Eclipse (H)** — The Moon's penumbral, umbral and antumbral shadows traverse different parts of Earth; eclipse appears either total or annular along different sections of its path; hybrid eclipses are also known as annular-total eclipses

The second character of the eclipse type is a qualifier defined as follows.

1. m = Middle eclipse of Saros series
2. n = Central eclipse with no northern limit
3. s = Central eclipse with no southern limit
4. + = Non-central eclipse with no northern limit
5. – = Non-central eclipse with no southern limit
6. 2 = Hybrid eclipse path begins total and ends annular
7. 3 = Hybrid eclipse path begins annular and ends total
8. b = Saros series begins (first eclipse in a Saros series)
9. e = Saros series ends (last eclipse in a Saros series)

Qualifiers 1 through 5 are used with annular, total or hybrid eclipses. Qualifiers 6 and 7 apply only to special classes of hybrid eclipses while qualifiers 8 and 9 are used exclusively with partial eclipses.

## 4.9 QLE (Quincena Lunar Eclipse Parameter)

A lunar eclipse always occurs within 15 days of a solar eclipse. The Quincena Lunar Eclipse parameter (QLE) identifies the type of the lunar eclipse and whether it precedes or succeeds a particular solar eclipse. There are three basic types of lunar eclipses:

1. **Penumbral Lunar Eclipse (n)** — The Moon passes partly or completely within Earth's penumbra
2. **Partial Lunar Eclipse (p)** — The Moon passes partly within Earth's umbra
3. **Total Lunar Eclipse (t)** — The Moon passes completely within Earth's umbra

The QLE consists of a two-character string. The characters identify the type of lunar eclipse preceding and succeeding a solar eclipse, respectively. In most instances, one of the two characters in the QLE is "–" indicating no lunar eclipse occurs. On some occasions, a double quincena occurs in which a solar eclipse is both preceded and succeeded by a lunar eclipse. The QLE then consists of two characters identifying the types of the two lunar eclipses. (Section 3.5).

## 4.10 Gamma

Gamma is the minimum distance from the lunar shadow axis to the center of Earth, in units of Earth's equatorial radius. This distance is positive or negative, depending on whether the axis of the shadow cone passes north or south of Earth's center. If gamma is between +0.997 and –0.997, the eclipse is a central one (either total, annular, or hybrid). The limiting value 0.997 differs from unity because of the flattening of Earth.

## 4.11 Ecl Mag (Eclipse Magnitude)

The eclipse magnitude is defined as the fraction of the Sun's diameter occulted by the Moon. For partial eclipses, the eclipse magnitude at the instant of greatest eclipse is given for the geographic position closest to the Moon's shadow axis. For total, annular, and hybrid eclipses, the eclipse magnitude is replaced by the ratio of the topocentric apparent diameters of the Moon and the Sun at greatest eclipse. The eclipse magnitude is always < 1.0 for partial and annular eclipses, but ≥ 1.0 for total and hybrid eclipses.

## 4.12 Lat Long (Latitude and Longitude)

The latitude and longitude correspond to the position intersected by the lunar shadow axis at greatest eclipse.

## 4.13 Sun Alt (Altitude of Sun)

The Sun's altitude at the geographic position intersected by the lunar shadow axis is given at the instant of greatest eclipse. For partial eclipses, the Sun's altitude is always 0° because the shadow axis misses Earth. In this case, the geographic position corresponds to the point closest to the shadow axis.

## 4.14 Sun Azm (Azimuth of Sun)

The Sun's azimuth at the geographic position intersected by the lunar shadow axis is given at the instant of greatest eclipse. The values 0°, 90°, 180°, and 270° correspond to the north, east, south and west, respectively.

## 4.15 Path Width

For total, annular, and hybrid eclipses, the width of the path of totality or annularity (kilometers) is given at the geographic position intersected by the lunar shadow axis at the instant of greatest eclipse.

## 4.16 Central Line Dur (Central Line Duration)

For total, annular, or hybrid eclipses, the central line duration of the total or annular phase (minutes and seconds) is given at the geographic position intersected by the lunar shadow axis at greatest eclipse.

For total and hybrid eclipses, this duration is close to the maximum of the total phase along the entire path. For annular eclipses, the duration at greatest eclipse may be near either the minimum or maximum duration of the annular phase along the path. If the annular phase duration exceeds approximately 2.3 min, then it is close to the maximum duration along the central line track. If the annular phase duration is less, however, then it corresponds to a minimum and the annular duration increases towards the ends of the path.

## 4.17 EclipseWise.com and the 21st Century Canon

Much of the content of the *21st Century Canon of Solar Eclipses* is based on the eclipse predictions website *www.EclipseWise.com*. A plain text file containing the entire solar eclipse catalog in *Appendix A* can be downloaded at: *www.EclipseWise.com/solar/SEpubs/21CCSE.txt*.

A web based version of the catalog is available at: *www.eclipsewise.com/solar/SEcatalog/21CCSEcat.html*

# Section 5: Explanation of Solar Eclipse Maps in Appendices B, C and D

## 5.1 Explanation of Small Global Eclipse Maps in Appendix B

Earth experiences 224 eclipses of the Sun during the 21st Century. An individual global map for each eclipse appears in *Appendix B*.

The geographic visibility of each eclipse is illustrated with an orthographic projection map of Earth showing the path of the Moon's penumbral (partial) and umbral/antumbral (total, hybrid, or annular) shadows with respect to the continental coastlines, political boundaries (circa 2016) and the Equator. North is to the top and the daylight terminator is drawn for the instant of greatest eclipse. An 'x' symbol marks the sub-solar point where the Sun appears directly overhead at that time. The salient features of the eclipse maps are identified in Figure 5–1, which serves as a key.

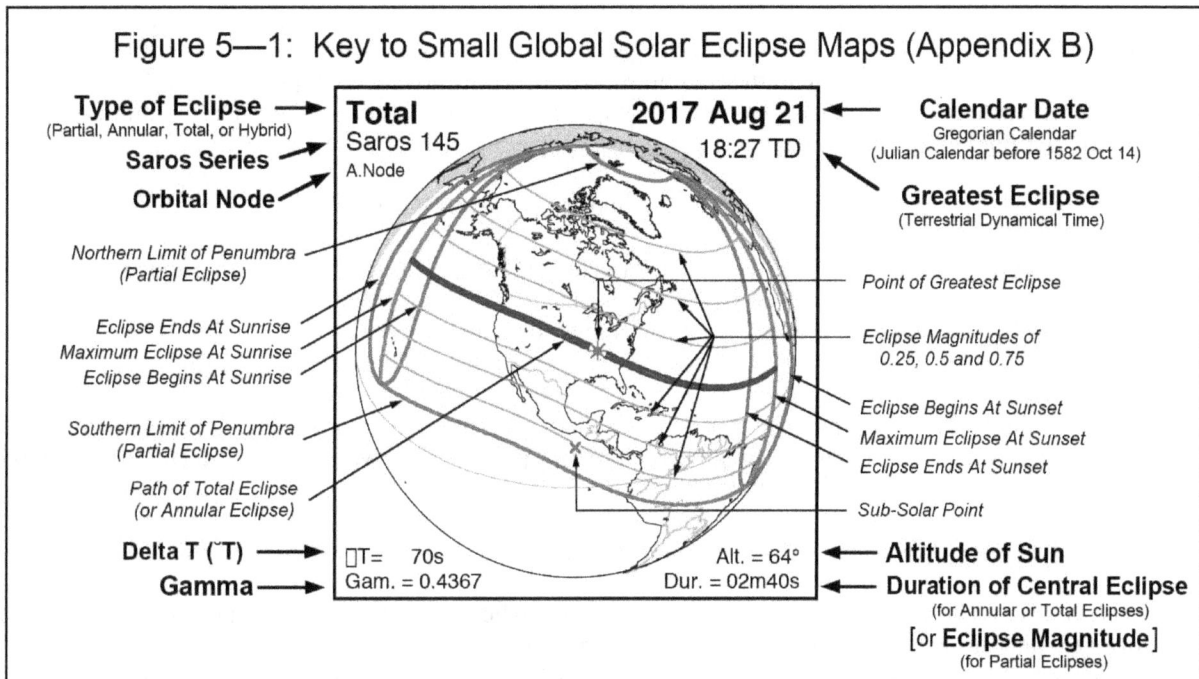

Figure 5—1: Key to Small Global Solar Eclipse Maps (Appendix B)

The limits of the Moon's penumbral shadow delineate the region of visibility of a partial solar eclipse. This irregular or saddle shaped region often covers more than half the daylight hemisphere of Earth and consists of several distinct zones or limits. At the northern and/or southern boundaries lie the limits of the penumbra's path. Partial eclipses have only one of these limits, as do central eclipses when the Moon's shadow axis falls no closer than about 0.45 radii from Earth's center. Great loops at the western and eastern extremes of the penumbra's path identify the areas where the eclipse begins/ends at sunrise and sunset, respectively. If the penumbra has both a northern and southern limit, the rising and setting curves form two separate, closed loops (e.g., 2017 Aug 21). Otherwise, the curves are connected in a distorted figure eight (e.g., 2019 Jul 02). Bisecting the *eclipse begins/ends at sunrise and sunset* loops is the curve of maximum eclipse at sunrise (western loop) and sunset (eastern loop).

The eclipse magnitude is defined as the fraction of the Sun's diameter occulted by the Moon. A curve of constant eclipse magnitude delineates the locus of all points where the local magnitude at maximum eclipse is equal to a constant value. The maps include *curves of constant eclipse magnitude* for values of 0.25, 0.5 and 0.75. These curves run exclusively between the curves of maximum eclipse at sunrise and sunset. They are approximately parallel to the northern/southern penumbral limits and the umbral/antumbral paths of central eclipses. The northern and southern limits of the penumbra may be thought of as curves of eclipse magnitude of 0.0. For total eclipses, the northern and southern limits of the umbra are curves of eclipse magnitude of 1.0.

Greatest eclipse is the instant when the axis of the Moon's shadow cone passes closest to Earth's center. Although greatest eclipse differs slightly from the instants of greatest magnitude and greatest duration (for total eclipses), the differences are relatively small. The point on Earth's surface intersected by the axis of the Moon's shadow cone at greatest eclipse is marked by an asterisk symbol '*'. For partial eclipses, the shadow axis misses Earth entirely, so the point of greatest eclipse lies on the day/night terminator and the Sun appears on the horizon.

Data relevant to an eclipse appear in the corners of each map. In the top left corner are the eclipse type (total, hybrid, annular, or partial), the Saros series of the eclipse, and the node of the Moon's orbit where the eclipse occurs. To the top right are the Gregorian calendar date, the time of greatest eclipse (Terrestrial Dynamical Time), and the value of Delta T ($\Delta T$).

The bottom left corner lists gamma (the minimum distance of the Moon's shadow cone axis from Earth's center, in Earth equatorial radii). The Sun's altitude at the geographic position of greatest eclipse is found to the lower right. The content of the final datum in the bottom right corner depends on the type of eclipse. If the eclipse is partial then the eclipse magnitude is given. If the eclipse is total, hybrid or annular, then the duration of the total or annular phase is given at the instant of greatest eclipse.

The list below briefly summarizes the parameters appearing in each global eclipse map.

**Solar Eclipse Type** – One of four basic types of solar eclipses: Partial, Annular Total or Hybrid (See Sect. 4.1).

**Saros Series** – The Saros series that the eclipse belongs to. Eclipses with an odd number take place at the Moon's ascending node, while those with an even number are at the descending node (See Sect. 4.7).

**Node** – The orbital node near which the eclipse takes place. The ascending node (A. Node) is the point where the Moon travels from south to north through Earth's orbital plane. Similarly, the descending node (D. Node) is the point where the Moon travels from north to south.

**Calendar Date** – Gregorian calendar date at greatest eclipse is given in the ISO order of YEAR-MONTH-DAY.

**Greatest Eclipse** – The instant (Terrestrial Dynamical Time) when the distance between the axis of the Moon's shadow cone and the center of Earth reaches a minimum (See Sect. 4.4).

**ΔT (Delta T)** – the arithmetic difference, in seconds, between Terrestrial Dynamical Time (TD) and Universal Time (UT1), (See Sect. 2.5).

**Gamma** – The minimum distance from the lunar shadow axis to the center of Earth, in units of Earth's equatorial radius (See Sect. 4.10).

**Altitude of Sun** – The Sun's altitude at the geographic position intersected by the lunar shadow axis is given at the instant of greatest eclipse (See Sect. 4.13).

**Duration of Central Eclipse** – The central line duration of the total or annular phase (in minutes and seconds) is given at the instant of greatest eclipse (See Sect. 4.16).

**Eclipse Magnitude** – The fraction of the Sun's diameter occulted by the Moon at the instant of greatest eclipse (See Sect. 4.11).

## 5.2 Explanation of Large Global Eclipse Maps in Appendix C

The heart of the *21st Century Canon of Solar Eclipses – Deluxe Edition* is the set of 224 full page global maps appearing in Appendix C — one map for every solar eclipse from 2001 through 2100. This feature is similar to the *Fifty Year Canon of Solar Eclipses: 1986 – 2035* (Espenak 1986) and makes *21st Century Canon* its modern successor.

The salient features of the large global maps are identified in Figure 5–2, which serves as a key.

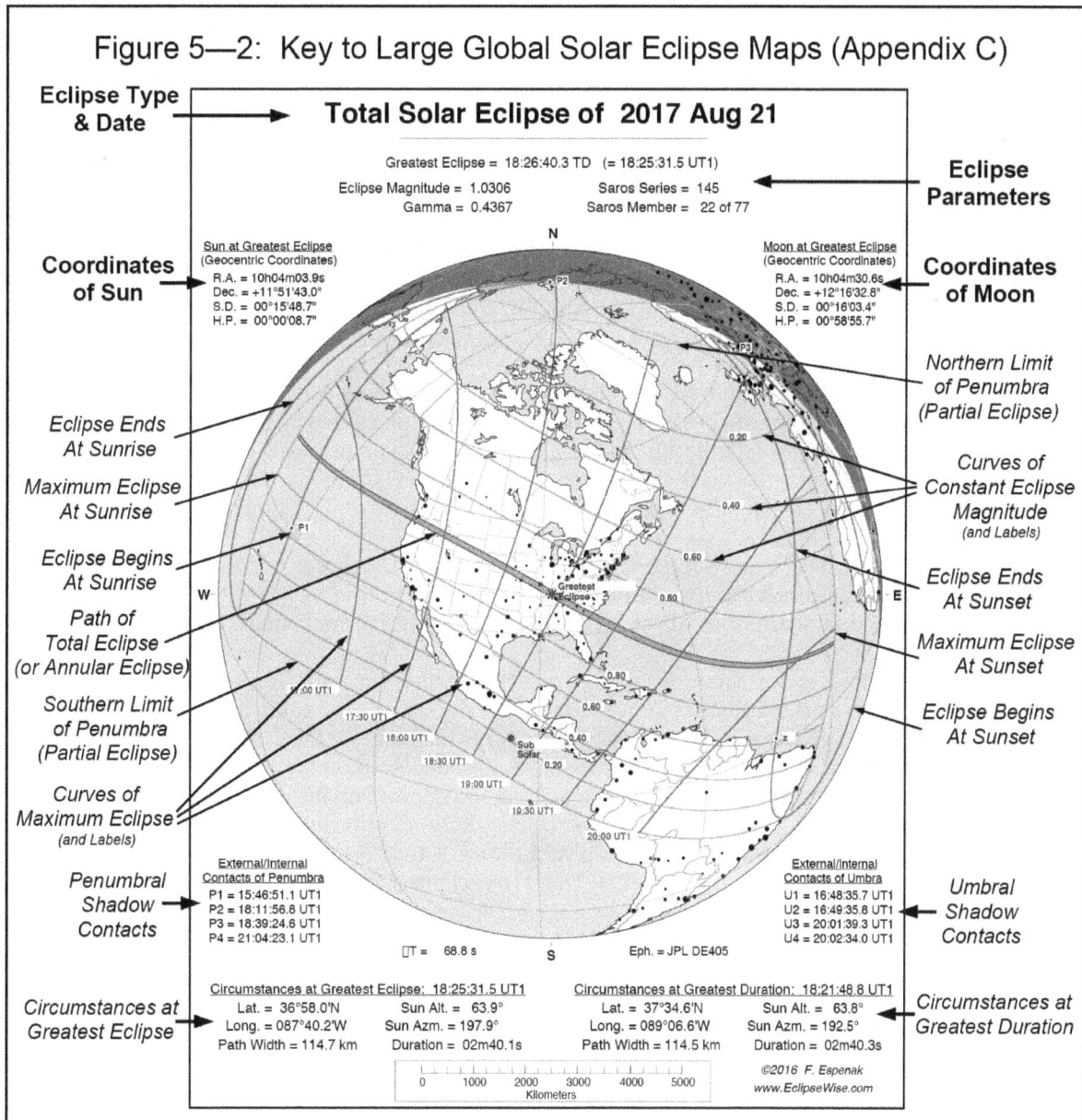

### Figure 5—2: Key to Large Global Solar Eclipse Maps (Appendix C)

Eclipse Type & Date → **Total Solar Eclipse of 2017 Aug 21**

Greatest Eclipse = 18:26:40.3 TD (= 18:25:31.5 UT1)
Eclipse Magnitude = 1.0306    Saros Series = 145
Gamma = 0.4367    Saros Member = 22 of 77

← Eclipse Parameters

Coordinates of Sun →
Sun at Greatest Eclipse (Geocentric Coordinates)
R.A. = 10h04m03.9s
Dec. = +11°51'43.0"
S.D. = 00°15'48.7"
H.P. = 00°00'08.7"

Moon at Greatest Eclipse (Geocentric Coordinates)
R.A. = 10h04m30.6s
Dec. = +12°16'32.8"
S.D. = 00°16'03.4"
H.P. = 00°58'55.7"
← Coordinates of Moon

Eclipse Ends At Sunrise
Maximum Eclipse At Sunrise
Eclipse Begins At Sunrise
Path of Total Eclipse (or Annular Eclipse)
Southern Limit of Penumbra (Partial Eclipse)
Curves of Maximum Eclipse (and Labels)

Northern Limit of Penumbra (Partial Eclipse)
Curves of Constant Eclipse Magnitude (and Labels)
Eclipse Ends At Sunset
Maximum Eclipse At Sunset
Eclipse Begins At Sunset

(map labels: 0.20, 0.40, 0.60, 0.80, Greatest Eclipse, Sub Solar, 17:00 UT1, 17:30 UT1, 18:00 UT1, 18:30 UT1, 19:00 UT1, 19:30 UT1, 20:00 UT1, N, S, E, W, P1, P2, P3)

Penumbral Shadow Contacts →
External/Internal Contacts of Penumbra
P1 = 15:46:51.1 UT1
P2 = 18:11:56.8 UT1
P3 = 18:39:24.6 UT1
P4 = 21:04:23.1 UT1

External/Internal Contacts of Umbra
U1 = 16:48:35.7 UT1
U2 = 16:49:35.8 UT1
U3 = 20:01:39.3 UT1
U4 = 20:02:34.0 UT1
← Umbral Shadow Contacts

ΔT = 68.8 s    Eph. = JPL DE405

Circumstances at Greatest Eclipse →
Circumstances at Greatest Eclipse: 18:25:31.5 UT1
Lat. = 36°58.0'N    Sun Alt. = 63.9°
Long. = 087°40.2'W    Sun Azm. = 197.9°
Path Width = 114.7 km    Duration = 02m40.1s

Circumstances at Greatest Duration: 18:21:48.8 UT1
Lat. = 37°34.6'N    Sun Alt. = 63.8°
Long. = 089°06.6'W    Sun Azm. = 192.5°
Path Width = 114.5 km    Duration = 02m40.3s
← Circumstances at Greatest Duration

0  1000  2000  3000  4000  5000
Kilometers

©2016 F. Espenak
www.EclipseWise.com

Each orthographic projection map shows the entire path of penumbral (partial) and umbral (total) or antumbral (annular) eclipse with respect to the continental coastlines, political boundaries (circa 2016), and major cities. North is to the top and the daylight terminator is plotted for the instant of greatest eclipse. An asterisk "*" indicates the sub-solar point on Earth (i.e., Sun in the Zenith).

The limits of the Moon's penumbral shadow delineate the region of visibility of the partial solar eclipse. This irregular or saddle shaped region often covers more than half of the daylight hemisphere of Earth and consists of several distinct zones or limits. At the northern and/or southern boundaries lie the limits of the penumbra's path. Partial eclipses have only one of these limits, as do central eclipses when the Moon's shadow axis falls no closer than about 0.45 radii from Earth's center.

Great loops at the western and eastern extremes of the penumbra's path identify the areas where the eclipse begins/ends at sunrise and sunset, respectively. If the penumbra has both a northern and southern limit, the rising and setting curves form two separate, closed loops. Otherwise, the curves are connected in a distorted figure eight. Bisecting the 'eclipse begins/ends at sunrise and sunset' loops is the curve of maximum eclipse at sunrise (western loop) and sunset (eastern loop. The points *P1* and *P4* mark the coordinates where the penumbral shadow first contacts (partial eclipse begins) and last contacts (partial eclipse ends) Earth's surface. If the penumbral path has both a northern and southern limit, then points *P2* and *P3* are also plotted. These correspond to the coordinates where the penumbral shadow cone is internally tangent to Earth's disk.

A curve of maximum eclipse is the locus of all points where the eclipse is at maximum at a given time. Curves of maximum eclipse are plotted at each half-hour Universal Time and are labeled with the time. They generally run between the penumbral limits in the north/south direction, or from the maximum eclipse at sunrise and sunset curves to one of the limits.

The curves of constant eclipse magnitude delineate the locus of all points where the magnitude at maximum eclipse is constant. These curves run between the curves of maximum eclipse at sunrise and sunset . They are roughly parallel to the northern/southern penumbral limits and the umbral paths of central eclipses (total, annular, and hybrid). In fact, the northern and southern limits of the penumbra are curves of constant magnitude of 0.0. The adjacent curves are for magnitudes of 0.2, 0.4, 0.6 and 0.8 and are labeled as such.

For total eclipses, the northern and southern limits of the umbra are curves of constant magnitude of 1.0. Antumbral path limits for annular eclipses are curves of maximum eclipse magnitude. The path of each total eclipse is darkly shaded. Similarly, the path of each annular eclipse is lightly shaded.

Greatest eclipse is defined as the instant when the axis of the Moon's shadow passes closest to Earth's center. Although greatest eclipse differs slightly from the instants of greatest magnitude and greatest duration (for total eclipses), the differences are small. An asterisk '*' marks the point on Earth's surface intersected by the lunar shadow axis at greatest eclipse. For partial eclipses, the shadow axis misses Earth entirely, so the point of greatest eclipse lies on the day/night terminator and the Sun appears on the horizon.

Data pertinent to the eclipse appear with each map. At the top is listed the instant of greatest eclipse, expressed in both Terrestrial Dynamical Time and Universal Time. The eclipse magnitude is defined as the fraction of the Sun's diameter obscured by the Moon at greatest eclipse. For central eclipses, the magnitude listed is actually the geocentric ratio of diameters of the Moon and the Sun. Gamma is the minimum distance (in Earth radii) of the Moon's shadow axis from Earth's center at greatest eclipse. The Saros series of the eclipse is listed, followed by the member position. The first number identifies the sequence position of the eclipse in the Saros, while the second number is the number of eclipses in the entire series.

In the upper left and right corners are the geocentric coordinates of the Sun and the Moon, respectively, at the instant of greatest eclipse. They are:

R.A. – Right Ascension
Dec. – Declination
S.D. – Apparent Semi-Diameter
H.P. – Horizontal Parallax

To the lower left are exterior/interior contact times of the Moon's penumbra with Earth, which are defined:

P1 – Instant of first exterior tangency of Penumbra with Earth's limb. (Partial Eclipse Begins)
P2 – Instant of first interior tangency of Penumbra with Earth's limb.
P3 – Instant of last interior tangency of Penumbra with Earth's limb.
P4 – Instant of last exterior tangency of Penumbra with Earth's limb. (Partial Eclipse Ends)

Not all eclipses have *P2* and *P3* penumbral contacts. They are only present in cases where the penumbral shadow falls completely within Earth's disk. For central eclipses, the lower right corner lists exterior/interior contact times of the Moon's umbral shadow with Earth's limb which are defined as follows:

U1 – Instant of first exterior tangency of Umbra with Earth's limb.
(Umbral [Total/Annular] Eclipse Begins)
U2 – Instant of first interior tangency of Umbra with Earth's limb.
U3 – Instant of last interior tangency of Umbra with Earth's limb.
U4 – Instant of last exterior tangency of Umbra with Earth's limb.
(Umbral [Total/Annular] Eclipse Ends)

At bottom left are the geographic coordinates of the position of greatest eclipse along with the local circumstances at that location (i.e., Sun altitude, Sun azimuth, path width and duration of totality/annularity). For central eclipses, similar data are also given at the bottom right for the position of greatest duration.

## 5.3 Explanation of Central Solar Eclipse Maps in Appendix D

The final series of 12 maps in Appendix D show the path of every central solar eclipse (total, annular and hybrid) with respect to continental coastlines, political boundaries (circa 2016), and major cities. The maps employ a transverse Mercator projection and are centered on three geographic regions:

1. North and South America
2. Europe and Africa
3. Asia and Australia

Furthermore, each map covers a 25-year period:

1. 2001–2025
2. 2026–2050
3. 2051–2075
4. 2076–2100

Each eclipse is identified by its date at the instant of greatest eclipse. An asterisk symbol marks the location of greatest eclipse.

The paths of total, annular, and hybrid eclipses can be identified as follows.

**Total Eclipses –** the path is shaded blue in (Color & Deluxe Editions);
the date appears in bold and is preceded by the letter **T**

**Annular Eclipses –** the path is shaded orange (Color & Deluxe Editions);
the date appears in italics and is preceded by the letter **A**

**Hybrid Eclipses –** the path is shaded blue along the total sections and
shaded orange along the annular sections;
the date appears in bold-italics and is preceded by the letter **H**

# References

*Astronomical Almanac for 1986,* Washington: US Government Printing Office; London: HM Stationery Office (1985).

*Astronomical Almanac for 2011,* Washington: US Government Printing Office; London: HM Stationery Office (2010).

Espenak, F., *Fifty Year Canon of Solar Eclipses: 1986–2035*, Sky Publishing Corp., Cambridge, Massuchusetts (1987).

Espenak, F., *Thousand Year Canon of Solar Eclipses: 1501 to 2500*, Astropixels Publishing, Portal, Arizona (2014).

Espenak, F., *Thousand Year Canon of Lunar Eclipses: 1501 to 2500*, Astropixels Publishing, Portal, Arizona (2014).

Espenak, F., and Meeus, J., *Five Millennium Canon of Solar Eclipses: –1999 to +3000 (2000 BCE to 3000 CE)*, NASA Tech. Pub. 2006–214141, NASA Goddard Space Flight Center, Greenbelt, Maryland (2006).

Espenak, F., and Meeus, J., *Five Millennium Canon of Lunar Eclipses: –1999 to +3000 (2000 BCE to 3000 CE)*, NASA Tech. Pub. 2006–214172, NASA Goddard Space Flight Center, Greenbelt, Maryland (2009a).

Espenak, F., and Meeus, J., *Five Millennium Catalog of Lunar Eclipses: –1999 to +3000 (2000 BCE to 3000 CE)*, NASA Tech. Pub. 2006–214173, NASA Goddard Space Flight Center, Greenbelt, Maryland (2009b).

Espenak, F., and Meeus, J., *Five Millennium Catalog of Solar Eclipses: –1999 to +3000 (2000 BCE to 3000 CE)*, NASA Tech. Pub. 2006–214174, NASA Goddard Space Flight Center, Greenbelt, Maryland (2009c).

*Explanatory Supplement to the Ephemeris*, H.M. Almanac Office, London (1974).

Littmann, M., Espenak, F., and Willcox, K., *Totality—Eclipses of the Sun*, 3rd Ed., Oxford University Press, New York (2008).

Meeus, J., *Mathematical Astronomy Morsels*, Willmann-Bell, pp. 56–62 (1997).

——, *More Mathematical Astronomy Morsels*, Willmann-Bell, pp. 70–72 (2002).

——, Grosjean, C.C., and Vanderleen, W., *Canon of Solar Eclipses*, Pergamon Press, Oxford, United Kingdom (1966).

Mucke, H., and Meeus, J., *Canon of Solar Eclipses: –2003 to +2526*, Astronomisches Büro, Vienna (1983).

Newcomb, S., "Tables of the Motion of the Earth on its Axis Around the Sun," *Astron. Papers Amer. Eph.,* Vol. 6, Part I (1895).

van den Bergh, *Periodicity and Variation of Solar (and Lunar) Eclipses*, Tjeenk Willink, and Haarlem, Netherlands (1955).

van der Sluys, M. , *http://hemel.waarnemen.com/Computing/deltat.html* (2010).

von Oppolzer, T.R., *Canon der Finsternisse*, Wien, (1887); Gingerich, O., (Translator) *Canon of Eclipses*, Dover Publications, New York (1962).

# Appendix A

## Solar Eclipse Catalog: 2001 to 2100

# Key to Solar Eclipse Catalog

**Cat Num** — sequential Catalog Number assigned to each eclipse from 1 to 2,389

**Calendar Date** — Gregorian date of Greatest Eclipse (Julian date prior to 1582 Oct 04)

**TD of Greatest Eclipse** — Terrestrial Dynamical Time of Greatest Eclipse

**ΔT** — arithmetic difference between Terrestrial Dynamical Time Universal Time (seconds)

**Luna Num** — number of synodic months, or lunations, since New Moon on 2000 Jan 06

**Saros Num** — Saros Series Number of eclipse

**Ecl Type** — Solar Eclipse Type

      P = Partial Solar Eclipse
      A = Annular Solar Eclipse
      T = Total Solar Eclipse
      H = Hybrid Solar Eclipse
      m = Middle eclipse of Saros series
      n = Central eclipse of with no northern limit
      s = Central eclipse of with no southern limit
      + = Non-central eclipse of with no northern limit
      – = Non-central eclipse of with no southern limit
      2 = Hybrid eclipse path begins total and ends annular
      3 = Hybrid eclipse path begins annular and ends total
      b = Saros series begins (first eclipse in a Saros series)
      e = Saros series ends (last eclipse in a Saros series)

**QLE** — Quincena Lunar Eclipse Parameter

      n = Penumbral Lunar Eclipse
      p = Partial Lunar Eclipse
      t = Total Lunar Eclipse

**Gamma** — minimum distance from the axis of the lunar shadow to the center of Earth

**Ecl Mag** — Eclipse Magnitude; fraction of the Sun's diameter obscured by the Moon

**Lat & Lng** — latitude and longitude where the Sun appears in zenith at greatest eclipse

**Sun Alt & Sun Azm** — altitude and azimuth of the Sun at greatest eclipse

**Path Width** — width of the central path (km) at greatest eclipse (total, annular & hybrid eclipses)

**Central Line Dur** — Central Line Duration (minutes. seconds) at greatest eclipse

| Cat Num | Calendar Date | TD of Greatest Eclipse | ΔT s | Luna Num | Saros Num | Ecl Type | QLE | Gamma | Ecl Mag | Lat ° | Long ° | Sun Alt ° | Sun Azm ° | Path Width km | Central Line Dur |
|---|---|---|---|---|---|---|---|---|---|---|---|---|---|---|---|
| 001 | 2001 Jun 21 | 12:04:46 | 64 | 18 | 127 | T | -p | -0.5701 | 1.0495 | 11.3S | 2.7E | 55 | 355 | 200 | 04m57s |
| 002 | 2001 Dec 14 | 20:53:01 | 64 | 24 | 132 | A | -n | 0.4089 | 0.9681 | 0.6N | 130.7W | 66 | 188 | 126 | 03m53s |
| 003 | 2002 Jun 10 | 23:45:22 | 64 | 30 | 137 | A | nn | 0.1993 | 0.9962 | 34.5N | 178.6W | 78 | 169 | 13 | 00m23s |
| 004 | 2002 Dec 04 | 07:32:16 | 64 | 36 | 142 | T | n- | -0.3020 | 1.0244 | 39.5S | 59.6E | 72 | 16 | 87 | 02m04s |
| 005 | 2003 May 31 | 04:09:23 | 64 | 42 | 147 | An | t- | 0.9960 | 0.9384 | 66.6N | 24.5W | 3 | 35 | - | 03m37s |
| 006 | 2003 Nov 23 | 22:50:22 | 65 | 48 | 152 | T | t- | -0.9638 | 1.0379 | 72.7S | 88.4E | 15 | 111 | 496 | 01m57s |
| 007 | 2004 Apr 19 | 13:35:05 | 65 | 53 | 119 | P | -t | -1.1335 | 0.7367 | 61.6S | 44.3E | 0 | 295 | | |
| 008 | 2004 Oct 14 | 03:00:23 | 65 | 59 | 124 | P | -t | 1.0348 | 0.9283 | 61.2N | 153.7W | 0 | 253 | | |
| 009 | 2005 Apr 08 | 20:36:51 | 65 | 65 | 129 | H | -n | -0.3473 | 1.0074 | 10.6S | 119.0W | 70 | 332 | 27 | 00m42s |
| 010 | 2005 Oct 03 | 10:32:47 | 65 | 71 | 134 | A | -p | 0.3306 | 0.9576 | 12.9N | 28.7E | 71 | 209 | 162 | 04m32s |
| 011 | 2006 Mar 29 | 10:12:23 | 65 | 77 | 139 | T | n- | 0.3843 | 1.0515 | 23.2N | 16.7E | 67 | 149 | 184 | 04m07s |
| 012 | 2006 Sep 22 | 11:41:16 | 65 | 83 | 144 | A | p- | -0.4062 | 0.9352 | 20.6S | 9.1W | 66 | 31 | 261 | 07m09s |
| 013 | 2007 Mar 19 | 02:32:58 | 65 | 89 | 149 | P | t- | 1.0728 | 0.8756 | 61.0N | 55.5E | 0 | 92 | | |
| 014 | 2007 Sep 11 | 12:32:24 | 65 | 95 | 154 | P | t- | -1.1255 | 0.7507 | 61.0S | 90.2W | 0 | 80 | | |
| 015 | 2008 Feb 07 | 03:56:10 | 65 | 100 | 121 | A | -t | -0.9570 | 0.9650 | 67.6S | 150.5W | 16 | 269 | 444 | 02m12s |
| 016 | 2008 Aug 01 | 10:22:12 | 66 | 106 | 126 | T | -p | 0.8307 | 1.0394 | 65.7N | 72.3E | 34 | 235 | 237 | 02m27s |
| 017 | 2009 Jan 26 | 07:59:45 | 66 | 112 | 131 | A | -n | -0.2820 | 0.9282 | 34.1S | 70.2E | 73 | 337 | 280 | 07m54s |
| 018 | 2009 Jul 22 | 02:36:25 | 66 | 118 | 136 | T | nn | 0.0698 | 1.0799 | 24.2N | 144.1E | 86 | 198 | 258 | 06m39s |
| 019 | 2010 Jan 15 | 07:07:39 | 66 | 124 | 141 | A | p- | 0.4002 | 0.9190 | 1.6N | 69.3E | 66 | 165 | 333 | 11m08s |
| 020 | 2010 Jul 11 | 19:34:38 | 66 | 130 | 146 | T | p- | -0.6788 | 1.0580 | 19.7S | 121.9W | 47 | 14 | 259 | 05m20s |
| 021 | 2011 Jan 04 | 08:51:42 | 66 | 136 | 151 | P | t- | 1.0626 | 0.8576 | 64.7N | 20.8E | 0 | 155 | | |
| 022 | 2011 Jun 01 | 21:17:18 | 66 | 141 | 118 | P | -t | 1.2130 | 0.6011 | 67.8N | 46.8E | 0 | 6 | | |
| 023 | 2011 Jul 01 | 08:39:30 | 66 | 142 | 156 | Pb | t- | -1.4917 | 0.0971 | 65.2S | 28.6E | 0 | 21 | | |
| 024 | 2011 Nov 25 | 06:21:25 | 67 | 147 | 123 | P | -t | -1.0536 | 0.9047 | 68.6S | 82.4W | 0 | 165 | | |
| 025 | 2012 May 20 | 23:53:54 | 67 | 153 | 128 | A | -p | 0.4828 | 0.9439 | 49.1N | 176.3E | 61 | 171 | 237 | 05m46s |
| 026 | 2012 Nov 13 | 22:12:55 | 67 | 159 | 133 | T | -n | -0.3719 | 1.0500 | 40.0S | 161.3W | 68 | 11 | 179 | 04m02s |
| 027 | 2013 May 10 | 00:26:20 | 67 | 165 | 138 | A | pn | -0.2694 | 0.9544 | 2.2N | 175.5E | 74 | 350 | 173 | 06m03s |
| 028 | 2013 Nov 03 | 12:47:36 | 67 | 171 | 143 | H3 | n- | 0.3272 | 1.0159 | 3.5N | 11.7W | 71 | 192 | 58 | 01m40s |
| 029 | 2014 Apr 29 | 06:04:33 | 67 | 177 | 148 | A- | t- | -1.0000 | 0.9868 | 70.6S | 131.3E | 0 | 319 | - | - |
| 030 | 2014 Oct 23 | 21:45:39 | 68 | 183 | 153 | P | t- | 1.0908 | 0.8114 | 71.2N | 97.2W | 0 | 231 | | |
| 031 | 2015 Mar 20 | 09:46:47 | 68 | 188 | 120 | T | -t | 0.9454 | 1.0446 | 64.4N | 6.6W | 18 | 135 | 463 | 02m47s |
| 032 | 2015 Sep 13 | 06:55:19 | 68 | 194 | 125 | P | -t | -1.1004 | 0.7875 | 72.1S | 2.3W | 0 | 77 | | |
| 033 | 2016 Mar 09 | 01:58:19 | 68 | 200 | 130 | T | -n | 0.2609 | 1.0450 | 10.1N | 148.8E | 75 | 162 | 155 | 04m09s |
| 034 | 2016 Sep 01 | 09:08:02 | 68 | 206 | 135 | A | -n | -0.3330 | 0.9736 | 10.7S | 37.8E | 70 | 16 | 100 | 03m06s |
| 035 | 2017 Feb 26 | 14:54:33 | 69 | 212 | 140 | A | n- | -0.4578 | 0.9922 | 34.7S | 31.2W | 63 | 340 | 31 | 00m44s |
| 036 | 2017 Aug 21 | 18:26:40 | 69 | 218 | 145 | T | p- | 0.4367 | 1.0306 | 37.0N | 87.7W | 64 | 198 | 115 | 02m40s |
| 037 | 2018 Feb 15 | 20:52:33 | 69 | 224 | 150 | P | t- | -1.2116 | 0.5991 | 71.0S | 0.6E | 0 | 228 | | |
| 038 | 2018 Jul 13 | 03:02:16 | 69 | 229 | 117 | P | -t | -1.3542 | 0.3365 | 67.9S | 127.4E | 0 | 8 | | |
| 039 | 2018 Aug 11 | 09:47:28 | 69 | 230 | 155 | P | t- | 1.1476 | 0.7368 | 70.4N | 174.5E | 0 | 321 | | |
| 040 | 2019 Jan 06 | 01:42:38 | 69 | 235 | 122 | P | -t | 1.1417 | 0.7146 | 67.4N | 153.6E | 0 | 178 | | |
| 041 | 2019 Jul 02 | 19:24:07 | 70 | 241 | 127 | T | -p | -0.6466 | 1.0459 | 17.4S | 109.0W | 50 | 359 | 201 | 04m33s |
| 042 | 2019 Dec 26 | 05:18:53 | 70 | 247 | 132 | A | -n | 0.4135 | 0.9701 | 1.0N | 102.2E | 66 | 184 | 118 | 03m39s |
| 043 | 2020 Jun 21 | 06:41:15 | 70 | 253 | 137 | Am | nn | 0.1209 | 0.9940 | 30.5N | 79.7E | 83 | 174 | 21 | 00m38s |
| 044 | 2020 Dec 14 | 16:14:39 | 70 | 259 | 142 | T | n- | -0.2939 | 1.0254 | 40.3S | 68.0W | 73 | 10 | 90 | 02m10s |
| 045 | 2021 Jun 10 | 10:43:07 | 70 | 265 | 147 | A | t- | 0.9152 | 0.9435 | 80.8N | 66.8W | 23 | 90 | 527 | 03m51s |
| 046 | 2021 Dec 04 | 07:34:38 | 71 | 271 | 152 | T | p- | -0.9526 | 1.0367 | 76.8S | 46.2W | 17 | 115 | 419 | 01m54s |
| 047 | 2022 Apr 30 | 20:42:37 | 71 | 276 | 119 | P | -t | -1.1901 | 0.6396 | 62.1S | 71.5W | 0 | 304 | | |
| 048 | 2022 Oct 25 | 11:01:20 | 71 | 282 | 124 | P | -t | 1.0701 | 0.8619 | 61.6N | 77.3E | 0 | 244 | | |
| 049 | 2023 Apr 20 | 04:17:56 | 71 | 288 | 129 | H | -n | -0.3952 | 1.0132 | 9.6S | 125.8E | 67 | 334 | 49 | 01m16s |
| 050 | 2023 Oct 14 | 18:00:41 | 71 | 294 | 134 | A | -p | 0.3753 | 0.9520 | 11.4N | 83.1W | 68 | 208 | 187 | 05m17s |
| 051 | 2024 Apr 08 | 18:18:29 | 71 | 300 | 139 | T | n- | 0.3431 | 1.0566 | 25.3N | 104.1W | 70 | 149 | 198 | 04m28s |
| 052 | 2024 Oct 02 | 18:46:13 | 72 | 306 | 144 | A | p- | -0.3509 | 0.9326 | 22.0S | 114.5W | 69 | 31 | 266 | 07m25s |
| 053 | 2025 Mar 29 | 10:48:36 | 72 | 312 | 149 | P | t- | 1.0405 | 0.9376 | 61.1N | 77.1W | 0 | 83 | | |
| 054 | 2025 Sep 21 | 19:43:04 | 72 | 318 | 154 | P | t- | -1.0651 | 0.8550 | 60.9S | 153.5E | 0 | 89 | | |
| 055 | 2026 Feb 17 | 12:13:06 | 72 | 323 | 121 | A | -t | -0.9743 | 0.9630 | 64.7S | 86.7E | 12 | 268 | 616 | 02m20s |
| 056 | 2026 Aug 12 | 17:47:06 | 72 | 329 | 126 | T | -p | 0.8977 | 1.0386 | 65.2N | 25.2W | 26 | 248 | 294 | 02m18s |
| 057 | 2027 Feb 06 | 16:00:48 | 73 | 335 | 131 | A | -n | -0.2952 | 0.9281 | 31.3S | 48.5W | 73 | 334 | 282 | 07m51s |
| 058 | 2027 Aug 02 | 10:07:50 | 73 | 341 | 136 | T | nn | 0.1421 | 1.0790 | 25.5N | 33.2E | 82 | 202 | 258 | 06m23s |
| 059 | 2028 Jan 26 | 15:08:59 | 73 | 347 | 141 | A | p- | 0.3901 | 0.9208 | 3.0N | 51.6W | 67 | 161 | 323 | 10m27s |
| 060 | 2028 Jul 22 | 02:56:40 | 73 | 353 | 146 | T | p- | -0.6056 | 1.0560 | 15.6S | 126.7E | 53 | 17 | 230 | 05m10s |

| Cat Num | Calendar Date | TD of Greatest Eclipse | ΔT s | Luna Num | Saros Num | Ecl Type | QLE | Gamma | Ecl Mag | Lat ° | Long ° | Sun Alt ° | Sun Azm ° | Path Width km | Central Line Dur |
|---|---|---|---|---|---|---|---|---|---|---|---|---|---|---|---|
| 061 | 2029 Jan 14 | 17:13:48 | 73 | 359 | 151 | P | t- | 1.0553 | 0.8714 | 63.7N | 114.2W | 0 | 145 | | |
| 062 | 2029 Jun 12 | 04:06:13 | 74 | 364 | 118 | P | -t | 1.2943 | 0.4576 | 66.8N | 66.2W | 0 | 355 | | |
| 063 | 2029 Jul 11 | 15:37:19 | 74 | 365 | 156 | P | t- | -1.4191 | 0.2303 | 64.3S | 85.6W | 0 | 30 | | |
| 064 | 2029 Dec 05 | 15:03:58 | 74 | 370 | 123 | P | -t | -1.0609 | 0.8911 | 67.5S | 135.6E | 0 | 177 | | |
| 065 | 2030 Jun 01 | 06:29:13 | 74 | 376 | 128 | A | -p | 0.5626 | 0.9443 | 56.5N | 80.1E | 55 | 176 | 250 | 05m21s |
| 066 | 2030 Nov 25 | 06:51:37 | 74 | 382 | 133 | T | -n | -0.3867 | 1.0468 | 43.6S | 71.2E | 67 | 7 | 169 | 03m44s |
| 067 | 2031 May 21 | 07:16:04 | 74 | 388 | 138 | A | nn | -0.1970 | 0.9589 | 8.9N | 71.7E | 79 | 354 | 152 | 05m26s |
| 068 | 2031 Nov 14 | 21:07:31 | 75 | 394 | 143 | H | n- | 0.3078 | 1.0106 | 0.6S | 137.6W | 72 | 189 | 38 | 01m08s |
| 069 | 2032 May 09 | 13:26:42 | 75 | 400 | 148 | A | t- | -0.9375 | 0.9957 | 51.3S | 7.1W | 20 | 345 | 44 | 00m22s |
| 070 | 2032 Nov 03 | 05:34:13 | 75 | 406 | 153 | P | t- | 1.0643 | 0.8554 | 70.4N | 132.6E | 0 | 218 | | |
| 071 | 2033 Mar 30 | 18:02:36 | 75 | 411 | 120 | T | -t | 0.9778 | 1.0462 | 71.3N | 155.8W | 11 | 111 | 781 | 02m37s |
| 072 | 2033 Sep 23 | 13:54:31 | 76 | 417 | 125 | P | -t | -1.1583 | 0.6890 | 72.2S | 121.3W | 0 | 91 | | |
| 073 | 2034 Mar 20 | 10:18:45 | 76 | 423 | 130 | T | -n | 0.2894 | 1.0458 | 16.1N | 22.2E | 73 | 162 | 159 | 04m09s |
| 074 | 2034 Sep 12 | 16:19:28 | 76 | 429 | 135 | A | -p | -0.3936 | 0.9736 | 18.2S | 72.6W | 67 | 18 | 102 | 02m58s |
| 075 | 2035 Mar 09 | 23:05:54 | 76 | 435 | 140 | A | n- | -0.4368 | 0.9919 | 29.0S | 155.0W | 64 | 340 | 31 | 00m48s |
| 076 | 2035 Sep 02 | 01:56:46 | 76 | 441 | 145 | T | p- | 0.3727 | 1.0320 | 29.1N | 158.0E | 68 | 199 | 116 | 02m54s |
| 077 | 2036 Feb 27 | 04:46:49 | 77 | 447 | 150 | P | t- | -1.1942 | 0.6286 | 71.6S | 131.5W | 0 | 242 | | |
| 078 | 2036 Jul 23 | 10:32:06 | 77 | 452 | 117 | P | -t | -1.4250 | 0.1992 | 68.9S | 3.5E | 0 | 19 | | |
| 079 | 2036 Aug 21 | 17:25:45 | 77 | 453 | 155 | P | t- | 1.0825 | 0.8622 | 71.1N | 47.0E | 0 | 309 | | |
| 080 | 2037 Jan 16 | 09:48:55 | 77 | 458 | 122 | P | -t | 1.1477 | 0.7049 | 68.5N | 20.8E | 0 | 166 | | |
| 081 | 2037 Jul 13 | 02:40:36 | 77 | 464 | 127 | T | -p | -0.7246 | 1.0413 | 24.8S | 139.1E | 43 | 3 | 201 | 03m58s |
| 082 | 2038 Jan 05 | 13:47:11 | 78 | 470 | 132 | A | -n | 0.4169 | 0.9728 | 2.1N | 25.5W | 65 | 179 | 107 | 03m18s |
| 083 | 2038 Jul 02 | 13:32:55 | 78 | 476 | 137 | A | nn | 0.0398 | 0.9911 | 25.4N | 21.9W | 88 | 179 | 31 | 01m00s |
| 084 | 2038 Dec 26 | 01:00:10 | 78 | 482 | 142 | T | n- | -0.2881 | 1.0269 | 40.3S | 163.9E | 73 | 5 | 95 | 02m18s |
| 085 | 2039 Jun 21 | 17:12:54 | 78 | 488 | 147 | A | p- | 0.8312 | 0.9454 | 78.9N | 102.1W | 33 | 153 | 365 | 04m05s |
| 086 | 2039 Dec 15 | 16:23:46 | 79 | 494 | 152 | T | p- | -0.9458 | 1.0356 | 80.9S | 172.7E | 18 | 123 | 380 | 01m51s |
| 087 | 2040 May 11 | 03:43:02 | 79 | 499 | 119 | P | -t | -1.2529 | 0.5306 | 62.8S | 174.4E | 0 | 313 | | |
| 088 | 2040 Nov 04 | 19:09:02 | 79 | 505 | 124 | P | -t | 1.0993 | 0.8074 | 62.2N | 53.4W | 0 | 234 | | |
| 089 | 2041 Apr 30 | 11:52:21 | 79 | 511 | 129 | T | -p | -0.4492 | 1.0189 | 9.6S | 12.2E | 63 | 337 | 72 | 01m51s |
| 090 | 2041 Oct 25 | 01:36:22 | 80 | 517 | 134 | A | -p | 0.4133 | 0.9467 | 9.9N | 162.8E | 66 | 206 | 213 | 06m07s |
| 091 | 2042 Apr 20 | 02:17:30 | 80 | 523 | 139 | T | n- | 0.2956 | 1.0614 | 27.0N | 137.3E | 73 | 151 | 210 | 04m51s |
| 092 | 2042 Oct 14 | 02:00:42 | 80 | 529 | 144 | A | n- | -0.3030 | 0.9301 | 23.7S | 137.8E | 72 | 30 | 273 | 07m44s |
| 093 | 2043 Apr 09 | 18:57:49 | 80 | 535 | 149 | T+ | t- | 1.0031 | 1.0096 | 61.3N | 151.9E | 0 | 74 | – | – |
| 094 | 2043 Oct 03 | 03:01:49 | 81 | 541 | 154 | A- | t- | -1.0102 | 0.9497 | 61.0S | 35.2E | 0 | 98 | – | – |
| 095 | 2044 Feb 28 | 20:24:40 | 81 | 546 | 121 | As | -t | -0.9954 | 0.9600 | 62.2S | 25.6W | 4 | 260 | – | 02m27s |
| 096 | 2044 Aug 23 | 01:17:02 | 81 | 552 | 126 | T | -t | 0.9613 | 1.0364 | 64.3N | 120.5W | 15 | 264 | 453 | 02m04s |
| 097 | 2045 Feb 16 | 23:56:07 | 81 | 558 | 131 | A | -n | -0.3125 | 0.9285 | 28.3S | 166.2W | 72 | 331 | 281 | 07m47s |
| 098 | 2045 Aug 12 | 17:42:39 | 82 | 564 | 136 | T | -n | 0.2116 | 1.0774 | 25.9N | 78.6W | 78 | 206 | 256 | 06m06s |
| 099 | 2046 Feb 05 | 23:06:26 | 82 | 570 | 141 | A | p- | 0.3765 | 0.9232 | 4.8N | 171.4W | 68 | 157 | 310 | 09m42s |
| 100 | 2046 Aug 02 | 10:21:13 | 82 | 576 | 146 | T | p- | -0.5350 | 1.0531 | 12.7S | 15.1E | 58 | 21 | 206 | 04m51s |
| 101 | 2047 Jan 26 | 01:33:18 | 82 | 582 | 151 | P | t- | 1.0450 | 0.8908 | 62.9N | 111.7E | 0 | 135 | | |
| 102 | 2047 Jun 23 | 10:52:31 | 83 | 587 | 118 | P | -t | 1.3766 | 0.3129 | 65.8N | 178.0W | 0 | 346 | | |
| 103 | 2047 Jul 22 | 22:36:17 | 83 | 588 | 156 | P | t- | -1.3477 | 0.3605 | 63.4S | 160.1E | 0 | 40 | | |
| 104 | 2047 Dec 16 | 23:50:12 | 83 | 593 | 123 | P | -t | -1.0661 | 0.8817 | 66.4S | 6.6W | 0 | 188 | | |
| 105 | 2048 Jun 11 | 12:58:53 | 83 | 599 | 128 | A | -p | 0.6468 | 0.9441 | 63.7N | 11.5W | 49 | 184 | 271 | 04m58s |
| 106 | 2048 Dec 05 | 15:35:27 | 83 | 605 | 133 | T | -n | -0.3973 | 1.0440 | 46.1S | 56.4W | 66 | 1 | 160 | 03m28s |
| 107 | 2049 May 31 | 13:59:59 | 84 | 611 | 138 | A | nn | -0.1187 | 0.9631 | 15.3N | 29.9W | 83 | 358 | 134 | 04m45s |
| 108 | 2049 Nov 25 | 05:33:48 | 84 | 617 | 143 | H | n- | 0.2943 | 1.0057 | 3.8S | 95.2E | 73 | 185 | 21 | 00m38s |
| 109 | 2050 May 20 | 20:42:50 | 84 | 623 | 148 | H | t- | -0.8688 | 1.0038 | 40.1S | 123.8W | 29 | 352 | 27 | 00m21s |
| 110 | 2050 Nov 14 | 13:30:53 | 85 | 629 | 153 | P | t- | 1.0447 | 0.8874 | 69.5N | 1.0E | 0 | 206 | | |
| 111 | 2051 Apr 11 | 02:10:39 | 85 | 634 | 120 | P | -t | 1.0169 | 0.9849 | 71.6N | 32.1E | 0 | 63 | | |
| 112 | 2051 Oct 04 | 21:02:15 | 85 | 640 | 125 | P | -t | -1.2094 | 0.6024 | 72.0S | 117.7E | 0 | 105 | | |
| 113 | 2052 Mar 30 | 18:31:53 | 85 | 646 | 130 | T | -n | 0.3239 | 1.0466 | 22.4N | 102.6W | 71 | 161 | 164 | 04m08s |
| 114 | 2052 Sep 22 | 23:39:10 | 86 | 652 | 135 | A | -p | -0.4480 | 0.9734 | 25.7S | 174.9E | 63 | 20 | 106 | 02m51s |
| 115 | 2053 Mar 20 | 07:08:19 | 86 | 658 | 140 | A | n- | -0.4089 | 0.9919 | 23.0S | 82.9E | 66 | 341 | 31 | 00m50s |
| 116 | 2053 Sep 12 | 09:34:09 | 86 | 664 | 145 | T | n- | 0.3140 | 1.0328 | 21.5N | 41.7E | 72 | 199 | 116 | 03m04s |
| 117 | 2054 Mar 09 | 12:33:40 | 87 | 670 | 150 | P | t- | -1.1711 | 0.6678 | 72.0S | 97.9E | 0 | 256 | | |
| 118 | 2054 Aug 03 | 18:04:02 | 87 | 675 | 117 | Pe | -t | -1.4941 | 0.0656 | 69.8S | 121.4W | 0 | 31 | | |
| 119 | 2054 Sep 02 | 01:09:34 | 87 | 676 | 155 | P | t- | 1.0215 | 0.9793 | 71.7N | 82.4W | 0 | 296 | | |
| 120 | 2055 Jan 27 | 17:54:05 | 87 | 681 | 122 | P | -t | 1.1550 | 0.6932 | 69.5N | 112.3W | 0 | 154 | | |

| Cat Num | Calendar Date | TD of Greatest Eclipse | ΔT s | Luna Num | Saros Num | Ecl Type | QLE | Gamma | Ecl Mag | Lat ° | Long ° | Sun Alt ° | Sun Azm ° | Path Width km | Central Line Dur |
|---|---|---|---|---|---|---|---|---|---|---|---|---|---|---|---|
| 121 | 2055 Jul 24 | 09:57:50 | 88 | 687 | 127 | T | -p | -0.8012 | 1.0359 | 33.3S | 25.7E | 37 | 8 | 202 | 03m17s |
| 122 | 2056 Jan 16 | 22:16:45 | 88 | 693 | 132 | A | -n | 0.4199 | 0.9760 | 3.9N | 153.6W | 65 | 175 | 95 | 02m52s |
| 123 | 2056 Jul 12 | 20:21:59 | 88 | 699 | 137 | A | nn | -0.0426 | 0.9878 | 19.4N | 123.8W | 88 | 3 | 43 | 01m26s |
| 124 | 2057 Jan 05 | 09:47:52 | 88 | 705 | 142 | T | n- | -0.2837 | 1.0287 | 39.2S | 35.1E | 73 | 359 | 102 | 02m29s |
| 125 | 2057 Jul 01 | 23:40:15 | 89 | 711 | 147 | A | p- | 0.7455 | 0.9464 | 71.5N | 176.3W | 41 | 177 | 298 | 04m22s |
| 126 | 2057 Dec 26 | 01:14:35 | 89 | 717 | 152 | T | p- | -0.9405 | 1.0348 | 84.9S | 21.7E | 19 | 141 | 355 | 01m50s |
| 127 | 2058 May 22 | 10:39:26 | 89 | 722 | 119 | P | -t | -1.3194 | 0.4141 | 63.5S | 61.1E | 0 | 322 | | |
| 128 | 2058 Jun 21 | 00:19:35 | 89 | 723 | 157 | Pb | t- | 1.4869 | 0.1261 | 65.9N | 9.8E | 0 | 13 | | |
| 129 | 2058 Nov 16 | 03:23:07 | 90 | 728 | 124 | P | -t | 1.1224 | 0.7644 | 62.9N | 174.1E | 0 | 225 | | |
| 130 | 2059 May 11 | 19:22:16 | 90 | 734 | 129 | T | -p | -0.5080 | 1.0242 | 10.7S | 100.5W | 59 | 340 | 95 | 02m23s |
| 131 | 2059 Nov 05 | 09:18:15 | 90 | 740 | 134 | A | -p | 0.4454 | 0.9417 | 8.7N | 47.0E | 63 | 203 | 238 | 07m00s |
| 132 | 2060 Apr 30 | 10:10:00 | 91 | 746 | 139 | T | n- | 0.2422 | 1.0660 | 28.0N | 20.8E | 76 | 154 | 222 | 05m15s |
| 133 | 2060 Oct 24 | 09:24:10 | 91 | 752 | 144 | A | nn | -0.2625 | 0.9277 | 25.8S | 28.0E | 75 | 28 | 281 | 08m06s |
| 134 | 2061 Apr 20 | 02:56:49 | 91 | 758 | 149 | T | t- | 0.9578 | 1.0476 | 64.5N | 59.1E | 16 | 97 | 559 | 02m37s |
| 135 | 2061 Oct 13 | 10:32:10 | 92 | 764 | 154 | A | t- | -0.9639 | 0.9469 | 62.1S | 54.5W | 15 | 79 | 743 | 03m41s |
| 136 | 2062 Mar 11 | 04:26:16 | 92 | 769 | 121 | P | -t | -1.0238 | 0.9331 | 61.0S | 147.2W | 0 | 263 | | |
| 137 | 2062 Sep 03 | 08:54:27 | 92 | 775 | 126 | P | -t | 1.0192 | 0.9749 | 61.3N | 150.2E | 0 | 286 | | |
| 138 | 2063 Feb 28 | 07:43:30 | 93 | 781 | 131 | A | -p | -0.3360 | 0.9293 | 25.2S | 77.6E | 70 | 329 | 280 | 07m41s |
| 139 | 2063 Aug 24 | 01:22:11 | 93 | 787 | 136 | T | -n | 0.2771 | 1.0750 | 25.6N | 168.3E | 74 | 209 | 252 | 05m49s |
| 140 | 2064 Feb 17 | 07:00:23 | 93 | 793 | 141 | A | p- | 0.3597 | 0.9262 | 7.0N | 69.6E | 69 | 154 | 295 | 08m56s |
| 141 | 2064 Aug 12 | 17:46:06 | 94 | 799 | 146 | T | p- | -0.4652 | 1.0495 | 10.9S | 96.1W | 62 | 24 | 184 | 04m28s |
| 142 | 2065 Feb 05 | 09:52:26 | 94 | 805 | 151 | P | t- | 1.0336 | 0.9123 | 62.2N | 22.0W | 0 | 125 | | |
| 143 | 2065 Jul 03 | 17:33:52 | 94 | 810 | 118 | P | -t | 1.4619 | 0.1639 | 64.8N | 71.7E | 0 | 336 | | |
| 144 | 2065 Aug 02 | 05:34:17 | 94 | 811 | 156 | P | t- | -1.2758 | 0.4903 | 62.7S | 46.4E | 0 | 49 | | |
| 145 | 2065 Dec 27 | 08:39:56 | 95 | 816 | 123 | P | -t | -1.0688 | 0.8769 | 65.4S | 149.3W | 0 | 198 | | |
| 146 | 2066 Jun 22 | 19:25:48 | 95 | 822 | 128 | A | -p | 0.7330 | 0.9435 | 70.1N | 96.5W | 43 | 198 | 309 | 04m40s |
| 147 | 2066 Dec 17 | 00:23:40 | 95 | 828 | 133 | T | -n | -0.4043 | 1.0416 | 47.4S | 175.6W | 66 | 355 | 152 | 03m14s |
| 148 | 2067 Jun 11 | 20:42:26 | 96 | 834 | 138 | A | nn | -0.0387 | 0.9670 | 21.0N | 130.3W | 88 | 2 | 119 | 04m05s |
| 149 | 2067 Dec 06 | 14:03:43 | 96 | 840 | 143 | H | n- | 0.2845 | 1.0011 | 6.0S | 32.5W | 74 | 181 | 4 | 00m08s |
| 150 | 2068 May 31 | 03:56:39 | 96 | 846 | 148 | T | p- | -0.7970 | 1.0110 | 31.0S | 123.1E | 37 | 357 | 63 | 01m06s |
| 151 | 2068 Nov 24 | 21:32:30 | 97 | 852 | 153 | P | t- | 1.0299 | 0.9109 | 68.5N | 131.2W | 0 | 194 | | |
| 152 | 2069 Apr 21 | 10:11:09 | 97 | 857 | 120 | P | -t | 1.0624 | 0.8992 | 71.0N | 101.4W | 0 | 50 | | |
| 153 | 2069 May 20 | 17:53:18 | 97 | 858 | 158 | Pb | t- | -1.4852 | 0.0879 | 68.8S | 70.1W | 0 | 342 | | |
| 154 | 2069 Oct 15 | 04:19:56 | 97 | 863 | 125 | P | -t | -1.2524 | 0.5298 | 71.6S | 5.6W | 0 | 119 | | |
| 155 | 2070 Apr 11 | 02:36:09 | 98 | 869 | 130 | T | -n | 0.3652 | 1.0472 | 29.1N | 134.9E | 68 | 162 | 168 | 04m04s |
| 156 | 2070 Oct 04 | 07:08:57 | 98 | 875 | 135 | A | -p | -0.4950 | 0.9731 | 32.8S | 60.2E | 60 | 21 | 110 | 02m44s |
| 157 | 2071 Mar 31 | 15:01:06 | 98 | 881 | 140 | A | n- | -0.3739 | 0.9919 | 16.7S | 37.2W | 68 | 342 | 31 | 00m52s |
| 158 | 2071 Sep 23 | 17:20:28 | 99 | 887 | 145 | T | n- | 0.2620 | 1.0333 | 14.2N | 76.9W | 75 | 198 | 116 | 03m11s |
| 159 | 2072 Mar 19 | 20:10:31 | 99 | 893 | 150 | P | t- | -1.1405 | 0.7199 | 72.2S | 30.5W | 0 | 270 | | |
| 160 | 2072 Sep 12 | 08:59:20 | 100 | 899 | 155 | T | t- | 0.9655 | 1.0558 | 69.8N | 101.8E | 14 | 240 | 732 | 03m13s |
| 161 | 2073 Feb 07 | 01:55:59 | 100 | 904 | 122 | P | -t | 1.1651 | 0.6768 | 70.5N | 114.7E | 0 | 141 | | |
| 162 | 2073 Aug 03 | 17:15:23 | 100 | 910 | 127 | T | -t | -0.8763 | 1.0294 | 43.2S | 89.6W | 28 | 14 | 206 | 02m29s |
| 163 | 2074 Jan 27 | 06:44:15 | 101 | 916 | 132 | A | -n | 0.4251 | 0.9798 | 6.6N | 78.6E | 65 | 171 | 79 | 02m21s |
| 164 | 2074 Jul 24 | 03:10:32 | 101 | 922 | 137 | A | nn | -0.1242 | 0.9838 | 12.8N | 133.5E | 83 | 7 | 58 | 01m57s |
| 165 | 2075 Jan 16 | 18:36:04 | 101 | 928 | 142 | T | n- | -0.2799 | 1.0311 | 37.2S | 94.3W | 74 | 354 | 110 | 02m42s |
| 166 | 2075 Jul 13 | 06:05:44 | 102 | 934 | 147 | A | p- | 0.6583 | 0.9467 | 63.1N | 95.0W | 49 | 186 | 262 | 04m45s |
| 167 | 2076 Jan 06 | 10:07:28 | 102 | 940 | 152 | T | p- | -0.9373 | 1.0342 | 87.2S | 173.9W | 20 | 203 | 340 | 01m49s |
| 168 | 2076 Jun 01 | 17:31:22 | 102 | 945 | 119 | P | -t | -1.3897 | 0.2897 | 64.4S | 51.4W | 0 | 331 | | |
| 169 | 2076 Jul 01 | 06:50:43 | 103 | 946 | 157 | P | t- | 1.4005 | 0.2746 | 67.0N | 98.3W | 0 | 3 | | |
| 170 | 2076 Nov 26 | 11:43:01 | 103 | 951 | 124 | P | -t | 1.1401 | 0.7315 | 63.7N | 39.9E | 0 | 215 | | |
| 171 | 2077 May 22 | 02:46:05 | 103 | 957 | 129 | T | -p | -0.5725 | 1.0290 | 13.1S | 148.1E | 55 | 343 | 119 | 02m54s |
| 172 | 2077 Nov 15 | 17:07:56 | 104 | 963 | 134 | A | -p | 0.4705 | 0.9371 | 7.8N | 71.0W | 62 | 199 | 262 | 07m54s |
| 173 | 2078 May 11 | 17:56:55 | 104 | 969 | 139 | T | n- | 0.1838 | 1.0701 | 28.1N | 93.9W | 79 | 158 | 232 | 05m40s |
| 174 | 2078 Nov 04 | 16:55:44 | 104 | 975 | 144 | A | nn | -0.2285 | 0.9255 | 27.8S | 83.5W | 77 | 25 | 287 | 08m29s |
| 175 | 2079 May 01 | 10:50:13 | 105 | 981 | 149 | T | p- | 0.9081 | 1.0512 | 66.2N | 46.5W | 24 | 108 | 406 | 02m55s |
| 176 | 2079 Oct 24 | 18:11:21 | 105 | 987 | 154 | A | t- | -0.9243 | 0.9484 | 63.4S | 160.8W | 22 | 72 | 495 | 03m39s |
| 177 | 2080 Mar 21 | 12:20:15 | 106 | 992 | 121 | P | -t | -1.0578 | 0.8734 | 60.9S | 85.7E | 0 | 271 | | |
| 178 | 2080 Sep 13 | 16:38:09 | 106 | 998 | 126 | P | -t | 1.0724 | 0.8743 | 61.1N | 25.6E | 0 | 277 | | |
| 179 | 2081 Mar 10 | 15:23:31 | 106 | 1004 | 131 | A | -p | -0.3653 | 0.9304 | 22.4S | 36.9W | 68 | 329 | 277 | 07m36s |
| 180 | 2081 Sep 03 | 09:07:31 | 107 | 1010 | 136 | T | -n | 0.3379 | 1.0720 | 24.6N | 53.4E | 70 | 211 | 247 | 05m33s |

| Cat Num | Calendar Date | TD of Greatest Eclipse | ΔT s | Luna Num | Saros Num | Ecl Type | QLE | Gamma | Ecl Mag | Lat ° | Long ° | Sun Alt ° | Sun Azm ° | Path Width km | Central Line Dur |
|---|---|---|---|---|---|---|---|---|---|---|---|---|---|---|---|
| 181 | 2082 Feb 27 | 14:47:00 | 107 | 1016 | 141 | A | p- | 0.3361 | 0.9298 | 9.4N | 47.3W | 70 | 152 | 277 | 08m12s |
| 182 | 2082 Aug 24 | 01:16:21 | 108 | 1022 | 146 | T | n- | -0.4004 | 1.0452 | 10.3S | 151.5E | 66 | 26 | 163 | 04m01s |
| 183 | 2083 Feb 16 | 18:06:36 | 108 | 1028 | 151 | P | t- | 1.0170 | 0.9433 | 61.6N | 154.3W | 0 | 116 | | |
| 184 | 2083 Jul 15 | 00:14:23 | 108 | 1033 | 118 | Pe | -t | 1.5464 | 0.0169 | 64.0N | 37.9W | 0 | 327 | | |
| 185 | 2083 Aug 13 | 12:34:41 | 108 | 1034 | 156 | P | t- | -1.2064 | 0.6146 | 62.1S | 67.7W | 0 | 58 | | |
| 186 | 2084 Jan 07 | 17:30:24 | 109 | 1039 | 123 | P | -t | -1.0715 | 0.8723 | 64.4S | 68.3E | 0 | 209 | | |
| 187 | 2084 Jul 03 | 01:50:26 | 109 | 1045 | 128 | A | -p | 0.8208 | 0.9421 | 75.0N | 169.3W | 35 | 222 | 377 | 04m25s |
| 188 | 2084 Dec 27 | 09:13:48 | 110 | 1051 | 133 | T | -n | -0.4094 | 1.0396 | 47.3S | 47.4E | 66 | 349 | 146 | 03m04s |
| 189 | 2085 Jun 22 | 03:21:16 | 110 | 1057 | 138 | A | nn | 0.0453 | 0.9704 | 26.2N | 131.0E | 87 | 186 | 106 | 03m29s |
| 190 | 2085 Dec 16 | 22:37:48 | 111 | 1063 | 143 | A | n- | 0.2786 | 0.9971 | 7.3S | 161.1W | 74 | 176 | 10 | 00m19s |
| 191 | 2086 Jun 11 | 11:07:14 | 111 | 1069 | 148 | T | p- | -0.7215 | 1.0174 | 23.2S | 12.2E | 44 | 2 | 86 | 01m48s |
| 192 | 2086 Dec 06 | 05:38:55 | 111 | 1075 | 153 | P | p- | 1.0194 | 0.9271 | 67.4N | 96.0E | 0 | 182 | | |
| 193 | 2087 May 02 | 18:04:42 | 112 | 1080 | 120 | P | -t | 1.1139 | 0.8011 | 70.3N | 127.3E | 0 | 37 | | |
| 194 | 2087 Jun 01 | 01:27:14 | 112 | 1081 | 158 | P | t- | -1.4186 | 0.2146 | 67.8S | 165.1E | 0 | 354 | | |
| 195 | 2087 Oct 26 | 11:46:57 | 112 | 1086 | 125 | P | -t | -1.2882 | 0.4696 | 71.0S | 130.8W | 0 | 132 | | |
| 196 | 2088 Apr 21 | 10:31:49 | 113 | 1092 | 130 | T | -p | 0.4135 | 1.0474 | 36.0N | 14.8E | 65 | 163 | 173 | 03m58s |
| 197 | 2088 Oct 14 | 14:48:05 | 113 | 1098 | 135 | A | -p | -0.5349 | 0.9727 | 39.7S | 56.3W | 57 | 21 | 115 | 02m38s |
| 198 | 2089 Apr 10 | 22:44:42 | 113 | 1104 | 140 | A | n- | -0.3319 | 0.9919 | 10.2S | 155.0W | 71 | 344 | 30 | 00m53s |
| 199 | 2089 Oct 04 | 01:15:23 | 114 | 1110 | 145 | T | n- | 0.2167 | 1.0333 | 7.4N | 162.5E | 77 | 197 | 115 | 03m14s |
| 200 | 2090 Mar 31 | 03:38:08 | 114 | 1116 | 150 | P | t- | -1.1028 | 0.7843 | 72.1S | 156.6W | 0 | 284 | | |
| 201 | 2090 Sep 23 | 16:56:36 | 115 | 1122 | 155 | T | t- | 0.9157 | 1.0562 | 60.7N | 40.8W | 23 | 218 | 463 | 03m36s |
| 202 | 2091 Feb 18 | 09:54:40 | 115 | 1127 | 122 | P | -t | 1.1779 | 0.6558 | 71.2N | 18.0W | 0 | 128 | | |
| 203 | 2091 Aug 15 | 00:34:43 | 116 | 1133 | 127 | T | -t | -0.9490 | 1.0216 | 55.6S | 150.2E | 18 | 23 | 237 | 01m38s |
| 204 | 2092 Feb 07 | 15:10:20 | 116 | 1139 | 132 | A | -n | 0.4322 | 0.9840 | 9.9N | 49.0W | 64 | 168 | 62 | 01m48s |
| 205 | 2092 Aug 03 | 09:59:33 | 116 | 1145 | 137 | A | nn | -0.2044 | 0.9794 | 5.6N | 30.0E | 78 | 10 | 75 | 02m31s |
| 206 | 2093 Jan 27 | 03:22:16 | 117 | 1151 | 142 | T | n- | -0.2737 | 1.0340 | 34.1S | 136.1E | 74 | 350 | 119 | 02m58s |
| 207 | 2093 Jul 23 | 12:32:04 | 117 | 1157 | 147 | A | p- | 0.5717 | 0.9463 | 54.6N | 1.0E | 55 | 191 | 241 | 05m11s |
| 208 | 2094 Jan 16 | 18:59:03 | 118 | 1163 | 152 | T | p- | -0.9333 | 1.0342 | 84.8S | 10.9W | 21 | 267 | 329 | 01m52s |
| 209 | 2094 Jun 13 | 00:22:11 | 118 | 1168 | 119 | P | -t | -1.4613 | 0.1618 | 65.3S | 163.9W | 0 | 341 | | |
| 210 | 2094 Jul 12 | 13:24:35 | 118 | 1169 | 157 | P | t- | 1.3149 | 0.4225 | 68.0N | 152.5E | 0 | 352 | | |
| 211 | 2094 Dec 07 | 20:05:56 | 119 | 1174 | 124 | P | -t | 1.1547 | 0.7046 | 64.7N | 95.3W | 0 | 205 | | |
| 212 | 2095 Jun 02 | 10:07:40 | 119 | 1180 | 129 | T | -p | -0.6396 | 1.0332 | 16.7S | 36.9E | 50 | 347 | 145 | 03m18s |
| 213 | 2095 Nov 27 | 01:02:57 | 120 | 1186 | 134 | A | -p | 0.4903 | 0.9330 | 7.2N | 169.5E | 61 | 195 | 285 | 08m47s |
| 214 | 2096 May 22 | 01:37:14 | 120 | 1192 | 139 | T | nn | 0.1196 | 1.0737 | 27.3N | 153.1E | 83 | 162 | 241 | 06m06s |
| 215 | 2096 Nov 15 | 00:36:15 | 121 | 1198 | 144 | A | nn | -0.2018 | 0.9237 | 29.7S | 163.0E | 78 | 22 | 294 | 08m53s |
| 216 | 2097 May 11 | 18:34:31 | 121 | 1204 | 149 | T | p- | 0.8516 | 1.0538 | 67.4N | 149.8W | 31 | 121 | 339 | 03m10s |
| 217 | 2097 Nov 04 | 02:01:25 | 121 | 1210 | 154 | A | t- | -0.8926 | 0.9494 | 65.8S | 86.5E | 26 | 68 | 411 | 03m36s |
| 218 | 2098 Apr 01 | 20:02:31 | 122 | 1215 | 121 | P | -t | -1.1005 | 0.7984 | 61.0S | 38.4W | 0 | 280 | | |
| 219 | 2098 Sep 25 | 00:31:16 | 122 | 1221 | 126 | P | -t | 1.1184 | 0.7871 | 61.1N | 101.3W | 0 | 268 | | |
| 220 | 2098 Oct 24 | 10:36:11 | 122 | 1222 | 164 | Pb | t- | -1.5407 | 0.0057 | 61.8S | 95.8W | 0 | 116 | | |
| 221 | 2099 Mar 21 | 22:54:32 | 123 | 1227 | 131 | A | -p | -0.4016 | 0.9318 | 20.0S | 149.4W | 66 | 329 | 275 | 07m32s |
| 222 | 2099 Sep 14 | 16:57:53 | 123 | 1233 | 136 | T | -n | 0.3942 | 1.0684 | 23.4N | 63.1W | 67 | 211 | 241 | 05m18s |
| 223 | 2100 Mar 10 | 22:28:11 | 124 | 1239 | 141 | A | n- | 0.3077 | 0.9338 | 12.0N | 162.8W | 72 | 151 | 257 | 07m29s |
| 224 | 2100 Sep 04 | 08:49:20 | 124 | 1245 | 146 | T | n- | -0.3384 | 1.0402 | 10.5S | 38.7E | 70 | 28 | 142 | 03m32s |

# Appendix B

# Small Global Solar Eclipse Maps: 2001 to 2100

# Key to Small Global Solar Eclipse Maps

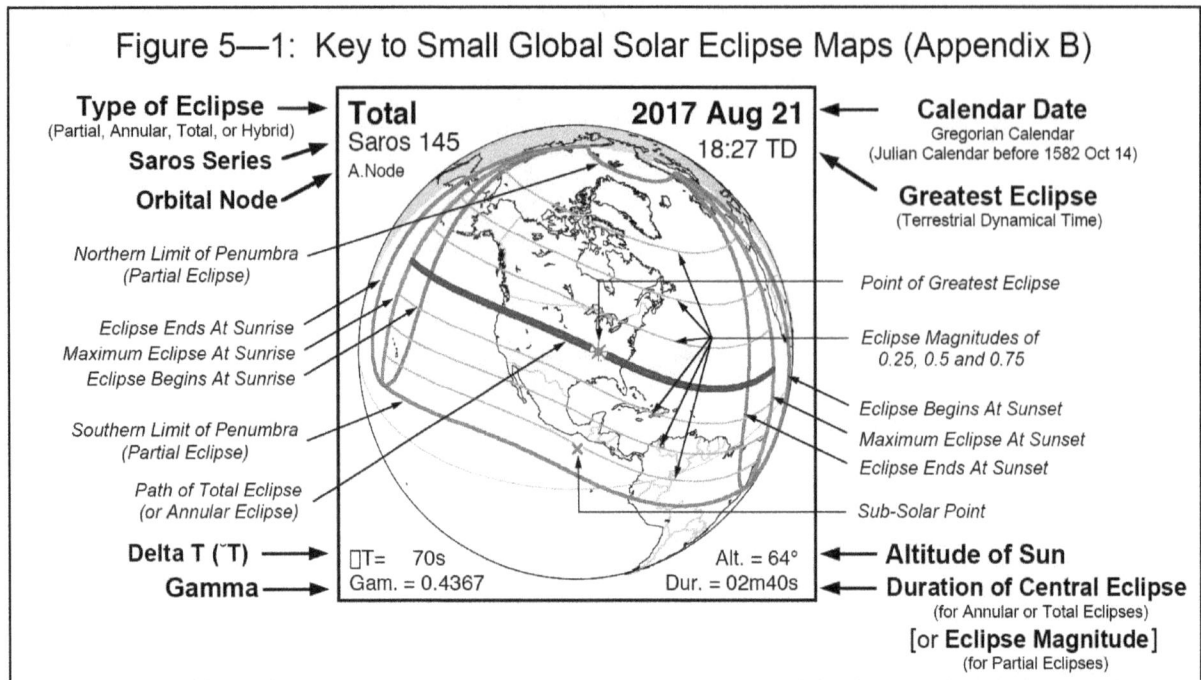

Figure 5—1: Key to Small Global Solar Eclipse Maps (Appendix B)

**Type of Eclipse** → (Partial, Annular, Total, or Hybrid)
**Saros Series** →
**Orbital Node** →

Total     2017 Aug 21
Saros 145    18:27 TD
A.Node

← **Calendar Date**
Gregorian Calendar
(Julian Calendar before 1582 Oct 14)
**Greatest Eclipse**
(Terrestrial Dynamical Time)

Northern Limit of Penumbra
(Partial Eclipse)

Eclipse Ends At Sunrise
Maximum Eclipse At Sunrise
Eclipse Begins At Sunrise

Southern Limit of Penumbra
(Partial Eclipse)

Path of Total Eclipse
(or Annular Eclipse)

Point of Greatest Eclipse

Eclipse Magnitudes of
0.25, 0.5 and 0.75

Eclipse Begins At Sunset
Maximum Eclipse At Sunset
Eclipse Ends At Sunset

Sub-Solar Point

**Delta T (˜T)** →
**Gamma** →

˜T= 70s
Gam. = 0.4367

Alt. = 64°
Dur. = 02m40s

← **Altitude of Sun**
**Duration of Central Eclipse**
(for Annular or Total Eclipses)
[or **Eclipse Magnitude**]
(for Partial Eclipses)

**Solar Eclipse Type** – One of four basic types of solar eclipses: Partial, Annular Total or Hybrid (See Sect. 4.1).

**Saros Series** – The Saros series that the eclipse belongs to. Eclipses with an odd number take place at the Moon's ascending node, while those with an even number are at the descending node (See Sect. 4.7).

**Node** – The orbital node near which the eclipse takes place. The ascending node (A. Node) is the point where the Moon travels from south to north through Earth's orbital plane. Similarly, the descending node (D. Node) is the point where the Moon travels from north to south.

**Calendar Date** – Gregorian calendar date at greatest eclipse is given in the ISO order of YEAR-MONTH-DAY.

**Greatest Eclipse** – The instant (Terrestrial Dynamical Time) when the distance between the axis of the Moon's shadow cone and the center of Earth reaches a minimum (See Sect. 4.4).

**ΔT (Delta T)** – the arithmetic difference, in seconds, between Terrestrial Dynamical Time (TD) and Universal Time (UT1), (See Sect. 2.5).

**Gamma** – The minimum distance from the lunar shadow axis to the center of Earth, in units of Earth's equatorial radius (See Sect. 4.10).

**Altitude of Sun** – The Sun's altitude at the geographic position intersected by the lunar shadow axis is given at the instant of greatest eclipse (See Sect. 4.13).

**Duration of Central Eclipse** – The central line duration of the total or annular phase (in minutes and seconds) is given at the instant of greatest eclipse (See Sect. 4.16).

**Eclipse Magnitude** – The fraction of the Sun's diameter occulted by the Moon at the instant of greatest eclipse (See Sect. 4.11).

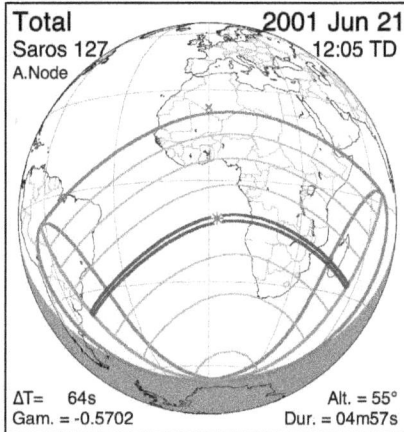

Total     2001 Jun 21
Saros 127     12:05 TD
A.Node
ΔT= 64s     Alt. = 55°
Gam. = -0.5702     Dur. = 04m57s

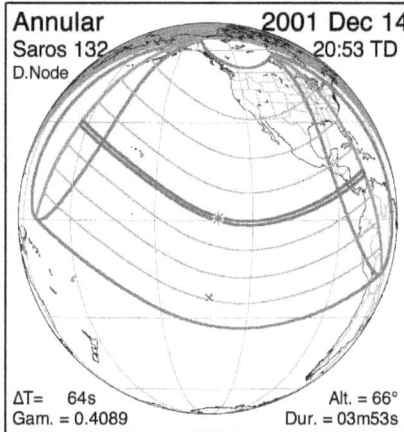

Annular     2001 Dec 14
Saros 132     20:53 TD
D.Node
ΔT= 64s     Alt. = 66°
Gam. = 0.4089     Dur. = 03m53s

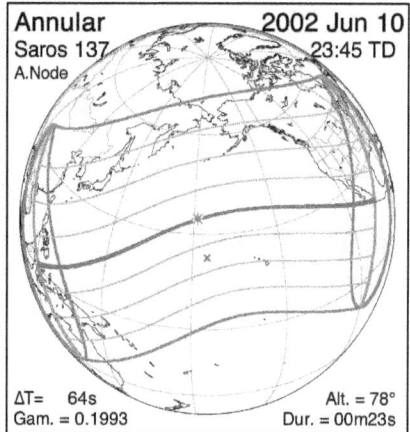

Annular     2002 Jun 10
Saros 137     23:45 TD
A.Node
ΔT= 64s     Alt. = 78°
Gam. = 0.1993     Dur. = 00m23s

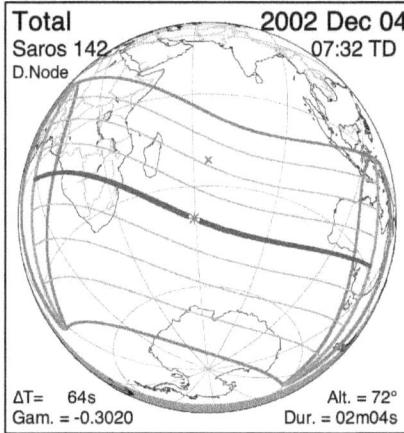

Total     2002 Dec 04
Saros 142     07:32 TD
D.Node
ΔT= 64s     Alt. = 72°
Gam. = -0.3020     Dur. = 02m04s

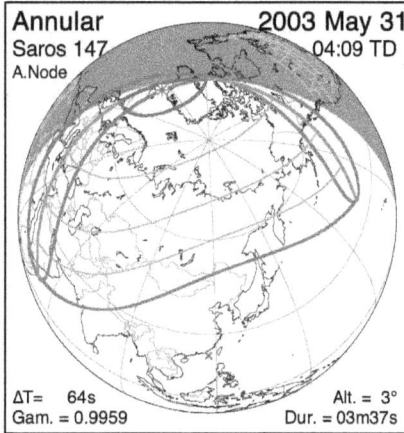

Annular     2003 May 31
Saros 147     04:09 TD
A.Node
ΔT= 64s     Alt. = 3°
Gam. = 0.9959     Dur. = 03m37s

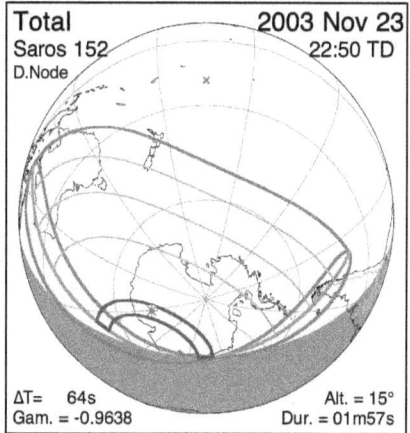

Total     2003 Nov 23
Saros 152     22:50 TD
D.Node
ΔT= 64s     Alt. = 15°
Gam. = -0.9638     Dur. = 01m57s

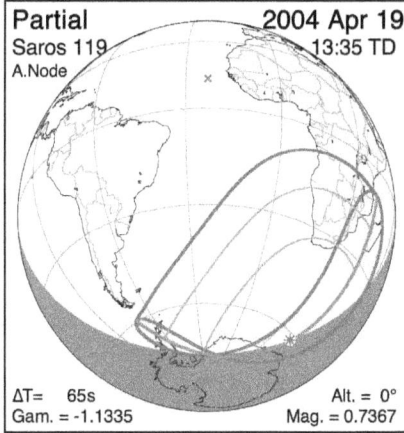

Partial     2004 Apr 19
Saros 119     13:35 TD
A.Node
ΔT= 65s     Alt. = 0°
Gam. = -1.1335     Mag. = 0.7367

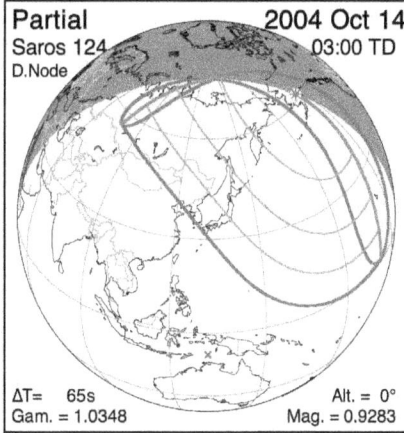

Partial     2004 Oct 14
Saros 124     03:00 TD
D.Node
ΔT= 65s     Alt. = 0°
Gam. = 1.0348     Mag. = 0.9283

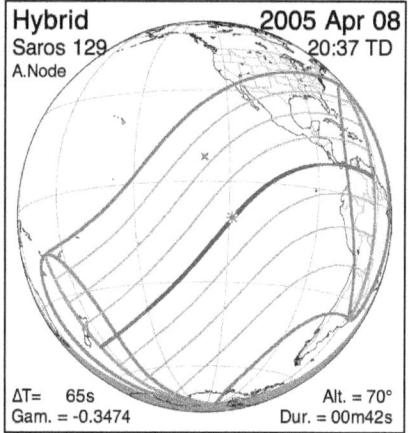

Hybrid     2005 Apr 08
Saros 129     20:37 TD
A.Node
ΔT= 65s     Alt. = 70°
Gam. = -0.3474     Dur. = 00m42s

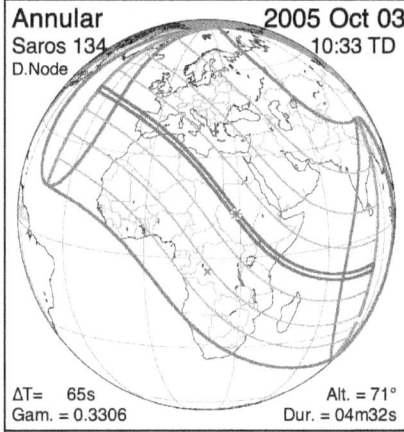

Annular     2005 Oct 03
Saros 134     10:33 TD
D.Node
ΔT= 65s     Alt. = 71°
Gam. = 0.3306     Dur. = 04m32s

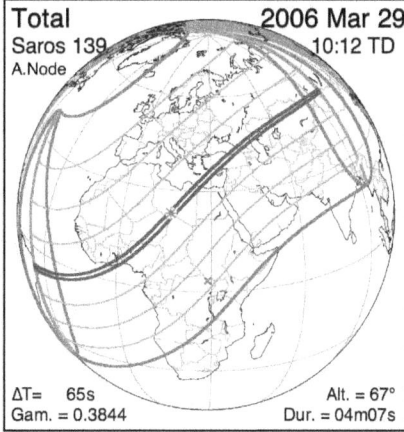

Total     2006 Mar 29
Saros 139     10:12 TD
A.Node
ΔT= 65s     Alt. = 67°
Gam. = 0.3844     Dur. = 04m07s

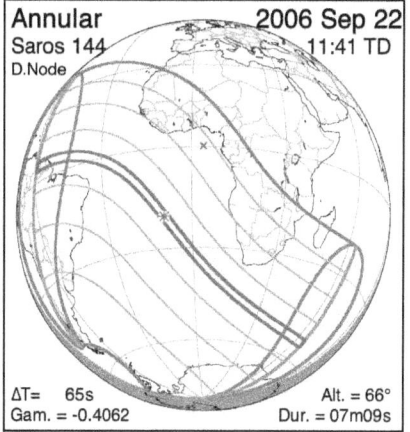

Annular     2006 Sep 22
Saros 144     11:41 TD
D.Node
ΔT= 65s     Alt. = 66°
Gam. = -0.4062     Dur. = 07m09s

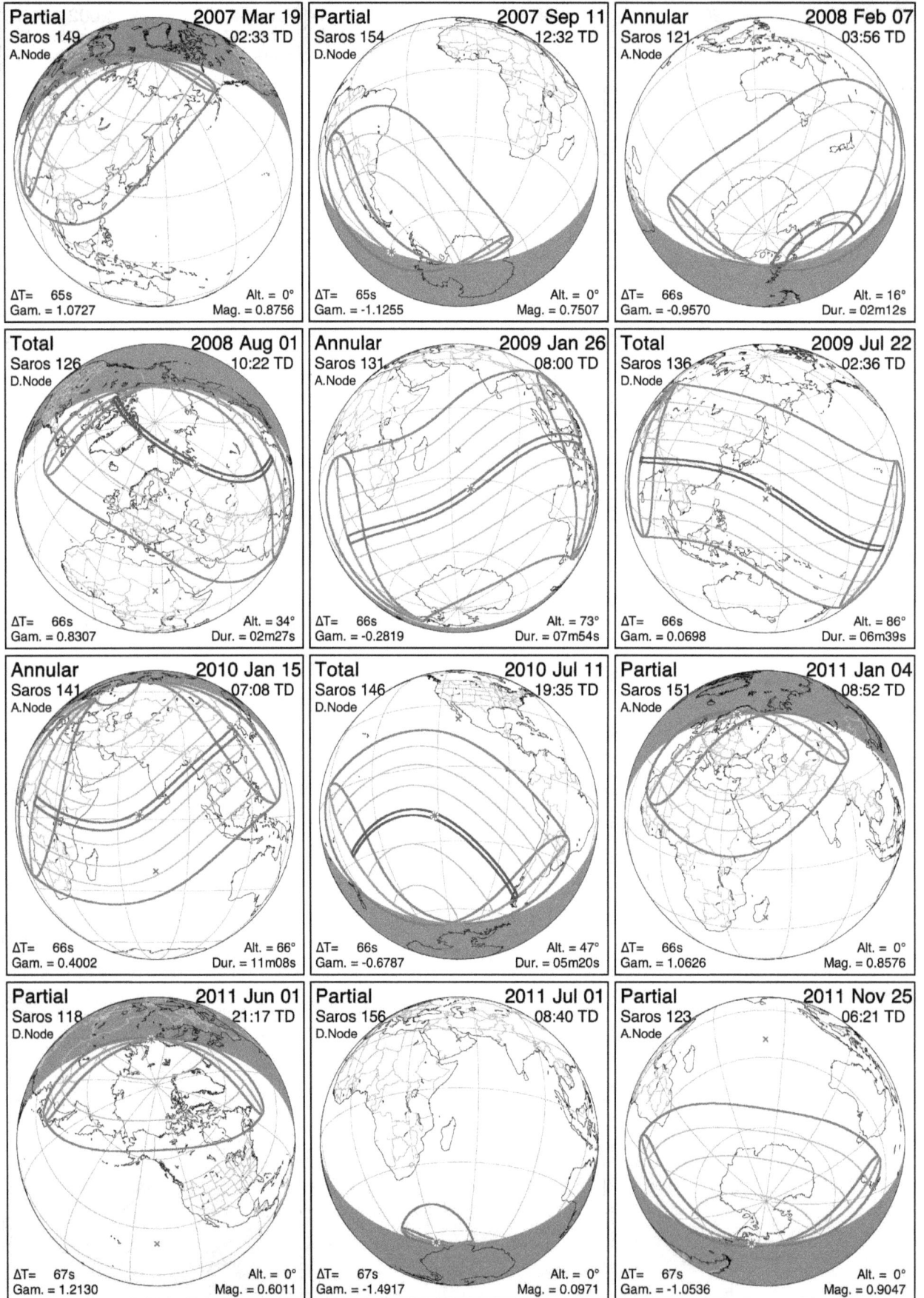

| Partial | 2007 Mar 19 |
|---|---|
| Saros 149 | 02:33 TD |
| A.Node | |
| ΔT= 65s | Alt. = 0° |
| Gam. = 1.0727 | Mag. = 0.8756 |

| Partial | 2007 Sep 11 |
|---|---|
| Saros 154 | 12:32 TD |
| D.Node | |
| ΔT= 65s | Alt. = 0° |
| Gam. = -1.1255 | Mag. = 0.7507 |

| Annular | 2008 Feb 07 |
|---|---|
| Saros 121 | 03:56 TD |
| A.Node | |
| ΔT= 66s | Alt. = 16° |
| Gam. = -0.9570 | Dur. = 02m12s |

| Total | 2008 Aug 01 |
|---|---|
| Saros 126 | 10:22 TD |
| D.Node | |
| ΔT= 66s | Alt. = 34° |
| Gam. = 0.8307 | Dur. = 02m27s |

| Annular | 2009 Jan 26 |
|---|---|
| Saros 131 | 08:00 TD |
| A.Node | |
| ΔT= 66s | Alt. = 73° |
| Gam. = -0.2819 | Dur. = 07m54s |

| Total | 2009 Jul 22 |
|---|---|
| Saros 136 | 02:36 TD |
| D.Node | |
| ΔT= 66s | Alt. = 86° |
| Gam. = 0.0698 | Dur. = 06m39s |

| Annular | 2010 Jan 15 |
|---|---|
| Saros 141 | 07:08 TD |
| A.Node | |
| ΔT= 66s | Alt. = 66° |
| Gam. = 0.4002 | Dur. = 11m08s |

| Total | 2010 Jul 11 |
|---|---|
| Saros 146 | 19:35 TD |
| D.Node | |
| ΔT= 66s | Alt. = 47° |
| Gam. = -0.6787 | Dur. = 05m20s |

| Partial | 2011 Jan 04 |
|---|---|
| Saros 151 | 08:52 TD |
| A.Node | |
| ΔT= 66s | Alt. = 0° |
| Gam. = 1.0626 | Mag. = 0.8576 |

| Partial | 2011 Jun 01 |
|---|---|
| Saros 118 | 21:17 TD |
| D.Node | |
| ΔT= 67s | Alt. = 0° |
| Gam. = 1.2130 | Mag. = 0.6011 |

| Partial | 2011 Jul 01 |
|---|---|
| Saros 156 | 08:40 TD |
| D.Node | |
| ΔT= 67s | Alt. = 0° |
| Gam. = -1.4917 | Mag. = 0.0971 |

| Partial | 2011 Nov 25 |
|---|---|
| Saros 123 | 06:21 TD |
| A.Node | |
| ΔT= 67s | Alt. = 0° |
| Gam. = -1.0536 | Mag. = 0.9047 |

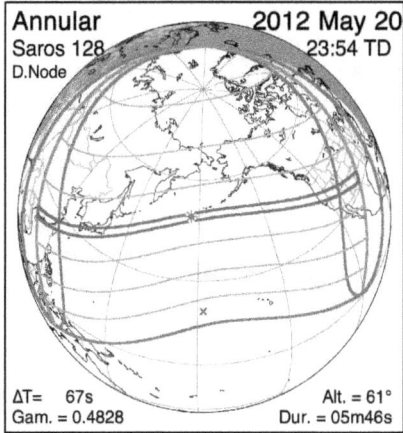

Annular      2012 May 20
Saros 128      23:54 TD
D.Node
ΔT= 67s      Alt. = 61°
Gam. = 0.4828      Dur. = 05m46s

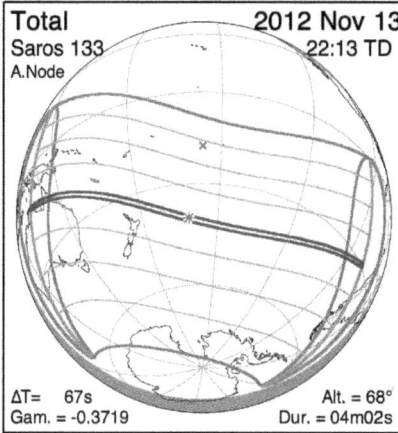

Total      2012 Nov 13
Saros 133      22:13 TD
A.Node
ΔT= 67s      Alt. = 68°
Gam. = -0.3719      Dur. = 04m02s

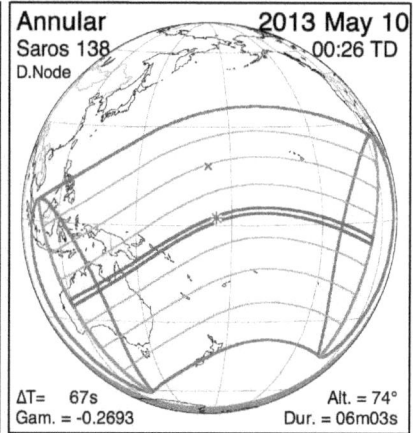

Annular      2013 May 10
Saros 138      00:26 TD
D.Node
ΔT= 67s      Alt. = 74°
Gam. = -0.2693      Dur. = 06m03s

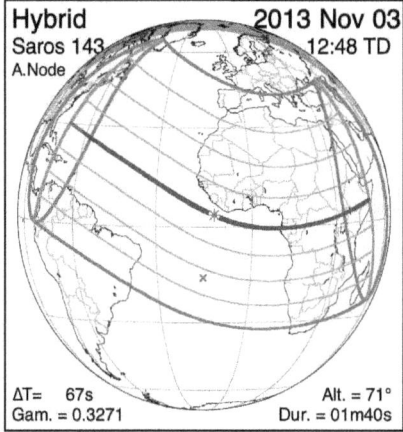

Hybrid      2013 Nov 03
Saros 143      12:48 TD
A.Node
ΔT= 67s      Alt. = 71°
Gam. = 0.3271      Dur. = 01m40s

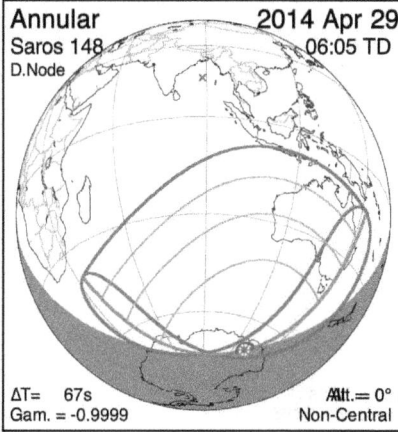

Annular      2014 Apr 29
Saros 148      06:05 TD
D.Node
ΔT= 67s      Alt. = 0°
Gam. = -0.9999      Non-Central

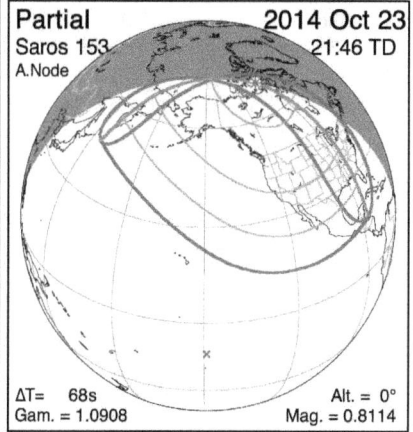

Partial      2014 Oct 23
Saros 153      21:46 TD
A.Node
ΔT= 68s      Alt. = 0°
Gam. = 1.0908      Mag. = 0.8114

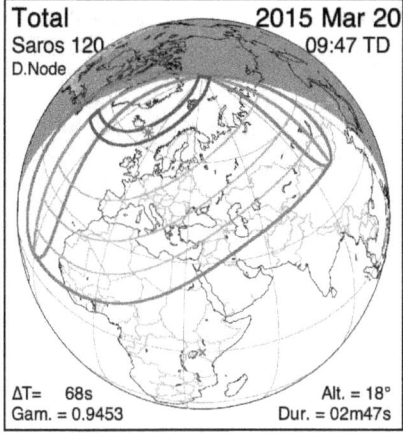

Total      2015 Mar 20
Saros 120      09:47 TD
D.Node
ΔT= 68s      Alt. = 18°
Gam. = 0.9453      Dur. = 02m47s

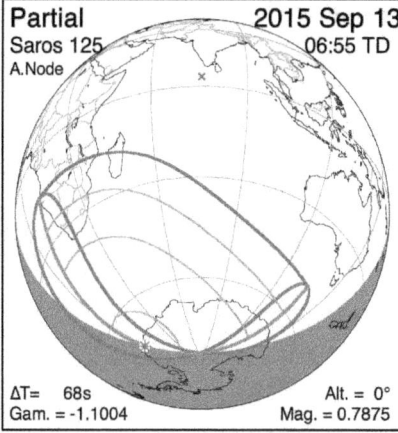

Partial      2015 Sep 13
Saros 125      06:55 TD
A.Node
ΔT= 68s      Alt. = 0°
Gam. = -1.1004      Mag. = 0.7875

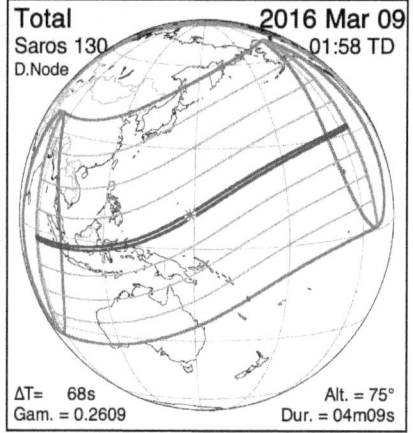

Total      2016 Mar 09
Saros 130      01:58 TD
D.Node
ΔT= 68s      Alt. = 75°
Gam. = 0.2609      Dur. = 04m09s

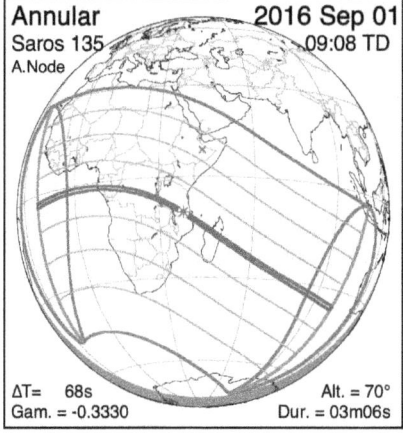

Annular      2016 Sep 01
Saros 135      09:08 TD
A.Node
ΔT= 68s      Alt. = 70°
Gam. = -0.3330      Dur. = 03m06s

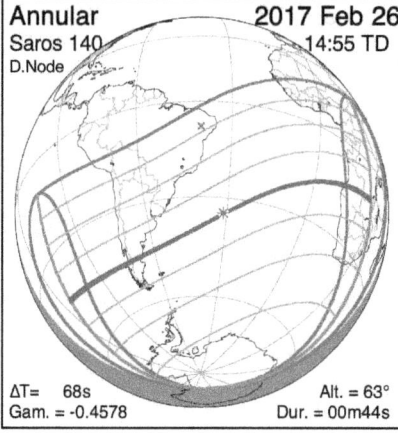

Annular      2017 Feb 26
Saros 140      14:55 TD
D.Node
ΔT= 68s      Alt. = 63°
Gam. = -0.4578      Dur. = 00m44s

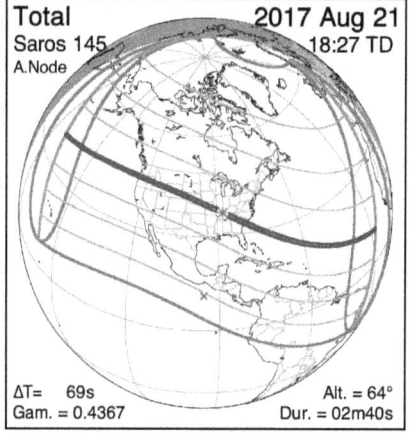

Total      2017 Aug 21
Saros 145      18:27 TD
A.Node
ΔT= 69s      Alt. = 64°
Gam. = 0.4367      Dur. = 02m40s

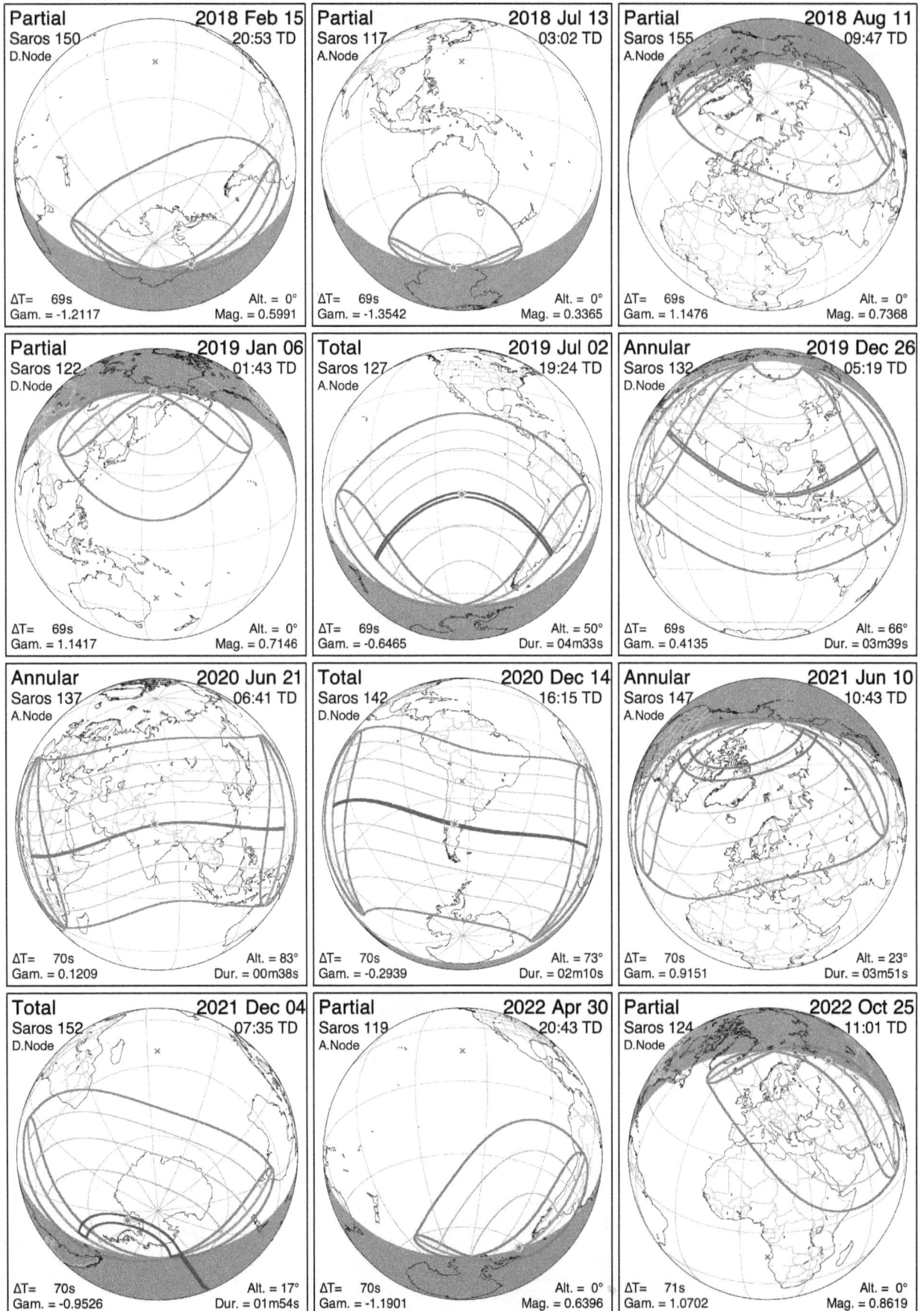

| Partial | 2018 Feb 15 |
|---|---|
| Saros 150 | 20:53 TD |
| D.Node | |
| ΔT= 69s | Alt. = 0° |
| Gam. = -1.2117 | Mag. = 0.5991 |

| Partial | 2018 Jul 13 |
|---|---|
| Saros 117 | 03:02 TD |
| A.Node | |
| ΔT= 69s | Alt. = 0° |
| Gam. = -1.3542 | Mag. = 0.3365 |

| Partial | 2018 Aug 11 |
|---|---|
| Saros 155 | 09:47 TD |
| A.Node | |
| ΔT= 69s | Alt. = 0° |
| Gam. = 1.1476 | Mag. = 0.7368 |

| Partial | 2019 Jan 06 |
|---|---|
| Saros 122 | 01:43 TD |
| D.Node | |
| ΔT= 69s | Alt. = 0° |
| Gam. = 1.1417 | Mag. = 0.7146 |

| Total | 2019 Jul 02 |
|---|---|
| Saros 127 | 19:24 TD |
| A.Node | |
| ΔT= 69s | Alt. = 50° |
| Gam. = -0.6465 | Dur. = 04m33s |

| Annular | 2019 Dec 26 |
|---|---|
| Saros 132 | 05:19 TD |
| D.Node | |
| ΔT= 69s | Alt. = 66° |
| Gam. = 0.4135 | Dur. = 03m39s |

| Annular | 2020 Jun 21 |
|---|---|
| Saros 137 | 06:41 TD |
| A.Node | |
| ΔT= 70s | Alt. = 83° |
| Gam. = 0.1209 | Dur. = 00m38s |

| Total | 2020 Dec 14 |
|---|---|
| Saros 142 | 16:15 TD |
| D.Node | |
| ΔT= 70s | Alt. = 73° |
| Gam. = -0.2939 | Dur. = 02m10s |

| Annular | 2021 Jun 10 |
|---|---|
| Saros 147 | 10:43 TD |
| A.Node | |
| ΔT= 70s | Alt. = 23° |
| Gam. = 0.9151 | Dur. = 03m51s |

| Total | 2021 Dec 04 |
|---|---|
| Saros 152 | 07:35 TD |
| D.Node | |
| ΔT= 70s | Alt. = 17° |
| Gam. = -0.9526 | Dur. = 01m54s |

| Partial | 2022 Apr 30 |
|---|---|
| Saros 119 | 20:43 TD |
| A.Node | |
| ΔT= 70s | Alt. = 0° |
| Gam. = -1.1901 | Mag. = 0.6396 |

| Partial | 2022 Oct 25 |
|---|---|
| Saros 124 | 11:01 TD |
| D.Node | |
| ΔT= 71s | Alt. = 0° |
| Gam. = 1.0702 | Mag. = 0.8619 |

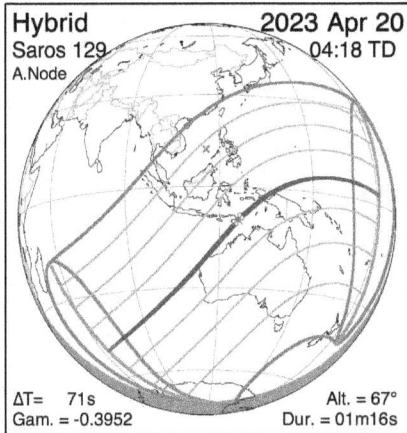

**Hybrid**     **2023 Apr 20**
Saros 129     04:18 TD
A.Node

ΔT= 71s     Alt. = 67°
Gam. = -0.3952     Dur. = 01m16s

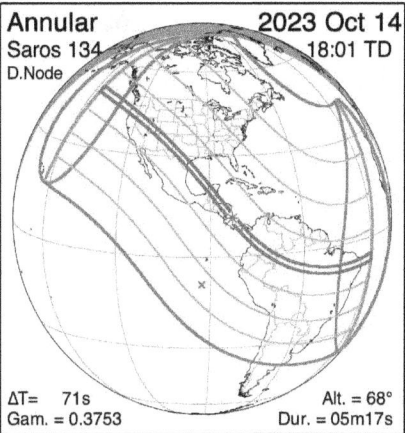

**Annular**     **2023 Oct 14**
Saros 134     18:01 TD
D.Node

ΔT= 71s     Alt. = 68°
Gam. = 0.3753     Dur. = 05m17s

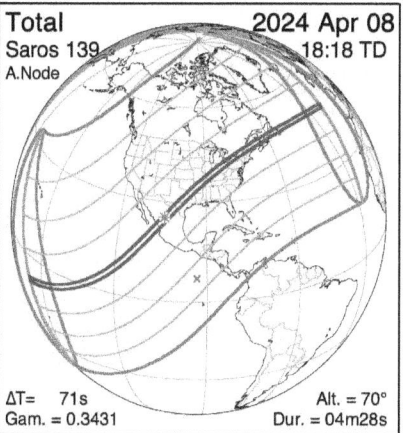

**Total**     **2024 Apr 08**
Saros 139     18:18 TD
A.Node

ΔT= 71s     Alt. = 70°
Gam. = 0.3431     Dur. = 04m28s

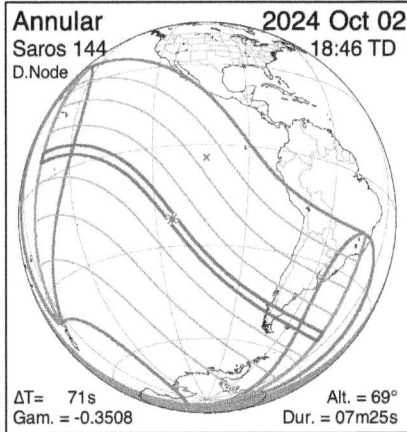

**Annular**     **2024 Oct 02**
Saros 144     18:46 TD
D.Node

ΔT= 71s     Alt. = 69°
Gam. = -0.3508     Dur. = 07m25s

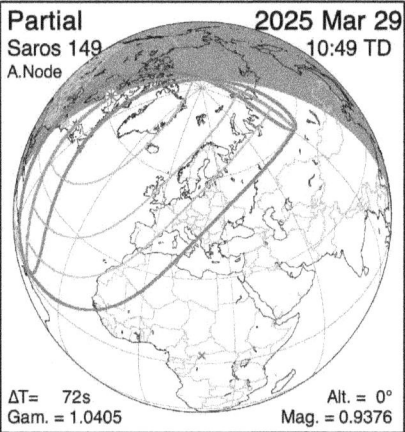

**Partial**     **2025 Mar 29**
Saros 149     10:49 TD
A.Node

ΔT= 72s     Alt. = 0°
Gam. = 1.0405     Mag. = 0.9376

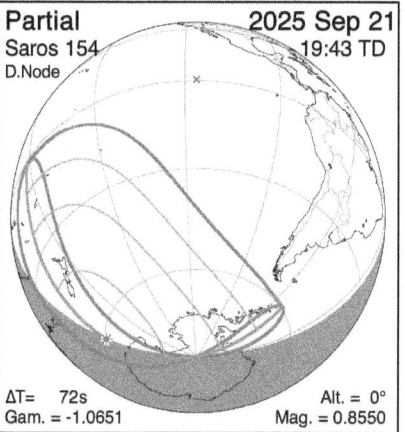

**Partial**     **2025 Sep 21**
Saros 154     19:43 TD
D.Node

ΔT= 72s     Alt. = 0°
Gam. = -1.0651     Mag. = 0.8550

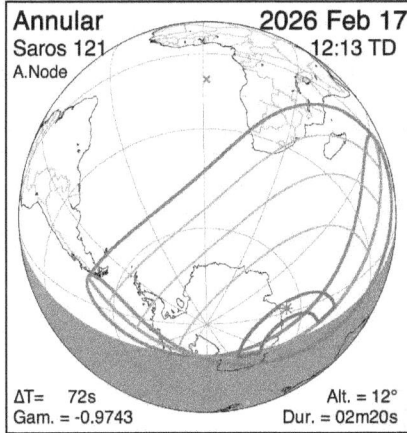

**Annular**     **2026 Feb 17**
Saros 121     12:13 TD
A.Node

ΔT= 72s     Alt. = 12°
Gam. = -0.9743     Dur. = 02m20s

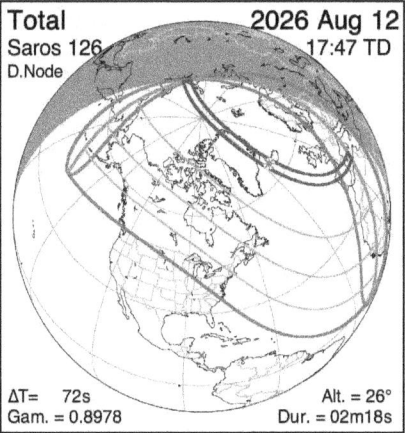

**Total**     **2026 Aug 12**
Saros 126     17:47 TD
D.Node

ΔT= 72s     Alt. = 26°
Gam. = 0.8978     Dur. = 02m18s

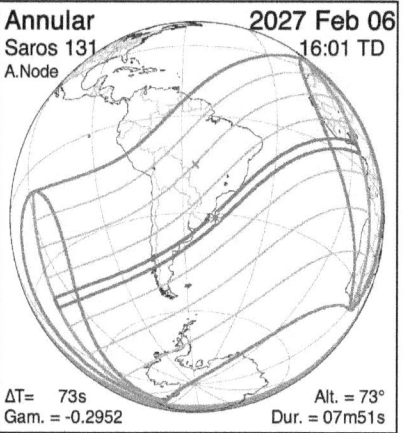

**Annular**     **2027 Feb 06**
Saros 131     16:01 TD
A.Node

ΔT= 73s     Alt. = 73°
Gam. = -0.2952     Dur. = 07m51s

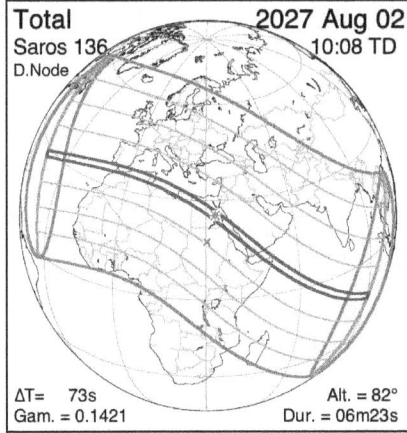

**Total**     **2027 Aug 02**
Saros 136     10:08 TD
D.Node

ΔT= 73s     Alt. = 82°
Gam. = 0.1421     Dur. = 06m23s

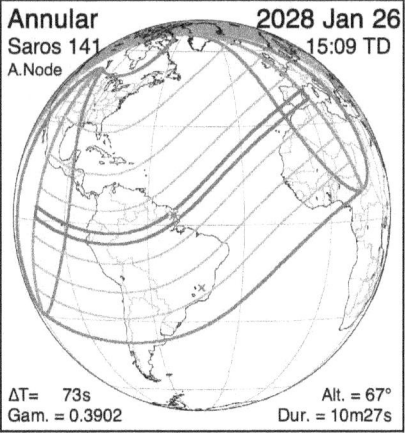

**Annular**     **2028 Jan 26**
Saros 141     15:09 TD
A.Node

ΔT= 73s     Alt. = 67°
Gam. = 0.3902     Dur. = 10m27s

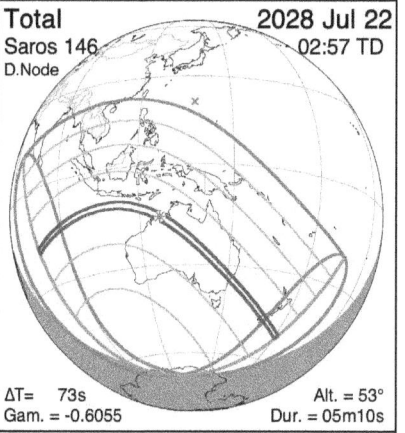

**Total**     **2028 Jul 22**
Saros 146     02:57 TD
D.Node

ΔT= 73s     Alt. = 53°
Gam. = -0.6055     Dur. = 05m10s

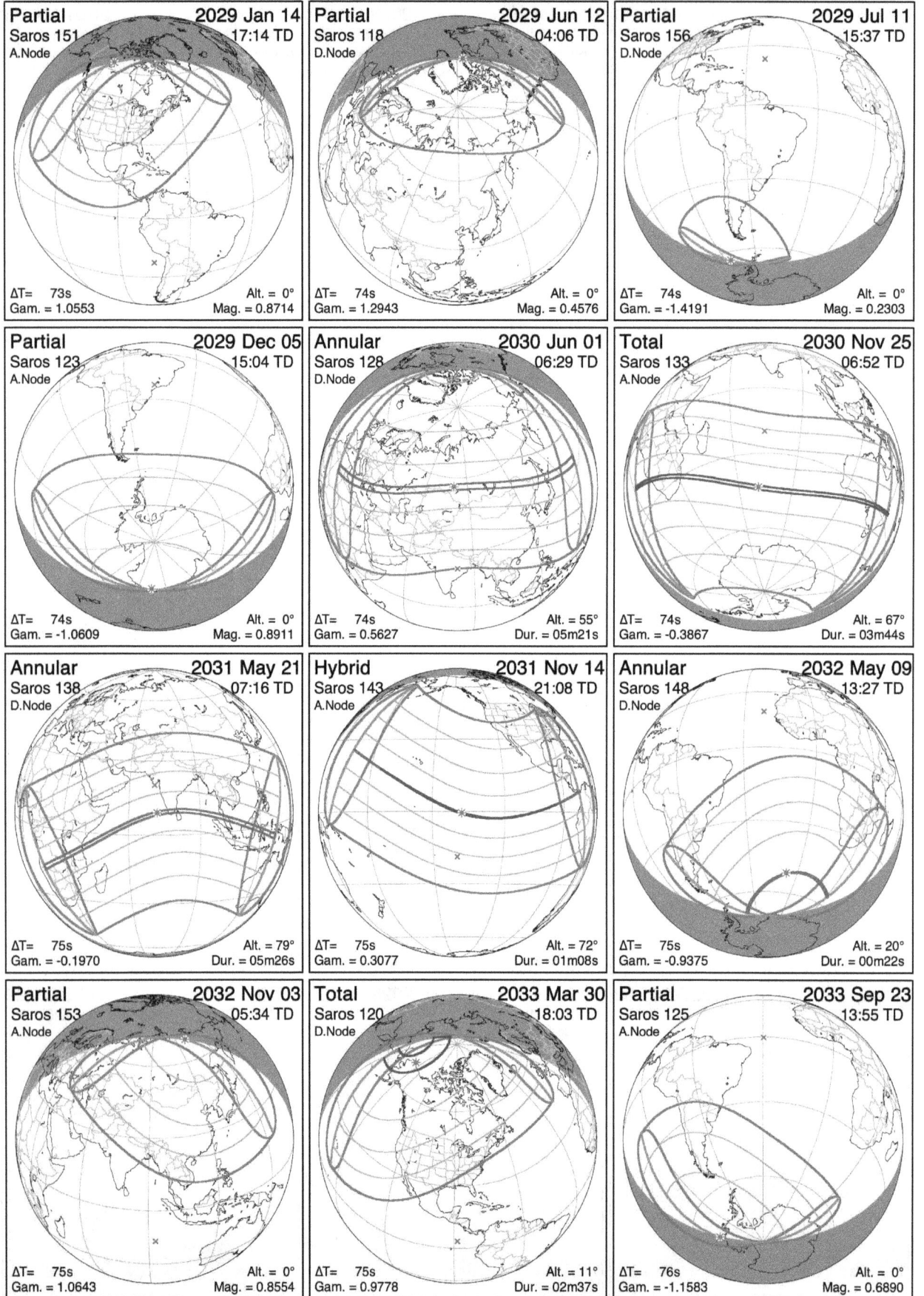

| Partial | 2029 Jan 14 |
|---|---|
| Saros 151 | 17:14 TD |
| A.Node | |

ΔT= 73s    Alt. = 0°
Gam. = 1.0553    Mag. = 0.8714

| Partial | 2029 Jun 12 |
|---|---|
| Saros 118 | 04:06 TD |
| D.Node | |

ΔT= 74s    Alt. = 0°
Gam. = 1.2943    Mag. = 0.4576

| Partial | 2029 Jul 11 |
|---|---|
| Saros 156 | 15:37 TD |
| D.Node | |

ΔT= 74s    Alt. = 0°
Gam. = -1.4191    Mag. = 0.2303

| Partial | 2029 Dec 05 |
|---|---|
| Saros 123 | 15:04 TD |
| A.Node | |

ΔT= 74s    Alt. = 0°
Gam. = -1.0609    Mag. = 0.8911

| Annular | 2030 Jun 01 |
|---|---|
| Saros 128 | 06:29 TD |
| D.Node | |

ΔT= 74s    Alt. = 55°
Gam. = 0.5627    Dur. = 05m21s

| Total | 2030 Nov 25 |
|---|---|
| Saros 133 | 06:52 TD |
| A.Node | |

ΔT= 74s    Alt. = 67°
Gam. = -0.3867    Dur. = 03m44s

| Annular | 2031 May 21 |
|---|---|
| Saros 138 | 07:16 TD |
| D.Node | |

ΔT= 75s    Alt. = 79°
Gam. = -0.1970    Dur. = 05m26s

| Hybrid | 2031 Nov 14 |
|---|---|
| Saros 143 | 21:08 TD |
| A.Node | |

ΔT= 75s    Alt. = 72°
Gam. = 0.3077    Dur. = 01m08s

| Annular | 2032 May 09 |
|---|---|
| Saros 148 | 13:27 TD |
| D.Node | |

ΔT= 75s    Alt. = 20°
Gam. = -0.9375    Dur. = 00m22s

| Partial | 2032 Nov 03 |
|---|---|
| Saros 153 | 05:34 TD |
| A.Node | |

ΔT= 75s    Alt. = 0°
Gam. = 1.0643    Mag. = 0.8554

| Total | 2033 Mar 30 |
|---|---|
| Saros 120 | 18:03 TD |
| D.Node | |

ΔT= 75s    Alt. = 11°
Gam. = 0.9778    Dur. = 02m37s

| Partial | 2033 Sep 23 |
|---|---|
| Saros 125 | 13:55 TD |
| A.Node | |

ΔT= 76s    Alt. = 0°
Gam. = -1.1583    Mag. = 0.6890

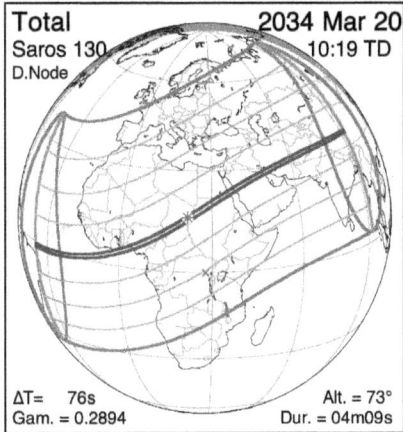

Total         2034 Mar 20
Saros 130       10:19 TD
D.Node

ΔT= 76s        Alt. = 73°
Gam. = 0.2894    Dur. = 04m09s

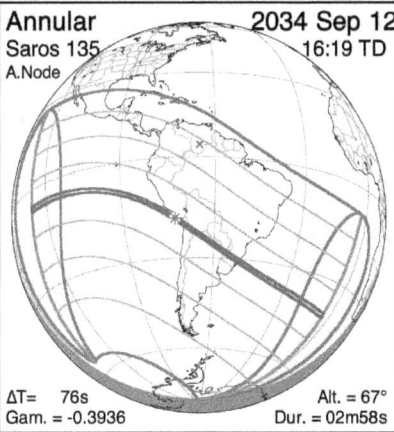

Annular       2034 Sep 12
Saros 135       16:19 TD
A.Node

ΔT= 76s        Alt. = 67°
Gam. = -0.3936    Dur. = 02m58s

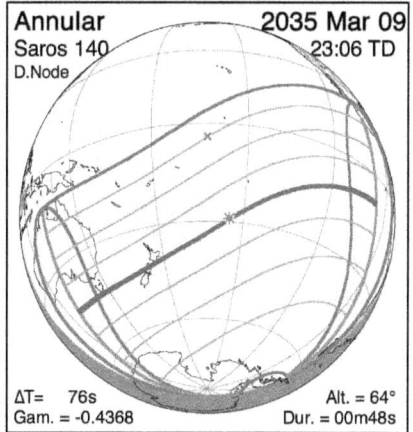

Annular       2035 Mar 09
Saros 140       23:06 TD
D.Node

ΔT= 76s        Alt. = 64°
Gam. = -0.4368    Dur. = 00m48s

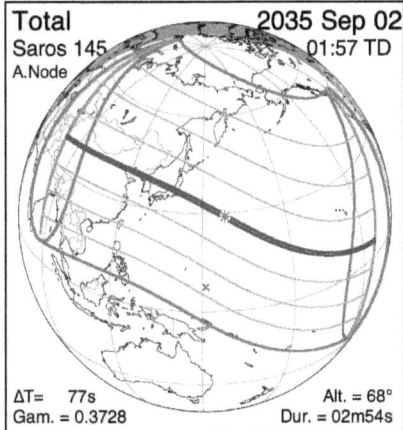

Total         2035 Sep 02
Saros 145       01:57 TD
A.Node

ΔT= 77s        Alt. = 68°
Gam. = 0.3728    Dur. = 02m54s

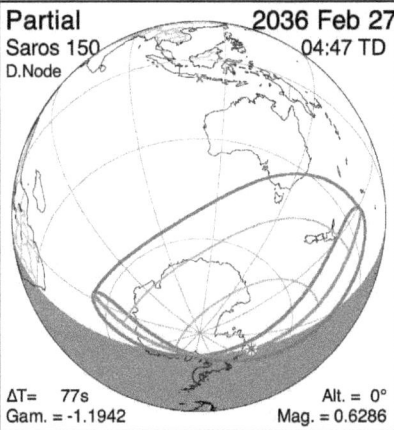

Partial        2036 Feb 27
Saros 150       04:47 TD
D.Node

ΔT= 77s        Alt. = 0°
Gam. = -1.1942    Mag. = 0.6286

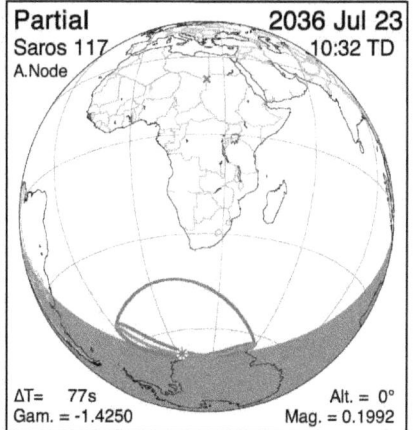

Partial        2036 Jul 23
Saros 117       10:32 TD
A.Node

ΔT= 77s        Alt. = 0°
Gam. = -1.4250    Mag. = 0.1992

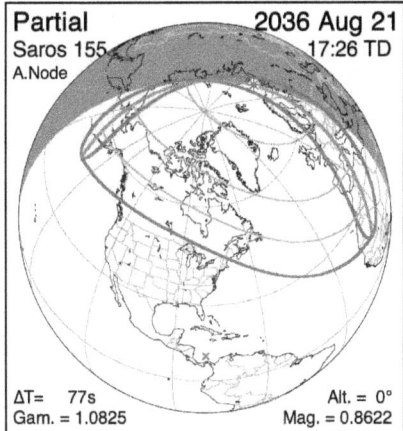

Partial        2036 Aug 21
Saros 155       17:26 TD
A.Node

ΔT= 77s        Alt. = 0°
Gam. = 1.0825    Mag. = 0.8622

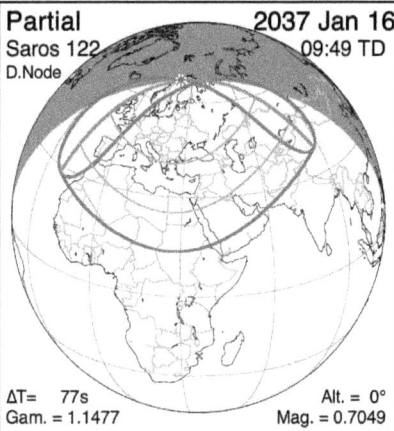

Partial        2037 Jan 16
Saros 122       09:49 TD
D.Node

ΔT= 77s        Alt. = 0°
Gam. = 1.1477    Mag. = 0.7049

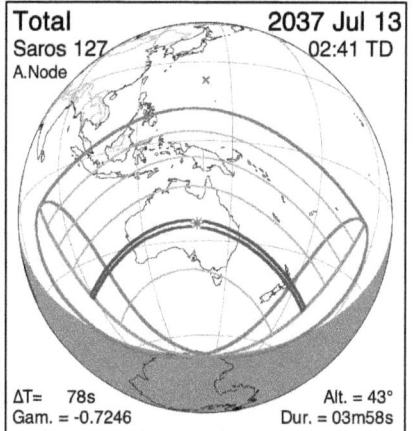

Total         2037 Jul 13
Saros 127       02:41 TD
A.Node

ΔT= 78s        Alt. = 43°
Gam. = -0.7246    Dur. = 03m58s

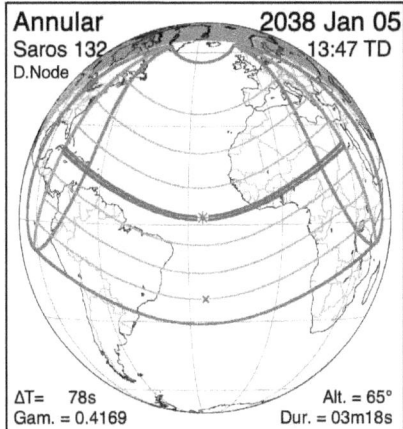

Annular       2038 Jan 05
Saros 132       13:47 TD
D.Node

ΔT= 78s        Alt. = 65°
Gam. = 0.4169    Dur. = 03m18s

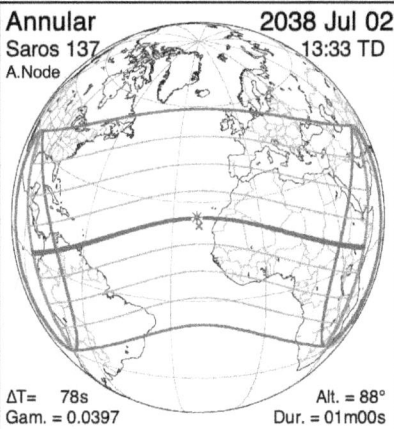

Annular       2038 Jul 02
Saros 137       13:33 TD
A.Node

ΔT= 78s        Alt. = 88°
Gam. = 0.0397    Dur. = 01m00s

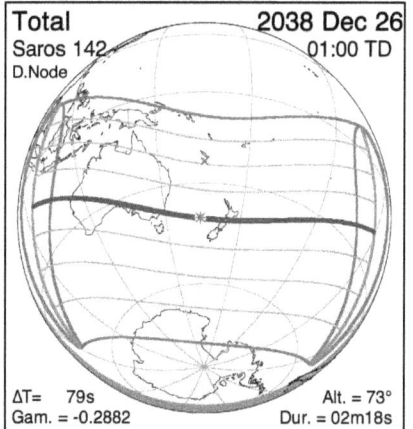

Total         2038 Dec 26
Saros 142       01:00 TD
D.Node

ΔT= 79s        Alt. = 73°
Gam. = -0.2882    Dur. = 02m18s

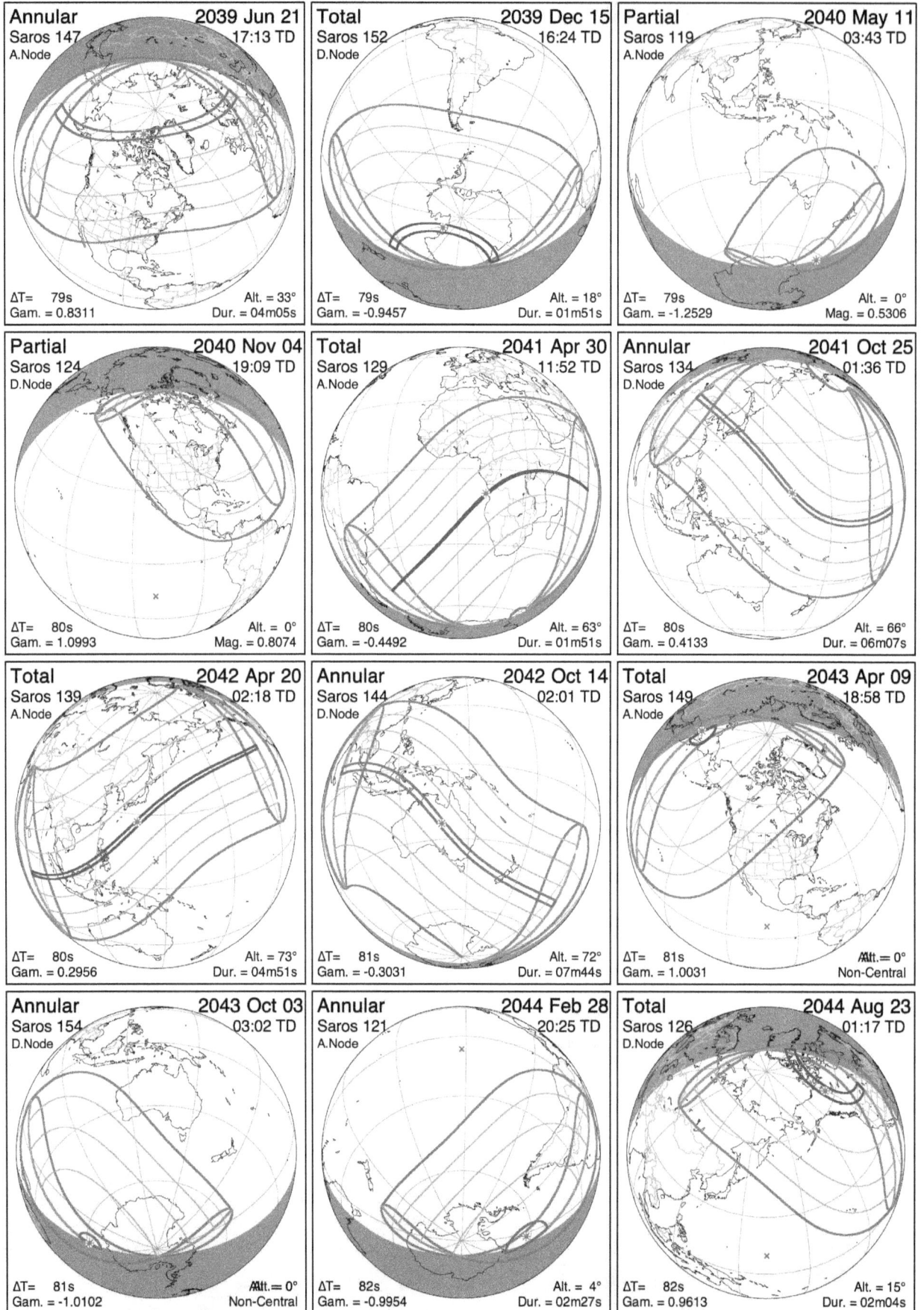

| Annular | 2039 Jun 21 |
|---|---|
| Saros 147 | 17:13 TD |
| A.Node | |
| ΔT= 79s | Alt. = 33° |
| Gam. = 0.8311 | Dur. = 04m05s |

| Total | 2039 Dec 15 |
|---|---|
| Saros 152 | 16:24 TD |
| D.Node | |
| ΔT= 79s | Alt. = 18° |
| Gam. = -0.9457 | Dur. = 01m51s |

| Partial | 2040 May 11 |
|---|---|
| Saros 119 | 03:43 TD |
| A.Node | |
| ΔT= 79s | Alt. = 0° |
| Gam. = -1.2529 | Mag. = 0.5306 |

| Partial | 2040 Nov 04 |
|---|---|
| Saros 124 | 19:09 TD |
| D.Node | |
| ΔT= 80s | Alt. = 0° |
| Gam. = 1.0993 | Mag. = 0.8074 |

| Total | 2041 Apr 30 |
|---|---|
| Saros 129 | 11:52 TD |
| A.Node | |
| ΔT= 80s | Alt. = 63° |
| Gam. = -0.4492 | Dur. = 01m51s |

| Annular | 2041 Oct 25 |
|---|---|
| Saros 134 | 01:36 TD |
| D.Node | |
| ΔT= 80s | Alt. = 66° |
| Gam. = 0.4133 | Dur. = 06m07s |

| Total | 2042 Apr 20 |
|---|---|
| Saros 139 | 02:18 TD |
| A.Node | |
| ΔT= 80s | Alt. = 73° |
| Gam. = 0.2956 | Dur. = 04m51s |

| Annular | 2042 Oct 14 |
|---|---|
| Saros 144 | 02:01 TD |
| D.Node | |
| ΔT= 81s | Alt. = 72° |
| Gam. = -0.3031 | Dur. = 07m44s |

| Total | 2043 Apr 09 |
|---|---|
| Saros 149 | 18:58 TD |
| A.Node | |
| ΔT= 81s | Alt.= 0° |
| Gam. = 1.0031 | Non-Central |

| Annular | 2043 Oct 03 |
|---|---|
| Saros 154 | 03:02 TD |
| D.Node | |
| ΔT= 81s | Alt.= 0° |
| Gam. = -1.0102 | Non-Central |

| Annular | 2044 Feb 28 |
|---|---|
| Saros 121 | 20:25 TD |
| A.Node | |
| ΔT= 82s | Alt. = 4° |
| Gam. = -0.9954 | Dur. = 02m27s |

| Total | 2044 Aug 23 |
|---|---|
| Saros 126 | 01:17 TD |
| D.Node | |
| ΔT= 82s | Alt. = 15° |
| Gam. = 0.9613 | Dur. = 02m04s |

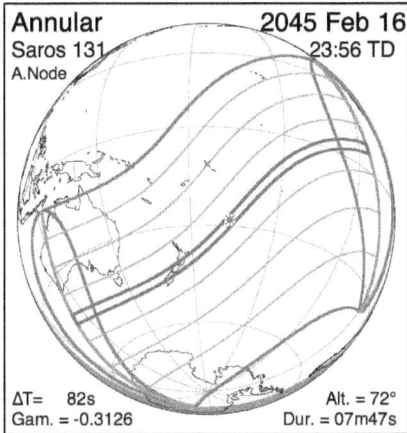

Annular     2045 Feb 16
Saros 131     23:56 TD
A.Node
ΔT= 82s     Alt. = 72°
Gam. = -0.3126     Dur. = 07m47s

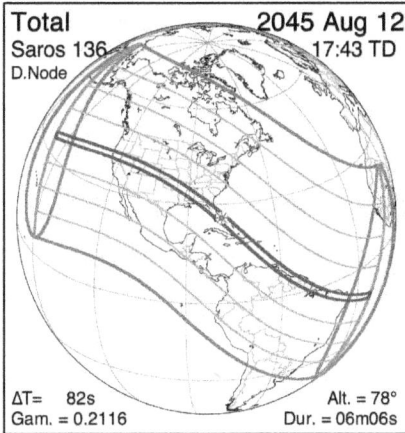

Total     2045 Aug 12
Saros 136     17:43 TD
D.Node
ΔT= 82s     Alt. = 78°
Gam. = 0.2116     Dur. = 06m06s

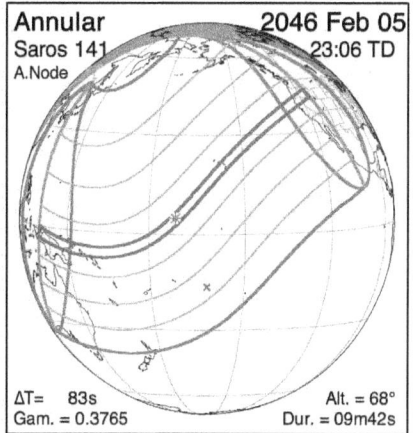

Annular     2046 Feb 05
Saros 141     23:06 TD
A.Node
ΔT= 83s     Alt. = 68°
Gam. = 0.3765     Dur. = 09m42s

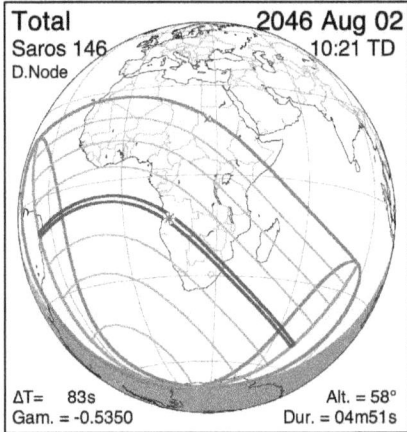

Total     2046 Aug 02
Saros 146     10:21 TD
D.Node
ΔT= 83s     Alt. = 58°
Gam. = -0.5350     Dur. = 04m51s

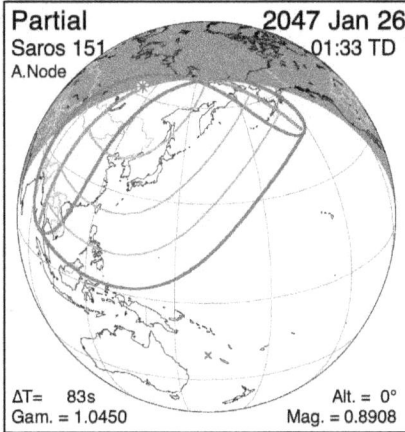

Partial     2047 Jan 26
Saros 151     01:33 TD
A.Node
ΔT= 83s     Alt. = 0°
Gam. = 1.0450     Mag. = 0.8908

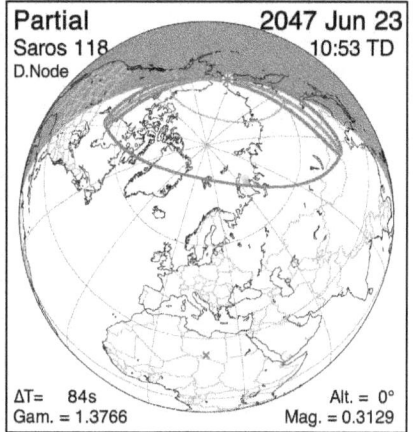

Partial     2047 Jun 23
Saros 118     10:53 TD
D.Node
ΔT= 84s     Alt. = 0°
Gam. = 1.3766     Mag. = 0.3129

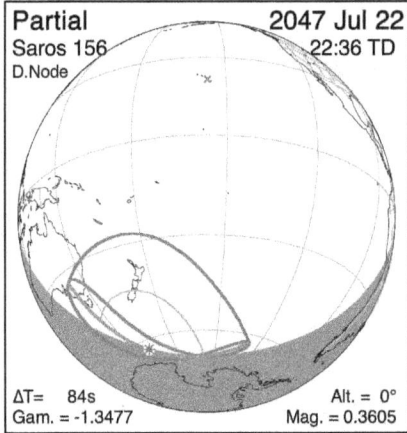

Partial     2047 Jul 22
Saros 156     22:36 TD
D.Node
ΔT= 84s     Alt. = 0°
Gam. = -1.3477     Mag. = 0.3605

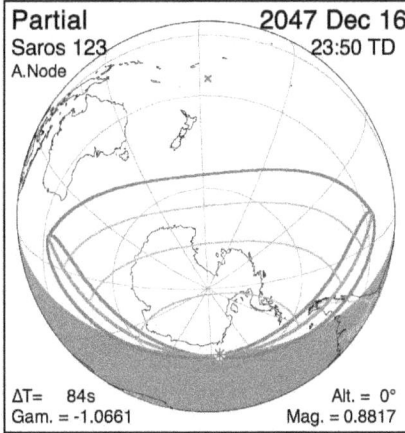

Partial     2047 Dec 16
Saros 123     23:50 TD
A.Node
ΔT= 84s     Alt. = 0°
Gam. = -1.0661     Mag. = 0.8817

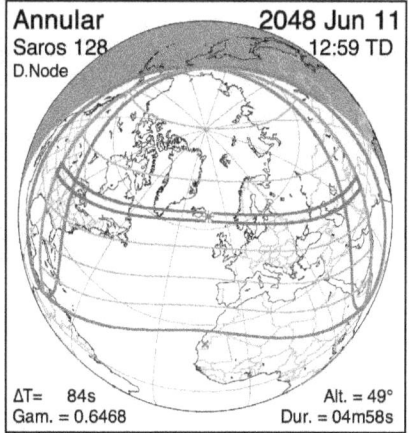

Annular     2048 Jun 11
Saros 128     12:59 TD
D.Node
ΔT= 84s     Alt. = 49°
Gam. = 0.6468     Dur. = 04m58s

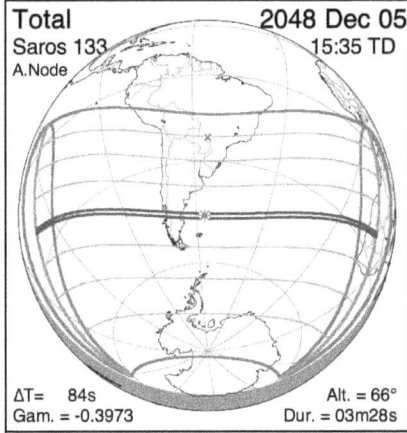

Total     2048 Dec 05
Saros 133     15:35 TD
A.Node
ΔT= 84s     Alt. = 66°
Gam. = -0.3973     Dur. = 03m28s

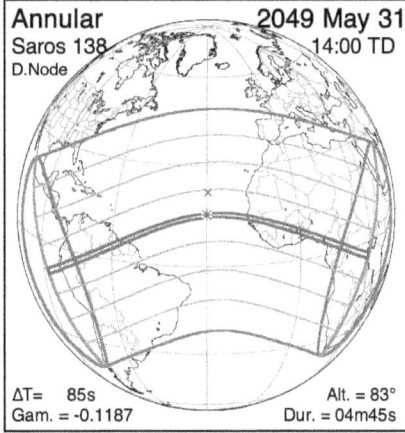

Annular     2049 May 31
Saros 138     14:00 TD
D.Node
ΔT= 85s     Alt. = 83°
Gam. = -0.1187     Dur. = 04m45s

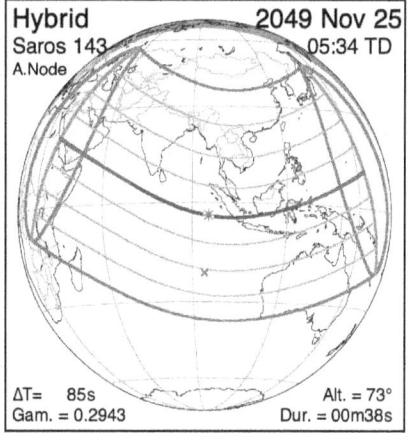

Hybrid     2049 Nov 25
Saros 143     05:34 TD
A.Node
ΔT= 85s     Alt. = 73°
Gam. = 0.2943     Dur. = 00m38s

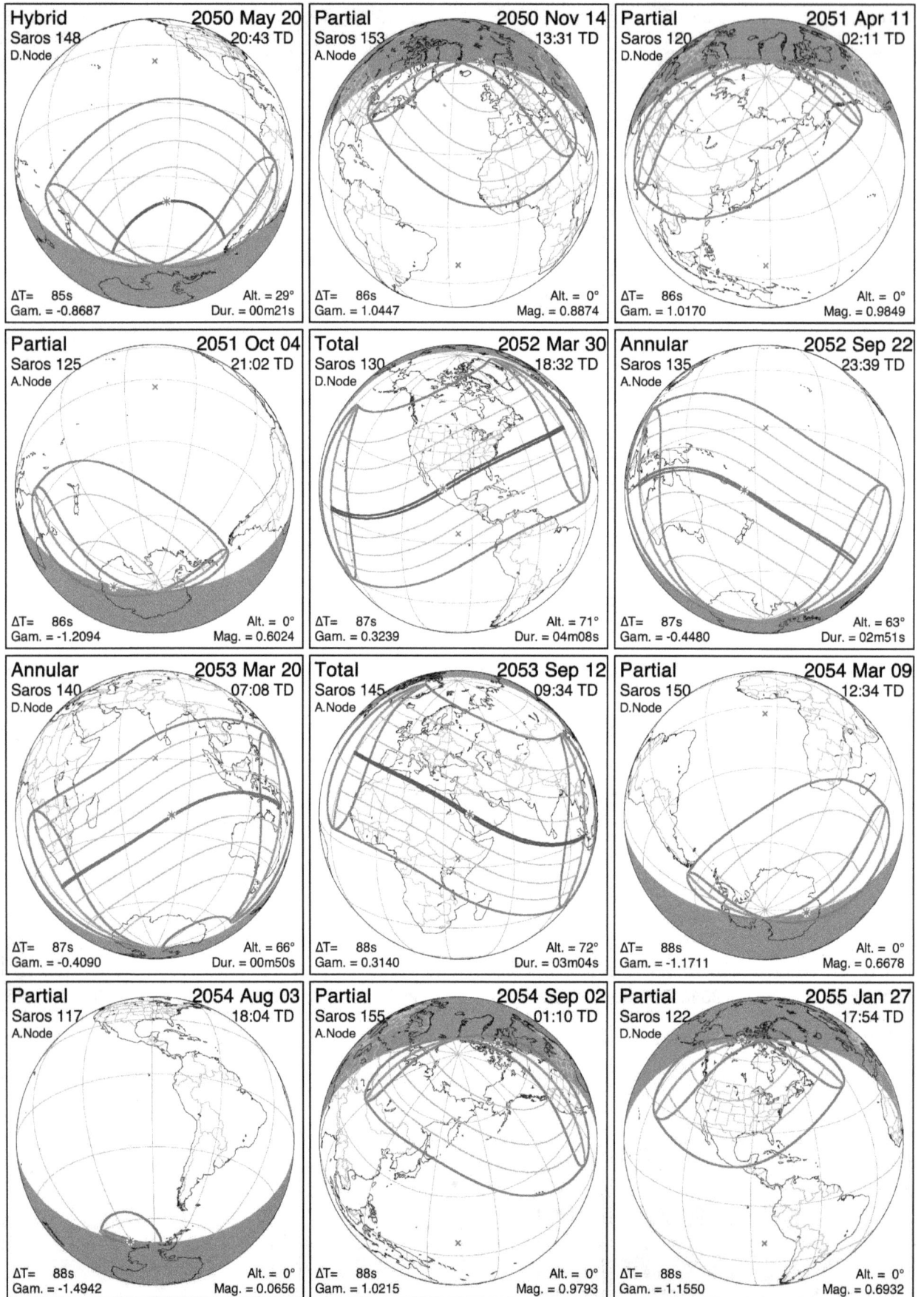

| Hybrid | 2050 May 20 |
|---|---|
| Saros 148 | 20:43 TD |
| D.Node | |
| ΔT= 85s | Alt. = 29° |
| Gam. = -0.8687 | Dur. = 00m21s |

| Partial | 2050 Nov 14 |
|---|---|
| Saros 153 | 13:31 TD |
| A.Node | |
| ΔT= 86s | Alt. = 0° |
| Gam. = 1.0447 | Mag. = 0.8874 |

| Partial | 2051 Apr 11 |
|---|---|
| Saros 120 | 02:11 TD |
| D.Node | |
| ΔT= 86s | Alt. = 0° |
| Gam. = 1.0170 | Mag. = 0.9849 |

| Partial | 2051 Oct 04 |
|---|---|
| Saros 125 | 21:02 TD |
| A.Node | |
| ΔT= 86s | Alt. = 0° |
| Gam. = -1.2094 | Mag. = 0.6024 |

| Total | 2052 Mar 30 |
|---|---|
| Saros 130 | 18:32 TD |
| D.Node | |
| ΔT= 87s | Alt. = 71° |
| Gam. = 0.3239 | Dur. = 04m08s |

| Annular | 2052 Sep 22 |
|---|---|
| Saros 135 | 23:39 TD |
| A.Node | |
| ΔT= 87s | Alt. = 63° |
| Gam. = -0.4480 | Dur. = 02m51s |

| Annular | 2053 Mar 20 |
|---|---|
| Saros 140 | 07:08 TD |
| D.Node | |
| ΔT= 87s | Alt. = 66° |
| Gam. = -0.4090 | Dur. = 00m50s |

| Total | 2053 Sep 12 |
|---|---|
| Saros 145 | 09:34 TD |
| A.Node | |
| ΔT= 88s | Alt. = 72° |
| Gam. = 0.3140 | Dur. = 03m04s |

| Partial | 2054 Mar 09 |
|---|---|
| Saros 150 | 12:34 TD |
| D.Node | |
| ΔT= 88s | Alt. = 0° |
| Gam. = -1.1711 | Mag. = 0.6678 |

| Partial | 2054 Aug 03 |
|---|---|
| Saros 117 | 18:04 TD |
| A.Node | |
| ΔT= 88s | Alt. = 0° |
| Gam. = -1.4942 | Mag. = 0.0656 |

| Partial | 2054 Sep 02 |
|---|---|
| Saros 155 | 01:10 TD |
| A.Node | |
| ΔT= 88s | Alt. = 0° |
| Gam. = 1.0215 | Mag. = 0.9793 |

| Partial | 2055 Jan 27 |
|---|---|
| Saros 122 | 17:54 TD |
| D.Node | |
| ΔT= 88s | Alt. = 0° |
| Gam. = 1.1550 | Mag. = 0.6932 |

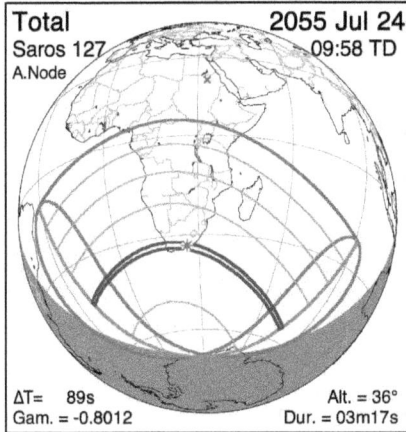

**Total**      **2055 Jul 24**
Saros 127      09:58 TD
A.Node

ΔT= 89s      Alt. = 36°
Gam. = -0.8012      Dur. = 03m17s

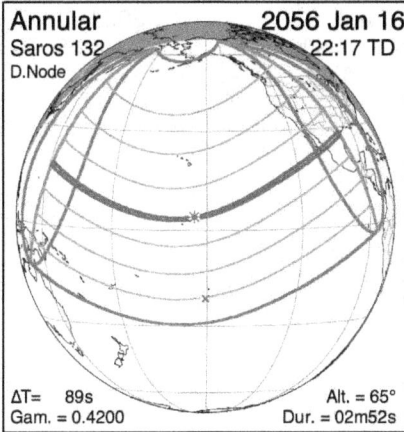

**Annular**      **2056 Jan 16**
Saros 132      22:17 TD
D.Node

ΔT= 89s      Alt. = 65°
Gam. = 0.4200      Dur. = 02m52s

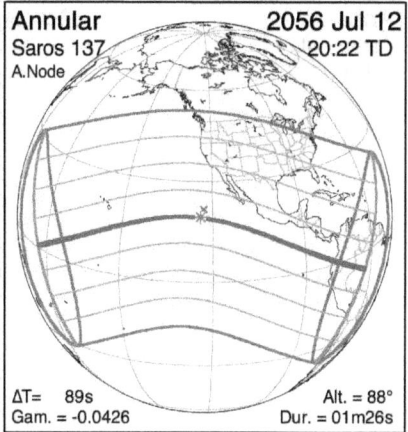

**Annular**      **2056 Jul 12**
Saros 137      20:22 TD
A.Node

ΔT= 89s      Alt. = 88°
Gam. = -0.0426      Dur. = 01m26s

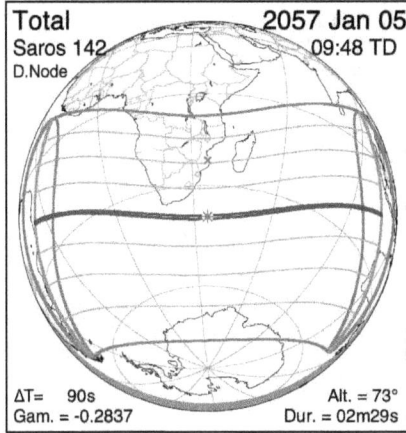

**Total**      **2057 Jan 05**
Saros 142      09:48 TD
D.Node

ΔT= 90s      Alt. = 73°
Gam. = -0.2837      Dur. = 02m29s

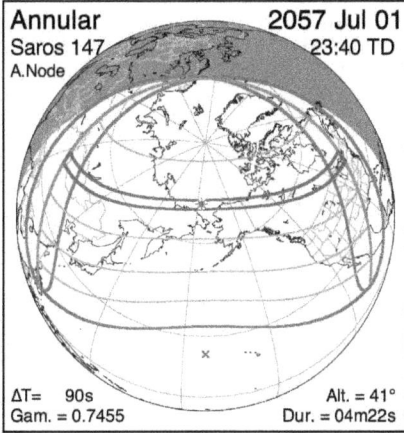

**Annular**      **2057 Jul 01**
Saros 147      23:40 TD
A.Node

ΔT= 90s      Alt. = 41°
Gam. = 0.7455      Dur. = 04m22s

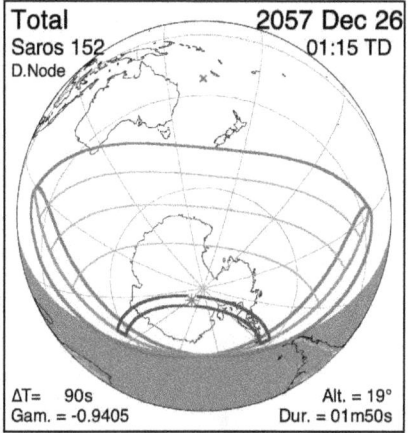

**Total**      **2057 Dec 26**
Saros 152      01:15 TD
D.Node

ΔT= 90s      Alt. = 19°
Gam. = -0.9405      Dur. = 01m50s

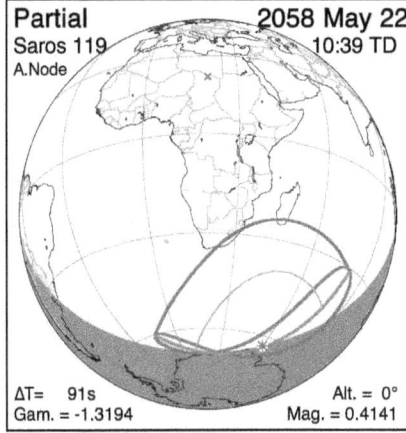

**Partial**      **2058 May 22**
Saros 119      10:39 TD
A.Node

ΔT= 91s      Alt. = 0°
Gam. = -1.3194      Mag. = 0.4141

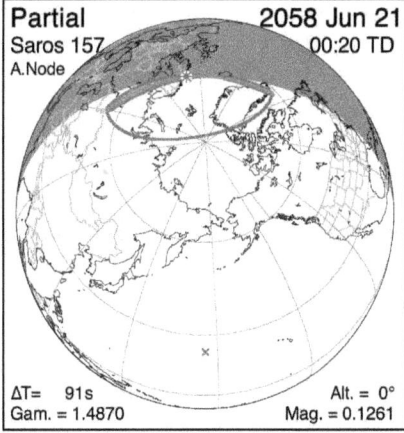

**Partial**      **2058 Jun 21**
Saros 157      00:20 TD
A.Node

ΔT= 91s      Alt. = 0°
Gam. = 1.4870      Mag. = 0.1261

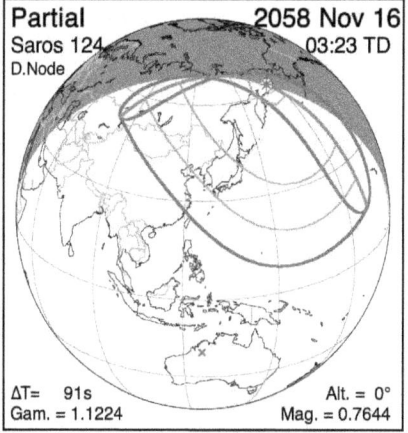

**Partial**      **2058 Nov 16**
Saros 124      03:23 TD
D.Node

ΔT= 91s      Alt. = 0°
Gam. = 1.1224      Mag. = 0.7644

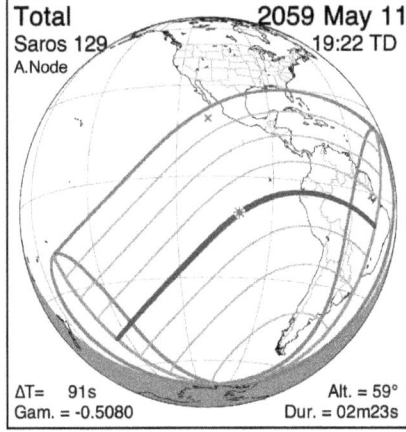

**Total**      **2059 May 11**
Saros 129      19:22 TD
A.Node

ΔT= 91s      Alt. = 59°
Gam. = -0.5080      Dur. = 02m23s

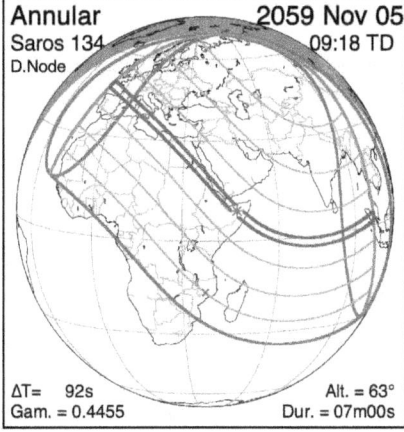

**Annular**      **2059 Nov 05**
Saros 134      09:18 TD
D.Node

ΔT= 92s      Alt. = 63°
Gam. = 0.4455      Dur. = 07m00s

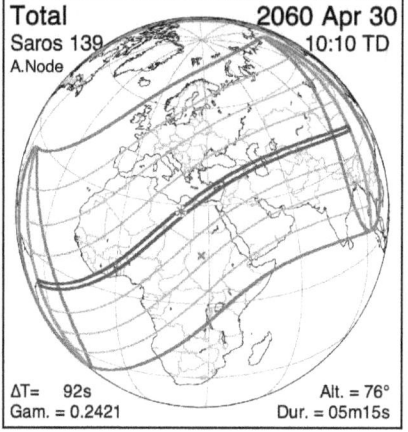

**Total**      **2060 Apr 30**
Saros 139      10:10 TD
A.Node

ΔT= 92s      Alt. = 76°
Gam. = 0.2421      Dur. = 05m15s

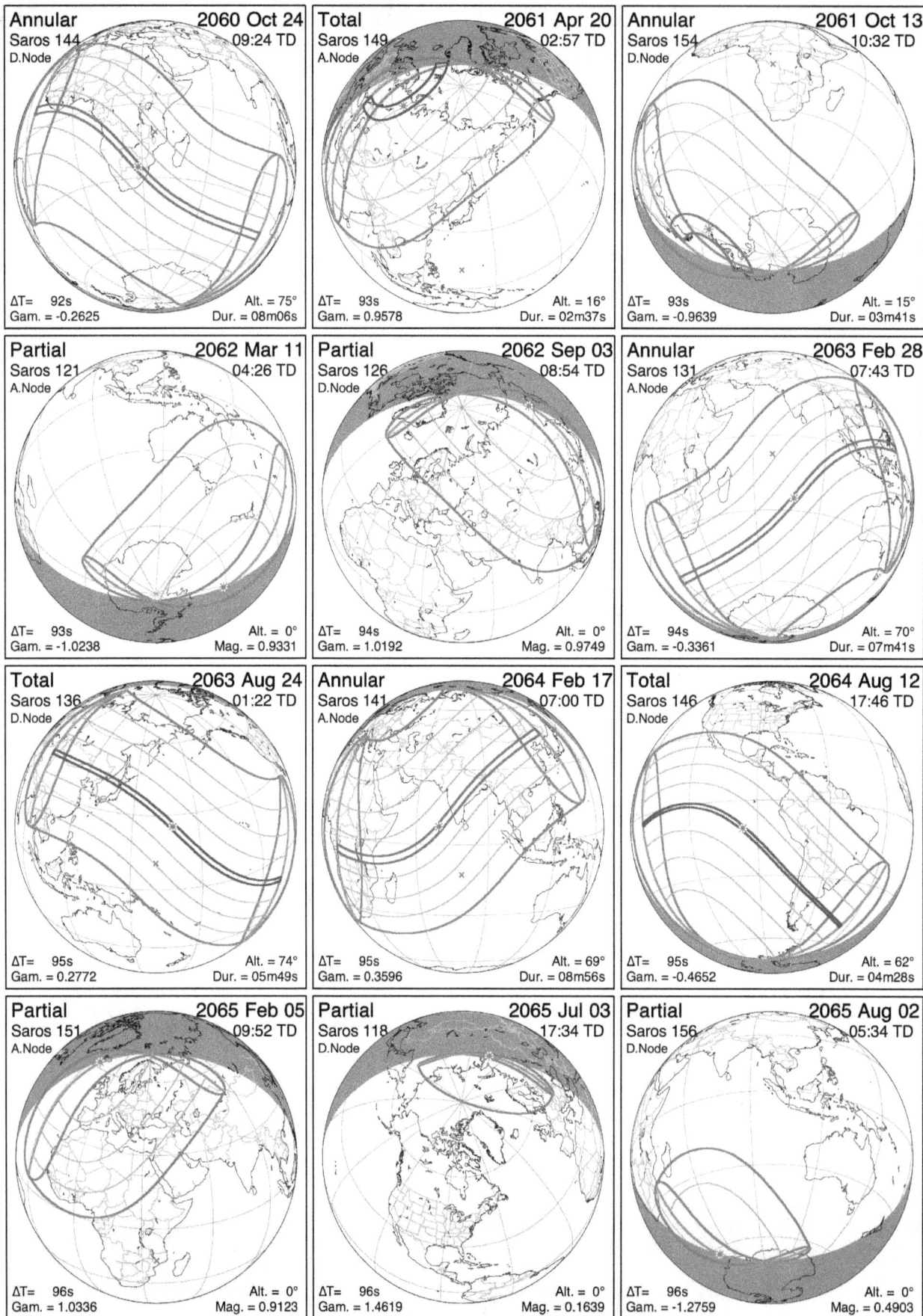

| Annular | 2060 Oct 24 |
|---|---|
| Saros 144 | 09:24 TD |
| D.Node | |
| ΔT= 92s | Alt. = 75° |
| Gam. = -0.2625 | Dur. = 08m06s |

| Total | 2061 Apr 20 |
|---|---|
| Saros 149 | 02:57 TD |
| A.Node | |
| ΔT= 93s | Alt. = 16° |
| Gam. = 0.9578 | Dur. = 02m37s |

| Annular | 2061 Oct 13 |
|---|---|
| Saros 154 | 10:32 TD |
| D.Node | |
| ΔT= 93s | Alt. = 15° |
| Gam. = -0.9639 | Dur. = 03m41s |

| Partial | 2062 Mar 11 |
|---|---|
| Saros 121 | 04:26 TD |
| A.Node | |
| ΔT= 93s | Alt. = 0° |
| Gam. = -1.0238 | Mag. = 0.9331 |

| Partial | 2062 Sep 03 |
|---|---|
| Saros 126 | 08:54 TD |
| D.Node | |
| ΔT= 94s | Alt. = 0° |
| Gam. = 1.0192 | Mag. = 0.9749 |

| Annular | 2063 Feb 28 |
|---|---|
| Saros 131 | 07:43 TD |
| A.Node | |
| ΔT= 94s | Alt. = 70° |
| Gam. = -0.3361 | Dur. = 07m41s |

| Total | 2063 Aug 24 |
|---|---|
| Saros 136 | 01:22 TD |
| D.Node | |
| ΔT= 95s | Alt. = 74° |
| Gam. = 0.2772 | Dur. = 05m49s |

| Annular | 2064 Feb 17 |
|---|---|
| Saros 141 | 07:00 TD |
| A.Node | |
| ΔT= 95s | Alt. = 69° |
| Gam. = 0.3596 | Dur. = 08m56s |

| Total | 2064 Aug 12 |
|---|---|
| Saros 146 | 17:46 TD |
| D.Node | |
| ΔT= 95s | Alt. = 62° |
| Gam. = -0.4652 | Dur. = 04m28s |

| Partial | 2065 Feb 05 |
|---|---|
| Saros 151 | 09:52 TD |
| A.Node | |
| ΔT= 96s | Alt. = 0° |
| Gam. = 1.0336 | Mag. = 0.9123 |

| Partial | 2065 Jul 03 |
|---|---|
| Saros 118 | 17:34 TD |
| D.Node | |
| ΔT= 96s | Alt. = 0° |
| Gam. = 1.4619 | Mag. = 0.1639 |

| Partial | 2065 Aug 02 |
|---|---|
| Saros 156 | 05:34 TD |
| D.Node | |
| ΔT= 96s | Alt. = 0° |
| Gam. = -1.2759 | Mag. = 0.4903 |

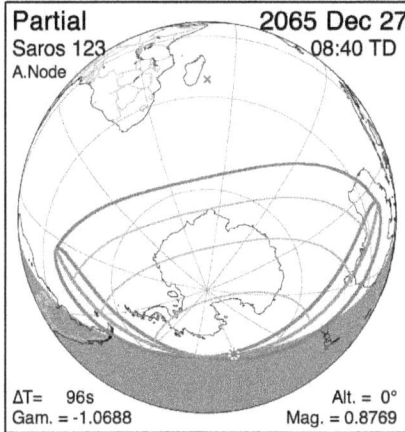

Partial      2065 Dec 27
Saros 123      08:40 TD
A.Node

ΔT= 96s      Alt. = 0°
Gam. = -1.0688      Mag. = 0.8769

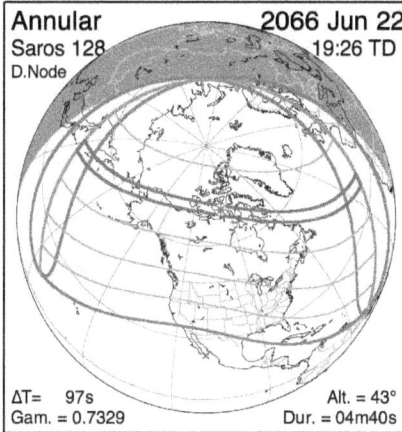

Annular      2066 Jun 22
Saros 128      19:26 TD
D.Node

ΔT= 97s      Alt. = 43°
Gam. = 0.7329      Mag. = 0.8769

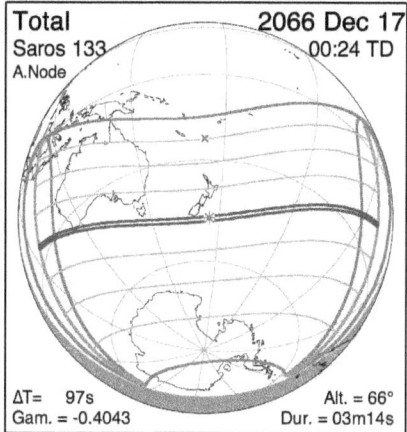

Total      2066 Dec 17
Saros 133      00:24 TD
A.Node

ΔT= 97s      Alt. = 66°
Gam. = -0.4043      Dur. = 03m14s

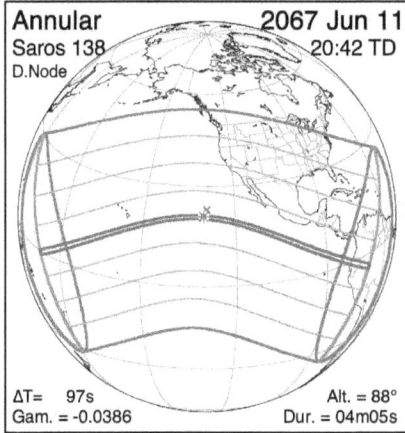

Annular      2067 Jun 11
Saros 138      20:42 TD
D.Node

ΔT= 97s      Alt. = 88°
Gam. = -0.0386      Dur. = 04m05s

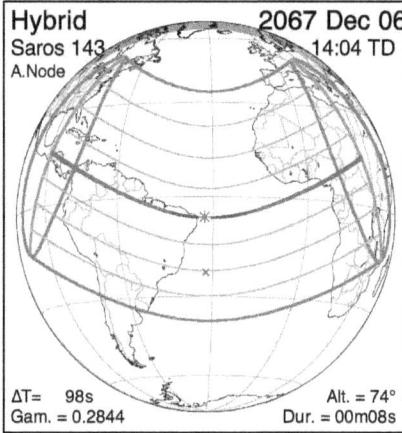

Hybrid      2067 Dec 06
Saros 143      14:04 TD
A.Node

ΔT= 98s      Alt. = 74°
Gam. = 0.2844      Dur. = 00m08s

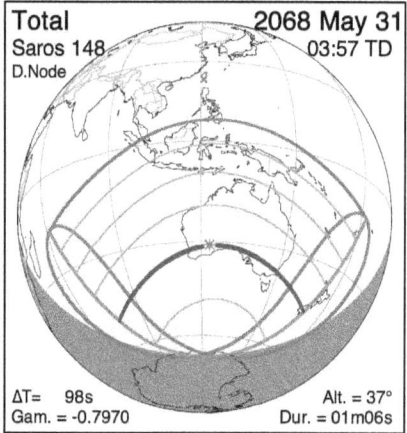

Total      2068 May 31
Saros 148      03:57 TD
D.Node

ΔT= 98s      Alt. = 37°
Gam. = -0.7970      Dur. = 01m06s

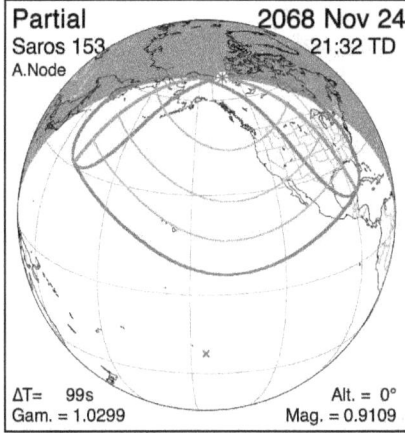

Partial      2068 Nov 24
Saros 153      21:32 TD
A.Node

ΔT= 99s      Alt. = 0°
Gam. = 1.0299      Mag. = 0.9109

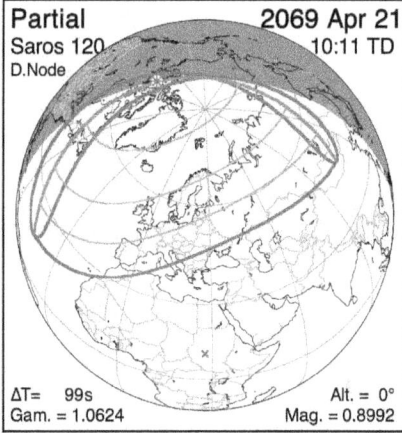

Partial      2069 Apr 21
Saros 120      10:11 TD
D.Node

ΔT= 99s      Alt. = 0°
Gam. = 1.0624      Mag. = 0.8992

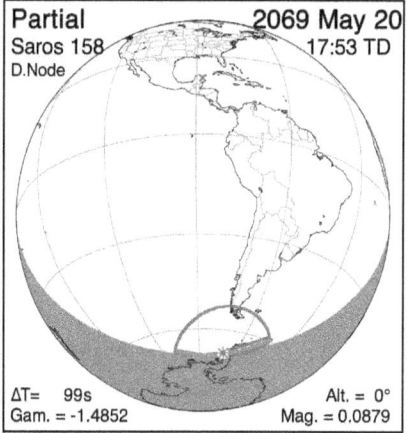

Partial      2069 May 20
Saros 158      17:53 TD
D.Node

ΔT= 99s      Alt. = 0°
Gam. = -1.4852      Mag. = 0.0879

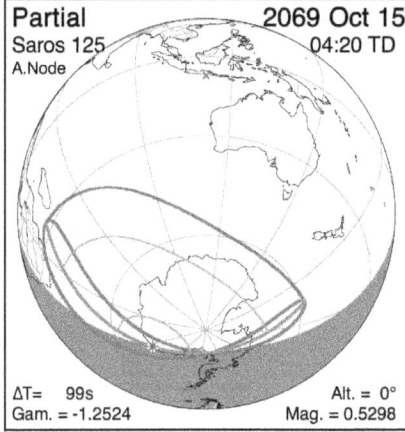

Partial      2069 Oct 15
Saros 125      04:20 TD
A.Node

ΔT= 99s      Alt. = 0°
Gam. = -1.2524      Mag. = 0.5298

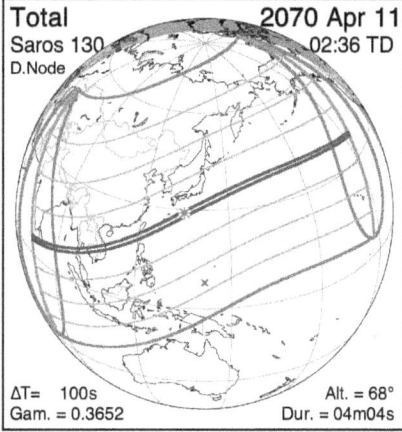

Total      2070 Apr 11
Saros 130      02:36 TD
D.Node

ΔT= 100s      Alt. = 68°
Gam. = 0.3652      Dur. = 04m04s

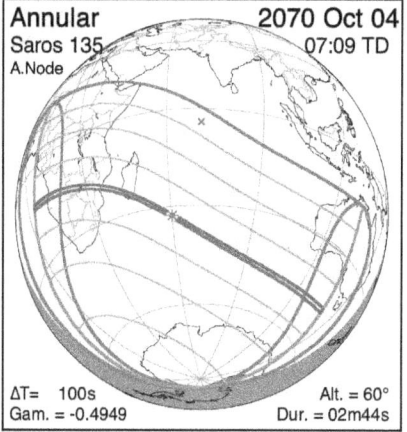

Annular      2070 Oct 04
Saros 135      07:09 TD
A.Node

ΔT= 100s      Alt. = 60°
Gam. = -0.4949      Dur. = 02m44s

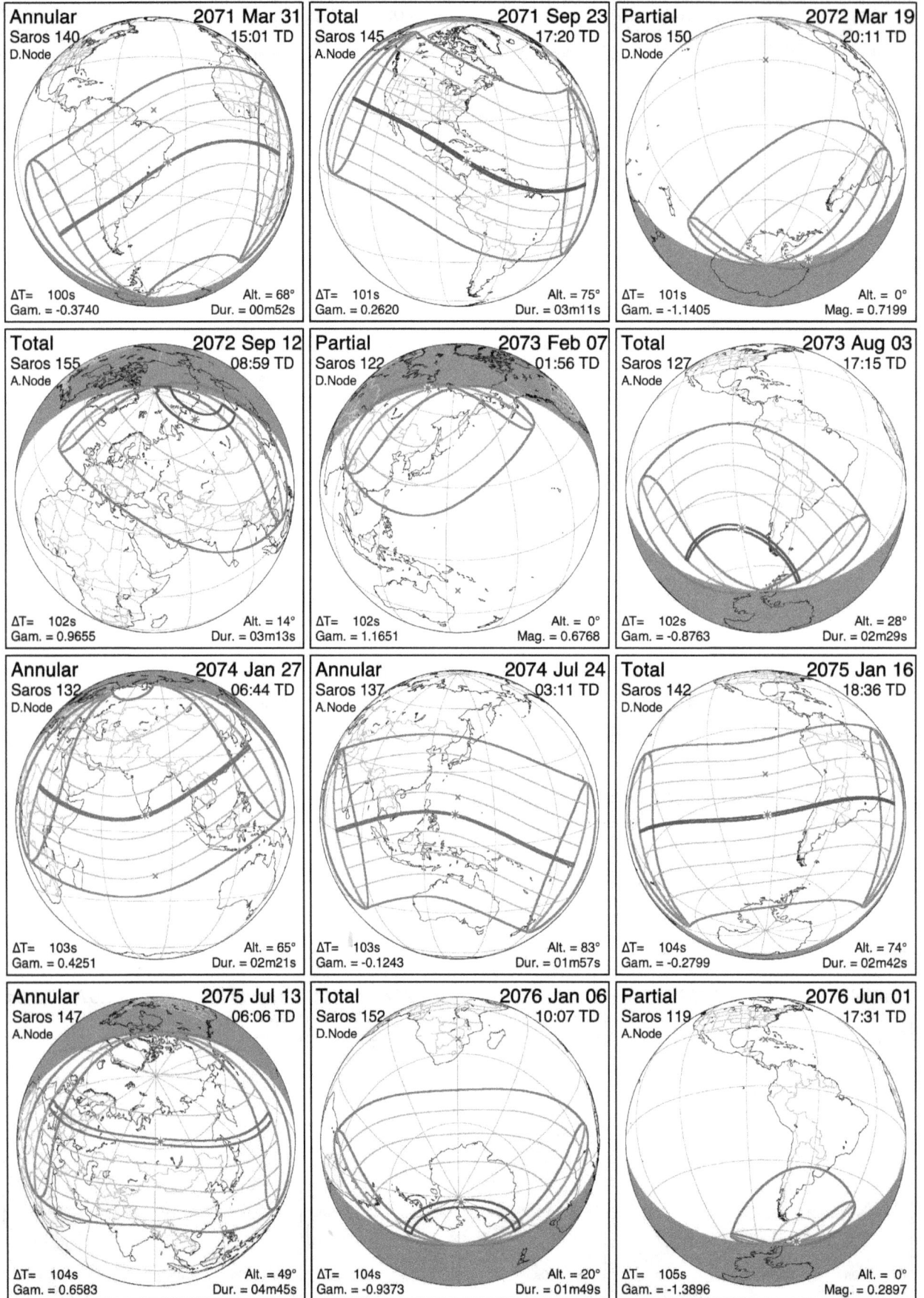

| Annular | 2071 Mar 31 |
|---|---|
| Saros 140 | 15:01 TD |
| D.Node | |
| ΔT= 100s | Alt. = 68° |
| Gam. = -0.3740 | Dur. = 00m52s |

| Total | 2071 Sep 23 |
|---|---|
| Saros 145 | 17:20 TD |
| A.Node | |
| ΔT= 101s | Alt. = 75° |
| Gam. = 0.2620 | Dur. = 03m11s |

| Partial | 2072 Mar 19 |
|---|---|
| Saros 150 | 20:11 TD |
| D.Node | |
| ΔT= 101s | Alt. = 0° |
| Gam. = -1.1405 | Mag. = 0.7199 |

| Total | 2072 Sep 12 |
|---|---|
| Saros 155 | 08:59 TD |
| A.Node | |
| ΔT= 102s | Alt. = 14° |
| Gam. = 0.9655 | Dur. = 03m13s |

| Partial | 2073 Feb 07 |
|---|---|
| Saros 122 | 01:56 TD |
| D.Node | |
| ΔT= 102s | Alt. = 0° |
| Gam. = 1.1651 | Mag. = 0.6768 |

| Total | 2073 Aug 03 |
|---|---|
| Saros 127 | 17:15 TD |
| A.Node | |
| ΔT= 102s | Alt. = 28° |
| Gam. = -0.8763 | Dur. = 02m29s |

| Annular | 2074 Jan 27 |
|---|---|
| Saros 132 | 06:44 TD |
| D.Node | |
| ΔT= 103s | Alt. = 65° |
| Gam. = 0.4251 | Dur. = 02m21s |

| Annular | 2074 Jul 24 |
|---|---|
| Saros 137 | 03:11 TD |
| A.Node | |
| ΔT= 103s | Alt. = 83° |
| Gam. = -0.1243 | Dur. = 01m57s |

| Total | 2075 Jan 16 |
|---|---|
| Saros 142 | 18:36 TD |
| D.Node | |
| ΔT= 104s | Alt. = 74° |
| Gam. = -0.2799 | Dur. = 02m42s |

| Annular | 2075 Jul 13 |
|---|---|
| Saros 147 | 06:06 TD |
| A.Node | |
| ΔT= 104s | Alt. = 49° |
| Gam. = 0.6583 | Dur. = 04m45s |

| Total | 2076 Jan 06 |
|---|---|
| Saros 152 | 10:07 TD |
| D.Node | |
| ΔT= 104s | Alt. = 20° |
| Gam. = -0.9373 | Dur. = 01m49s |

| Partial | 2076 Jun 01 |
|---|---|
| Saros 119 | 17:31 TD |
| A.Node | |
| ΔT= 105s | Alt. = 0° |
| Gam. = -1.3896 | Mag. = 0.2897 |

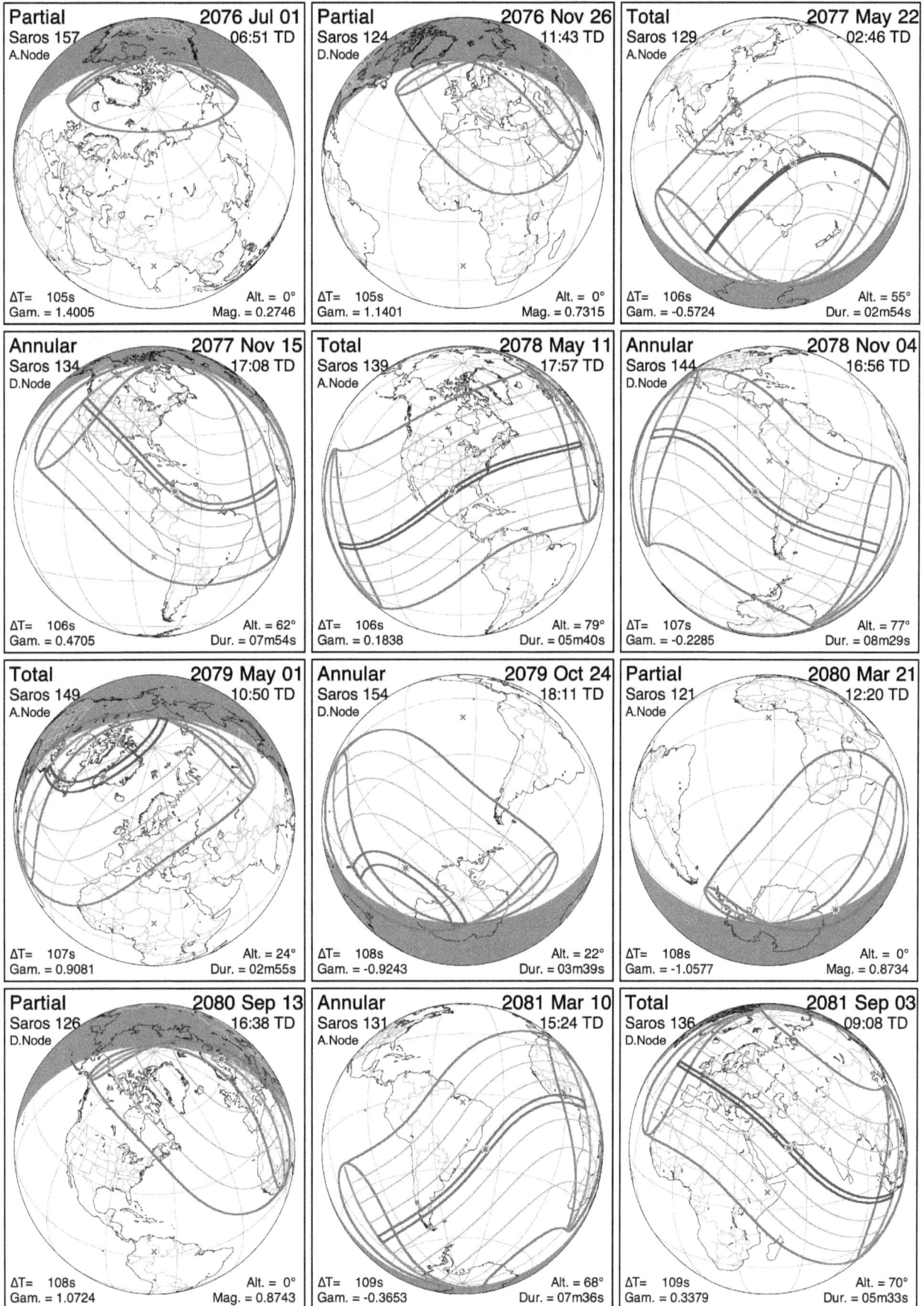

| Partial | 2076 Jul 01 |
|---|---|
| Saros 157 | 06:51 TD |
| A.Node | |
| ΔT= 105s | Alt. = 0° |
| Gam. = 1.4005 | Mag. = 0.2746 |

| Partial | 2076 Nov 26 |
|---|---|
| Saros 124 | 11:43 TD |
| D.Node | |
| ΔT= 105s | Alt. = 0° |
| Gam. = 1.1401 | Mag. = 0.7315 |

| Total | 2077 May 22 |
|---|---|
| Saros 129 | 02:46 TD |
| A.Node | |
| ΔT= 106s | Alt. = 55° |
| Gam. = -0.5724 | Dur. = 02m54s |

| Annular | 2077 Nov 15 |
|---|---|
| Saros 134 | 17:08 TD |
| D.Node | |
| ΔT= 106s | Alt. = 62° |
| Gam. = 0.4705 | Dur. = 07m54s |

| Total | 2078 May 11 |
|---|---|
| Saros 139 | 17:57 TD |
| A.Node | |
| ΔT= 106s | Alt. = 79° |
| Gam. = 0.1838 | Dur. = 05m40s |

| Annular | 2078 Nov 04 |
|---|---|
| Saros 144 | 16:56 TD |
| D.Node | |
| ΔT= 107s | Alt. = 77° |
| Gam. = -0.2285 | Dur. = 08m29s |

| Total | 2079 May 01 |
|---|---|
| Saros 149 | 10:50 TD |
| A.Node | |
| ΔT= 107s | Alt. = 24° |
| Gam. = 0.9081 | Dur. = 02m55s |

| Annular | 2079 Oct 24 |
|---|---|
| Saros 154 | 18:11 TD |
| D.Node | |
| ΔT= 108s | Alt. = 22° |
| Gam. = -0.9243 | Dur. = 03m39s |

| Partial | 2080 Mar 21 |
|---|---|
| Saros 121 | 12:20 TD |
| A.Node | |
| ΔT= 108s | Alt. = 0° |
| Gam. = -1.0577 | Mag. = 0.8734 |

| Partial | 2080 Sep 13 |
|---|---|
| Saros 126 | 16:38 TD |
| D.Node | |
| ΔT= 108s | Alt. = 0° |
| Gam. = 1.0724 | Mag. = 0.8743 |

| Annular | 2081 Mar 10 |
|---|---|
| Saros 131 | 15:24 TD |
| A.Node | |
| ΔT= 109s | Alt. = 68° |
| Gam. = -0.3653 | Dur. = 07m36s |

| Total | 2081 Sep 03 |
|---|---|
| Saros 136 | 09:08 TD |
| D.Node | |
| ΔT= 109s | Alt. = 70° |
| Gam. = 0.3379 | Dur. = 05m33s |

49

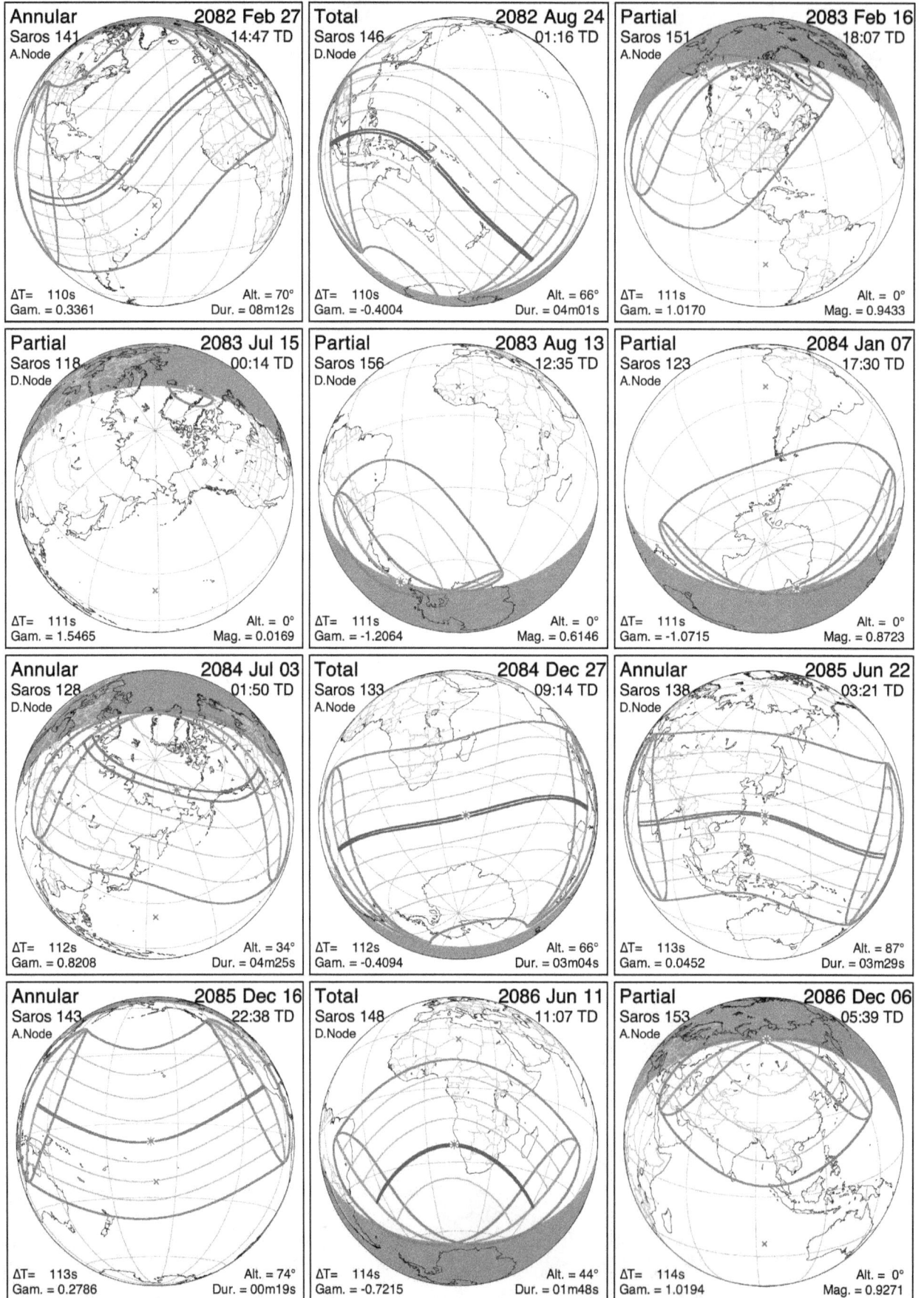

| Annular | 2082 Feb 27 |
|---|---|
| Saros 141 | 14:47 TD |
| A.Node | |

| ΔT= 110s | Alt. = 70° |
|---|---|
| Gam. = 0.3361 | Dur. = 08m12s |

| Total | 2082 Aug 24 |
|---|---|
| Saros 146 | 01:16 TD |
| D.Node | |

| ΔT= 110s | Alt. = 66° |
|---|---|
| Gam. = -0.4004 | Dur. = 04m01s |

| Partial | 2083 Feb 16 |
|---|---|
| Saros 151 | 18:07 TD |
| A.Node | |

| ΔT= 111s | Alt. = 0° |
|---|---|
| Gam. = 1.0170 | Mag. = 0.9433 |

| Partial | 2083 Jul 15 |
|---|---|
| Saros 118 | 00:14 TD |
| D.Node | |

| ΔT= 111s | Alt. = 0° |
|---|---|
| Gam. = 1.5465 | Mag. = 0.0169 |

| Partial | 2083 Aug 13 |
|---|---|
| Saros 156 | 12:35 TD |
| D.Node | |

| ΔT= 111s | Alt. = 0° |
|---|---|
| Gam. = -1.2064 | Mag. = 0.6146 |

| Partial | 2084 Jan 07 |
|---|---|
| Saros 123 | 17:30 TD |
| A.Node | |

| ΔT= 111s | Alt. = 0° |
|---|---|
| Gam. = -1.0715 | Mag. = 0.8723 |

| Annular | 2084 Jul 03 |
|---|---|
| Saros 128 | 01:50 TD |
| D.Node | |

| ΔT= 112s | Alt. = 34° |
|---|---|
| Gam. = 0.8208 | Dur. = 04m25s |

| Total | 2084 Dec 27 |
|---|---|
| Saros 133 | 09:14 TD |
| A.Node | |

| ΔT= 112s | Alt. = 66° |
|---|---|
| Gam. = -0.4094 | Dur. = 03m04s |

| Annular | 2085 Jun 22 |
|---|---|
| Saros 138 | 03:21 TD |
| D.Node | |

| ΔT= 113s | Alt. = 87° |
|---|---|
| Gam. = 0.0452 | Dur. = 03m29s |

| Annular | 2085 Dec 16 |
|---|---|
| Saros 143 | 22:38 TD |
| A.Node | |

| ΔT= 113s | Alt. = 74° |
|---|---|
| Gam. = 0.2786 | Dur. = 00m19s |

| Total | 2086 Jun 11 |
|---|---|
| Saros 148 | 11:07 TD |
| D.Node | |

| ΔT= 114s | Alt. = 44° |
|---|---|
| Gam. = -0.7215 | Dur. = 01m48s |

| Partial | 2086 Dec 06 |
|---|---|
| Saros 153 | 05:39 TD |
| A.Node | |

| ΔT= 114s | Alt. = 0° |
|---|---|
| Gam. = 1.0194 | Mag. = 0.9271 |

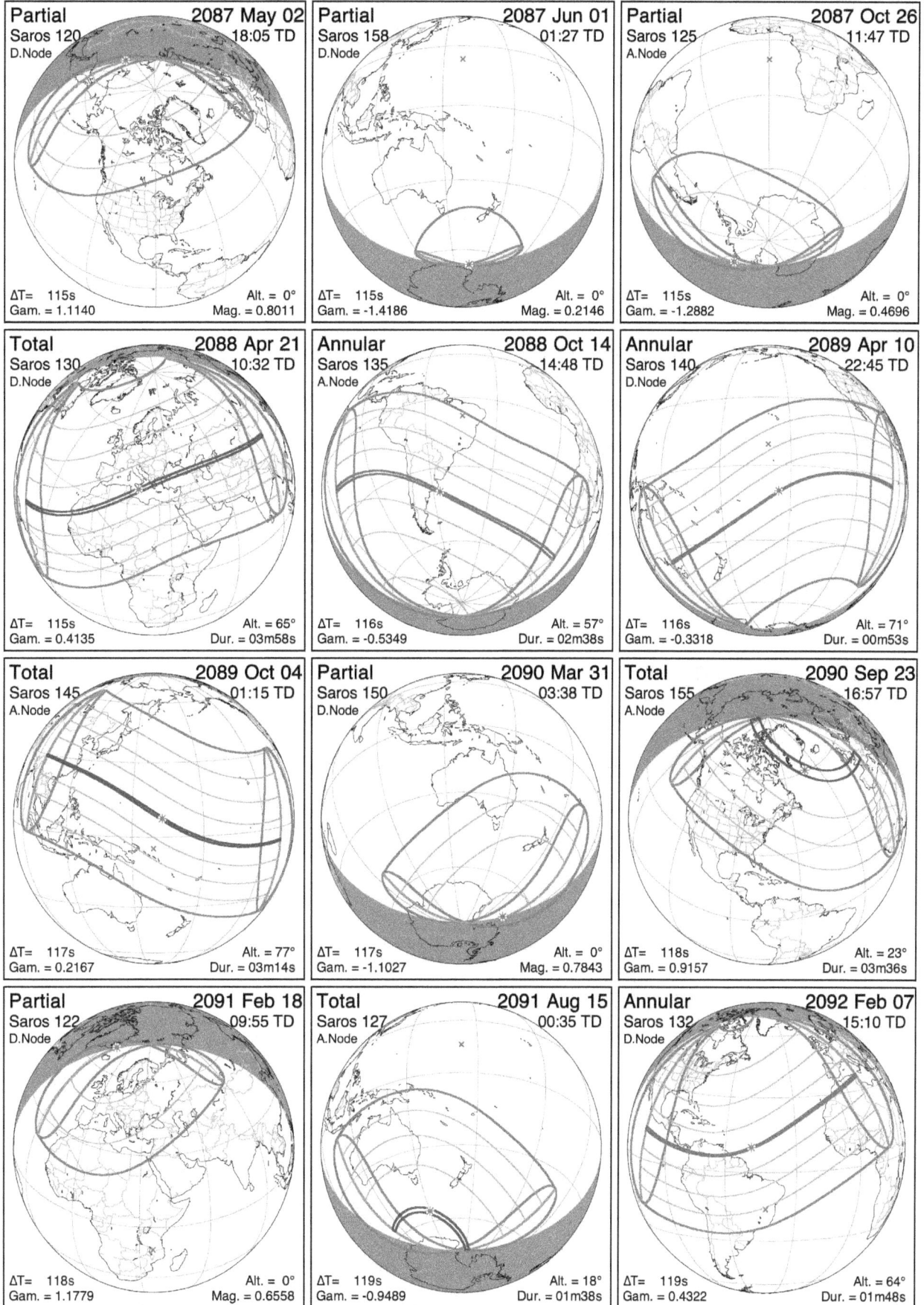

| Partial | 2087 May 02 |
|---|---|
| Saros 120 | 18:05 TD |
| D.Node | |

| ΔT= 115s | Alt. = 0° |
|---|---|
| Gam. = 1.1140 | Mag. = 0.8011 |

| Partial | 2087 Jun 01 |
|---|---|
| Saros 158 | 01:27 TD |
| D.Node | |

| ΔT= 115s | Alt. = 0° |
|---|---|
| Gam. = -1.4186 | Mag. = 0.2146 |

| Partial | 2087 Oct 26 |
|---|---|
| Saros 125 | 11:47 TD |
| A.Node | |

| ΔT= 115s | Alt. = 0° |
|---|---|
| Gam. = -1.2882 | Mag. = 0.4696 |

| Total | 2088 Apr 21 |
|---|---|
| Saros 130 | 10:32 TD |
| D.Node | |

| ΔT= 115s | Alt. = 65° |
|---|---|
| Gam. = 0.4135 | Dur. = 03m58s |

| Annular | 2088 Oct 14 |
|---|---|
| Saros 135 | 14:48 TD |
| A.Node | |

| ΔT= 116s | Alt. = 57° |
|---|---|
| Gam. = -0.5349 | Dur. = 02m38s |

| Annular | 2089 Apr 10 |
|---|---|
| Saros 140 | 22:45 TD |
| D.Node | |

| ΔT= 116s | Alt. = 71° |
|---|---|
| Gam. = -0.3318 | Dur. = 00m53s |

| Total | 2089 Oct 04 |
|---|---|
| Saros 145 | 01:15 TD |
| A.Node | |

| ΔT= 117s | Alt. = 77° |
|---|---|
| Gam. = 0.2167 | Dur. = 03m14s |

| Partial | 2090 Mar 31 |
|---|---|
| Saros 150 | 03:38 TD |
| D.Node | |

| ΔT= 117s | Alt. = 0° |
|---|---|
| Gam. = -1.1027 | Mag. = 0.7843 |

| Total | 2090 Sep 23 |
|---|---|
| Saros 155 | 16:57 TD |
| A.Node | |

| ΔT= 118s | Alt. = 23° |
|---|---|
| Gam. = 0.9157 | Dur. = 03m36s |

| Partial | 2091 Feb 18 |
|---|---|
| Saros 122 | 09:55 TD |
| D.Node | |

| ΔT= 118s | Alt. = 0° |
|---|---|
| Gam. = 1.1779 | Mag. = 0.6558 |

| Total | 2091 Aug 15 |
|---|---|
| Saros 127 | 00:35 TD |
| A.Node | |

| ΔT= 119s | Alt. = 18° |
|---|---|
| Gam. = -0.9489 | Dur. = 01m38s |

| Annular | 2092 Feb 07 |
|---|---|
| Saros 132 | 15:10 TD |
| D.Node | |

| ΔT= 119s | Alt. = 64° |
|---|---|
| Gam. = 0.4322 | Dur. = 01m48s |

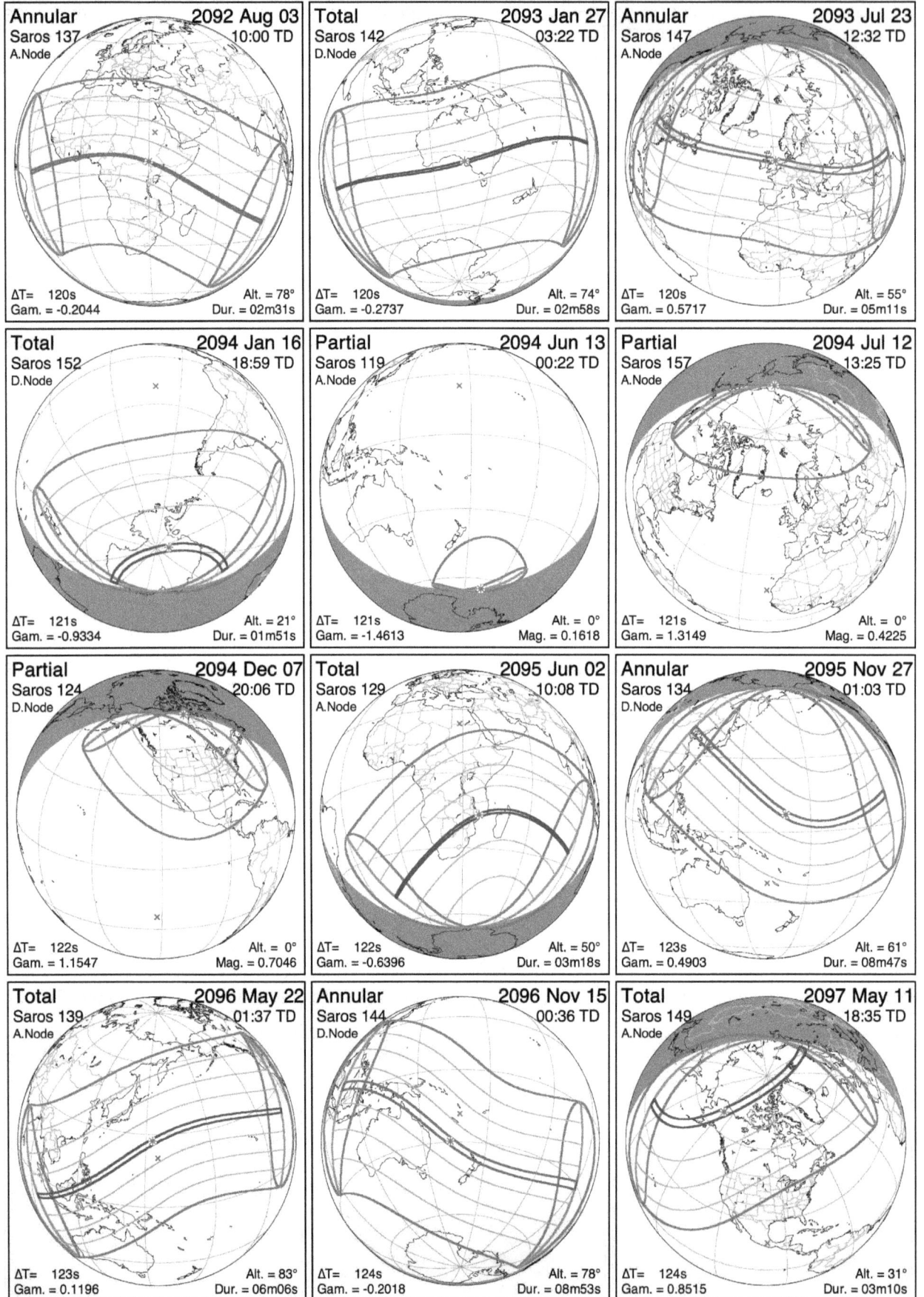

| Annular | 2092 Aug 03 |
|---|---|
| Saros 137 | 10:00 TD |
| A.Node | |

| ΔT= 120s | Alt. = 78° |
|---|---|
| Gam. = -0.2044 | Dur. = 02m31s |

| Total | 2093 Jan 27 |
|---|---|
| Saros 142 | 03:22 TD |
| D.Node | |

| ΔT= 120s | Alt. = 74° |
|---|---|
| Gam. = -0.2737 | Dur. = 02m58s |

| Annular | 2093 Jul 23 |
|---|---|
| Saros 147 | 12:32 TD |
| A.Node | |

| ΔT= 120s | Alt. = 55° |
|---|---|
| Gam. = 0.5717 | Dur. = 05m11s |

| Total | 2094 Jan 16 |
|---|---|
| Saros 152 | 18:59 TD |
| D.Node | |

| ΔT= 121s | Alt. = 21° |
|---|---|
| Gam. = -0.9334 | Dur. = 01m51s |

| Partial | 2094 Jun 13 |
|---|---|
| Saros 119 | 00:22 TD |
| A.Node | |

| ΔT= 121s | Alt. = 0° |
|---|---|
| Gam. = -1.4613 | Mag. = 0.1618 |

| Partial | 2094 Jul 12 |
|---|---|
| Saros 157 | 13:25 TD |
| A.Node | |

| ΔT= 121s | Alt. = 0° |
|---|---|
| Gam. = 1.3149 | Mag. = 0.4225 |

| Partial | 2094 Dec 07 |
|---|---|
| Saros 124 | 20:06 TD |
| D.Node | |

| ΔT= 122s | Alt. = 0° |
|---|---|
| Gam. = 1.1547 | Mag. = 0.7046 |

| Total | 2095 Jun 02 |
|---|---|
| Saros 129 | 10:08 TD |
| A.Node | |

| ΔT= 122s | Alt. = 50° |
|---|---|
| Gam. = -0.6396 | Dur. = 03m18s |

| Annular | 2095 Nov 27 |
|---|---|
| Saros 134 | 01:03 TD |
| D.Node | |

| ΔT= 123s | Alt. = 61° |
|---|---|
| Gam. = 0.4903 | Dur. = 08m47s |

| Total | 2096 May 22 |
|---|---|
| Saros 139 | 01:37 TD |
| A.Node | |

| ΔT= 123s | Alt. = 83° |
|---|---|
| Gam. = 0.1196 | Dur. = 06m06s |

| Annular | 2096 Nov 15 |
|---|---|
| Saros 144 | 00:36 TD |
| D.Node | |

| ΔT= 124s | Alt. = 78° |
|---|---|
| Gam. = -0.2018 | Dur. = 08m53s |

| Total | 2097 May 11 |
|---|---|
| Saros 149 | 18:35 TD |
| A.Node | |

| ΔT= 124s | Alt. = 31° |
|---|---|
| Gam. = 0.8515 | Dur. = 03m10s |

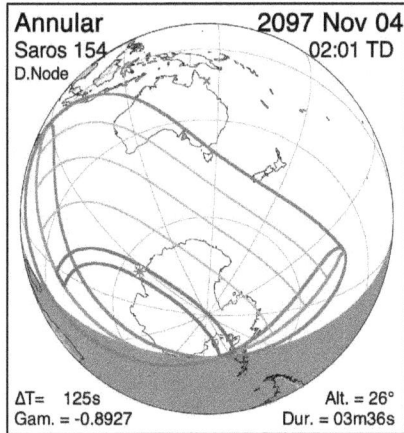

Annular       2097 Nov 04
Saros 154       02:01 TD
D.Node

ΔT= 125s       Alt. = 26°
Gam. = -0.8927       Dur. = 03m36s

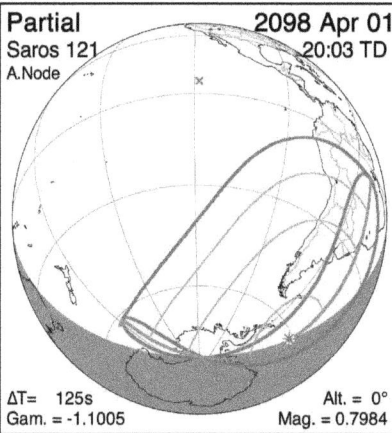

Partial       2098 Apr 01
Saros 121       20:03 TD
A.Node

ΔT= 125s       Alt. = 0°
Gam. = -1.1005       Mag. = 0.7984

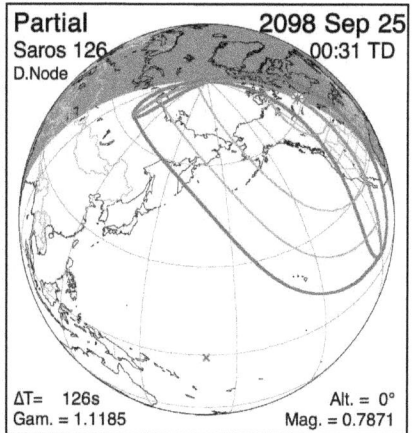

Partial       2098 Sep 25
Saros 126       00:31 TD
D.Node

ΔT= 126s       Alt. = 0°
Gam. = 1.1185       Mag. = 0.7871

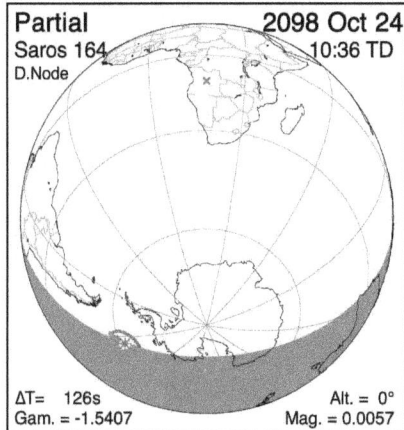

Partial       2098 Oct 24
Saros 164       10:36 TD
D.Node

ΔT= 126s       Alt. = 0°
Gam. = -1.5407       Mag. = 0.0057

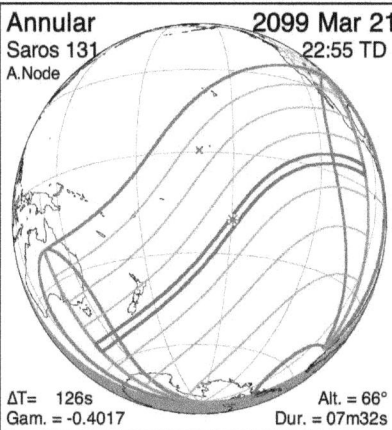

Annular       2099 Mar 21
Saros 131       22:55 TD
A.Node

ΔT= 126s       Alt. = 66°
Gam. = -0.4017       Dur. = 07m32s

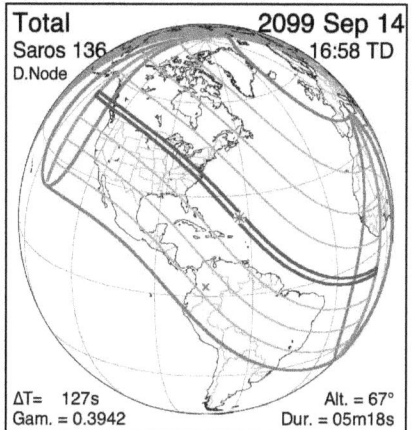

Total       2099 Sep 14
Saros 136       16:58 TD
D.Node

ΔT= 127s       Alt. = 67°
Gam. = 0.3942       Dur. = 05m18s

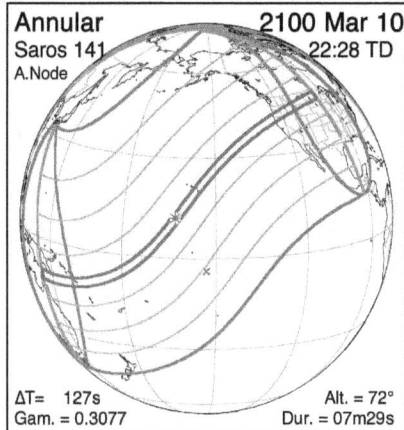

Annular       2100 Mar 10
Saros 141       22:28 TD
A.Node

ΔT= 127s       Alt. = 72°
Gam. = 0.3077       Dur. = 07m29s

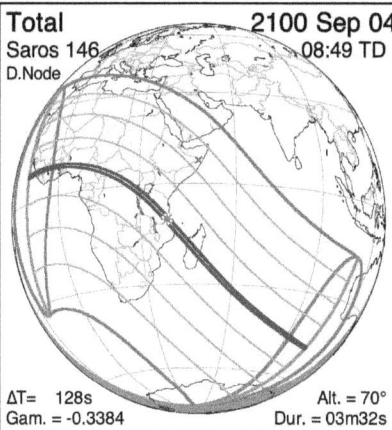

Total       2100 Sep 04
Saros 146       08:49 TD
D.Node

ΔT= 128s       Alt. = 70°
Gam. = -0.3384       Dur. = 03m32s

# Appendix C

## Large Global Solar Eclipse Maps: 2001 to 2100

# Key to Large Global Solar Eclipse Maps

## Figure 5—2: Key to Large Global Solar Eclipse Maps (Appendix C)

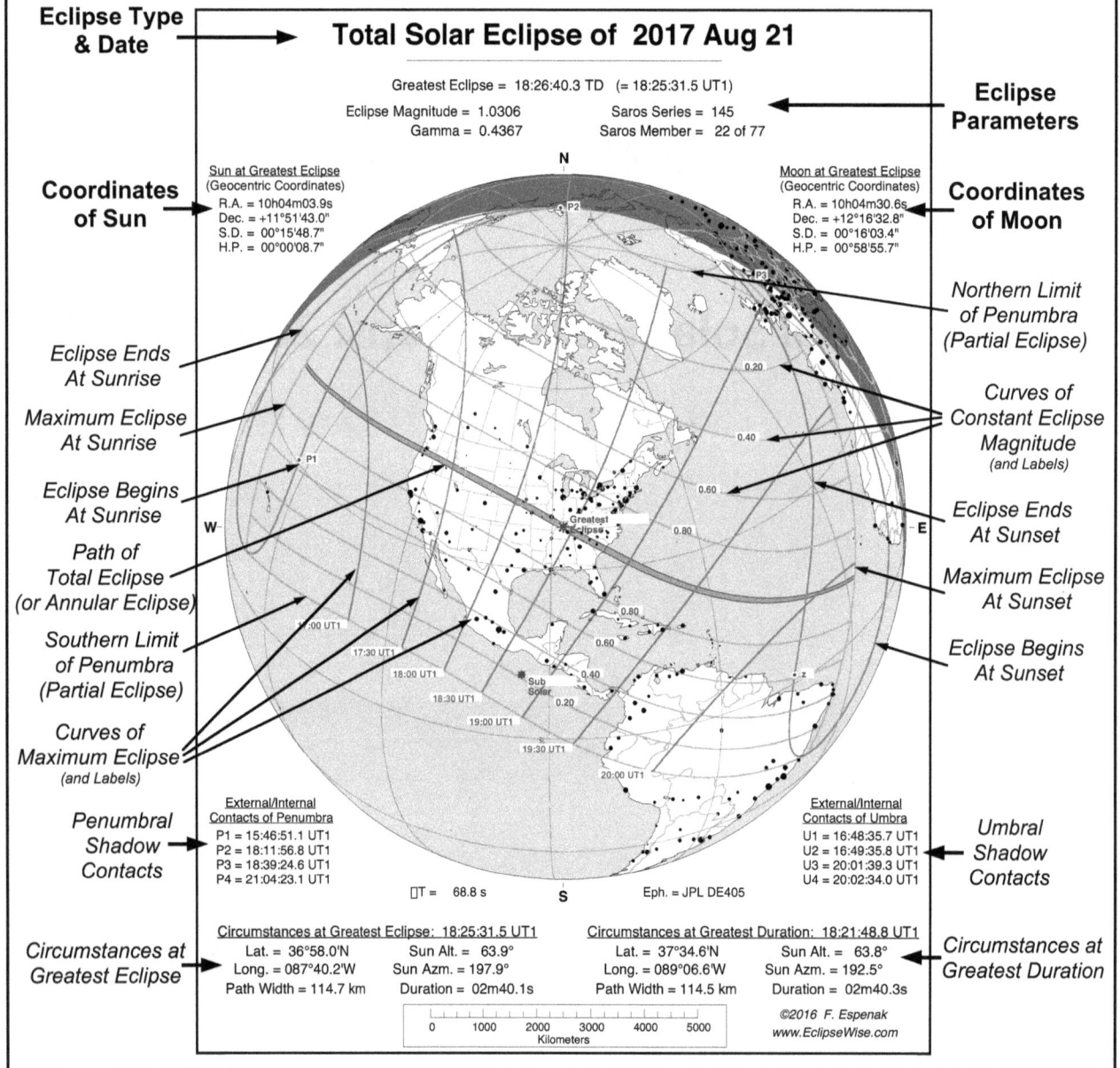

**Eclipse Type & Date** →

### Total Solar Eclipse of 2017 Aug 21

Greatest Eclipse = 18:26:40.3 TD (= 18:25:31.5 UT1)

Eclipse Magnitude = 1.0306          Saros Series = 145
Gamma = 0.4367                      Saros Member = 22 of 77

← **Eclipse Parameters**

**Coordinates of Sun** →

Sun at Greatest Eclipse
(Geocentric Coordinates)
R.A. = 10h04m03.9s
Dec. = +11°51'43.0"
S.D. = 00°15'48.7"
H.P. = 00°00'08.7"

Moon at Greatest Eclipse
(Geocentric Coordinates)
R.A. = 10h04m30.6s
Dec. = +12°16'32.8"
S.D. = 00°16'03.4"
H.P. = 00°58'55.7"

← **Coordinates of Moon**

*Eclipse Ends At Sunrise*

*Maximum Eclipse At Sunrise*

*Eclipse Begins At Sunrise*

*Path of Total Eclipse (or Annular Eclipse)*

*Southern Limit of Penumbra (Partial Eclipse)*

*Curves of Maximum Eclipse (and Labels)*

*Northern Limit of Penumbra (Partial Eclipse)*

*Curves of Constant Eclipse Magnitude (and Labels)*

*Eclipse Ends At Sunset*

*Maximum Eclipse At Sunset*

*Eclipse Begins At Sunset*

*Penumbral Shadow Contacts* →

External/Internal
Contacts of Penumbra
P1 = 15:46:51.1 UT1
P2 = 18:11:56.8 UT1
P3 = 18:39:24.6 UT1
P4 = 21:04:23.1 UT1

External/Internal
Contacts of Umbra
U1 = 16:48:35.7 UT1
U2 = 16:49:35.8 UT1
U3 = 20:01:39.3 UT1
U4 = 20:02:34.0 UT1

← *Umbral Shadow Contacts*

ΔT = 68.8 s          Eph. = JPL DE405

*Circumstances at Greatest Eclipse*

Circumstances at Greatest Eclipse: 18:25:31.5 UT1
Lat. = 36°58.0'N          Sun Alt. = 63.9°
Long. = 087°40.2'W       Sun Azm. = 197.9°
Path Width = 114.7 km    Duration = 02m40.1s

Circumstances at Greatest Duration: 18:21:48.8 UT1
Lat. = 37°34.6'N          Sun Alt. = 63.8°
Long. = 089°06.6'W       Sun Azm. = 192.5°
Path Width = 114.5 km    Duration = 02m40.3s

*Circumstances at Greatest Duration*

0   1000   2000   3000   4000   5000
Kilometers

©2016 F. Espenak
www.EclipseWise.com

# Total Solar Eclipse of 2001 Jun 21

Greatest Eclipse = 12:04:46.3 TD  (= 12:03:42.2 UT1)

| | |
|---|---|
| Eclipse Magnitude = 1.0495 | Saros Series = 127 |
| Gamma = -0.5701 | Saros Member = 57 of 82 |

Sun at Greatest Eclipse
(Geocentric Coordinates)
R.A. = 06h00m46.1s
Dec. = +23°26'18.2"
S.D. = 00°15'44.3"
H.P. = 00°00'08.7"

Moon at Greatest Eclipse
(Geocentric Coordinates)
R.A. = 06h01m00.5s
Dec. = +22°52'27.2"
S.D. = 00°16'17.6"
H.P. = 00°59'47.9"

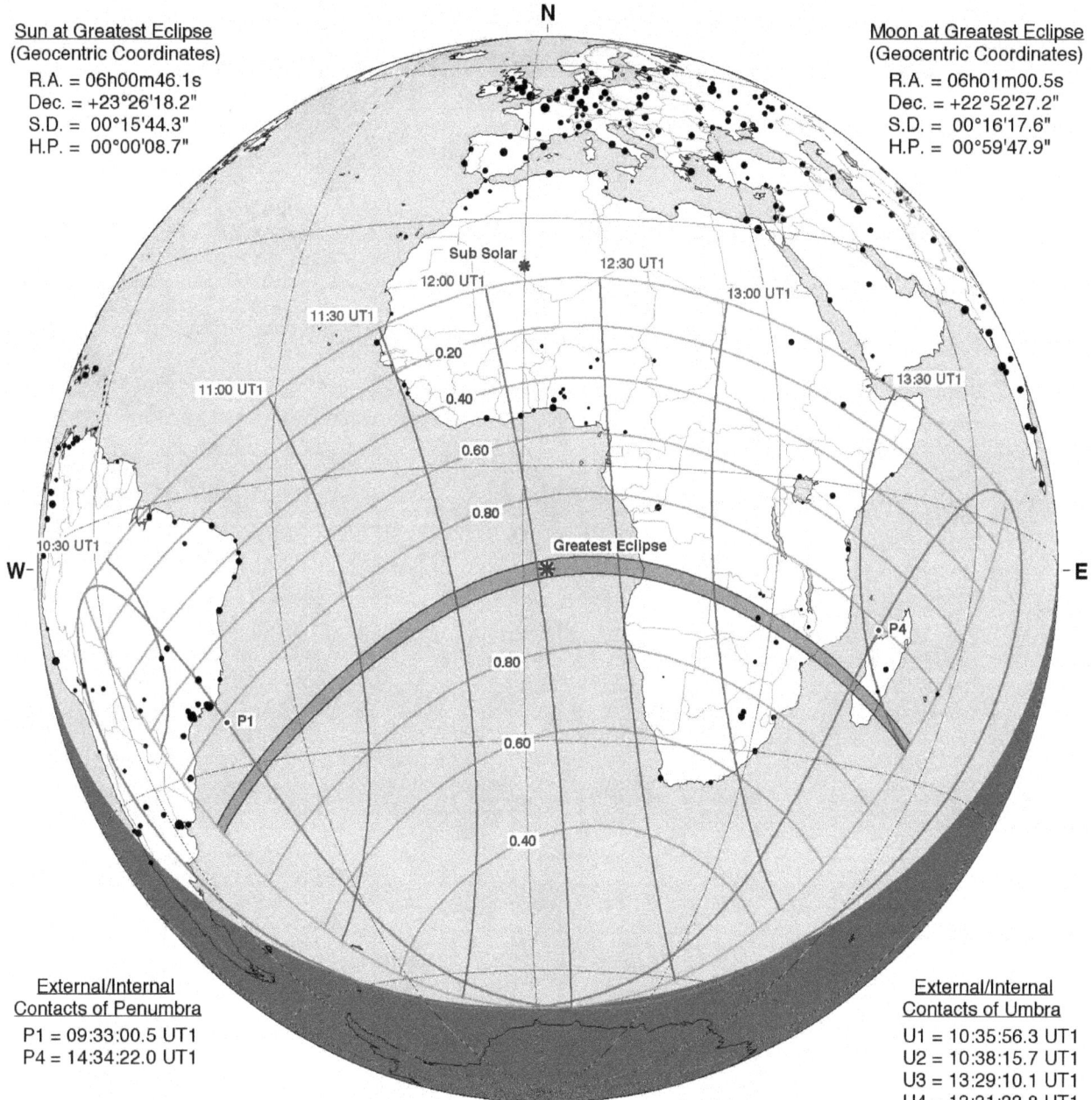

External/Internal
Contacts of Penumbra
P1 = 09:33:00.5 UT1
P4 = 14:34:22.0 UT1

External/Internal
Contacts of Umbra
U1 = 10:35:56.3 UT1
U2 = 10:38:15.7 UT1
U3 = 13:29:10.1 UT1
U4 = 13:31:33.8 UT1

ΔT = 64.2 s          Eph. = JPL DE405

Circumstances at Greatest Eclipse: 12:03:42.2 UT1
Lat. = 11°15.2'S          Sun Alt. = 55.2°
Long. = 002°44.8'E        Sun Azm. = 354.8°
Path Width = 200.0 km     Duration = 04m56.5s

Circumstances at Greatest Duration: 12:06:07.4 UT1
Lat. = 11°08.5'S          Sun Alt. = 55.1°
Long. = 003°28.5'E        Sun Azm. = 352.7°
Path Width = 200.3 km     Duration = 04m56.7s

©2016 F. Espenak
www.EclipseWise.com

# Annular Solar Eclipse of 2001 Dec 14

Greatest Eclipse = 20:53:01.4 TD  (= 20:51:57.1 UT1)

| | |
|---|---|
| Eclipse Magnitude = 0.9681 | Saros Series = 132 |
| Gamma = 0.4089 | Saros Member = 45 of 71 |

**Sun at Greatest Eclipse**
(Geocentric Coordinates)
R.A. = 17h29m14.2s
Dec. = -23°15'02.0"
S.D. = 00°16'14.9"
H.P. = 00°00'08.9"

**Moon at Greatest Eclipse**
(Geocentric Coordinates)
R.A. = 17h29m29.5s
Dec. = -22°52'05.9"
S.D. = 00°15'30.3"
H.P. = 00°56'54.2"

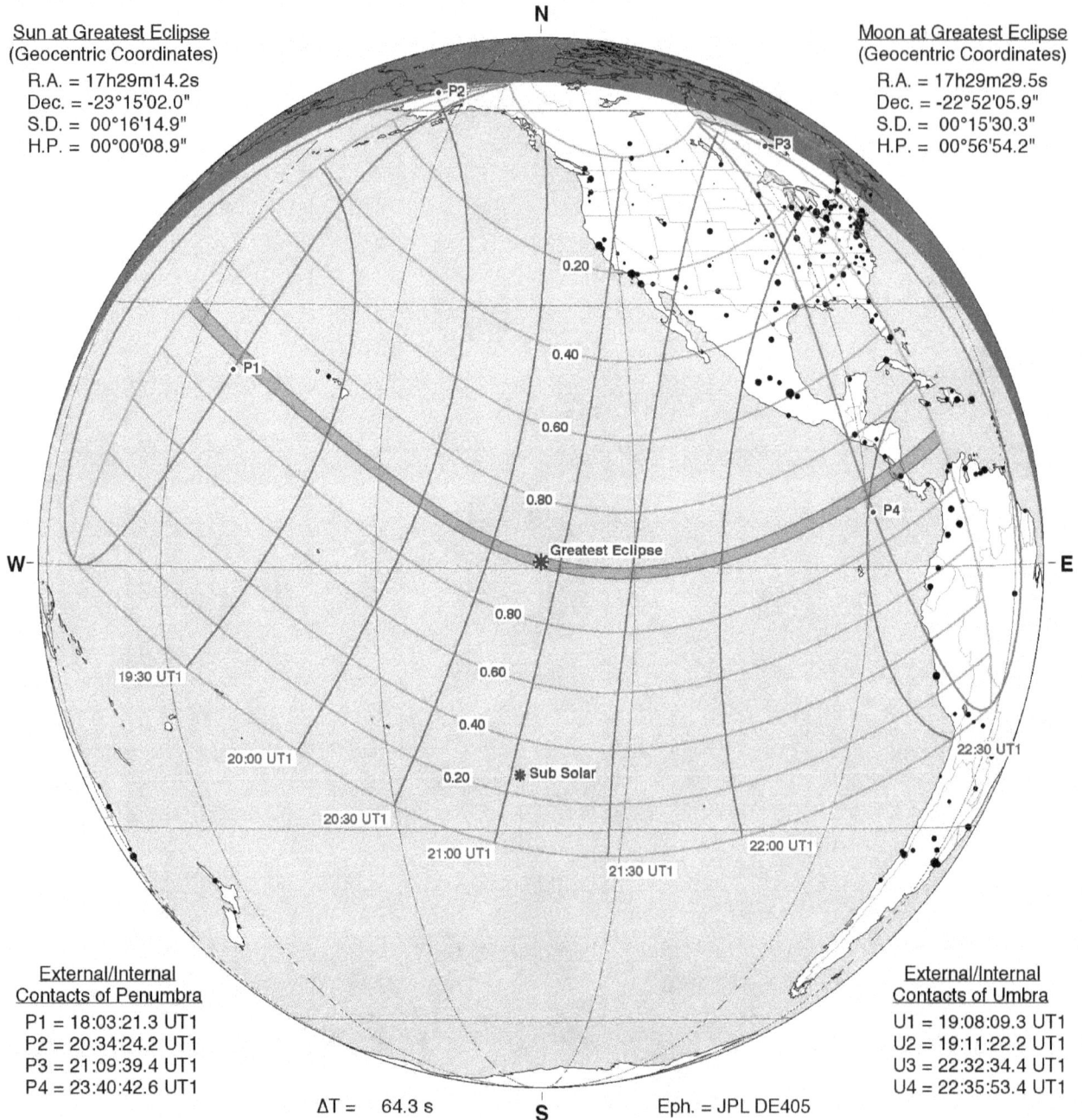

**External/Internal
Contacts of Penumbra**
P1 = 18:03:21.3 UT1
P2 = 20:34:24.2 UT1
P3 = 21:09:39.4 UT1
P4 = 23:40:42.6 UT1

**External/Internal
Contacts of Umbra**
U1 = 19:08:09.3 UT1
U2 = 19:11:22.2 UT1
U3 = 22:32:34.4 UT1
U4 = 22:35:53.4 UT1

ΔT =   64.3 s          Eph. = JPL DE405

**Circumstances at Greatest Eclipse: 20:51:57.1 UT1**

| | |
|---|---|
| Lat. = 00°38.1'N | Sun Alt. =  65.9° |
| Long. = 130°42.2'W | Sun Azm. = 188.0° |
| Path Width = 125.6 km | Duration = 03m53.0s |

**Circumstances at Greatest Duration: 21:07:23.8 UT1**

| | |
|---|---|
| Lat. = 00°14.2'S | Sun Alt. =  64.4° |
| Long. = 126°41.3'W | Sun Azm. = 205.0° |
| Path Width = 127.4 km | Duration = 03m54.1s |

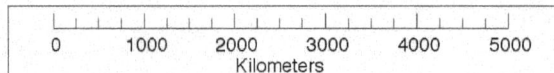

©2016  F. Espenak
www.EclipseWise.com

# Annular Solar Eclipse of 2002 Jun 10

Greatest Eclipse = 23:45:22.2 TD  (= 23:44:17.9 UT1)

Eclipse Magnitude = 0.9962          Saros Series = 137
Gamma = 0.1993                      Saros Member = 35 of 70

Sun at Greatest Eclipse
(Geocentric Coordinates)
R.A. = 05h16m04.1s
Dec. = +23°03'18.9"
S.D. = 00°15'45.1"
H.P. = 00°00'08.7"

Moon at Greatest Eclipse
(Geocentric Coordinates)
R.A. = 05h15m55.6s
Dec. = +23°14'25.0"
S.D. = 00°15'27.1"
H.P. = 00°56'42.5"

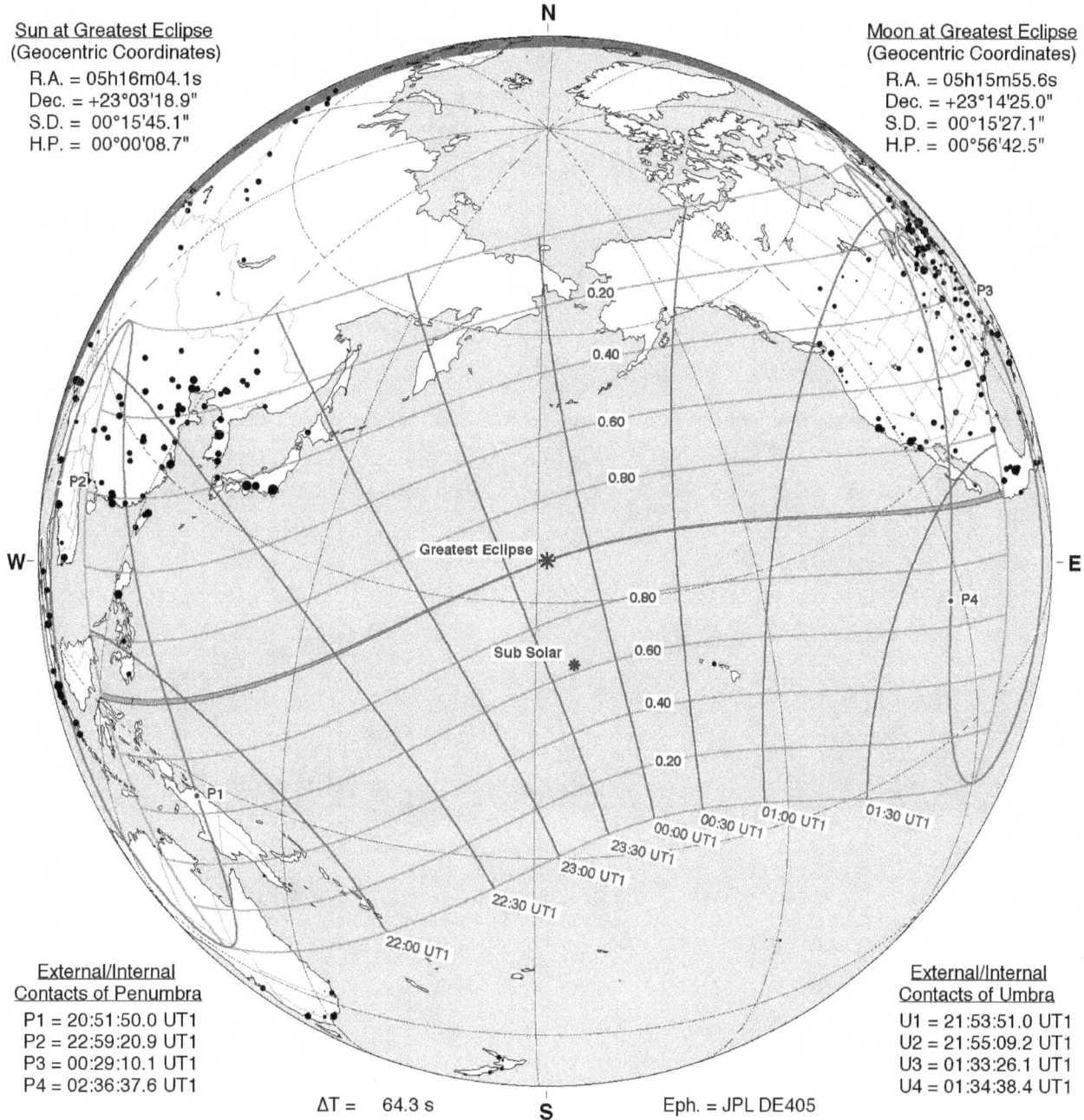

External/Internal
Contacts of Penumbra
P1 = 20:51:50.0 UT1
P2 = 22:59:20.9 UT1
P3 = 00:29:10.1 UT1
P4 = 02:36:37.6 UT1

External/Internal
Contacts of Umbra
U1 = 21:53:51.0 UT1
U2 = 21:55:09.2 UT1
U3 = 01:33:26.1 UT1
U4 = 01:34:38.4 UT1

ΔT = 64.3 s          Eph. = JPL DE405

Circumstances at Greatest Eclipse: 23:44:17.9 UT1
Lat. = 34°33.0'N        Sun Alt. = 78.3°
Long. = 178°37.2'W      Sun Azm. = 169.0°
Path Width = 13.5 km    Duration = 00m22.8s

Circumstances at Greatest Duration: 21:54:30.1 UT1
Lat. = 01°19.8'N        Sun Alt. = 0.0°
Long. = 120°40.5'E      Sun Azm. = 66.9°
Path Width = 78.3 km    Duration = 01m12.6s

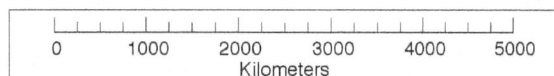

0   1000   2000   3000   4000   5000
Kilometers

©2016  F. Espenak
www.EclipseWise.com

59

# Total Solar Eclipse of 2002 Dec 04

Greatest Eclipse = 07:32:15.7 TD (= 07:31:11.3 UT1)

| | |
|---|---|
| Eclipse Magnitude = 1.0244 | Saros Series = 142 |
| Gamma = -0.3020 | Saros Member = 22 of 72 |

**Sun at Greatest Eclipse**
(Geocentric Coordinates)
R.A. = 16h41m50.9s
Dec. = -22°13'29.2"
S.D. = 00°16'13.6"
H.P. = 00°00'08.9"

**Moon at Greatest Eclipse**
(Geocentric Coordinates)
R.A. = 16h41m32.9s
Dec. = -22°31'05.2"
S.D. = 00°16'21.5"
H.P. = 01°00'02.3"

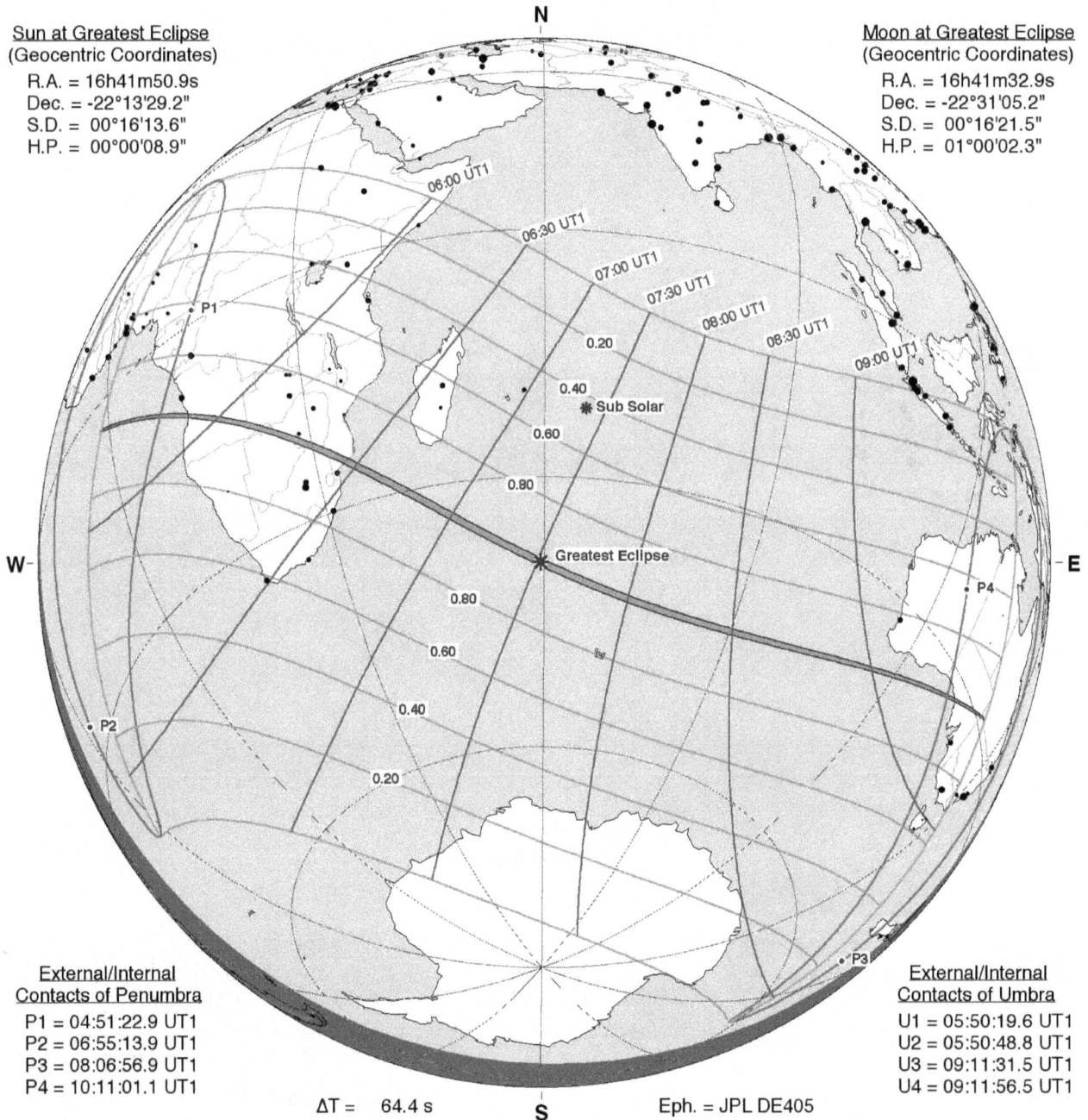

**External/Internal Contacts of Penumbra**
P1 = 04:51:22.9 UT1
P2 = 06:55:13.9 UT1
P3 = 08:06:56.9 UT1
P4 = 10:11:01.1 UT1

ΔT = 64.4 s

Eph. = JPL DE405

**External/Internal Contacts of Umbra**
U1 = 05:50:19.6 UT1
U2 = 05:50:48.8 UT1
U3 = 09:11:31.5 UT1
U4 = 09:11:56.5 UT1

**Circumstances at Greatest Eclipse: 07:31:11.3 UT1**

| | |
|---|---|
| Lat. = 39°27.4'S | Sun Alt. = 72.2° |
| Long. = 059°33.3'E | Sun Azm. = 15.8° |
| Path Width = 87.0 km | Duration = 02m03.8s |

**Circumstances at Greatest Duration: 07:31:56.7 UT1**

| | |
|---|---|
| Lat. = 39°34.2'S | Sun Alt. = 72.2° |
| Long. = 059°52.6'E | Sun Azm. = 14.2° |
| Path Width = 86.9 km | Duration = 02m03.8s |

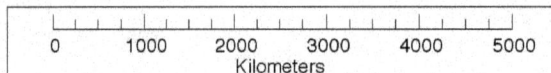

©2016 F. Espenak
www.EclipseWise.com

# Annular Solar Eclipse of 2003 May 31

Greatest Eclipse = 04:09:22.5 TD  (= 04:08:18.0 UT1)

Eclipse Magnitude = 0.9384          Saros Series = 147
Gamma = 0.9960          Saros Member = 22 of 80

Sun at Greatest Eclipse
(Geocentric Coordinates)
R.A. = 04h30m33.5s
Dec. = +21°50'57.1"
S.D. = 00°15'46.5"
H.P. = 00°00'08.7"

Moon at Greatest Eclipse
(Geocentric Coordinates)
R.A. = 04h29m35.5s
Dec. = +22°43'13.1"
S.D. = 00°14'48.1"
H.P. = 00°54'19.3"

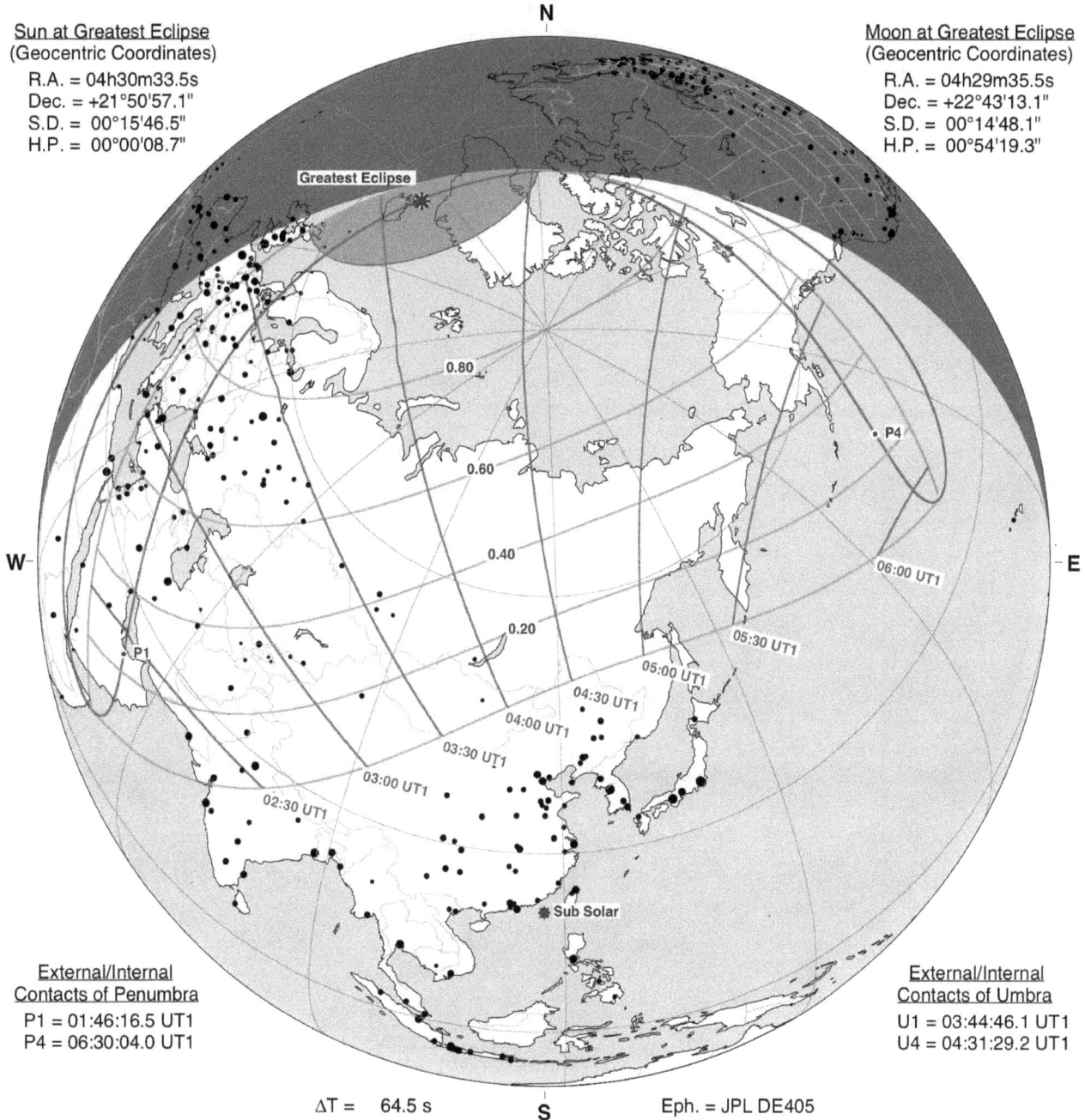

Greatest Eclipse

0.80
0.60
0.40
0.20

P1
P4

06:00 UT1
05:30 UT1
05:00 UT1
04:30 UT1
04:00 UT1
03:30 UT1
03:00 UT1
02:30 UT1

N
W — — E
S

Sub Solar

External/Internal
Contacts of Penumbra
P1 = 01:46:16.5 UT1
P4 = 06:30:04.0 UT1

External/Internal
Contacts of Umbra
U1 = 03:44:46.1 UT1
U4 = 04:31:29.2 UT1

ΔT = 64.5 s          Eph. = JPL DE405

Circumstances at Greatest Eclipse: 04:08:18.0 UT1
Lat. = 66°33.4'N          Sun Alt. = 2.9°
Long. = 024°29.5'W          Sun Azm. = 35.1°
Path Width = 4638.4 km          Duration = 03m36.8s

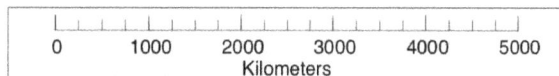

Circumstances at Greatest Duration: 04:07:54.1 UT1
Lat. = 66°26.6'N          Sun Alt. = 2.9°
Long. = 023°59.7'W          Sun Azm. = 35.4°
Path Width = 4659.3 km          Duration = 03m36.8s

©2016 F. Espenak
www.EclipseWise.com

0   1000   2000   3000   4000   5000
Kilometers

# Total Solar Eclipse of 2003 Nov 23

Greatest Eclipse = 22:50:21.7 TD   (= 22:49:17.2 UT1)

| | |
|---|---|
| Eclipse Magnitude = 1.0379 | Saros Series = 152 |
| Gamma = -0.9638 | Saros Member = 12 of 70 |

**Sun at Greatest Eclipse**
(Geocentric Coordinates)
R.A. = 15h56m23.2s
Dec. = -20°24'22.8"
S.D. = 00°16'11.8"
H.P. = 00°00'08.9"

**Moon at Greatest Eclipse**
(Geocentric Coordinates)
R.A. = 15h55m07.5s
Dec. = -21°20'45.7"
S.D. = 00°16'44.7"
H.P. = 01°01'27.3"

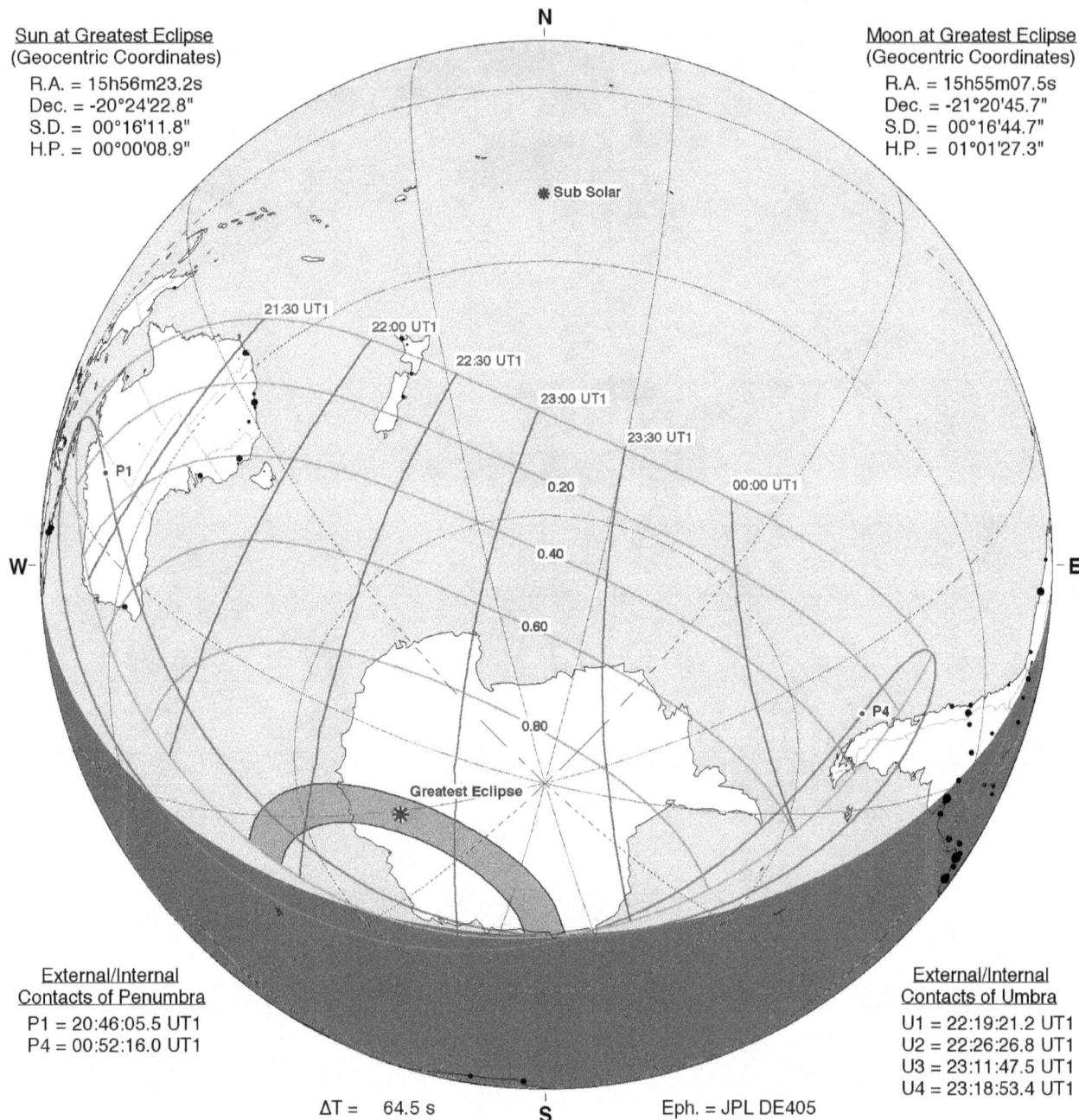

N

☀ Sub Solar

21:30 UT1
22:00 UT1
22:30 UT1
23:00 UT1
23:30 UT1
0.20
00:00 UT1
0.40
0.60
0.80

P1

P4

W — —— E

Greatest Eclipse
☀

S

**External/Internal Contacts of Penumbra**
P1 = 20:46:05.5 UT1
P4 = 00:52:16.0 UT1

**External/Internal Contacts of Umbra**
U1 = 22:19:21.2 UT1
U2 = 22:26:26.8 UT1
U3 = 23:11:47.5 UT1
U4 = 23:18:53.4 UT1

ΔT = 64.5 s

Eph. = JPL DE405

**Circumstances at Greatest Eclipse: 22:49:17.2 UT1**

| | |
|---|---|
| Lat. = 72°40.0'S | Sun Alt. = 14.9° |
| Long. = 088°23.7'E | Sun Azm. = 111.2° |
| Path Width = 495.5 km | Duration = 01m57.3s |

**Circumstances at Greatest Duration: 22:49:14.2 UT1**

| | |
|---|---|
| Lat. = 72°38.5'S | Sun Alt. = 14.9° |
| Long. = 088°26.5'E | Sun Azm. = 111.1° |
| Path Width = 495.6 km | Duration = 01m57.3s |

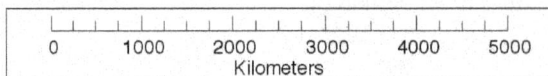

©2016  F. Espenak
www.EclipseWise.com

0    1000    2000    3000    4000    5000
Kilometers

# Partial Solar Eclipse of 2004 Apr 19

Greatest Eclipse = 13:35:05.3 TD   (= 13:34:00.7 UT1)

Eclipse Magnitude = 0.7367

Gamma = -1.1335

Saros Series = 119

Saros Member = 65 of 71

**Sun at Greatest Eclipse**
(Geocentric Coordinates)
R.A. = 01h50m58.6s
Dec. = +11°24'41.2"
S.D. = 00°15'55.2"
H.P. = 00°00'08.8"

**Moon at Greatest Eclipse**
(Geocentric Coordinates)
R.A. = 01h52m50.5s
Dec. = +10°28'42.9"
S.D. = 00°15'01.7"
H.P. = 00°55'09.3"

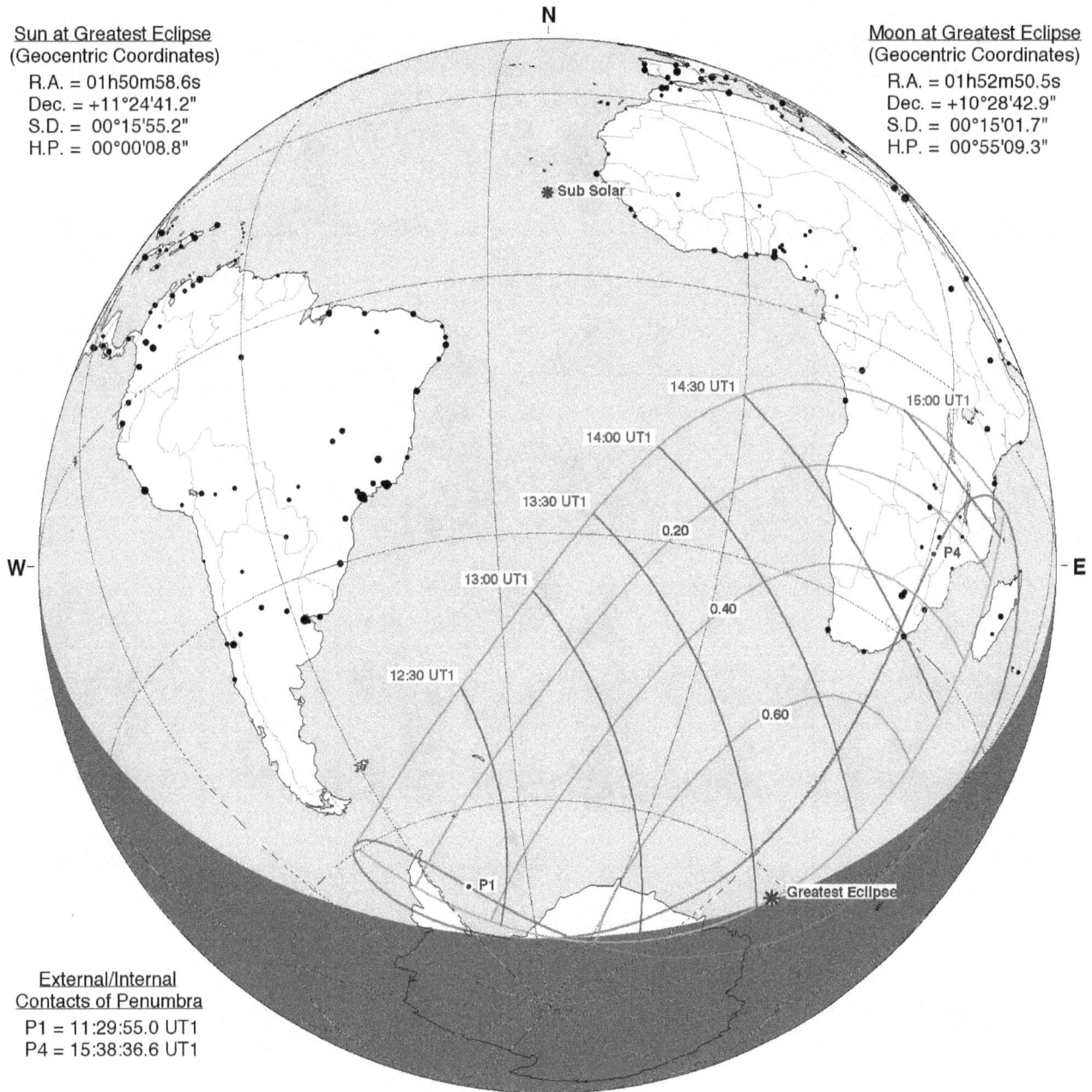

**External/Internal Contacts of Penumbra**
P1 = 11:29:55.0 UT1
P4 = 15:38:36.6 UT1

ΔT = 64.6 s

Eph. = JPL DE405

**Circumstances at Greatest Eclipse: 13:34:00.7 UT1**
Lat. = 61°35.6'S
Long. = 044°15.6'E
Sun Alt. = 0.0°
Sun Azm. = 294.6°

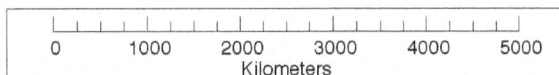

```
0    1000   2000   3000   4000   5000
Kilometers
```

©2016  F. Espenak
www.EclipseWise.com

# Partial Solar Eclipse of 2004 Oct 14

Greatest Eclipse = 03:00:23.0 TD (= 02:59:18.4 UT1)

Eclipse Magnitude = 0.9283

Gamma = 1.0348

Saros Series = 124

Saros Member = 54 of 73

**Sun at Greatest Eclipse**
(Geocentric Coordinates)
R.A. = 13h18m00.5s
Dec. = -08°14'10.7"
S.D. = 00°16'02.2"
H.P. = 00°00'08.8"

**Moon at Greatest Eclipse**
(Geocentric Coordinates)
R.A. = 13h19m53.4s
Dec. = -07°20'43.5"
S.D. = 00°15'55.4"
H.P. = 00°58'26.4"

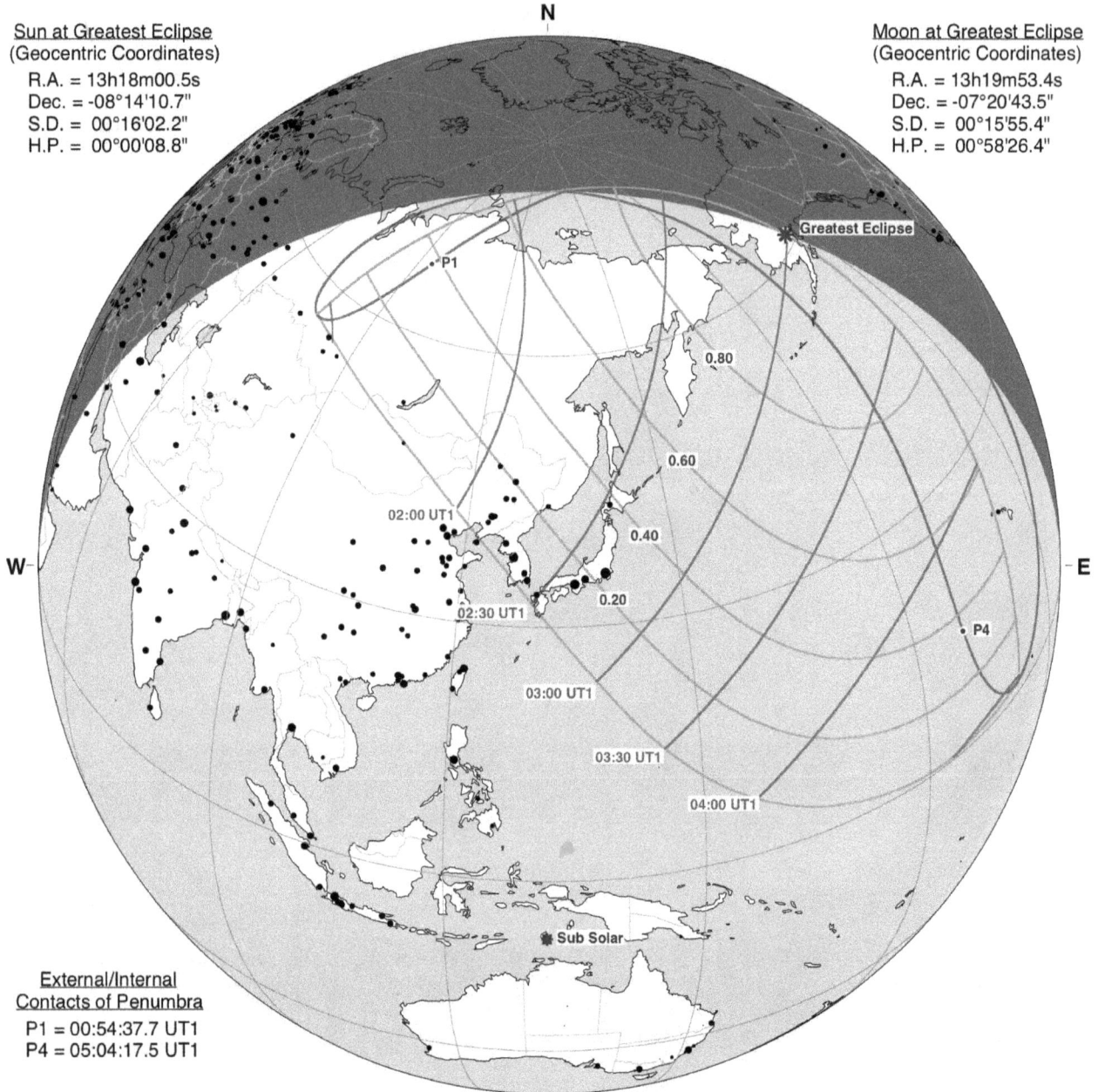

N

Greatest Eclipse

P1

0.80

0.60

0.40

02:00 UT1

0.20

W

E

02:30 UT1

P4

03:00 UT1

03:30 UT1

04:00 UT1

**External/Internal
Contacts of Penumbra**
P1 = 00:54:37.7 UT1
P4 = 05:04:17.5 UT1

Sub Solar

ΔT = 64.6 s

S

Eph. = JPL DE405

**Circumstances at Greatest Eclipse: 02:59:18.4 UT1**
Lat. = 61°15.0'N
Long. = 153°42.1'W
Sun Alt. = 0.0°
Sun Azm. = 252.7°

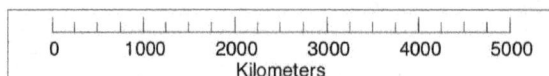

0    1000    2000    3000    4000    5000
Kilometers

# Hybrid Solar Eclipse of 2005 Apr 08

Greatest Eclipse = 20:36:50.5 TD (= 20:35:45.9 UT1)

| | |
|---|---|
| Eclipse Magnitude = 1.0074 | Saros Series = 129 |
| Gamma = -0.3473 | Saros Member = 51 of 80 |

**Sun at Greatest Eclipse**
(Geocentric Coordinates)
R.A. = 01h10m30.3s
Dec. = +07°28'47.2"
S.D. = 00°15'58.1"
H.P. = 00°00'08.8"

**Moon at Greatest Eclipse**
(Geocentric Coordinates)
R.A. = 01h11m08.3s
Dec. = +07°10'59.2"
S.D. = 00°15'50.7"
H.P. = 00°58'09.0"

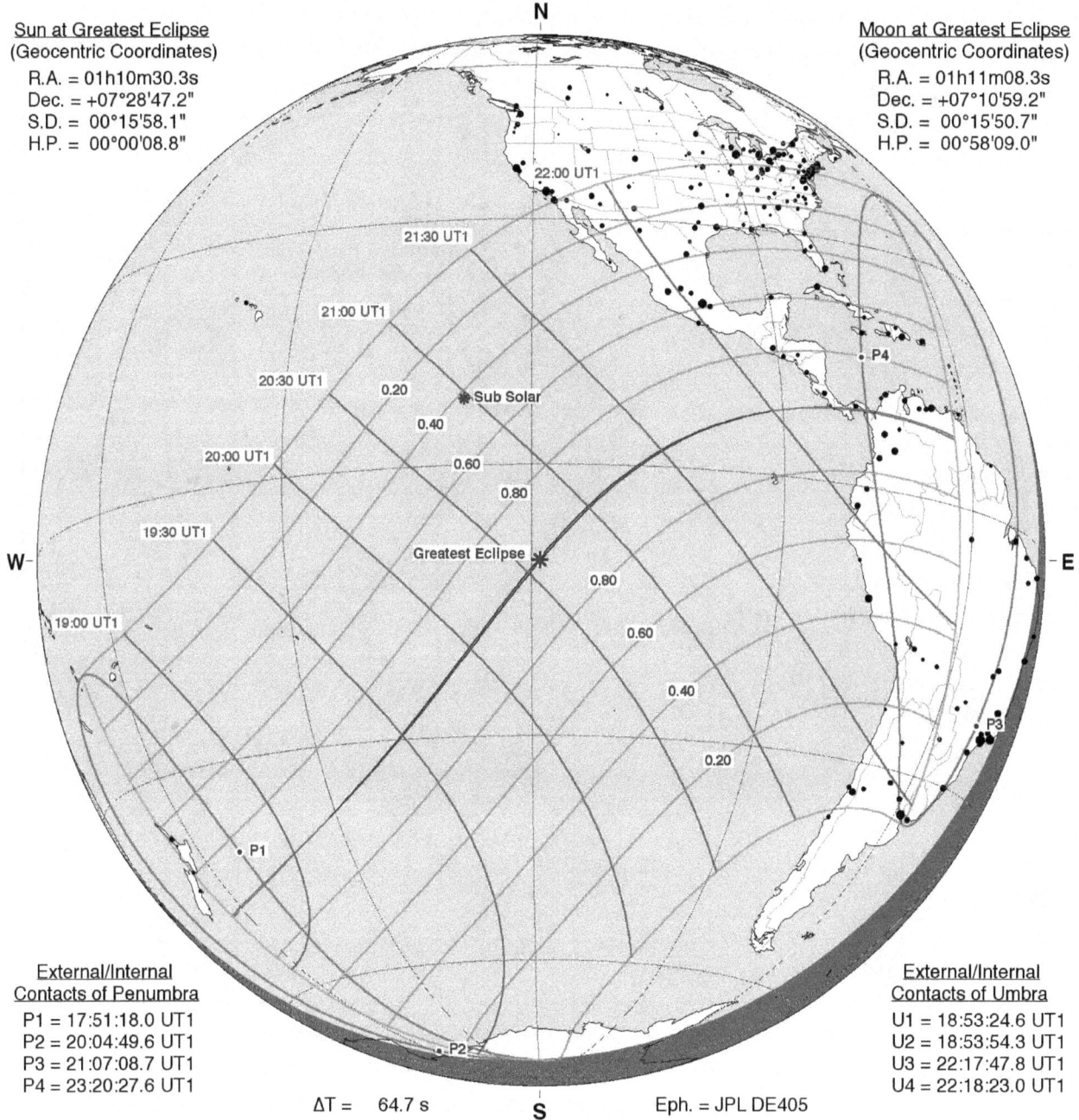

**External/Internal Contacts of Penumbra**
P1 = 17:51:18.0 UT1
P2 = 20:04:49.6 UT1
P3 = 21:07:08.7 UT1
P4 = 23:20:27.6 UT1

**External/Internal Contacts of Umbra**
U1 = 18:53:24.6 UT1
U2 = 18:53:54.3 UT1
U3 = 22:17:47.8 UT1
U4 = 22:18:23.0 UT1

ΔT = 64.7 s

Eph. = JPL DE405

**Circumstances at Greatest Eclipse: 20:35:45.9 UT1**

| | |
|---|---|
| Lat. = 10°34.0'S | Sun Alt. = 69.6° |
| Long. = 118°59.1'W | Sun Azm. = 331.9° |
| Path Width = 27.0 km | Duration = 00m42.1s |

**Circumstances at Greatest Duration: 20:33:14.1 UT1**

| | |
|---|---|
| Lat. = 11°12.1'S | Sun Alt. = 69.6° |
| Long. = 119°32.6'W | Sun Azm. = 335.7° |
| Path Width = 27.0 km | Duration = 00m42.1s |

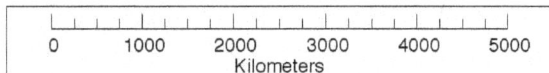

©2016 F. Espenak
www.EclipseWise.com

# Annular Solar Eclipse of 2005 Oct 03

Greatest Eclipse = 10:32:47.3 TD  (= 10:31:42.6 UT1)

Eclipse Magnitude = 0.9576
Gamma = 0.3306

Saros Series = 134
Saros Member = 43 of 71

**Sun at Greatest Eclipse**
(Geocentric Coordinates)
R.A. = 12h37m55.0s
Dec. = -04°05'04.2"
S.D. = 00°15'59.1"
H.P. = 00°00'08.8"

**Moon at Greatest Eclipse**
(Geocentric Coordinates)
R.A. = 12h38m30.3s
Dec. = -03°49'04.7"
S.D. = 00°15'05.2"
H.P. = 00°55'22.1"

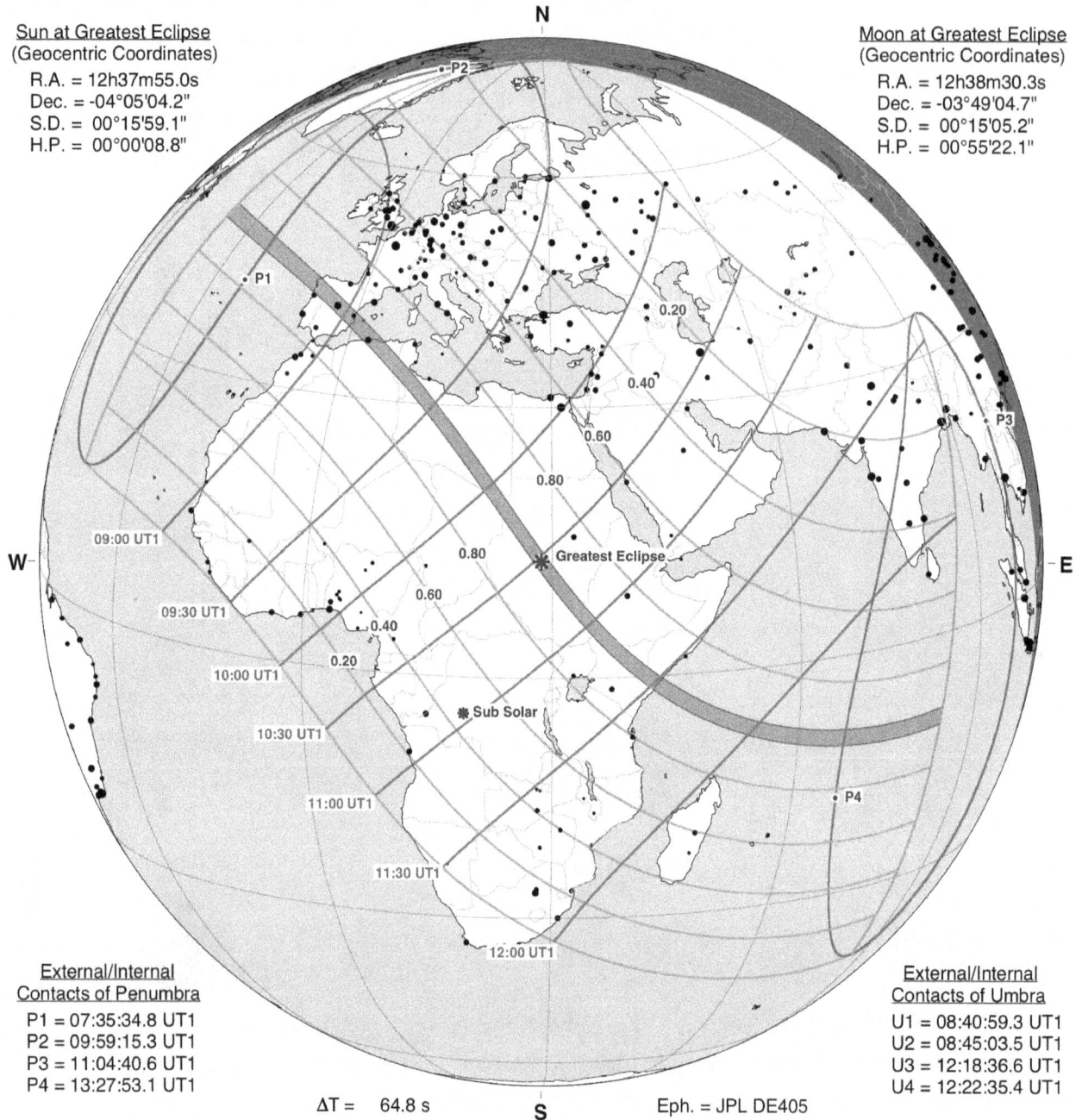

N

P2

P1

0.20

0.40

0.60

0.80

P3

09:00 UT1

W —

0.80

0.60

— E

09:30 UT1

0.40

0.20

10:00 UT1

10:30 UT1

* Greatest Eclipse

* Sub Solar

11:00 UT1

11:30 UT1

P4

12:00 UT1

**External/Internal
Contacts of Penumbra**
P1 = 07:35:34.8 UT1
P2 = 09:59:15.3 UT1
P3 = 11:04:40.6 UT1
P4 = 13:27:53.1 UT1

**External/Internal
Contacts of Umbra**
U1 = 08:40:59.3 UT1
U2 = 08:45:03.5 UT1
U3 = 12:18:36.6 UT1
U4 = 12:22:35.4 UT1

ΔT =   64.8 s

S

Eph. = JPL DE405

**Circumstances at Greatest Eclipse: 10:31:42.6 UT1**

Lat. = 12°53.4'N
Long. = 028°44.1'E
Path Width = 162.2 km

Sun Alt. =  70.6°
Sun Azm. = 209.4°
Duration = 04m31.5s

**Circumstances at Greatest Duration: 10:37:00.5 UT1**

Lat. = 11°35.3'N
Long. = 029°45.4'E
Path Width = 163.1 km

Sun Alt. =  70.4°
Sun Azm. = 217.4°
Duration = 04m31.6s

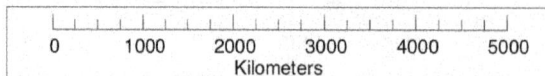

©2016  F. Espenak
www.EclipseWise.com

0      1000     2000     3000     4000     5000
Kilometers

# Total Solar Eclipse of 2006 Mar 29

Greatest Eclipse = 10:12:22.7 TD   (= 10:11:17.8 UT1)

Eclipse Magnitude = 1.0515          Saros Series = 139
Gamma = 0.3843          Saros Member = 29 of 71

Sun at Greatest Eclipse
(Geocentric Coordinates)
R.A. = 00h31m31.7s
Dec. = +03°24'10.3"
S.D. = 00°16'01.1"
H.P. = 00°00'08.8"

Moon at Greatest Eclipse
(Geocentric Coordinates)
R.A. = 00h30m46.6s
Dec. = +03°44'36.2"
S.D. = 00°16'34.9"
H.P. = 01°00'51.4"

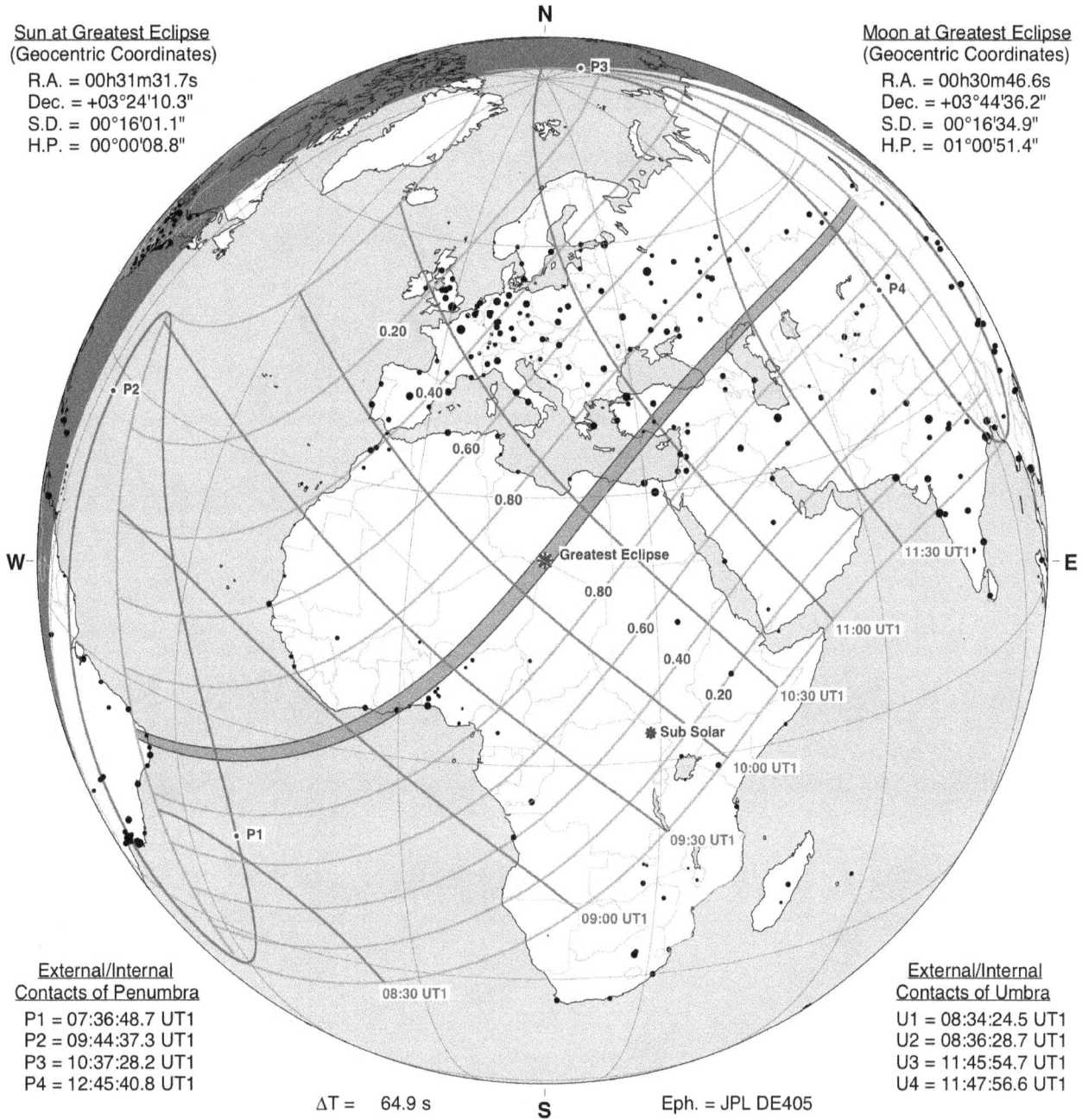

N

P3
P4
0.20
P2
0.40
0.60
0.80
W
E
Greatest Eclipse
0.80
0.60
11:30 UT1
0.40
11:00 UT1
0.20
10:30 UT1
Sub Solar
10:00 UT1
09:30 UT1
P1
09:00 UT1
08:30 UT1
S

External/Internal
Contacts of Penumbra
P1 = 07:36:48.7 UT1
P2 = 09:44:37.3 UT1
P3 = 10:37:28.2 UT1
P4 = 12:45:40.8 UT1

ΔT = 64.9 s

Eph. = JPL DE405

External/Internal
Contacts of Umbra
U1 = 08:34:24.5 UT1
U2 = 08:36:28.7 UT1
U3 = 11:45:54.7 UT1
U4 = 11:47:56.6 UT1

Circumstances at Greatest Eclipse: 10:11:17.8 UT1

| | |
|---|---|
| Lat. = 23°09.1'N | Sun Alt. = 67.3° |
| Long. = 016°44.8'E | Sun Azm. = 148.6° |
| Path Width = 183.5 km | Duration = 04m06.7s |

Circumstances at Greatest Duration: 10:11:40.5 UT1

| | |
|---|---|
| Lat. = 23°15.7'N | Sun Alt. = 67.3° |
| Long. = 016°50.8'E | Sun Azm. = 149.2° |
| Path Width = 183.5 km | Duration = 04m06.7s |

0   1000   2000   3000   4000   5000
Kilometers

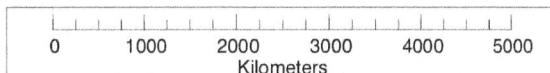

©2016  F. Espenak
www.EclipseWise.com

# Annular Solar Eclipse of 2006 Sep 22

Greatest Eclipse = 11:41:16.4 TD   (= 11:40:11.4 UT1)

| | |
|---|---|
| Eclipse Magnitude = 0.9352 | Saros Series = 144 |
| Gamma = -0.4062 | Saros Member = 16 of 70 |

Sun at Greatest Eclipse
(Geocentric Coordinates)
R.A. = 11h57m32.9s
Dec. = +00°15'56.9"
S.D. = 00°15'56.1"
H.P. = 00°00'08.8"

Moon at Greatest Eclipse
(Geocentric Coordinates)
R.A. = 11h56m50.2s
Dec. = -00°03'07.8"
S.D. = 00°14'41.9"
H.P. = 00°53'56.7"

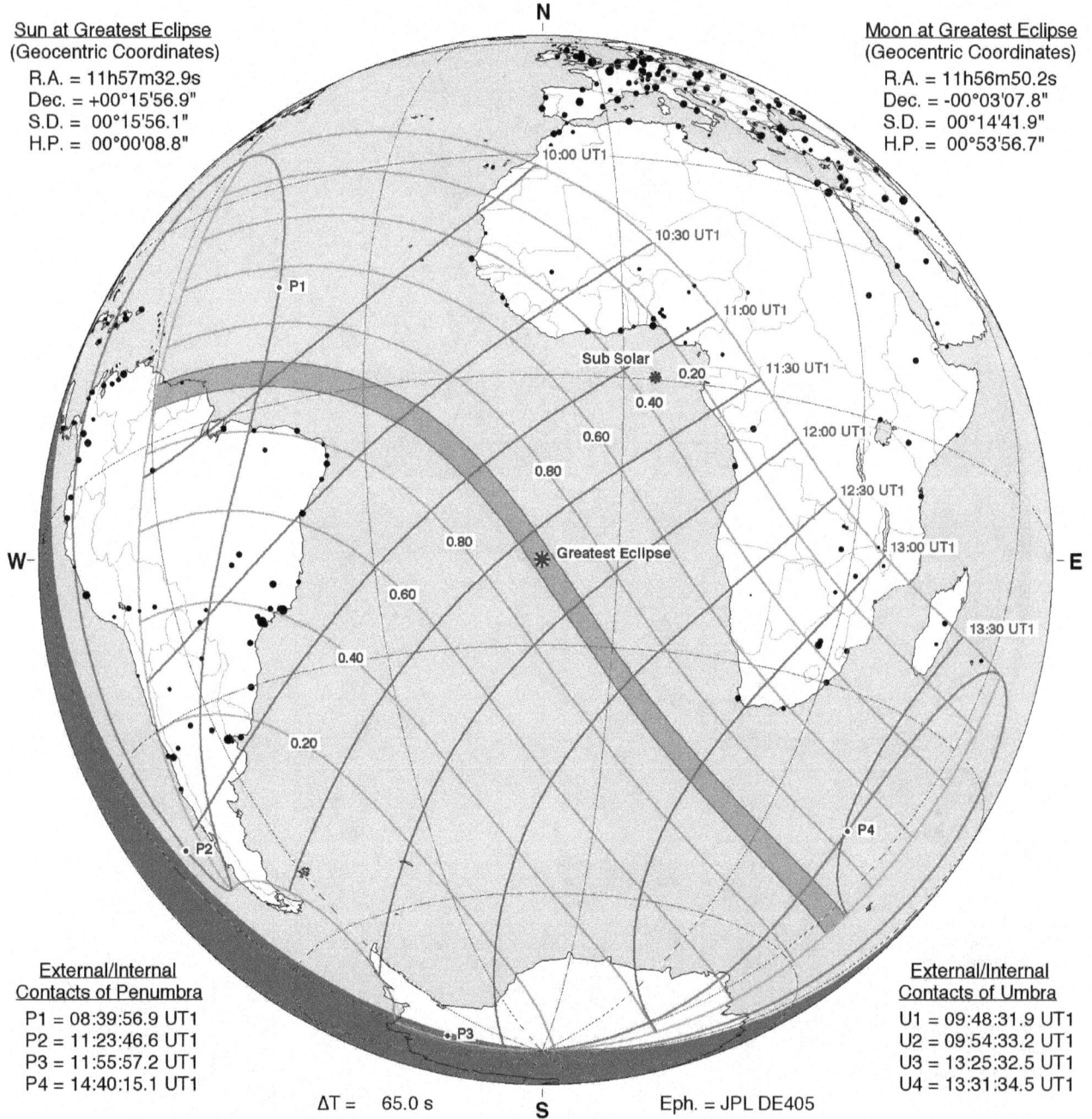

External/Internal
Contacts of Penumbra
P1 = 08:39:56.9 UT1
P2 = 11:23:46.6 UT1
P3 = 11:55:57.2 UT1
P4 = 14:40:15.1 UT1

External/Internal
Contacts of Umbra
U1 = 09:48:31.9 UT1
U2 = 09:54:33.2 UT1
U3 = 13:25:32.5 UT1
U4 = 13:31:34.5 UT1

ΔT = 65.0 s

Eph. = JPL DE405

Circumstances at Greatest Eclipse: 11:40:11.4 UT1
Lat. = 20°38.8'S          Sun Alt. = 65.9°
Long. = 009°04.6'W        Sun Azm. = 31.2°
Path Width = 261.0 km     Duration = 07m09.3s

Circumstances at Greatest Duration: 11:39:23.5 UT1
Lat. = 20°26.7'S          Sun Alt. = 65.9°
Long. = 009°13.7'W        Sun Azm. = 32.2°
Path Width = 261.3 km     Duration = 07m09.3s

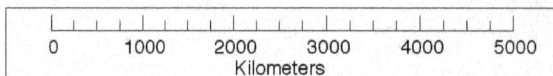

©2016 F. Espenak
www.EclipseWise.com

# Partial Solar Eclipse of 2007 Mar 19

Greatest Eclipse = 02:32:57.5 TD   (= 02:31:52.3 UT1)

Eclipse Magnitude = 0.8756          Saros Series = 149
Gamma = 1.0728          Saros Member = 20 of 71

**Sun at Greatest Eclipse**
(Geocentric Coordinates)
R.A. = 23h53m04.0s
Dec. = -00°45'04.8"
S.D. = 00°16'04.0"
H.P. = 00°00'08.8"

**Moon at Greatest Eclipse**
(Geocentric Coordinates)
R.A. = 23h50m57.2s
Dec. = +00°12'14.7"
S.D. = 00°16'40.7"
H.P. = 01°01'12.5"

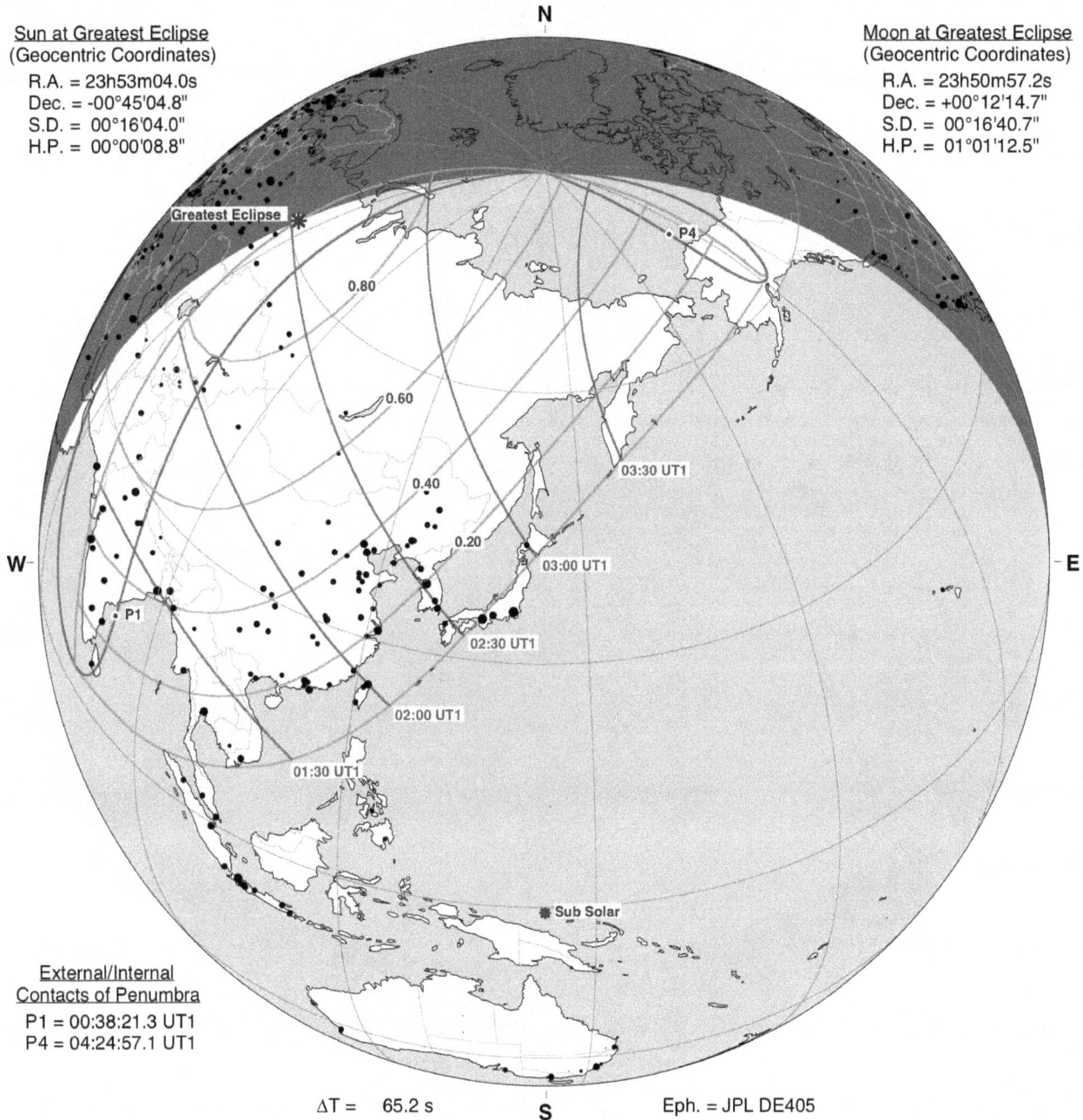

N

Greatest Eclipse

P4

0.80

0.60

0.40

03:30 UT1

0.20

03:00 UT1

W

E

02:30 UT1

P1

02:00 UT1

01:30 UT1

Sub Solar

**External/Internal
Contacts of Penumbra**
P1 = 00:38:21.3 UT1
P4 = 04:24:57.1 UT1

ΔT = 65.2 s          S          Eph. = JPL DE405

Circumstances at Greatest Eclipse: 02:31:52.3 UT1
Lat. = 61°02.9'N          Sun Alt. = 0.0°
Long. = 055°28.1'E          Sun Azm. = 91.6°

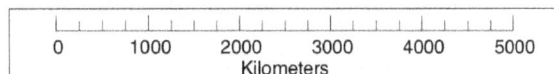

0     1000     2000     3000     4000     5000
Kilometers

# Partial Solar Eclipse of 2007 Sep 11

Greatest Eclipse = 12:32:24.5 TD   (= 12:31:19.2 UT1)

| | |
|---|---|
| Eclipse Magnitude = 0.7507 | Saros Series = 154 |
| Gamma = -1.1255 | Saros Member = 6 of 71 |

**Sun at Greatest Eclipse**
(Geocentric Coordinates)
R.A. = 11h17m20.8s
Dec. = +04°35'13.3"
S.D. = 00°15'53.2"
H.P. = 00°00'08.7"

**Moon at Greatest Eclipse**
(Geocentric Coordinates)
R.A. = 11h15m21.9s
Dec. = +03°40'57.3"
S.D. = 00°15'00.5"
H.P. = 00°55'05.0"

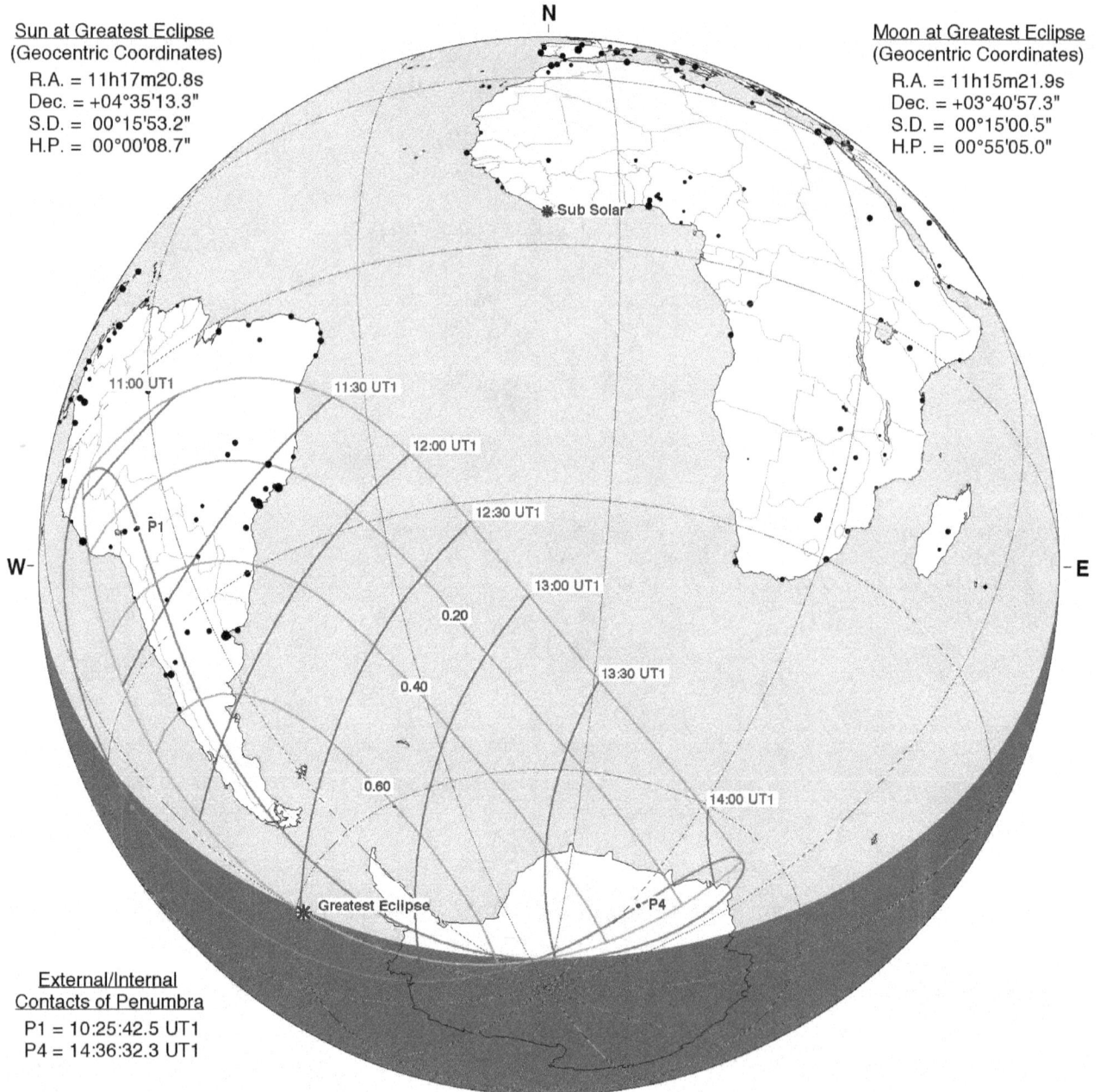

N

Sub Solar

11:00 UT1

11:30 UT1

12:00 UT1

12:30 UT1

P1

13:00 UT1

W

0.20

0.40

13:30 UT1

0.60

14:00 UT1

Greatest Eclipse

P4

E

**External/Internal
Contacts of Penumbra**
P1 = 10:25:42.5 UT1
P4 = 14:36:32.3 UT1

ΔT = 65.3 s

S

Eph. = JPL DE405

**Circumstances at Greatest Eclipse: 12:31:19.2 UT1**

| | |
|---|---|
| Lat. = 61°00.3'S | Sun Alt. = 0.0° |
| Long. = 090°14.1'W | Sun Azm. = 80.5° |

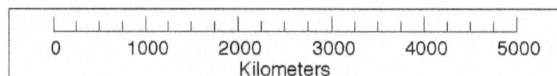

©2016 F. Espenak
www.EclipseWise.com

| 0 | 1000 | 2000 | 3000 | 4000 | 5000 |
|---|---|---|---|---|---|

Kilometers

# Annular Solar Eclipse of 2008 Feb 07

Greatest Eclipse = 03:56:10.5 TD  (= 03:55:05.0 UT1)

| | |
|---|---|
| Eclipse Magnitude = 0.9650 | Saros Series = 121 |
| Gamma = -0.9570 | Saros Member = 60 of 71 |

**Sun at Greatest Eclipse**
(Geocentric Coordinates)
R.A. = 21h20m44.7s
Dec. = -15°30'56.2"
S.D. = 00°16'13.1"
H.P. = 00°00'08.9"

**Moon at Greatest Eclipse**
(Geocentric Coordinates)
R.A. = 21h22m15.3s
Dec. = -16°21'00.5"
S.D. = 00°15'35.2"
H.P. = 00°57'12.3"

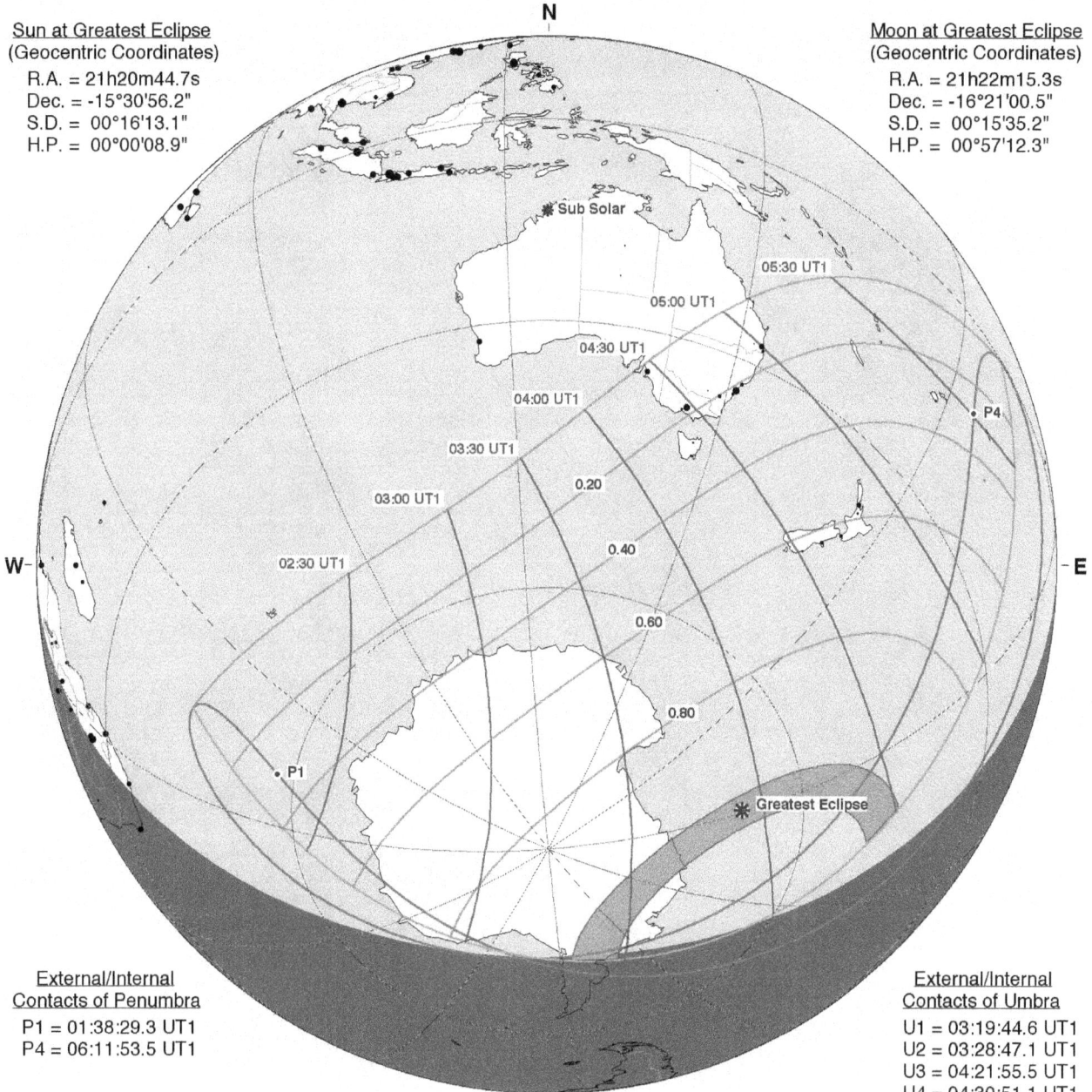

N

Sub Solar

05:30 UT1
05:00 UT1
04:30 UT1
04:00 UT1
03:30 UT1
03:00 UT1
02:30 UT1

P4

0.20
0.40
0.60
0.80

W

E

P1

Greatest Eclipse

**External/Internal Contacts of Penumbra**
P1 = 01:38:29.3 UT1
P4 = 06:11:53.5 UT1

**External/Internal Contacts of Umbra**
U1 = 03:19:44.6 UT1
U2 = 03:28:47.1 UT1
U3 = 04:21:55.5 UT1
U4 = 04:30:51.1 UT1

ΔT = 65.4 s

S

Eph. = JPL DE405

**Circumstances at Greatest Eclipse: 03:55:05.0 UT1**

| | |
|---|---|
| Lat. = 67°34.6'S | Sun Alt. = 16.3° |
| Long. = 150°30.1'W | Sun Azm. = 268.8° |
| Path Width = 444.4 km | Duration = 02m11.7s |

**Circumstances at Greatest Duration: 03:23:57.9 UT1**

| | |
|---|---|
| Lat. = 72°57.4'S | Sun Alt. = 0.0° |
| Long. = 072°31.9'W | Sun Azm. = 204.1° |
| Path Width = 581.2 km | Duration = 02m14.1s |

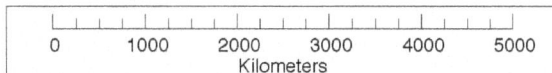

0   1000   2000   3000   4000   5000
Kilometers

# Total Solar Eclipse of 2008 Aug 01

Greatest Eclipse = 10:22:12.3 TD  (= 10:21:06.7 UT1)

| | |
|---|---|
| Eclipse Magnitude = 1.0394 | Saros Series = 126 |
| Gamma = 0.8307 | Saros Member = 47 of 72 |

**N**

**Sun at Greatest Eclipse**
(Geocentric Coordinates)
R.A. = 08h47m54.1s
Dec. = +17°51'56.4"
S.D. = 00°15'45.5"
H.P. = 00°00'08.7"

**Moon at Greatest Eclipse**
(Geocentric Coordinates)
R.A. = 08h49m08.8s
Dec. = +18°38'01.6"
S.D. = 00°16'14.1"
H.P. = 00°59'34.8"

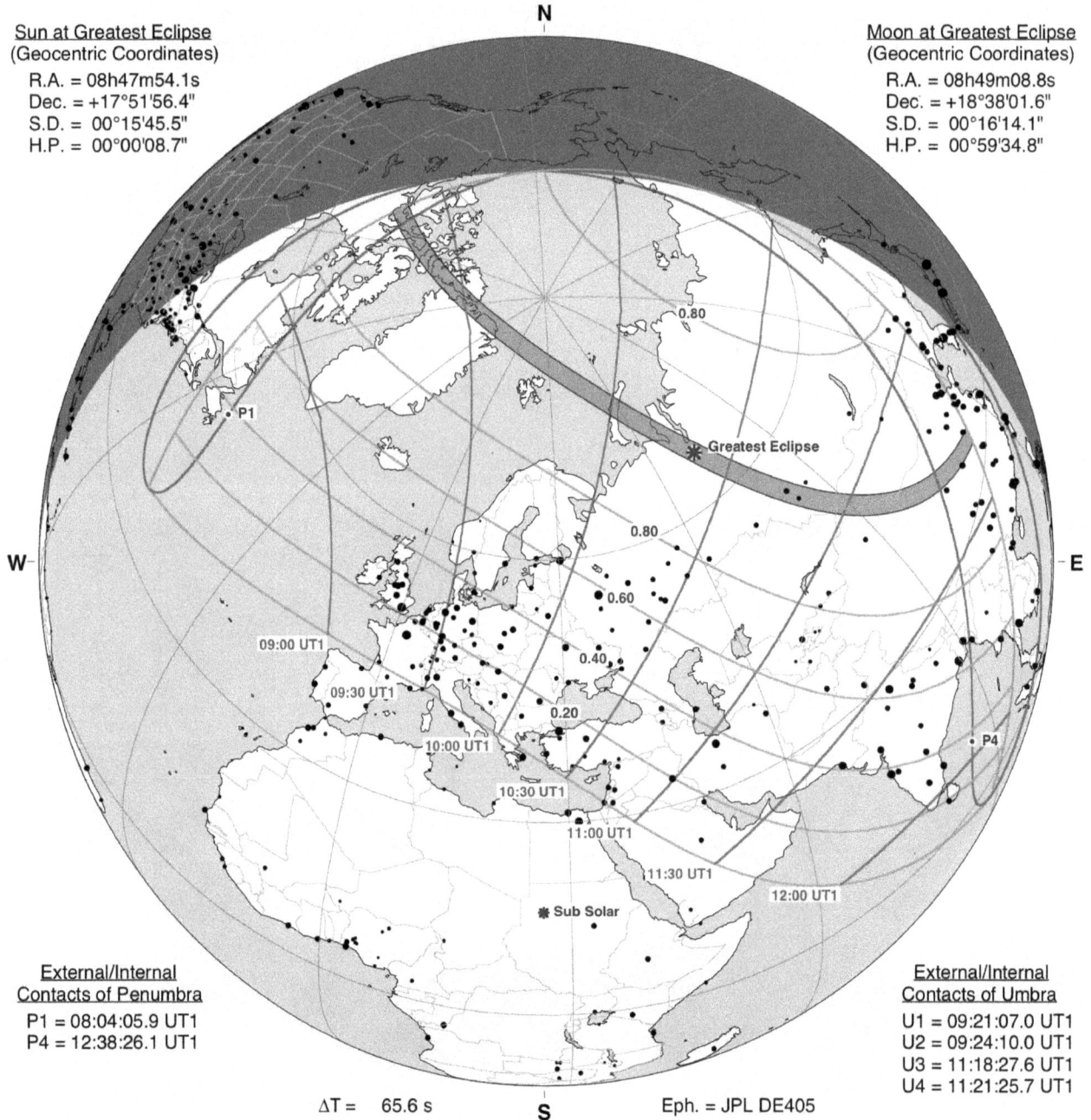

0.80

Greatest Eclipse

**W**

0.80

0.60

09:00 UT1

0.40

09:30 UT1

0.20

10:00 UT1

**E**

P4

10:30 UT1

11:00 UT1

11:30 UT1

12:00 UT1

Sub Solar

**External/Internal Contacts of Penumbra**
P1 = 08:04:05.9 UT1
P4 = 12:38:26.1 UT1

**External/Internal Contacts of Umbra**
U1 = 09:21:07.0 UT1
U2 = 09:24:10.0 UT1
U3 = 11:18:27.6 UT1
U4 = 11:21:25.7 UT1

ΔT = 65.6 s

**S**

Eph. = JPL DE405

| Circumstances at Greatest Eclipse: 10:21:06.7 UT1 | | Circumstances at Greatest Duration: 10:19:11.5 UT1 | |
|---|---|---|---|
| Lat. = 65°39.2'N | Sun Alt. = 33.5° | Lat. = 66°30.6'N | Sun Alt. = 33.5° |
| Long. = 072°17.8'E | Sun Azm. = 235.2° | Long. = 071°20.0'E | Sun Azm. = 233.2° |
| Path Width = 236.9 km | Duration = 02m27.1s | Path Width = 236.1 km | Duration = 02m27.2s |

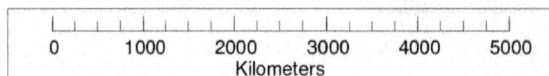

0   1000   2000   3000   4000   5000
Kilometers

# Annular Solar Eclipse of 2009 Jan 26

Greatest Eclipse = 07:59:44.5 TD  (= 07:58:38.8 UT1)

Eclipse Magnitude = 0.9282          Saros Series = 131
Gamma = -0.2820          Saros Member = 50 of 70

Sun at Greatest Eclipse
(Geocentric Coordinates)
R.A. = 20h35m32.8s
Dec. = -18°38'55.0"
S.D. = 00°16'14.6"
H.P. = 00°00'08.9"

Moon at Greatest Eclipse
(Geocentric Coordinates)
R.A. = 20h35m55.2s
Dec. = -18°53'18.2"
S.D. = 00°14'51.6"
H.P. = 00°54'32.2"

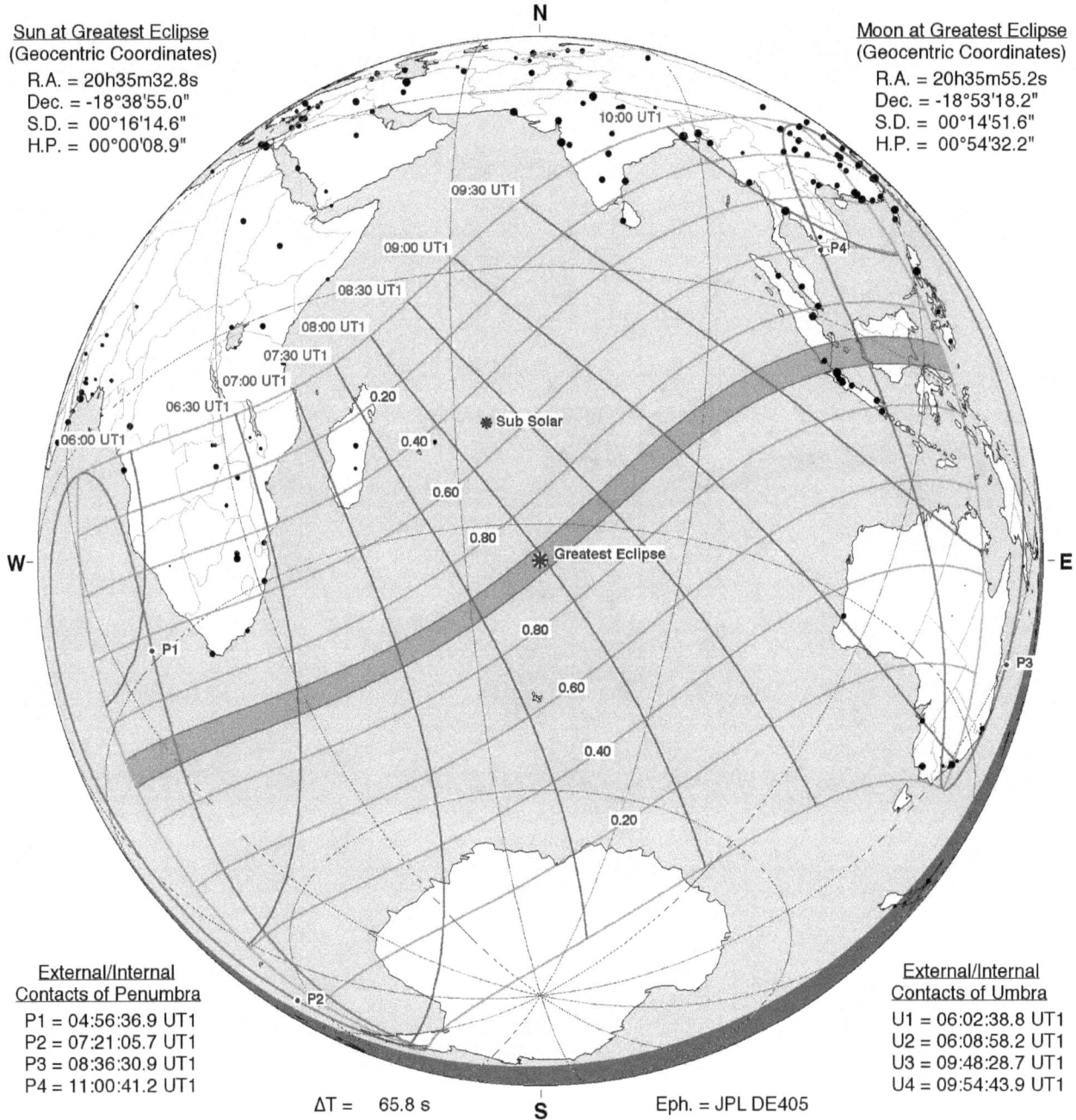

External/Internal
Contacts of Penumbra
P1 = 04:56:36.9 UT1
P2 = 07:21:05.7 UT1
P3 = 08:36:30.9 UT1
P4 = 11:00:41.2 UT1

External/Internal
Contacts of Umbra
U1 = 06:02:38.8 UT1
U2 = 06:08:58.2 UT1
U3 = 09:48:28.7 UT1
U4 = 09:54:43.9 UT1

ΔT =  65.8 s          Eph. = JPL DE405

Circumstances at Greatest Eclipse: 07:58:38.8 UT1

| | |
|---|---|
| Lat. = 34°04.5'S | Sun Alt. = 73.4° |
| Long. = 070°13.9'E | Sun Azm. = 337.0° |
| Path Width = 280.2 km | Duration = 07m53.6s |

Circumstances at Greatest Duration: 07:42:17.9 UT1

| | |
|---|---|
| Lat. = 37°03.7'S | Sun Alt. = 71.5° |
| Long. = 065°12.8'E | Sun Azm. = 7.0° |
| Path Width = 279.0 km | Duration = 07m56.0s |

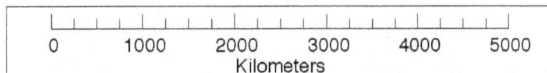

©2016  F. Espenak
www.EclipseWise.com

# Total Solar Eclipse of 2009 Jul 22

Greatest Eclipse = 02:36:24.6 TD (= 02:35:18.7 UT1)

Eclipse Magnitude = 1.0799          Saros Series = 136
Gamma = 0.0698          Saros Member = 37 of 71

Sun at Greatest Eclipse
(Geocentric Coordinates)
R.A. = 08h06m24.1s
Dec. = +20°16'03.0"
S.D. = 00°15'44.5"
H.P. = 00°00'08.7"

Moon at Greatest Eclipse
(Geocentric Coordinates)
R.A. = 08h06m29.6s
Dec. = +20°20'07.0"
S.D. = 00°16'42.7"
H.P. = 01°01'19.8"

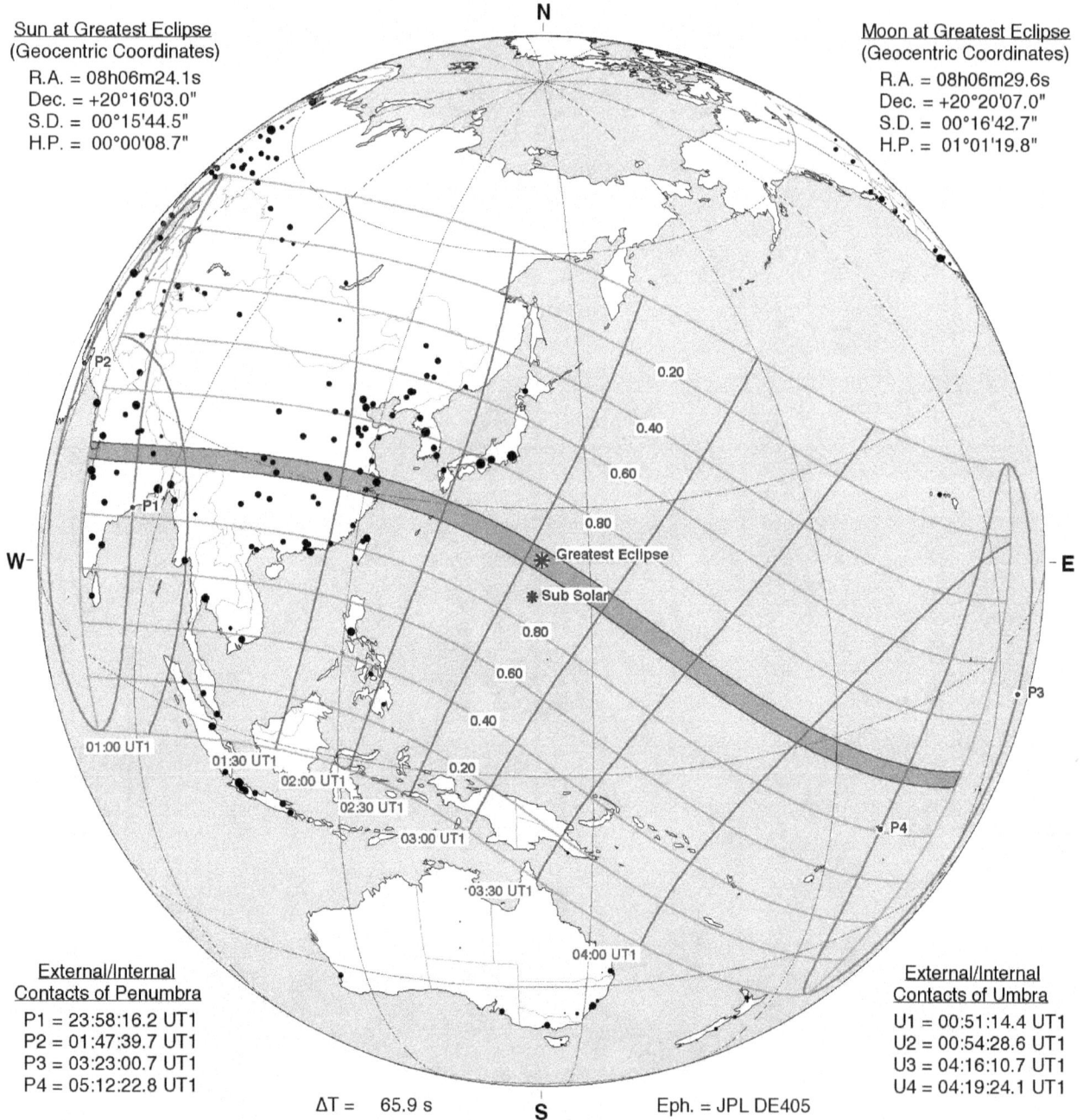

External/Internal
Contacts of Penumbra
P1 = 23:58:16.2 UT1
P2 = 01:47:39.7 UT1
P3 = 03:23:00.7 UT1
P4 = 05:12:22.8 UT1

ΔT = 65.9 s

Eph. = JPL DE405

External/Internal
Contacts of Umbra
U1 = 00:51:14.4 UT1
U2 = 00:54:28.6 UT1
U3 = 04:16:10.7 UT1
U4 = 04:19:24.1 UT1

Circumstances at Greatest Eclipse: 02:35:18.7 UT1

| Lat. = 24°13.2'N | Sun Alt. = 85.9° |
| Long. = 144°06.9'E | Sun Azm. = 197.6° |
| Path Width = 258.5 km | Duration = 06m38.9s |

Circumstances at Greatest Duration: 02:29:16.6 UT1

| Lat. = 25°14.5'N | Sun Alt. = 84.6° |
| Long. = 142°05.5'E | Sun Azm. = 157.4° |
| Path Width = 258.2 km | Duration = 06m39.5s |

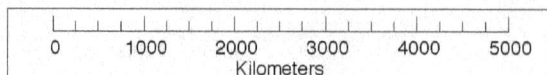

©2016 F. Espenak
www.EclipseWise.com

# Annular Solar Eclipse of 2010 Jan 15

Greatest Eclipse = 07:07:39.2 TD  (= 07:06:33.2 UT1)

| Eclipse Magnitude = 0.9190 | Saros Series = 141 |
|---|---|
| Gamma = 0.4002 | Saros Member = 23 of 70 |

Sun at Greatest Eclipse
(Geocentric Coordinates)
R.A. = 19h47m51.0s
Dec. = -21°07'38.7"
S.D. = 00°16'15.5"
H.P. = 00°00'08.9"

Moon at Greatest Eclipse
(Geocentric Coordinates)
R.A. = 19h47m25.3s
Dec. = -20°46'54.8"
S.D. = 00°14'44.3"
H.P. = 00°54'05.4"

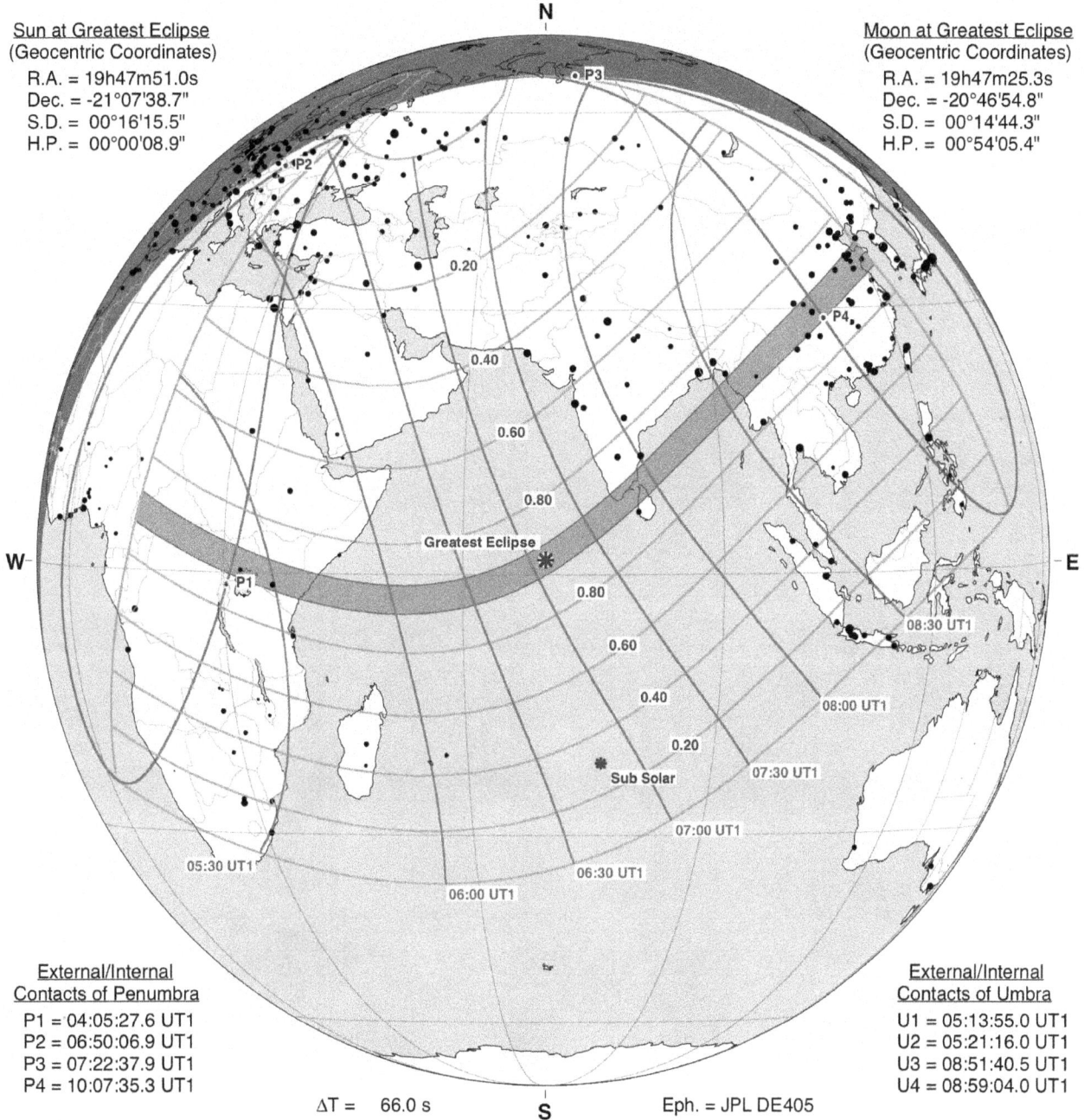

N

P3

P2

0.20

0.40

P4

0.60

0.80

Greatest Eclipse

0.80

W — — E

0.60

P1

08:30 UT1

0.40

08:00 UT1

0.20

07:30 UT1

Sub Solar

07:00 UT1

05:30 UT1

06:30 UT1

06:00 UT1

External/Internal
Contacts of Penumbra
P1 = 04:05:27.6 UT1
P2 = 06:50:06.9 UT1
P3 = 07:22:37.9 UT1
P4 = 10:07:35.3 UT1

External/Internal
Contacts of Umbra
U1 = 05:13:55.0 UT1
U2 = 05:21:16.0 UT1
U3 = 08:51:40.5 UT1
U4 = 08:59:04.0 UT1

$\Delta T$ = 66.0 s

S

Eph. = JPL DE405

### Circumstances at Greatest Eclipse: 07:06:33.2 UT1

| Lat. = 01°37.4'N | Sun Alt. = 66.4° |
|---|---|
| Long. = 069°17.4'E | Sun Azm. = 164.9° |
| Path Width = 333.1 km | Duration = 11m07.8s |

### Circumstances at Greatest Duration: 06:54:29.6 UT1

| Lat. = 00°13.2'N | Sun Alt. = 65.6° |
|---|---|
| Long. = 066°39.1'E | Sun Azm. = 151.8° |
| Path Width = 337.2 km | Duration = 11m10.7s |

©2016  F. Espenak
www.EclipseWise.com

| 0 | 1000 | 2000 | 3000 | 4000 | 5000 |
|---|---|---|---|---|---|

Kilometers

# Total Solar Eclipse of 2010 Jul 11

Greatest Eclipse = 19:34:37.9 TD  (= 19:33:31.7 UT1)

| | |
|---|---|
| Eclipse Magnitude = 1.0580 | Saros Series = 146 |
| Gamma = -0.6788 | Saros Member = 27 of 76 |

**Sun at Greatest Eclipse**
(Geocentric Coordinates)
R.A. = 07h23m57.6s
Dec. = +22°02'11.0"
S.D. = 00°15'43.9"
H.P. = 00°00'08.7"

**Moon at Greatest Eclipse**
(Geocentric Coordinates)
R.A. = 07h23m15.8s
Dec. = +21°22'29.3"
S.D. = 00°16'26.6"
H.P. = 01°00'20.9"

N

Sub Solar

18:30 UT1
19:00 UT1
19:30 UT1
20:00 UT1
20:30 UT1
0.20
0.40
0.60
0.80
18:00 UT1
P1
W
Greatest Eclipse
E
21:00 UT1
0.80
P4
0.60

**External/Internal Contacts of Penumbra**
P1 = 17:09:37.8 UT1
P4 = 21:57:14.6 UT1

**External/Internal Contacts of Umbra**
U1 = 18:15:12.3 UT1
U2 = 18:18:30.1 UT1
U3 = 20:48:19.8 UT1
U4 = 20:51:41.3 UT1

ΔT = 66.2 s      S      Eph. = JPL DE405

**Circumstances at Greatest Eclipse: 19:33:31.7 UT1**

| | |
|---|---|
| Lat. = 19°44.8'S | Sun Alt. = 47.1° |
| Long. = 121°52.5'W | Sun Azm. = 13.5° |
| Path Width = 258.6 km | Duration = 05m20.2s |

**Circumstances at Greatest Duration: 19:31:26.0 UT1**

| | |
|---|---|
| Lat. = 19°29.0'S | Sun Alt. = 47.1° |
| Long. = 122°32.2'W | Sun Azm. = 15.1° |
| Path Width = 259.5 km | Duration = 05m20.3s |

0  1000  2000  3000  4000  5000
Kilometers

©2016  F. Espenak
www.EclipseWise.com

# Partial Solar Eclipse of 2011 Jan 04

Greatest Eclipse = 08:51:42.4 TD  (= 08:50:36.1 UT1)

Eclipse Magnitude = 0.8576          Saros Series = 151

Gamma = 1.0626          Saros Member = 14 of 72

Sun at Greatest Eclipse
(Geocentric Coordinates)
R.A. = 18h59m14.9s
Dec. = -22°44'21.1"
S.D. = 00°16'15.9"
H.P. = 00°00'08.9"

Moon at Greatest Eclipse
(Geocentric Coordinates)
R.A. = 18h58m23.8s
Dec. = -21°46'01.2"
S.D. = 00°15'18.1"
H.P. = 00°56'09.6"

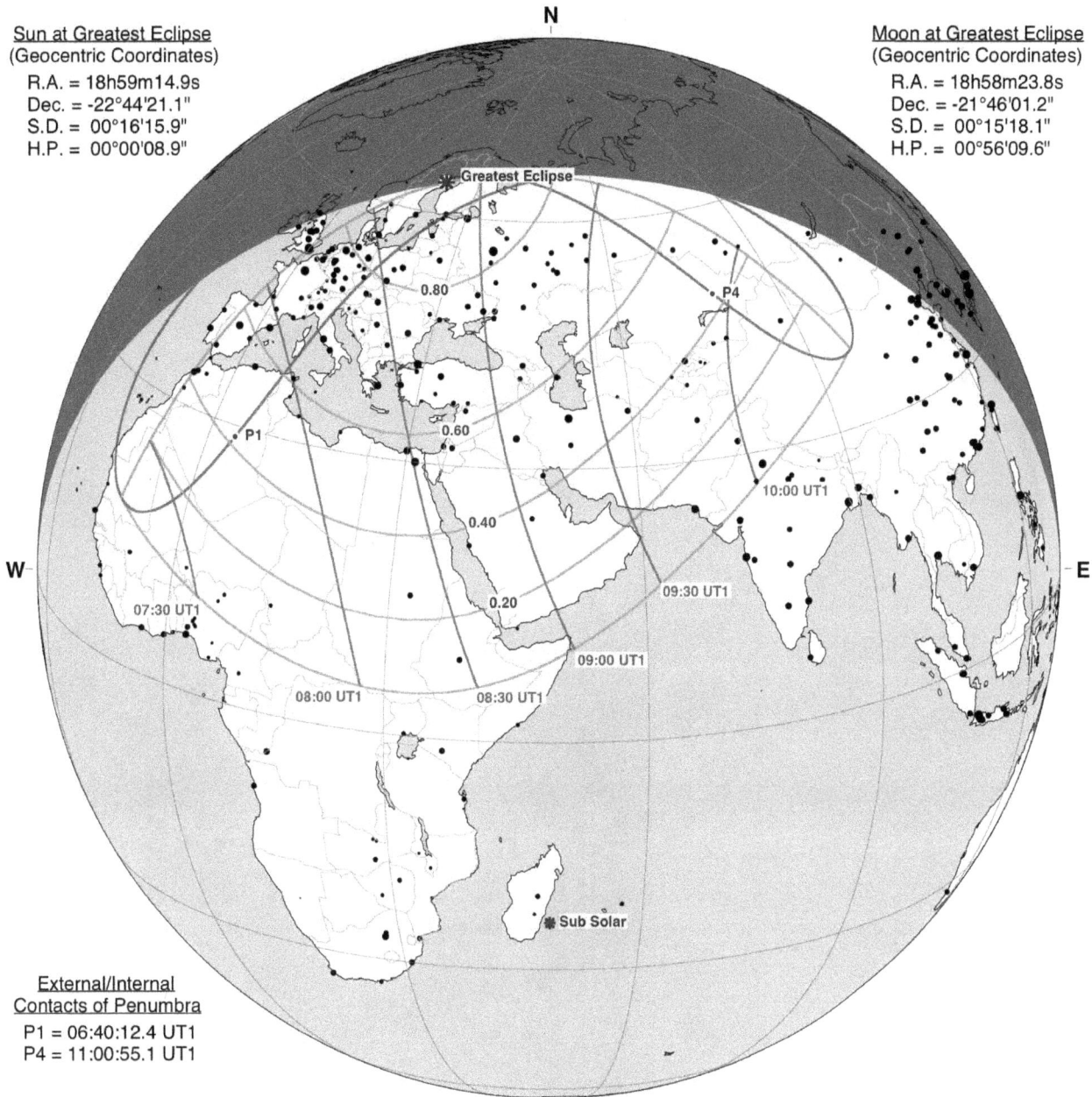

N

* Greatest Eclipse

0.80

P4

0.60

P1

0.40

10:00 UT1

W

0.20

09:30 UT1

E

07:30 UT1

09:00 UT1

08:00 UT1          08:30 UT1

* Sub Solar

External/Internal
Contacts of Penumbra
P1 = 06:40:12.4 UT1
P4 = 11:00:55.1 UT1

ΔT = 66.3 s          S          Eph. = JPL DE405

Circumstances at Greatest Eclipse: 08:50:36.1 UT1

Lat. = 64°39.1'N          Sun Alt. =   0.0°
Long. = 020°47.9'E          Sun Azm. = 154.5°

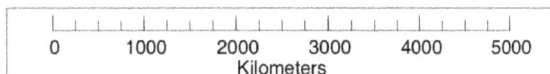

0     1000    2000    3000    4000    5000
Kilometers

©2016  F. Espenak
www.EclipseWise.com

# Partial Solar Eclipse of 2011 Jun 01

Greatest Eclipse = 21:17:18.4 TD   (= 21:16:12.0 UT1)

Eclipse Magnitude = 0.6011          Saros Series = 118
Gamma = 1.2130                      Saros Member = 68 of 72

Sun at Greatest Eclipse
(Geocentric Coordinates)
R.A. = 04h37m53.4s
Dec. = +22°05'47.3"
S.D. = 00°15'46.3"
H.P. = 00°00'08.7"

Moon at Greatest Eclipse
(Geocentric Coordinates)
R.A. = 04h37m41.0s
Dec. = +23°13'19.4"
S.D. = 00°15'13.4"
H.P. = 00°55'52.1"

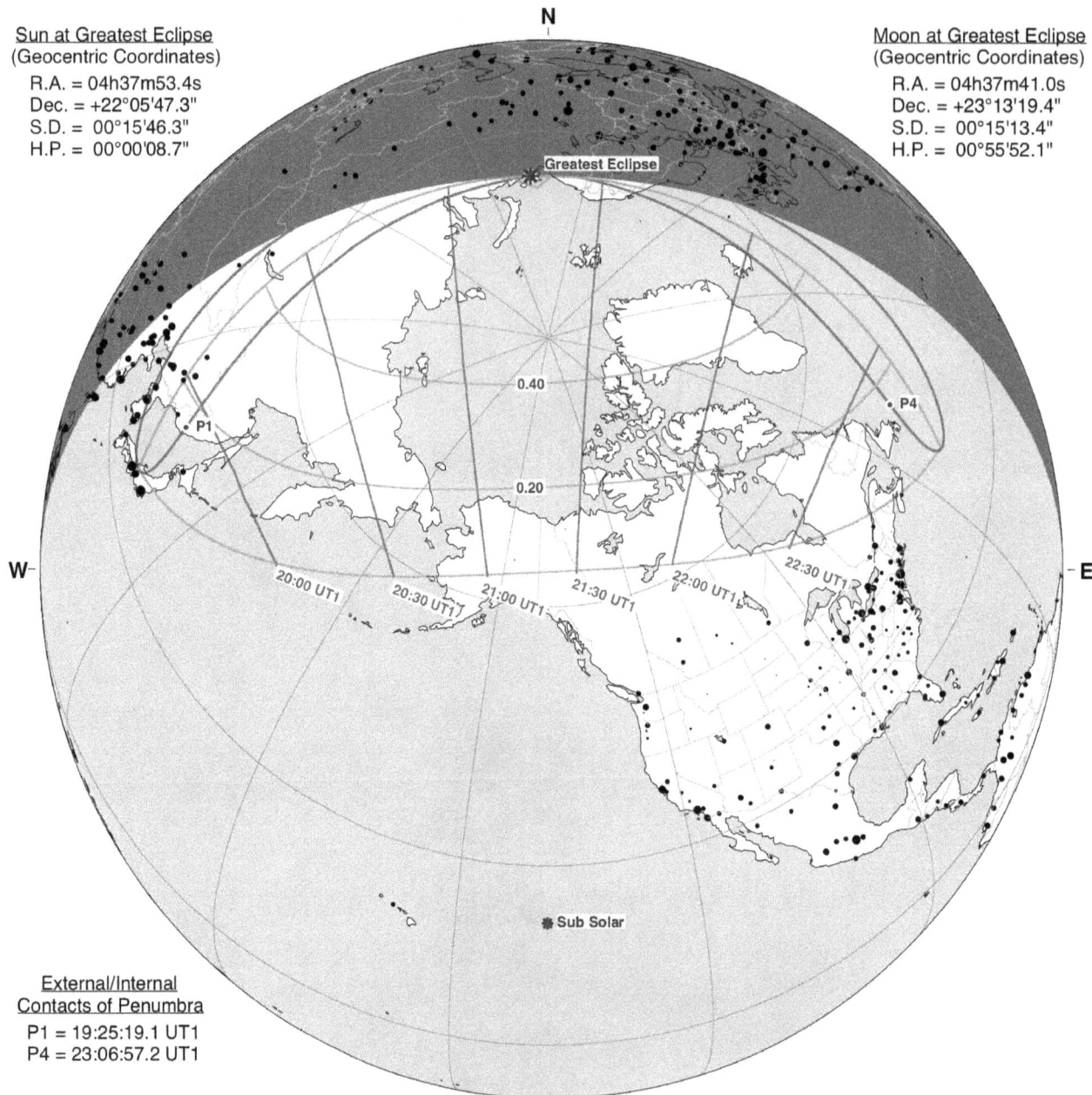

N

Greatest Eclipse

0.40

0.20

P1

P4

W

E

20:00 UT1   20:30 UT1   21:00 UT1   21:30 UT1   22:00 UT1   22:30 UT1

Sub Solar

External/Internal
Contacts of Penumbra
P1 = 19:25:19.1 UT1
P4 = 23:06:57.2 UT1

ΔT = 66.4 s          S          Eph. = JPL DE405

Circumstances at Greatest Eclipse: 21:16:12.0 UT1
Lat. = 67°47.0'N          Sun Alt. = 0.0°
Long. = 046°47.2'E        Sun Azm. = 5.9°

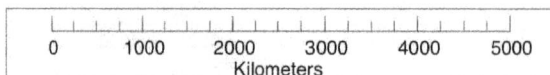

0   1000   2000   3000   4000   5000
Kilometers

©2016  F. Espenak
www.EclipseWise.com

# Partial Solar Eclipse of 2011 Jul 01

Greatest Eclipse = 08:39:30.3 TD   (= 08:38:23.9 UT1)

| | |
|---|---|
| Eclipse Magnitude = 0.0971 | Saros Series = 156 |
| Gamma = -1.4917 | Saros Member = 1 of 69 |

N

**Sun at Greatest Eclipse**
(Geocentric Coordinates)
R.A. = 06h40m01.7s
Dec. = +23°07'05.9"
S.D. = 00°15'43.9"
H.P. = 00°00'08.6"

**Moon at Greatest Eclipse**
(Geocentric Coordinates)
R.A. = 06h39m02.0s
Dec. = +21°42'47.5"
S.D. = 00°15'38.6"
H.P. = 00°57'24.6"

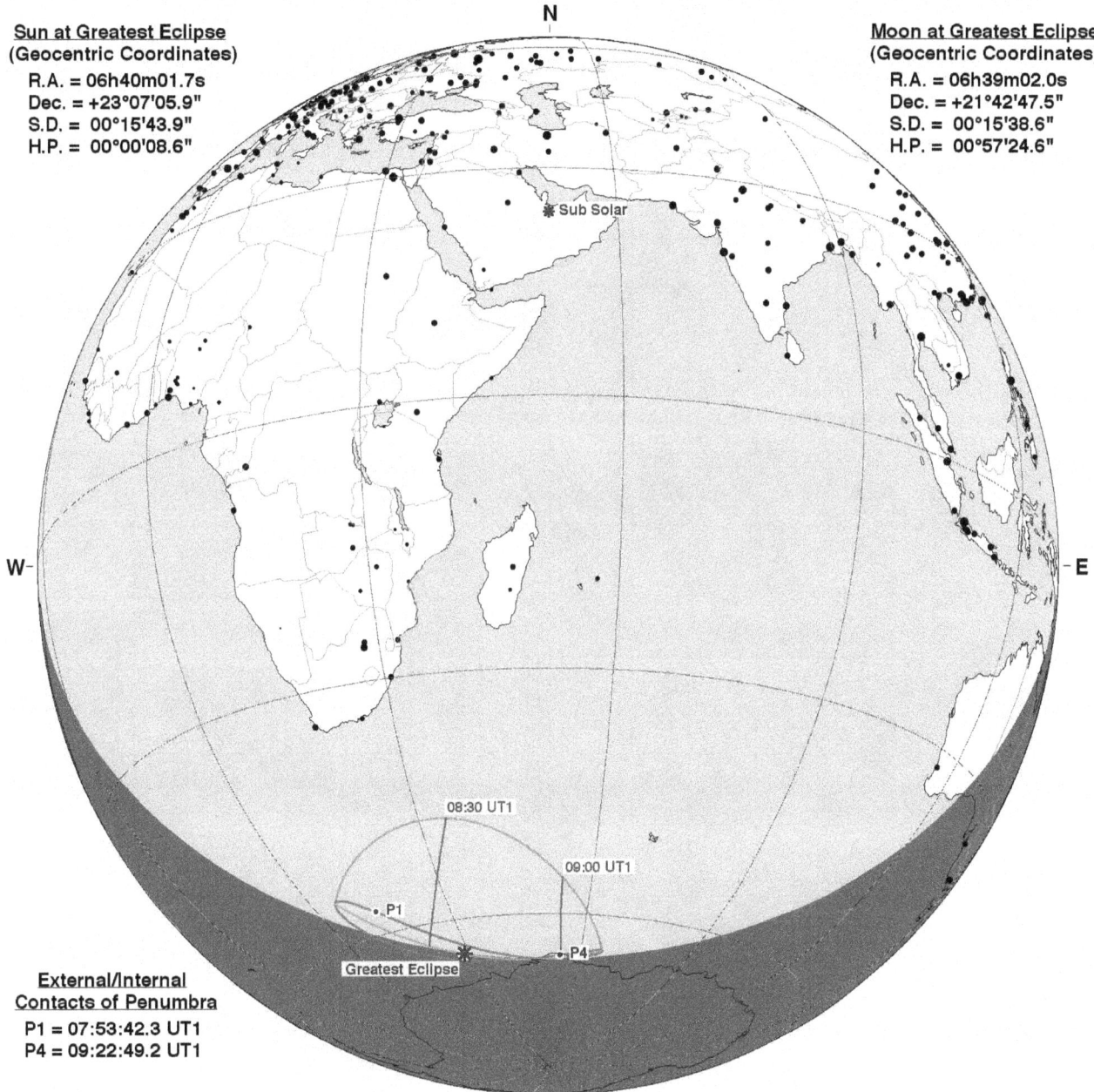

* Sub Solar

W—

—E

08:30 UT1

09:00 UT1

P1

Greatest Eclipse

P4

**External/Internal**
**Contacts of Penumbra**
P1 = 07:53:42.3 UT1
P4 = 09:22:49.2 UT1

ΔT = 66.4 s

S

Eph. = JPL DE405

**Circumstances at Greatest Eclipse: 08:38:23.9 UT1**

| | |
|---|---|
| Lat. = 65°09.5'S | Sun Alt. = 0.0° |
| Long. = 028°38.2'E | Sun Azm. = 20.8° |

| 0 | 1000 | 2000 | 3000 | 4000 | 5000 |
|---|---|---|---|---|---|

Kilometers

©2016 F. Espenak
www.EclipseWise.com

# Partial Solar Eclipse of 2011 Nov 25

Greatest Eclipse = 06:21:24.5 TD  (= 06:20:18.0 UT1)

| | |
|---|---|
| Eclipse Magnitude = 0.9047 | Saros Series = 123 |
| Gamma = -1.0536 | Saros Member = 53 of 70 |

**N**

**Sun at Greatest Eclipse**
(Geocentric Coordinates)
R.A. = 16h02m13.7s
Dec. = -20°40'56.2"
S.D. = 00°16'12.1"
H.P. = 00°00'08.9"

**Moon at Greatest Eclipse**
(Geocentric Coordinates)
R.A. = 16h01m46.2s
Dec. = -21°44'25.4"
S.D. = 00°16'32.6"
H.P. = 01°00'42.7"

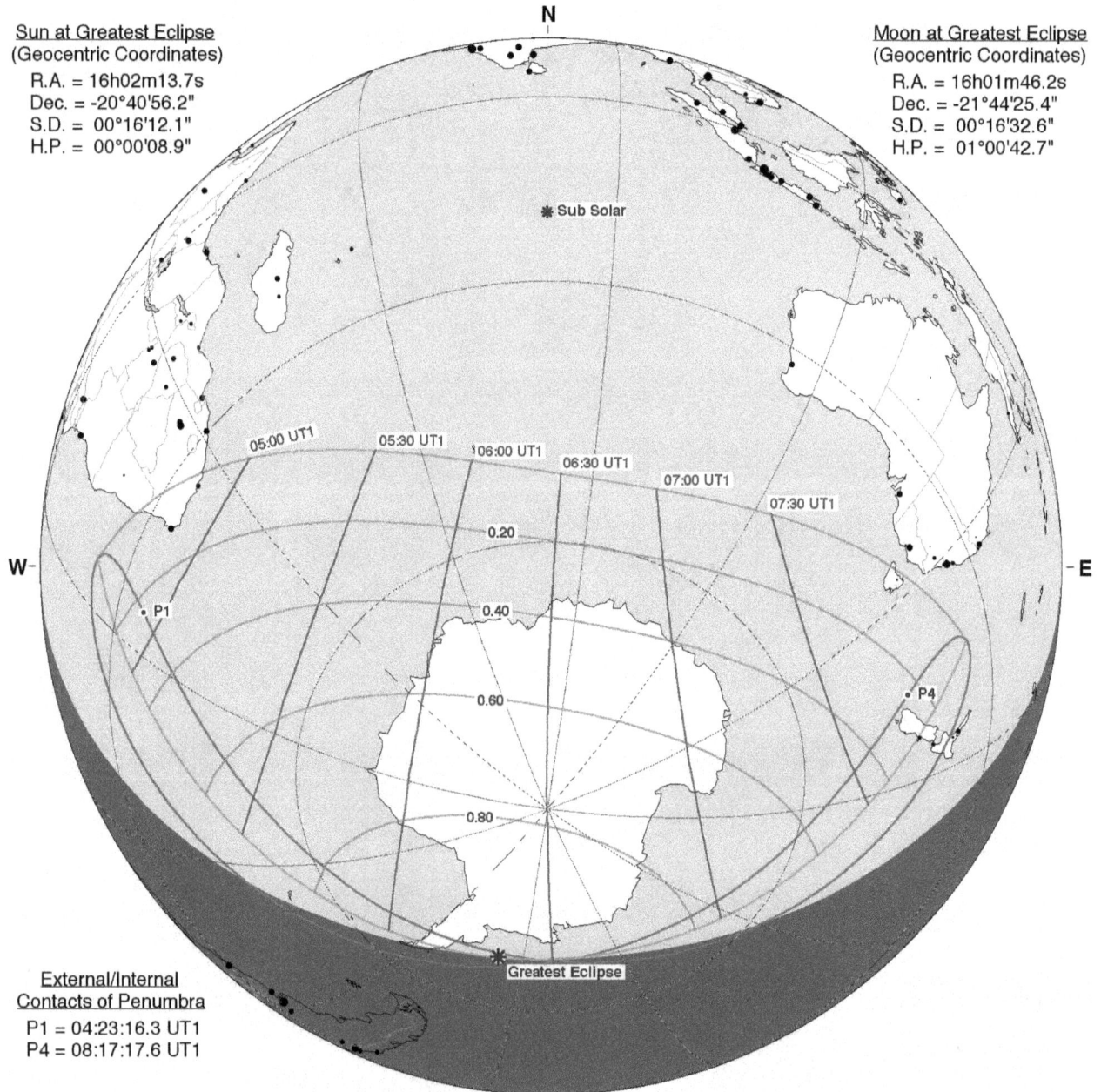

✷ Sub Solar

05:00 UT1   05:30 UT1   06:00 UT1   06:30 UT1   07:00 UT1   07:30 UT1

0.20

**W**

0.40

P1

0.60

P4

0.80

✷ Greatest Eclipse

**External/Internal Contacts of Penumbra**
P1 = 04:23:16.3 UT1
P4 = 08:17:17.6 UT1

ΔT = 66.5 s       **S**       Eph. = JPL DE405

**Circumstances at Greatest Eclipse: 06:20:18.0 UT1**

| | |
|---|---|
| Lat. = 68°34.1'S | Sun Alt. = 0.0° |
| Long. = 082°25.9'W | Sun Azm. = 165.1° |

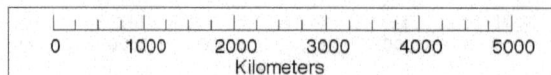

**E**

0    1000    2000    3000    4000    5000
Kilometers

# Annular Solar Eclipse of 2012 May 20

Greatest Eclipse = 23:53:53.6 TD (= 23:52:47.0 UT1)

| | |
|---|---|
| Eclipse Magnitude = 0.9439 | Saros Series = 128 |
| Gamma = 0.4828 | Saros Member = 58 of 73 |

**Sun at Greatest Eclipse**
(Geocentric Coordinates)
R.A. = 03h52m43.0s
Dec. = +20°13'15.1"
S.D. = 00°15'48.1"
H.P. = 00°00'08.7"

**Moon at Greatest Eclipse**
(Geocentric Coordinates)
R.A. = 03h52m30.7s
Dec. = +20°39'06.3"
S.D. = 00°14'43.3"
H.P. = 00°54'01.7"

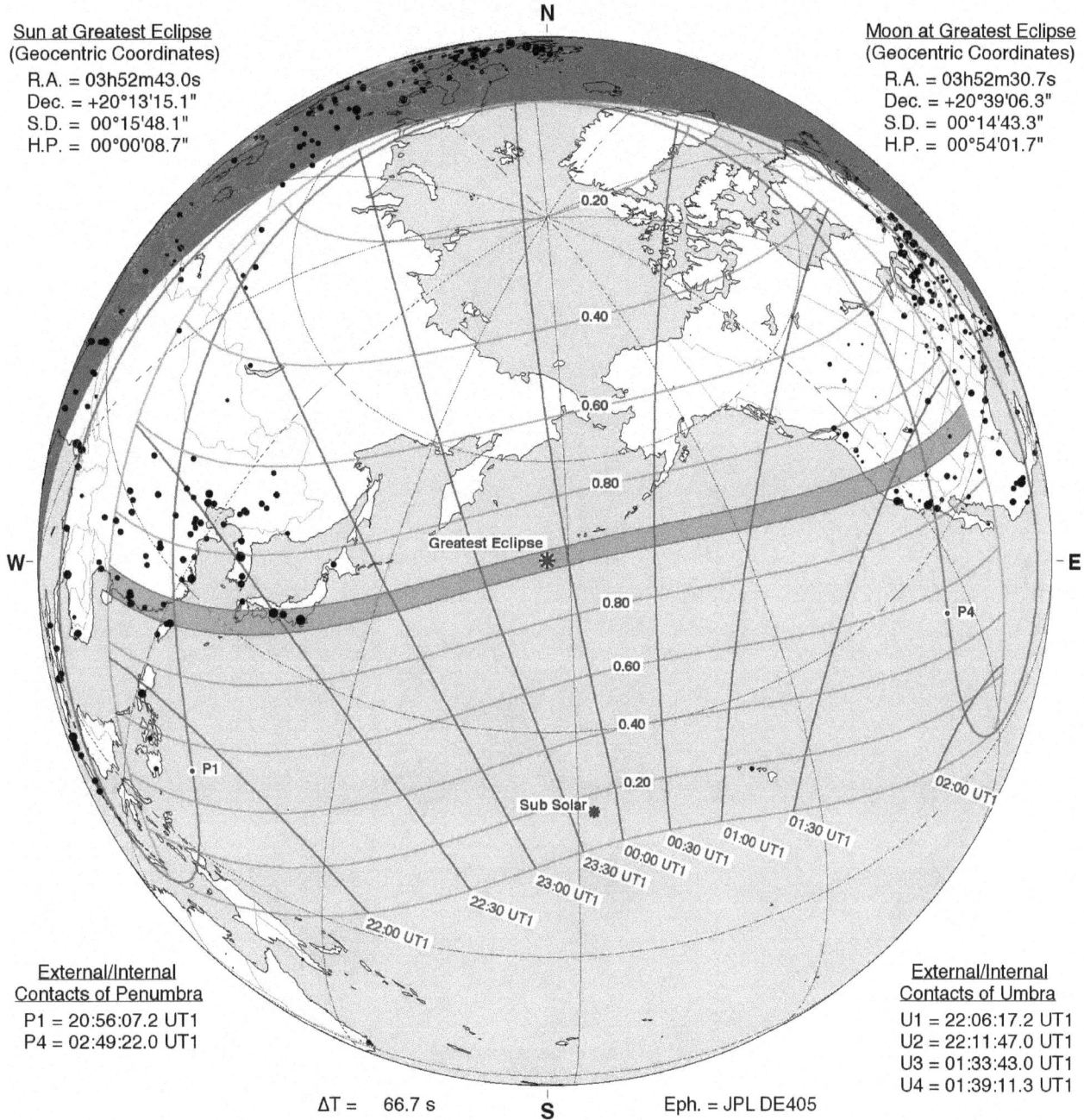

N
W — — E
S

0.20
0.40
0.60
0.80
Greatest Eclipse
0.80
0.60
0.40
0.20
Sub Solar
P1
P4

02:00 UT1
01:30 UT1
01:00 UT1
00:30 UT1
00:00 UT1
23:30 UT1
23:00 UT1
22:30 UT1
22:00 UT1

**External/Internal**
**Contacts of Penumbra**
P1 = 20:56:07.2 UT1
P4 = 02:49:22.0 UT1

**External/Internal**
**Contacts of Umbra**
U1 = 22:06:17.2 UT1
U2 = 22:11:47.0 UT1
U3 = 01:33:43.0 UT1
U4 = 01:39:11.3 UT1

ΔT = 66.7 s        Eph. = JPL DE405

**Circumstances at Greatest Eclipse: 23:52:47.0 UT1**

| | |
|---|---|
| Lat. = 49°05.6'N | Sun Alt. = 60.9° |
| Long. = 176°16.4'E | Sun Azm. = 171.0° |
| Path Width = 236.9 km | Duration = 05m46.3s |

**Circumstances at Greatest Duration: 23:55:53.1 UT1**

| | |
|---|---|
| Lat. = 49°19.3'N | Sun Alt. = 60.8° |
| Long. = 177°45.8'E | Sun Azm. = 175.4° |
| Path Width = 236.7 km | Duration = 05m46.3s |

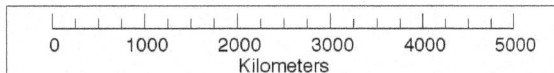

0   1000   2000   3000   4000   5000
Kilometers

©2016 F. Espenak
www.EclipseWise.com

# Total Solar Eclipse of 2012 Nov 13

Greatest Eclipse = 22:12:55.2 TD  (= 22:11:48.4 UT1)

| | |
|---|---|
| Eclipse Magnitude = 1.0500 | Saros Series = 133 |
| Gamma = -0.3719 | Saros Member = 45 of 72 |

**N**

Sun at Greatest Eclipse
(Geocentric Coordinates)
R.A. = 15h18m06.7s
Dec. = -18°15'02.6"
S.D. = 00°16'09.9"
H.P. = 00°00'08.9"

Moon at Greatest Eclipse
(Geocentric Coordinates)
R.A. = 15h17m51.2s
Dec. = -18°37'29.5"
S.D. = 00°16'42.4"
H.P. = 01°01'19.0"

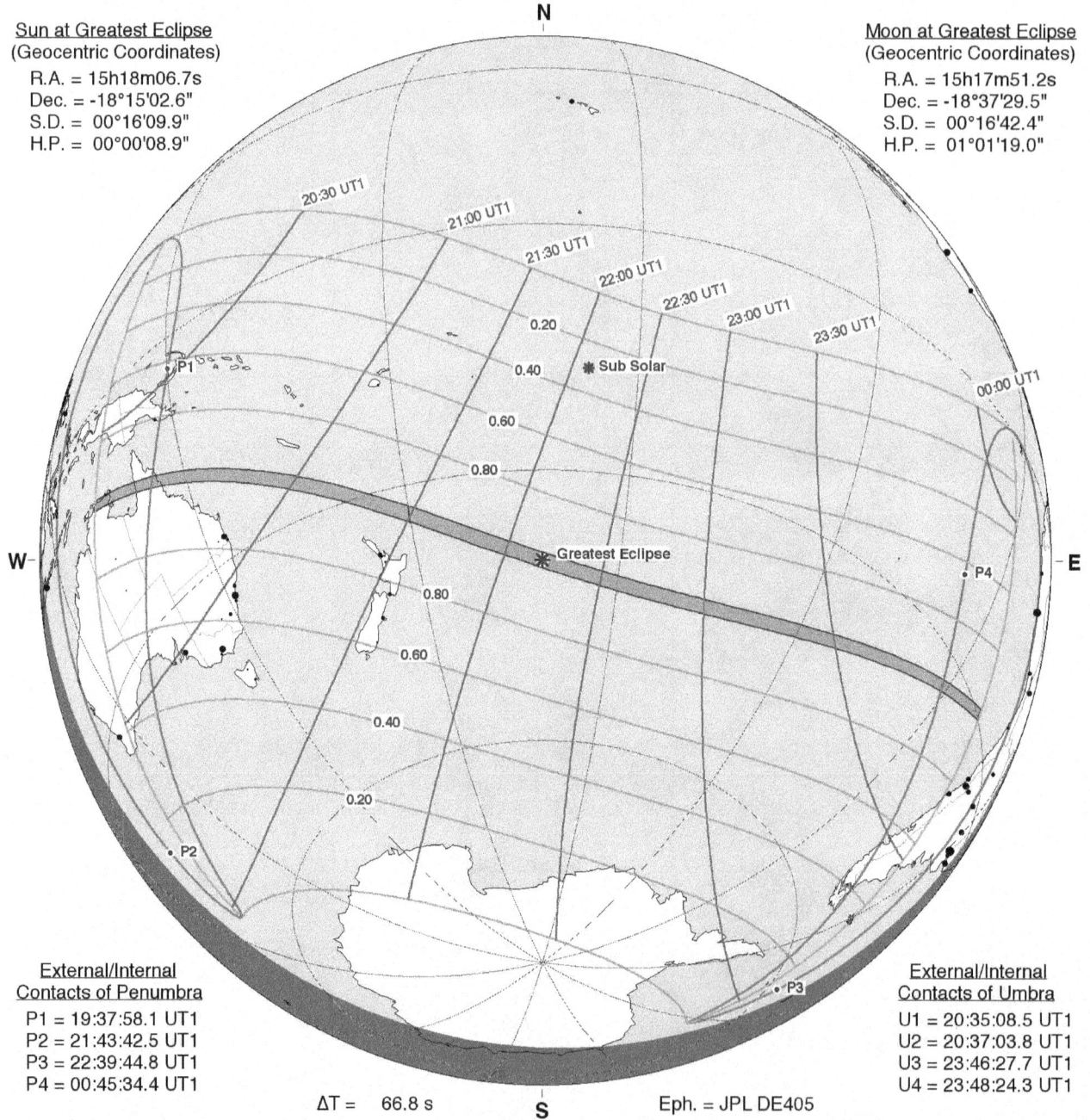

20:30 UT1
21:00 UT1
21:30 UT1
22:00 UT1
22:30 UT1
23:00 UT1
23:30 UT1
00:00 UT1

0.20
0.40
Sub Solar
0.60
0.80
P1
Greatest Eclipse
0.80
0.60
0.40
0.20
P2
P3
P4

**W**

**E**

External/Internal
Contacts of Penumbra
P1 = 19:37:58.1 UT1
P2 = 21:43:42.5 UT1
P3 = 22:39:44.8 UT1
P4 = 00:45:34.4 UT1

External/Internal
Contacts of Umbra
U1 = 20:35:08.5 UT1
U2 = 20:37:03.8 UT1
U3 = 23:46:27.7 UT1
U4 = 23:48:24.3 UT1

ΔT = 66.8 s

**S**

Eph. = JPL DE405

| Circumstances at Greatest Eclipse: 22:11:48.4 UT1 | | Circumstances at Greatest Duration: 22:14:00.1 UT1 | |
|---|---|---|---|
| Lat. = 39°57.4'S | Sun Alt. = 68.0° | Lat. = 40°11.7'S | Sun Alt. = 67.9° |
| Long. = 161°20.2'W | Sun Azm. = 11.4° | Long. = 160°19.4'W | Sun Azm. = 7.4° |
| Path Width = 179.0 km | Duration = 04m02.2s | Path Width = 178.8 km | Duration = 04m02.2s |

0   1000   2000   3000   4000   5000
Kilometers

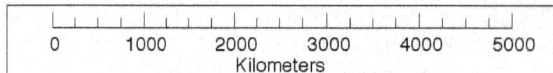

©2016 F. Espenak
www.EclipseWise.com

# Annular Solar Eclipse of 2013 May 10

Greatest Eclipse = 00:26:20.3 TD   (= 00:25:13.3 UT1)

Eclipse Magnitude = 0.9544          Saros Series = 138
Gamma = -0.2694            Saros Member = 31 of 70

Sun at Greatest Eclipse
(Geocentric Coordinates)
R.A. = 03h08m17.4s
Dec. = +17°36'34.3"
S.D. = 00°15'50.4"
H.P. = 00°00'08.7"

Moon at Greatest Eclipse
(Geocentric Coordinates)
R.A. = 03h08m28.1s
Dec. = +17°22'06.3"
S.D. = 00°14'53.8"
H.P. = 00°54'40.4"

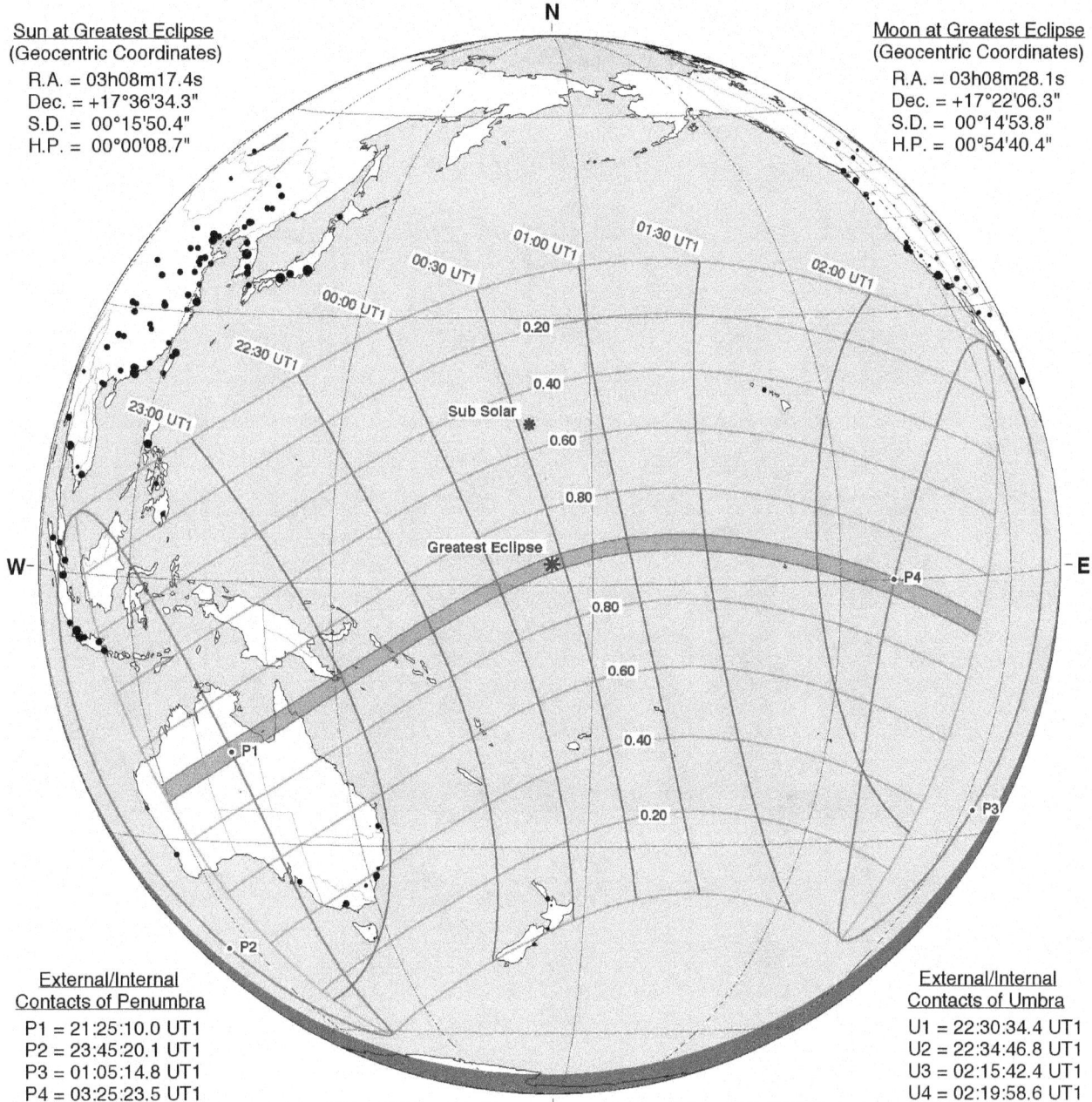

N

01:00 UT1
01:30 UT1
00:30 UT1
02:00 UT1
00:00 UT1
0.20
22:30 UT1
0.40
23:00 UT1
Sub Solar
0.60
0.80
Greatest Eclipse
W
E
P4
0.80
0.60
0.40
P1
0.20
P3
P2
S

External/Internal
Contacts of Penumbra
P1 = 21:25:10.0 UT1
P2 = 23:45:20.1 UT1
P3 = 01:05:14.8 UT1
P4 = 03:25:23.5 UT1

ΔT = 67.0 s

Eph. = JPL DE405

External/Internal
Contacts of Umbra
U1 = 22:30:34.4 UT1
U2 = 22:34:46.8 UT1
U3 = 02:15:42.4 UT1
U4 = 02:19:58.6 UT1

Circumstances at Greatest Eclipse: 00:25:13.3 UT1
Lat. = 02°13.2'N        Sun Alt. = 74.4°
Long. = 175°27.9'E      Sun Azm. = 350.5°
Path Width = 172.6 km   Duration = 06m03.4s

Circumstances at Greatest Duration: 00:35:20.7 UT1
Lat. = 02°58.7'N        Sun Alt. = 73.6°
Long. = 177°50.9'E      Sun Azm. = 333.6°
Path Width = 173.6 km   Duration = 06m04.3s

| 0 | 1000 | 2000 | 3000 | 4000 | 5000 |
Kilometers

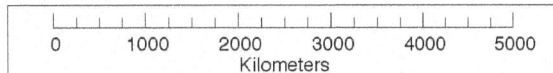

# Hybrid Solar Eclipse of 2013 Nov 03

Greatest Eclipse = 12:47:36.1 TD  (= 12:46:28.9 UT1)

| | |
|---|---|
| Eclipse Magnitude = 1.0159 | Saros Series = 143 |
| Gamma = 0.3272 | Saros Member = 23 of 72 |

**Sun at Greatest Eclipse**
(Geocentric Coordinates)
R.A. = 14h35m19.9s
Dec. = -15°12'22.5"
S.D. = 00°16'07.4"
H.P. = 00°00'08.9"

**Moon at Greatest Eclipse**
(Geocentric Coordinates)
R.A. = 14h35m37.0s
Dec. = -14°53'30.7"
S.D. = 00°16'07.6"
H.P. = 00°59'11.0"

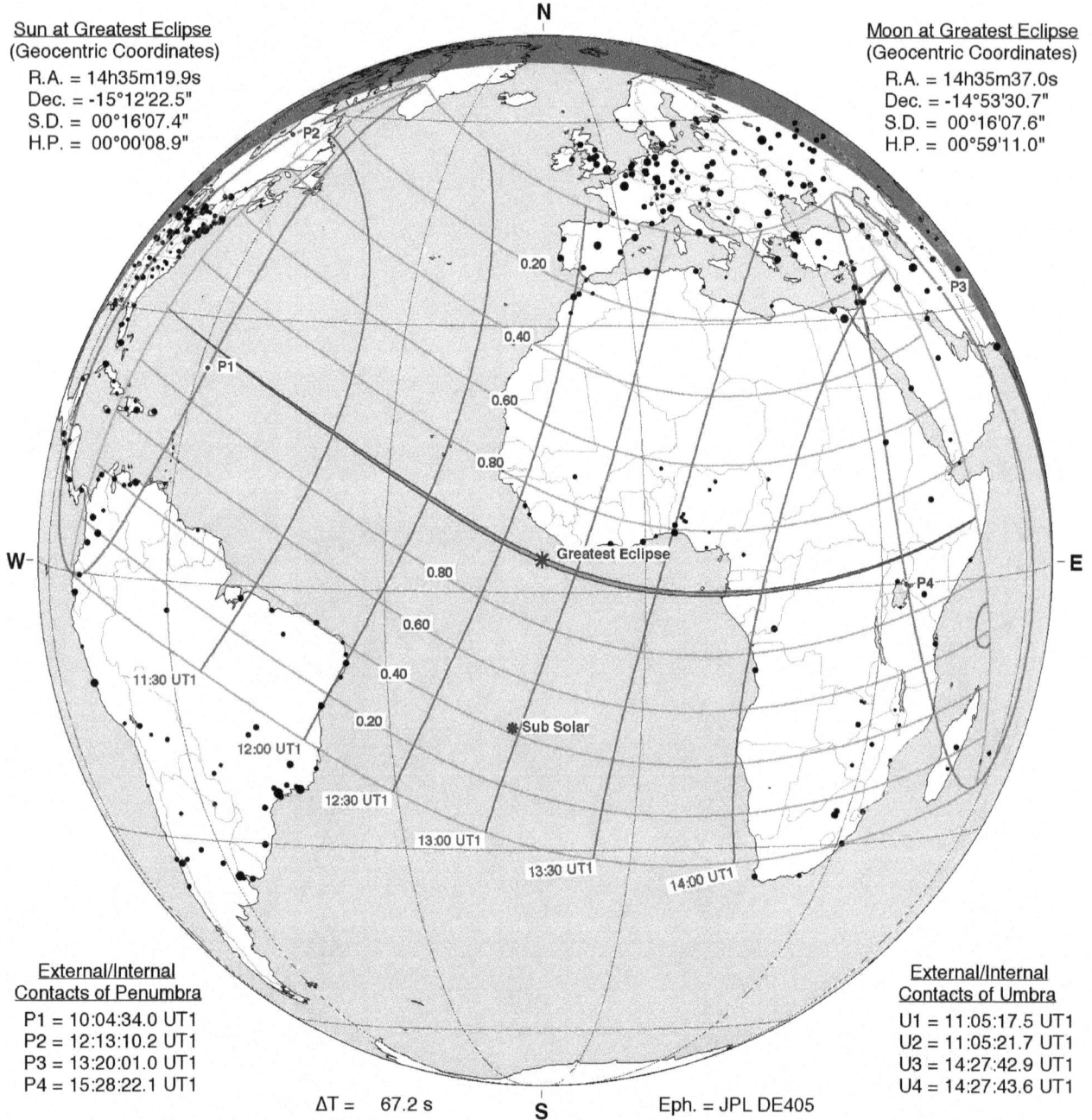

P2

P1

P3

0.20

0.40

0.60

0.80

N

W

E

* Greatest Eclipse

* Sub Solar

P4

0.80

0.60

0.40

0.20

11:30 UT1

12:00 UT1

12:30 UT1

13:00 UT1

13:30 UT1

14:00 UT1

S

**External/Internal
Contacts of Penumbra**
P1 = 10:04:34.0 UT1
P2 = 12:13:10.2 UT1
P3 = 13:20:01.0 UT1
P4 = 15:28:22.1 UT1

**External/Internal
Contacts of Umbra**
U1 = 11:05:17.5 UT1
U2 = 11:05:21.7 UT1
U3 = 14:27:42.9 UT1
U4 = 14:27:43.6 UT1

ΔT = 67.2 s          Eph. = JPL DE405

**Circumstances at Greatest Eclipse: 12:46:28.9 UT1**

| | |
|---|---|
| Lat. = 03°29.4'N | Sun Alt. = 70.9° |
| Long. = 011°41.9'W | Sun Azm. = 192.0° |
| Path Width = 57.5 km | Duration = 01m39.6s |

**Circumstances at Greatest Duration: 12:50:51.1 UT1**

| | |
|---|---|
| Lat. = 03°00.9'N | Sun Alt. = 70.7° |
| Long. = 010°29.9'W | Sun Azm. = 198.8° |
| Path Width = 57.7 km | Duration = 01m39.7s |

0        1000      2000      3000      4000      5000
Kilometers

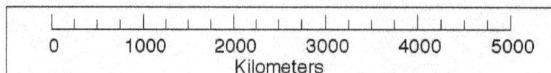

©2016 F. Espenak
www.EclipseWise.com

# Annular Solar Eclipse of 2014 Apr 29

Greatest Eclipse = 06:04:32.9 TD  (= 06:03:25.5 UT1)

| | |
|---|---|
| Eclipse Magnitude = 0.9868 | Saros Series = 148 |
| Gamma = -1.0000 | Saros Member = 21 of 75 |

**Sun at Greatest Eclipse**
(Geocentric Coordinates)
R.A. = 02h25m52.9s
Dec. = +14°26'54.2"
S.D. = 00°15'52.9"
H.P. = 00°00'08.7"

**Moon at Greatest Eclipse**
(Geocentric Coordinates)
R.A. = 02h26m46.0s
Dec. = +13°31'06.8"
S.D. = 00°15'38.4"
H.P. = 00°57'24.1"

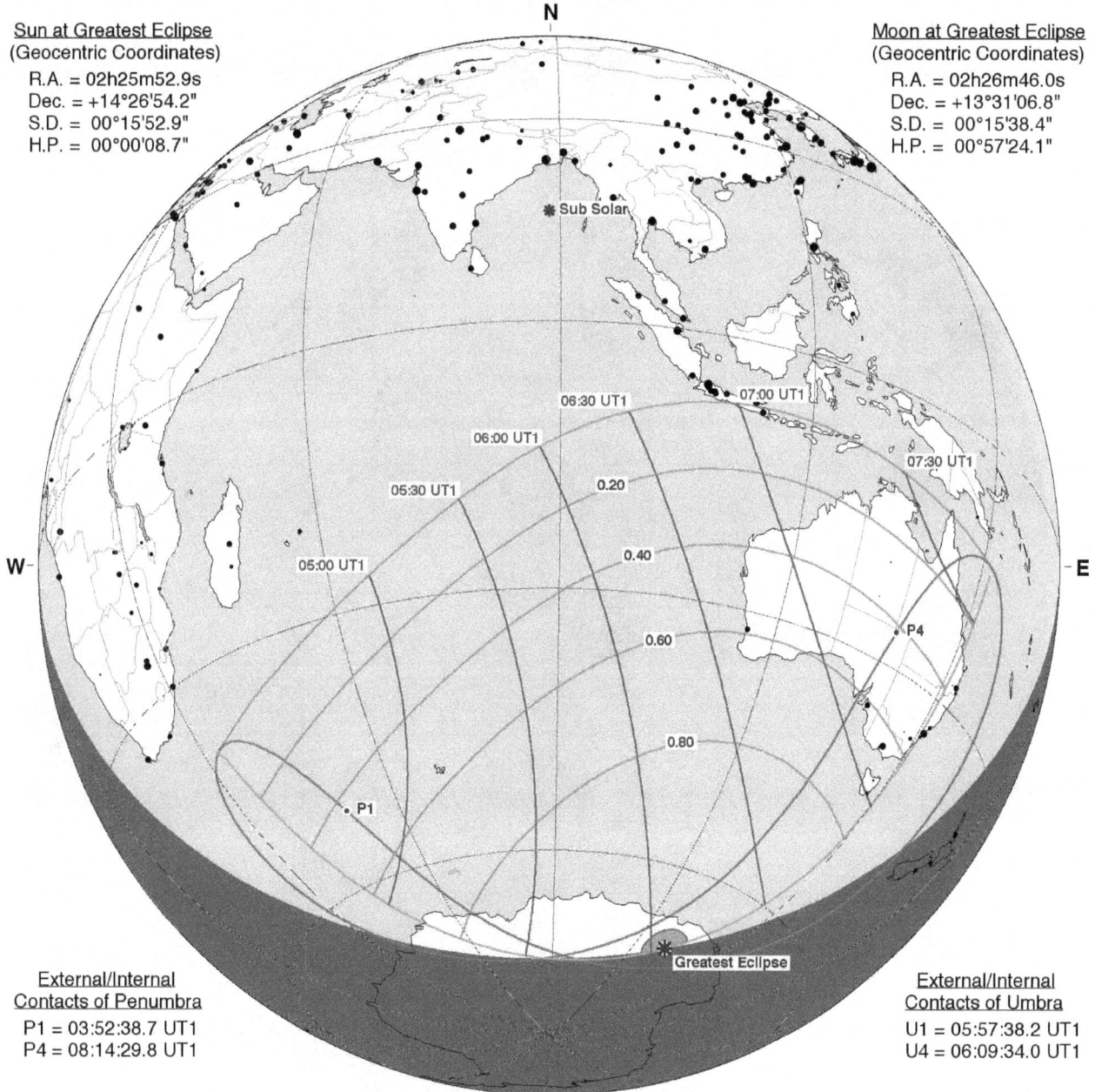

N

Sub Solar

06:30 UT1    07:00 UT1

06:00 UT1

07:30 UT1

05:30 UT1    0.20

05:00 UT1    0.40

W — — E

0.60

P4

0.80

P1

Greatest Eclipse

S

**External/Internal**
**Contacts of Penumbra**
P1 = 03:52:38.7 UT1
P4 = 08:14:29.8 UT1

**External/Internal**
**Contacts of Umbra**
U1 = 05:57:38.2 UT1
U4 = 06:09:34.0 UT1

ΔT = 67.3 s          Eph. = JPL DE405

**Circumstances at Greatest Eclipse: 06:03:25.5 UT1**

| | |
|---|---|
| Lat. = 70°38.7'S | Sun Alt. = 0.0° |
| Long. = 131°15.4'E | Sun Azm. = 318.8° |
| Path Width = 0.0 km | Duration = 00m00.0s |

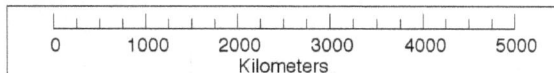

0    1000    2000    3000    4000    5000
Kilometers

*©2016 F. Espenak*
*www.EclipseWise.com*

# Partial Solar Eclipse of 2014 Oct 23

Greatest Eclipse = 21:45:39.2 TD (= 21:44:31.7 UT1)

| | |
|---|---|
| Eclipse Magnitude = 0.8114 | Saros Series = 153 |
| Gamma = 1.0908 | Saros Member = 9 of 70 |

<u>Sun at Greatest Eclipse</u>
(Geocentric Coordinates)
R.A. = 13h53m11.9s
Dec. = -11°36'45.1"
S.D. = 00°16'04.6"
H.P. = 00°00'08.8"

<u>Moon at Greatest Eclipse</u>
(Geocentric Coordinates)
R.A. = 13h54m15.8s
Dec. = -10°37'52.6"
S.D. = 00°15'15.5"
H.P. = 00°55'59.9"

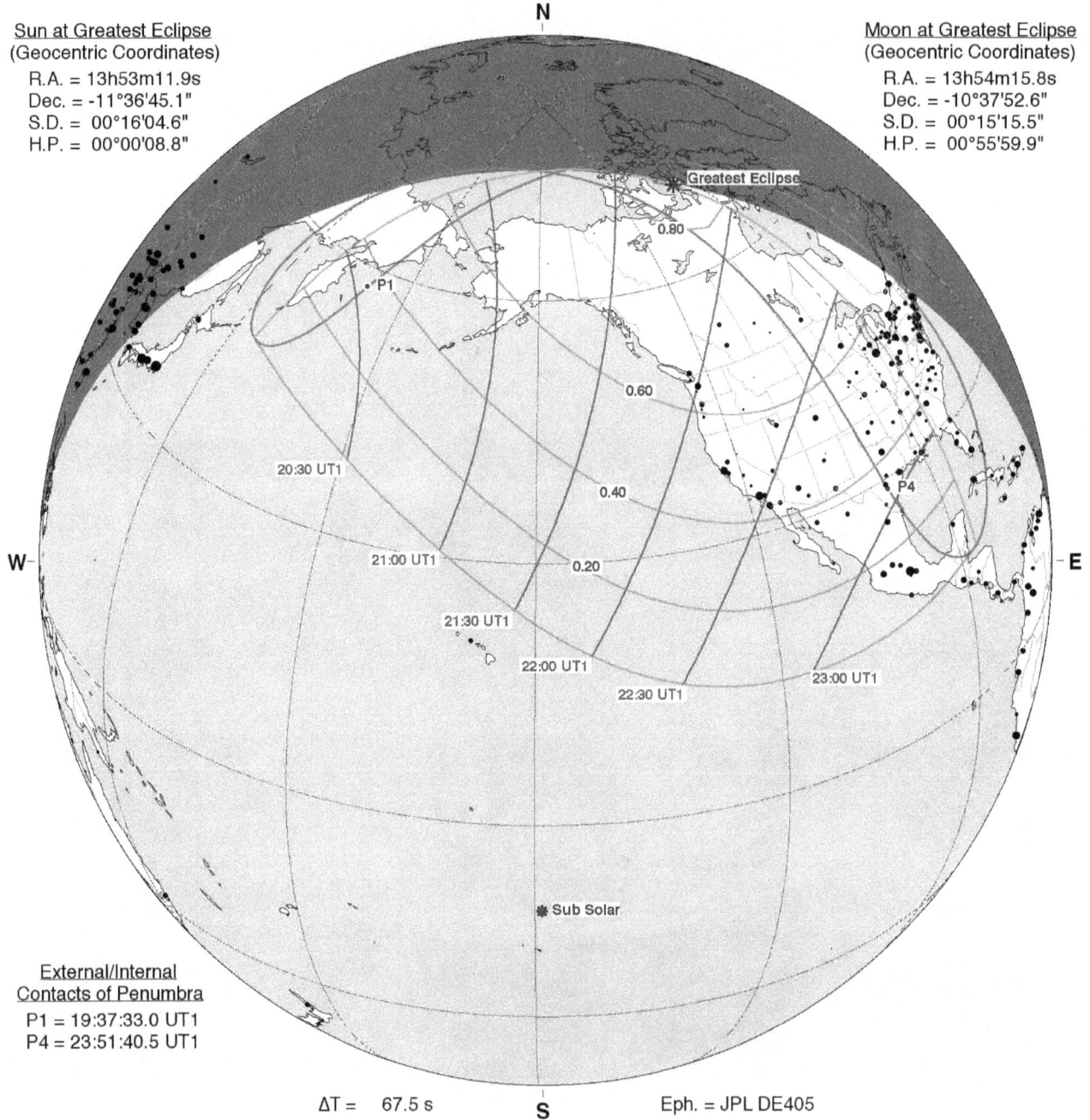

N

Greatest Eclipse

0.80
0.60
0.40
0.20

P1
P4

20:30 UT1
21:00 UT1
21:30 UT1
22:00 UT1
22:30 UT1
23:00 UT1

W

E

Sub Solar

<u>External/Internal</u>
<u>Contacts of Penumbra</u>
P1 = 19:37:33.0 UT1
P4 = 23:51:40.5 UT1

ΔT = 67.5 s

S

Eph. = JPL DE405

<u>Circumstances at Greatest Eclipse: 21:44:31.7 UT1</u>
Lat. = 71°10.1'N          Sun Alt. = 0.0°
Long. = 097°10.3'W     Sun Azm. = 231.4°

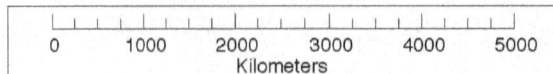

0    1000   2000   3000   4000   5000
Kilometers

©2016 F. Espenak
www.EclipseWise.com

# Total Solar Eclipse of 2015 Mar 20

Greatest Eclipse = 09:46:46.8 TD   (= 09:45:39.2 UT1)

| | |
|---|---|
| Eclipse Magnitude = 1.0446 | Saros Series = 120 |
| Gamma = 0.9454 | Saros Member = 61 of 71 |

**N**

Sun at Greatest Eclipse
(Geocentric Coordinates)
R.A. = 23h58m01.5s
Dec. = -00°12'50.4"
S.D. = 00°16'03.7"
H.P. = 00°00'08.8"

Moon at Greatest Eclipse
(Geocentric Coordinates)
R.A. = 23h56m50.5s
Dec. = +00°42'08.7"
S.D. = 00°16'41.6"
H.P. = 01°01'15.8"

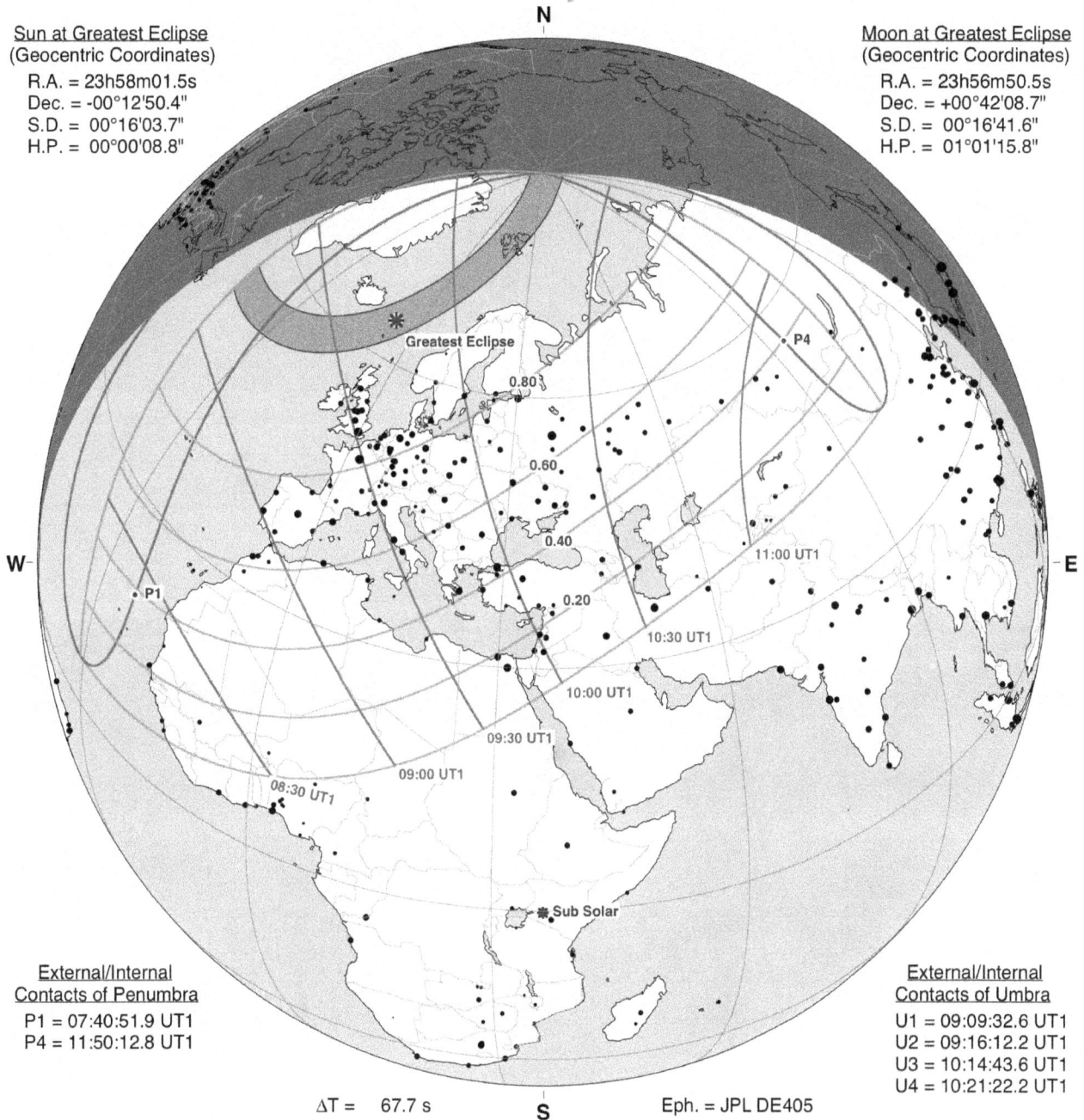

Greatest Eclipse

P4

0.80

0.60

0.40

11:00 UT1

**W**

0.20

10:30 UT1

P1

10:00 UT1

09:30 UT1

09:00 UT1

08:30 UT1

**E**

Sub Solar

External/Internal
Contacts of Penumbra
P1 = 07:40:51.9 UT1
P4 = 11:50:12.8 UT1

External/Internal
Contacts of Umbra
U1 = 09:09:32.6 UT1
U2 = 09:16:12.2 UT1
U3 = 10:14:43.6 UT1
U4 = 10:21:22.2 UT1

ΔT = 67.7 s         **S**         Eph. = JPL DE405

| Circumstances at Greatest Eclipse: 09:45:39.2 UT1 | | Circumstances at Greatest Duration: 09:45:16.5 UT1 | |
|---|---|---|---|
| Lat. = 64°25.9'N | Sun Alt. = 18.5° | Lat. = 64°17.0'N | Sun Alt. = 18.5° |
| Long. = 006°38.8'W | Sun Azm. = 135.0° | Long. = 006°53.6'W | Sun Azm. = 134.6° |
| Path Width = 462.6 km | Duration = 02m46.9s | Path Width = 463.3 km | Duration = 02m46.9s |

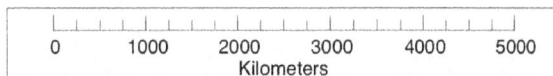

| 0 | 1000 | 2000 | 3000 | 4000 | 5000 |
|---|---|---|---|---|---|

Kilometers

©2016  F. Espenak
www.EclipseWise.com

# Partial Solar Eclipse of 2015 Sep 13

Greatest Eclipse = 06:55:19.2 TD   (= 06:54:11.4 UT1)

Eclipse Magnitude = 0.7875          Saros Series = 125
Gamma = -1.1004            Saros Member = 54 of 73

**Sun at Greatest Eclipse**
(Geocentric Coordinates)
R.A. = 11h23m54.6s
Dec. = +03°53'20.1"
S.D. = 00°15'53.6"
H.P. = 00°00'08.7"

**Moon at Greatest Eclipse**
(Geocentric Coordinates)
R.A. = 11h22m43.3s
Dec. = +02°56'47.8"
S.D. = 00°14'43.0"
H.P. = 00°54'00.6"

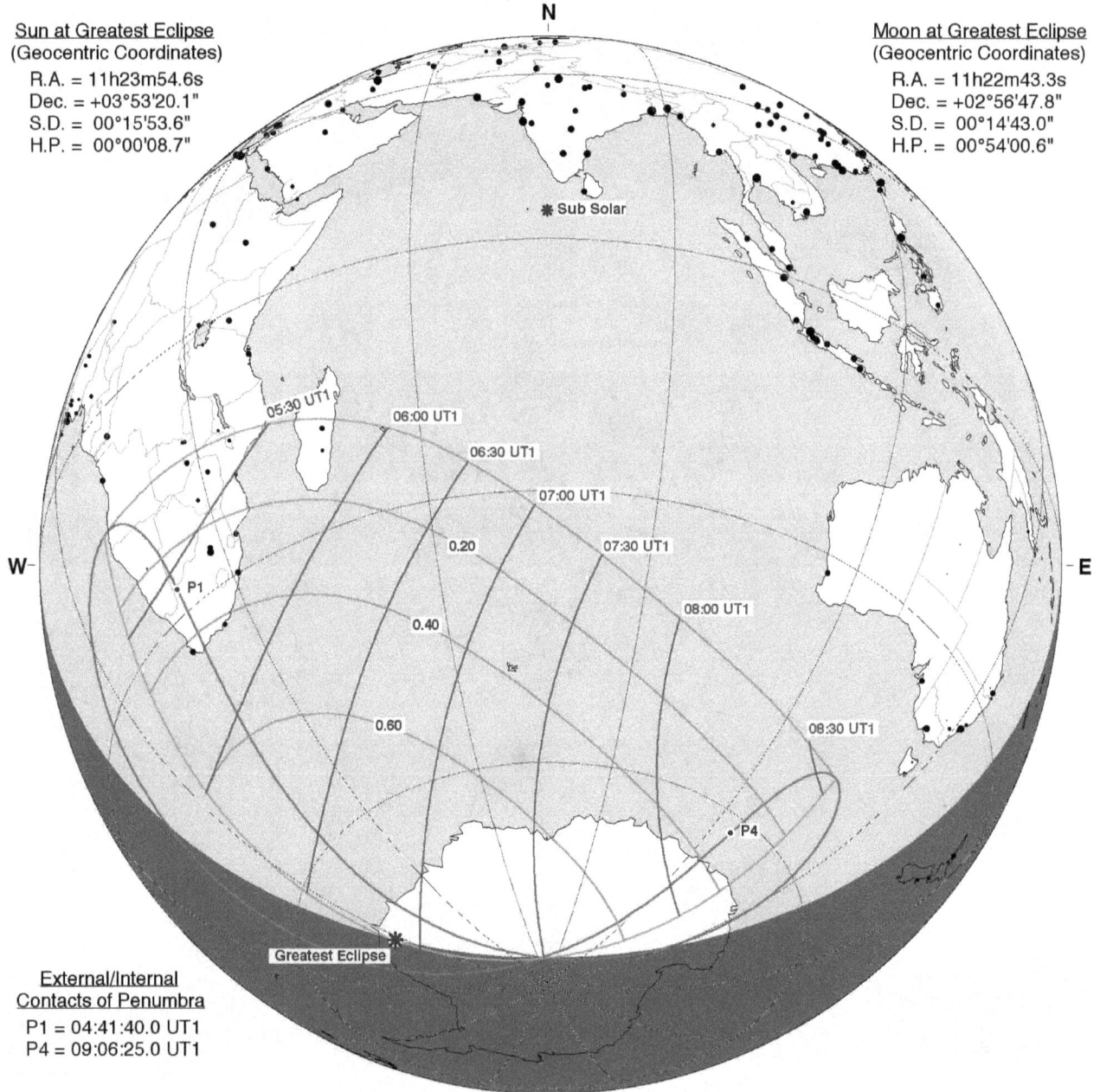

N

✳ Sub Solar

05:30 UT1
06:00 UT1
06:30 UT1
07:00 UT1
07:30 UT1
0.20
08:00 UT1
0.40
08:30 UT1
0.60

W

E

P1

P4

✳ Greatest Eclipse

**External/Internal Contacts of Penumbra**
P1 = 04:41:40.0 UT1
P4 = 09:06:25.0 UT1

ΔT =   67.8 s          S          Eph. = JPL DE405

**Circumstances at Greatest Eclipse: 06:54:11.4 UT1**
Lat. = 72°06.4'S          Sun Alt. =   0.0°
Long. = 002°18.2'W          Sun Azm. = 77.2°

| 0 | 1000 | 2000 | 3000 | 4000 | 5000 |
Kilometers

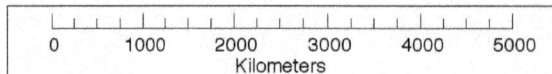

# Total Solar Eclipse of 2016 Mar 09

Greatest Eclipse = 01:58:19.5 TD (= 01:57:11.4 UT1)

| | |
|---|---|
| Eclipse Magnitude = 1.0450 | Saros Series = 130 |
| Gamma = 0.2609 | Saros Member = 52 of 73 |

**Sun at Greatest Eclipse**
(Geocentric Coordinates)
R.A. = 23h19m17.6s
Dec. = -04°22'46.4"
S.D. = 00°16'06.5"
H.P. = 00°00'08.9"

**Moon at Greatest Eclipse**
(Geocentric Coordinates)
R.A. = 23h18m58.7s
Dec. = -04°07'40.6"
S.D. = 00°16'33.5"
H.P. = 01°00'46.2"

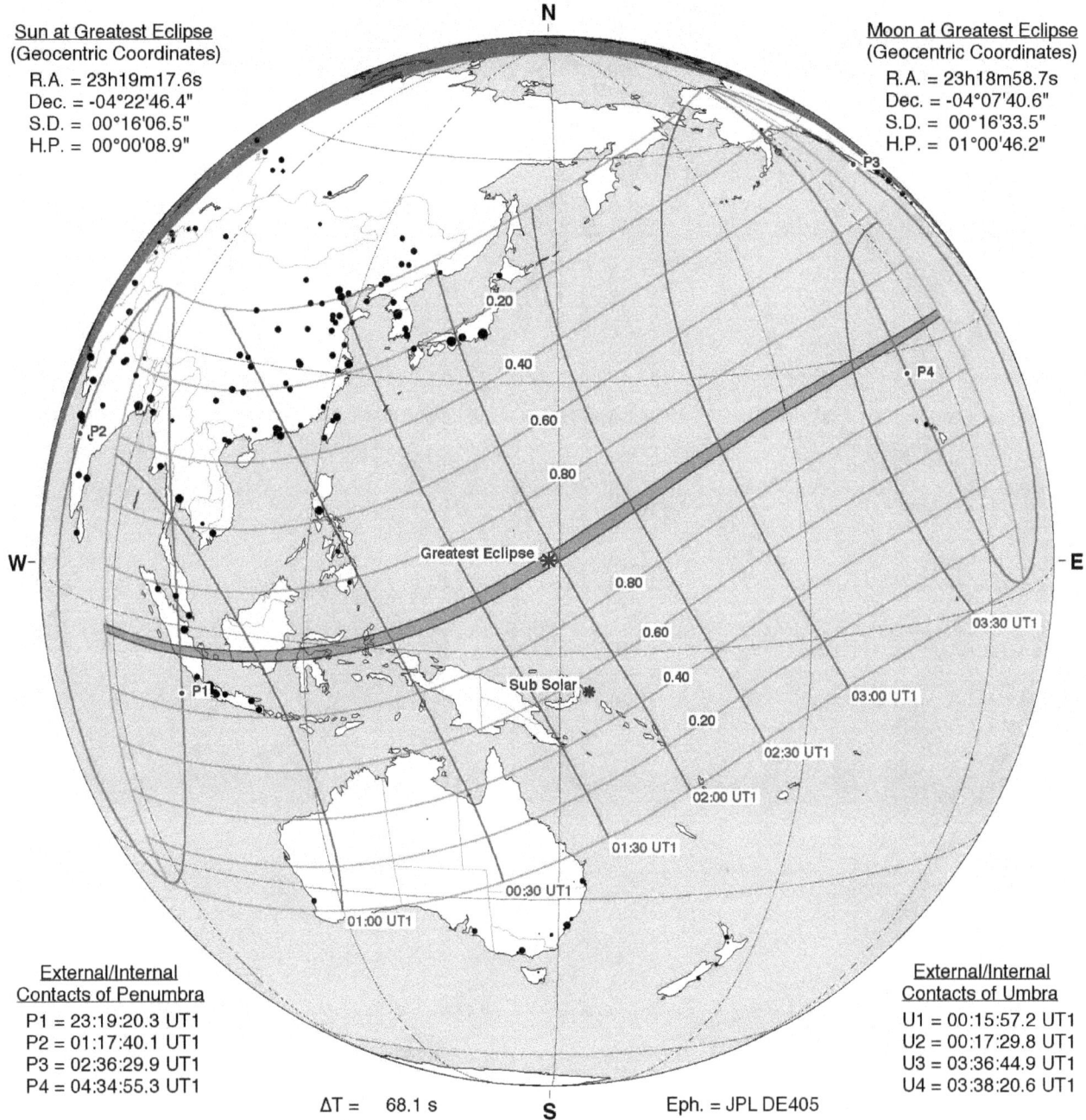

N

0.20
0.40
0.60
0.80
Greatest Eclipse
0.80
0.60
0.40
0.20

P2
P1
P3
P4

Sub Solar

03:30 UT1
03:00 UT1
02:30 UT1
02:00 UT1
01:30 UT1
00:30 UT1
01:00 UT1

W — — E

**External/Internal**
**Contacts of Penumbra**
P1 = 23:19:20.3 UT1
P2 = 01:17:40.1 UT1
P3 = 02:36:29.9 UT1
P4 = 04:34:55.3 UT1

ΔT = 68.1 s

S

Eph. = JPL DE405

**External/Internal**
**Contacts of Umbra**
U1 = 00:15:57.2 UT1
U2 = 00:17:29.8 UT1
U3 = 03:36:44.9 UT1
U4 = 03:38:20.6 UT1

**Circumstances at Greatest Eclipse: 01:57:11.4 UT1**

| | |
|---|---|
| Lat. = 10°07.3'N | Sun Alt. = 74.8° |
| Long. = 148°47.6'E | Sun Azm. = 162.5° |
| Path Width = 155.1 km | Duration = 04m09.5s |

**Circumstances at Greatest Duration: 01:56:51.8 UT1**

| | |
|---|---|
| Lat. = 10°04.0'N | Sun Alt. = 74.8° |
| Long. = 148°42.1'E | Sun Azm. = 161.8° |
| Path Width = 155.1 km | Duration = 04m09.5s |

0   1000   2000   3000   4000   5000
Kilometers

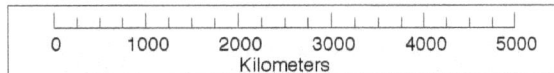

©2016 F. Espenak
www.EclipseWise.com

# Annular Solar Eclipse of 2016 Sep 01

Greatest Eclipse = 09:08:02.0 TD   (= 09:06:53.7 UT1)

Eclipse Magnitude = 0.9736          Saros Series = 135
Gamma = -0.3330          Saros Member = 39 of 71

Sun at Greatest Eclipse
(Geocentric Coordinates)
R.A. = 10h43m43.3s
Dec. = +08°03'38.0"
S.D. = 00°15'51.0"
H.P. = 00°00'08.7"

Moon at Greatest Eclipse
(Geocentric Coordinates)
R.A. = 10h43m22.2s
Dec. = +07°45'51.0"
S.D. = 00°15'12.4"
H.P. = 00°55'48.6"

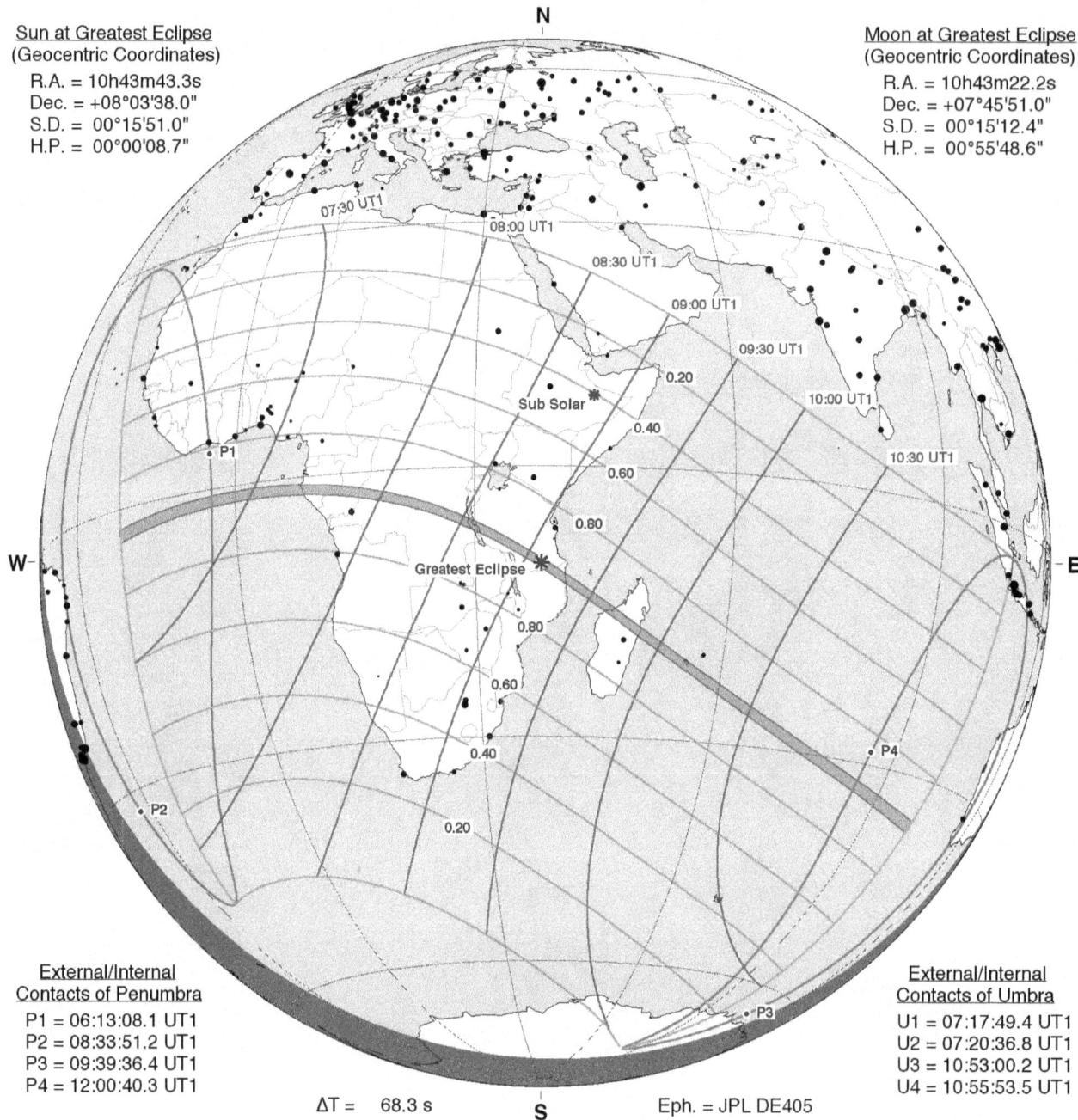

N

07:30 UT1
08:00 UT1
08:30 UT1
09:00 UT1
09:30 UT1
10:00 UT1
10:30 UT1

0.20
Sub Solar
0.40
0.60
0.80

W

Greatest Eclipse

0.80
0.60
0.40
0.20

E

P1
P2
P3
P4

External/Internal
Contacts of Penumbra
P1 = 06:13:08.1 UT1
P2 = 08:33:51.2 UT1
P3 = 09:39:36.4 UT1
P4 = 12:00:40.3 UT1

ΔT =   68.3 s

S

Eph. = JPL DE405

External/Internal
Contacts of Umbra
U1 = 07:17:49.4 UT1
U2 = 07:20:36.8 UT1
U3 = 10:53:00.2 UT1
U4 = 10:55:53.5 UT1

Circumstances at Greatest Eclipse: 09:06:53.7 UT1

| | |
|---|---|
| Lat. = 10°40.9'S | Sun Alt. = 70.5° |
| Long. = 037°45.7'E | Sun Azm. = 16.4° |
| Path Width = 99.7 km | Duration = 03m05.6s |

Circumstances at Greatest Duration: 09:05:09.4 UT1

| | |
|---|---|
| Lat. = 10°25.9'S | Sun Alt. = 70.5° |
| Long. = 037°20.7'E | Sun Azm. = 19.1° |
| Path Width = 99.8 km | Duration = 03m05.6s |

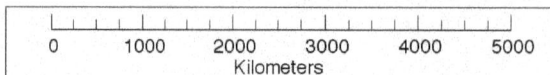

©2016  F. Espenak
www.EclipseWise.com

0   1000   2000   3000   4000   5000
Kilometers

# Annular Solar Eclipse of 2017 Feb 26

Greatest Eclipse = 14:54:32.8 TD  (= 14:53:24.3 UT1)

| | |
|---|---|
| Eclipse Magnitude = 0.9922 | Saros Series = 140 |
| Gamma = -0.4578 | Saros Member = 29 of 71 |

**Sun at Greatest Eclipse**
(Geocentric Coordinates)
R.A. = 22h39m23.1s
Dec. = -08°29'38.8"
S.D. = 00°16'09.0"
H.P. = 00°00'08.9"

**Moon at Greatest Eclipse**
(Geocentric Coordinates)
R.A. = 22h39m53.2s
Dec. = -08°55'03.6"
S.D. = 00°15'47.8"
H.P. = 00°57'58.6"

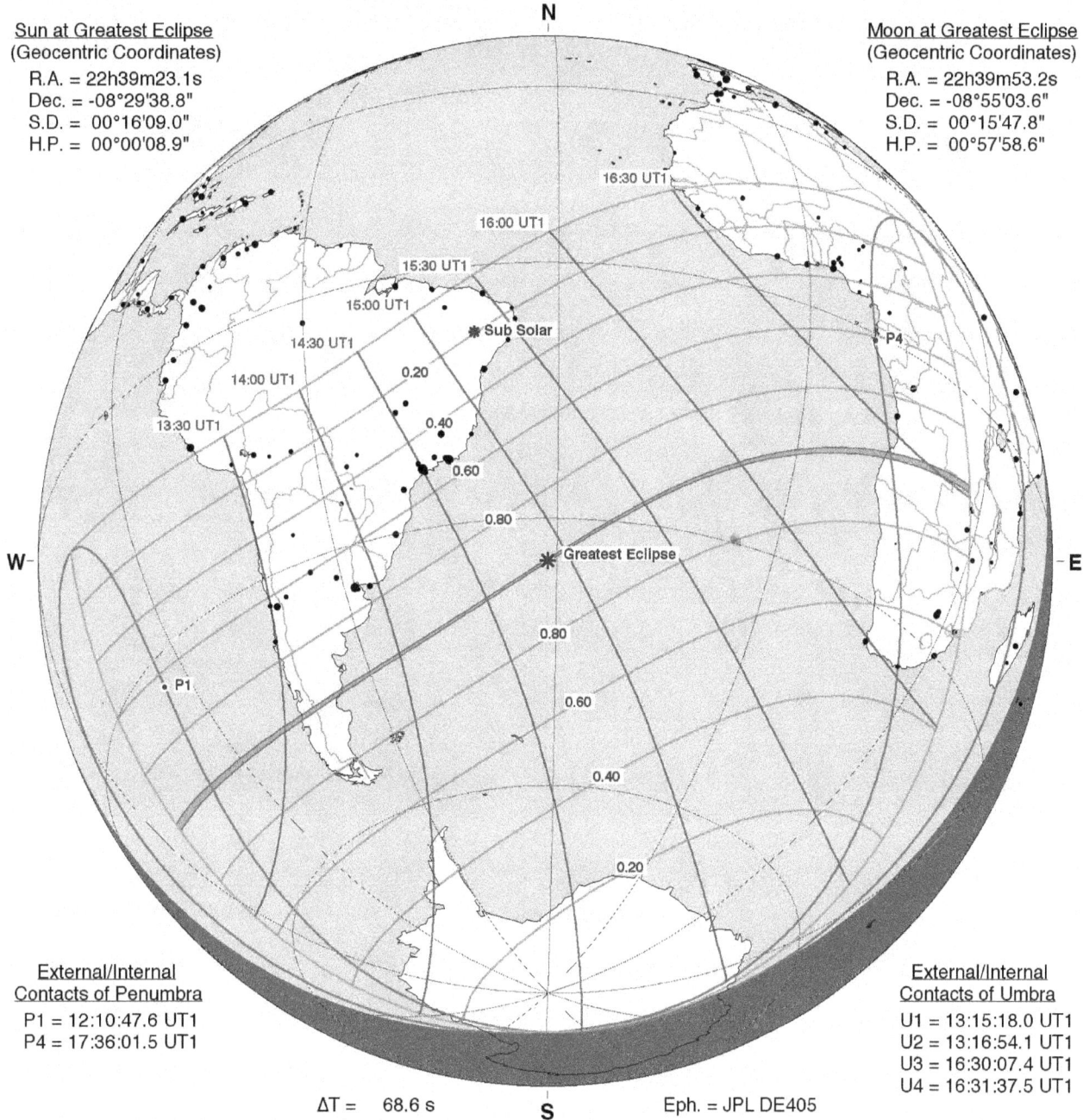

N

16:30 UT1
16:00 UT1
15:30 UT1
15:00 UT1
14:30 UT1
14:00 UT1
13:30 UT1

Sub Solar

P4

0.20
0.40
0.60
0.80

W

Greatest Eclipse

E

0.80
0.60
0.40

P1

0.20

S

**External/Internal
Contacts of Penumbra**
P1 = 12:10:47.6 UT1
P4 = 17:36:01.5 UT1

**External/Internal
Contacts of Umbra**
U1 = 13:15:18.0 UT1
U2 = 13:16:54.1 UT1
U3 = 16:30:07.4 UT1
U4 = 16:31:37.5 UT1

ΔT =  68.6 s

Eph. = JPL DE405

**Circumstances at Greatest Eclipse: 14:53:24.3 UT1**

| | |
|---|---|
| Lat. = 34°40.8'S | Sun Alt. =  62.6° |
| Long. = 031°11.4'W | Sun Azm. = 340.5° |
| Path Width =  30.6 km | Duration = 00m44.0s |

**Circumstances at Greatest Duration: 13:16:06.0 UT1**

| | |
|---|---|
| Lat. =  43°07.5'S | Sun Alt. =   0.0° |
| Long. = 113°52.9'W | Sun Azm. = 101.7° |
| Path Width =  96.3 km | Duration = 01m22.4s |

0    1000    2000    3000    4000    5000
Kilometers

©2016  F. Espenak
www.EclipseWise.com

# Total Solar Eclipse of 2017 Aug 21

Greatest Eclipse = 18:26:40.3 TD  (= 18:25:31.5 UT1)

| | |
|---|---|
| Eclipse Magnitude = 1.0306 | Saros Series = 145 |
| Gamma = 0.4367 | Saros Member = 22 of 77 |

**Sun at Greatest Eclipse**
(Geocentric Coordinates)
R.A. = 10h04m03.9s
Dec. = +11°51'43.0"
S.D. = 00°15'48.7"
H.P. = 00°00'08.7"

**Moon at Greatest Eclipse**
(Geocentric Coordinates)
R.A. = 10h04m30.6s
Dec. = +12°16'32.8"
S.D. = 00°16'03.4"
H.P. = 00°58'55.7"

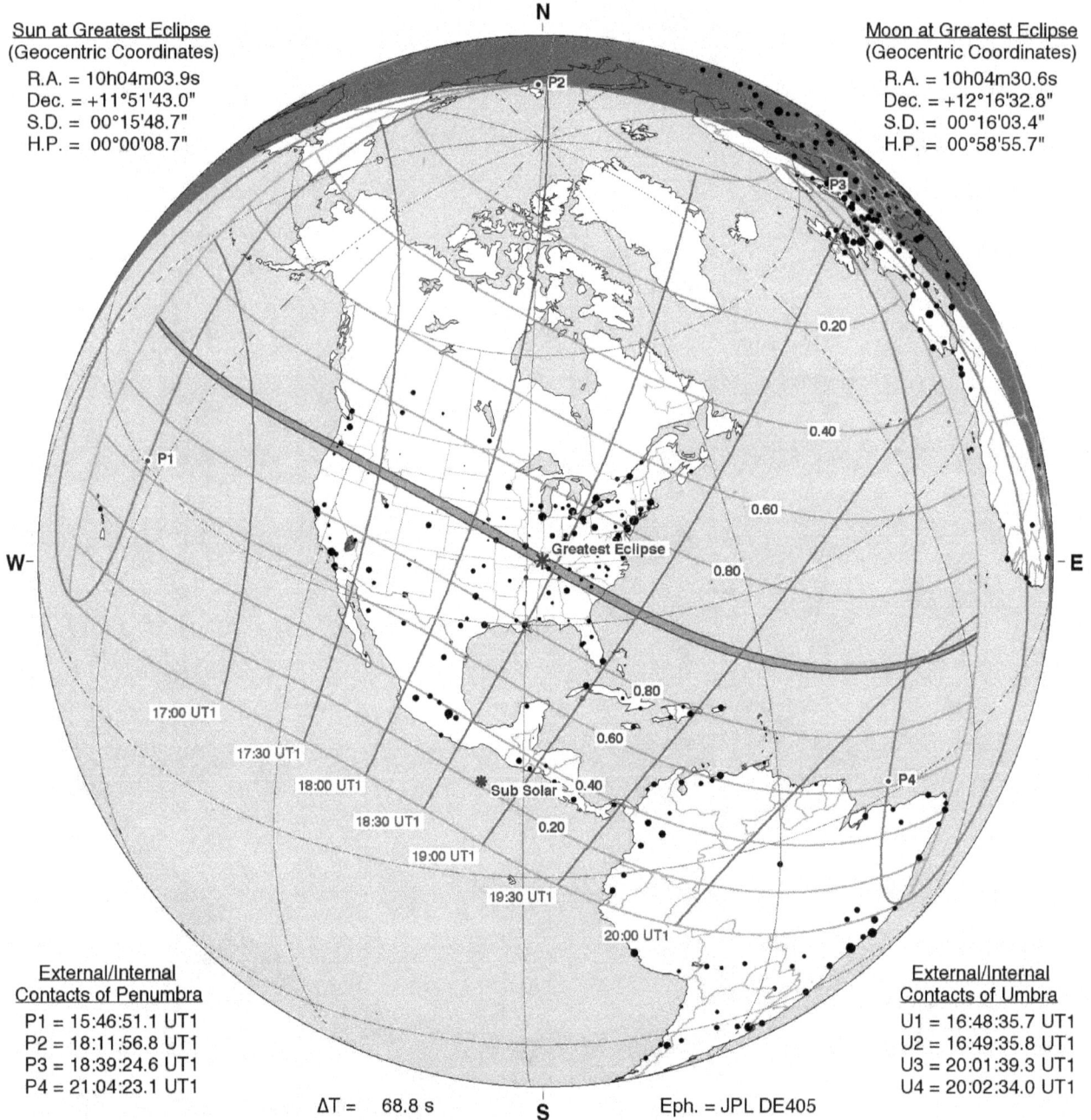

**External/Internal Contacts of Penumbra**
P1 = 15:46:51.1 UT1
P2 = 18:11:56.8 UT1
P3 = 18:39:24.6 UT1
P4 = 21:04:23.1 UT1

**External/Internal Contacts of Umbra**
U1 = 16:48:35.7 UT1
U2 = 16:49:35.8 UT1
U3 = 20:01:39.3 UT1
U4 = 20:02:34.0 UT1

ΔT = 68.8 s

Eph. = JPL DE405

**Circumstances at Greatest Eclipse: 18:25:31.5 UT1**

| | |
|---|---|
| Lat. = 36°58.0'N | Sun Alt. = 63.9° |
| Long. = 087°40.2'W | Sun Azm. = 197.9° |
| Path Width = 114.7 km | Duration = 02m40.1s |

**Circumstances at Greatest Duration: 18:21:48.8 UT1**

| | |
|---|---|
| Lat. = 37°34.6'N | Sun Alt. = 63.8° |
| Long. = 089°06.6'W | Sun Azm. = 192.5° |
| Path Width = 114.5 km | Duration = 02m40.3s |

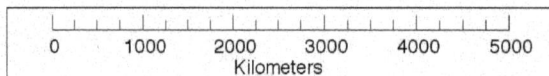

©2016 F. Espenak
www.EclipseWise.com

0 1000 2000 3000 4000 5000
Kilometers

# Partial Solar Eclipse of 2018 Feb 15

Greatest Eclipse = 20:52:33.3 TD   (= 20:51:24.3 UT1)

Eclipse Magnitude = 0.5991          Saros Series = 150
Gamma = -1.2116                     Saros Member = 17 of 71

Sun at Greatest Eclipse
(Geocentric Coordinates)
R.A. = 21h57m18.8s
Dec. = -12°28'07.3"
S.D. = 00°16'11.4"
H.P. = 00°00'08.9"

Moon at Greatest Eclipse
(Geocentric Coordinates)
R.A. = 21h58m26.9s
Dec. = -13°32'29.9"
S.D. = 00°14'59.4"
H.P. = 00°55'00.9"

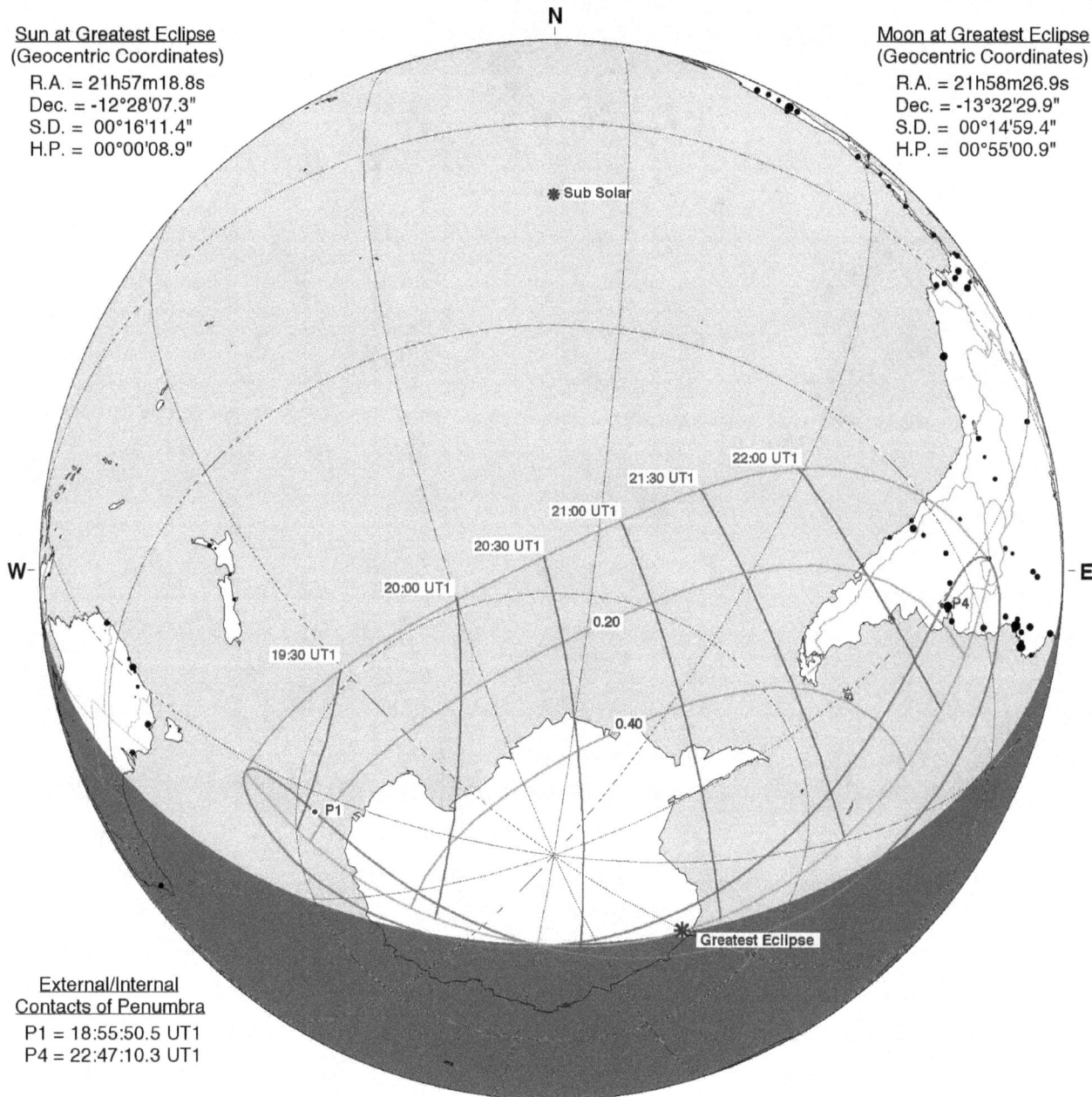

N

* Sub Solar

22:00 UT1
21:30 UT1
21:00 UT1
20:30 UT1
20:00 UT1
19:30 UT1

P4

0.20

0.40

W — 

— E

P1

* Greatest Eclipse

External/Internal
Contacts of Penumbra
P1 = 18:55:50.5 UT1
P4 = 22:47:10.3 UT1

ΔT =   69.0 s          S          Eph. = JPL DE405

Circumstances at Greatest Eclipse: 20:51:24.3 UT1
Lat. = 71°01.6'S          Sun Alt. =   0.0°
Long. = 000°38.5'E        Sun Azm. = 228.4°

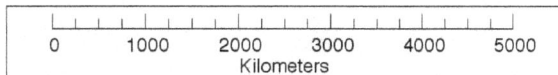

| | | | | | |
|---|---|---|---|---|---|
| 0 | 1000 | 2000 | 3000 | 4000 | 5000 |

Kilometers

©2016  F. Espenak
www.EclipseWise.com

# Partial Solar Eclipse of 2018 Jul 13

Greatest Eclipse = 03:02:16.1 TD   (= 03:01:07.0 UT1)

Eclipse Magnitude = 0.3365          Saros Series = 117
Gamma = -1.3542                     Saros Member = 69 of 71

Sun at Greatest Eclipse
(Geocentric Coordinates)
R.A. = 07h29m31.1s
Dec. = +21°50'30.6"
S.D. = 00°15'44.0"
H.P. = 00°00'08.7"

Moon at Greatest Eclipse
(Geocentric Coordinates)
R.A. = 07h29m10.9s
Dec. = +20°27'46.1"
S.D. = 00°16'42.8"
H.P. = 01°01'20.4"

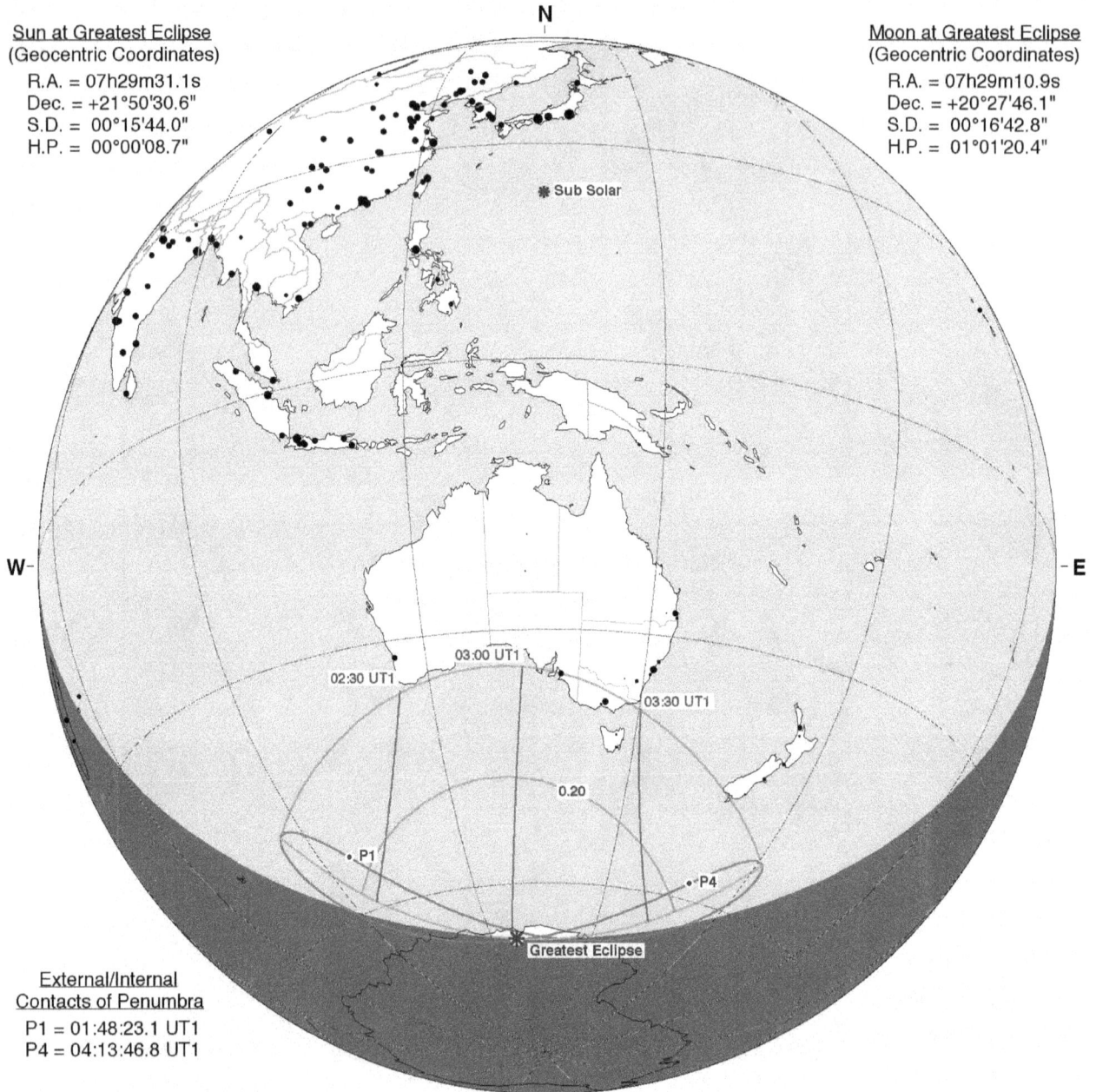

N

* Sub Solar

W

E

03:00 UT1
02:30 UT1
03:30 UT1

0.20

P1
P4

Greatest Eclipse

External/Internal
Contacts of Penumbra
P1 = 01:48:23.1 UT1
P4 = 04:13:46.8 UT1

ΔT = 69.2 s          S          Eph. = JPL DE405

Circumstances at Greatest Eclipse: 03:01:07.0 UT1
Lat. = 67°55.4'S        Sun Alt. =   0.0°
Long. = 127°25.8'E      Sun Azm. =   8.1°

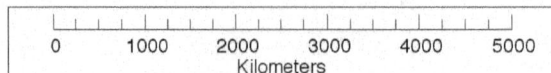

| | | | | | |
|---|---|---|---|---|---|
| 0 | 1000 | 2000 | 3000 | 4000 | 5000 |

Kilometers

©2016  F. Espenak
www.EclipseWise.com

# Partial Solar Eclipse of 2018 Aug 11

Greatest Eclipse = 09:47:28.0 TD   (= 09:46:18.8 UT1)

Eclipse Magnitude = 0.7368          Saros Series = 155
Gamma = 1.1476                      Saros Member = 6 of 71

Sun at Greatest Eclipse
(Geocentric Coordinates)
R.A. = 09h24m28.1s
Dec. = +15°13'19.1"
S.D. = 00°15'46.8"
H.P. = 00°00'08.7"

Moon at Greatest Eclipse
(Geocentric Coordinates)
R.A. = 09h25m31.3s
Dec. = +16°21'40.4"
S.D. = 00°16'40.0"
H.P. = 01°01'10.1"

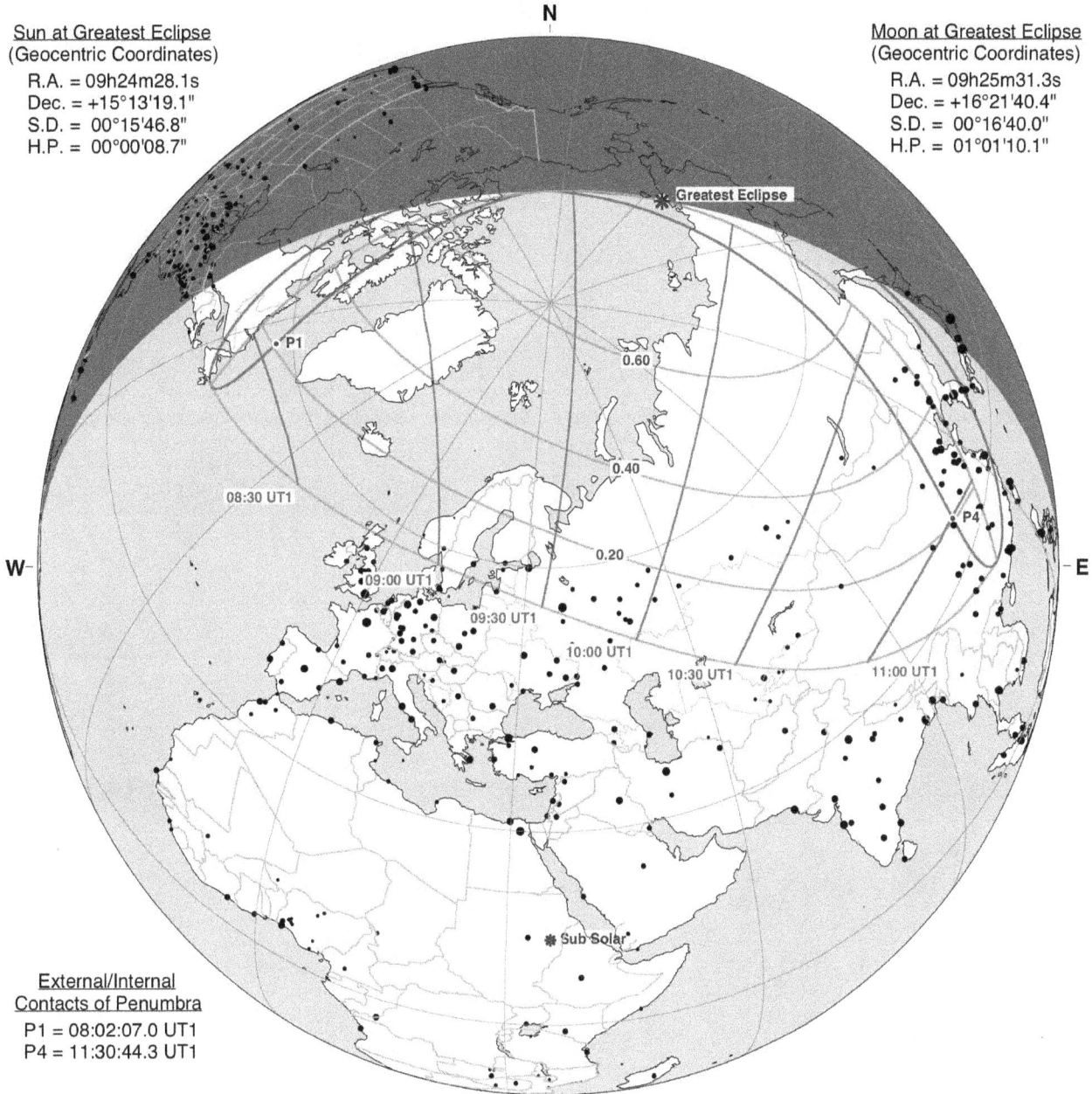

N

Greatest Eclipse

P1

0.60

0.40

08:30 UT1

W

0.20

09:00 UT1

P4

09:30 UT1

10:00 UT1

10:30 UT1     11:00 UT1

E

Sub Solar

External/Internal
Contacts of Penumbra
P1 = 08:02:07.0 UT1
P4 = 11:30:44.3 UT1

ΔT =   69.2 s        S          Eph. = JPL DE405

Circumstances at Greatest Eclipse: 09:46:18.8 UT1
Lat. = 70°23.3'N          Sun Alt. =   0.0°
Long. = 174°28.1'E        Sun Azm. = 321.5°

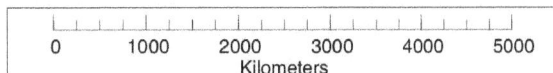

| | | | | | |
|---|---|---|---|---|---|
| 0 | 1000 | 2000 | 3000 | 4000 | 5000 |

Kilometers

©2016  F. Espenak
www.EclipseWise.com

# Partial Solar Eclipse of 2019 Jan 06

Greatest Eclipse = 01:42:37.7 TD  (= 01:41:28.3 UT1)

| | |
|---|---|
| Eclipse Magnitude = 0.7146 | Saros Series = 122 |
| Gamma = 1.1417 | Saros Member = 58 of 70 |

**Sun at Greatest Eclipse**
(Geocentric Coordinates)
R.A. = 19h06m57.4s
Dec. = -22°32'36.6"
S.D. = 00°16'15.9"
H.P. = 00°00'08.9"

**Moon at Greatest Eclipse**
(Geocentric Coordinates)
R.A. = 19h06m53.0s
Dec. = -21°30'36.4"
S.D. = 00°14'50.4"
H.P. = 00°54'27.6"

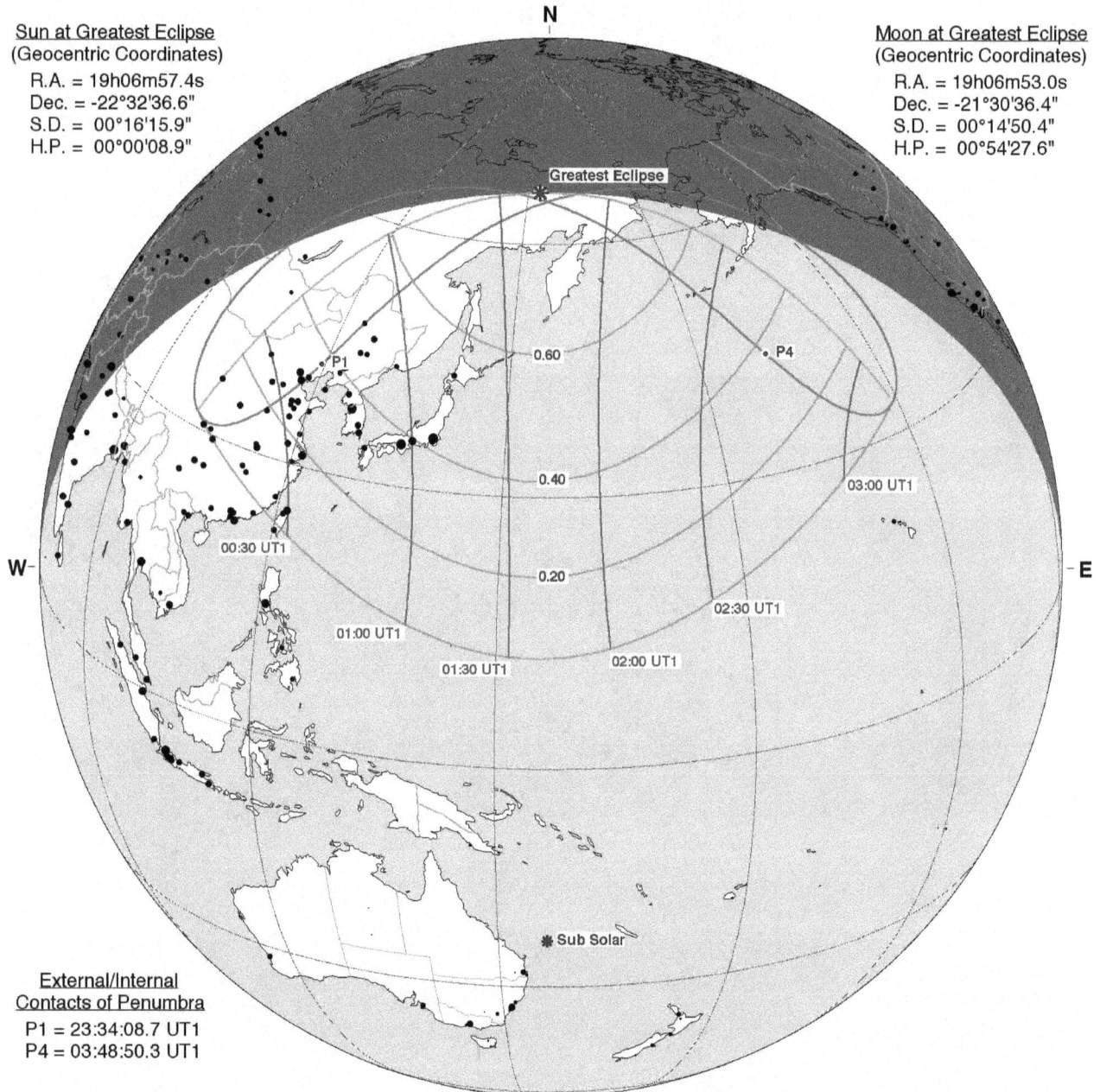

N

Greatest Eclipse

0.60

P1    P4

0.40

03:00 UT1

00:30 UT1

W    0.20    E

02:30 UT1

01:00 UT1

01:30 UT1    02:00 UT1

Sub Solar

**External/Internal
Contacts of Penumbra**
P1 = 23:34:08.7 UT1
P4 = 03:48:50.3 UT1

ΔT =   69.4 s    S    Eph. = JPL DE405

Circumstances at Greatest Eclipse: 01:41:28.3 UT1

| | |
|---|---|
| Lat. = 67°26.1'N | Sun Alt. =   0.0° |
| Long. = 153°33.7'E | Sun Azm. = 177.7° |

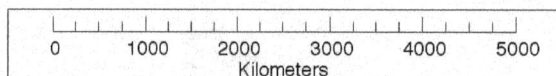

0    1000    2000    3000    4000    5000
Kilometers

©2016  F. Espenak
www.EclipseWise.com

# Total Solar Eclipse of 2019 Jul 02

Greatest Eclipse = 19:24:07.5 TD   (= 19:22:57.9 UT1)

| | |
|---|---|
| Eclipse Magnitude = 1.0459 | Saros Series = 127 |
| Gamma = -0.6466 | Saros Member = 58 of 82 |

**Sun at Greatest Eclipse**
(Geocentric Coordinates)
R.A. = 06h46m14.8s
Dec. = +23°00'36.4"
S.D. = 00°15'43.8"
H.P. = 00°00'08.6"

**Moon at Greatest Eclipse**
(Geocentric Coordinates)
R.A. = 06h46m17.9s
Dec. = +22°22'09.2"
S.D. = 00°16'14.9"
H.P. = 00°59'37.8"

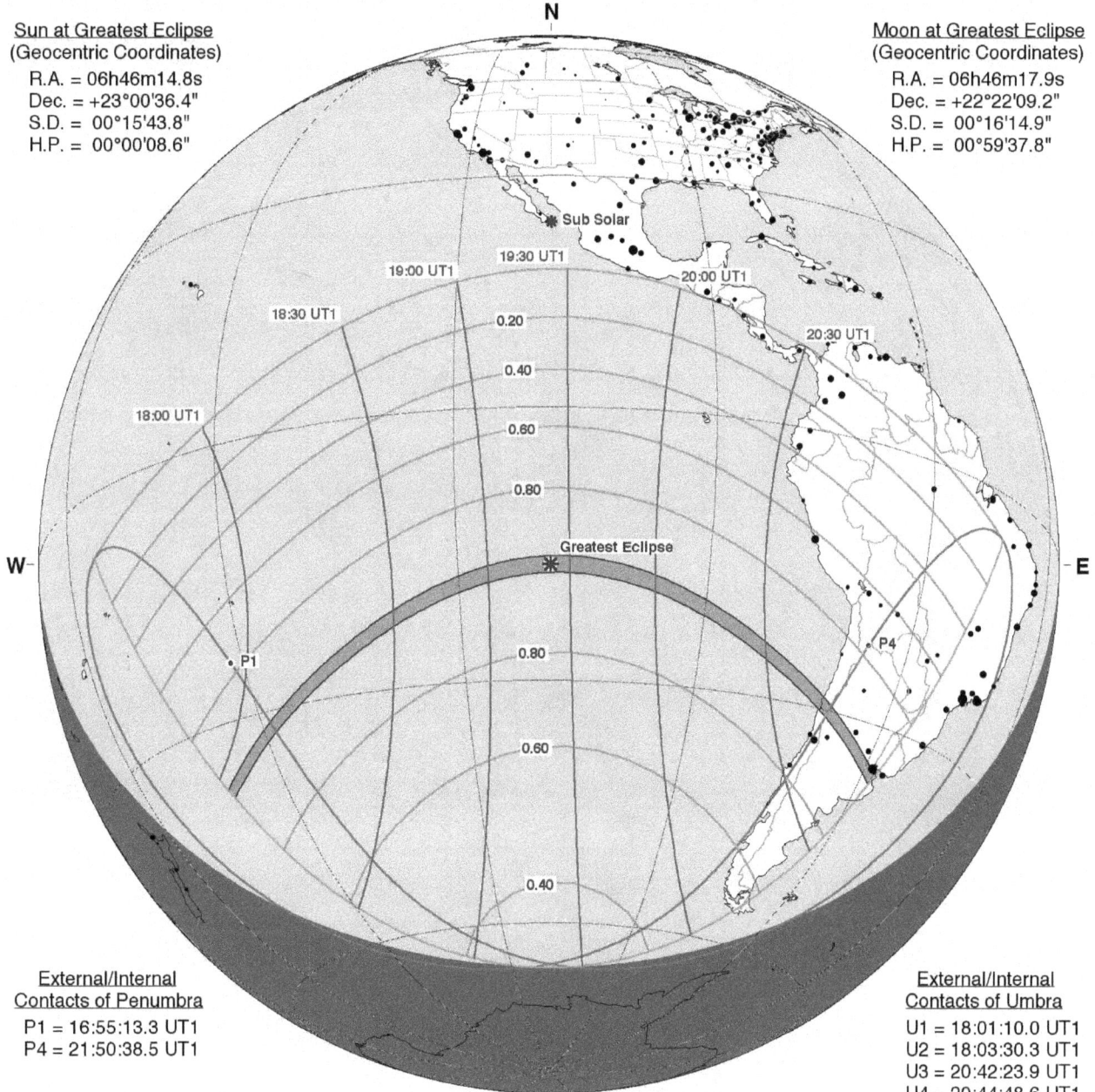

N

Sub Solar

19:00 UT1
19:30 UT1
20:00 UT1
18:30 UT1
20:30 UT1
0.20
0.40
18:00 UT1
0.60
0.80
W

Greatest Eclipse

E

P1
P4
0.80
0.60
0.40

**External/Internal
Contacts of Penumbra**
P1 = 16:55:13.3 UT1
P4 = 21:50:38.5 UT1

**External/Internal
Contacts of Umbra**
U1 = 18:01:10.0 UT1
U2 = 18:03:30.3 UT1
U3 = 20:42:23.9 UT1
U4 = 20:44:48.6 UT1

ΔT =   69.6 s

S

Eph. = JPL DE405

**Circumstances at Greatest Eclipse: 19:22:57.9 UT1**

| | |
|---|---|
| Lat. = 17°23.3'S | Sun Alt. = 49.6° |
| Long. = 108°59.9'W | Sun Azm. = 359.0° |
| Path Width = 200.6 km | Duration = 04m32.8s |

**Circumstances at Greatest Duration: 19:24:09.3 UT1**

| | |
|---|---|
| Lat. = 17°22.8'S | Sun Alt. = 49.6° |
| Long. = 108°37.3'W | Sun Azm. = 358.0° |
| Path Width = 200.7 km | Duration = 04m32.8s |

0   1000   2000   3000   4000   5000
Kilometers

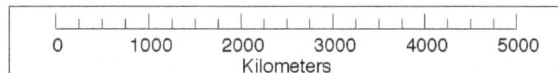

# Annular Solar Eclipse of 2019 Dec 26

Greatest Eclipse = 05:18:53.1 TD   (= 05:17:43.3 UT1)

Eclipse Magnitude = 0.9701          Saros Series = 132
Gamma = 0.4135                      Saros Member = 46 of 71

**Sun at Greatest Eclipse**
(Geocentric Coordinates)
R.A. = 18h17m56.7s
Dec. = -23°22'19.2"
S.D. = 00°16'15.7"
H.P. = 00°00'08.9"

**Moon at Greatest Eclipse**
(Geocentric Coordinates)
R.A. = 18h18m03.7s
Dec. = -22°58'50.4"
S.D. = 00°15'33.0"
H.P. = 00°57'04.0"

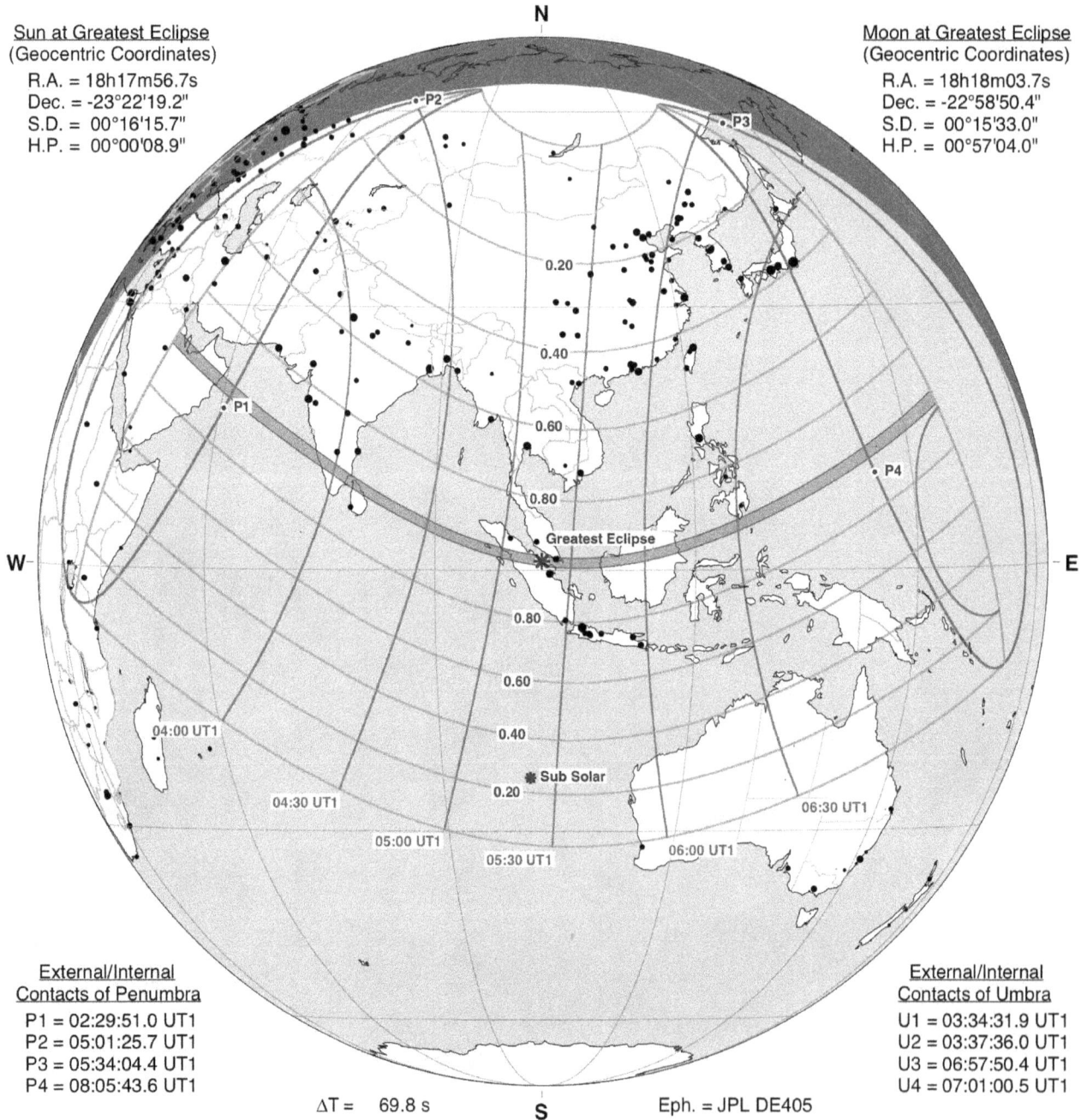

N

P2

P3

0.20

0.40

0.60

0.80

P1

P4

Greatest Eclipse

W

E

0.80

0.60

0.40

04:00 UT1

0.20

Sub Solar

04:30 UT1

06:30 UT1

05:00 UT1

05:30 UT1

06:00 UT1

**External/Internal
Contacts of Penumbra**
P1 = 02:29:51.0 UT1
P2 = 05:01:25.7 UT1
P3 = 05:34:04.4 UT1
P4 = 08:05:43.6 UT1

**External/Internal
Contacts of Umbra**
U1 = 03:34:31.9 UT1
U2 = 03:37:36.0 UT1
U3 = 06:57:50.4 UT1
U4 = 07:01:00.5 UT1

ΔT = 69.8 s          S          Eph. = JPL DE405

**Circumstances at Greatest Eclipse: 05:17:43.3 UT1**

| | |
|---|---|
| Lat. = 01°00.5'N | Sun Alt. = 65.6° |
| Long. = 102°14.9'E | Sun Azm. = 183.6° |
| Path Width = 117.9 km | Duration = 03m39.5s |

**Circumstances at Greatest Duration: 05:28:29.5 UT1**

| | |
|---|---|
| Lat. = 00°46.7'N | Sun Alt. = 64.9° |
| Long. = 105°05.1'E | Sun Azm. = 195.6° |
| Path Width = 118.6 km | Duration = 03m40.0s |

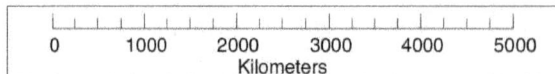

©2016 F. Espenak
www.EclipseWise.com

0   1000   2000   3000   4000   5000
Kilometers

# Annular Solar Eclipse of 2020 Jun 21

Greatest Eclipse = 06:41:15.4 TD  (= 06:40:05.4 UT1)

Eclipse Magnitude = 0.9940          Saros Series = 137

Gamma = 0.1209          Saros Member = 36 of 70

**Sun at Greatest Eclipse**
(Geocentric Coordinates)
R.A. = 06h01m33.0s
Dec. = +23°26'09.7"
S.D. = 00°15'44.2"
H.P. = 00°00'08.7"

**Moon at Greatest Eclipse**
(Geocentric Coordinates)
R.A. = 06h01m30.2s
Dec. = +23°32'56.7"
S.D. = 00°15'24.0"
H.P. = 00°56'31.1"

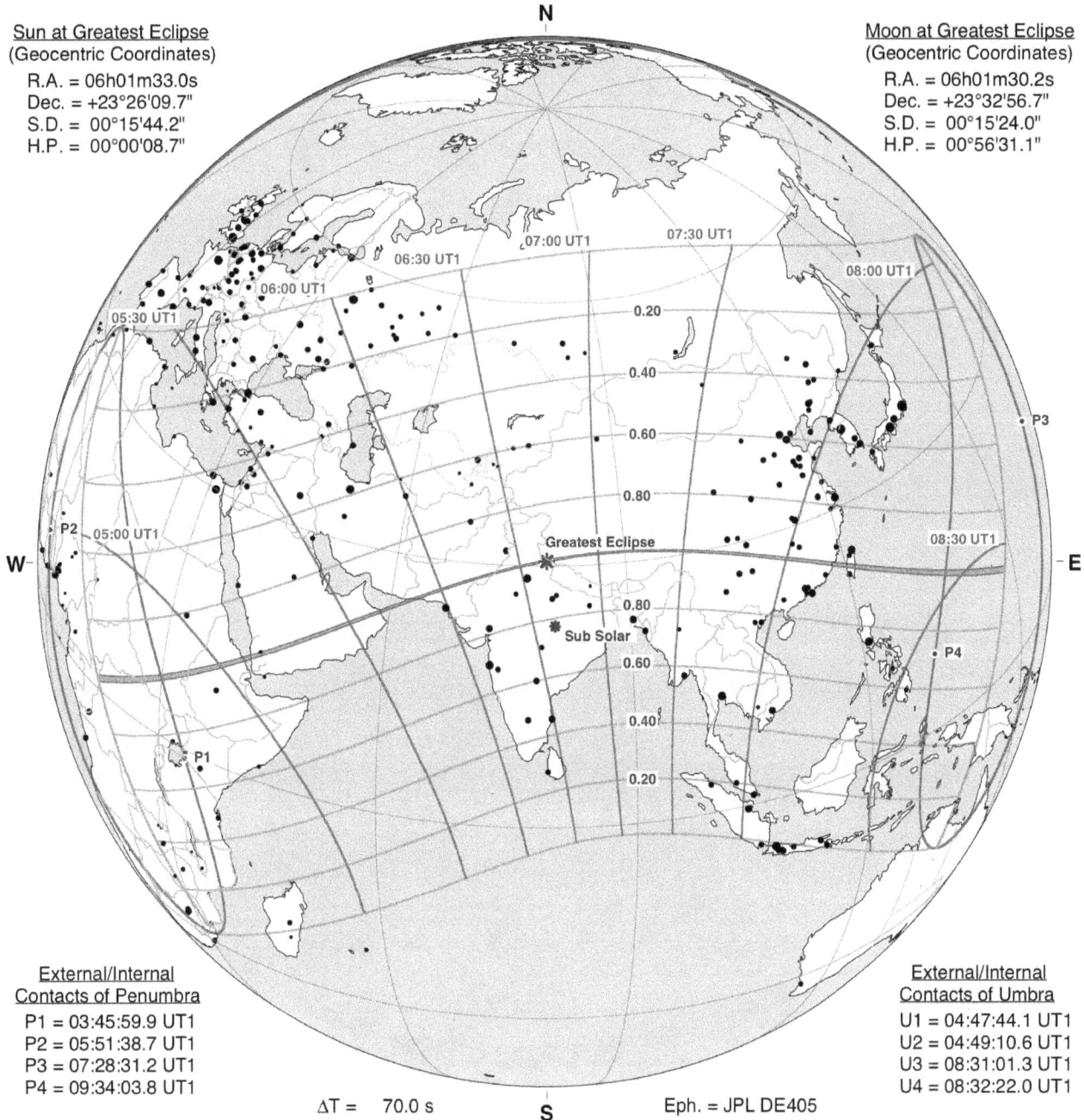

N

07:00 UT1
07:30 UT1
06:30 UT1
08:00 UT1
06:00 UT1
05:30 UT1
0.20
0.40
0.60
0.80
P3
P2
05:00 UT1
W
Greatest Eclipse
0.80
08:30 UT1
E
Sub Solar
0.60
0.40
P4
0.20
P1
S

**External/Internal Contacts of Penumbra**
P1 = 03:45:59.9 UT1
P2 = 05:51:38.7 UT1
P3 = 07:28:31.2 UT1
P4 = 09:34:03.8 UT1

**External/Internal Contacts of Umbra**
U1 = 04:47:44.1 UT1
U2 = 04:49:10.6 UT1
U3 = 08:31:01.3 UT1
U4 = 08:32:22.0 UT1

ΔT = 70.0 s          Eph. = JPL DE405

**Circumstances at Greatest Eclipse: 06:40:05.4 UT1**
Lat. = 30°31.2'N          Sun Alt. = 82.9°
Long. = 079°40.0'E          Sun Azm. = 174.2°
Path Width = 21.2 km          Duration = 00m38.2s

**Circumstances at Greatest Duration: 04:48:27.4 UT1**
Lat. = 01°16.1'N          Sun Alt. = 0.0°
Long. = 017°47.9'E          Sun Azm. = 66.6°
Path Width = 85.2 km          Duration = 01m22.4s

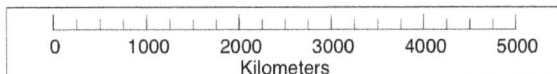

©2016 F. Espenak
www.EclipseWise.com

0      1000    2000    3000    4000    5000
Kilometers

# Total Solar Eclipse of 2020 Dec 14

Greatest Eclipse = 16:14:39.4 TD   (= 16:13:29.1 UT1)

| | |
|---|---|
| Eclipse Magnitude = 1.0254 | Saros Series = 142 |
| Gamma = -0.2939 | Saros Member = 23 of 72 |

**Sun at Greatest Eclipse**
(Geocentric Coordinates)
R.A. = 17h30m05.9s
Dec. = -23°15'32.3"
S.D. = 00°16'14.9"
H.P. = 00°00'08.9"

**Moon at Greatest Eclipse**
(Geocentric Coordinates)
R.A. = 17h29m54.3s
Dec. = -23°32'58.8"
S.D. = 00°16'23.7"
H.P. = 01°00'10.4"

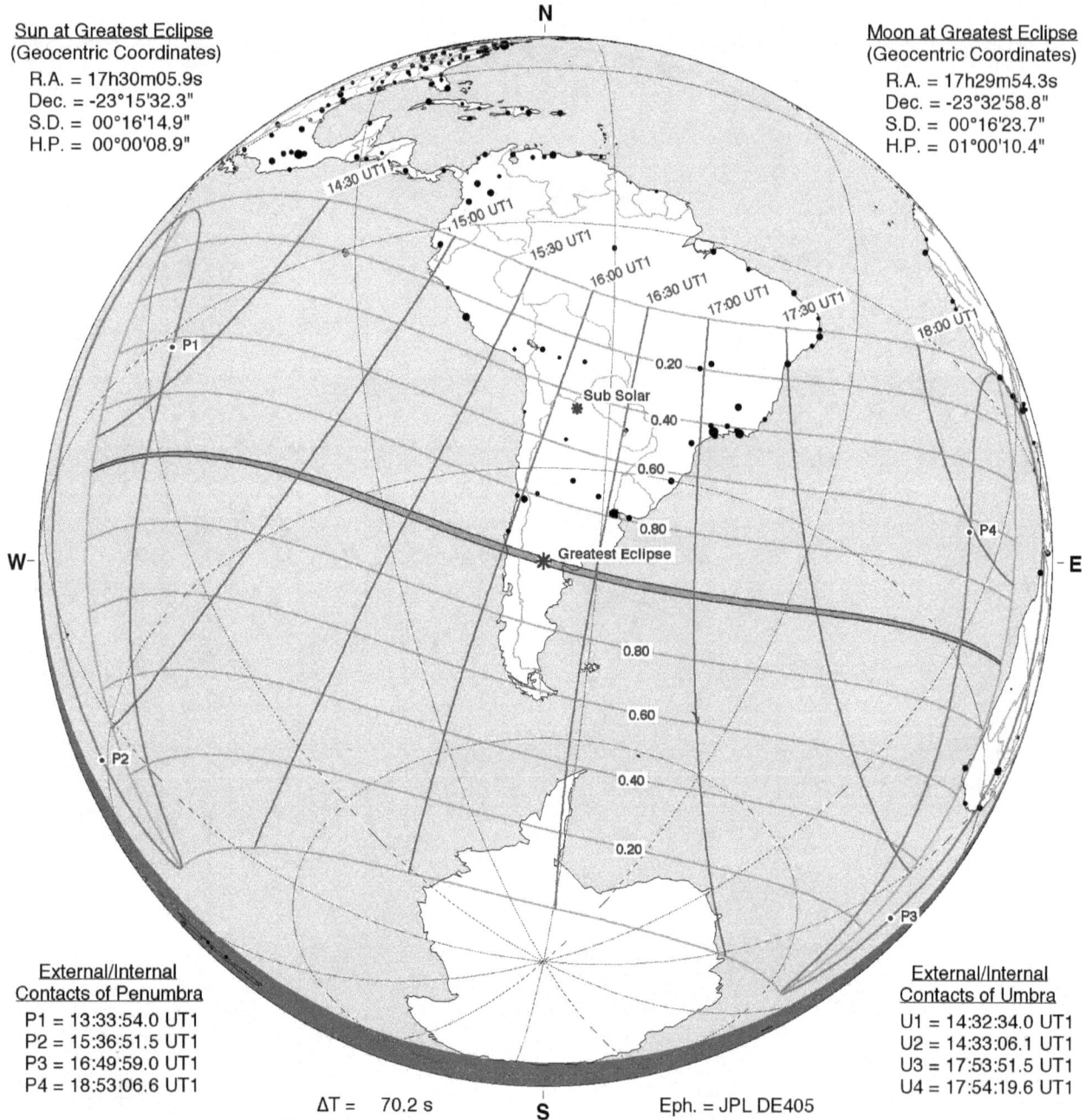

**External/Internal**
**Contacts of Penumbra**
P1 = 13:33:54.0 UT1
P2 = 15:36:51.5 UT1
P3 = 16:49:59.0 UT1
P4 = 18:53:06.6 UT1

**External/Internal**
**Contacts of Umbra**
U1 = 14:32:34.0 UT1
U2 = 14:33:06.1 UT1
U3 = 17:53:51.5 UT1
U4 = 17:54:19.6 UT1

ΔT =   70.2 s          Eph. = JPL DE405

Circumstances at Greatest Eclipse:  16:13:29.1 UT1

| | |
|---|---|
| Lat. = 40°20.1'S | Sun Alt. = 72.7° |
| Long. = 067°57.4'W | Sun Azm. = 10.3° |
| Path Width = 90.2 km | Duration = 02m09.7s |

Circumstances at Greatest Duration:  16:13:34.9 UT1

| | |
|---|---|
| Lat. = 40°20.7'S | Sun Alt. = 72.7° |
| Long. = 067°54.8'W | Sun Azm. = 10.1° |
| Path Width = 90.2 km | Duration = 02m09.7s |

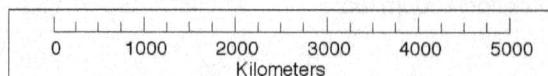

©2016  F. Espenak
www.EclipseWise.com

| 0 | 1000 | 2000 | 3000 | 4000 | 5000 |
|---|---|---|---|---|---|

Kilometers

# Annular Solar Eclipse of 2021 Jun 10

Greatest Eclipse = 10:43:06.7 TD  (= 10:41:56.3 UT1)

Eclipse Magnitude = 0.9435          Saros Series = 147
Gamma = 0.9152          Saros Member = 23 of 80

**Sun at Greatest Eclipse**
(Geocentric Coordinates)
R.A. = 05h15m31.4s
Dec. = +23°02'37.1"
S.D. = 00°15'45.2"
H.P. = 00°00'08.7"

**Moon at Greatest Eclipse**
(Geocentric Coordinates)
R.A. = 05h14m53.6s
Dec. = +23°51'21.6"
S.D. = 00°14'46.8"
H.P. = 00°54'14.5"

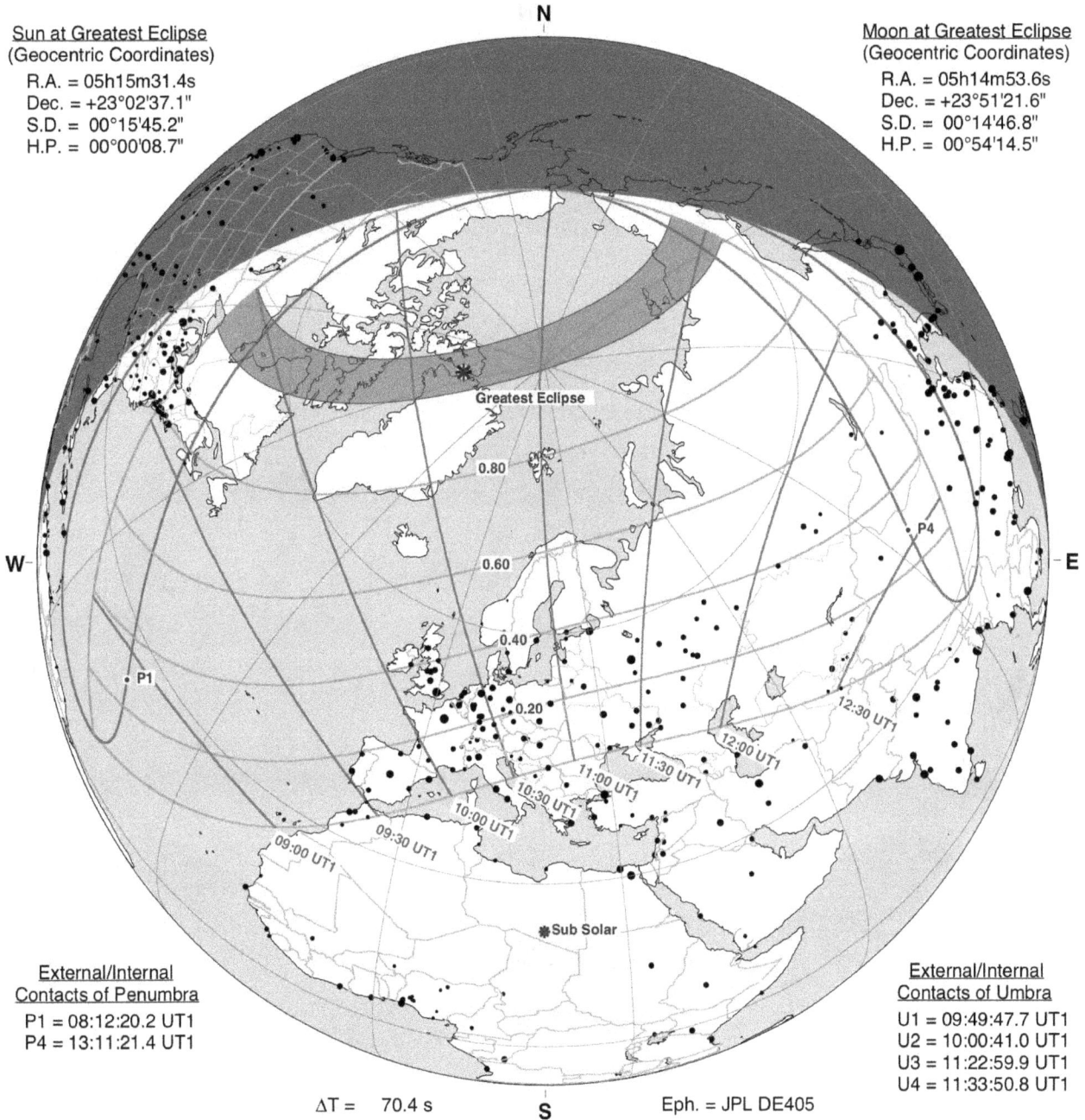

N

Greatest Eclipse

0.80

0.60

0.40

0.20

P4

12:30 UT1
12:00 UT1
11:30 UT1
11:00 UT1
10:30 UT1
10:00 UT1
09:30 UT1
09:00 UT1

W

E

P1

Sub Solar

**External/Internal**
**Contacts of Penumbra**
P1 = 08:12:20.2 UT1
P4 = 13:11:21.4 UT1

**External/Internal**
**Contacts of Umbra**
U1 = 09:49:47.7 UT1
U2 = 10:00:41.0 UT1
U3 = 11:22:59.9 UT1
U4 = 11:33:50.8 UT1

ΔT = 70.4 s          S          Eph. = JPL DE405

**Circumstances at Greatest Eclipse: 10:41:56.3 UT1**

| | |
|---|---|
| Lat. = 80°48.9'N | Sun Alt. = 23.3° |
| Long. = 066°46.1'W | Sun Azm. = 89.9° |
| Path Width = 526.8 km | Duration = 03m51.2s |

**Circumstances at Greatest Duration: 10:41:57.4 UT1**

| | |
|---|---|
| Lat. = 80°49.4'N | Sun Alt. = 23.3° |
| Long. = 066°46.4'W | Sun Azm. = 89.9° |
| Path Width = 526.8 km | Duration = 03m51.2s |

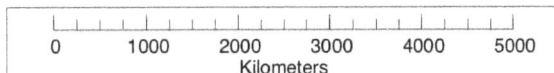

©2016 F. Espenak
www.EclipseWise.com

Kilometers
0  1000  2000  3000  4000  5000

# Total Solar Eclipse of 2021 Dec 04

Greatest Eclipse = 07:34:37.9 TD  (= 07:33:27.3 UT1)

| | |
|---|---|
| Eclipse Magnitude = 1.0367 | Saros Series = 152 |
| Gamma = -0.9526 | Saros Member = 13 of 70 |

Sun at Greatest Eclipse
(Geocentric Coordinates)
R.A. = 16h43m32.4s
Dec. = -22°16'29.4"
S.D. = 00°16'13.6"
H.P. = 00°00'08.9"

Moon at Greatest Eclipse
(Geocentric Coordinates)
R.A. = 16h42m35.0s
Dec. = -23°13'22.3"
S.D. = 00°16'44.7"
H.P. = 01°01'27.3"

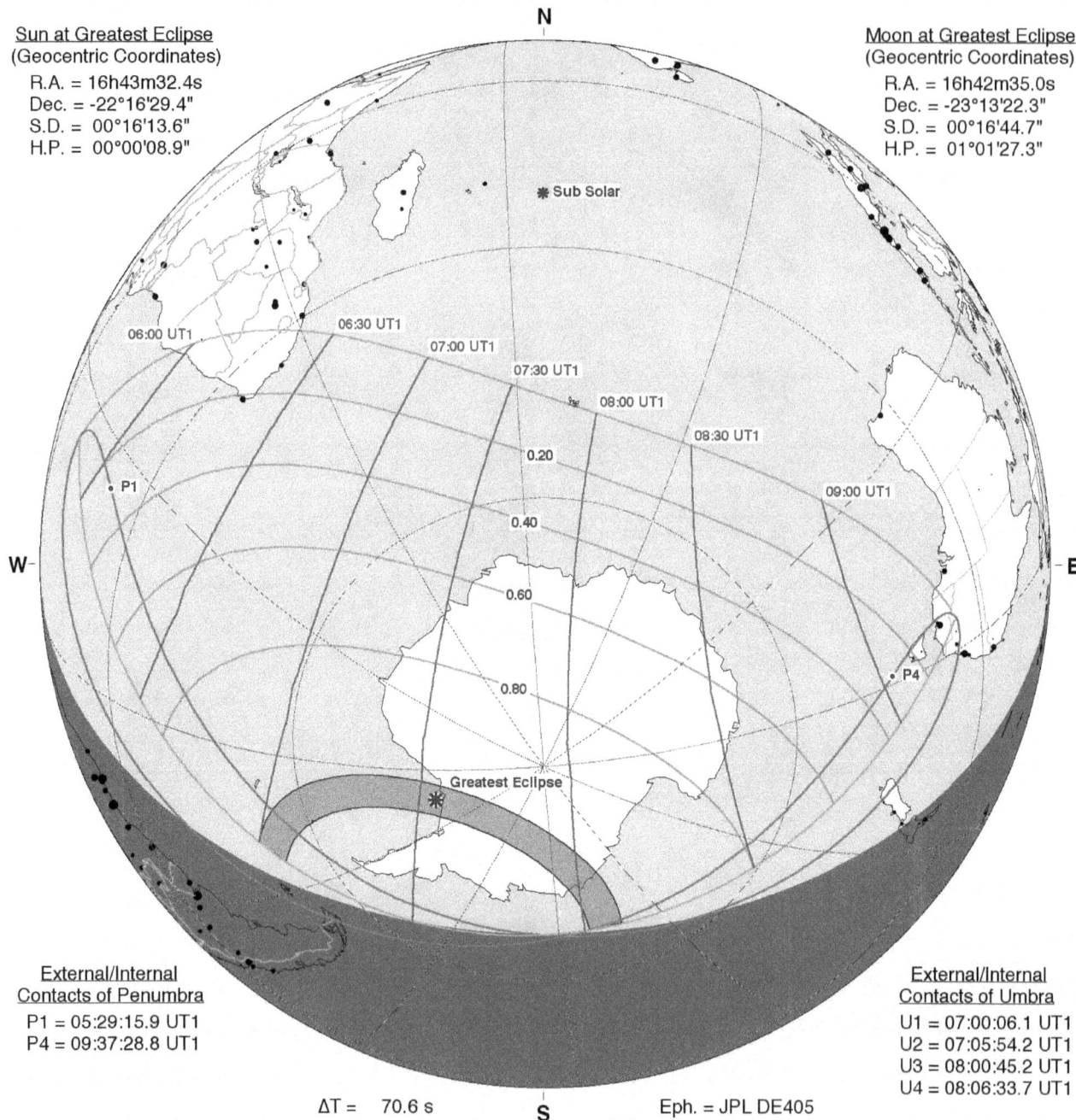

External/Internal
Contacts of Penumbra
P1 = 05:29:15.9 UT1
P4 = 09:37:28.8 UT1

External/Internal
Contacts of Umbra
U1 = 07:00:06.1 UT1
U2 = 07:05:54.2 UT1
U3 = 08:00:45.2 UT1
U4 = 08:06:33.7 UT1

ΔT = 70.6 s          Eph. = JPL DE405

Circumstances at Greatest Eclipse: 07:33:27.3 UT1

| | |
|---|---|
| Lat. = 76°46.6'S | Sun Alt. = 17.2° |
| Long. = 046°13.7'W | Sun Azm. = 114.8° |
| Path Width = 418.8 km | Duration = 01m54.4s |

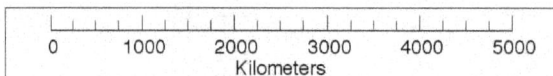

Circumstances at Greatest Duration: 07:33:29.6 UT1

| | |
|---|---|
| Lat. = 76°47.7'S | Sun Alt. = 17.2° |
| Long. = 046°16.7'W | Sun Azm. = 114.8° |
| Path Width = 418.7 km | Duration = 01m54.4s |

©2016  F. Espenak
www.EclipseWise.com

# Partial Solar Eclipse of 2022 Apr 30

Greatest Eclipse = 20:42:36.5 TD  (= 20:41:25.8 UT1)

Eclipse Magnitude = 0.6396　　　Saros Series = 119
Gamma = -1.1901　　　Saros Member = 66 of 71

Sun at Greatest Eclipse
(Geocentric Coordinates)
R.A. = 02h32m15.6s
Dec. = +14°57'53.5"
S.D. = 00°15'52.6"
H.P. = 00°00'08.7"

Moon at Greatest Eclipse
(Geocentric Coordinates)
R.A. = 02h34m04.8s
Dec. = +13°57'48.8"
S.D. = 00°15'04.0"
H.P. = 00°55'17.7"

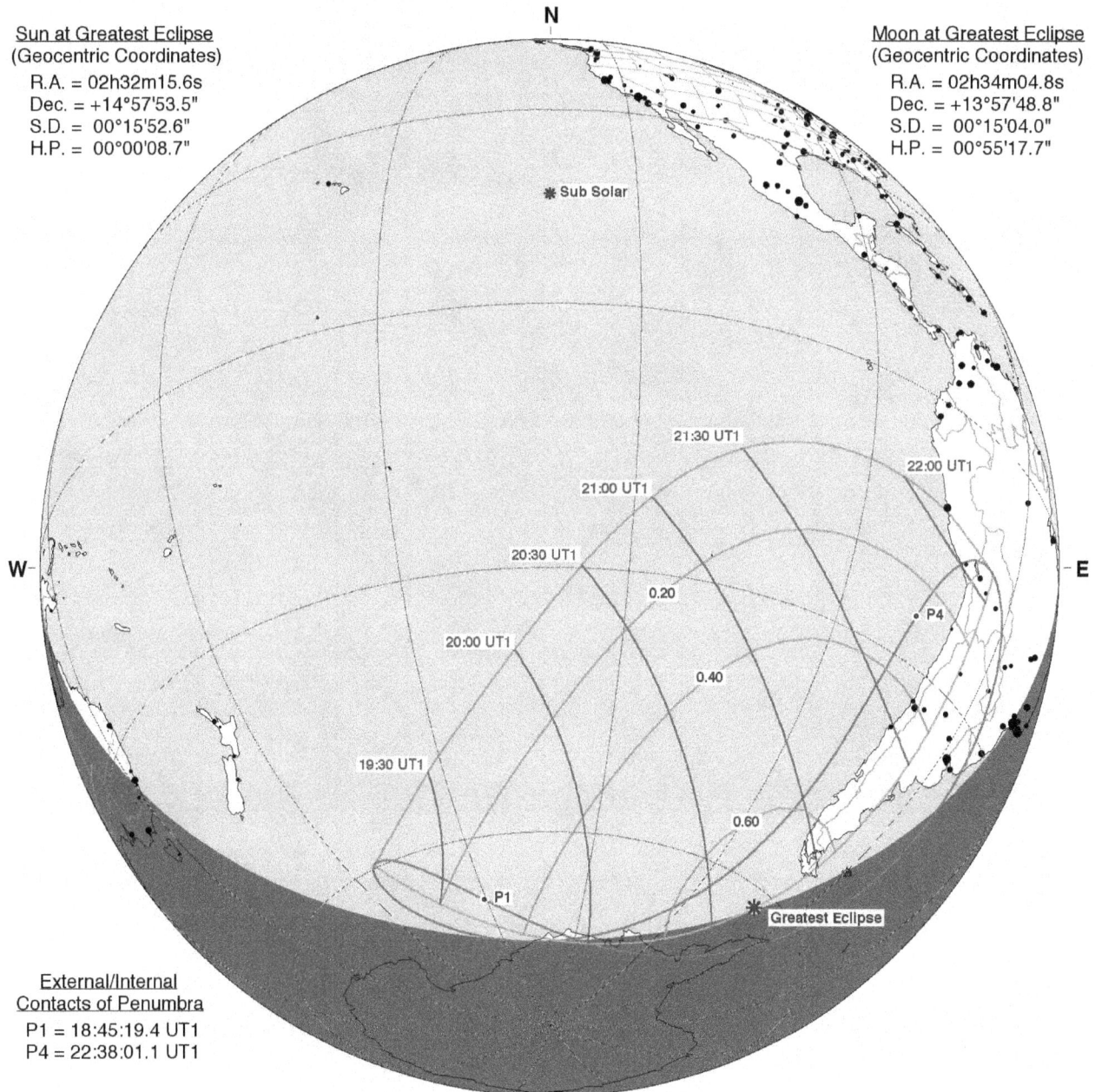

N

* Sub Solar

21:30 UT1

22:00 UT1

21:00 UT1

20:30 UT1

0.20

W

P4

E

20:00 UT1

0.40

19:30 UT1

0.60

P1

* Greatest Eclipse

External/Internal
Contacts of Penumbra
P1 = 18:45:19.4 UT1
P4 = 22:38:01.1 UT1

ΔT = 70.7 s　　S　　Eph. = JPL DE405

Circumstances at Greatest Eclipse: 20:41:25.8 UT1
Lat. = 62°07.0'S　　Sun Alt. = 0.0°
Long. = 071°28.7'W　　Sun Azm. = 303.5°

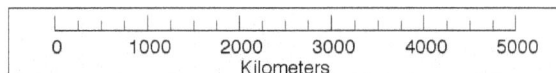

0　1000　2000　3000　4000　5000
Kilometers

©2016  F. Espenak
www.EclipseWise.com

# Partial Solar Eclipse of 2022 Oct 25

Greatest Eclipse = 11:01:20.0 TD  (= 11:00:09.1 UT1)

Eclipse Magnitude = 0.8619      Saros Series = 124
Gamma = 1.0701      Saros Member = 55 of 73

**Sun at Greatest Eclipse**
(Geocentric Coordinates)
R.A. = 13h59m20.5s
Dec. = -12°10'17.0"
S.D. = 00°16'05.0"
H.P. = 00°00'08.8"

**Moon at Greatest Eclipse**
(Geocentric Coordinates)
R.A. = 14h01m10.9s
Dec. = -11°14'16.0"
S.D. = 00°15'52.6"
H.P. = 00°58'16.0"

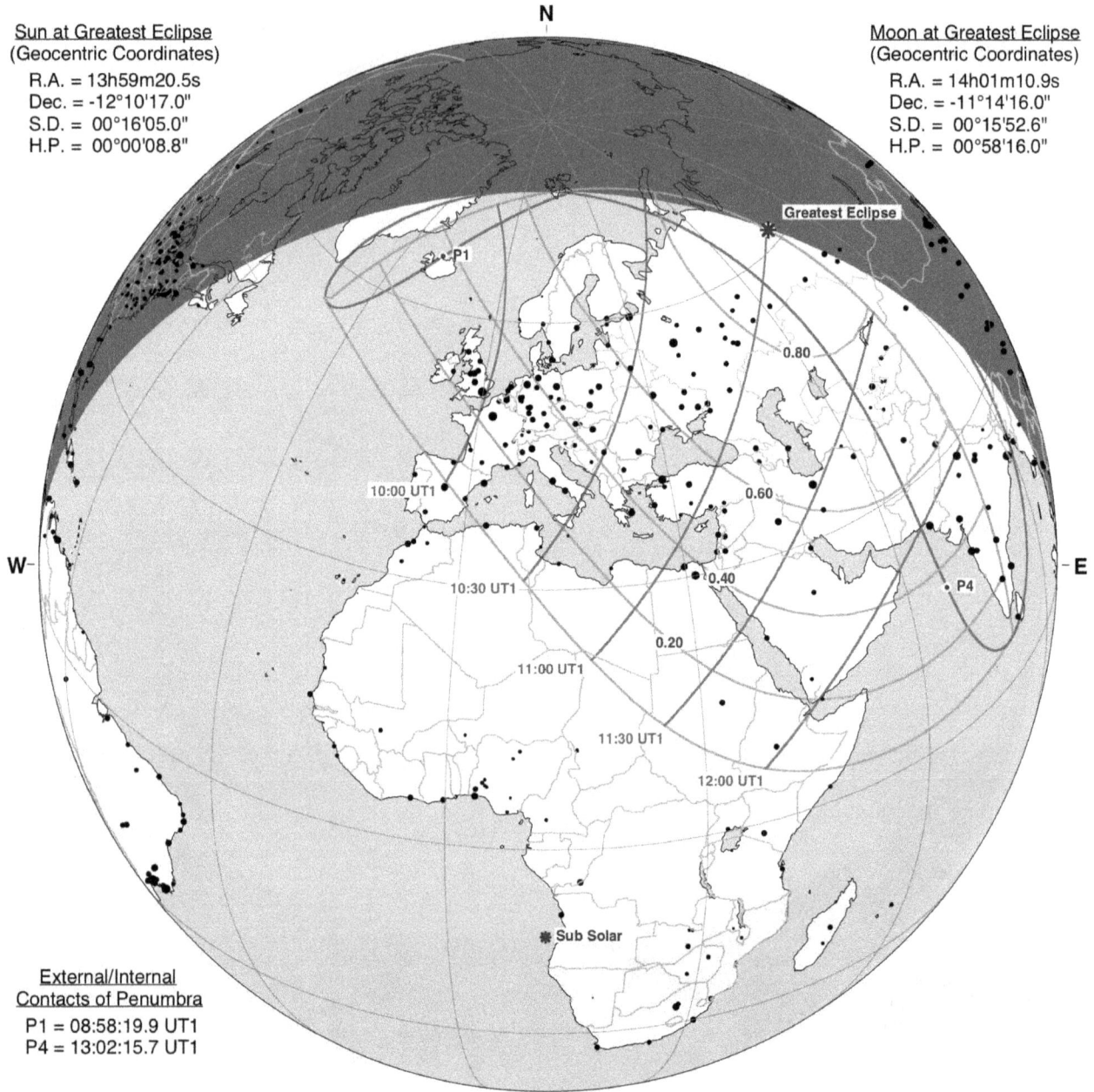

N

Greatest Eclipse

P1

0.80

0.60

10:00 UT1

W — E

10:30 UT1

0.40

P4

11:00 UT1

0.20

11:30 UT1

12:00 UT1

Sub Solar

**External/Internal**
**Contacts of Penumbra**
P1 = 08:58:19.9 UT1
P4 = 13:02:15.7 UT1

ΔT = 70.9 s     S     Eph. = JPL DE405

**Circumstances at Greatest Eclipse: 11:00:09.1 UT1**
Lat. = 61°38.9'N     Sun Alt. = 0.0°
Long. = 077°20.8'E     Sun Azm. = 243.6°

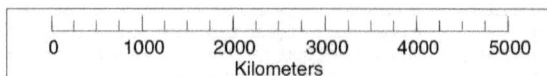

| | | | | | |
|---|---|---|---|---|---|
| 0 | 1000 | 2000 | 3000 | 4000 | 5000 |

Kilometers

©2016 F. Espenak
www.EclipseWise.com

# Hybrid Solar Eclipse of 2023 Apr 20

Greatest Eclipse = 04:17:56.0 TD  (= 04:16:44.9 UT1)

Eclipse Magnitude = 1.0132          Saros Series = 129
Gamma = -0.3952          Saros Member = 52 of 80

Sun at Greatest Eclipse
(Geocentric Coordinates)
R.A. = 01h51m01.7s
Dec. = +11°24'54.1"
S.D. = 00°15'55.4"
H.P. = 00°00'08.8"

Moon at Greatest Eclipse
(Geocentric Coordinates)
R.A. = 01h51m43.2s
Dec. = +11°04'16.7"
S.D. = 00°15'53.6"
H.P. = 00°58'19.9"

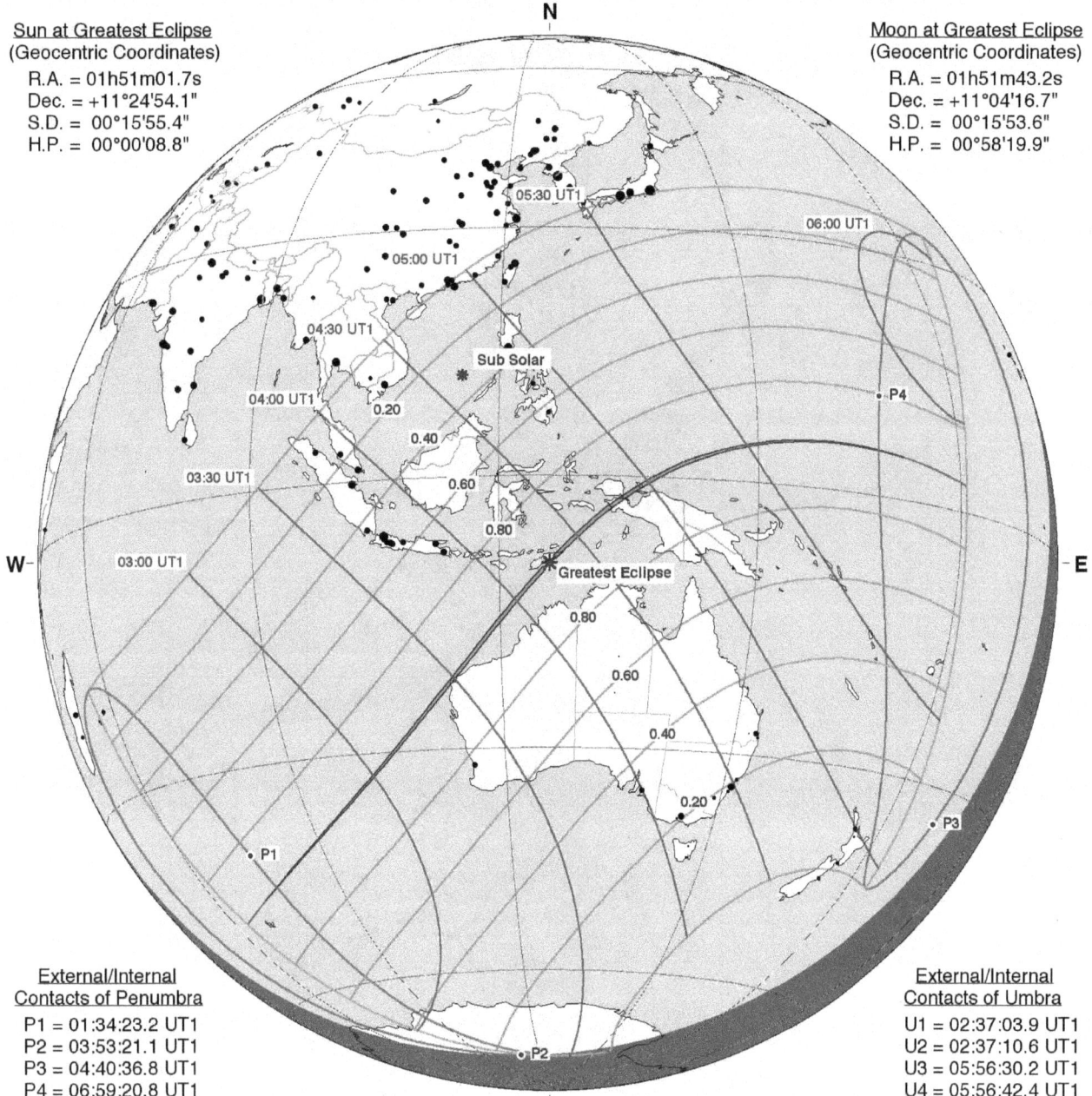

N

05:30 UT1
06:00 UT1
05:00 UT1
04:30 UT1
Sub Solar
P4
04:00 UT1
0.20
0.40
03:30 UT1
0.60
0.80
W
03:00 UT1
Greatest Eclipse
0.80
0.60
0.40
0.20
P3
E
P1
P2
S

External/Internal
Contacts of Penumbra
P1 = 01:34:23.2 UT1
P2 = 03:53:21.1 UT1
P3 = 04:40:36.8 UT1
P4 = 06:59:20.8 UT1

ΔT =   71.1 s

Eph. = JPL DE405

External/Internal
Contacts of Umbra
U1 = 02:37:03.9 UT1
U2 = 02:37:10.6 UT1
U3 = 05:56:30.2 UT1
U4 = 05:56:42.4 UT1

Circumstances at Greatest Eclipse: 04:16:44.9 UT1
Lat. =  09°35.7'S          Sun Alt. =  66.7°
Long. = 125°46.8'E          Sun Azm. = 334.0°
Path Width =  49.0 km          Duration = 01m16.1s

Circumstances at Greatest Duration: 04:16:15.6 UT1
Lat. =  09°42.5'S          Sun Alt. =  66.7°
Long. = 125°40.1'E          Sun Azm. = 334.6°
Path Width =  48.9 km          Duration = 01m16.1s

| 0 | 1000 | 2000 | 3000 | 4000 | 5000 |
|---|---|---|---|---|---|

Kilometers

©2016  F. Espenak
www.EclipseWise.com

# Annular Solar Eclipse of 2023 Oct 14

Greatest Eclipse = 18:00:40.6 TD  (= 17:59:29.3 UT1)

| | |
|---|---|
| Eclipse Magnitude = 0.9520 | Saros Series = 134 |
| Gamma = 0.3753 | Saros Member = 44 of 71 |

**Sun at Greatest Eclipse**
(Geocentric Coordinates)
R.A. = 13h18m05.4s
Dec. = -08°14'36.7"
S.D. = 00°16'02.0"
H.P. = 00°00'08.8"

**Moon at Greatest Eclipse**
(Geocentric Coordinates)
R.A. = 13h18m44.3s
Dec. = -07°56'18.9"
S.D. = 00°15'02.9"
H.P. = 00°55'13.8"

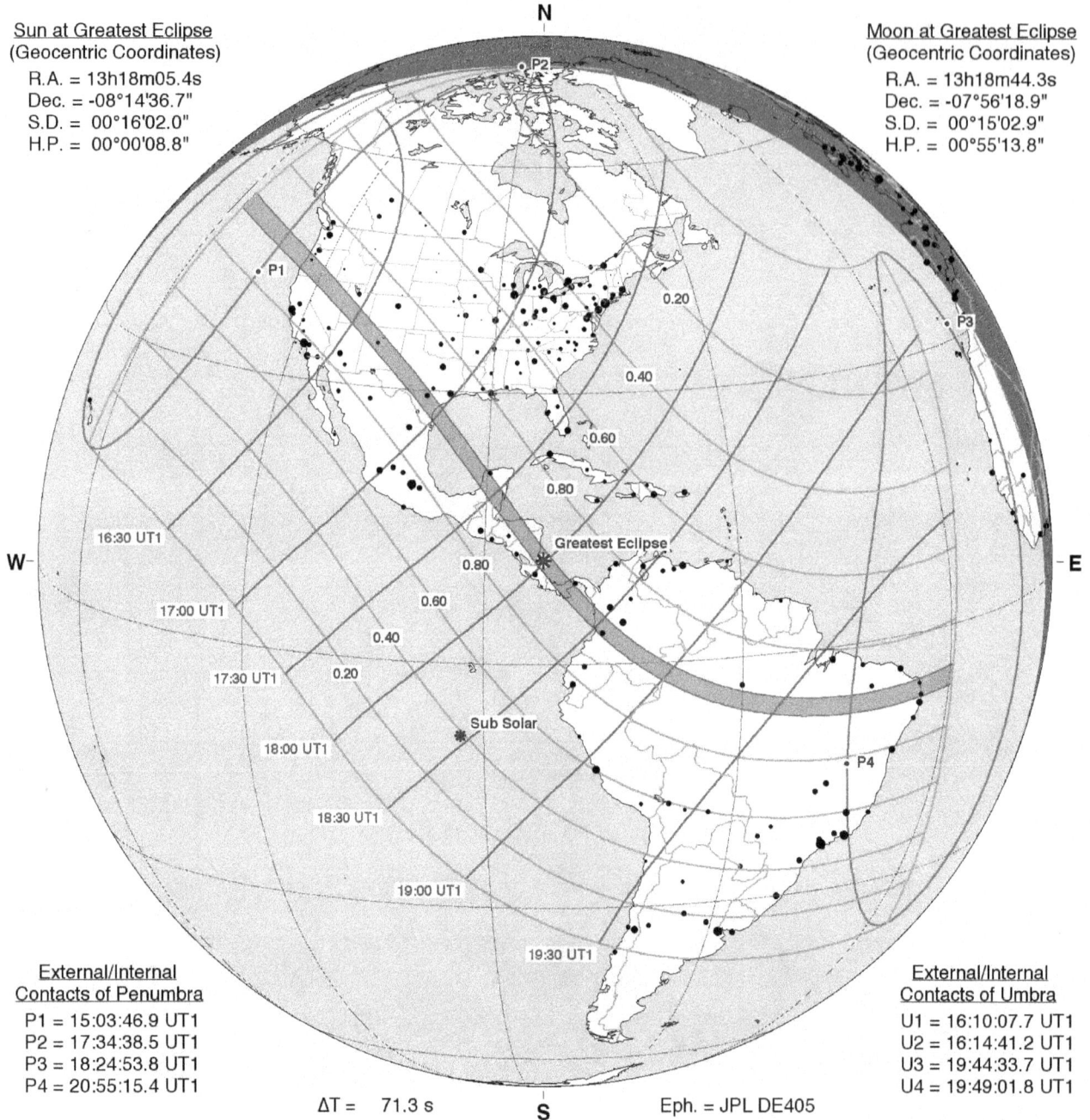

**External/Internal Contacts of Penumbra**
P1 = 15:03:46.9 UT1
P2 = 17:34:38.5 UT1
P3 = 18:24:53.8 UT1
P4 = 20:55:15.4 UT1

**External/Internal Contacts of Umbra**
U1 = 16:10:07.7 UT1
U2 = 16:14:41.2 UT1
U3 = 19:44:33.7 UT1
U4 = 19:49:01.8 UT1

ΔT = 71.3 s

Eph. = JPL DE405

**Circumstances at Greatest Eclipse: 17:59:29.3 UT1**

| | |
|---|---|
| Lat. = 11°22.1'N | Sun Alt. = 67.9° |
| Long. = 083°06.1'W | Sun Azm. = 208.0° |
| Path Width = 187.4 km | Duration = 05m17.2s |

**Circumstances at Greatest Duration: 18:13:09.2 UT1**

| | |
|---|---|
| Lat. = 08°14.6'N | Sun Alt. = 66.8° |
| Long. = 080°24.1'W | Sun Azm. = 225.1° |
| Path Width = 191.1 km | Duration = 05m17.8s |

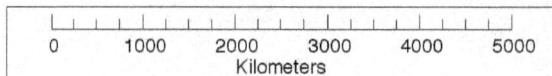

©2016  F. Espenak
www.EclipseWise.com

# Total Solar Eclipse of 2024 Apr 08

Greatest Eclipse = 18:18:29.4 TD   (= 18:17:17.9 UT1)

| | |
|---|---|
| Eclipse Magnitude = 1.0566 | Saros Series = 139 |
| Gamma = 0.3431 | Saros Member = 30 of 71 |

**Sun at Greatest Eclipse**
(Geocentric Coordinates)
R.A. = 01h11m36.9s
Dec. = +07°35'29.4"
S.D. = 00°15'58.2"
H.P. = 00°00'08.8"

**Moon at Greatest Eclipse**
(Geocentric Coordinates)
R.A. = 01h10m57.5s
Dec. = +07°53'55.5"
S.D. = 00°16'36.3"
H.P. = 01°00'56.6"

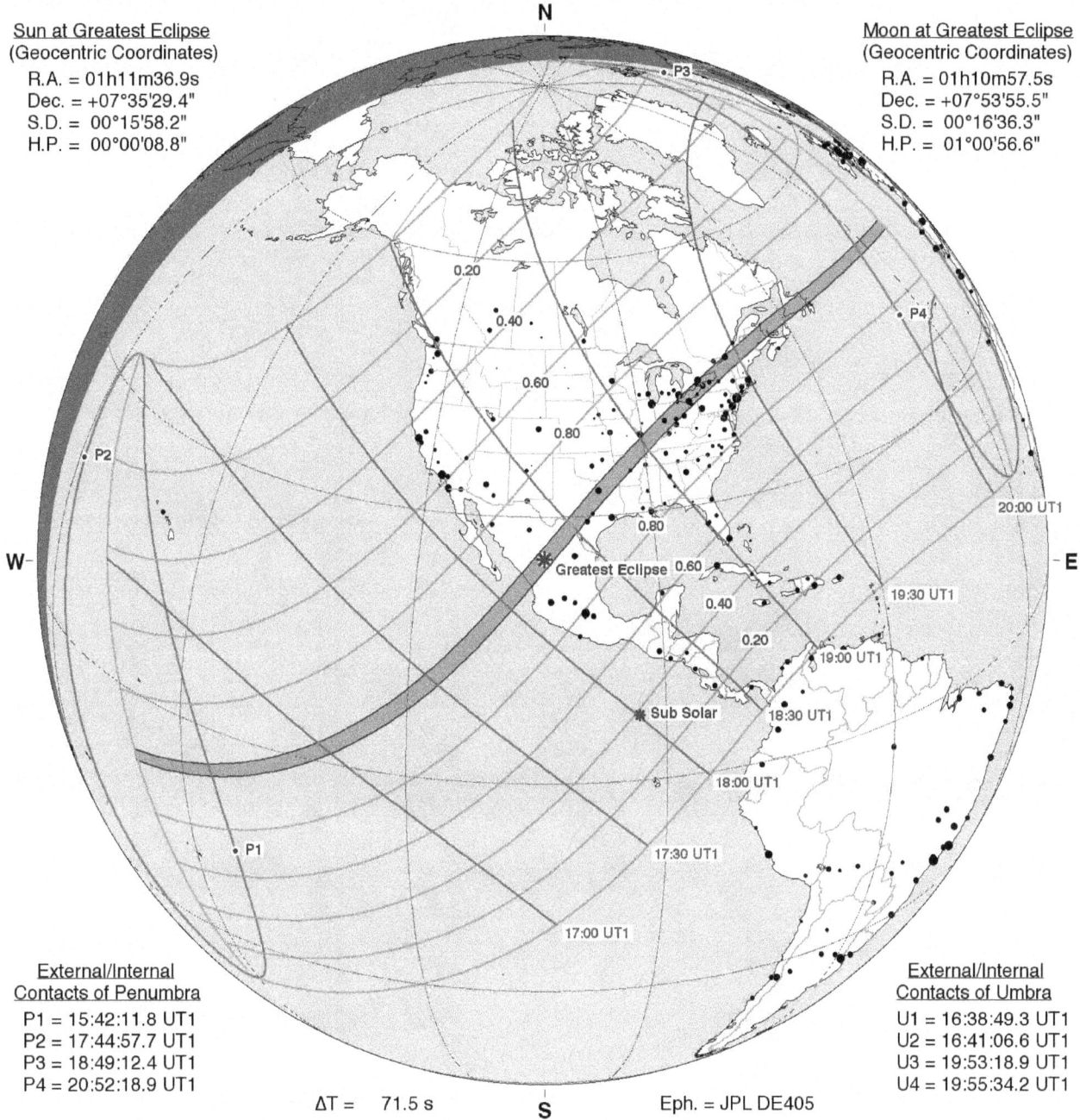

**External/Internal Contacts of Penumbra**
P1 = 15:42:11.8 UT1
P2 = 17:44:57.7 UT1
P3 = 18:49:12.4 UT1
P4 = 20:52:18.9 UT1

ΔT = 71.5 s

Eph. = JPL DE405

**External/Internal Contacts of Umbra**
U1 = 16:38:49.3 UT1
U2 = 16:41:06.6 UT1
U3 = 19:53:18.9 UT1
U4 = 19:55:34.2 UT1

**Circumstances at Greatest Eclipse: 18:17:17.9 UT1**

| | |
|---|---|
| Lat. = 25°17.4'N | Sun Alt. = 69.8° |
| Long. = 104°08.3'W | Sun Azm. = 149.4° |
| Path Width = 197.5 km | Duration = 04m28.1s |

**Circumstances at Greatest Duration: 18:19:32.7 UT1**

| | |
|---|---|
| Lat. = 25°55.8'N | Sun Alt. = 69.7° |
| Long. = 103°31.1'W | Sun Azm. = 153.3° |
| Path Width = 197.0 km | Duration = 04m28.2s |

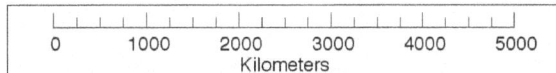

©2016 F. Espenak
www.EclipseWise.com

# Annular Solar Eclipse of 2024 Oct 02

Greatest Eclipse = 18:46:13.3 TD  (= 18:45:01.6 UT1)

| | |
|---|---|
| Eclipse Magnitude = 0.9326 | Saros Series = 144 |
| Gamma = -0.3509 | Saros Member = 17 of 70 |

Sun at Greatest Eclipse
(Geocentric Coordinates)
R.A. = 12h36m58.9s
Dec. = -03°59'03.9"
S.D. = 00°15'58.9"
H.P. = 00°00'08.8"

Moon at Greatest Eclipse
(Geocentric Coordinates)
R.A. = 12h36m22.3s
Dec. = -04°15'35.4"
S.D. = 00°14'41.8"
H.P. = 00°53'56.4"

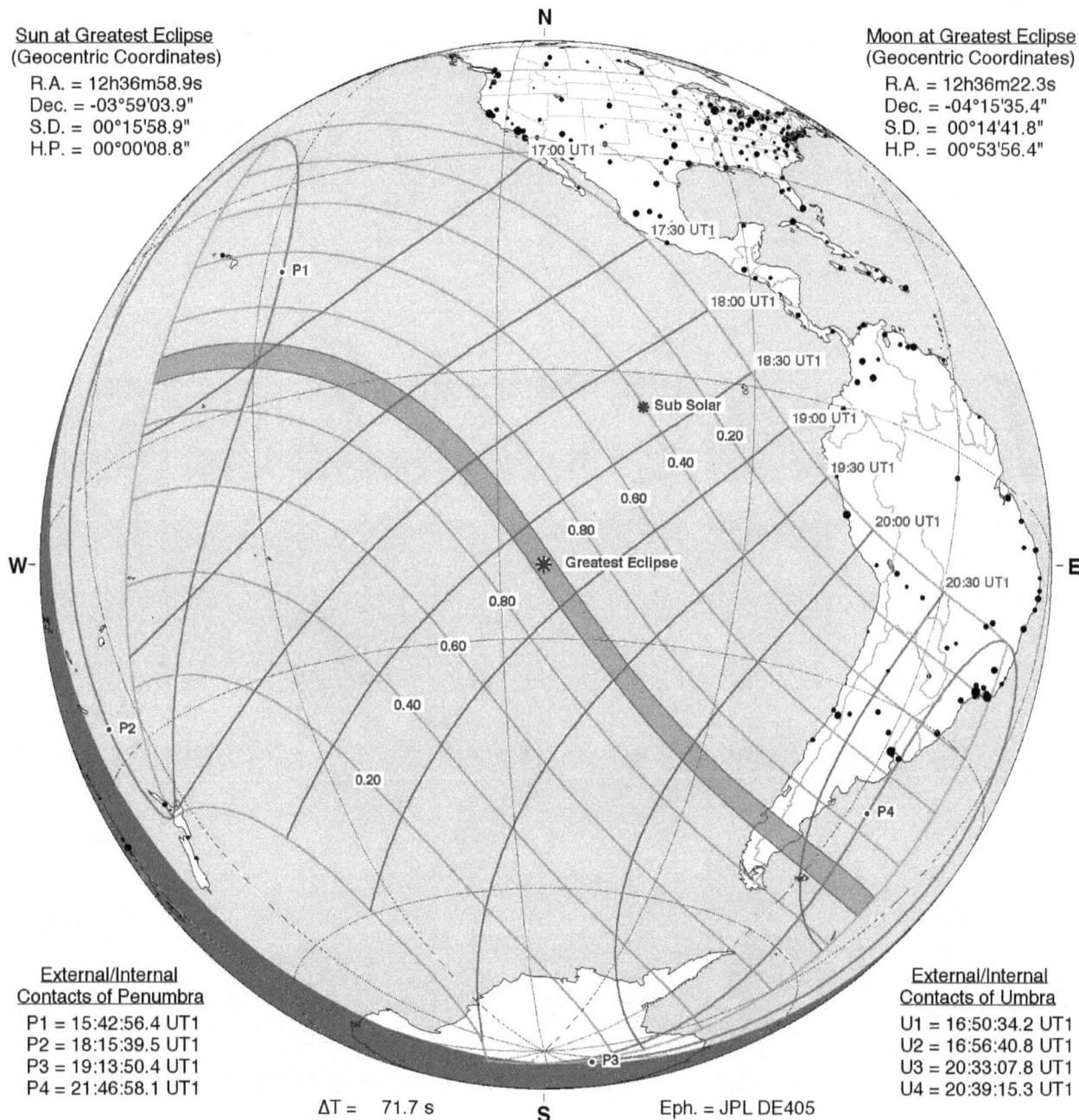

External/Internal
Contacts of Penumbra
P1 = 15:42:56.4 UT1
P2 = 18:15:39.5 UT1
P3 = 19:13:50.4 UT1
P4 = 21:46:58.1 UT1

External/Internal
Contacts of Umbra
U1 = 16:50:34.2 UT1
U2 = 16:56:40.8 UT1
U3 = 20:33:07.8 UT1
U4 = 20:39:15.3 UT1

ΔT = 71.7 s        Eph. = JPL DE405

Circumstances at Greatest Eclipse: 18:45:01.6 UT1

| | |
|---|---|
| Lat. = 21°57.2'S | Sun Alt. = 69.3° |
| Long. = 114°30.5'W | Sun Azm. = 31.1° |
| Path Width = 266.5 km | Duration = 07m25.1s |

Circumstances at Greatest Duration: 18:53:00.1 UT1

| | |
|---|---|
| Lat. = 23°57.8'S | Sun Alt. = 68.9° |
| Long. = 112°56.3'W | Sun Azm. = 19.6° |
| Path Width = 264.4 km | Duration = 07m25.4s |

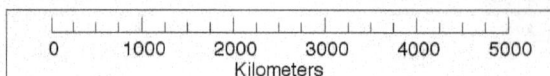

©2016  F. Espenak
www.EclipseWise.com

# Partial Solar Eclipse of 2025 Mar 29

Greatest Eclipse = 10:48:36.1 TD  (= 10:47:24.2 UT1)

| Eclipse Magnitude = 0.9376 | Saros Series = 149 |
|---|---|
| Gamma = 1.0405 | Saros Member = 21 of 71 |

Sun at Greatest Eclipse
(Geocentric Coordinates)
R.A. = 00h33m03.1s
Dec. = +03°33'55.0"
S.D. = 00°16'01.1"
H.P. = 00°00'08.8"

Moon at Greatest Eclipse
(Geocentric Coordinates)
R.A. = 00h31m00.8s
Dec. = +04°29'34.1"
S.D. = 00°16'39.4"
H.P. = 01°01'07.8"

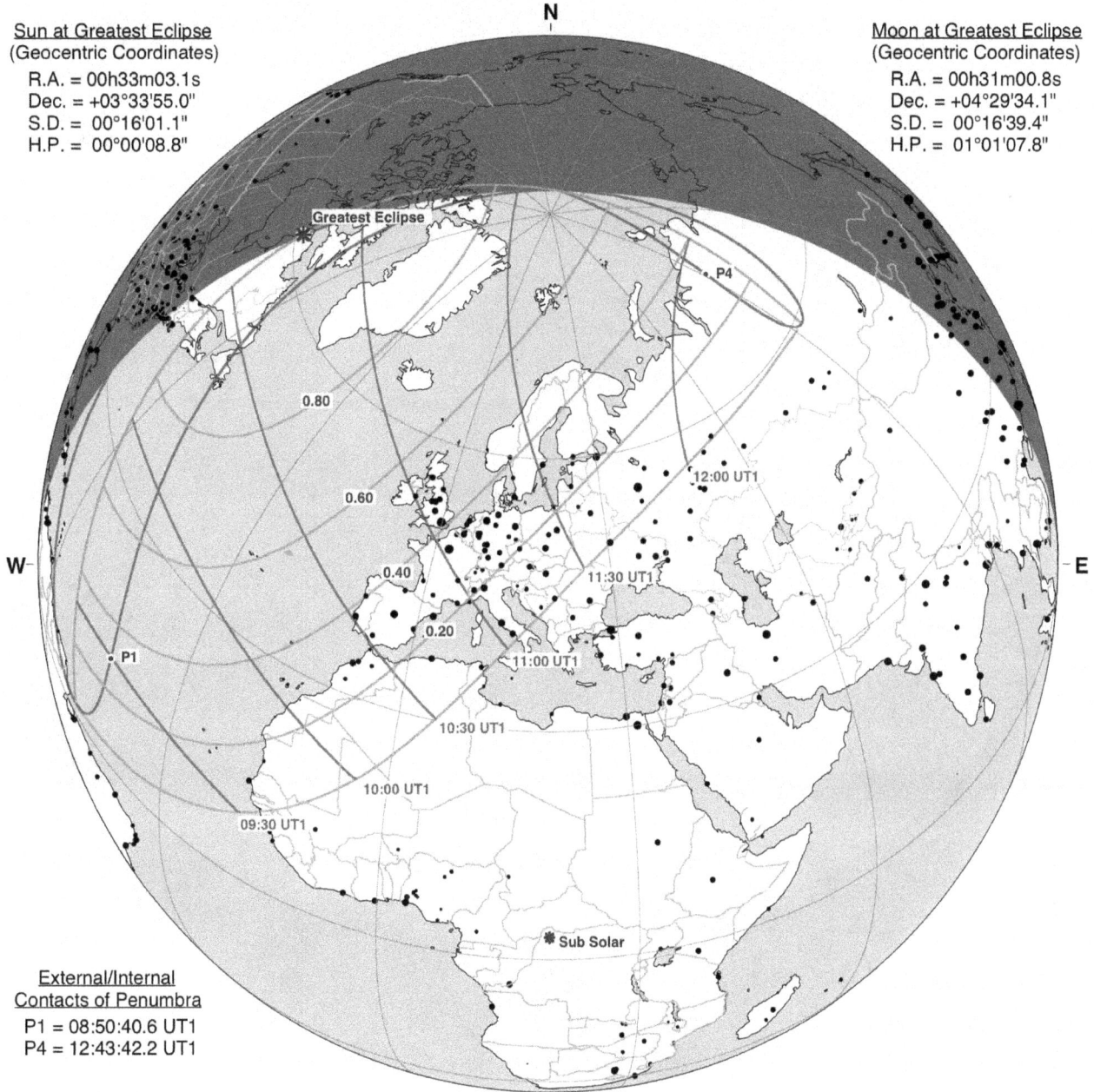

Greatest Eclipse

P4

0.80

0.60

0.40

0.20

12:00 UT1

11:30 UT1

11:00 UT1

10:30 UT1

10:00 UT1

09:30 UT1

P1

N

S

W

E

Sub Solar

External/Internal
Contacts of Penumbra
P1 = 08:50:40.6 UT1
P4 = 12:43:42.2 UT1

ΔT = 71.9 s

Eph. = JPL DE405

Circumstances at Greatest Eclipse: 10:47:24.2 UT1

| Lat. = 61°06.0'N | Sun Alt. =   0.0° |
|---|---|
| Long. = 077°05.1'W | Sun Azm. = 82.6° |

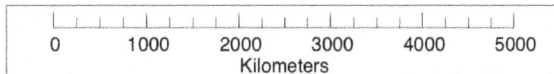

| 0 | 1000 | 2000 | 3000 | 4000 | 5000 |
|---|---|---|---|---|---|

Kilometers

©2016  F. Espenak
www.EclipseWise.com

# Partial Solar Eclipse of 2025 Sep 21

Greatest Eclipse = 19:43:04.2 TD   (= 19:41:52.2 UT1)

| | |
|---|---|
| Eclipse Magnitude = 0.8550 | Saros Series = 154 |
| Gamma = -1.0651 | Saros Member = 7 of 71 |

Sun at Greatest Eclipse
(Geocentric Coordinates)
R.A. = 11h56m36.9s
Dec. = +00°22'00.7"
S.D. = 00°15'55.9"
H.P. = 00°00'08.8"

Moon at Greatest Eclipse
(Geocentric Coordinates)
R.A. = 11h54m42.8s
Dec. = -00°29'14.7"
S.D. = 00°15'02.8"
H.P. = 00°55'13.2"

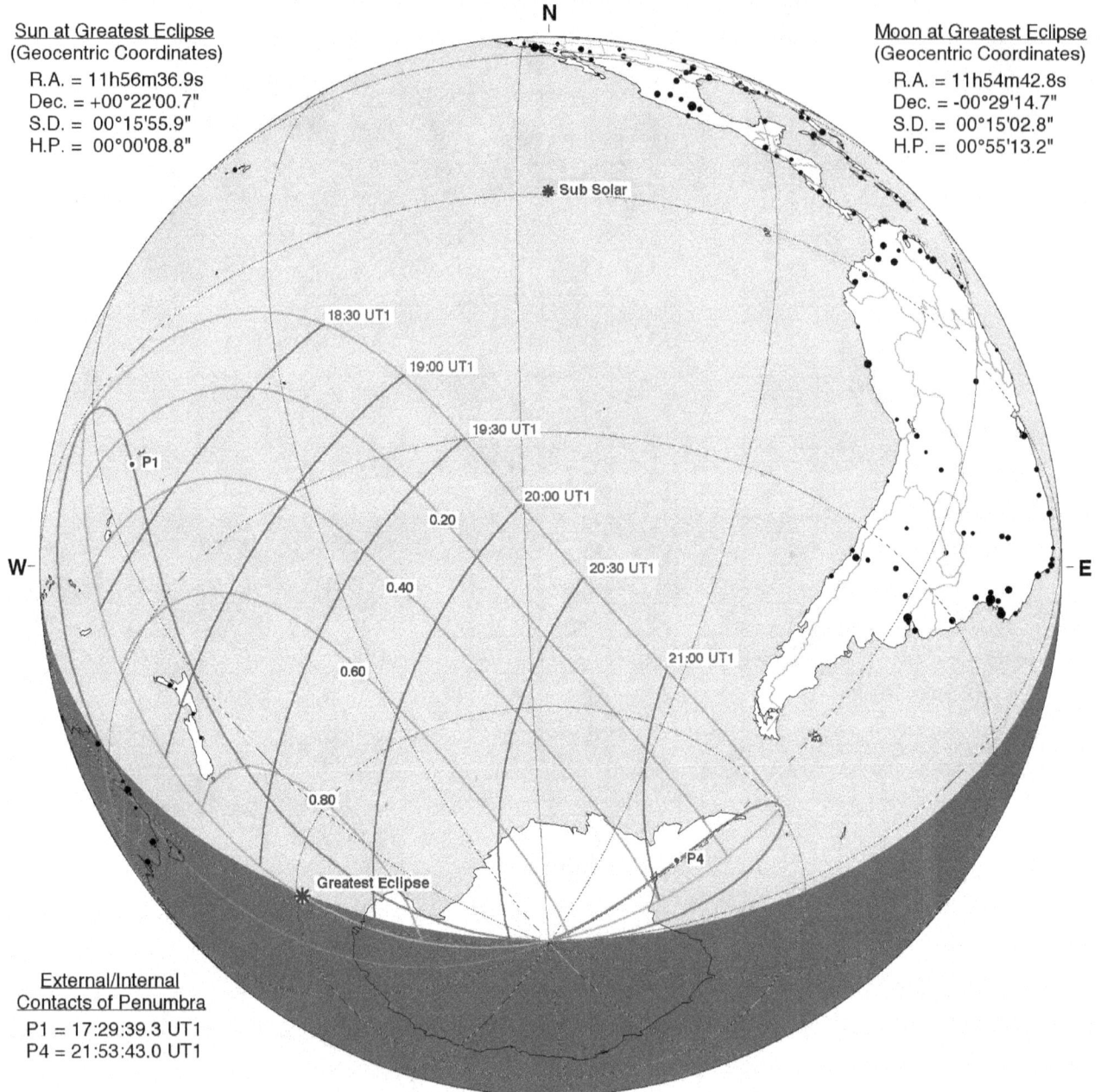

N

* Sub Solar

18:30 UT1
19:00 UT1
19:30 UT1
20:00 UT1
20:30 UT1
21:00 UT1

P1

0.20
0.40
0.60
0.80

W

E

Greatest Eclipse

P4

External/Internal
Contacts of Penumbra
P1 = 17:29:39.3 UT1
P4 = 21:53:43.0 UT1

ΔT = 72.1 s

S

Eph. = JPL DE405

Circumstances at Greatest Eclipse: 19:41:52.2 UT1

| | |
|---|---|
| Lat. = 60°54.1'S | Sun Alt. = 0.0° |
| Long. = 153°29.9'E | Sun Azm. = 89.2° |

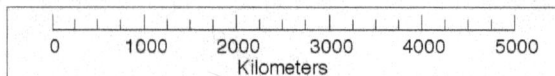

0   1000   2000   3000   4000   5000
Kilometers

©2016 F. Espenak
www.EclipseWise.com

# Annular Solar Eclipse of 2026 Feb 17

Greatest Eclipse = 12:13:05.8 TD   (= 12:11:53.6 UT1)

| | |
|---|---|
| Eclipse Magnitude = 0.9630 | Saros Series = 121 |
| Gamma = -0.9743 | Saros Member = 61 of 71 |

**Sun at Greatest Eclipse**
(Geocentric Coordinates)
R.A. = 22h03m54.3s
Dec. = -11°52'42.3"
S.D. = 00°16'11.1"
H.P. = 00°00'08.9"

**Moon at Greatest Eclipse**
(Geocentric Coordinates)
R.A. = 22h05m34.0s
Dec. = -12°42'29.5"
S.D. = 00°15'32.4"
H.P. = 00°57'02.0"

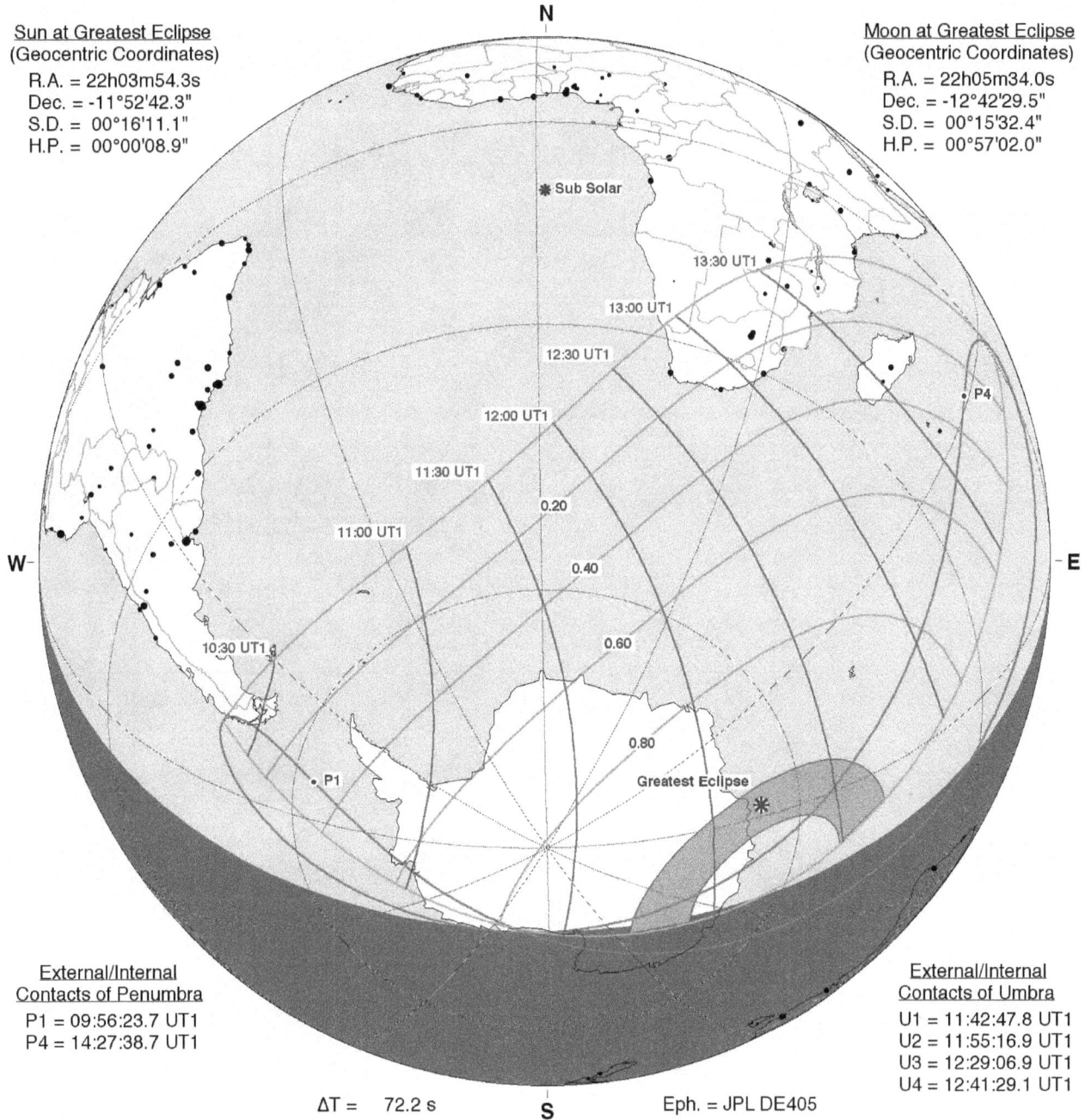

N

* Sub Solar

13:30 UT1
13:00 UT1
12:30 UT1
12:00 UT1
11:30 UT1
11:00 UT1
10:30 UT1

0.20
0.40
0.60
0.80

P4
P1

Greatest Eclipse

W — — E

**External/Internal**
**Contacts of Penumbra**
P1 = 09:56:23.7 UT1
P4 = 14:27:38.7 UT1

**External/Internal**
**Contacts of Umbra**
U1 = 11:42:47.8 UT1
U2 = 11:55:16.9 UT1
U3 = 12:29:06.9 UT1
U4 = 12:41:29.1 UT1

ΔT = 72.2 s          S          Eph. = JPL DE405

**Circumstances at Greatest Eclipse: 12:11:53.6 UT1**

| | |
|---|---|
| Lat. = 64°43.1'S | Sun Alt. = 12.3° |
| Long. = 086°44.4'E | Sun Azm. = 268.3° |
| Path Width = 616.2 km | Duration = 02m19.6s |

**Circumstances at Greatest Duration: 11:48:15.2 UT1**

| | |
|---|---|
| Lat. = 71°57.6'S | Sun Alt. = 0.0° |
| Long. = 136°39.8'E | Sun Azm. = 228.4° |
| Path Width = 765.7 km | Duration = 02m20.9s |

0  1000  2000  3000  4000  5000
Kilometers

# Total Solar Eclipse of 2026 Aug 12

Greatest Eclipse = 17:47:05.8 TD (= 17:45:53.3 UT1)

| | |
|---|---|
| Eclipse Magnitude = 1.0386 | Saros Series = 126 |
| Gamma = 0.8977 | Saros Member = 48 of 72 |

**Sun at Greatest Eclipse**
(Geocentric Coordinates)
R.A. = 09h29m47.3s
Dec. = +14°48'04.5"
S.D. = 00°15'47.0"
H.P. = 00°00'08.7"

**Moon at Greatest Eclipse**
(Geocentric Coordinates)
R.A. = 09h31m17.3s
Dec. = +15°36'58.5"
S.D. = 00°16'16.9"
H.P. = 00°59'45.1"

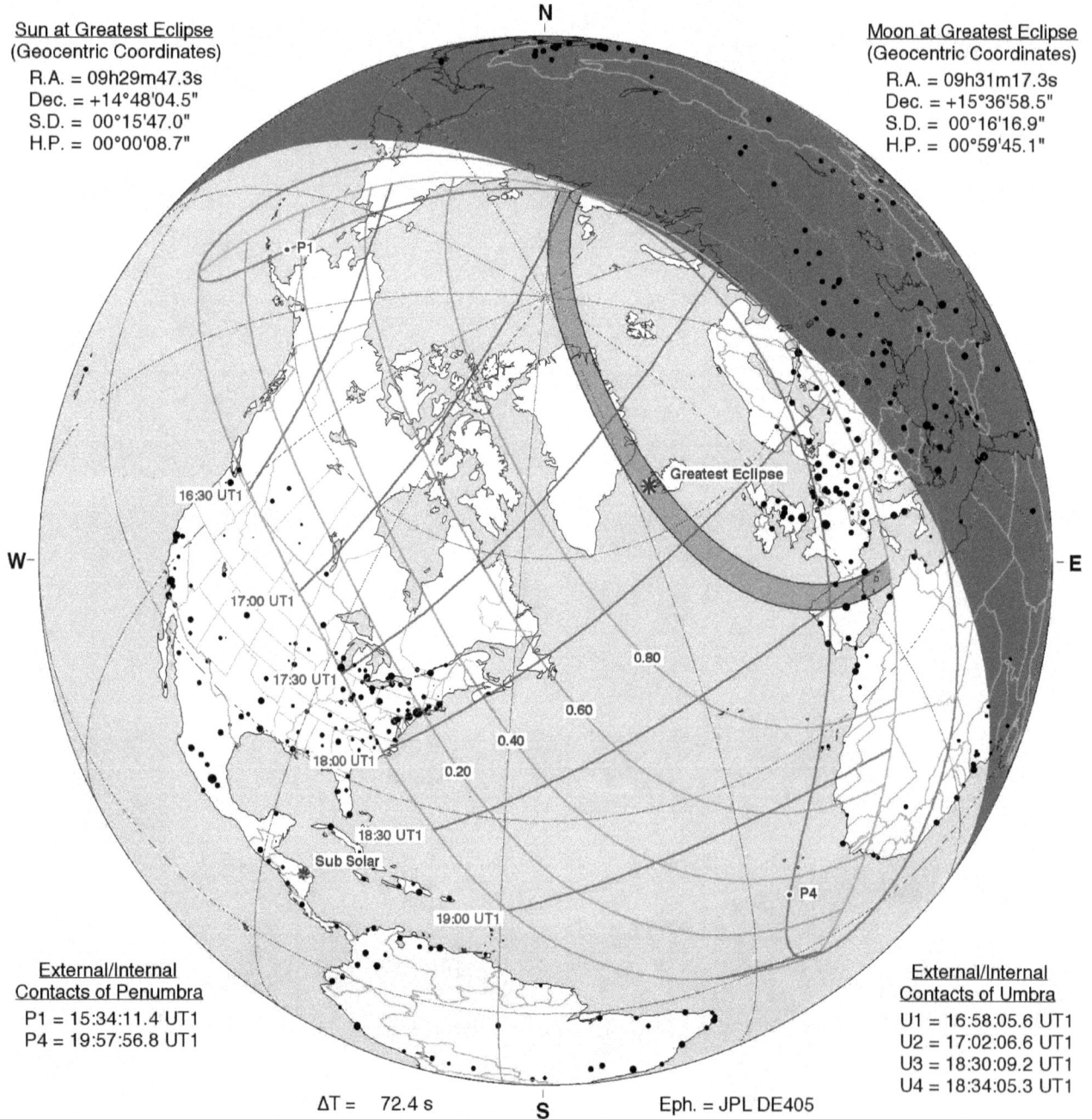

**External/Internal Contacts of Penumbra**
P1 = 15:34:11.4 UT1
P4 = 19:57:56.8 UT1

**External/Internal Contacts of Umbra**
U1 = 16:58:05.6 UT1
U2 = 17:02:06.6 UT1
U3 = 18:30:09.2 UT1
U4 = 18:34:05.3 UT1

ΔT = 72.4 s          Eph. = JPL DE405

**Circumstances at Greatest Eclipse: 17:45:53.3 UT1**

| | |
|---|---|
| Lat. = 65°13.4'N | Sun Alt. = 25.8° |
| Long. = 025°13.7'W | Sun Azm. = 248.4° |
| Path Width = 294.0 km | Duration = 02m18.2s |

**Circumstances at Greatest Duration: 17:44:41.5 UT1**

| | |
|---|---|
| Lat. = 65°49.8'N | Sun Alt. = 25.7° |
| Long. = 025°29.1'W | Sun Azm. = 247.6° |
| Path Width = 292.9 km | Duration = 02m18.2s |

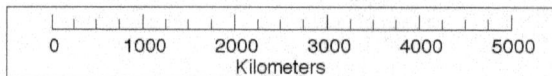

©2016 F. Espenak
www.EclipseWise.com

Kilometers
0   1000   2000   3000   4000   5000

# Annular Solar Eclipse of 2027 Feb 06

Greatest Eclipse = 16:00:47.7 TD   (= 15:59:35.1 UT1)

| | |
|---|---|
| Eclipse Magnitude = 0.9281 | Saros Series = 131 |
| Gamma = -0.2952 | Saros Member = 51 of 70 |

**N**

Sun at Greatest Eclipse
(Geocentric Coordinates)
R.A. = 21h20m17.6s
Dec. = -15°32'54.5"
S.D. = 00°16'13.1"
H.P. = 00°00'08.9"

Moon at Greatest Eclipse
(Geocentric Coordinates)
R.A. = 21h20m44.2s
Dec. = -15°47'36.0"
S.D. = 00°14'50.2"
H.P. = 00°54'27.0"

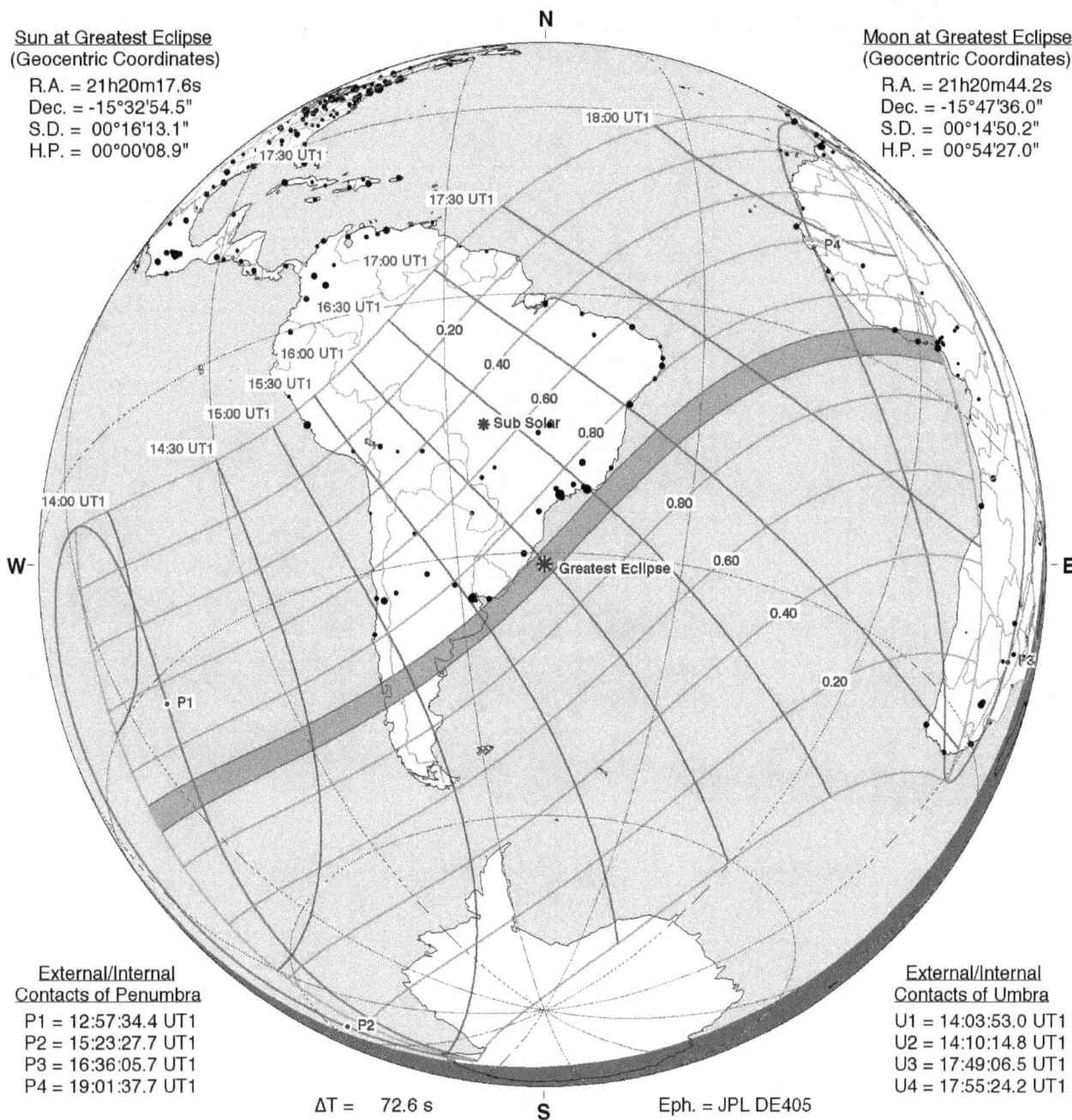

18:00 UT1

17:30 UT1

17:30 UT1

P4

17:00 UT1

16:30 UT1

0.20

16:00 UT1

0.40

15:30 UT1

0.60

15:00 UT1

* Sub Solar

0.80

14:30 UT1

0.80

14:00 UT1

0.60

* Greatest Eclipse

**W**

0.40

**E**

0.20

P1

P3

P2

External/Internal
Contacts of Penumbra
P1 = 12:57:34.4 UT1
P2 = 15:23:27.7 UT1
P3 = 16:36:05.7 UT1
P4 = 19:01:37.7 UT1

External/Internal
Contacts of Umbra
U1 = 14:03:53.0 UT1
U2 = 14:10:14.8 UT1
U3 = 17:49:06.5 UT1
U4 = 17:55:24.2 UT1

ΔT = 72.6 s

**S**

Eph. = JPL DE405

### Circumstances at Greatest Eclipse: 15:59:35.1 UT1

| | |
|---|---|
| Lat. = 31°18.2'S | Sun Alt. = 72.7° |
| Long. = 048°28.0'W | Sun Azm. = 333.5° |
| Path Width = 281.5 km | Duration = 07m50.9s |

### Circumstances at Greatest Duration: 15:41:47.6 UT1

| | |
|---|---|
| Lat. = 35°04.2'S | Sun Alt. = 70.4° |
| Long. = 053°22.3'W | Sun Azm. = 4.1° |
| Path Width = 279.7 km | Duration = 07m53.5s |

0   1000   2000   3000   4000   5000
Kilometers

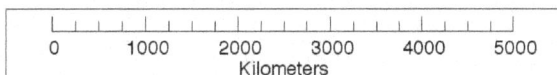

# Total Solar Eclipse of 2027 Aug 02

Greatest Eclipse = 10:07:50.2 TD   (= 10:06:37.4 UT1)

| | |
|---|---|
| Eclipse Magnitude = 1.0790 | Saros Series = 136 |
| Gamma = 0.1421 | Saros Member = 38 of 71 |

**N**

Sun at Greatest Eclipse
(Geocentric Coordinates)
R.A. = 08h49m26.9s
Dec. = +17°45'41.3"
S.D. = 00°15'45.5"
H.P. = 00°00'08.7"

Moon at Greatest Eclipse
(Geocentric Coordinates)
R.A. = 08h49m40.1s
Dec. = +17°53'47.8"
S.D. = 00°16'43.1"
H.P. = 01°01'21.4"

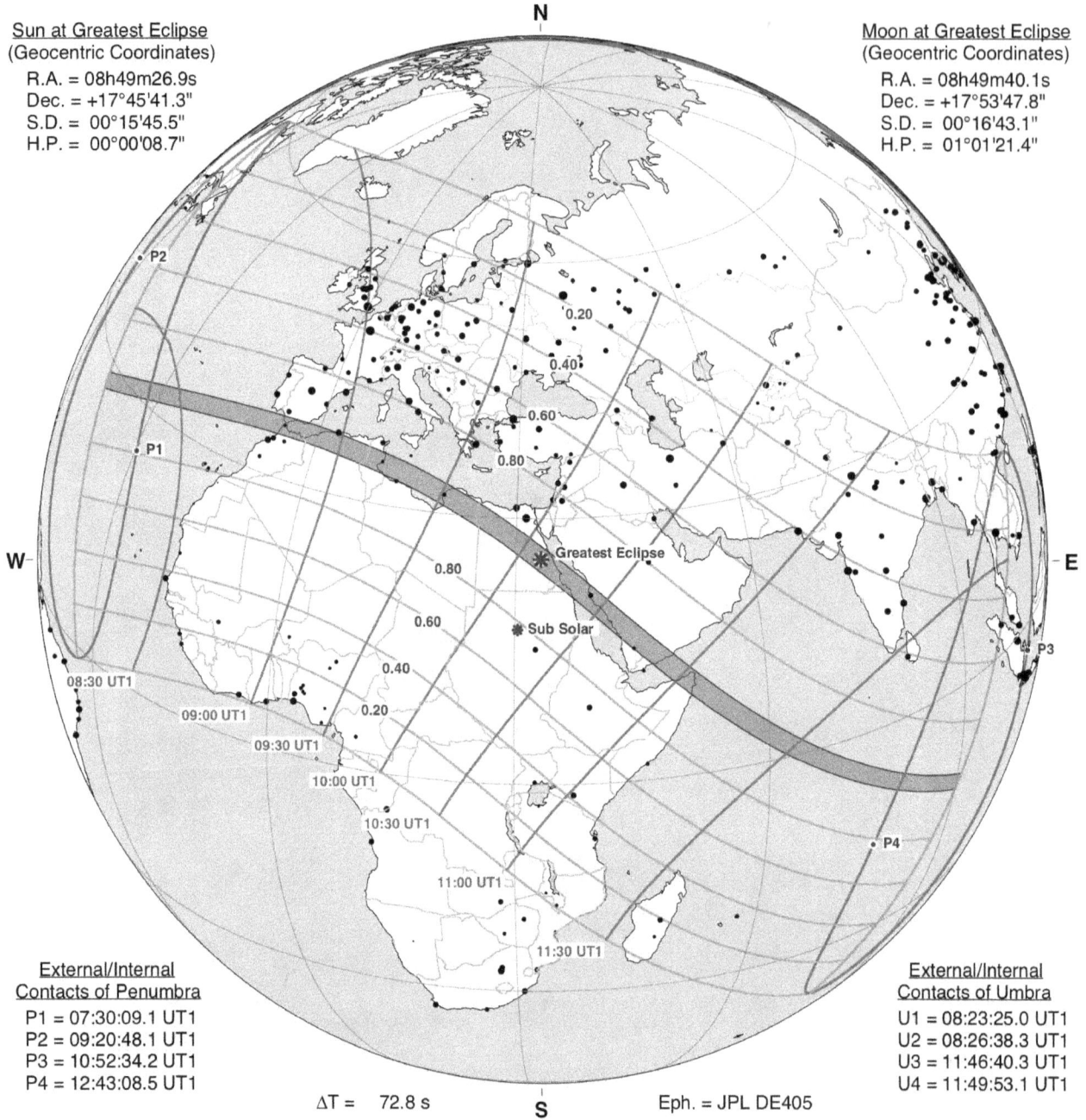

0.20
0.40
0.60
0.80

**W** —

— **E**

0.80
Greatest Eclipse

0.60
Sub Solar

0.40

0.20

08:30 UT1
09:00 UT1
09:30 UT1
10:00 UT1
10:30 UT1
11:00 UT1
11:30 UT1

P1
P2
P3
P4

External/Internal
Contacts of Penumbra
P1 = 07:30:09.1 UT1
P2 = 09:20:48.1 UT1
P3 = 10:52:34.2 UT1
P4 = 12:43:08.5 UT1

External/Internal
Contacts of Umbra
U1 = 08:23:25.0 UT1
U2 = 08:26:38.3 UT1
U3 = 11:46:40.3 UT1
U4 = 11:49:53.1 UT1

ΔT = 72.8 s

**S**

Eph. = JPL DE405

Circumstances at Greatest Eclipse: 10:06:37.4 UT1

| | |
|---|---|
| Lat. = 25°30.2'N | Sun Alt. = 81.7° |
| Long. = 033°11.0'E | Sun Azm. = 202.0° |
| Path Width = 257.7 km | Duration = 06m22.6s |

Circumstances at Greatest Duration: 10:00:20.9 UT1

| | |
|---|---|
| Lat. = 26°48.9'N | Sun Alt. = 80.9° |
| Long. = 031°07.9'E | Sun Azm. = 177.8° |
| Path Width = 257.2 km | Duration = 06m23.2s |

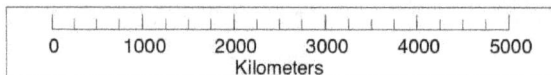

©2016 F. Espenak
www.EclipseWise.com

0   1000   2000   3000   4000   5000
Kilometers

# Annular Solar Eclipse of 2028 Jan 26

Greatest Eclipse = 15:08:58.8 TD  (= 15:07:45.8 UT1)

| | |
|---|---|
| Eclipse Magnitude = 0.9208 | Saros Series = 141 |
| Gamma = 0.3901 | Saros Member = 24 of 70 |

**Sun at Greatest Eclipse**
(Geocentric Coordinates)
R.A. = 20h34m14.2s
Dec. = -18°43'33.0"
S.D. = 00°16'14.6"
H.P. = 00°00'08.9"

**Moon at Greatest Eclipse**
(Geocentric Coordinates)
R.A. = 20h33m43.7s
Dec. = -18°23'46.3"
S.D. = 00°14'45.1"
H.P. = 00°54'08.3"

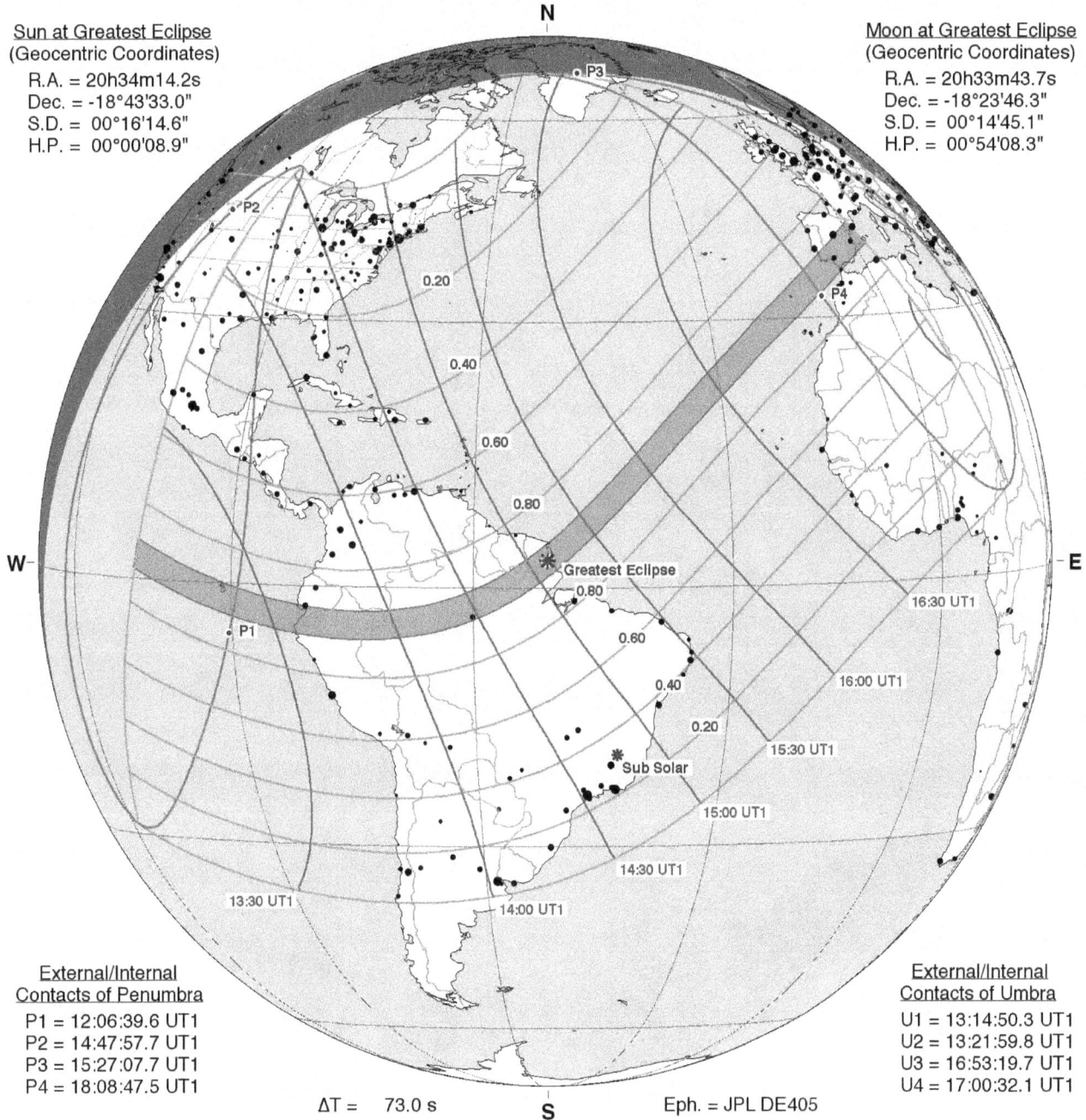

N

P3

P2

0.20

0.40

0.60

0.80

P4

W — E

0.80
Greatest Eclipse

0.60

0.40
16:30 UT1

P1

0.20
16:00 UT1

15:30 UT1

Sub Solar
15:00 UT1

14:30 UT1

13:30 UT1
14:00 UT1

**External/Internal
Contacts of Penumbra**
P1 = 12:06:39.6 UT1
P2 = 14:47:57.7 UT1
P3 = 15:27:07.7 UT1
P4 = 18:08:47.5 UT1

**External/Internal
Contacts of Umbra**
U1 = 13:14:50.3 UT1
U2 = 13:21:59.8 UT1
U3 = 16:53:19.7 UT1
U4 = 17:00:32.1 UT1

ΔT = 73.0 s     S     Eph. = JPL DE405

**Circumstances at Greatest Eclipse: 15:07:45.8 UT1**

| | |
|---|---|
| Lat. = 02°57.6'N | Sun Alt. = 67.0° |
| Long. = 051°33.6'W | Sun Azm. = 161.0° |
| Path Width = 323.0 km | Duration = 10m27.1s |

**Circumstances at Greatest Duration: 14:53:07.6 UT1**

| | |
|---|---|
| Lat. = 00°47.6'N | Sun Alt. = 65.8° |
| Long. = 054°39.0'W | Sun Azm. = 144.6° |
| Path Width = 328.7 km | Duration = 10m30.6s |

0  1000  2000  3000  4000  5000
Kilometers

# Total Solar Eclipse of 2028 Jul 22

Greatest Eclipse = 02:56:39.6 TD   (= 02:55:26.4 UT1)

| | |
|---|---|
| Eclipse Magnitude = 1.0560 | Saros Series = 146 |
| Gamma = -0.6056 | Saros Member = 28 of 76 |

**Sun at Greatest Eclipse**
(Geocentric Coordinates)
R.A. = 08h08m03.8s
Dec. = +20°10'53.0"
S.D. = 00°15'44.5"
H.P. = 00°00'08.7"

**Moon at Greatest Eclipse**
(Geocentric Coordinates)
R.A. = 08h07m16.7s
Dec. = +19°36'14.4"
S.D. = 00°16'24.3"
H.P. = 01°00'12.3"

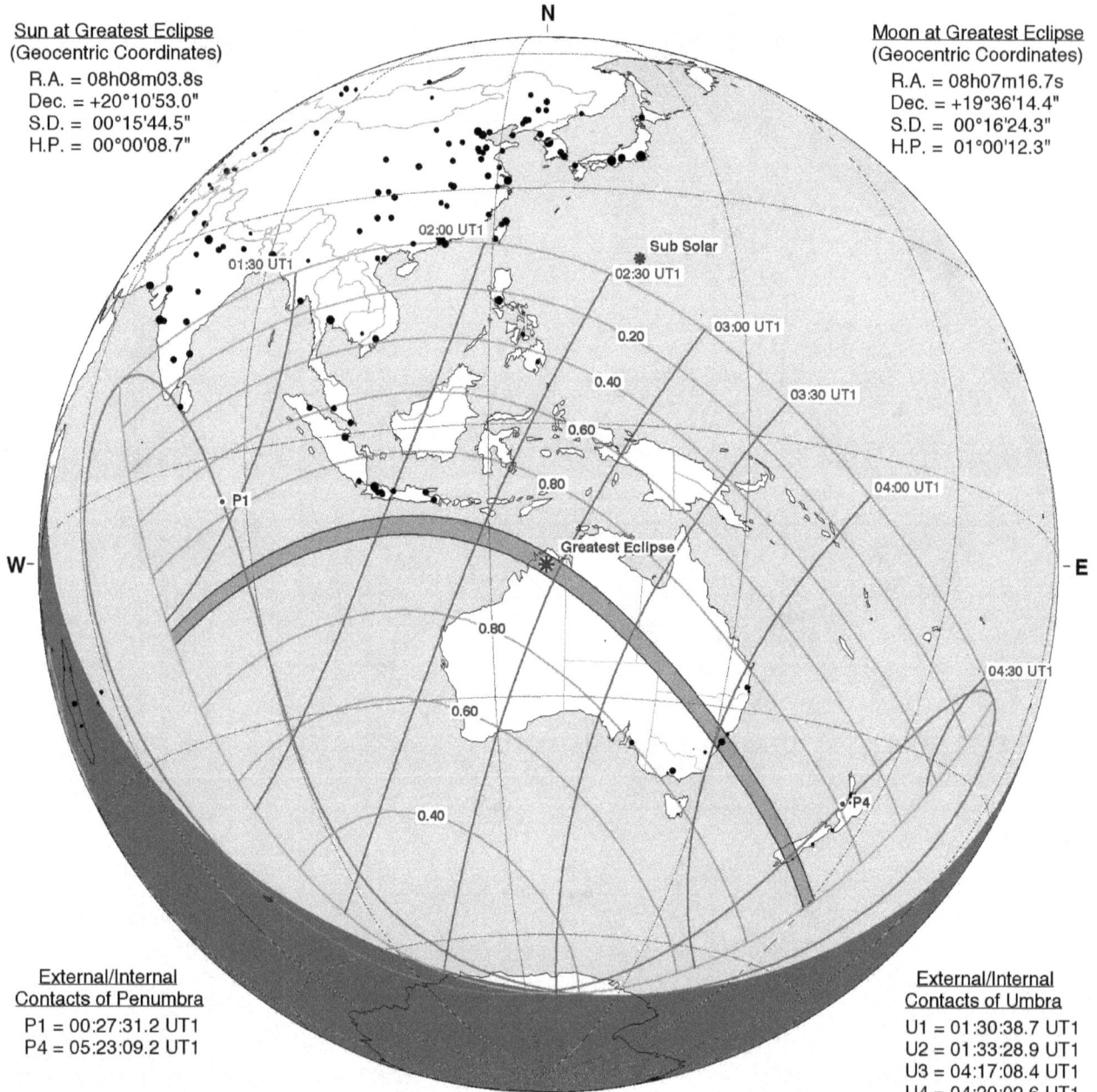

N

Sub Solar
02:30 UT1

02:00 UT1
01:30 UT1
03:00 UT1
0.20
0.40
03:30 UT1
0.60
04:00 UT1
0.80
P1
Greatest Eclipse
0.80
04:30 UT1
0.60
P4
0.40

W — E

S

**External/Internal**
**Contacts of Penumbra**
P1 = 00:27:31.2 UT1
P4 = 05:23:09.2 UT1

**External/Internal**
**Contacts of Umbra**
U1 = 01:30:38.7 UT1
U2 = 01:33:28.9 UT1
U3 = 04:17:08.4 UT1
U4 = 04:20:02.6 UT1

ΔT = 73.2 s          Eph. = JPL DE405

**Circumstances at Greatest Eclipse: 02:55:26.4 UT1**

| | |
|---|---|
| Lat. = 15°34.7'S | Sun Alt. = 52.6° |
| Long. = 126°42.4'E | Sun Azm. = 17.2° |
| Path Width = 230.2 km | Duration = 05m09.7s |

**Circumstances at Greatest Duration: 02:52:17.7 UT1**

| | |
|---|---|
| Lat. = 15°04.6'S | Sun Alt. = 52.6° |
| Long. = 125°47.3'E | Sun Azm. = 19.9° |
| Path Width = 231.4 km | Duration = 05m09.9s |

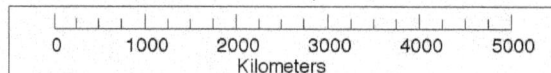

0   1000   2000   3000   4000   5000
Kilometers

©2016  F. Espenak
www.EclipseWise.com

# Partial Solar Eclipse of 2029 Jan 14

Greatest Eclipse = 17:13:47.5 TD   (= 17:12:34.1 UT1)

| | |
|---|---|
| Eclipse Magnitude = 0.8714 | Saros Series = 151 |
| Gamma = 1.0553 | Saros Member = 15 of 72 |

**N**

Sun at Greatest Eclipse
(Geocentric Coordinates)
R.A. = 19h47m03.1s
Dec. = -21°09'31.8"
S.D. = 00°16'15.6"
H.P. = 00°00'08.9"

Moon at Greatest Eclipse
(Geocentric Coordinates)
R.A. = 19h45m53.5s
Dec. = -20°12'32.3"
S.D. = 00°15'20.6"
H.P. = 00°56'18.7"

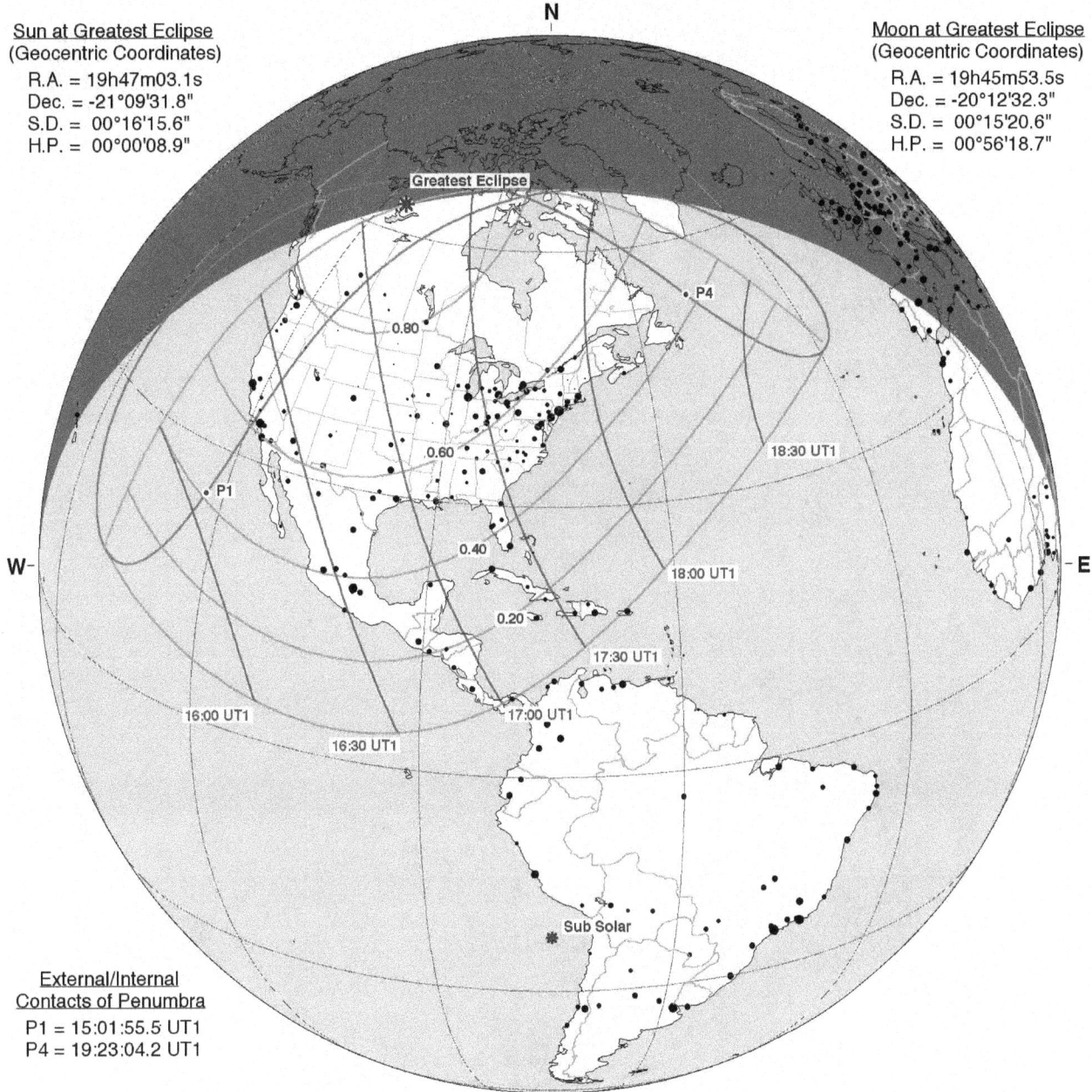

Greatest Eclipse

P4

0.80

P1

18:30 UT1

0.60

0.40

18:00 UT1

0.20

17:30 UT1

16:00 UT1

17:00 UT1

16:30 UT1

Sub Solar

**W**

**E**

External/Internal
Contacts of Penumbra
P1 = 15:01:55.5 UT1
P4 = 19:23:04.2 UT1

ΔT = 73.4 s

**S**

Eph. = JPL DE405

Circumstances at Greatest Eclipse: 17:12:34.1 UT1

| | |
|---|---|
| Lat. = 63°42.1'N | Sun Alt. = 0.0° |
| Long. = 114°13.5'W | Sun Azm. = 144.6° |

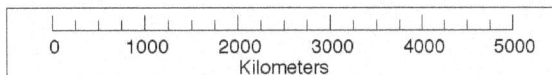

| | | | | | |
|---|---|---|---|---|---|
| 0 | 1000 | 2000 | 3000 | 4000 | 5000 |

Kilometers

# Partial Solar Eclipse of 2029 Jun 12

Greatest Eclipse = 04:06:13.0 TD   (= 04:04:59.5 UT1)

Eclipse Magnitude = 0.4576

Gamma = 1.2943

Saros Series = 118

Saros Member = 69 of 72

**Sun at Greatest Eclipse**
(Geocentric Coordinates)
R.A. = 05h22m58.2s
Dec. = +23°09'45.7"
S.D. = 00°15'45.0"
H.P. = 00°00'08.7"

**Moon at Greatest Eclipse**
(Geocentric Coordinates)
R.A. = 05h23m08.9s
Dec. = +24°21'37.7"
S.D. = 00°15'10.6"
H.P. = 00°55'42.0"

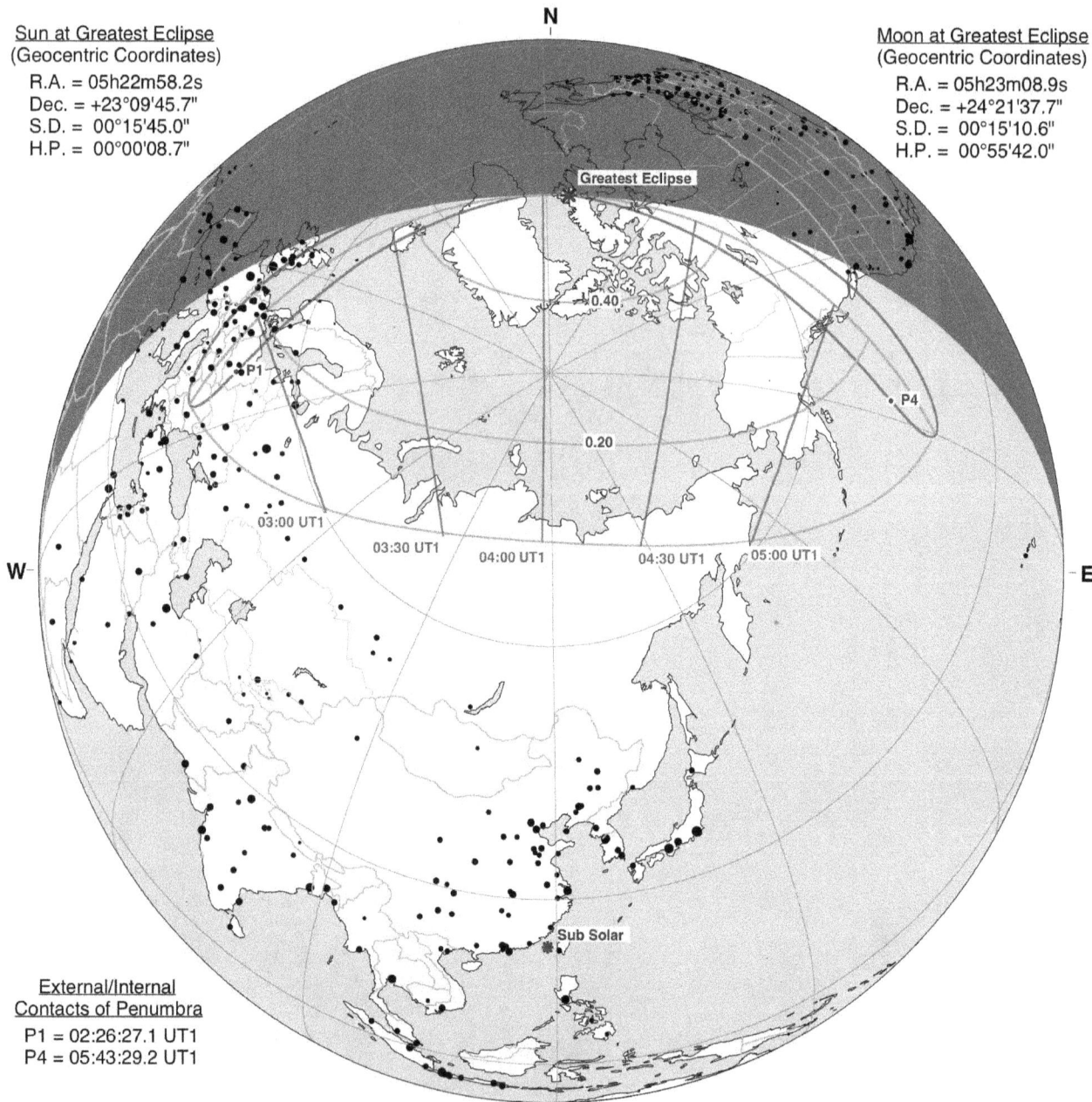

N

Greatest Eclipse

0.40

0.20

03:00 UT1

03:30 UT1

04:00 UT1

04:30 UT1

05:00 UT1

P1

P4

W

E

Sub Solar

**External/Internal
Contacts of Penumbra**
P1 = 02:26:27.1 UT1
P4 = 05:43:29.2 UT1

ΔT =   73.6 s

S

Eph. = JPL DE405

**Circumstances at Greatest Eclipse: 04:04:59.5 UT1**
Lat. = 66°45.9'N
Long. = 066°11.0'W
Sun Alt. =   0.0°
Sun Azm. = 355.5°

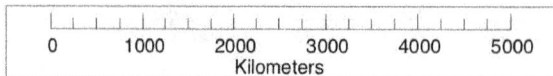

0   1000   2000   3000   4000   5000
Kilometers

©2016 F. Espenak
www.EclipseWise.com

# Partial Solar Eclipse of 2029 Jul 11

Greatest Eclipse = 15:37:18.9 TD   (= 15:36:05.3 UT1)

Eclipse Magnitude = 0.2303          Saros Series = 156
Gamma = -1.4191                      Saros Member = 2 of 69

Sun at Greatest Eclipse
(Geocentric Coordinates)
R.A. = 07h24m55.6s
Dec. = +22°00'04.3"
S.D. = 00°15'43.9"
H.P. = 00°00'08.7"

Moon at Greatest Eclipse
(Geocentric Coordinates)
R.A. = 07h23m33.7s
Dec. = +20°41'22.0"
S.D. = 00°15'35.3"
H.P. = 00°57'12.6"

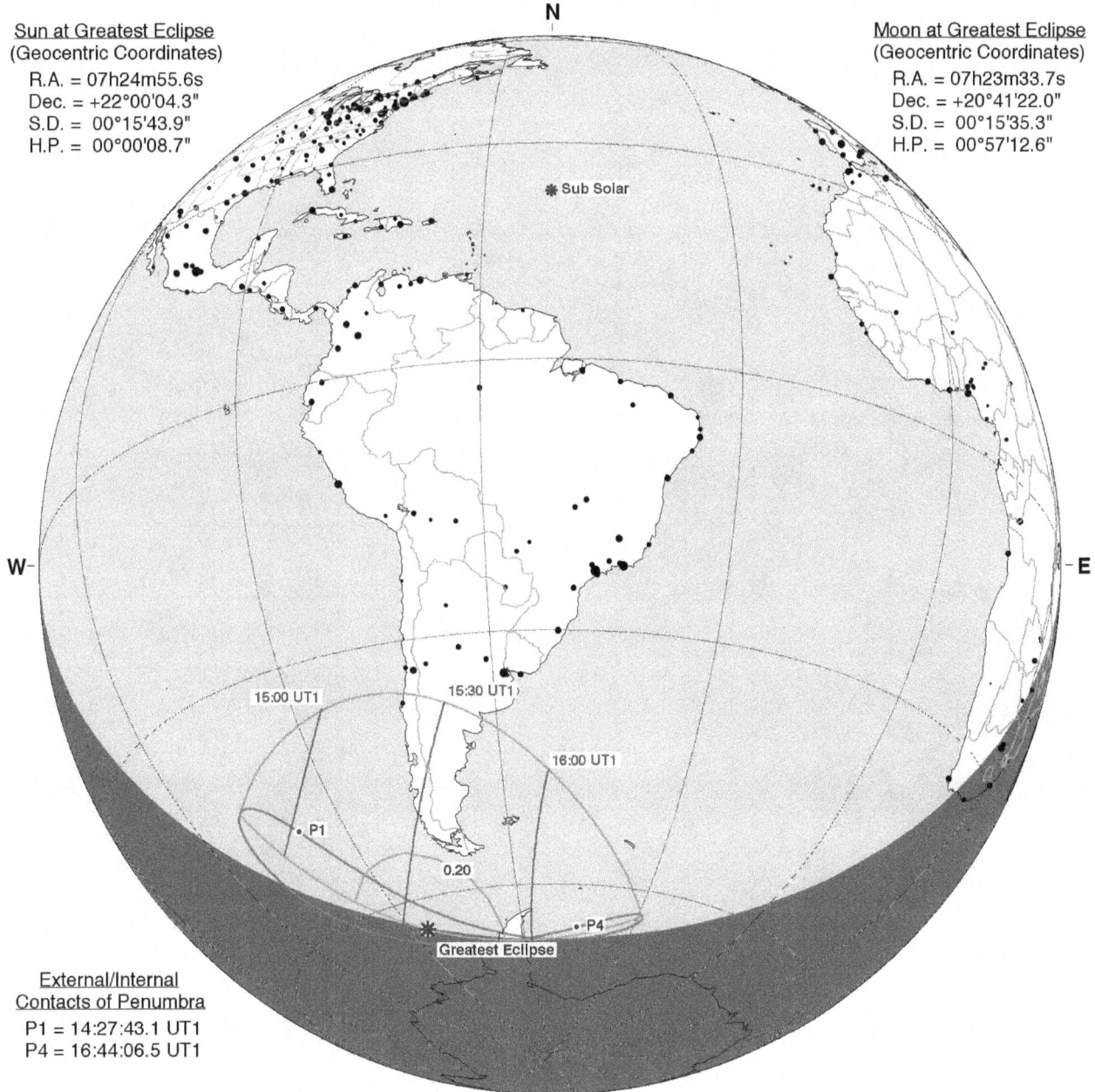

N

* Sub Solar

W

E

15:00 UT1    15:30 UT1

16:00 UT1

P1

0.20

*
Greatest Eclipse

P4

External/Internal
Contacts of Penumbra
P1 = 14:27:43.1 UT1
P4 = 16:44:06.5 UT1

ΔT = 73.6 s          S          Eph. = JPL DE405

Circumstances at Greatest Eclipse: 15:36:05.3 UT1
Lat. = 64°15.8'S          Sun Alt. = 0.0°
Long. = 085°37.4'W       Sun Azm. = 30.4°

0   1000   2000   3000   4000   5000
Kilometers

©2016  F. Espenak
www.EclipseWise.com

# Partial Solar Eclipse of 2029 Dec 05

Greatest Eclipse = 15:03:58.0 TD   (= 15:02:44.2 UT1)

| | |
|---|---|
| Eclipse Magnitude = 0.8911 | Saros Series = 123 |
| Gamma = -1.0609 | Saros Member = 54 of 70 |

Sun at Greatest Eclipse
(Geocentric Coordinates)
R.A. = 16h49m34.2s
Dec. = -22°26'54.3"
S.D. = 00°16'13.8"
H.P. = 00°00'08.9"

Moon at Greatest Eclipse
(Geocentric Coordinates)
R.A. = 16h49m27.4s
Dec. = -23°31'15.0"
S.D. = 00°16'34.3"
H.P. = 01°00'49.1"

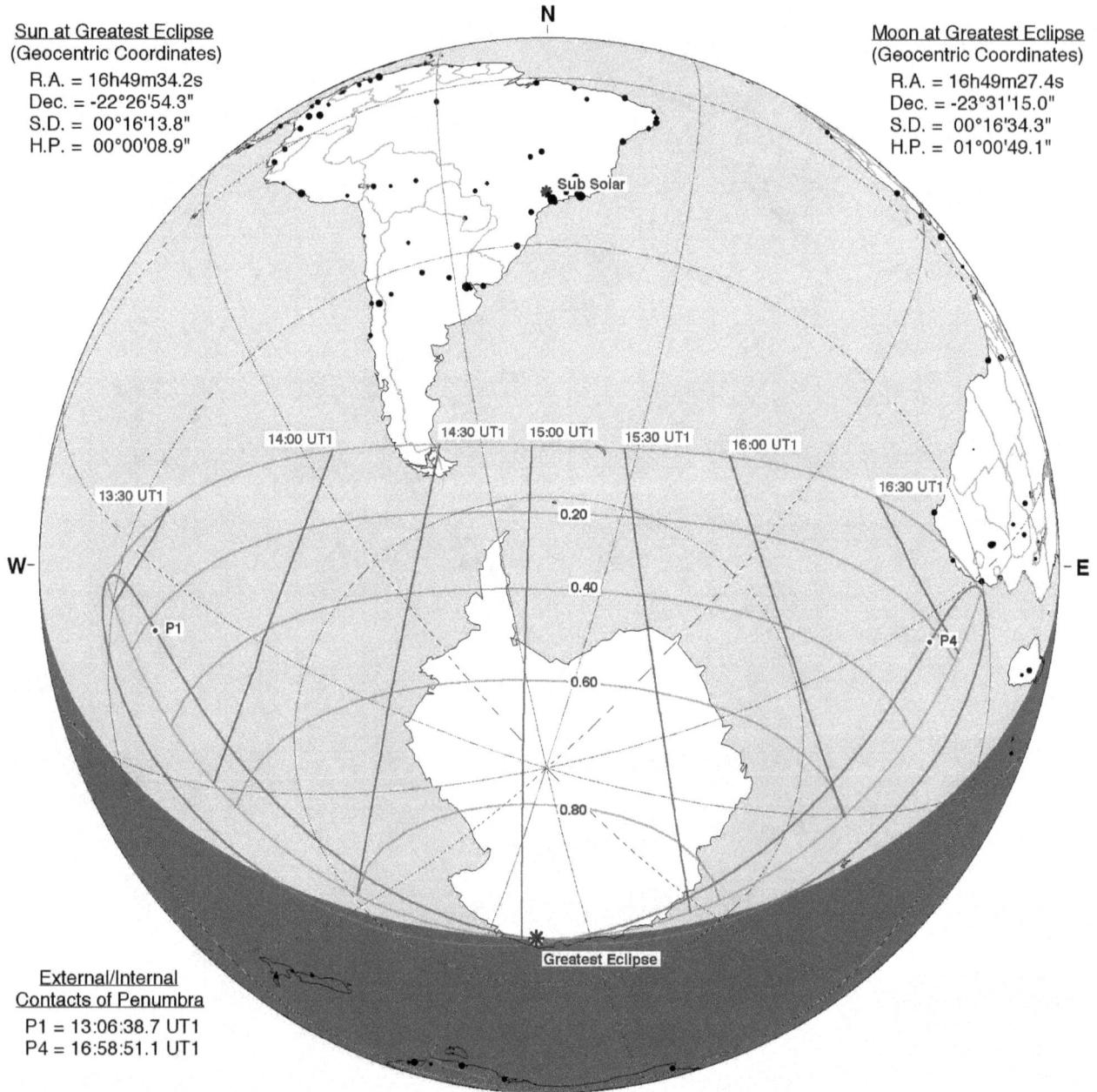

N

Sub Solar

14:00 UT1   14:30 UT1   15:00 UT1   15:30 UT1   16:00 UT1

13:30 UT1

16:30 UT1

W

0.20

0.40

P1

0.60

P4

0.80

E

*
Greatest Eclipse

External/Internal
Contacts of Penumbra
P1 = 13:06:38.7 UT1
P4 = 16:58:51.1 UT1

ΔT = 73.8 s

S

Eph. = JPL DE405

Circumstances at Greatest Eclipse: 15:02:44.2 UT1

| | |
|---|---|
| Lat. = 67°30.8'S | Sun Alt. = 0.0° |
| Long. = 135°38.7'E | Sun Azm. = 176.6° |

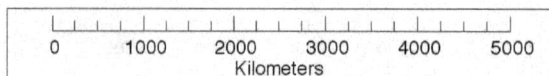

| 0 | 1000 | 2000 | 3000 | 4000 | 5000 |
|---|---|---|---|---|---|

Kilometers

©2016  F. Espenak
www.EclipseWise.com

# Annular Solar Eclipse of 2030 Jun 01

Greatest Eclipse = 06:29:12.9 TD  (= 06:27:58.8 UT1)

Eclipse Magnitude = 0.9443          Saros Series = 128
Gamma = 0.5626          Saros Member = 59 of 73

**Sun at Greatest Eclipse**
(Geocentric Coordinates)
R.A. = 04h37m01.2s
Dec. = +22°03'55.3"
S.D. = 00°15'46.4"
H.P. = 00°00'08.7"

**Moon at Greatest Eclipse**
(Geocentric Coordinates)
R.A. = 04h36m55.8s
Dec. = +22°34'11.5"
S.D. = 00°14'42.7"
H.P. = 00°53'59.6"

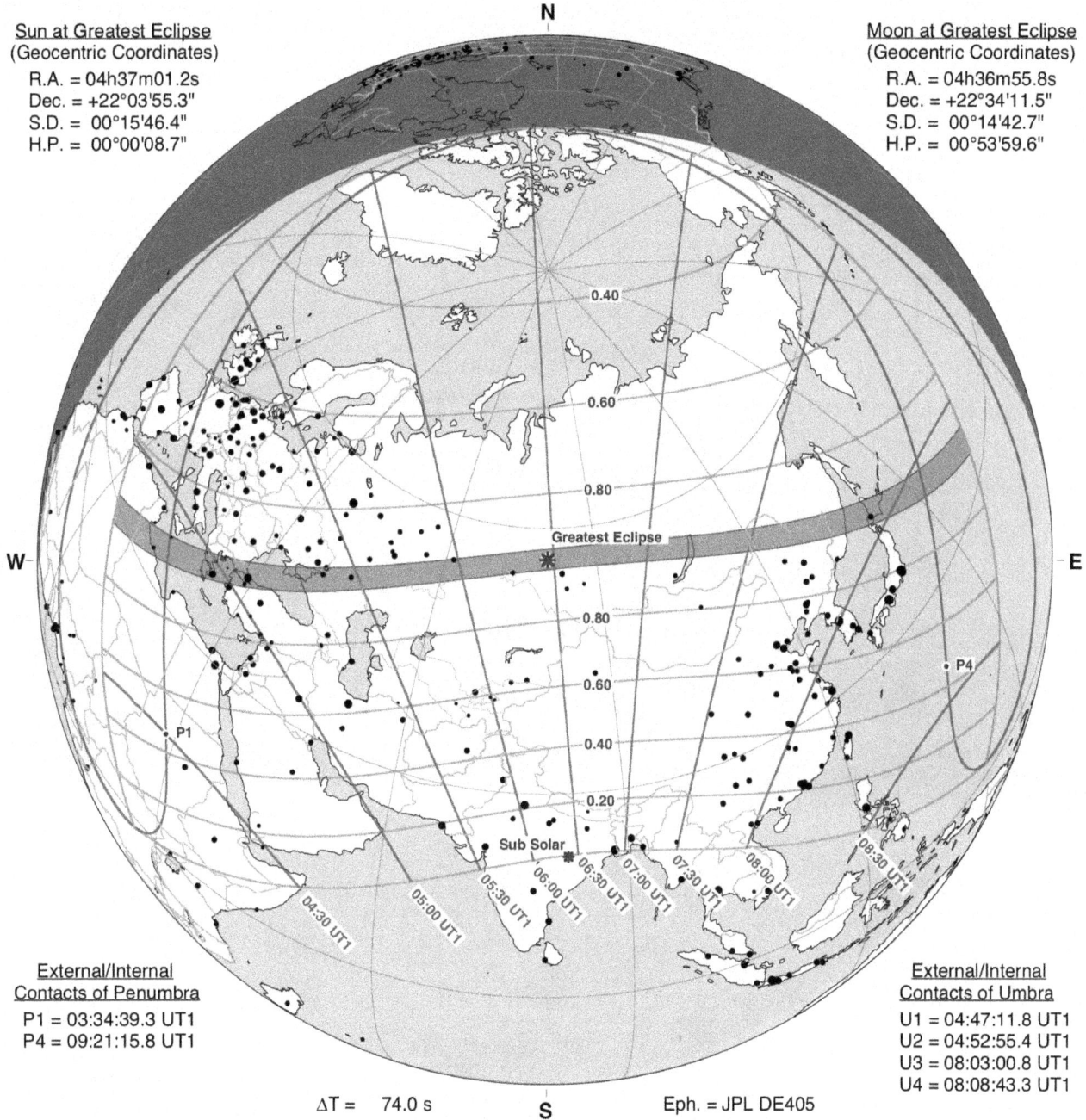

N

0.40

0.60

0.80

Greatest Eclipse

W

E

0.80

0.60

P4

0.40

0.20

Sub Solar

04:30 UT1
05:00 UT1
05:30 UT1
06:00 UT1
06:30 UT1
07:00 UT1
07:30 UT1
08:00 UT1
08:30 UT1

P1

**External/Internal**
**Contacts of Penumbra**
P1 = 03:34:39.3 UT1
P4 = 09:21:15.8 UT1

**External/Internal**
**Contacts of Umbra**
U1 = 04:47:11.8 UT1
U2 = 04:52:55.4 UT1
U3 = 08:03:00.8 UT1
U4 = 08:08:43.3 UT1

ΔT = 74.0 s          S          Eph. = JPL DE405

**Circumstances at Greatest Eclipse: 06:27:58.8 UT1**
Lat. = 56°31.5'N          Sun Alt. = 55.5°
Long. = 080°03.7'E          Sun Azm. = 176.1°
Path Width = 249.6 km          Duration = 05m20.8s

**Circumstances at Greatest Duration: 06:28:41.2 UT1**
Lat. = 56°32.9'N          Sun Alt. = 55.5°
Long. = 080°30.1'E          Sun Azm. = 177.1°
Path Width = 249.6 km          Duration = 05m20.8s

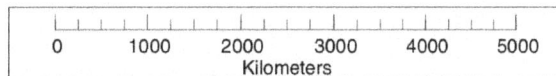

0    1000   2000   3000   4000   5000
Kilometers

©2016 F. Espenak
www.EclipseWise.com

# Total Solar Eclipse of 2030 Nov 25

Greatest Eclipse = 06:51:36.9 TD (= 06:50:22.7 UT1)

| | |
|---|---|
| Eclipse Magnitude = 1.0468 | Saros Series = 133 |
| Gamma = -0.3867 | Saros Member = 46 of 72 |

**Sun at Greatest Eclipse**
(Geocentric Coordinates)
R.A. = 16h03m58.7s
Dec. = -20°45'39.0"
S.D. = 00°16'12.1"
H.P. = 00°00'08.9"

**Moon at Greatest Eclipse**
(Geocentric Coordinates)
R.A. = 16h03m49.1s
Dec. = -21°09'10.6"
S.D. = 00°16'41.7"
H.P. = 01°01'16.4"

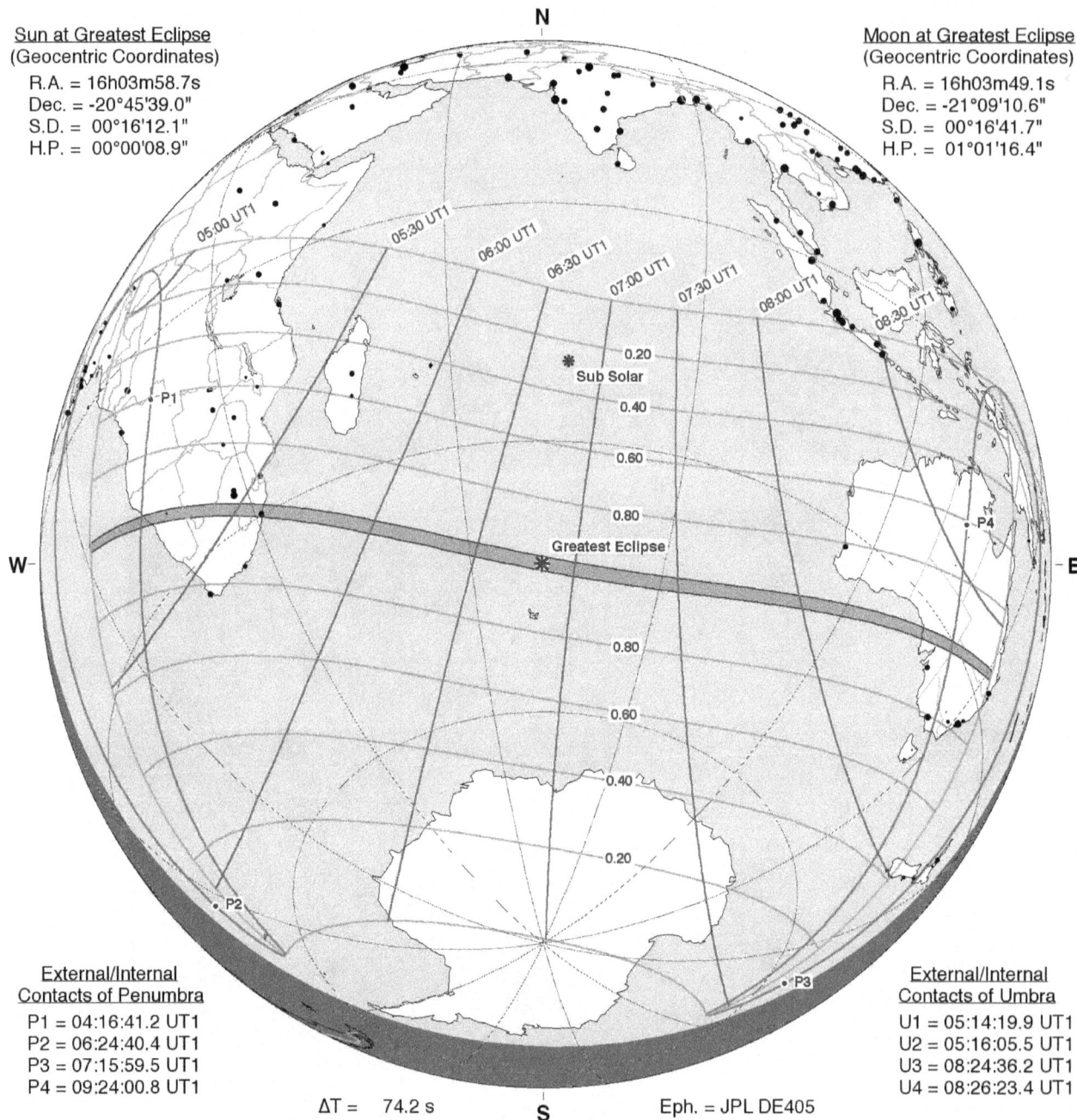

**External/Internal Contacts of Penumbra**
P1 = 04:16:41.2 UT1
P2 = 06:24:40.4 UT1
P3 = 07:15:59.5 UT1
P4 = 09:24:00.8 UT1

**External/Internal Contacts of Umbra**
U1 = 05:14:19.9 UT1
U2 = 05:16:05.5 UT1
U3 = 08:24:36.2 UT1
U4 = 08:26:23.4 UT1

ΔT = 74.2 s

Eph. = JPL DE405

**Circumstances at Greatest Eclipse: 06:50:22.7 UT1**
Lat. = 43°36.6'S
Long. = 071°13.7'E
Path Width = 169.3 km
Sun Alt. = 67.0°
Sun Azm. = 7.0°
Duration = 03m43.5s

**Circumstances at Greatest Duration: 06:51:56.0 UT1**
Lat. = 43°42.7'S
Long. = 072°01.4'E
Path Width = 169.2 km
Sun Alt. = 67.0°
Sun Azm. = 4.1°
Duration = 03m43.6s

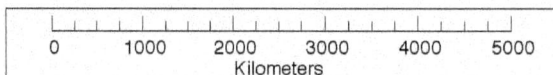

0 1000 2000 3000 4000 5000
Kilometers

©2016 F. Espenak
www.EclipseWise.com

# Annular Solar Eclipse of 2031 May 21

Greatest Eclipse = 07:16:04.3 TD   (= 07:14:49.9 UT1)

| | |
|---|---|
| Eclipse Magnitude = 0.9589 | Saros Series = 138 |
| Gamma = -0.1970 | Saros Member = 32 of 70 |

**N**

**Sun at Greatest Eclipse**
(Geocentric Coordinates)
R.A. = 03h51m34.6s
Dec. = +20°09'39.2"
S.D. = 00°15'48.2"
H.P. = 00°00'08.7"

**Moon at Greatest Eclipse**
(Geocentric Coordinates)
R.A. = 03h51m39.8s
Dec. = +19°58'57.5"
S.D. = 00°14'55.8"
H.P. = 00°54'47.5"

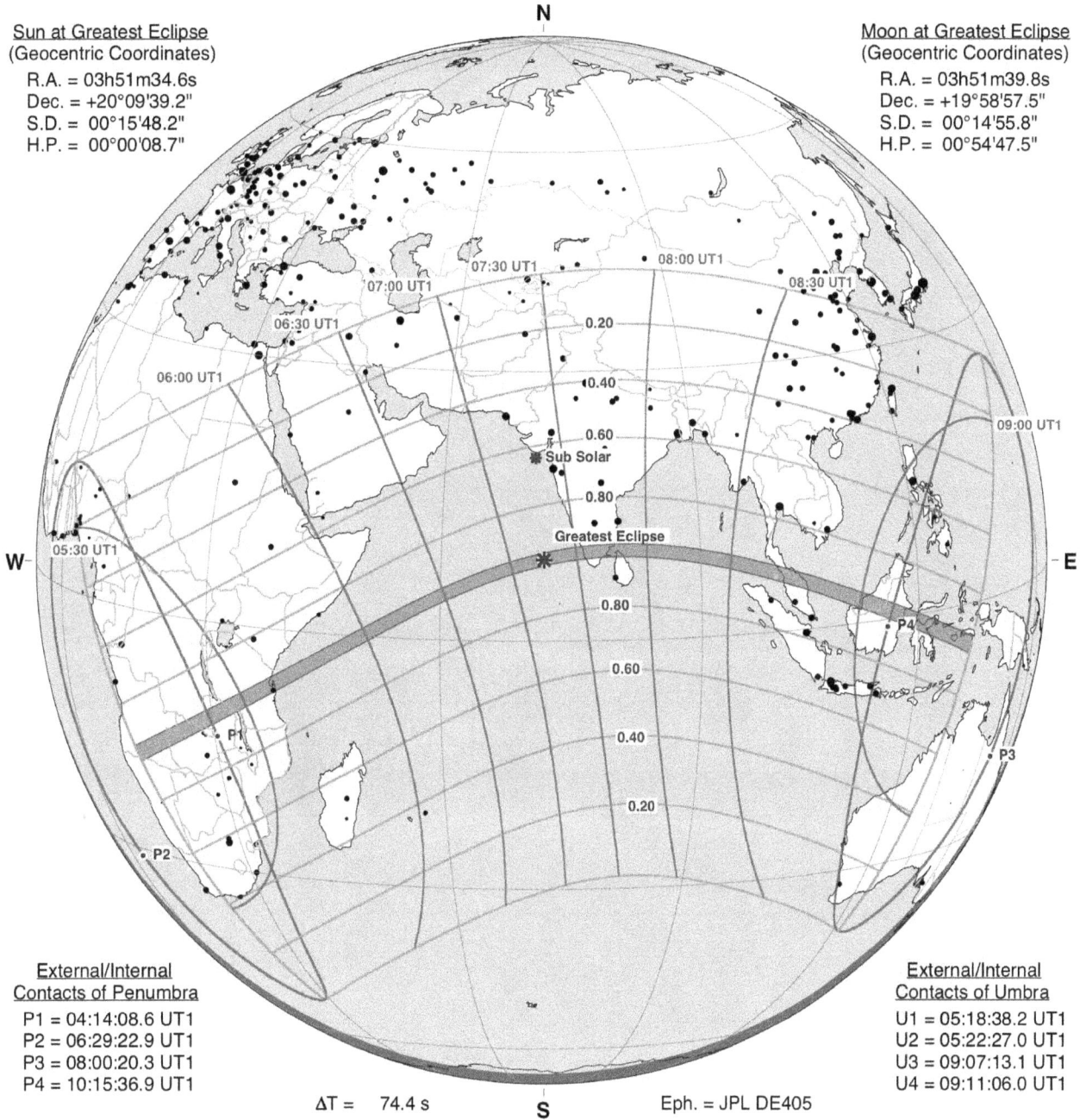

08:00 UT1
07:30 UT1
08:30 UT1
07:00 UT1
06:30 UT1
0.20
06:00 UT1
0.40
09:00 UT1
0.60
Sub Solar
0.80
Greatest Eclipse
05:30 UT1
**W**
0.80
**E**
0.60
P4
0.40
P1
0.20
P3
P2

**External/Internal Contacts of Penumbra**
P1 = 04:14:08.6 UT1
P2 = 06:29:22.9 UT1
P3 = 08:00:20.3 UT1
P4 = 10:15:36.9 UT1

**External/Internal Contacts of Umbra**
U1 = 05:18:38.2 UT1
U2 = 05:22:27.0 UT1
U3 = 09:07:13.1 UT1
U4 = 09:11:06.0 UT1

ΔT = 74.4 s

**S**

Eph. = JPL DE405

| **Circumstances at Greatest Eclipse: 07:14:49.9 UT1** | | **Circumstances at Greatest Duration: 07:23:14.4 UT1** | |
|---|---|---|---|
| Lat. = 08°55.6'N | Sun Alt. = 78.7° | Lat. = 09°19.1'N | Sun Alt. = 77.9° |
| Long. = 071°43.6'E | Sun Azm. = 353.8° | Long. = 073°47.6'E | Sun Azm. = 334.7° |
| Path Width = 152.2 km | Duration = 05m25.5s | Path Width = 152.7 km | Duration = 05m26.1s |

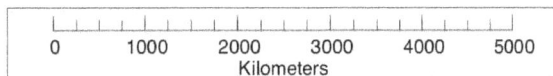

©2016  F. Espenak
www.EclipseWise.com

0   1000   2000   3000   4000   5000
Kilometers

# Hybrid Solar Eclipse of 2031 Nov 14

Greatest Eclipse = 21:07:30.7 TD  (= 21:06:16.0 UT1)

| | |
|---|---|
| Eclipse Magnitude = 1.0106 | Saros Series = 143 |
| Gamma = 0.3078 | Saros Member = 24 of 72 |

**Sun at Greatest Eclipse**
(Geocentric Coordinates)
R.A. = 15h19m31.2s
Dec. = -18°20'14.5"
S.D. = 00°16'09.9"
H.P. = 00°00'08.9"

**Moon at Greatest Eclipse**
(Geocentric Coordinates)
R.A. = 15h19m43.3s
Dec. = -18°02'21.3"
S.D. = 00°16'05.0"
H.P. = 00°59'01.4"

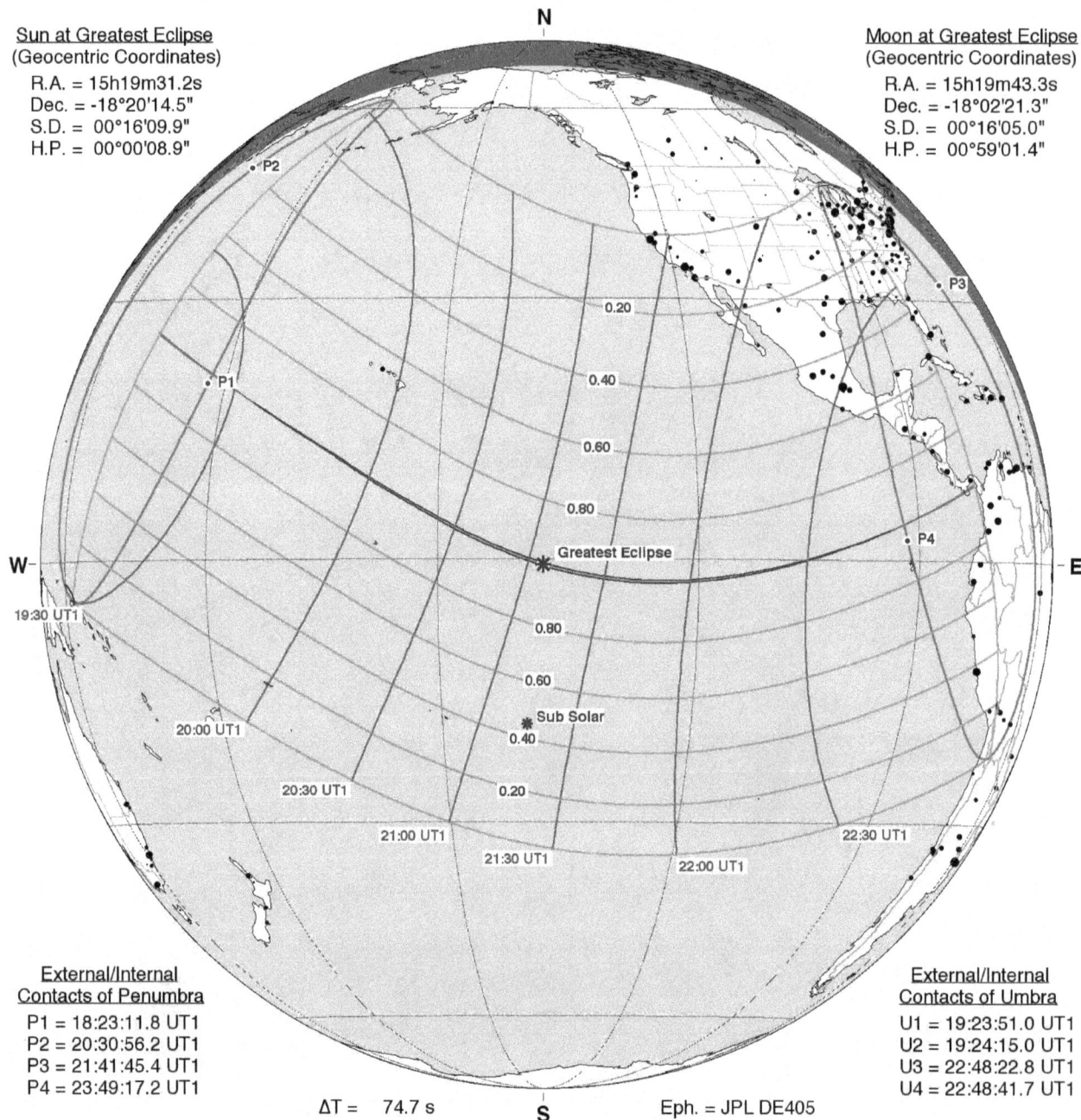

**External/Internal Contacts of Penumbra**
P1 = 18:23:11.8 UT1
P2 = 20:30:56.2 UT1
P3 = 21:41:45.4 UT1
P4 = 23:49:17.2 UT1

**External/Internal Contacts of Umbra**
U1 = 19:23:51.0 UT1
U2 = 19:24:15.0 UT1
U3 = 22:48:22.8 UT1
U4 = 22:48:41.7 UT1

ΔT = 74.7 s          Eph. = JPL DE405

**Circumstances at Greatest Eclipse: 21:06:16.0 UT1**

| | |
|---|---|
| Lat. = 00°37.9'S | Sun Alt. = 72.1° |
| Long. = 137°38.7'W | Sun Azm. = 188.7° |
| Path Width = 38.3 km | Duration = 01m08.3s |

**Circumstances at Greatest Duration: 21:10:29.2 UT1**

| | |
|---|---|
| Lat. = 00°57.7'S | Sun Alt. = 71.9° |
| Long. = 136°28.4'W | Sun Azm. = 195.6° |
| Path Width = 38.4 km | Duration = 01m08.4s |

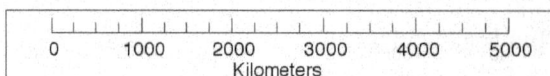

©2016  F. Espenak
www.EclipseWise.com

0   1000   2000   3000   4000   5000
Kilometers

# Annular Solar Eclipse of  2032 May 09

Greatest Eclipse =  13:26:42.4 TD   (= 13:25:27.5 UT1)

| | |
|---|---|
| Eclipse Magnitude =  0.9957 | Saros Series =  148 |
| Gamma = -0.9375 | Saros Member =  22 of 75 |

N

**Sun at Greatest Eclipse**
(Geocentric Coordinates)
R.A. = 03h08m06.7s
Dec. = +17°35'43.5"
S.D. =  00°15'50.4"
H.P. =  00°00'08.7"

**Moon at Greatest Eclipse**
(Geocentric Coordinates)
R.A. = 03h08m46.1s
Dec. = +16°42'42.0"
S.D. =  00°15'41.5"
H.P. =  00°57'35.4"

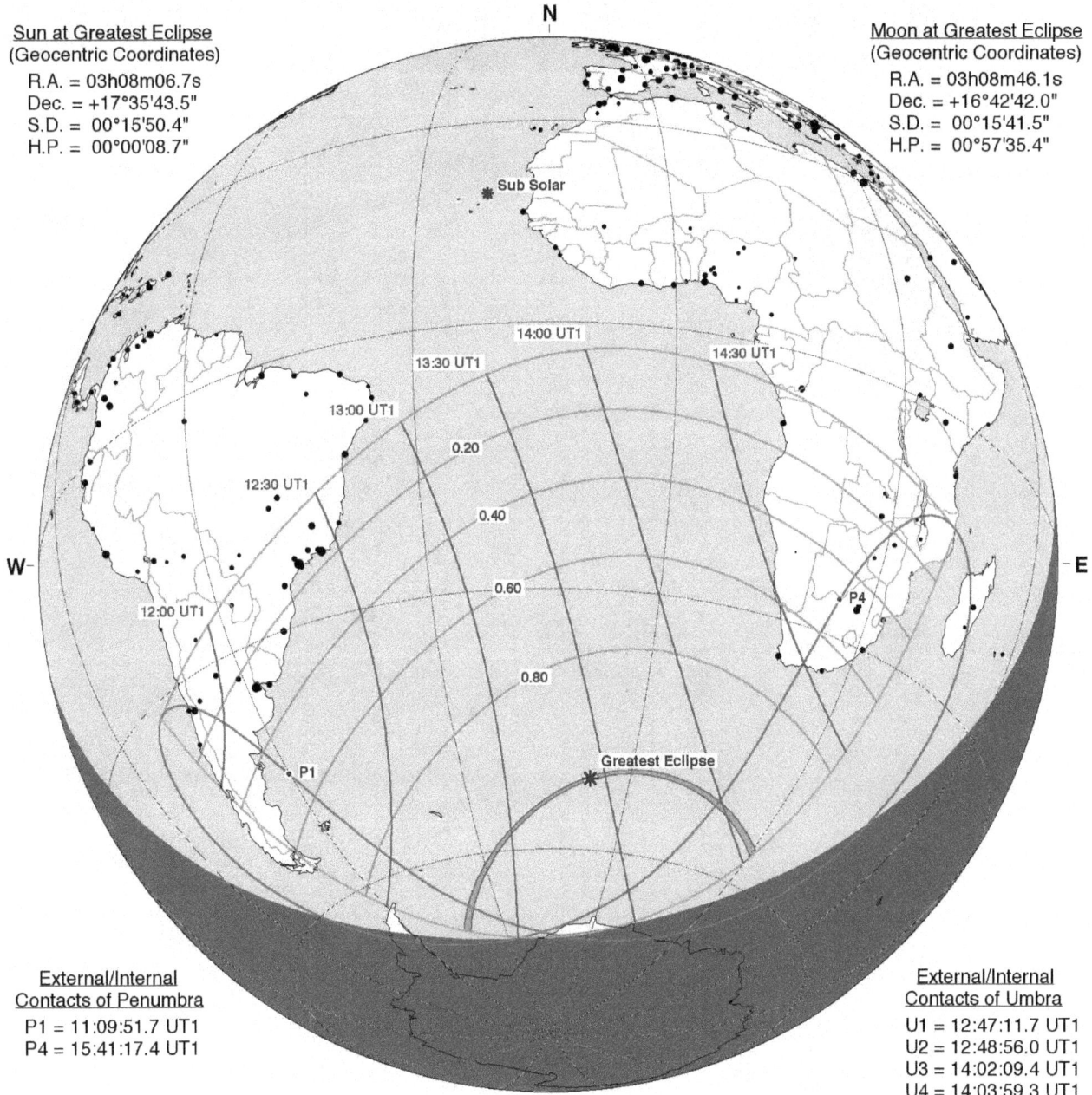

Sub Solar

14:00 UT1

13:30 UT1

14:30 UT1

13:00 UT1

0.20

12:30 UT1

0.40

0.60

P4

12:00 UT1

0.80

W

E

P1

Greatest Eclipse

**External/Internal
Contacts of Penumbra**
P1 = 11:09:51.7 UT1
P4 = 15:41:17.4 UT1

**External/Internal
Contacts of Umbra**
U1 = 12:47:11.7 UT1
U2 = 12:48:56.0 UT1
U3 = 14:02:09.4 UT1
U4 = 14:03:59.3 UT1

ΔT =    74.9 s         S         Eph. = JPL DE405

**Circumstances at Greatest Eclipse:  13:25:27.5 UT1**

| | |
|---|---|
| Lat. =  51°15.6'S | Sun Alt. =  19.9° |
| Long. = 007°04.4'W | Sun Azm. = 344.6° |
| Path Width =  43.7 km | Duration =  00m22.0s |

**Circumstances at Greatest Duration:  12:48:03.3 UT1**

| | |
|---|---|
| Lat. =  69°57.7'S | Sun Alt. =   0.0° |
| Long. = 042°31.9'W | Sun Azm. =  28.1° |
| Path Width =  86.7 km | Duration =  00m40.8s |

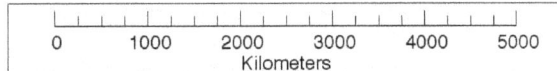

©2016  F. Espenak
www.EclipseWise.com

0    1000    2000    3000    4000    5000
Kilometers

# Partial Solar Eclipse of 2032 Nov 03

Greatest Eclipse = 05:34:12.9 TD  (= 05:32:57.8 UT1)

Eclipse Magnitude = 0.8554          Saros Series = 153
Gamma = 1.0643          Saros Member = 10 of 70

Sun at Greatest Eclipse
(Geocentric Coordinates)
R.A. = 14h35m40.9s
Dec. = -15°13'54.9"
S.D. = 00°16'07.4"
H.P. = 00°00'08.9"

Moon at Greatest Eclipse
(Geocentric Coordinates)
R.A. = 14h36m33.6s
Dec. = -14°16'01.1"
S.D. = 00°15'13.0"
H.P. = 00°55'50.8"

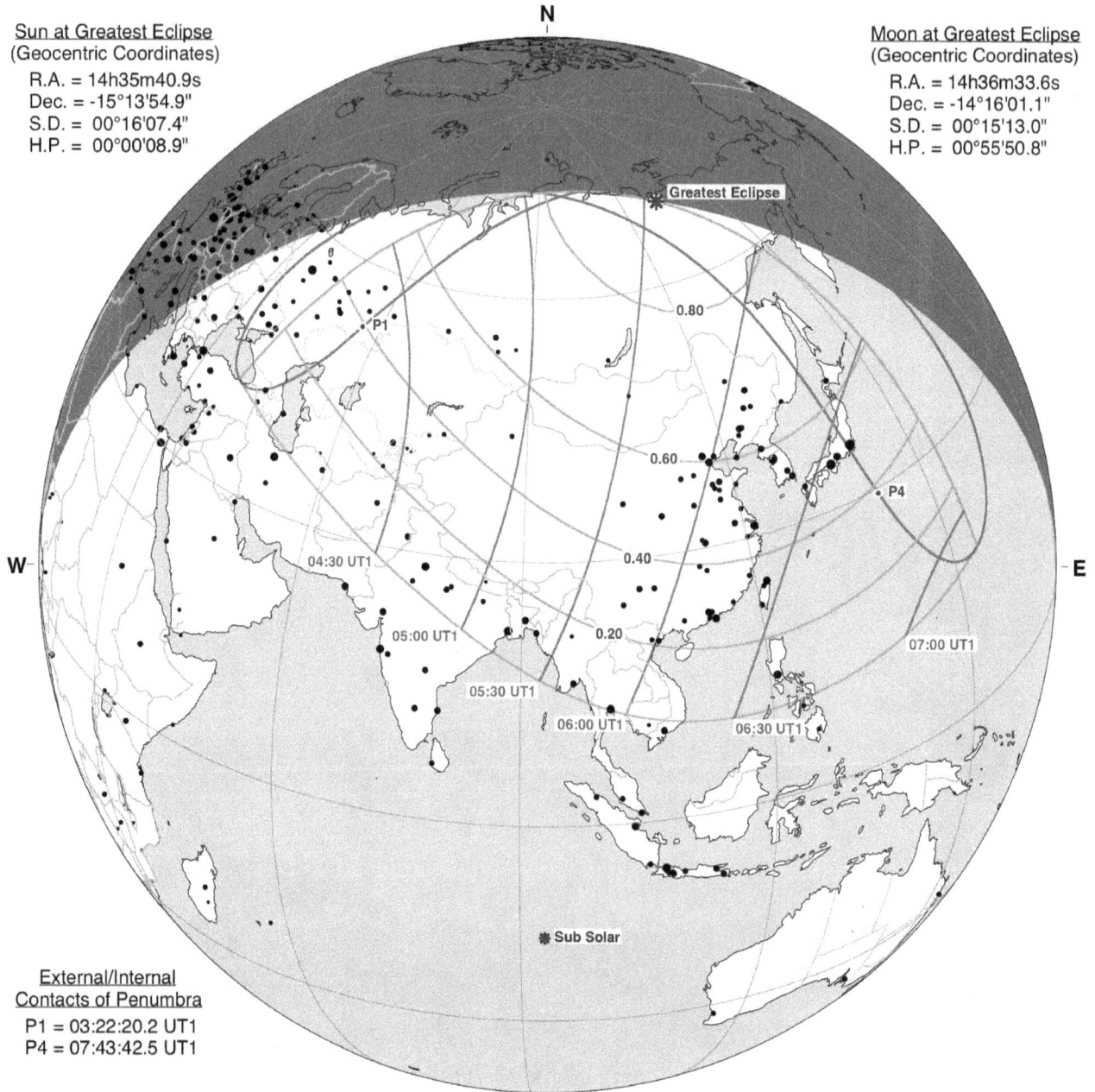

N

Greatest Eclipse

0.80

P1

0.60

P4

W

04:30 UT1

0.40

E

05:00 UT1

0.20

07:00 UT1

05:30 UT1

06:00 UT1          06:30 UT1

Sub Solar

External/Internal
Contacts of Penumbra
P1 = 03:22:20.2 UT1
P4 = 07:43:42.5 UT1

ΔT = 75.1 s          S          Eph. = JPL DE405

Circumstances at Greatest Eclipse: 05:32:57.8 UT1
Lat. = 70°25.7'N          Sun Alt. =  0.0°
Long. = 132°36.5'E          Sun Azm. = 218.3°

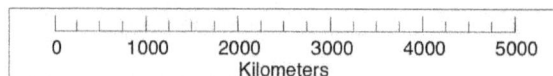

0  1000  2000  3000  4000  5000
Kilometers

©2016  F. Espenak
www.EclipseWise.com

# Total Solar Eclipse of 2033 Mar 30

Greatest Eclipse = 18:02:35.7 TD   (= 18:01:20.4 UT1)

Eclipse Magnitude = 1.0462          Saros Series = 120
Gamma = 0.9778                      Saros Member = 62 of 71

**N**

Sun at Greatest Eclipse
(Geocentric Coordinates)
R.A. = 00h38m02.8s
Dec. = +04°05'47.8"
S.D. = 00°16'00.8"
H.P. = 00°00'08.8"

Moon at Greatest Eclipse
(Geocentric Coordinates)
R.A. = 00h36m50.4s
Dec. = +05°02'48.6"
S.D. = 00°16'42.2"
H.P. = 01°01'18.3"

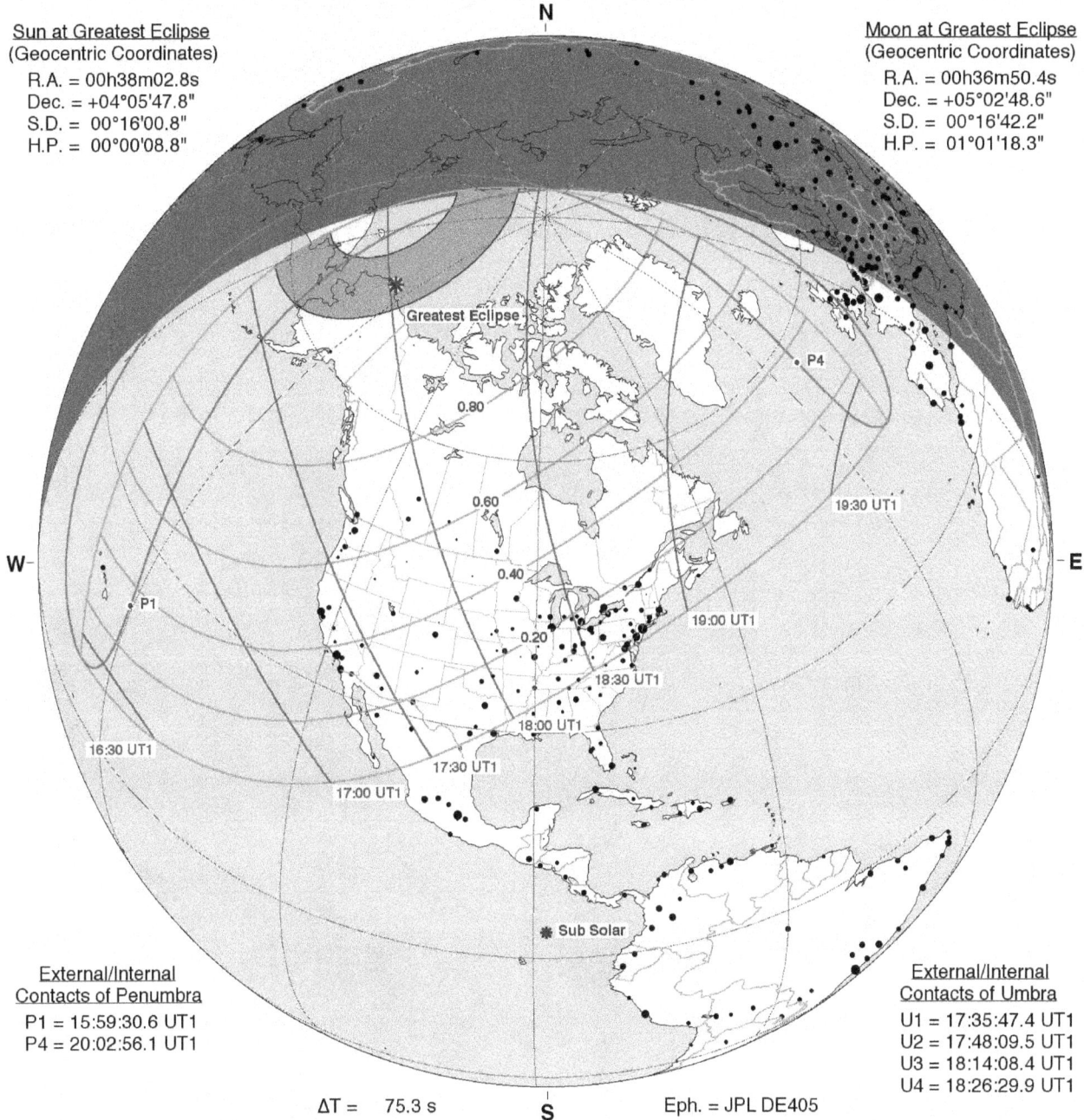

Greatest Eclipse

P4

0.80

0.60

19:30 UT1

**W**                                                                    **E**

P1

0.40

19:00 UT1

0.20

18:30 UT1

18:00 UT1

16:30 UT1

17:30 UT1

17:00 UT1

Sub Solar

External/Internal
Contacts of Penumbra
P1 = 15:59:30.6 UT1
P4 = 20:02:56.1 UT1

External/Internal
Contacts of Umbra
U1 = 17:35:47.4 UT1
U2 = 17:48:09.5 UT1
U3 = 18:14:08.4 UT1
U4 = 18:26:29.9 UT1

ΔT = 75.3 s          **S**          Eph. = JPL DE405

Circumstances at Greatest Eclipse: 18:01:20.4 UT1

| | |
|---|---|
| Lat. = 71°18.9'N | Sun Alt. = 11.2° |
| Long. = 155°50.1'W | Sun Azm. = 111.1° |
| Path Width = 781.2 km | Duration = 02m37.0s |

Circumstances at Greatest Duration: 18:01:04.2 UT1

| | |
|---|---|
| Lat. = 71°10.5'N | Sun Alt. = 11.2° |
| Long. = 155°55.9'W | Sun Azm. = 110.9° |
| Path Width = 782.6 km | Duration = 02m37.0s |

©2016 F. Espenak
www.EclipseWise.com

0    1000    2000    3000    4000    5000
Kilometers

# Partial Solar Eclipse of  2033 Sep 23

Greatest Eclipse =  13:54:31.2 TD   (= 13:53:15.7 UT1)

Eclipse Magnitude =  0.6890          Saros Series =  125
Gamma = -1.1583                      Saros Member =  55 of 73

**Sun at Greatest Eclipse**
(Geocentric Coordinates)
R.A. = 12h03m08.9s
Dec. = -00°20'27.7"
S.D. = 00°15'56.3"
H.P. = 00°00'08.8"

**Moon at Greatest Eclipse**
(Geocentric Coordinates)
R.A. = 12h01m52.5s
Dec. = -01°19'54.7"
S.D. = 00°14'43.6"
H.P. = 00°54'03.0"

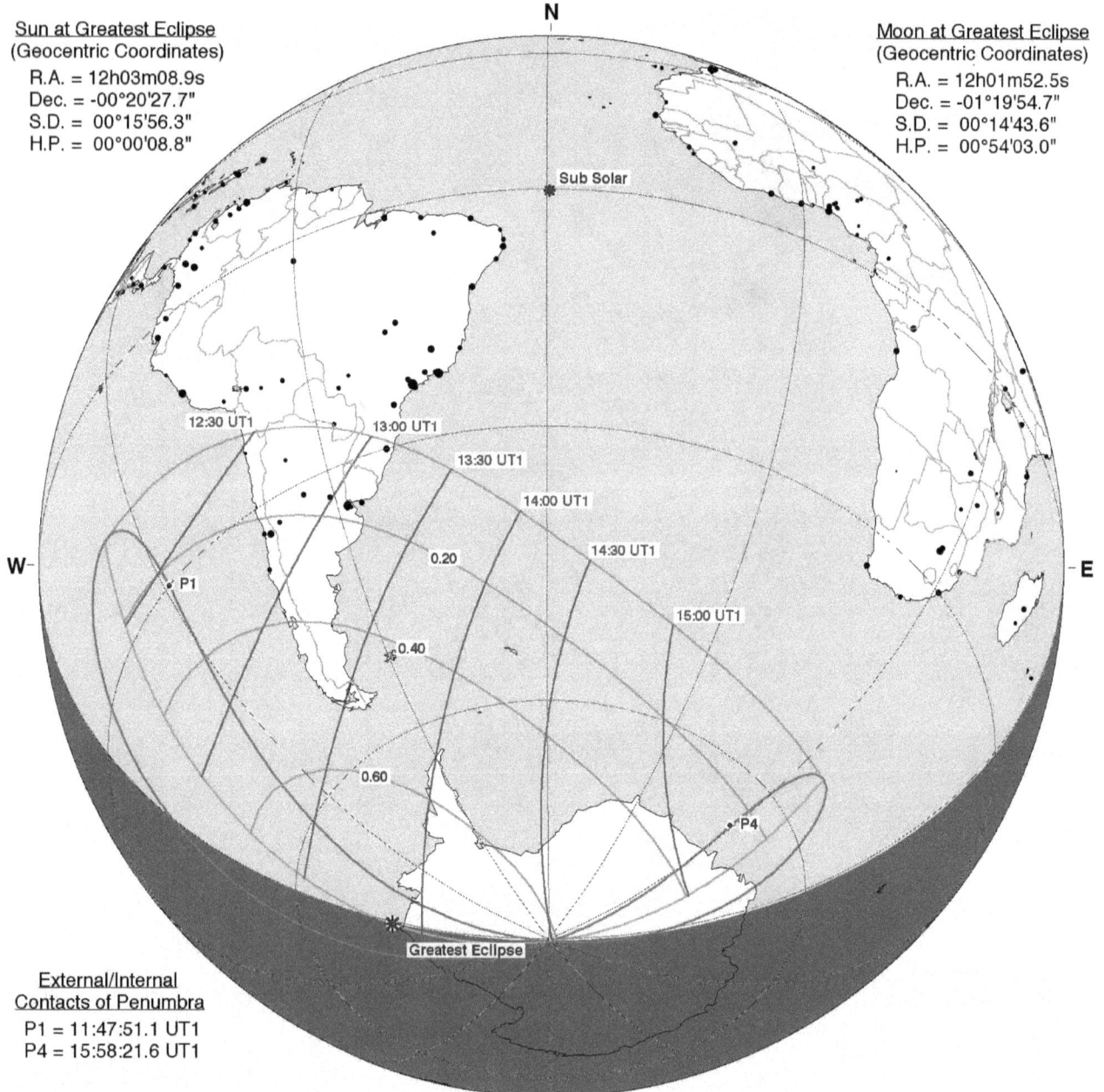

N

Sub Solar

12:30 UT1      13:00 UT1

13:30 UT1

14:00 UT1

0.20        14:30 UT1

W —                                          — E

0.40          15:00 UT1

P1

0.60

P4

* Greatest Eclipse

**External/Internal
Contacts of Penumbra**
P1 = 11:47:51.1 UT1
P4 = 15:58:21.6 UT1

ΔT =   75.5 s          S          Eph. = JPL DE405

Circumstances at Greatest Eclipse:  13:53:15.7 UT1
Lat. = 72°12.2'S          Sun Alt. =   0.0°
Long. = 121°15.2'W       Sun Azm. =  91.1°

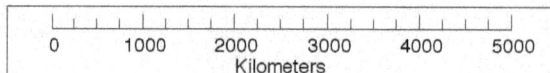

| | | | | | |
|---|---|---|---|---|---|
| 0 | 1000 | 2000 | 3000 | 4000 | 5000 |

Kilometers

©2016  F. Espenak
www.EclipseWise.com

# Total Solar Eclipse of 2034 Mar 20

Greatest Eclipse = 10:18:45.2 TD   (= 10:17:29.5 UT1)

| Eclipse Magnitude = 1.0458 | Saros Series = 130 |
|---|---|
| Gamma = 0.2894 | Saros Member = 53 of 73 |

Sun at Greatest Eclipse
(Geocentric Coordinates)
R.A. = 23h59m32.7s
Dec. = -00°02'58.0"
S.D. = 00°16'03.7"
H.P. = 00°00'08.8"

Moon at Greatest Eclipse
(Geocentric Coordinates)
R.A. = 23h59m11.3s
Dec. = +00°13'42.6"
S.D. = 00°16'31.6"
H.P. = 01°00'39.3"

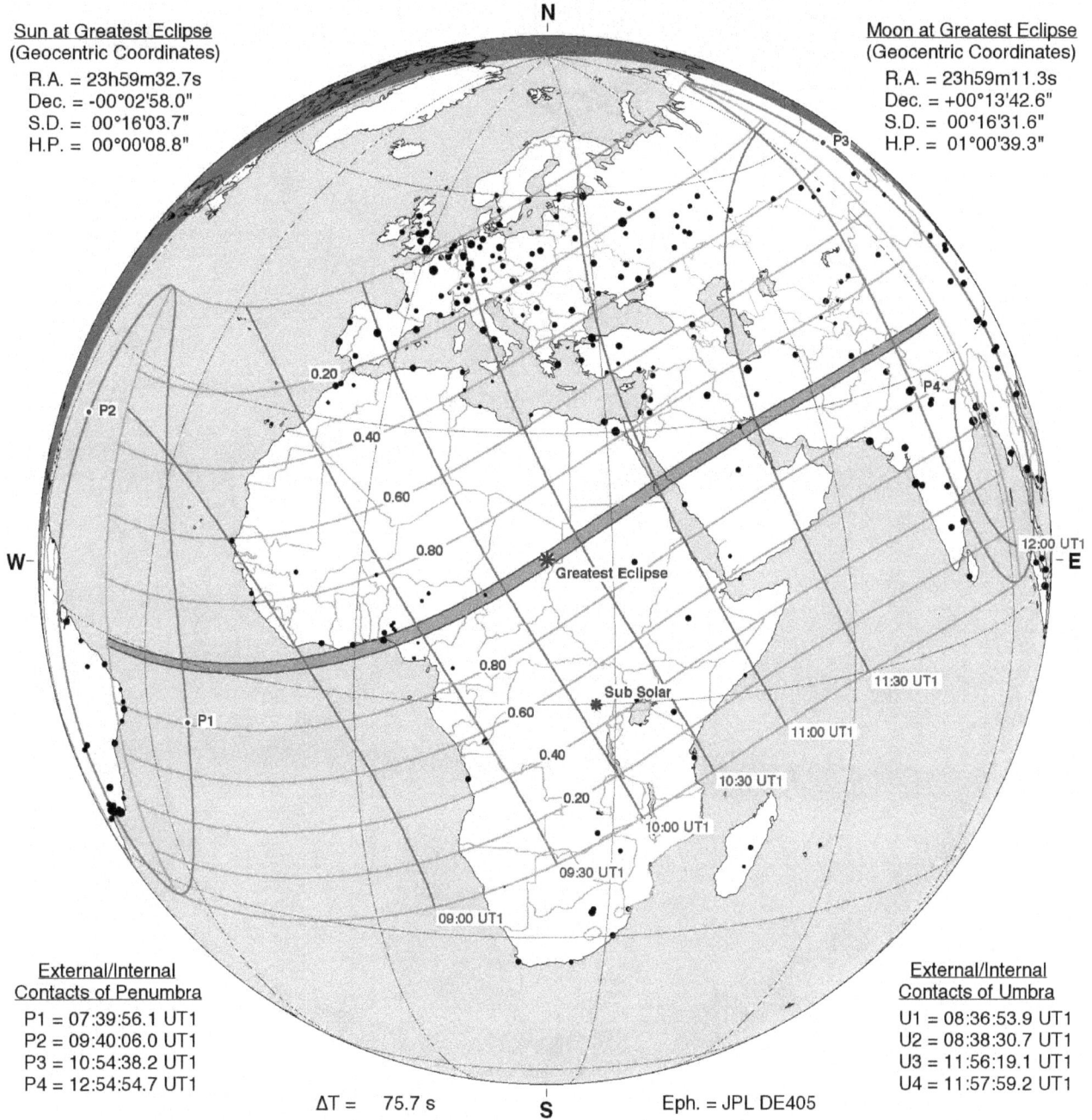

External/Internal
Contacts of Penumbra
P1 = 07:39:56.1 UT1
P2 = 09:40:06.0 UT1
P3 = 10:54:38.2 UT1
P4 = 12:54:54.7 UT1

External/Internal
Contacts of Umbra
U1 = 08:36:53.9 UT1
U2 = 08:38:30.7 UT1
U3 = 11:56:19.1 UT1
U4 = 11:57:59.2 UT1

ΔT = 75.7 s          Eph. = JPL DE405

Circumstances at Greatest Eclipse: 10:17:29.5 UT1

| Lat. = 16°03.3'N | Sun Alt. = 73.1° |
|---|---|
| Long. = 022°13.1'E | Sun Azm. = 161.6° |
| Path Width = 159.1 km | Duration = 04m09.3s |

Circumstances at Greatest Duration: 10:18:26.1 UT1

| Lat. = 16°13.3'N | Sun Alt. = 73.1° |
|---|---|
| Long. = 022°29.6'E | Sun Azm. = 163.5° |
| Path Width = 159.0 km | Duration = 04m09.4s |

©2016  F. Espenak
www.EclipseWise.com

# Annular Solar Eclipse of 2034 Sep 12

Greatest Eclipse = 16:19:27.5 TD   (= 16:18:11.6 UT1)

| | |
|---|---|
| Eclipse Magnitude = 0.9736 | Saros Series = 135 |
| Gamma = -0.3936 | Saros Member = 40 of 71 |

**N**

Sun at Greatest Eclipse
(Geocentric Coordinates)
R.A. = 11h23m10.9s
Dec. = +03°57'57.5"
S.D. = 00°15'53.5"
H.P. = 00°00'08.7"

Moon at Greatest Eclipse
(Geocentric Coordinates)
R.A. = 11h22m44.5s
Dec. = +03°36'59.6"
S.D. = 00°15'15.1"
H.P. = 00°55'58.6"

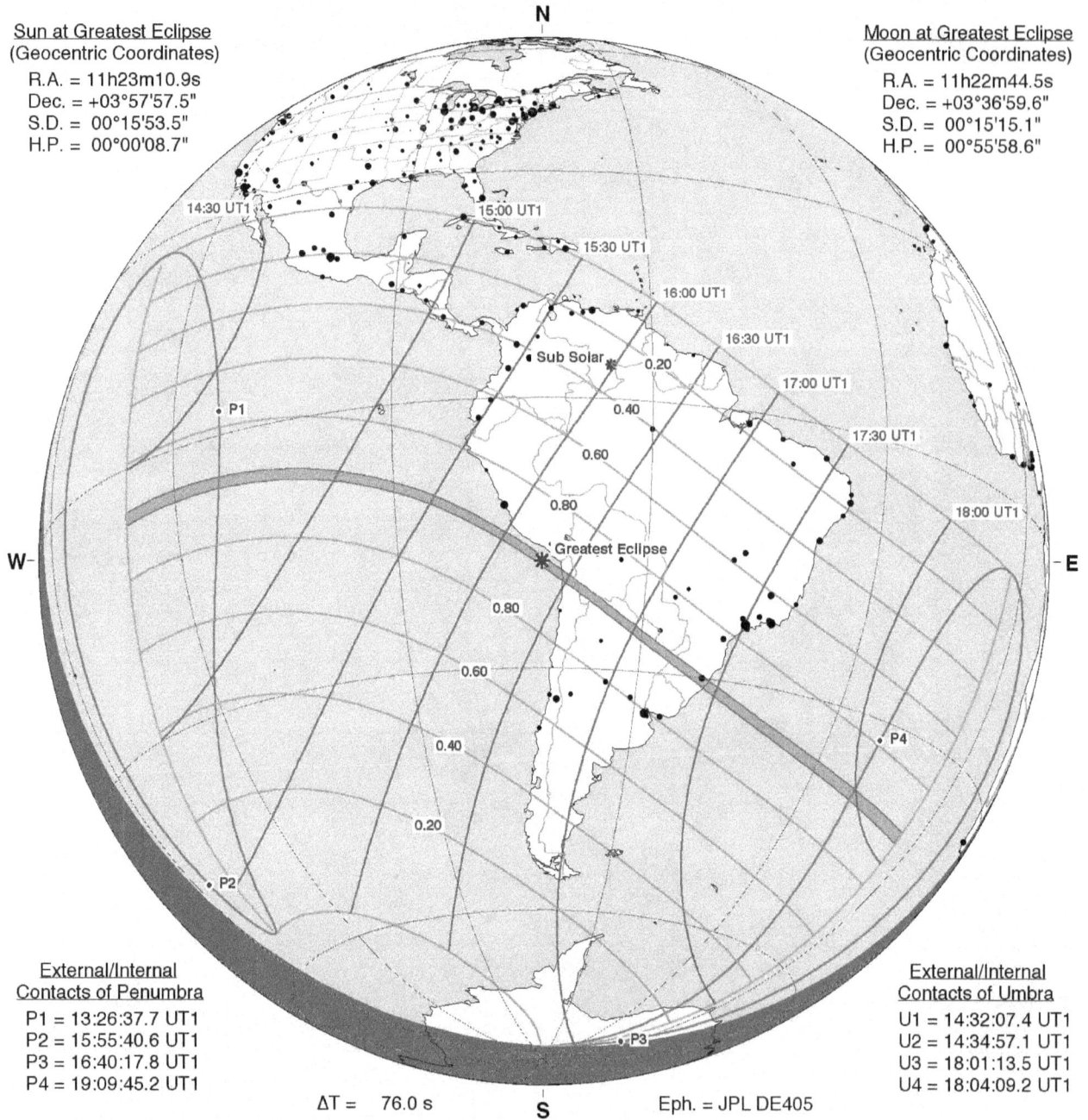

14:30 UT1
15:00 UT1
15:30 UT1
16:00 UT1
16:30 UT1
17:00 UT1
17:30 UT1
18:00 UT1

Sub Solar
0.20
P1
0.40
0.60
0.80
Greatest Eclipse
0.80
0.60
0.40
0.20
P2
P4
P3

**W**

**E**

External/Internal
Contacts of Penumbra
P1 = 13:26:37.7 UT1
P2 = 15:55:40.6 UT1
P3 = 16:40:17.8 UT1
P4 = 19:09:45.2 UT1

External/Internal
Contacts of Umbra
U1 = 14:32:07.4 UT1
U2 = 14:34:57.1 UT1
U3 = 18:01:13.5 UT1
U4 = 18:04:09.2 UT1

ΔT = 76.0 s

**S**

Eph. = JPL DE405

Circumstances at Greatest Eclipse: 16:18:11.6 UT1

| | |
|---|---|
| Lat. = 18°14.5'S | Sun Alt. = 66.7° |
| Long. = 072°37.4'W | Sun Azm. = 18.2° |
| Path Width = 102.1 km | Duration = 02m57.7s |

Circumstances at Greatest Duration: 16:29:28.2 UT1

| | |
|---|---|
| Lat. = 20°04.3'S | Sun Alt. = 65.9° |
| Long. = 069°44.5'W | Sun Azm. = 3.5° |
| Path Width = 101.7 km | Duration = 02m57.8s |

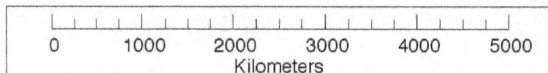

0   1000   2000   3000   4000   5000
Kilometers

©2016  F. Espenak
www.EclipseWise.com

# Annular Solar Eclipse of 2035 Mar 09

Greatest Eclipse = 23:05:53.6 TD (= 23:04:37.4 UT1)

| Eclipse Magnitude = 0.9919 | Saros Series = 140 |
|---|---|
| Gamma = -0.4368 | Saros Member = 30 of 71 |

**Sun at Greatest Eclipse**
(Geocentric Coordinates)
R.A. = 23h20m17.6s
Dec. = -04°16'22.2"
S.D. = 00°16'06.5"
H.P. = 00°00'08.9"

**Moon at Greatest Eclipse**
(Geocentric Coordinates)
R.A. = 23h20m47.9s
Dec. = -04°40'23.8"
S.D. = 00°15'44.9"
H.P. = 00°57'47.9"

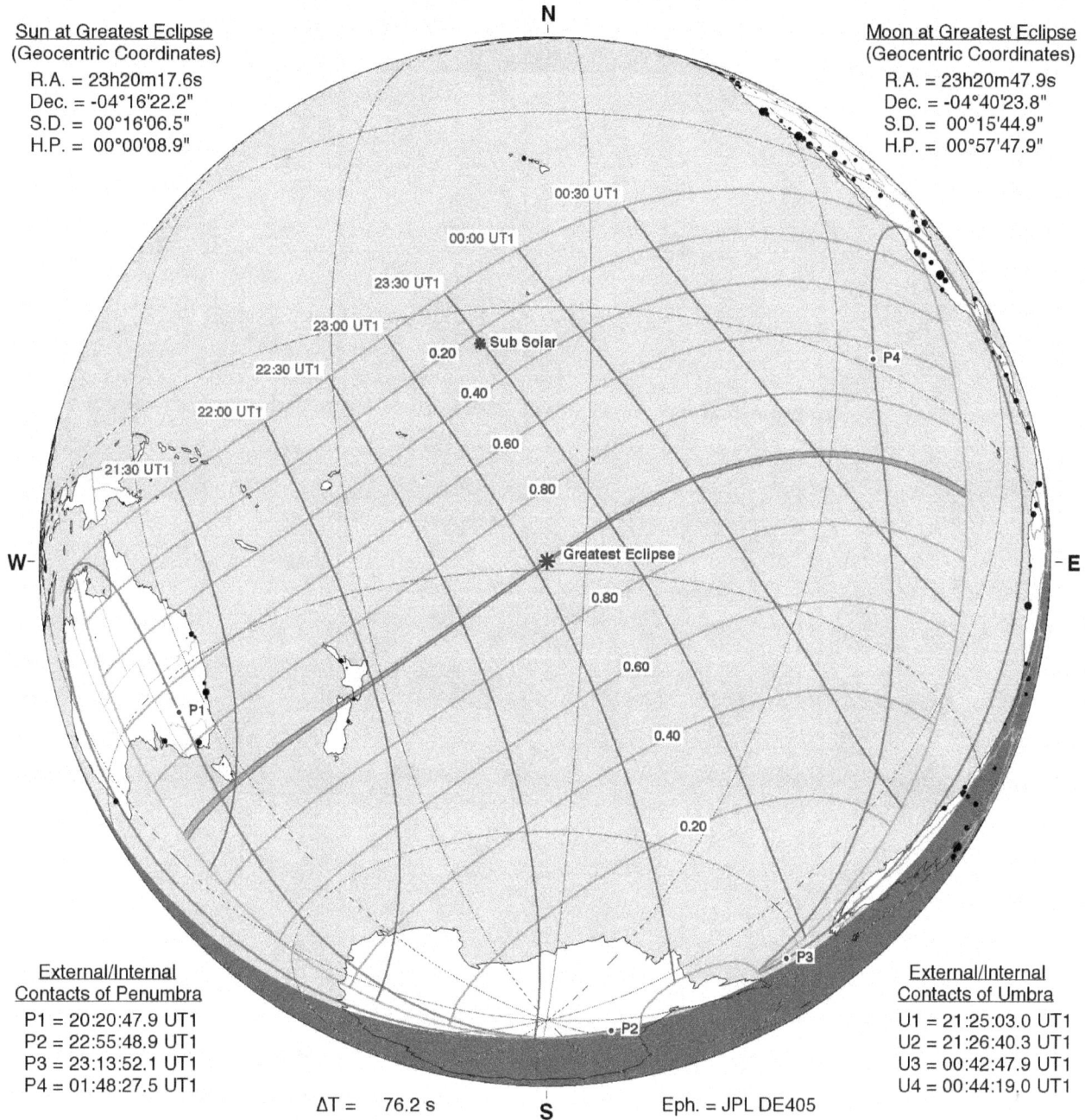

N

00:30 UT1
00:00 UT1
23:30 UT1
23:00 UT1
22:30 UT1
22:00 UT1
21:30 UT1

Sub Solar
0.20
0.40
0.60
0.80

P4

W — Greatest Eclipse — E
0.80
0.60
0.40
0.20

P1

P3
P2

S

**External/Internal Contacts of Penumbra**
P1 = 20:20:47.9 UT1
P2 = 22:55:48.9 UT1
P3 = 23:13:52.1 UT1
P4 = 01:48:27.5 UT1

ΔT = 76.2 s

Eph. = JPL DE405

**External/Internal Contacts of Umbra**
U1 = 21:25:03.0 UT1
U2 = 21:26:40.3 UT1
U3 = 00:42:47.9 UT1
U4 = 00:44:19.0 UT1

**Circumstances at Greatest Eclipse: 23:04:37.4 UT1**

| Lat. = 29°02.6'S | Sun Alt. = 63.9° |
|---|---|
| Long. = 154°58.1'W | Sun Azm. = 340.2° |
| Path Width = 31.5 km | Duration = 00m47.5s |

**Circumstances at Greatest Duration: 21:25:51.6 UT1**

| Lat. = 43°22.2'S | Sun Alt. = 0.0° |
|---|---|
| Long. = 127°04.5'E | Sun Azm. = 95.9° |
| Path Width = 95.5 km | Duration = 01m25.9s |

0   1000   2000   3000   4000   5000
Kilometers

# Total Solar Eclipse of 2035 Sep 02

Greatest Eclipse = 01:56:46.3 TD  (= 01:55:29.9 UT1)

| | |
|---|---|
| Eclipse Magnitude = 1.0320 | Saros Series = 145 |
| Gamma = 0.3727 | Saros Member = 23 of 77 |

**Sun at Greatest Eclipse**
(Geocentric Coordinates)
R.A. = 10h44m07.3s
Dec. = +08°01'09.8"
S.D. = 00°15'50.9"
H.P. = 00°00'08.7"

**Moon at Greatest Eclipse**
(Geocentric Coordinates)
R.A. = 10h44m32.4s
Dec. = +08°22'14.7"
S.D. = 00°16'06.4"
H.P. = 00°59'06.9"

**External/Internal Contacts of Penumbra**
P1 = 23:15:29.4 UT1
P2 = 01:27:32.2 UT1
P3 = 02:23:45.3 UT1
P4 = 04:35:41.4 UT1

**External/Internal Contacts of Umbra**
U1 = 00:15:49.1 UT1
U2 = 00:16:50.3 UT1
U3 = 03:34:21.2 UT1
U4 = 03:35:17.3 UT1

ΔT = 76.4 s

Eph. = JPL DE405

**Circumstances at Greatest Eclipse: 01:55:29.9 UT1**

| | |
|---|---|
| Lat. = 29°05.6'N | Sun Alt. = 67.9° |
| Long. = 158°00.9'E | Sun Azm. = 198.5° |
| Path Width = 116.3 km | Duration = 02m54.2s |

**Circumstances at Greatest Duration: 01:52:01.1 UT1**

| | |
|---|---|
| Lat. = 29°41.4'N | Sun Alt. = 67.8° |
| Long. = 156°51.6'E | Sun Azm. = 192.9° |
| Path Width = 116.1 km | Duration = 02m54.4s |

©2016 F. Espenak
www.EclipseWise.com

# Partial Solar Eclipse of 2036 Feb 27

Greatest Eclipse = 04:46:49.0 TD  (= 04:45:32.4 UT1)

| | |
|---|---|
| Eclipse Magnitude = 0.6286 | Saros Series = 150 |
| Gamma = -1.1942 | Saros Member = 18 of 71 |

**Sun at Greatest Eclipse**
(Geocentric Coordinates)
R.A. = 22h39m15.4s
Dec. = -08°30'21.2"
S.D. = 00°16'09.1"
H.P. = 00°00'08.9"

**Moon at Greatest Eclipse**
(Geocentric Coordinates)
R.A. = 22h40m29.9s
Dec. = -09°33'05.6"
S.D. = 00°14'57.5"
H.P. = 00°54'53.9"

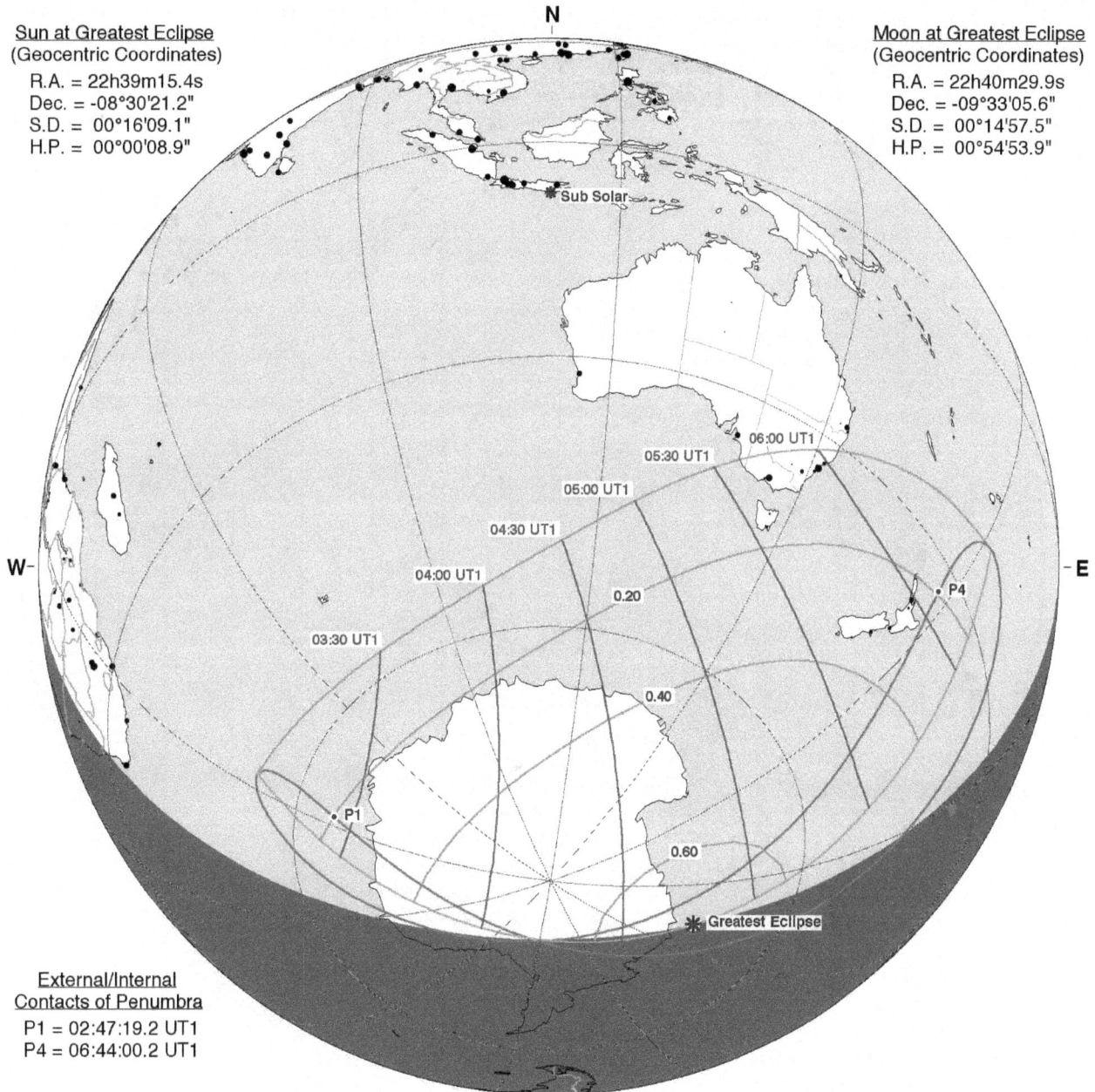

N

Sub Solar

06:00 UT1
05:30 UT1
05:00 UT1
04:30 UT1
04:00 UT1
03:30 UT1

0.20
0.40
0.60

P1
P4

W

E

Greatest Eclipse

**External/Internal
Contacts of Penumbra**
P1 = 02:47:19.2 UT1
P4 = 06:44:00.2 UT1

ΔT = 76.7 s

S

Eph. = JPL DE405

**Circumstances at Greatest Eclipse: 04:45:32.4 UT1**

| | |
|---|---|
| Lat. = 71°38.9'S | Sun Alt. = 0.0° |
| Long. = 131°27.3'W | Sun Azm. = 242.0° |

0   1000   2000   3000   4000   5000
Kilometers

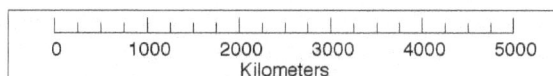

# Partial Solar Eclipse of 2036 Jul 23

Greatest Eclipse = 10:32:06.5 TD  (= 10:30:49.6 UT1)

Eclipse Magnitude = 0.1992          Saros Series = 117

Gamma = -1.4250          Saros Member = 70 of 71

Sun at Greatest Eclipse
(Geocentric Coordinates)
R.A. = 08h13m32.5s
Dec. = +19°53'41.2"
S.D. = 00°15'44.6"
H.P. = 00°00'08.7"

Moon at Greatest Eclipse
(Geocentric Coordinates)
R.A. = 08h12m46.3s
Dec. = +18°27'12.2"
S.D. = 00°16'42.4"
H.P. = 01°01'18.7"

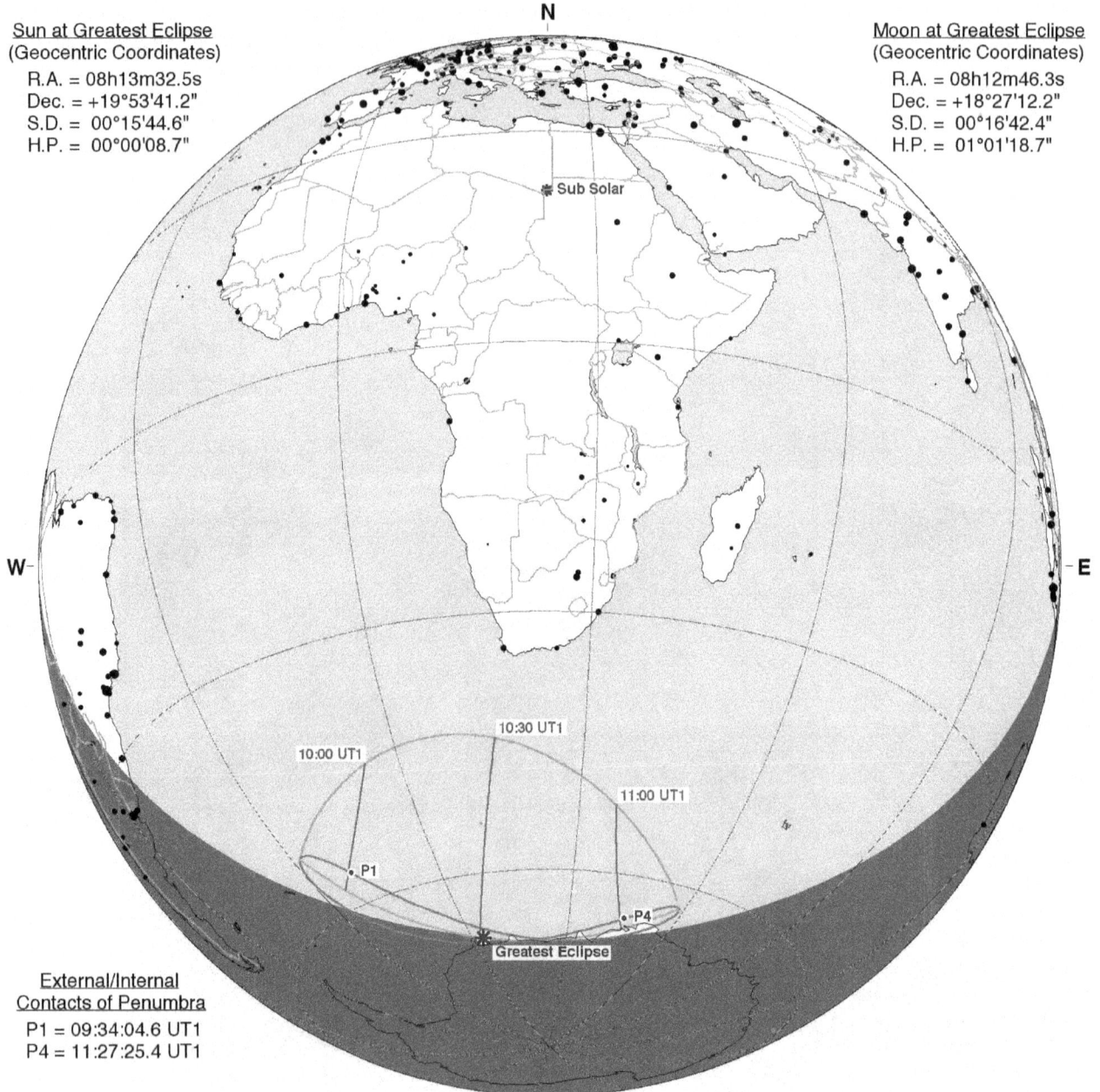

N

Sub Solar

W

E

10:30 UT1
10:00 UT1
11:00 UT1
P1
P4
Greatest Eclipse

External/Internal
Contacts of Penumbra
P1 = 09:34:04.6 UT1
P4 = 11:27:25.4 UT1

ΔT = 76.8 s          S          Eph. = JPL DE405

Circumstances at Greatest Eclipse: 10:30:49.6 UT1
Lat. = 68°53.0'S          Sun Alt. = 0.0°
Long. = 003°32.3'E          Sun Azm. = 19.1°

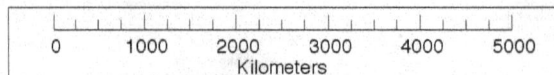

| 0 | 1000 | 2000 | 3000 | 4000 | 5000 |
Kilometers

©2016  F. Espenak
www.EclipseWise.com

# Partial Solar Eclipse of 2036 Aug 21

Greatest Eclipse = 17:25:45.4 TD   (= 17:24:28.5 UT1)

| Eclipse Magnitude = 0.8622 | Saros Series = 155 |
| Gamma = 1.0825 | Saros Member = 7 of 71 |

**N**

Sun at Greatest Eclipse
(Geocentric Coordinates)
R.A. = 10h05m24.9s
Dec. = +11°44'16.4"
S.D. = 00°15'48.7"
H.P. = 00°00'08.7"

Moon at Greatest Eclipse
(Geocentric Coordinates)
R.A. = 10h06m34.6s
Dec. = +12°48'10.2"
S.D. = 00°16'41.1"
H.P. = 01°01'14.1"

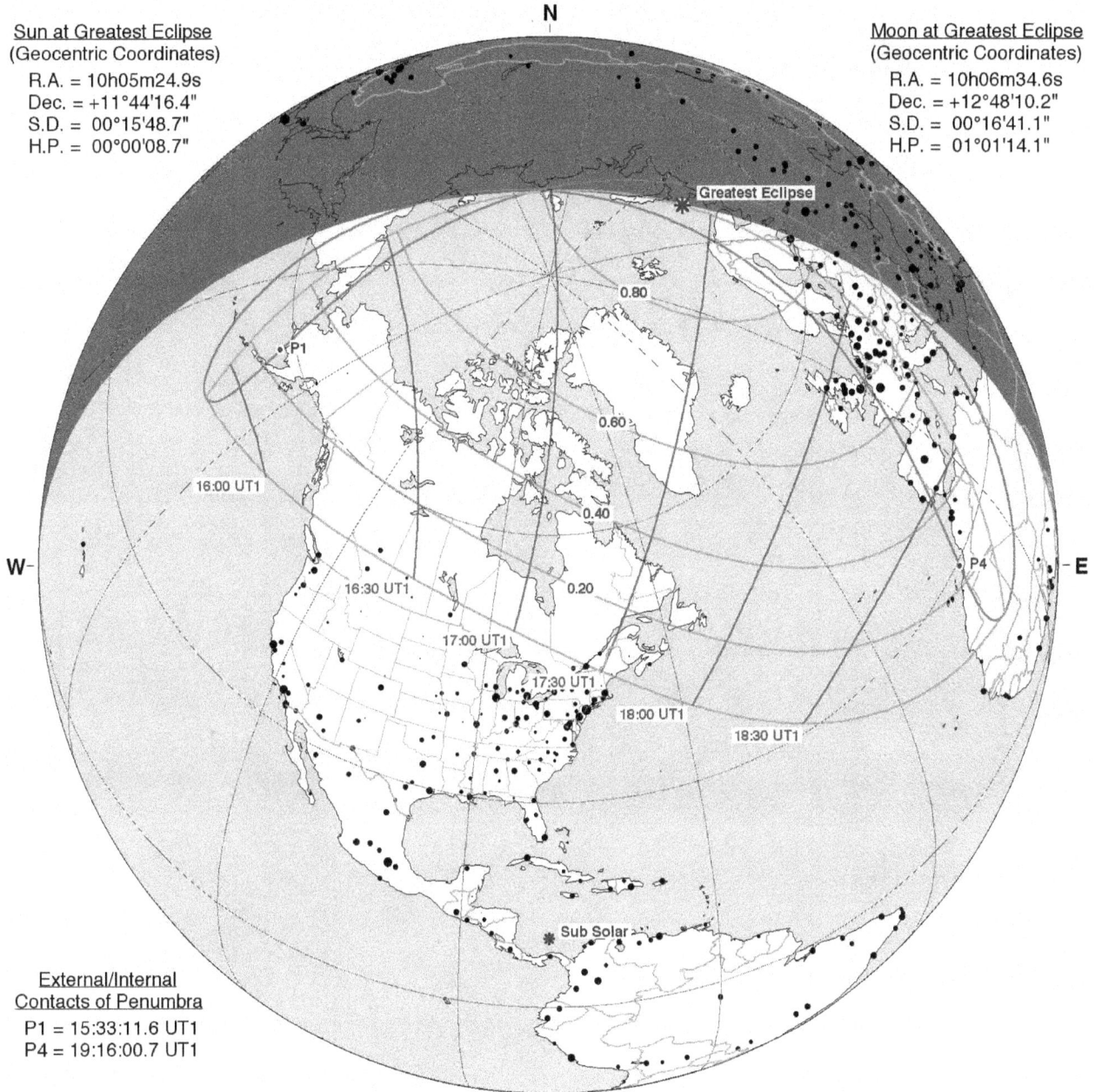

Greatest Eclipse

0.80

P1

0.60

16:00 UT1

0.40

16:30 UT1

0.20

17:00 UT1

17:30 UT1

18:00 UT1

P4

18:30 UT1

**W** —

— **E**

External/Internal
Contacts of Penumbra
P1 = 15:33:11.6 UT1
P4 = 19:16:00.7 UT1

Sub Solar

ΔT = 76.9 s      **S**      Eph. = JPL DE405

Circumstances at Greatest Eclipse: 17:24:28.5 UT1

| Lat. = 71°07.2'N | Sun Alt. = 0.0° |
| Long. = 046°59.1'E | Sun Azm. = 308.9° |

| 0 | 1000 | 2000 | 3000 | 4000 | 5000 |
Kilometers

©2016  F. Espenak
www.EclipseWise.com

135

# Partial Solar Eclipse of 2037 Jan 16

Greatest Eclipse = 09:48:55.1 TD (= 09:47:38.0 UT1)

Eclipse Magnitude = 0.7049     Saros Series = 122
Gamma = 1.1477     Saros Member = 59 of 70

**Sun at Greatest Eclipse**
(Geocentric Coordinates)
R.A. = 19h54m30.1s
Dec. = -20°49'43.5"
S.D. = 00°16'15.5"
H.P. = 00°00'08.9"

**Moon at Greatest Eclipse**
(Geocentric Coordinates)
R.A. = 19h54m05.4s
Dec. = -19°47'33.7"
S.D. = 00°14'51.8"
H.P. = 00°54'32.8"

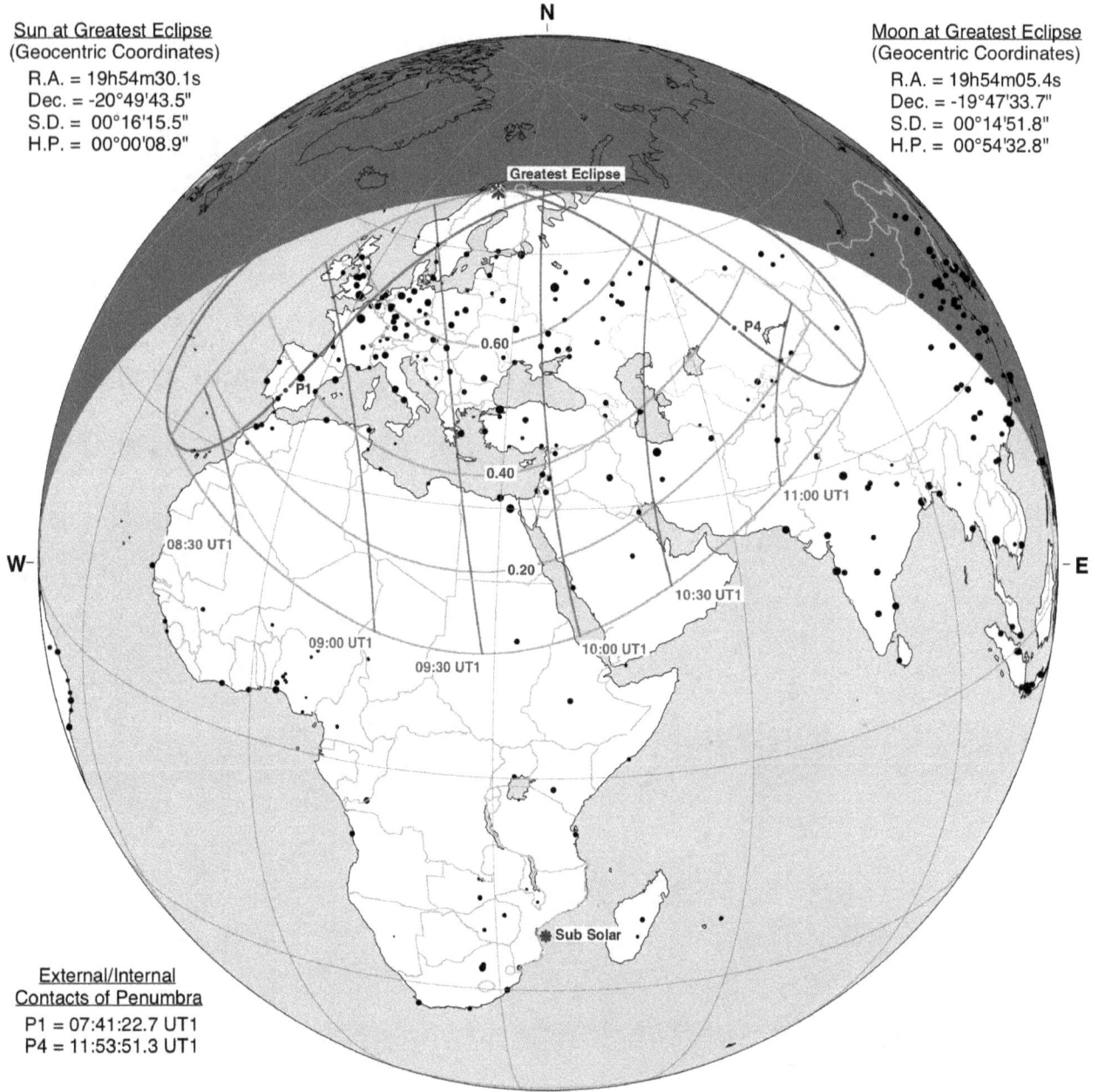

N

Greatest Eclipse

0.60

P1

P4

0.40

11:00 UT1

08:30 UT1

W

0.20

E

10:30 UT1

09:00 UT1

10:00 UT1

09:30 UT1

Sub Solar

**External/Internal
Contacts of Penumbra**
P1 = 07:41:22.7 UT1
P4 = 11:53:51.3 UT1

ΔT = 77.1 s     S     Eph. = JPL DE405

**Circumstances at Greatest Eclipse: 09:47:38.0 UT1**
Lat. = 68°31.3'N     Sun Alt. = 0.0°
Long. = 020°47.6'E     Sun Azm. = 166.2°

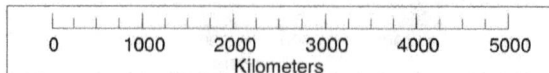

0   1000   2000   3000   4000   5000
Kilometers

©2016 F. Espenak
www.EclipseWise.com

# Total Solar Eclipse of 2037 Jul 13

Greatest Eclipse = 02:40:35.9 TD  (= 02:39:18.5 UT1)

Eclipse Magnitude = 1.0413      Saros Series = 127

Gamma = -0.7246      Saros Member = 59 of 82

**Sun at Greatest Eclipse**
(Geocentric Coordinates)
R.A. = 07h31m06.7s
Dec. = +21°46'57.5"
S.D. = 00°15'44.0"
H.P. = 00°00'08.7"

**Moon at Greatest Eclipse**
(Geocentric Coordinates)
R.A. = 07h30m56.4s
Dec. = +21°04'03.1"
S.D. = 00°16'12.0"
H.P. = 00°59'27.3"

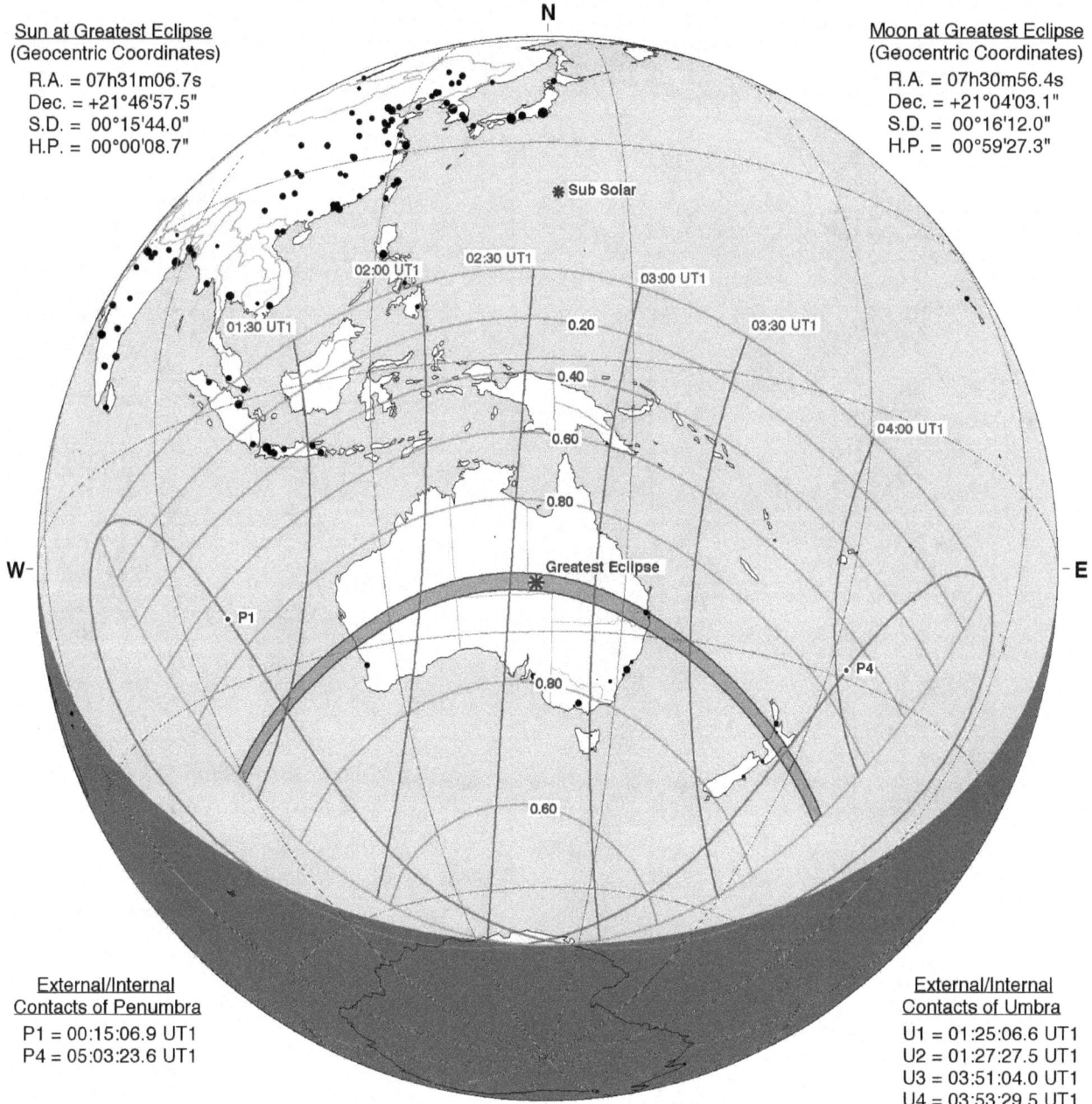

N

＊ Sub Solar

02:00 UT1   02:30 UT1   03:00 UT1
01:30 UT1   03:30 UT1
0.20
0.40
04:00 UT1
0.60
0.80
W
Greatest Eclipse ＊
E
P1
0.80
P4
0.60

**External/Internal Contacts of Penumbra**
P1 = 00:15:06.9 UT1
P4 = 05:03:23.6 UT1

**External/Internal Contacts of Umbra**
U1 = 01:25:06.6 UT1
U2 = 01:27:27.5 UT1
U3 = 03:51:04.0 UT1
U4 = 03:53:29.5 UT1

ΔT = 77.3 s    S    Eph. = JPL DE405

**Circumstances at Greatest Eclipse: 02:39:18.5 UT1**
Lat. = 24°45.8'S    Sun Alt. = 43.4°
Long. = 139°03.2'E    Sun Azm. = 3.3°
Path Width = 200.7 km    Duration = 03m58.4s

**Circumstances at Greatest Duration: 02:39:45.9 UT1**
Lat. = 24°46.7'S    Sun Alt. = 43.4°
Long. = 139°12.5'E    Sun Azm. = 2.9°
Path Width = 200.6 km    Duration = 03m58.4s

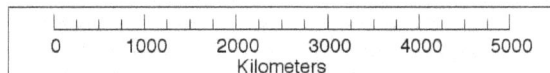

©2016 F. Espenak
www.EclipseWise.com

0   1000   2000   3000   4000   5000
Kilometers

# Annular Solar Eclipse of 2038 Jan 05

Greatest Eclipse = 13:47:10.9 TD   (= 13:45:53.4 UT1)

| | |
|---|---|
| Eclipse Magnitude = 0.9728 | Saros Series = 132 |
| Gamma = 0.4169 | Saros Member = 47 of 71 |

**Sun at Greatest Eclipse**
(Geocentric Coordinates)
R.A. = 19h06m27.4s
Dec. = -22°33'17.3"
S.D. = 00°16'15.9"
H.P. = 00°00'08.9"

**Moon at Greatest Eclipse**
(Geocentric Coordinates)
R.A. = 19h06m25.9s
Dec. = -22°09'29.7"
S.D. = 00°15'35.7"
H.P. = 00°57'13.9"

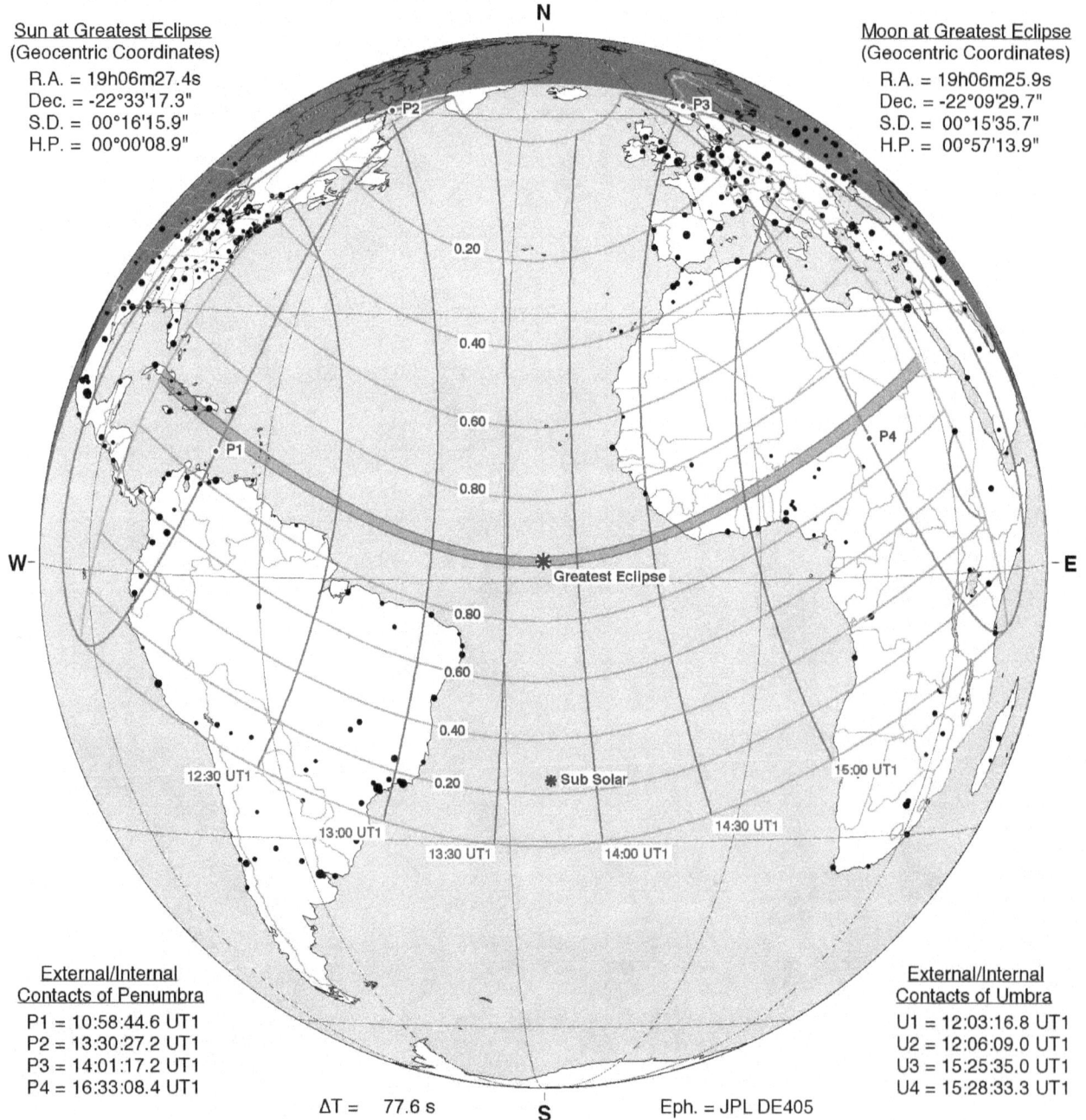

N

Greatest Eclipse

W — E

Sub Solar

12:30 UT1   13:00 UT1   13:30 UT1   14:00 UT1   14:30 UT1   15:00 UT1

**External/Internal Contacts of Penumbra**
P1 = 10:58:44.6 UT1
P2 = 13:30:27.2 UT1
P3 = 14:01:17.2 UT1
P4 = 16:33:08.4 UT1

**External/Internal Contacts of Umbra**
U1 = 12:03:16.8 UT1
U2 = 12:06:09.0 UT1
U3 = 15:25:35.0 UT1
U4 = 15:28:33.3 UT1

ΔT = 77.6 s

S

Eph. = JPL DE405

**Circumstances at Greatest Eclipse: 13:45:53.4 UT1**

| | |
|---|---|
| Lat. = 02°05.7'N | Sun Alt. = 65.3° |
| Long. = 025°28.0'W | Sun Azm. = 179.2° |
| Path Width = 107.2 km | Duration = 03m18.5s |

**Circumstances at Greatest Duration: 13:52:36.3 UT1**

| | |
|---|---|
| Lat. = 02°11.2'N | Sun Alt. = 65.1° |
| Long. = 023°41.3'W | Sun Azm. = 186.8° |
| Path Width = 107.4 km | Duration = 03m18.6s |

0   1000   2000   3000   4000   5000
Kilometers

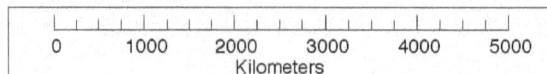

©2016 F. Espenak
www.EclipseWise.com

# Annular Solar Eclipse of  2038 Jul 02

Greatest Eclipse =  13:32:55.0 TD   (= 13:31:37.2 UT1)

| | |
|---|---|
| Eclipse Magnitude =  0.9911 | Saros Series =  137 |
| Gamma =  0.0398 | Saros Member =  37 of 70 |

**Sun at Greatest Eclipse**
(Geocentric Coordinates)
R.A. = 06h46m55.4s
Dec. = +22°59'44.2"
S.D. = 00°15'43.9"
H.P. = 00°00'08.6"

**Moon at Greatest Eclipse**
(Geocentric Coordinates)
R.A. = 06h46m55.2s
Dec. = +23°01'58.2"
S.D. = 00°15'20.9"
H.P. = 00°56'19.9"

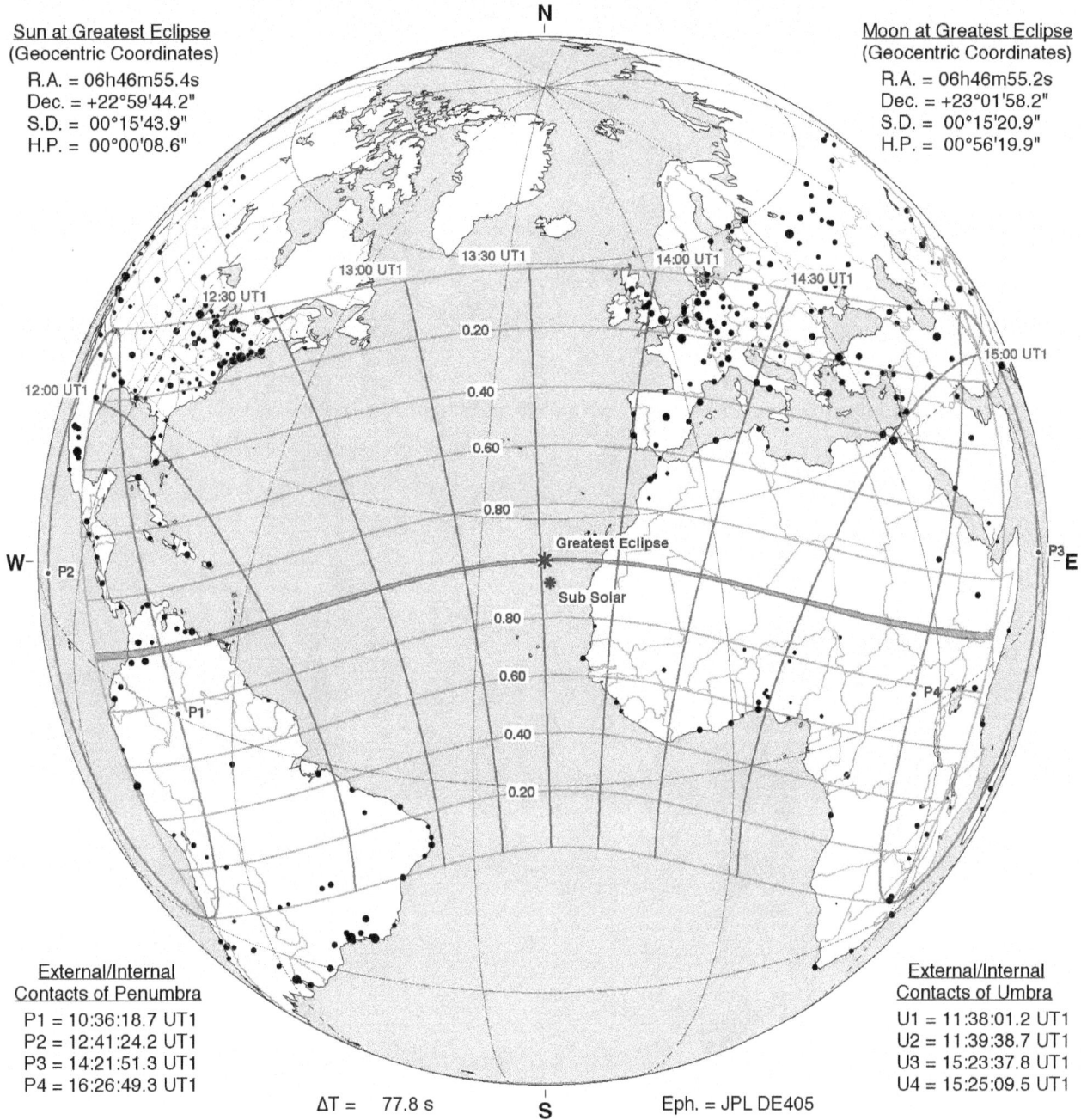

N

13:30 UT1
13:00 UT1
14:00 UT1
12:30 UT1
14:30 UT1
0.20
12:00 UT1
15:00 UT1
0.40
0.60
0.80
Greatest Eclipse
Sub Solar
W
P2
P3
E
0.80
0.60
P4
0.40
P1
0.20

S

**External/Internal**
**Contacts of Penumbra**
P1 = 10:36:18.7 UT1
P2 = 12:41:24.2 UT1
P3 = 14:21:51.3 UT1
P4 = 16:26:49.3 UT1

**External/Internal**
**Contacts of Umbra**
U1 = 11:38:01.2 UT1
U2 = 11:39:38.7 UT1
U3 = 15:23:37.8 UT1
U4 = 15:25:09.5 UT1

ΔT =   77.8 s

Eph. = JPL DE405

**Circumstances at Greatest Eclipse:  13:31:37.2 UT1**

| | |
|---|---|
| Lat. =  25°25.4'N | Sun Alt. =  87.6° |
| Long. = 021°54.8'W | Sun Azm. =  179.0° |
| Path Width =  31.2 km | Duration =  00m59.7s |

**Circumstances at Greatest Duration:  11:38:50.0 UT1**

| | |
|---|---|
| Lat. =  01°05.8'N | Sun Alt. =   0.0° |
| Long. = 084°08.2'W | Sun Azm. =  67.0° |
| Path Width =  94.2 km | Duration =  01m34.9s |

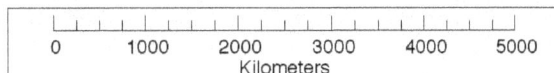

©2016  F. Espenak
www.EclipseWise.com

0    1000    2000    3000    4000    5000
Kilometers

# Total Solar Eclipse of 2038 Dec 26

Greatest Eclipse = 01:00:09.7 TD  (= 00:58:51.6 UT1)

| | |
|---|---|
| Eclipse Magnitude = 1.0269 | Saros Series = 142 |
| Gamma = -0.2881 | Saros Member = 24 of 72 |

**Sun at Greatest Eclipse**
(Geocentric Coordinates)
R.A. = 18h18m51.7s
Dec. = -23°21'47.8"
S.D. = 00°16'15.7"
H.P. = 00°00'08.9"

**Moon at Greatest Eclipse**
(Geocentric Coordinates)
R.A. = 18h18m46.7s
Dec. = -23°39'05.4"
S.D. = 00°16'25.8"
H.P. = 01°00'18.1"

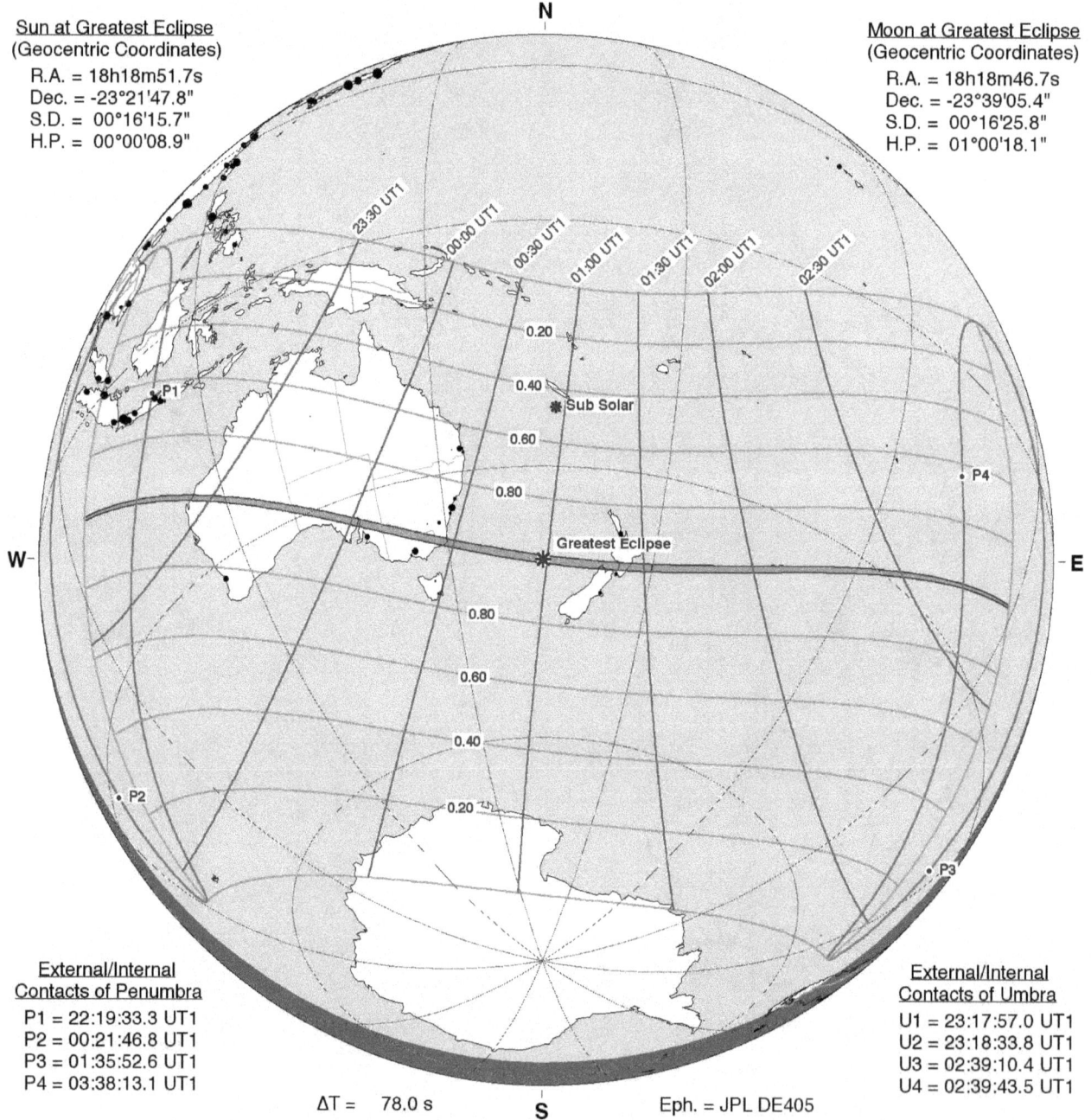

**External/Internal Contacts of Penumbra**
P1 = 22:19:33.3 UT1
P2 = 00:21:46.8 UT1
P3 = 01:35:52.6 UT1
P4 = 03:38:13.1 UT1

**External/Internal Contacts of Umbra**
U1 = 23:17:57.0 UT1
U2 = 23:18:33.8 UT1
U3 = 02:39:10.4 UT1
U4 = 02:39:43.5 UT1

ΔT = 78.0 s

Eph. = JPL DE405

**Circumstances at Greatest Eclipse: 00:58:51.6 UT1**

| | |
|---|---|
| Lat. = 40°16.7'S | Sun Alt. = 73.0° |
| Long. = 163°55.6'E | Sun Azm. = 4.5° |
| Path Width = 95.3 km | Duration = 02m18.2s |

**Circumstances at Greatest Duration: 00:58:08.2 UT1**

| | |
|---|---|
| Lat. = 40°14.8'S | Sun Alt. = 73.0° |
| Long. = 163°35.6'E | Sun Azm. = 6.2° |
| Path Width = 95.3 km | Duration = 02m18.2s |

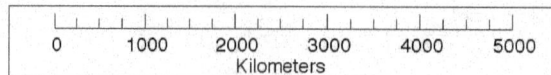

*©2016  F. Espenak*
*www.EclipseWise.com*

# Annular Solar Eclipse of 2039 Jun 21

Greatest Eclipse = 17:12:53.8 TD   (= 17:11:35.5 UT1)

| | |
|---|---|
| Eclipse Magnitude = 0.9454 | Saros Series = 147 |
| Gamma = 0.8312 | Saros Member = 24 of 80 |

**N**

Sun at Greatest Eclipse
(Geocentric Coordinates)
R.A. = 06h00m54.5s
Dec. = +23°26'03.6"
S.D. = 00°15'44.3"
H.P. = 00°00'08.7"

Moon at Greatest Eclipse
(Geocentric Coordinates)
R.A. = 06h00m35.3s
Dec. = +24°10'44.9"
S.D. = 00°14'45.6"
H.P. = 00°54'10.2"

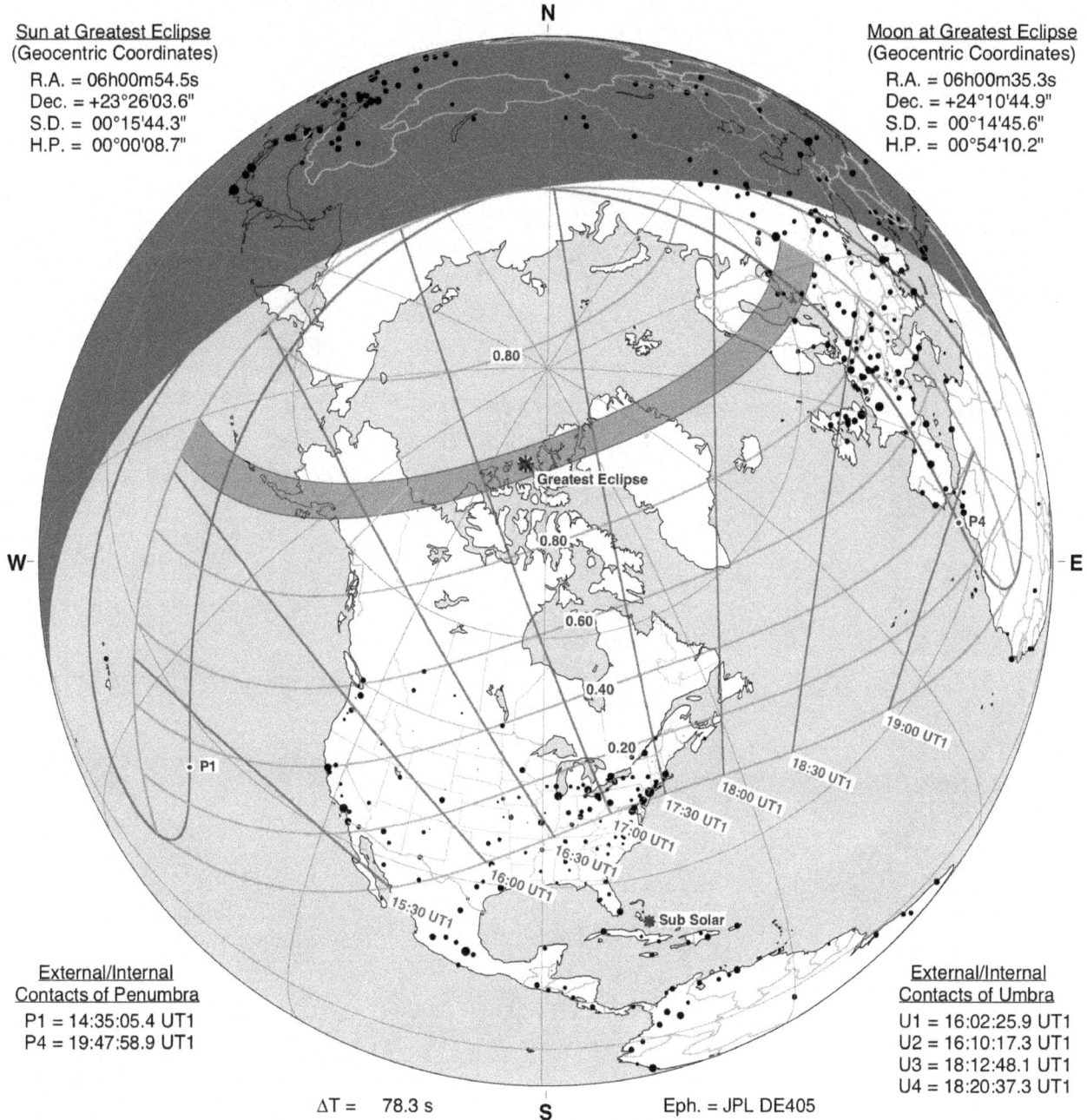

0.80

Greatest Eclipse

P4

0.80

0.60

0.40

0.20

19:00 UT1

18:30 UT1

18:00 UT1

P1

17:30 UT1

17:00 UT1

16:30 UT1

16:00 UT1

15:30 UT1

Sub Solar

**W** —                                                                — **E**

External/Internal
Contacts of Penumbra
P1 = 14:35:05.4 UT1
P4 = 19:47:58.9 UT1

External/Internal
Contacts of Umbra
U1 = 16:02:25.9 UT1
U2 = 16:10:17.3 UT1
U3 = 18:12:48.1 UT1
U4 = 18:20:37.3 UT1

ΔT =   78.3 s         **S**         Eph. = JPL DE405

| Circumstances at Greatest Eclipse: 17:11:35.5 UT1 | | Circumstances at Greatest Duration: 17:11:45.6 UT1 | |
|---|---|---|---|
| Lat. = 78°52.5'N | Sun Alt. =  33.4° | Lat. = 78°54.8'N | Sun Alt. =  33.4° |
| Long. = 102°07.9'W | Sun Azm. = 152.6° | Long. = 101°47.9'W | Sun Azm. = 153.1° |
| Path Width = 365.3 km | Duration = 04m04.9s | Path Width = 365.2 km | Duration = 04m04.9s |

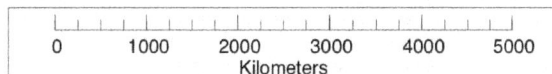

©2016  F. Espenak
www.EclipseWise.com

Kilometers
0   1000   2000   3000   4000   5000

# Total Solar Eclipse of 2039 Dec 15

Greatest Eclipse = 16:23:45.9 TD  (= 16:22:27.4 UT1)

| Eclipse Magnitude = 1.0356 | Saros Series = 152 |
|---|---|
| Gamma = -0.9458 | Saros Member = 14 of 70 |

Sun at Greatest Eclipse
(Geocentric Coordinates)
R.A. = 17h31m51.4s
Dec. = -23°16'37.6"
S.D. = 00°16'14.9"
H.P. = 00°00'08.9"

Moon at Greatest Eclipse
(Geocentric Coordinates)
R.A. = 17h31m14.4s
Dec. = -24°13'58.8"
S.D. = 00°16'44.6"
H.P. = 01°01'26.8"

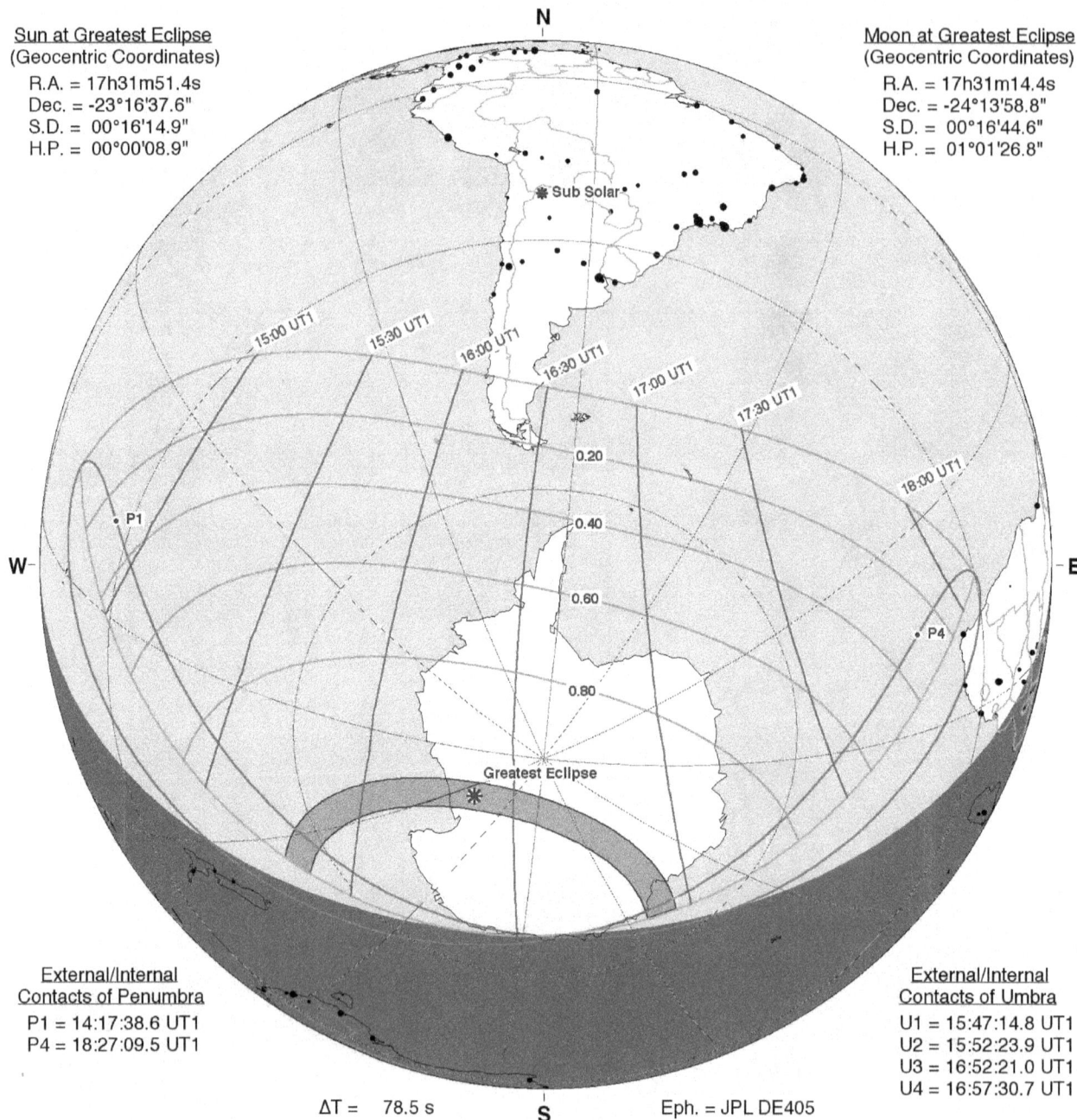

External/Internal
Contacts of Penumbra
P1 = 14:17:38.6 UT1
P4 = 18:27:09.5 UT1

External/Internal
Contacts of Umbra
U1 = 15:47:14.8 UT1
U2 = 15:52:23.9 UT1
U3 = 16:52:21.0 UT1
U4 = 16:57:30.7 UT1

ΔT = 78.5 s          Eph. = JPL DE405

Circumstances at Greatest Eclipse: 16:22:27.4 UT1

| Lat. = 80°51.4'S | Sun Alt. = 18.4° |
|---|---|
| Long. = 172°44.9'E | Sun Azm. = 123.4° |
| Path Width = 379.8 km | Duration = 01m51.4s |

Circumstances at Greatest Duration: 16:22:32.9 UT1

| Lat. = 80°54.0'S | Sun Alt. = 18.4° |
|---|---|
| Long. = 172°32.7'E | Sun Azm. = 123.5° |
| Path Width = 379.7 km | Duration = 01m51.4s |

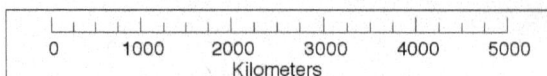

©2016  F. Espenak
www.EclipseWise.com

# Partial Solar Eclipse of 2040 May 11

Greatest Eclipse = 03:43:02.1 TD   (= 03:41:43.4 UT1)

| Eclipse Magnitude = 0.5306 | Saros Series = 119 |
|---|---|
| Gamma = -1.2529 | Saros Member = 67 of 71 |

**Sun at Greatest Eclipse**
(Geocentric Coordinates)
R.A. = 03h14m33.6s
Dec. = +18°01'19.7"
S.D. = 00°15'50.1"
H.P. = 00°00'08.7"

**Moon at Greatest Eclipse**
(Geocentric Coordinates)
R.A. = 03h16m16.3s
Dec. = +16°56'30.8"
S.D. = 00°15'06.4"
H.P. = 00°55'26.7"

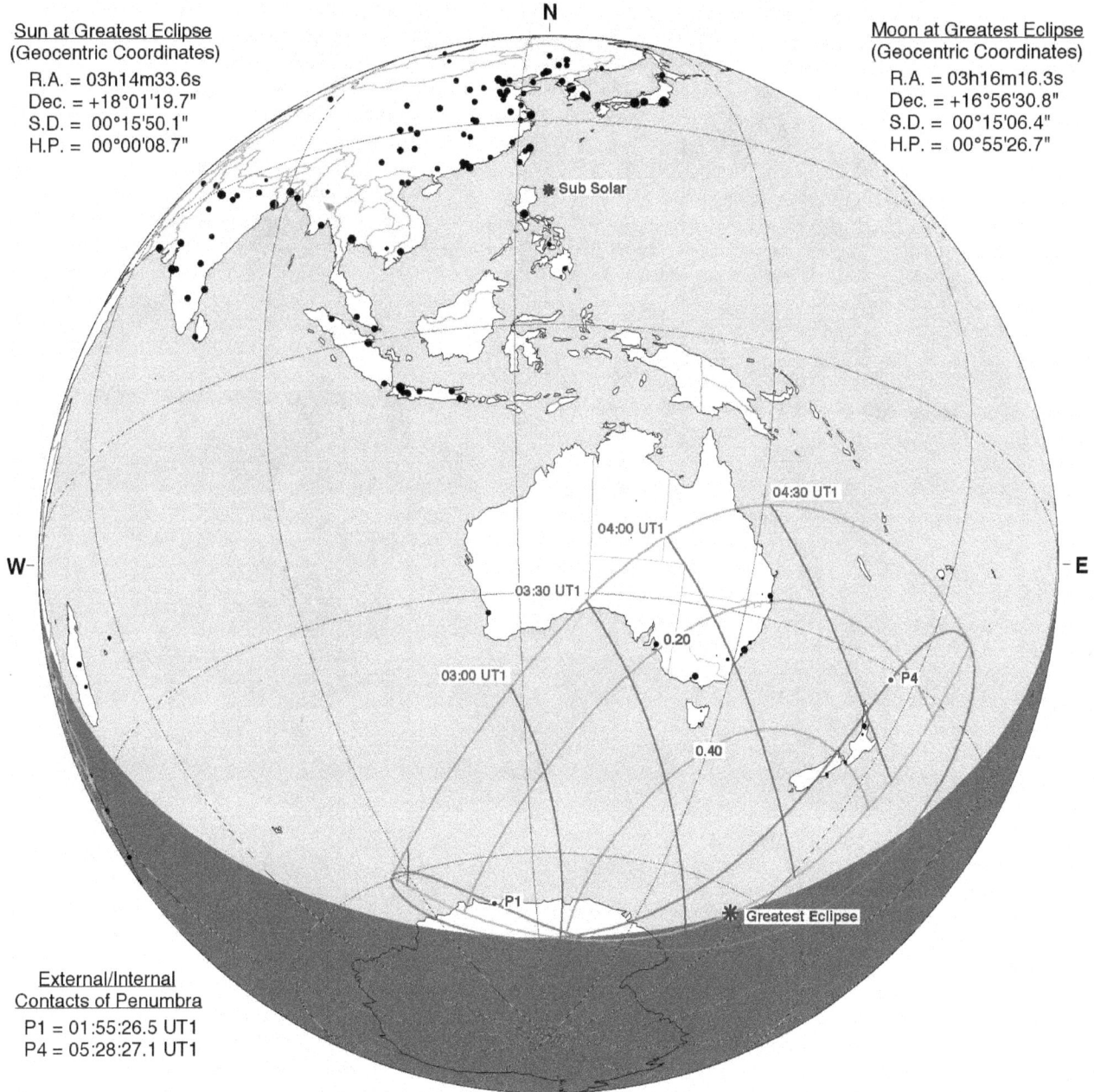

N

Sub Solar

04:30 UT1
04:00 UT1
03:30 UT1
0.20
03:00 UT1
0.40
P4
P1
Greatest Eclipse

W — E

**External/Internal
Contacts of Penumbra**
P1 = 01:55:26.5 UT1
P4 = 05:28:27.1 UT1

ΔT = 78.8 s   S   Eph. = JPL DE405

Circumstances at Greatest Eclipse: 03:41:43.4 UT1

| Lat. = 62°46.1'S | Sun Alt. = 0.0° |
|---|---|
| Long. = 174°23.3'E | Sun Azm. = 312.5° |

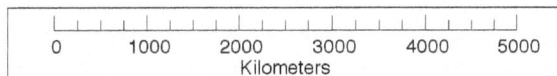

0   1000   2000   3000   4000   5000
Kilometers

# Partial Solar Eclipse of 2040 Nov 04

Greatest Eclipse = 19:09:02.0 TD (= 19:07:43.0 UT1)

Eclipse Magnitude = 0.8074          Saros Series = 124
Gamma = 1.0993                      Saros Member = 56 of 73

**Sun at Greatest Eclipse**
(Geocentric Coordinates)
R.A. = 14h42m06.9s
Dec. = -15°43'53.8"
S.D. = 00°16'07.7"
H.P. = 00°00'08.9"

**Moon at Greatest Eclipse**
(Geocentric Coordinates)
R.A. = 14h43m50.8s
Dec. = -14°45'19.8"
S.D. = 00°15'49.8"
H.P. = 00°58'05.7"

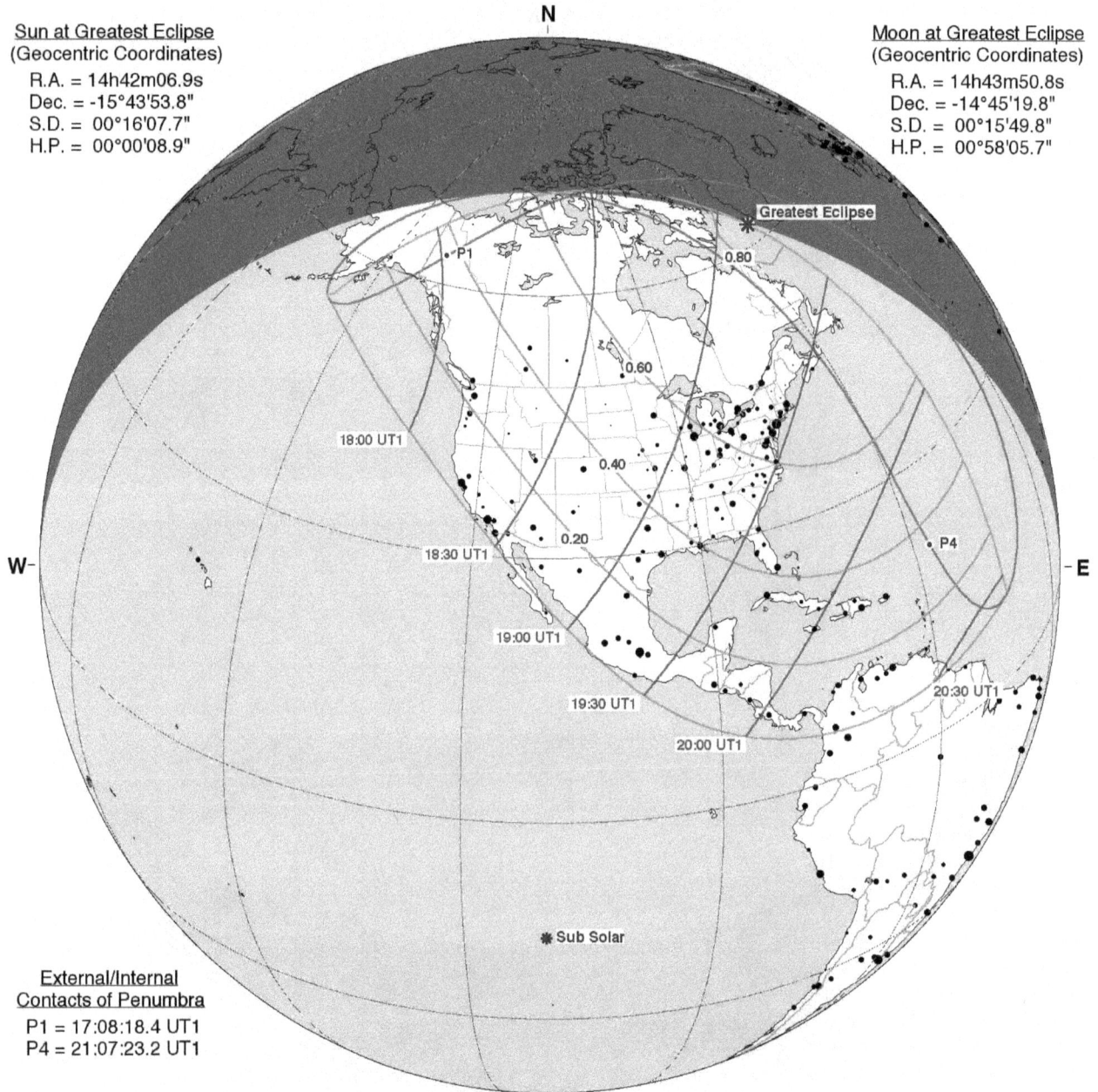

N

Greatest Eclipse
0.80
P1
0.60
18:00 UT1
0.40
18:30 UT1
0.20
P4
19:00 UT1
19:30 UT1
20:30 UT1
20:00 UT1

W

E

* Sub Solar

**External/Internal
Contacts of Penumbra**
P1 = 17:08:18.4 UT1
P4 = 21:07:23.2 UT1

ΔT = 79.0 s          S          Eph. = JPL DE405

**Circumstances at Greatest Eclipse: 19:07:43.0 UT1**
Lat. = 62°12.1'N          Sun Alt. = 0.0°
Long. = 053°24.2'W        Sun Azm. = 234.4°

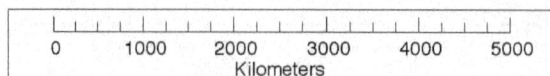

| 0 | 1000 | 2000 | 3000 | 4000 | 5000 |
Kilometers

©2016 F. Espenak
www.EclipseWise.com

# Total Solar Eclipse of  2041 Apr 30

Greatest Eclipse =  11:52:20.8 TD   (= 11:51:01.5 UT1)

Eclipse Magnitude =  1.0189          Saros Series =  129

Gamma = -0.4492          Saros Member =  53 of 80

Sun at Greatest Eclipse
(Geocentric Coordinates)
R.A. = 02h32m22.2s
Dec. = +14°58'18.8"
S.D. = 00°15'52.6"
H.P. = 00°00'08.7"

Moon at Greatest Eclipse
(Geocentric Coordinates)
R.A. = 02h33m06.0s
Dec. = +14°34'20.1"
S.D. = 00°15'56.6"
H.P. = 00°58'30.8"

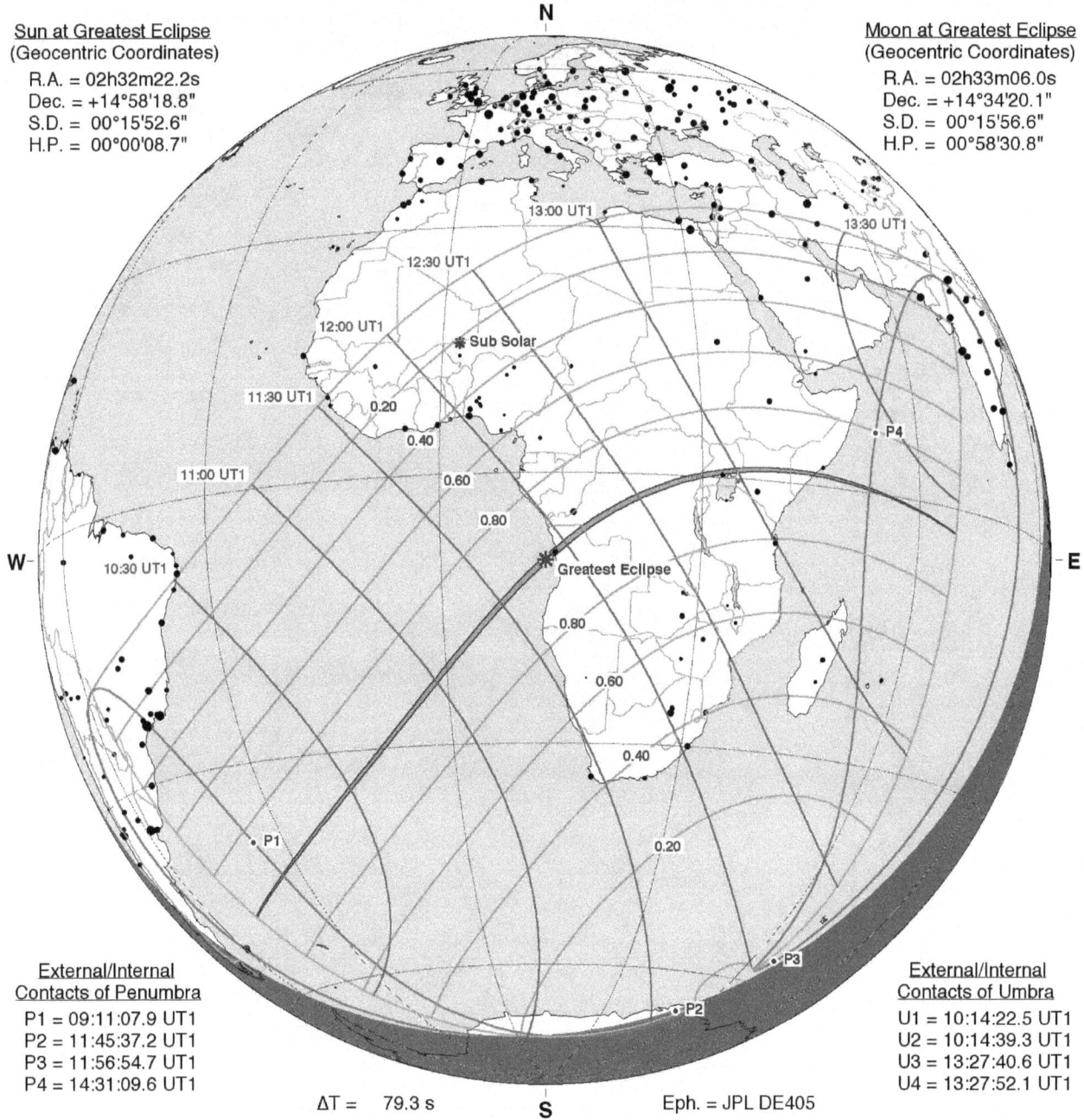

External/Internal
Contacts of Penumbra
P1 = 09:11:07.9 UT1
P2 = 11:45:37.2 UT1
P3 = 11:56:54.7 UT1
P4 = 14:31:09.6 UT1

External/Internal
Contacts of Umbra
U1 = 10:14:22.5 UT1
U2 = 10:14:39.3 UT1
U3 = 13:27:40.6 UT1
U4 = 13:27:52.1 UT1

ΔT =   79.3 s          Eph. = JPL DE405

Circumstances at Greatest Eclipse:  11:51:01.5 UT1

| | |
|---|---|
| Lat. =  09°37.2'S | Sun Alt. =  63.2° |
| Long. = 012°09.7'E | Sun Azm. = 336.7° |
| Path Width =  71.7 km | Duration =  01m50.6s |

Circumstances at Greatest Duration:  11:52:09.5 UT1

| | |
|---|---|
| Lat. =  09°22.9'S | Sun Alt. =  63.2° |
| Long. = 012°26.2'E | Sun Azm. = 335.4° |
| Path Width =  71.7 km | Duration =  01m50.6s |

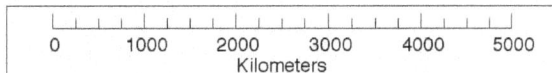

©2016  F. Espenak
www.EclipseWise.com

145

# Annular Solar Eclipse of 2041 Oct 25

Greatest Eclipse = 01:36:21.7 TD  (= 01:35:02.2 UT1)

| | |
|---|---|
| Eclipse Magnitude = 0.9467 | Saros Series = 134 |
| Gamma = 0.4133 | Saros Member = 45 of 71 |

**Sun at Greatest Eclipse**
(Geocentric Coordinates)
R.A. = 13h59m22.0s
Dec. = -12°10'20.1"
S.D. = 00°16'04.9"
H.P. = 00°00'08.8"

**Moon at Greatest Eclipse**
(Geocentric Coordinates)
R.A. = 14h00m02.5s
Dec. = -11°49'54.3"
S.D. = 00°15'00.8"
H.P. = 00°55'06.0"

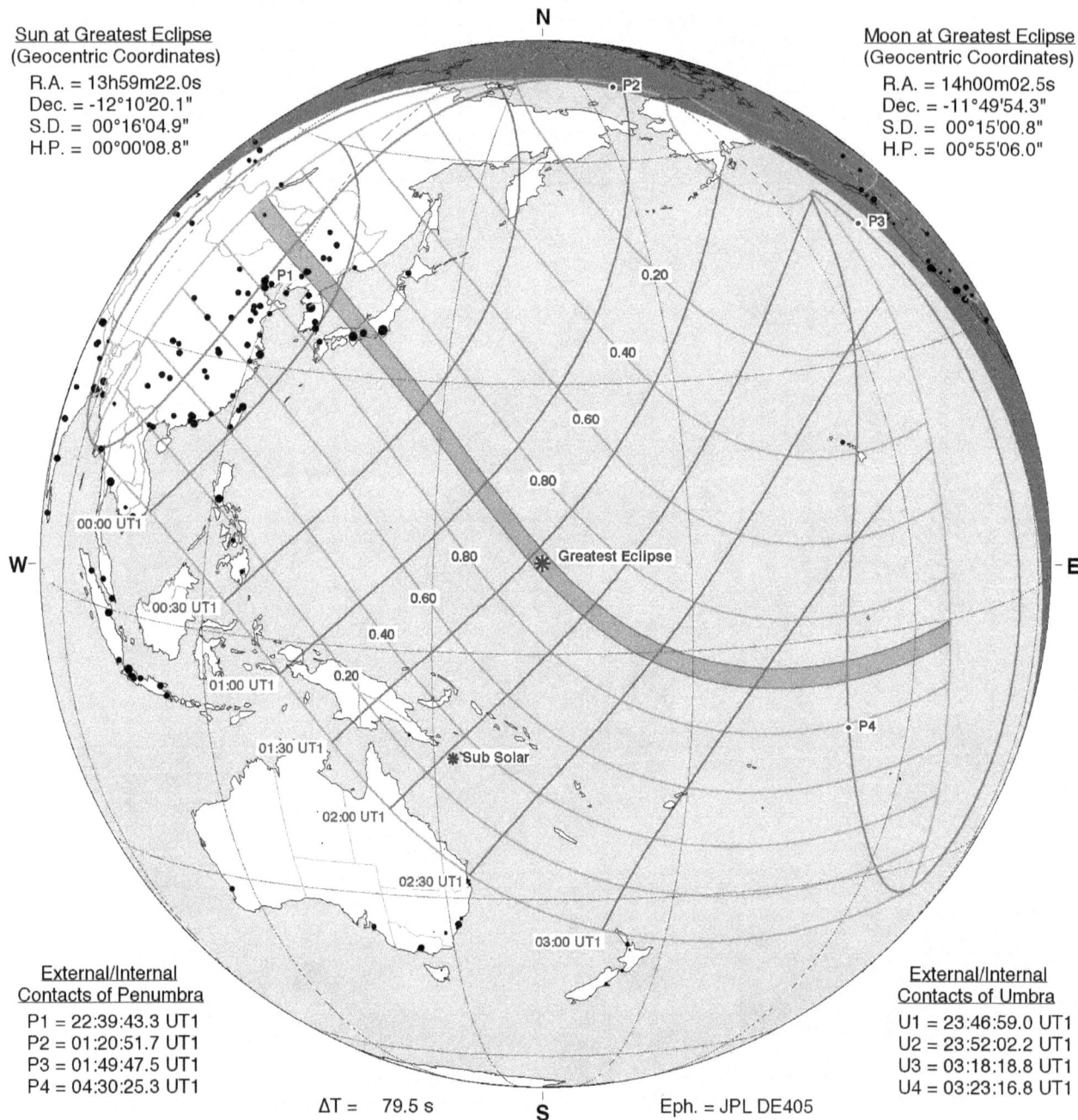

**External/Internal Contacts of Penumbra**
P1 = 22:39:43.3 UT1
P2 = 01:20:51.7 UT1
P3 = 01:49:47.5 UT1
P4 = 04:30:25.3 UT1

**External/Internal Contacts of Umbra**
U1 = 23:46:59.0 UT1
U2 = 23:52:02.2 UT1
U3 = 03:18:18.8 UT1
U4 = 03:23:16.8 UT1

ΔT = 79.5 s          Eph. = JPL DE405

**Circumstances at Greatest Eclipse: 01:35:02.2 UT1**

| | |
|---|---|
| Lat. = 09°56.0'N | Sun Alt. = 65.5° |
| Long. = 162°49.8'E | Sun Azm. = 205.6° |
| Path Width = 213.1 km | Duration = 06m07.1s |

**Circumstances at Greatest Duration: 01:50:35.9 UT1**

| | |
|---|---|
| Lat. = 06°42.8'N | Sun Alt. = 64.2° |
| Long. = 166°01.4'E | Sun Azm. = 223.0° |
| Path Width = 218.3 km | Duration = 06m08.4s |

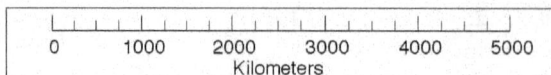

©2016 F. Espenak
www.EclipseWise.com

# Total Solar Eclipse of 2042 Apr 20

Greatest Eclipse = 02:17:30.1 TD  (= 02:16:10.4 UT1)

| | |
|---|---|
| Eclipse Magnitude = 1.0614 | Saros Series = 139 |
| Gamma = 0.2956 | Saros Member = 31 of 71 |

**Sun at Greatest Eclipse**
(Geocentric Coordinates)
R.A. = 01h52m12.4s
Dec. = +11°31'19.4"
S.D. = 00°15'55.3"
H.P. = 00°00'08.8"

**Moon at Greatest Eclipse**
(Geocentric Coordinates)
R.A. = 01h51m39.9s
Dec. = +11°47'27.9"
S.D. = 00°16'37.6"
H.P. = 01°01'01.4"

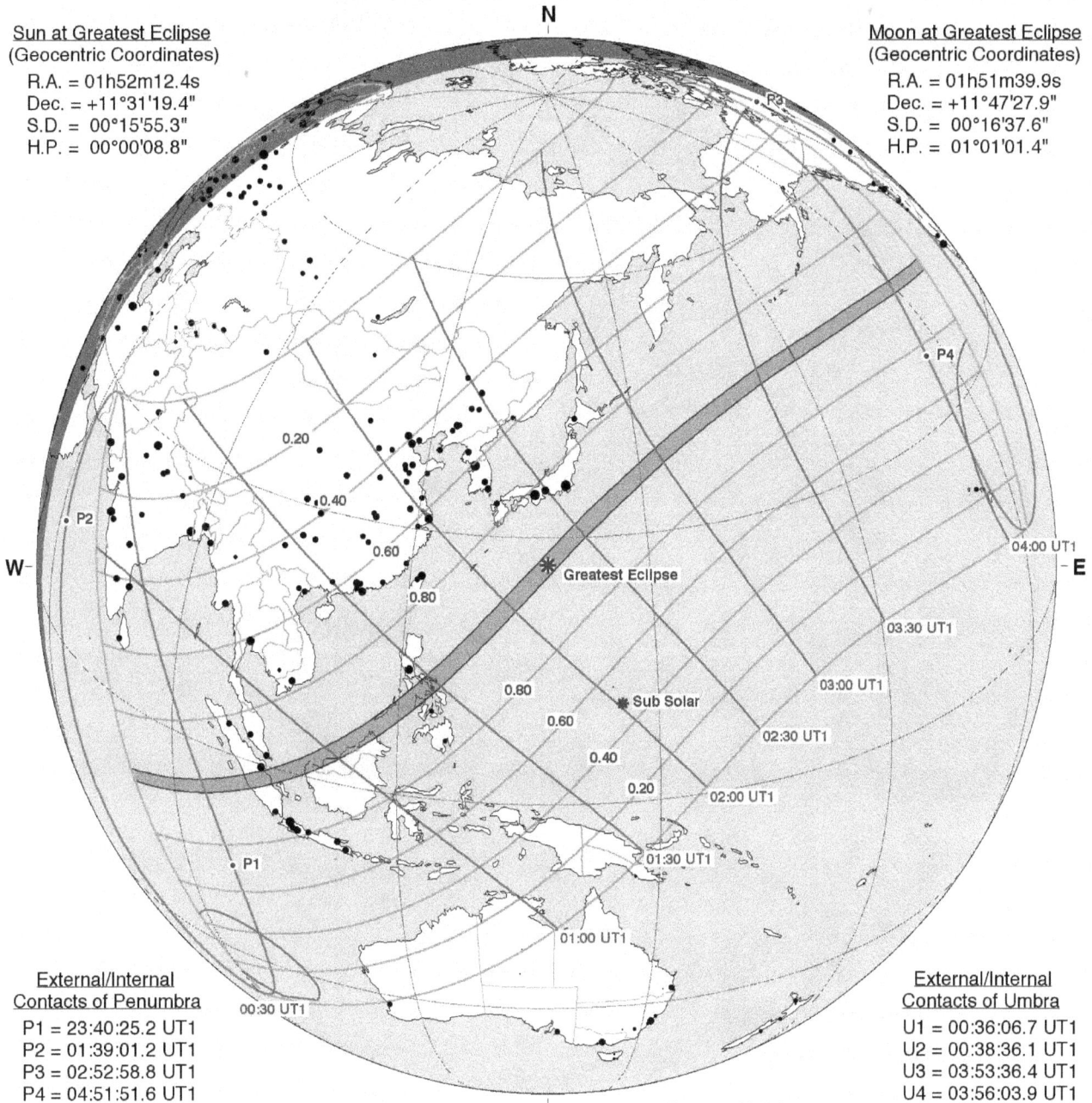

N

P3

P4

0.20

0.40

0.60

0.80

P2

W — 

— E

04:00 UT1

Greatest Eclipse

03:30 UT1

0.80

03:00 UT1

0.60

Sub Solar

02:30 UT1

0.40

0.20

02:00 UT1

P1

01:30 UT1

01:00 UT1

00:30 UT1

**External/Internal**
**Contacts of Penumbra**
P1 = 23:40:25.2 UT1
P2 = 01:39:01.2 UT1
P3 = 02:52:58.8 UT1
P4 = 04:51:51.6 UT1

ΔT =   79.8 s

S

Eph. = JPL DE405

**External/Internal**
**Contacts of Umbra**
U1 = 00:36:06.7 UT1
U2 = 00:38:36.1 UT1
U3 = 03:53:36.4 UT1
U4 = 03:56:03.9 UT1

**Circumstances at Greatest Eclipse:  02:16:10.4 UT1**

| | |
|---|---|
| Lat. = 26°57.2'N | Sun Alt. =  72.7° |
| Long. = 137°17.0'E | Sun Azm. = 151.2° |
| Path Width = 210.4 km | Duration = 04m51.0s |

**Circumstances at Greatest Duration:  02:20:00.4 UT1**

| | |
|---|---|
| Lat. = 27°59.1'N | Sun Alt. =  72.5° |
| Long. = 138°25.2'E | Sun Azm. = 159.0° |
| Path Width = 209.7 km | Duration = 04m51.2s |

©2016  F. Espenak
www.EclipseWise.com

| | | | | | |
|---|---|---|---|---|---|
| 0 | 1000 | 2000 | 3000 | 4000 | 5000 |

Kilometers

# Annular Solar Eclipse of 2042 Oct 14

Greatest Eclipse = 02:00:41.9 TD (= 01:59:21.8 UT1)

| | |
|---|---|
| Eclipse Magnitude = 0.9301 | Saros Series = 144 |
| Gamma = -0.3030 | Saros Member = 18 of 70 |

**Sun at Greatest Eclipse**
(Geocentric Coordinates)
R.A. = 13h17m05.8s
Dec. = -08°08'35.1"
S.D. = 00°16'01.9"
H.P. = 00°00'08.8"

**Moon at Greatest Eclipse**
(Geocentric Coordinates)
R.A. = 13h16m35.0s
Dec. = -08°23'00.1"
S.D. = 00°14'41.9"
H.P. = 00°53'56.6"

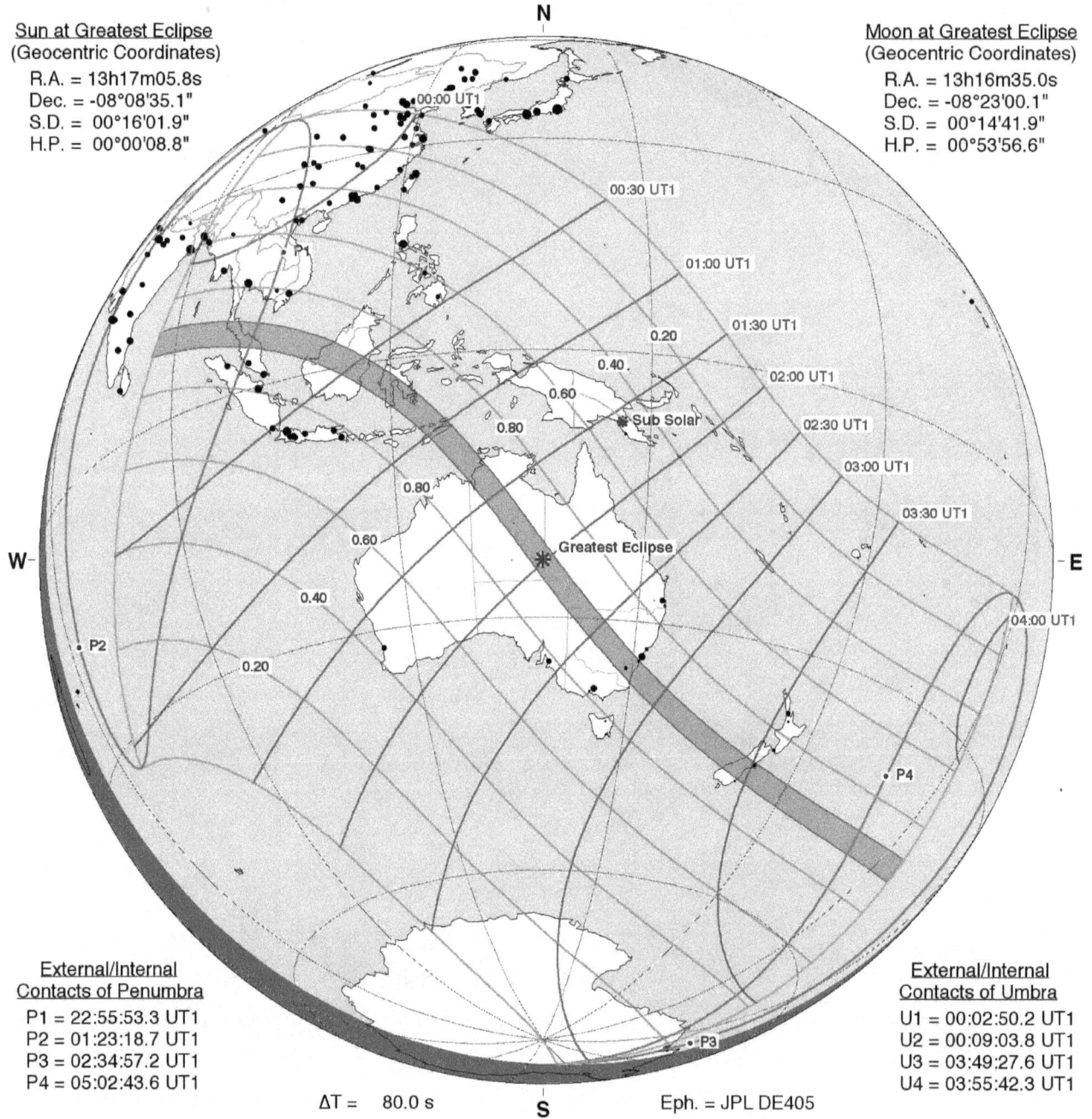

**External/Internal Contacts of Penumbra**
P1 = 22:55:53.3 UT1
P2 = 01:23:18.7 UT1
P3 = 02:34:57.2 UT1
P4 = 05:02:43.6 UT1

**External/Internal Contacts of Umbra**
U1 = 00:02:50.2 UT1
U2 = 00:09:03.8 UT1
U3 = 03:49:27.6 UT1
U4 = 03:55:42.3 UT1

ΔT = 80.0 s     Eph. = JPL DE405

**Circumstances at Greatest Eclipse: 01:59:21.8 UT1**

| | |
|---|---|
| Lat. = 23°44.7'S | Sun Alt. = 72.2° |
| Long. = 137°47.4'E | Sun Azm. = 30.1° |
| Path Width = 273.3 km | Duration = 07m44.2s |

**Circumstances at Greatest Duration: 02:13:39.4 UT1**

| | |
|---|---|
| Lat. = 27°13.1'S | Sun Alt. = 70.8° |
| Long. = 140°49.9'E | Sun Azm. = 6.8° |
| Path Width = 270.8 km | Duration = 07m45.4s |

0    1000    2000    3000    4000    5000
Kilometers

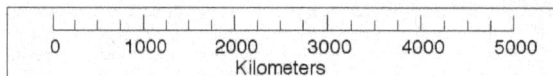

©2016  F. Espenak
www.EclipseWise.com

# Total Solar Eclipse of 2043 Apr 09

Greatest Eclipse = 18:57:49.4 TD (= 18:56:29.1 UT1)

| | |
|---|---|
| Eclipse Magnitude = 1.0096 | Saros Series = 149 |
| Gamma = 1.0031 | Saros Member = 22 of 71 |

Sun at Greatest Eclipse
(Geocentric Coordinates)
R.A. = 01h13m12.2s
Dec. = +07°45'05.1"
S.D. = 00°15'58.1"
H.P. = 00°00'08.8"

Moon at Greatest Eclipse
(Geocentric Coordinates)
R.A. = 01h11m17.3s
Dec. = +08°39'09.1"
S.D. = 00°16'38.0"
H.P. = 01°01'02.7"

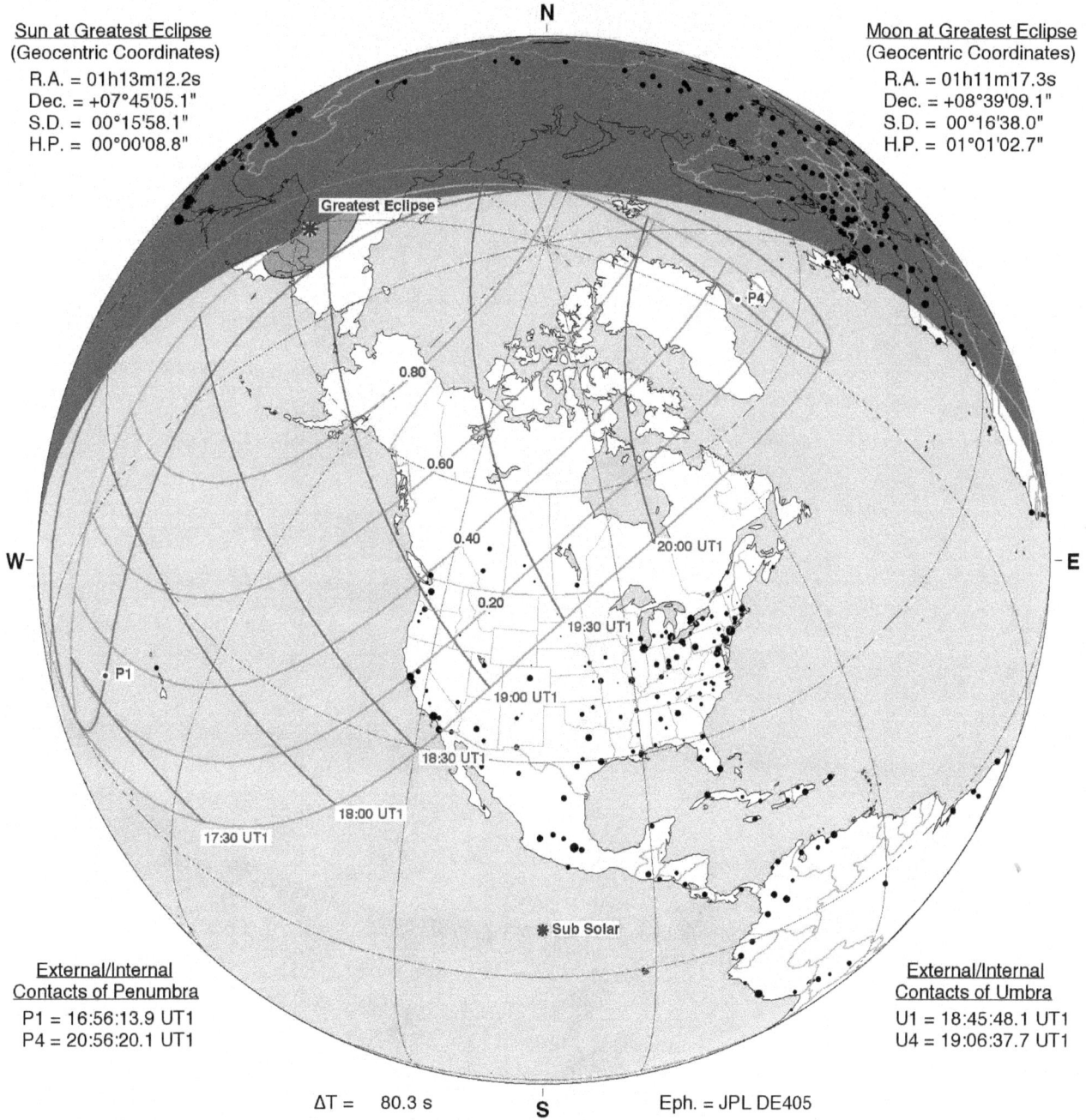

N

Greatest Eclipse

0.80
0.60
0.40
0.20

20:00 UT1
19:30 UT1
19:00 UT1
18:30 UT1
18:00 UT1
17:30 UT1

P4
P1

W — — E

Sub Solar

External/Internal
Contacts of Penumbra
P1 = 16:56:13.9 UT1
P4 = 20:56:20.1 UT1

External/Internal
Contacts of Umbra
U1 = 18:45:48.1 UT1
U4 = 19:06:37.7 UT1

ΔT = 80.3 s

S

Eph. = JPL DE405

Circumstances at Greatest Eclipse: 18:56:29.1 UT1

| | |
|---|---|
| Lat. = 61°18.7'N | Sun Alt. = 0.0° |
| Long. = 151°56.2'E | Sun Azm. = 73.7° |
| Path Width = 0.0 km | Duration = 00m00.0s |

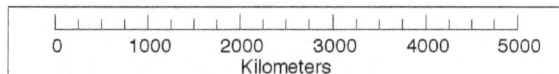

0 1000 2000 3000 4000 5000
Kilometers

©2016 F. Espenak
www.EclipseWise.com

149

# Annular Solar Eclipse of 2043 Oct 03

Greatest Eclipse = 03:01:48.9 TD  (= 03:00:28.4 UT1)

| | |
|---|---|
| Eclipse Magnitude = 0.9497 | Saros Series = 154 |
| Gamma = -1.0102 | Saros Member = 8 of 71 |

**Sun at Greatest Eclipse**
(Geocentric Coordinates)
R.A. = 12h36m02.9s
Dec. = -03°53'04.6"
S.D. = 00°15'58.8"
H.P. = 00°00'08.8"

**Moon at Greatest Eclipse**
(Geocentric Coordinates)
R.A. = 12h34m15.0s
Dec. = -04°41'56.9"
S.D. = 00°15'05.1"
H.P. = 00°55'21.7"

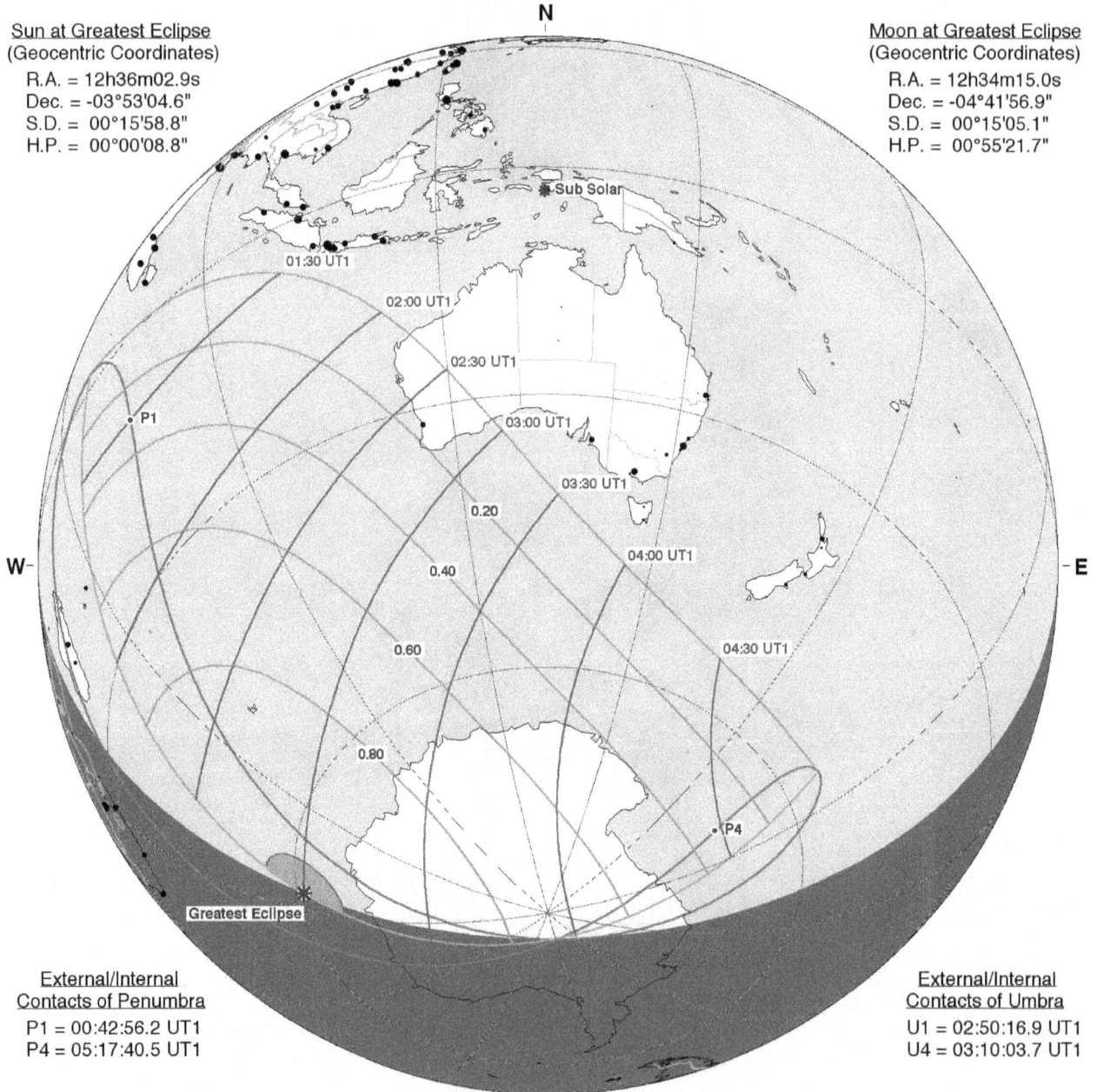

N

Sub Solar

01:30 UT1
02:00 UT1
02:30 UT1
03:00 UT1
03:30 UT1
04:00 UT1
04:30 UT1

P1

0.20
0.40
0.60
0.80

P4

W
E

Greatest Eclipse

**External/Internal**
**Contacts of Penumbra**
P1 = 00:42:56.2 UT1
P4 = 05:17:40.5 UT1

**External/Internal**
**Contacts of Umbra**
U1 = 02:50:16.9 UT1
U4 = 03:10:03.7 UT1

ΔT = 80.5 s

S

Eph. = JPL DE405

**Circumstances at Greatest Eclipse: 03:00:28.4 UT1**

| | |
|---|---|
| Lat. = 60°57.6'S | Sun Alt. = 0.0° |
| Long. = 035°14.0'E | Sun Azm. = 98.0° |
| Path Width = 0.0 km | Duration = 00m00.0s |

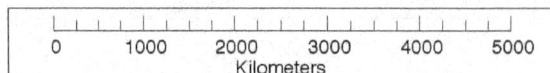

0 1000 2000 3000 4000 5000
Kilometers

# Annular Solar Eclipse of 2044 Feb 28

Greatest Eclipse = 20:24:39.5 TD   (= 20:23:18.7 UT1)

| Eclipse Magnitude = 0.9600 | Saros Series = 121 |
|---|---|
| Gamma = -0.9954 | Saros Member = 62 of 71 |

Sun at Greatest Eclipse
(Geocentric Coordinates)
R.A. = 22h45m44.1s
Dec. = -07°51'30.6"
S.D. = 00°16'08.8"
H.P. = 00°00'08.9"

Moon at Greatest Eclipse
(Geocentric Coordinates)
R.A. = 22h47m30.6s
Dec. = -08°41'25.7"
S.D. = 00°15'29.6"
H.P. = 00°56'51.8"

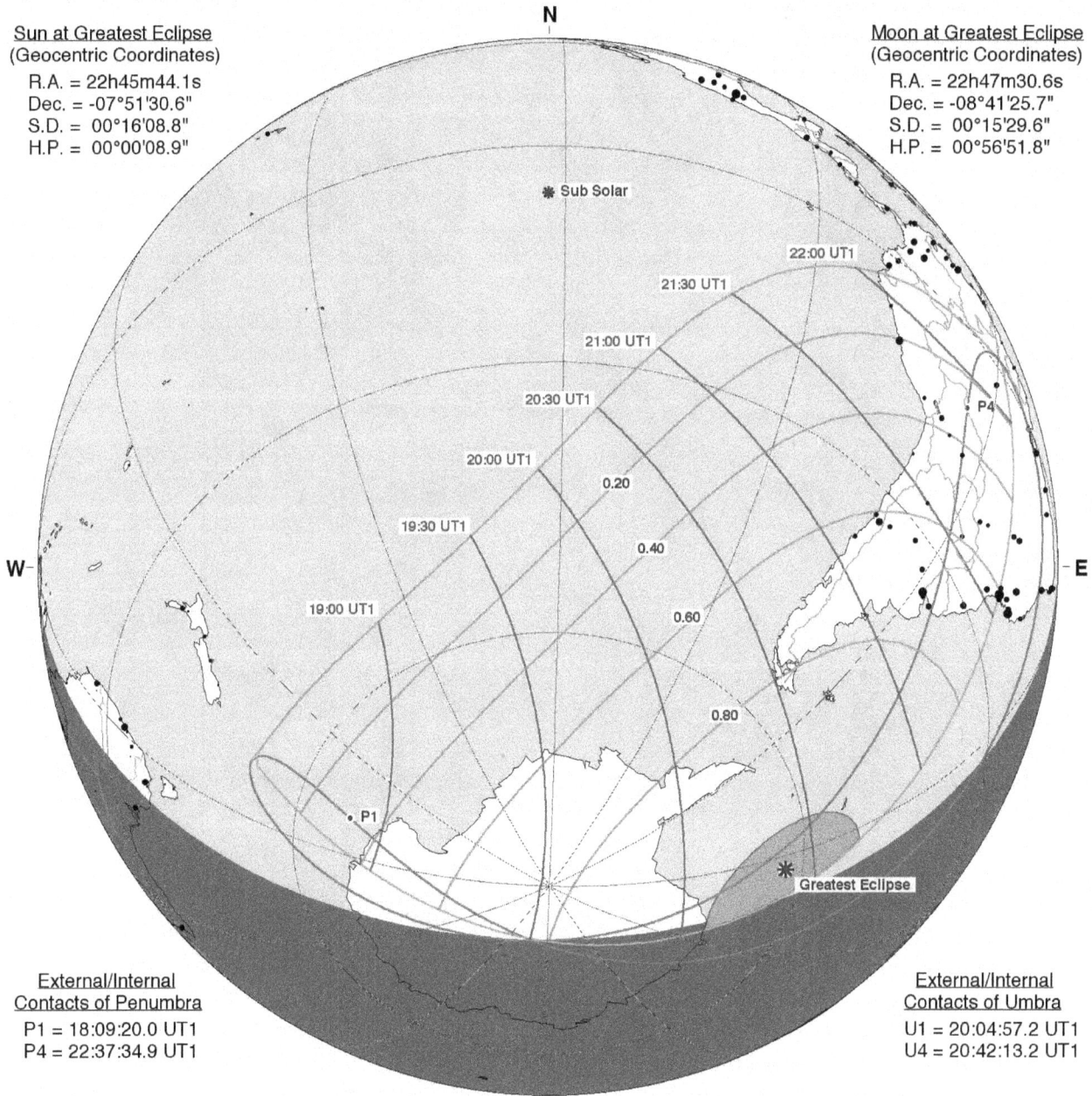

N

☀ Sub Solar

22:00 UT1
21:30 UT1
21:00 UT1
20:30 UT1
20:00 UT1
19:30 UT1
19:00 UT1

0.20
0.40
0.60
0.80

W — — E

P4

P1

☀ Greatest Eclipse

S

External/Internal
Contacts of Penumbra
P1 = 18:09:20.0 UT1
P4 = 22:37:34.9 UT1

External/Internal
Contacts of Umbra
U1 = 20:04:57.2 UT1
U4 = 20:42:13.2 UT1

ΔT = 80.8 s          Eph. = JPL DE405

Circumstances at Greatest Eclipse: 20:23:18.7 UT1

| Lat. = 62°10.9'S | Sun Alt. = 3.7° |
|---|---|
| Long. = 025°38.6'W | Sun Azm. = 260.1° |
| Path Width = 2237.9 km | Duration = 02m27.2s |

Circumstances at Greatest Duration: 20:16:25.2 UT1

| Lat. = 64°41.7'S | Sun Alt. = 0.0° |
|---|---|
| Long. = 014°00.7'W | Sun Azm. = 251.4° |
| Path Width = 2582.2 km | Duration = 02m27.4s |

©2016 F. Espenak
www.EclipseWise.com

| 0 | 1000 | 2000 | 3000 | 4000 | 5000 |
|---|---|---|---|---|---|

Kilometers

# Total Solar Eclipse of 2044 Aug 23

Greatest Eclipse = 01:17:01.7 TD  (= 01:15:40.6 UT1)

| | |
|---|---|
| Eclipse Magnitude = 1.0364 | Saros Series = 126 |
| Gamma = 0.9613 | Saros Member = 49 of 72 |

**Sun at Greatest Eclipse**
(Geocentric Coordinates)
R.A. = 10h10m33.4s
Dec. = +11°16'02.2"
S.D. = 00°15'48.9"
H.P. = 00°00'08.7"

**Moon at Greatest Eclipse**
(Geocentric Coordinates)
R.A. = 10h12m17.2s
Dec. = +12°07'34.4"
S.D. = 00°16'19.6"
H.P. = 00°59'55.1"

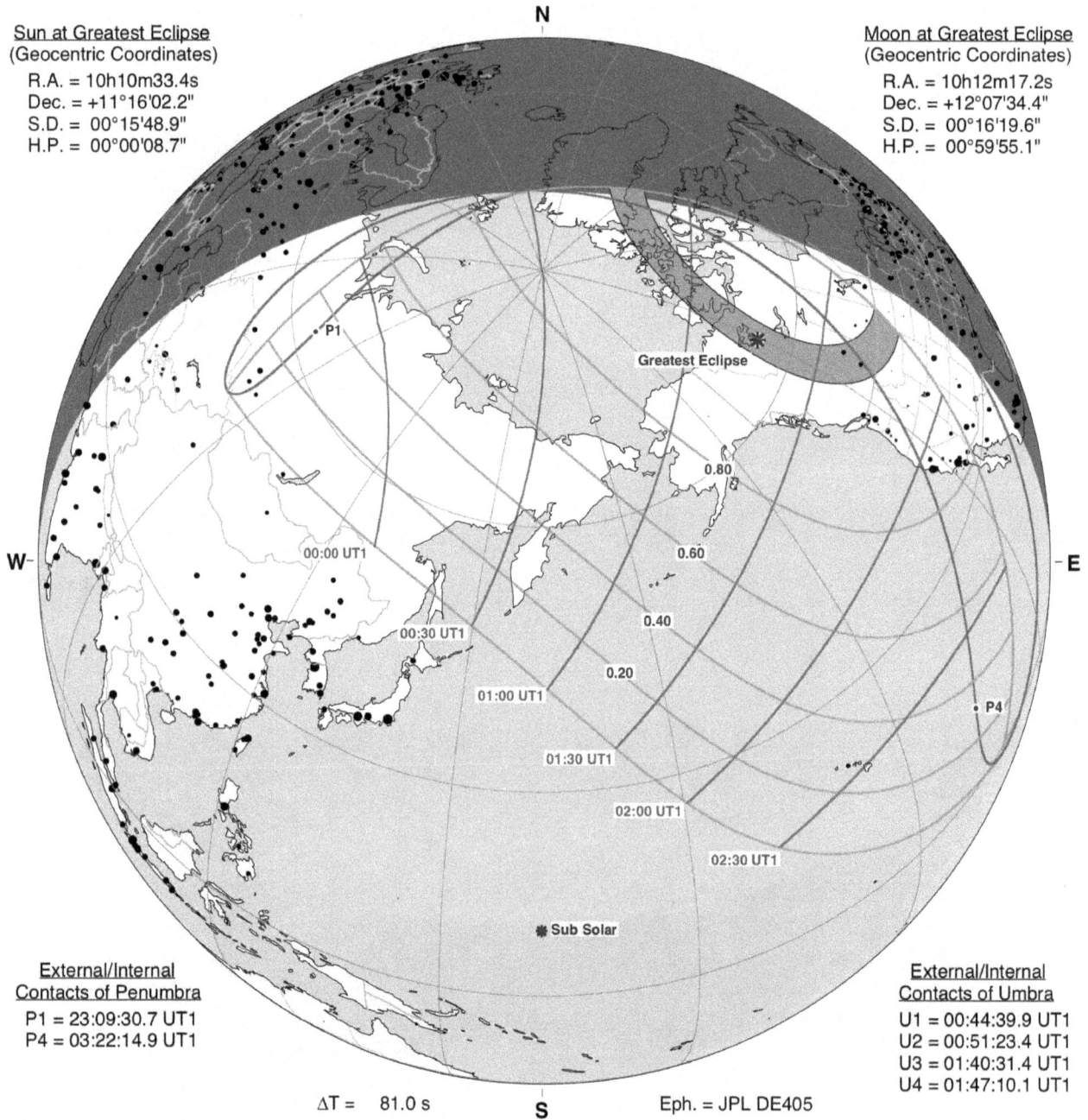

N

P1

Greatest Eclipse

0.80

0.60

0.40

0.20

00:00 UT1

00:30 UT1

01:00 UT1

01:30 UT1

02:00 UT1

02:30 UT1

W

E

P4

* Sub Solar

**External/Internal Contacts of Penumbra**
P1 = 23:09:30.7 UT1
P4 = 03:22:14.9 UT1

**External/Internal Contacts of Umbra**
U1 = 00:44:39.9 UT1
U2 = 00:51:23.4 UT1
U3 = 01:40:31.4 UT1
U4 = 01:47:10.1 UT1

ΔT = 81.0 s

S

Eph. = JPL DE405

**Circumstances at Greatest Eclipse: 01:15:40.6 UT1**

| | |
|---|---|
| Lat. = 64°20.1'N | Sun Alt. = 15.4° |
| Long. = 120°28.6'W | Sun Azm. = 263.9° |
| Path Width = 452.7 km | Duration = 02m03.9s |

**Circumstances at Greatest Duration: 01:15:14.8 UT1**

| | |
|---|---|
| Lat. = 64°34.1'N | Sun Alt. = 15.4° |
| Long. = 120°25.0'W | Sun Azm. = 263.8° |
| Path Width = 451.6 km | Duration = 02m03.9s |

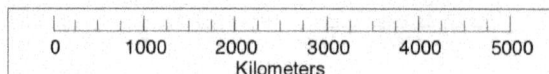

©2016  F. Espenak
www.EclipseWise.com

| 0 | 1000 | 2000 | 3000 | 4000 | 5000 |
|---|---|---|---|---|---|

Kilometers

# Annular Solar Eclipse of 2045 Feb 16

Greatest Eclipse = 23:56:06.6 TD  (= 23:54:45.3 UT1)

Eclipse Magnitude = 0.9285          Saros Series = 131
Gamma = -0.3125          Saros Member = 52 of 70

Sun at Greatest Eclipse
(Geocentric Coordinates)
R.A. = 22h03m27.1s
Dec. = -11°55'04.8"
S.D. = 00°16'11.2"
H.P. = 00°00'08.9"

Moon at Greatest Eclipse
(Geocentric Coordinates)
R.A. = 22h03m57.6s
Dec. = -12°10'17.7"
S.D. = 00°14'48.9"
H.P. = 00°54'22.2"

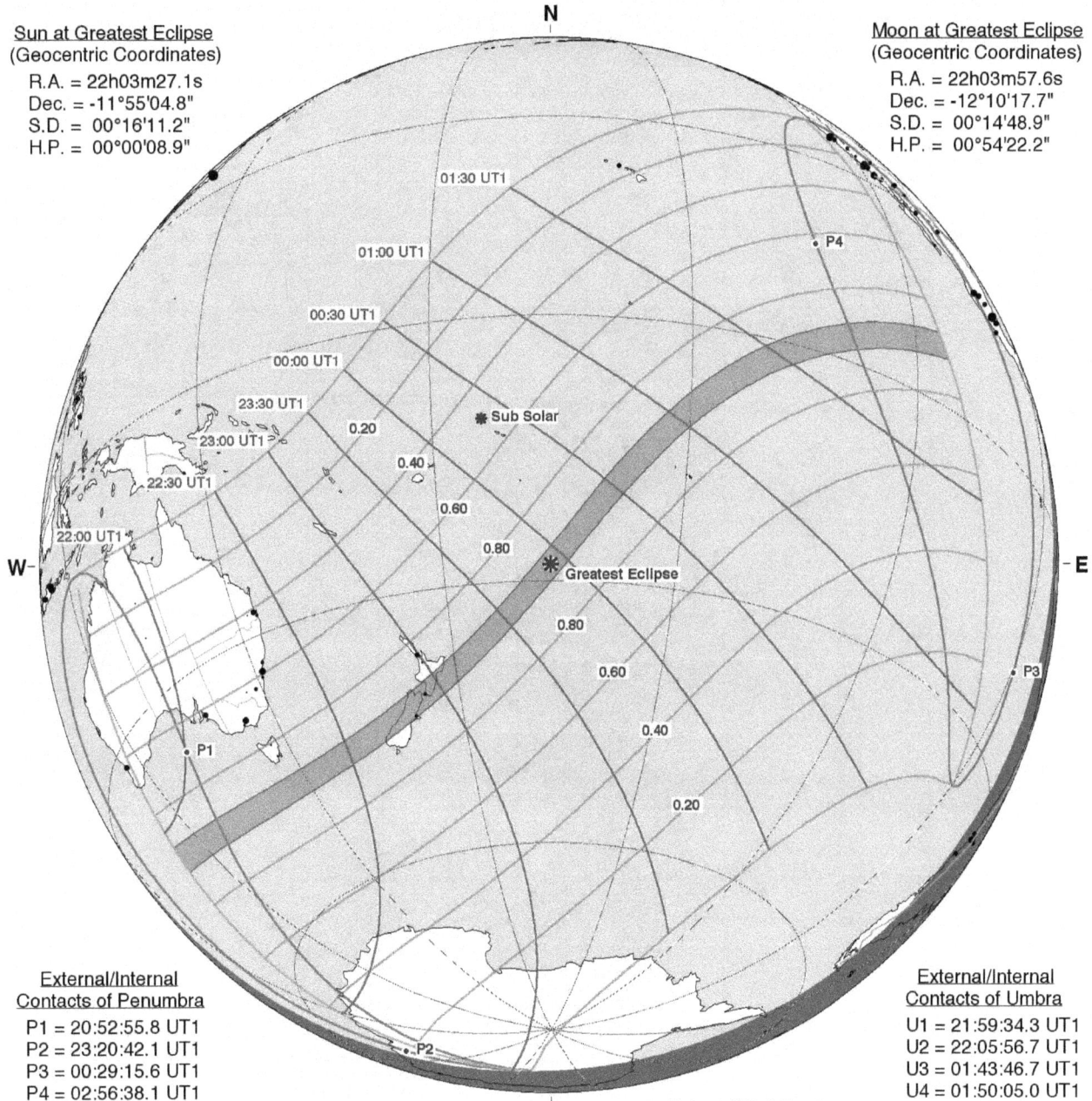

External/Internal
Contacts of Penumbra
P1 = 20:52:55.8 UT1
P2 = 23:20:42.1 UT1
P3 = 00:29:15.6 UT1
P4 = 02:56:38.1 UT1

External/Internal
Contacts of Umbra
U1 = 21:59:34.3 UT1
U2 = 22:05:56.7 UT1
U3 = 01:43:46.7 UT1
U4 = 01:50:05.0 UT1

ΔT = 81.3 s

Eph. = JPL DE405

Circumstances at Greatest Eclipse: 23:54:45.3 UT1
Lat. = 28°15.4'S          Sun Alt. = 71.6°
Long. = 166°13.8'W          Sun Azm. = 331.0°
Path Width = 281.2 km          Duration = 07m46.5s

Circumstances at Greatest Duration: 23:37:15.9 UT1
Lat. = 32°20.8'S          Sun Alt. = 69.6°
Long. = 170°31.5'W          Sun Azm. = 359.1°
Path Width = 278.7 km          Duration = 07m48.7s

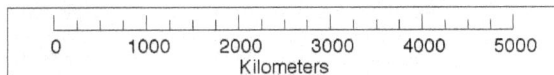

©2016 F. Espenak
www.EclipseWise.com

0    1000    2000    3000    4000    5000
Kilometers

# Total Solar Eclipse of 2045 Aug 12

Greatest Eclipse = 17:42:39.1 TD (= 17:41:17.5 UT1)

| | |
|---|---|
| Eclipse Magnitude = 1.0774 | Saros Series = 136 |
| Gamma = 0.2116 | Saros Member = 39 of 71 |

**Sun at Greatest Eclipse**
(Geocentric Coordinates)
R.A. = 09h31m17.7s
Dec. = +14°40'40.5"
S.D. = 00°15'47.0"
H.P. = 00°00'08.7"

**Moon at Greatest Eclipse**
(Geocentric Coordinates)
R.A. = 09h31m39.7s
Dec. = +14°52'29.9"
S.D. = 00°16'43.3"
H.P. = 01°01'22.3"

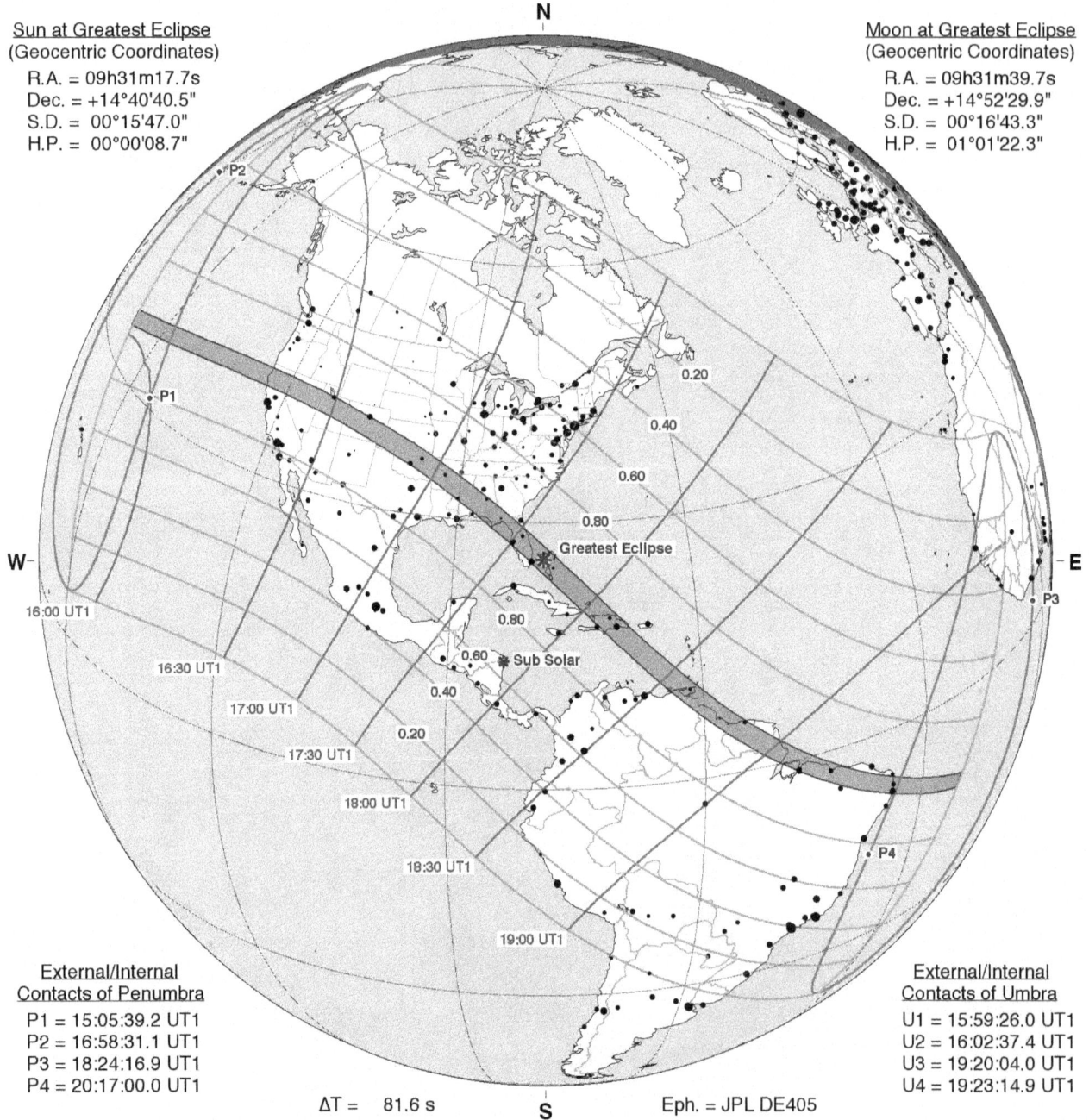

**External/Internal Contacts of Penumbra**
P1 = 15:05:39.2 UT1
P2 = 16:58:31.1 UT1
P3 = 18:24:16.9 UT1
P4 = 20:17:00.0 UT1

**External/Internal Contacts of Umbra**
U1 = 15:59:26.0 UT1
U2 = 16:02:37.4 UT1
U3 = 19:20:04.0 UT1
U4 = 19:23:14.9 UT1

ΔT = 81.6 s          Eph. = JPL DE405

**Circumstances at Greatest Eclipse: 17:41:17.5 UT1**
Lat. = 25°54.6'N          Sun Alt. = 77.6°
Long. = 078°34.1'W        Sun Azm. = 205.7°
Path Width = 255.6 km     Duration = 06m05.7s

**Circumstances at Greatest Duration: 17:35:29.1 UT1**
Lat. = 27°19.3'N          Sun Alt. = 77.2°
Long. = 080°22.9'W        Sun Azm. = 189.9°
Path Width = 254.8 km     Duration = 06m06.2s

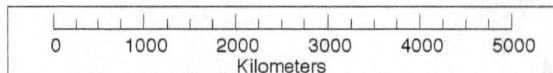

©2016 F. Espenak
www.EclipseWise.com

0   1000   2000   3000   4000   5000
Kilometers

# Annular Solar Eclipse of 2046 Feb 05

Greatest Eclipse = 23:06:26.2 TD   (= 23:05:04.3 UT1)

| Eclipse Magnitude = 0.9232 | Saros Series = 141 |
|---|---|
| Gamma = 0.3765 | Saros Member = 25 of 70 |

**N**

Sun at Greatest Eclipse
(Geocentric Coordinates)
R.A. = 21h19m00.8s
Dec. = -15°38'42.4"
S.D. = 00°16'13.2"
H.P. = 00°00'08.9"

Moon at Greatest Eclipse
(Geocentric Coordinates)
R.A. = 21h18m27.2s
Dec. = -15°20'02.1"
S.D. = 00°14'46.0"
H.P. = 00°54'11.7"

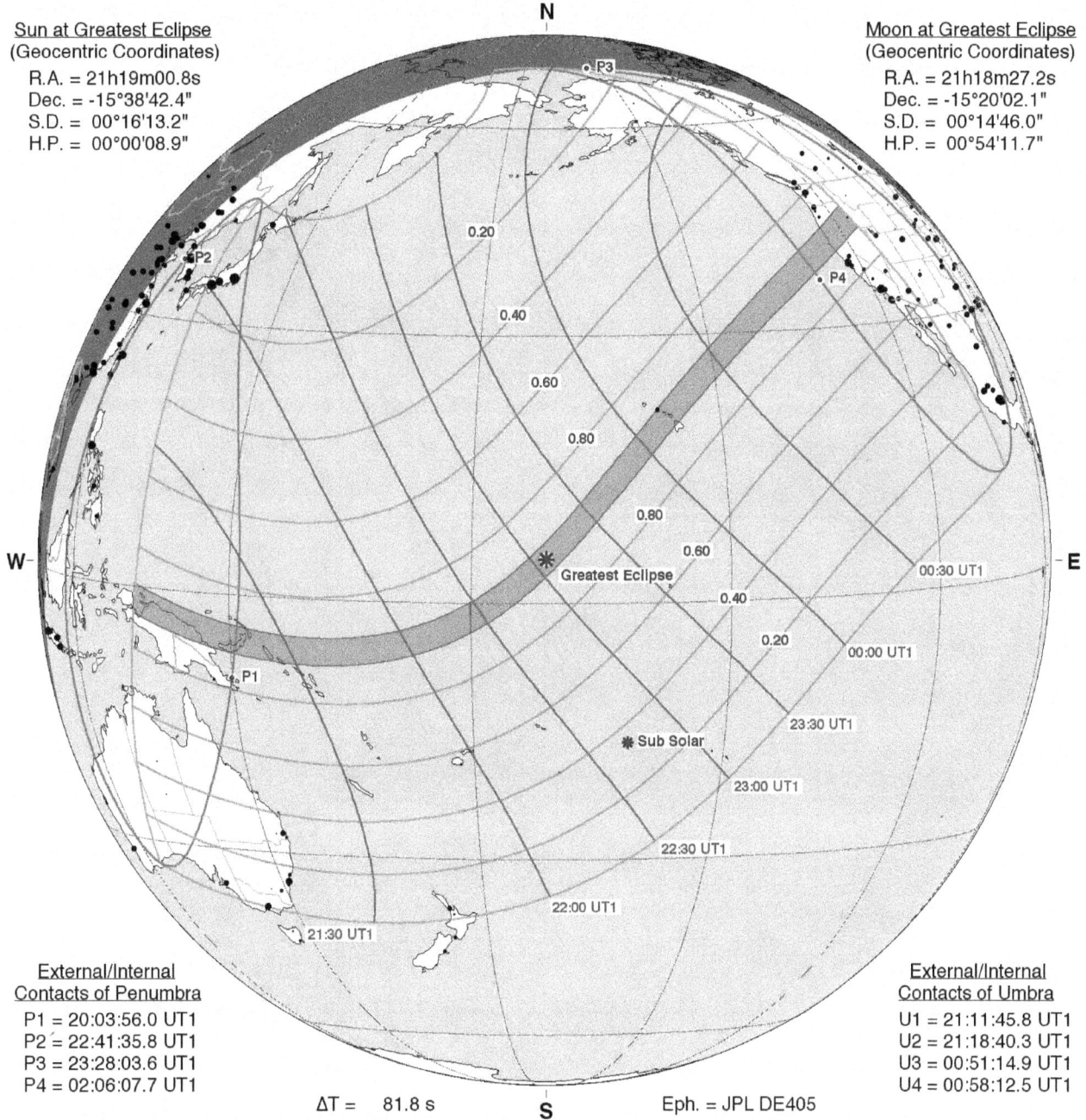

P3

0.20

0.40

0.60

0.80

P2

P4

0.80

0.60

**W**

Greatest Eclipse

00:30 UT1

**E**

0.40

0.20

00:00 UT1

P1

00:30 UT1

23:30 UT1

Sub Solar

23:00 UT1

22:30 UT1

22:00 UT1

21:30 UT1

External/Internal
Contacts of Penumbra
P1 = 20:03:56.0 UT1
P2 = 22:41:35.8 UT1
P3 = 23:28:03.6 UT1
P4 = 02:06:07.7 UT1

External/Internal
Contacts of Umbra
U1 = 21:11:45.8 UT1
U2 = 21:18:40.3 UT1
U3 = 00:51:14.9 UT1
U4 = 00:58:12.5 UT1

ΔT =    81.8 s

**S**

Eph. = JPL DE405

| Circumstances at Greatest Eclipse: 23:05:04.3 UT1 | | Circumstances at Greatest Duration: 22:49:00.6 UT1 | |
|---|---|---|---|
| Lat. = 04°46.9'N | Sun Alt. = 67.9° | Lat. = 01°56.1'N | Sun Alt. = 66.4° |
| Long. = 171°25.6'W | Sun Azm. = 157.4° | Long. = 174°40.8'W | Sun Azm. = 138.7° |
| Path Width = 310.1 km | Duration = 09m42.4s | Path Width = 316.7 km | Duration = 09m45.6s |

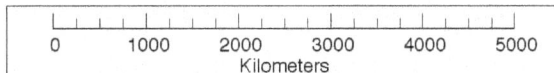

Kilometers
0    1000    2000    3000    4000    5000

©2016  F. Espenak
www.EclipseWise.com

155

# Total Solar Eclipse of 2046 Aug 02

Greatest Eclipse = 10:21:13.4 TD  (= 10:19:51.3 UT1)

Eclipse Magnitude = 1.0531          Saros Series = 146

Gamma = -0.5350          Saros Member = 29 of 76

**Sun at Greatest Eclipse**
(Geocentric Coordinates)
R.A. = 08h51m04.7s
Dec. = +17°39'03.1"
S.D. = 00°15'45.5"
H.P. = 00°00'08.7"

**Moon at Greatest Eclipse**
(Geocentric Coordinates)
R.A. = 08h50m16.1s
Dec. = +17°09'10.7"
S.D. = 00°16'21.8"
H.P. = 01°00'03.4"

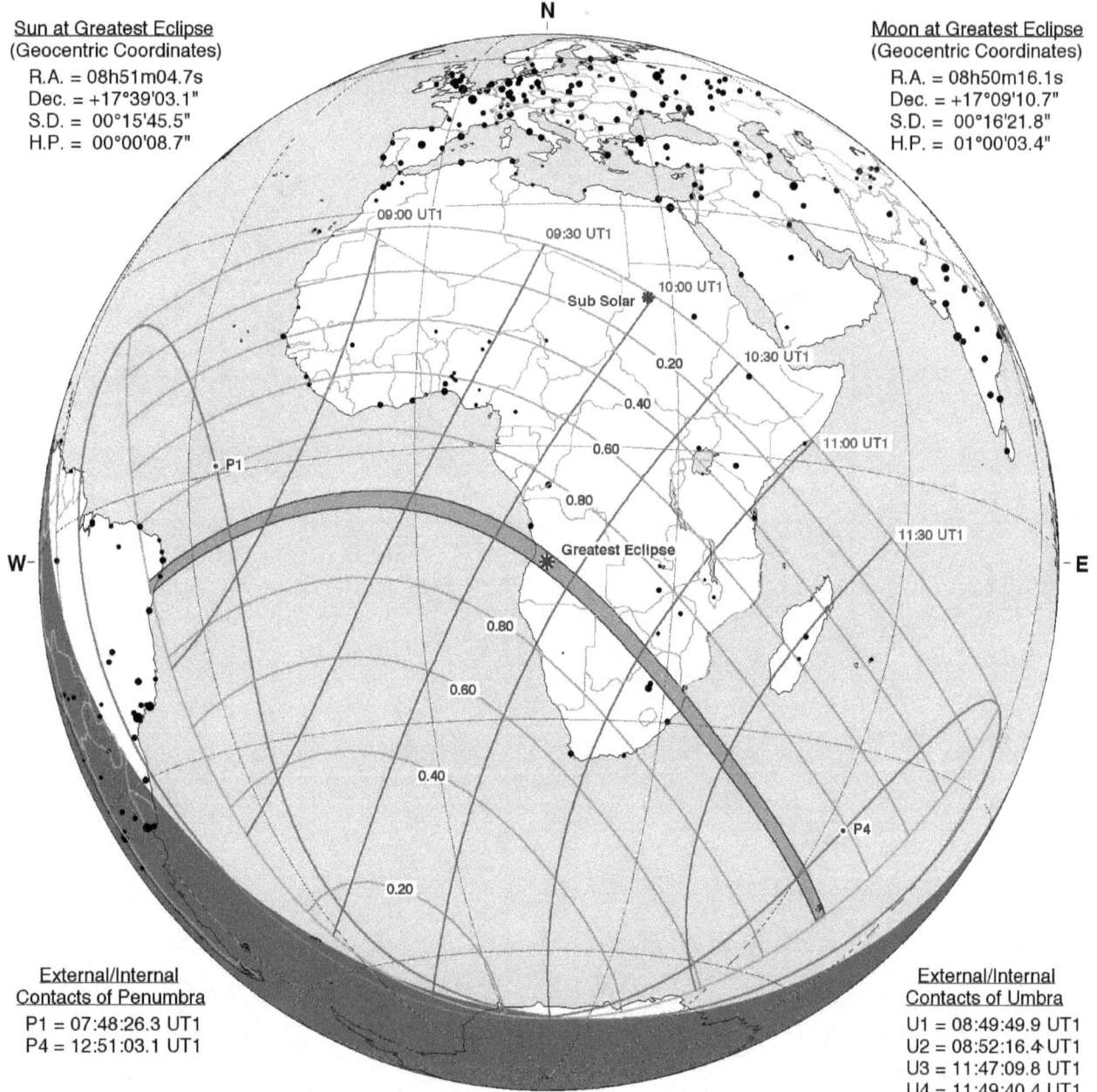

**External/Internal
Contacts of Penumbra**
P1 = 07:48:26.3 UT1
P4 = 12:51:03.1 UT1

**External/Internal
Contacts of Umbra**
U1 = 08:49:49.9 UT1
U2 = 08:52:16.4 UT1
U3 = 11:47:09.8 UT1
U4 = 11:49:40.4 UT1

ΔT = 82.1 s          Eph. = JPL DE405

**Circumstances at Greatest Eclipse: 10:19:51.3 UT1**

| | |
|---|---|
| Lat. = 12°43.8'S | Sun Alt. = 57.6° |
| Long. = 015°08.8'E | Sun Azm. = 20.7° |
| Path Width = 206.0 km | Duration = 04m51.4s |

**Circumstances at Greatest Duration: 10:16:03.2 UT1**

| | |
|---|---|
| Lat. = 12°00.3'S | Sun Alt. = 57.5° |
| Long. = 014°06.4'E | Sun Azm. = 24.4° |
| Path Width = 207.2 km | Duration = 04m51.6s |

0    1000    2000    3000    4000    5000
Kilometers

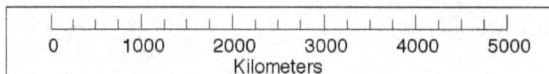

©2016 F. Espenak
www.EclipseWise.com

# Partial Solar Eclipse of 2047 Jan 26

Greatest Eclipse = 01:33:17.8 TD   (= 01:31:55.4 UT1)

Eclipse Magnitude = 0.8908          Saros Series = 151
Gamma = 1.0450                      Saros Member = 16 of 72

Sun at Greatest Eclipse
(Geocentric Coordinates)
R.A. = 20h33m28.4s
Dec. = -18°46'10.9"
S.D. = 00°16'14.7"
H.P. = 00°00'08.9"

Moon at Greatest Eclipse
(Geocentric Coordinates)
R.A. = 20h32m04.0s
Dec. = -17°50'50.8"
S.D. = 00°15'23.2"
H.P. = 00°56'28.0"

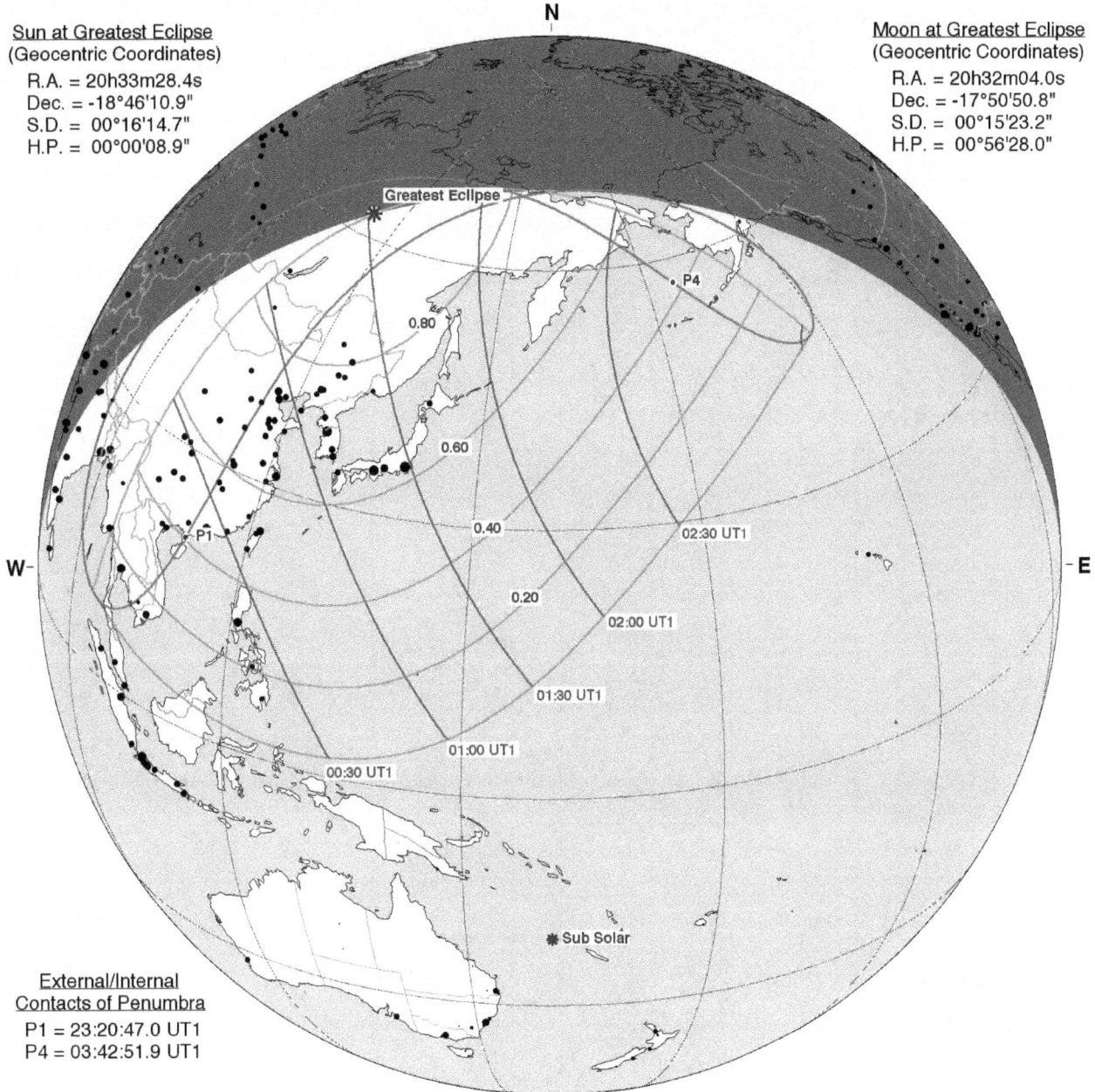

N

Greatest Eclipse

P4

0.80

0.60

0.40        02:30 UT1

W -                                          - E

0.20

02:00 UT1

P1

01:30 UT1

01:00 UT1

00:30 UT1

Sub Solar

External/Internal
Contacts of Penumbra
P1 = 23:20:47.0 UT1
P4 = 03:42:51.9 UT1

ΔT = 82.4 s        S        Eph. = JPL DE405

Circumstances at Greatest Eclipse: 01:31:55.4 UT1
Lat. = 62°52.0'N        Sun Alt. = 0.0°
Long. = 111°42.3'E      Sun Azm. = 134.9°

0    1000   2000   3000   4000   5000
Kilometers

# Partial Solar Eclipse of 2047 Jun 23

Greatest Eclipse = 10:52:30.6 TD  (= 10:51:08.0 UT1)

Eclipse Magnitude = 0.3129          Saros Series = 118
Gamma = 1.3766          Saros Member = 70 of 72

**N**

Sun at Greatest Eclipse
(Geocentric Coordinates)
R.A. = 06h08m27.7s
Dec. = +23°25'10.2"
S.D. = 00°15'44.2"
H.P. = 00°00'08.7"

Moon at Greatest Eclipse
(Geocentric Coordinates)
R.A. = 06h09m05.2s
Dec. = +24°40'56.6"
S.D. = 00°15'07.9"
H.P. = 00°55'32.1"

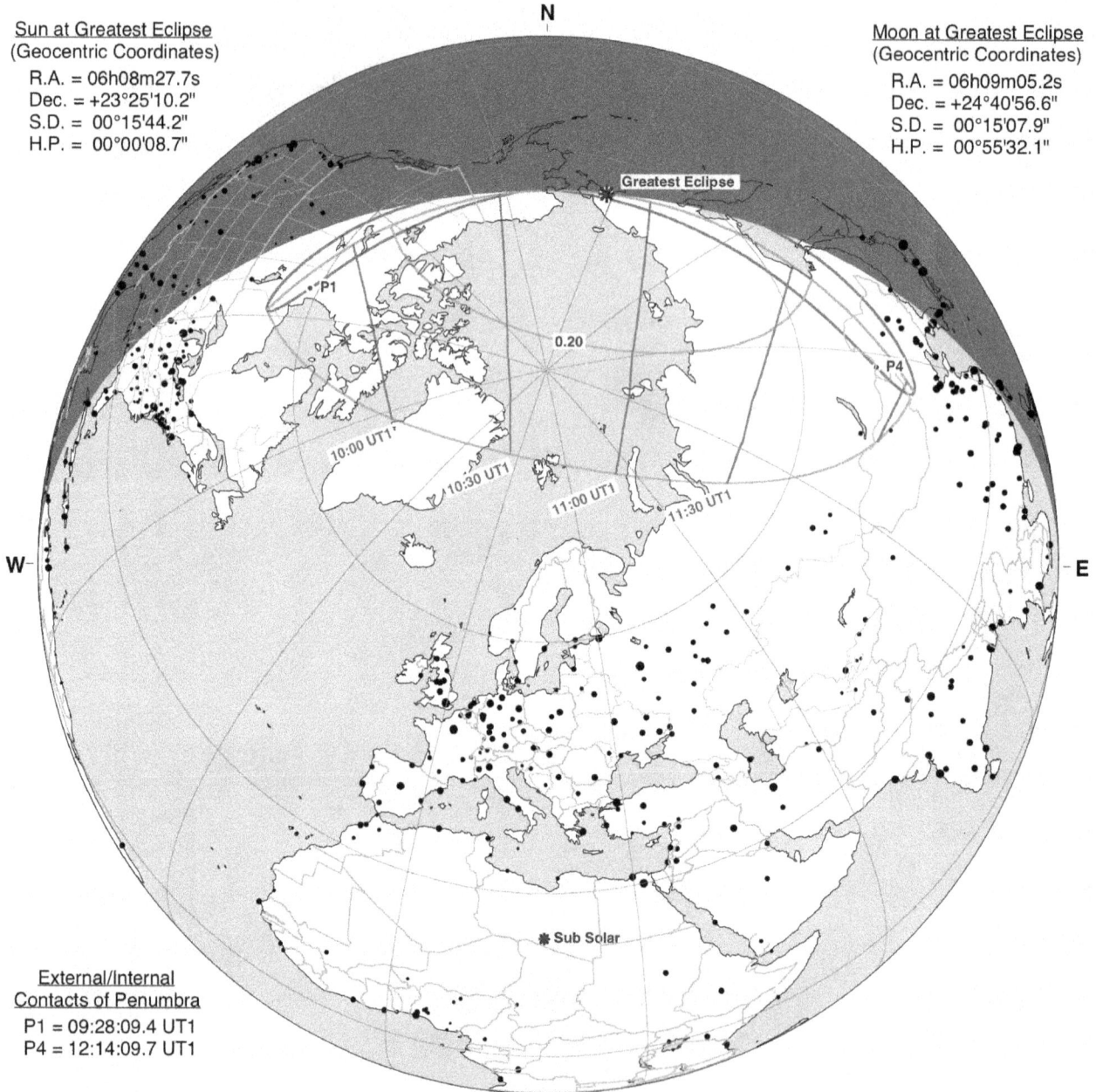

Greatest Eclipse

0.20

P1

P4

10:00 UT1
10:30 UT1
11:00 UT1
11:30 UT1

**W**

**E**

Sub Solar

External/Internal
Contacts of Penumbra
P1 = 09:28:09.4 UT1
P4 = 12:14:09.7 UT1

ΔT = 82.6 s          **S**          Eph. = JPL DE405

Circumstances at Greatest Eclipse: 10:51:08.0 UT1
Lat. = 65°46.0'N          Sun Alt. = 0.0°
Long. = 178°02.4'W          Sun Azm. = 345.5°

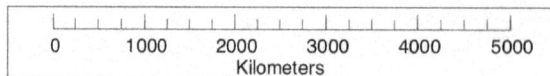

0    1000    2000    3000    4000    5000
Kilometers

©2016  F. Espenak
www.EclipseWise.com

# Partial Solar Eclipse of 2047 Jul 22

Greatest Eclipse = 22:36:17.4 TD  (= 22:34:54.7 UT1)

| Eclipse Magnitude = 0.3605 | Saros Series = 156 |
|---|---|
| Gamma = -1.3477 | Saros Member = 3 of 69 |

Sun at Greatest Eclipse
(Geocentric Coordinates)
R.A. = 08h08m59.7s
Dec. = +20°07'53.9"
S.D. = 00°15'44.5"
H.P. = 00°00'08.7"

Moon at Greatest Eclipse
(Geocentric Coordinates)
R.A. = 08h07m21.2s
Dec. = +18°54'51.1"
S.D. = 00°15'32.1"
H.P. = 00°57'00.9"

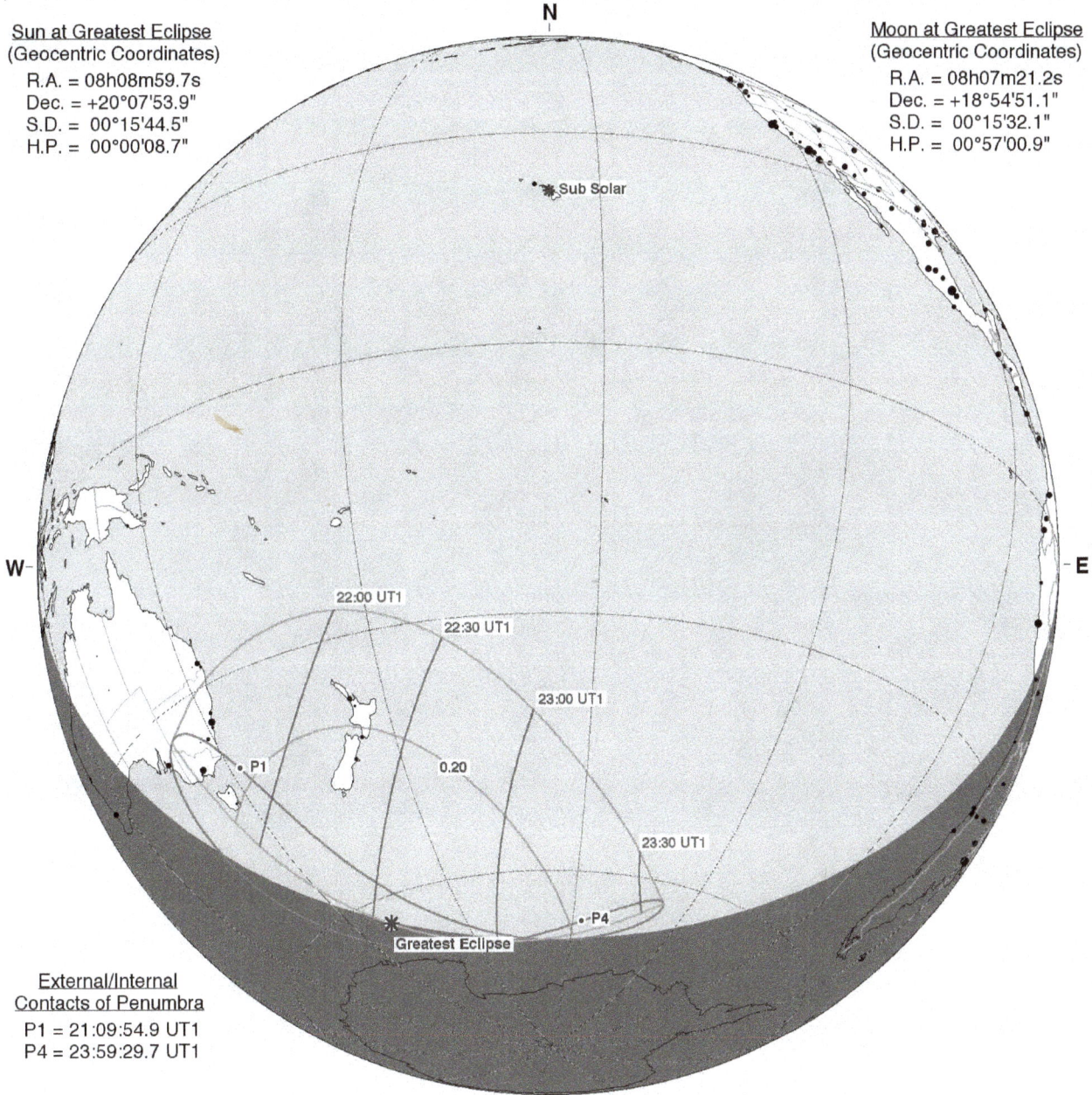

N

Sub Solar

W

E

22:00 UT1
22:30 UT1
23:00 UT1
0.20
23:30 UT1
P1
P4
Greatest Eclipse

External/Internal
Contacts of Penumbra
P1 = 21:09:54.9 UT1
P4 = 23:59:29.7 UT1

ΔT = 82.7 s

S

Eph. = JPL DE405

Circumstances at Greatest Eclipse: 22:34:54.7 UT1

| Lat. = 63°26.5'S | Sun Alt. = 0.0° |
|---|---|
| Long. = 160°08.3'E | Sun Azm. = 39.7° |

0      1000    2000    3000    4000    5000
Kilometers

©2016  F. Espenak
www.EclipseWise.com

159

# Partial Solar Eclipse of 2047 Dec 16

Greatest Eclipse = 23:50:12.3 TD   (= 23:48:49.4 UT1)

Eclipse Magnitude = 0.8817          Saros Series = 123
Gamma = -1.0661                     Saros Member = 55 of 70

Sun at Greatest Eclipse
(Geocentric Coordinates)
R.A. = 17h37m56.6s
Dec. = -23°20'10.9"
S.D. = 00°16'15.0"
H.P. = 00°00'08.9"

Moon at Greatest Eclipse
(Geocentric Coordinates)
R.A. = 17h38m13.1s
Dec. = -24°24'51.1"
S.D. = 00°16'35.9"
H.P. = 01°00'54.9"

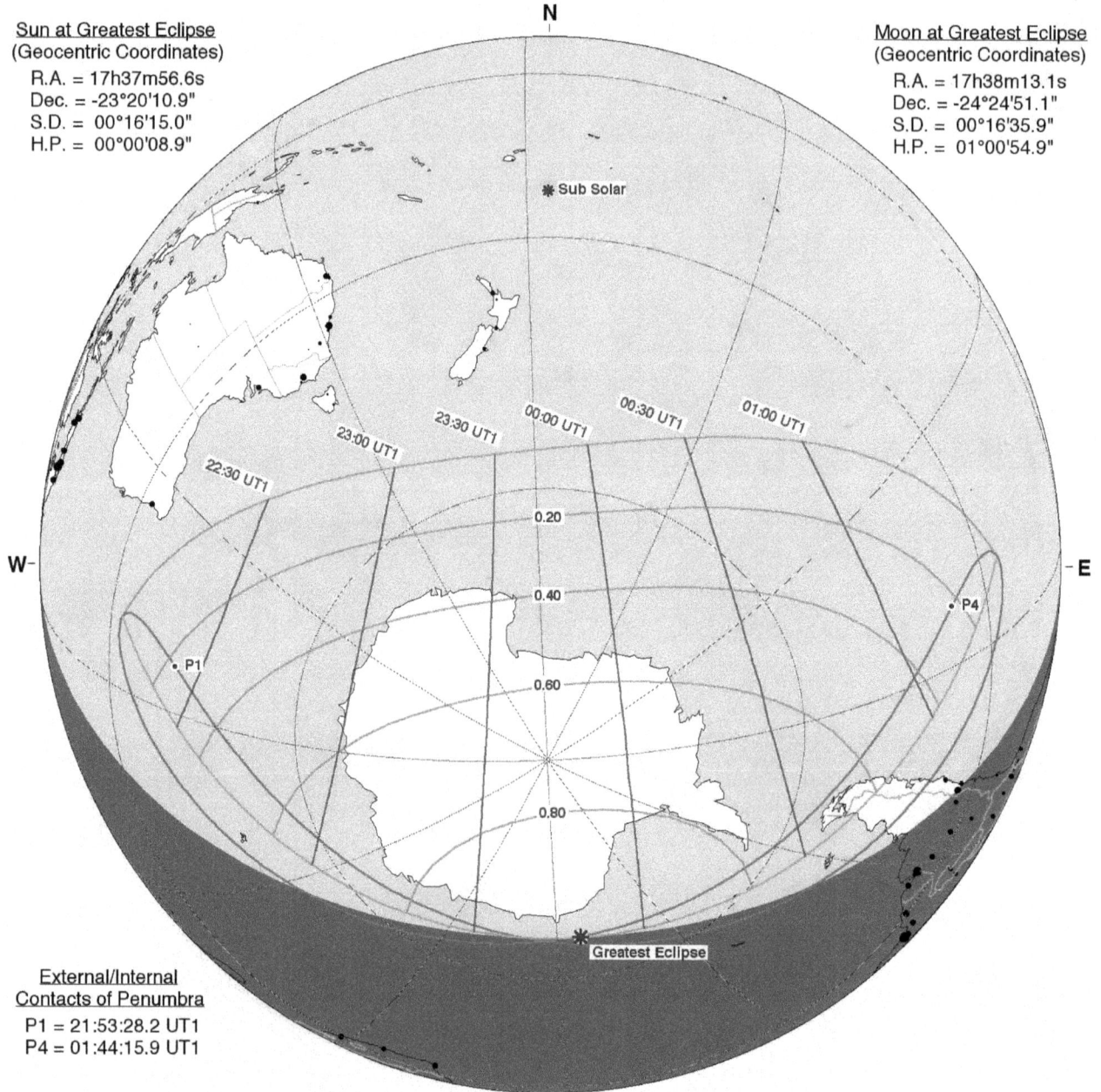

External/Internal
Contacts of Penumbra
P1 = 21:53:28.2 UT1
P4 = 01:44:15.9 UT1

ΔT = 82.9 s          Eph. = JPL DE405

Circumstances at Greatest Eclipse: 23:48:49.4 UT1
Lat. = 66°26.6'S          Sun Alt. = 0.0°
Long. = 006°38.2'W        Sun Azm. = 187.7°

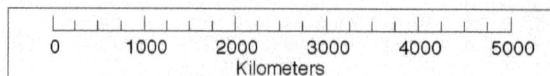

# Annular Solar Eclipse of 2048 Jun 11

Greatest Eclipse = 12:58:52.8 TD  (= 12:57:29.6 UT1)

Eclipse Magnitude = 0.9441  Saros Series = 128

Gamma = 0.6468  Saros Member = 60 of 73

Sun at Greatest Eclipse
(Geocentric Coordinates)
R.A. = 05h22m03.9s
Dec. = +23°08'47.0"
S.D. = 00°15'45.1"
H.P. = 00°00'08.7"

Moon at Greatest Eclipse
(Geocentric Coordinates)
R.A. = 05h22m09.1s
Dec. = +23°43'34.6"
S.D. = 00°14'42.3"
H.P. = 00°53'58.0"

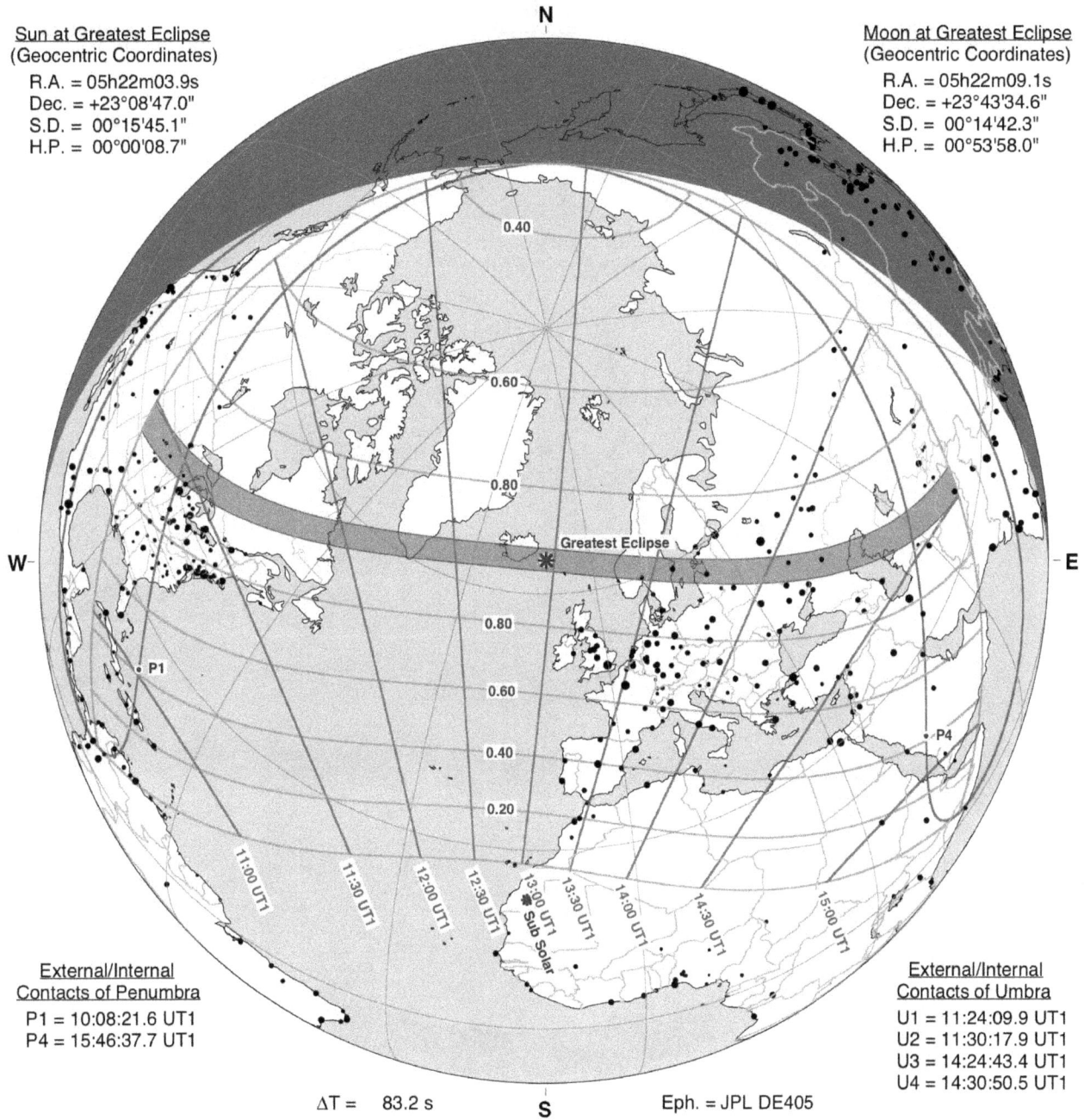

External/Internal
Contacts of Penumbra
P1 = 10:08:21.6 UT1
P4 = 15:46:37.7 UT1

External/Internal
Contacts of Umbra
U1 = 11:24:09.9 UT1
U2 = 11:30:17.9 UT1
U3 = 14:24:43.4 UT1
U4 = 14:30:50.5 UT1

ΔT = 83.2 s  Eph. = JPL DE405

Circumstances at Greatest Eclipse: 12:57:29.6 UT1

Lat. = 63°41.3'N  Sun Alt. = 49.4°
Long. = 011°32.5'W  Sun Azm. = 184.0°
Path Width = 271.5 km  Duration = 04m58.3s

Circumstances at Greatest Duration: 12:56:04.3 UT1

Lat. = 63°43.9'N  Sun Alt. = 49.4°
Long. = 012°43.9'W  Sun Azm. = 181.9°
Path Width = 271.5 km  Duration = 04m58.3s

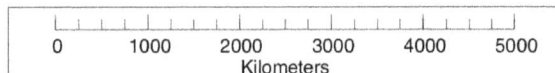

0  1000  2000  3000  4000  5000
Kilometers

©2016 F. Espenak
www.EclipseWise.com

# Total Solar Eclipse of 2048 Dec 05

Greatest Eclipse = 15:35:26.7 TD  (= 15:34:03.2 UT1)

| Eclipse Magnitude = 1.0440 | Saros Series = 133 |
|---|---|
| Gamma = -0.3973 | Saros Member = 47 of 72 |

**Sun at Greatest Eclipse**
(Geocentric Coordinates)
R.A. = 16h51m20.5s
Dec. = -22°29'40.9"
S.D. = 00°16'13.8"
H.P. = 00°00'08.9"

**Moon at Greatest Eclipse**
(Geocentric Coordinates)
R.A. = 16h51m18.6s
Dec. = -22°53'56.4"
S.D. = 00°16'40.9"
H.P. = 01°01'13.3"

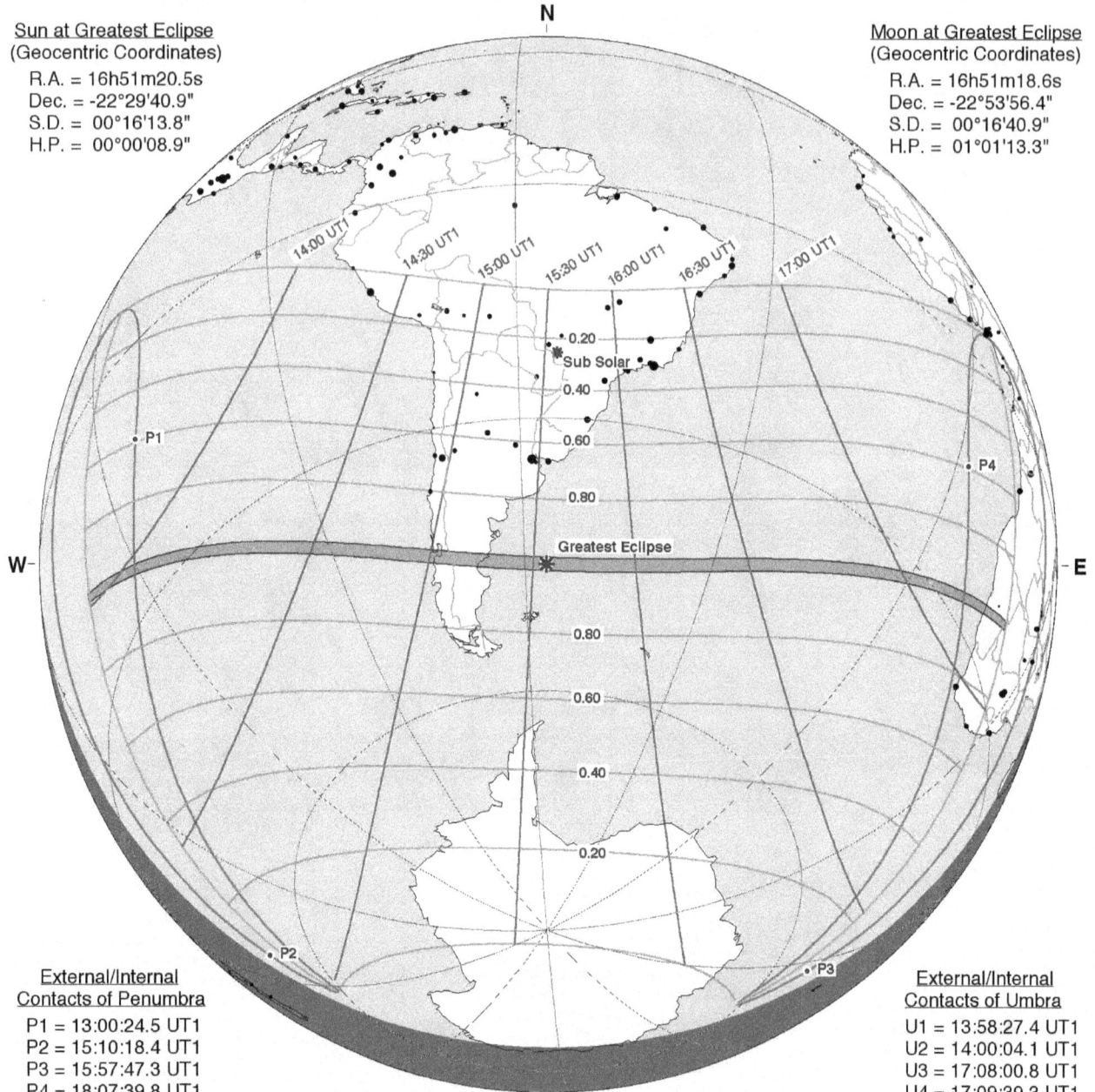

**External/Internal Contacts of Penumbra**
P1 = 13:00:24.5 UT1
P2 = 15:10:18.4 UT1
P3 = 15:57:47.3 UT1
P4 = 18:07:39.8 UT1

**External/Internal Contacts of Umbra**
U1 = 13:58:27.4 UT1
U2 = 14:00:04.1 UT1
U3 = 17:08:00.8 UT1
U4 = 17:09:39.3 UT1

ΔT = 83.5 s          Eph. = JPL DE405

**Circumstances at Greatest Eclipse: 15:34:03.2 UT1**

| Lat. = 46°07.9'S | Sun Alt. = 66.4° |
|---|---|
| Long. = 056°24.4'W | Sun Azm. = 1.4° |
| Path Width = 160.3 km | Duration = 03m27.6s |

**Circumstances at Greatest Duration: 15:34:47.8 UT1**

| Lat. = 46°08.5'S | Sun Alt. = 66.4° |
|---|---|
| Long. = 056°00.0'W | Sun Azm. = 0.0° |
| Path Width = 160.3 km | Duration = 03m27.6s |

0    1000    2000    3000    4000    5000
Kilometers

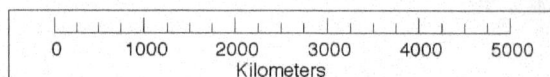

©2016  F. Espenak
www.EclipseWise.com

# Annular Solar Eclipse of 2049 May 31

Greatest Eclipse = 13:59:58.8 TD   (= 13:58:35.0 UT1)

| Eclipse Magnitude = 0.9631 | Saros Series = 138 |
|---|---|
| Gamma = -0.1187 | Saros Member = 33 of 70 |

**Sun at Greatest Eclipse**
(Geocentric Coordinates)
R.A. = 04h35m51.4s
Dec. = +22°01'26.4"
S.D. = 00°15'46.5"
H.P. = 00°00'08.7"

**Moon at Greatest Eclipse**
(Geocentric Coordinates)
R.A. = 04h35m52.6s
Dec. = +21°54'56.7"
S.D. = 00°14'57.8"
H.P. = 00°54'55.1"

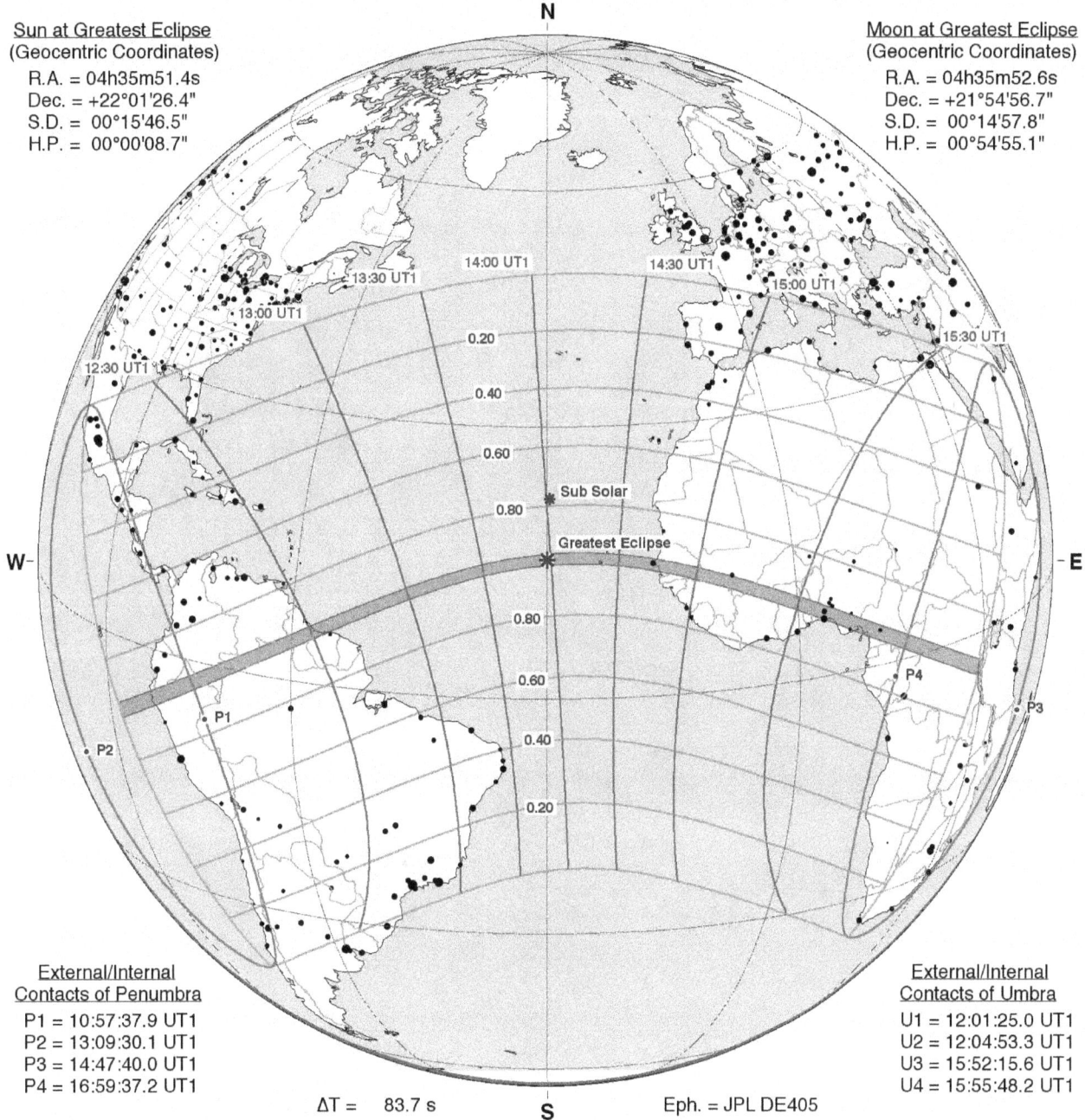

**External/Internal**
**Contacts of Penumbra**
P1 = 10:57:37.9 UT1
P2 = 13:09:30.1 UT1
P3 = 14:47:40.0 UT1
P4 = 16:59:37.2 UT1

ΔT = 83.7 s

Eph. = JPL DE405

**External/Internal**
**Contacts of Umbra**
U1 = 12:01:25.0 UT1
U2 = 12:04:53.3 UT1
U3 = 15:52:15.6 UT1
U4 = 15:55:48.2 UT1

**Circumstances at Greatest Eclipse: 13:58:35.0 UT1**

| Lat. = 15°18.6'N | Sun Alt. = 83.3° |
|---|---|
| Long. = 029°53.3'W | Sun Azm. = 357.6° |
| Path Width = 134.4 km | Duration = 04m45.3s |

**Circumstances at Greatest Duration: 14:04:03.2 UT1**

| Lat. = 15°23.8'N | Sun Alt. = 82.8° |
|---|---|
| Long. = 028°27.6'W | Sun Azm. = 336.6° |
| Path Width = 134.6 km | Duration = 04m45.5s |

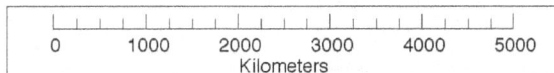

©2016 F. Espenak
www.EclipseWise.com

# Hybrid Solar Eclipse of 2049 Nov 25

Greatest Eclipse = 05:33:47.9 TD (= 05:32:23.9 UT1)

| | |
|---|---|
| Eclipse Magnitude = 1.0057 | Saros Series = 143 |
| Gamma = 0.2943 | Saros Member = 25 of 72 |

**Sun at Greatest Eclipse**
(Geocentric Coordinates)
R.A. = 16h05m24.9s
Dec. = -20°49'25.8"
S.D. = 00°16'12.0"
H.P. = 00°00'08.9"

**Moon at Greatest Eclipse**
(Geocentric Coordinates)
R.A. = 16h05m31.7s
Dec. = -20°32'13.6"
S.D. = 00°16'02.3"
H.P. = 00°58'51.9"

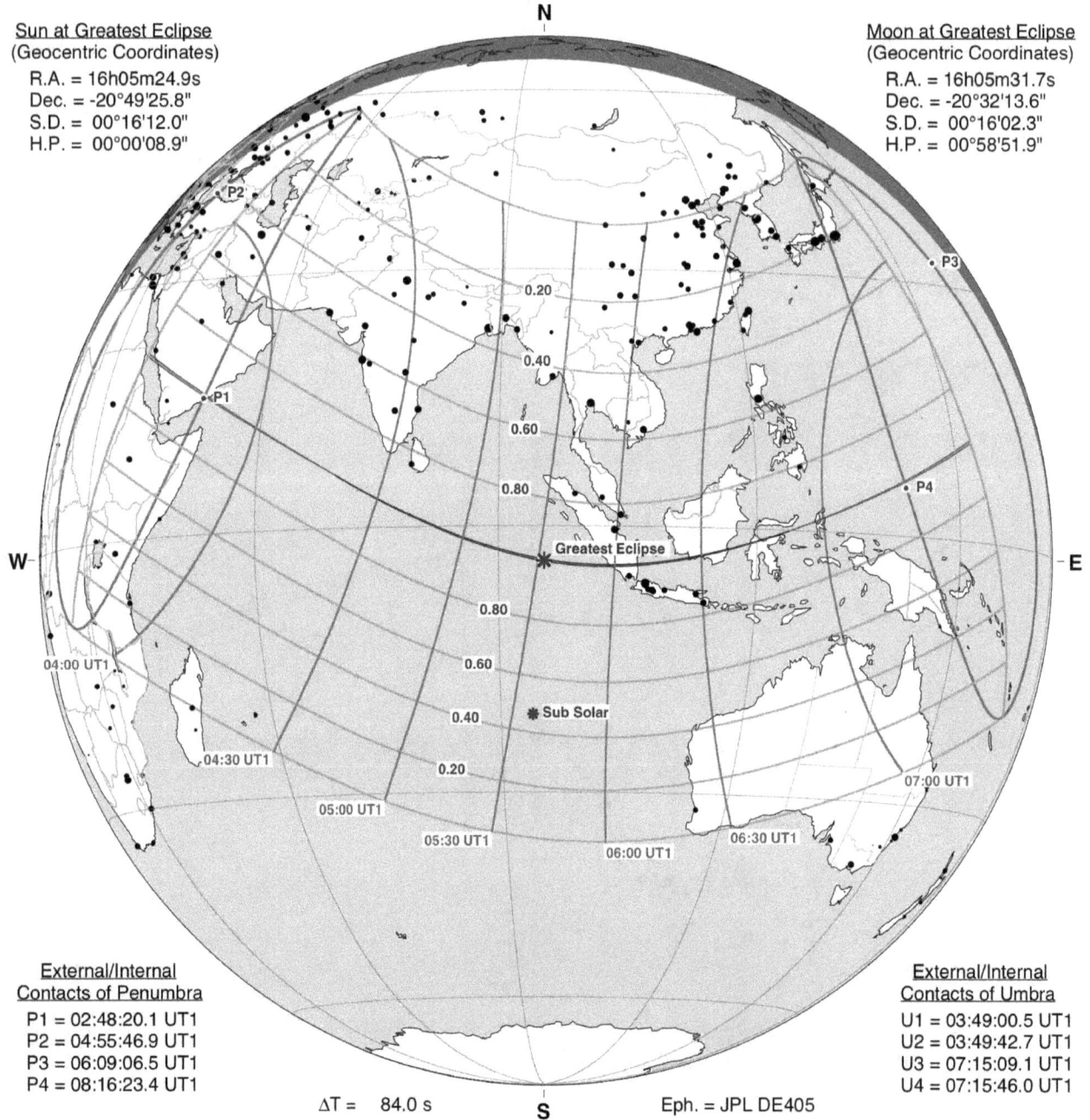

N

P2

P3

0.20

0.40

0.60

0.80

P1

P4

W — Greatest Eclipse — E

0.80

0.60

04:00 UT1

0.40    Sub Solar

04:30 UT1

0.20

05:00 UT1

07:00 UT1

05:30 UT1    06:30 UT1

06:00 UT1

**External/Internal**
**Contacts of Penumbra**
P1 = 02:48:20.1 UT1
P2 = 04:55:46.9 UT1
P3 = 06:09:06.5 UT1
P4 = 08:16:23.4 UT1

**External/Internal**
**Contacts of Umbra**
U1 = 03:49:00.5 UT1
U2 = 03:49:42.7 UT1
U3 = 07:15:09.1 UT1
U4 = 07:15:46.0 UT1

ΔT = 84.0 s     S     Eph. = JPL DE405

**Circumstances at Greatest Eclipse: 05:32:23.9 UT1**

| | |
|---|---|
| Lat. = 03°48.2'S | Sun Alt. = 72.9° |
| Long. = 095°12.7'E | Sun Azm. = 185.0° |
| Path Width = 20.7 km | Duration = 00m37.5s |

**Circumstances at Greatest Duration: 03:49:21.6 UT1**

| | |
|---|---|
| Lat. = 21°02.0'N | Sun Alt. = 0.0° |
| Long. = 037°48.4'E | Sun Azm. = 112.4° |
| Path Width = 40.0 km | Duration = 00m41.8s |

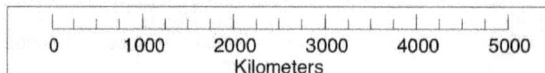

| 0 | 1000 | 2000 | 3000 | 4000 | 5000 |
|---|---|---|---|---|---|

Kilometers

# Hybrid Solar Eclipse of 2050 May 20

Greatest Eclipse = 20:42:50.2 TD  (= 20:41:25.9 UT1)

Eclipse Magnitude = 1.0038          Saros Series = 148
Gamma = -0.8688          Saros Member = 23 of 75

Sun at Greatest Eclipse
(Geocentric Coordinates)
R.A. = 03h51m25.4s
Dec. = +20°09'01.9"
S.D. = 00°15'48.3"
H.P. = 00°00'08.7"

Moon at Greatest Eclipse
(Geocentric Coordinates)
R.A. = 03h51m49.6s
Dec. = +19°19'17.1"
S.D. = 00°15'44.7"
H.P. = 00°57'47.0"

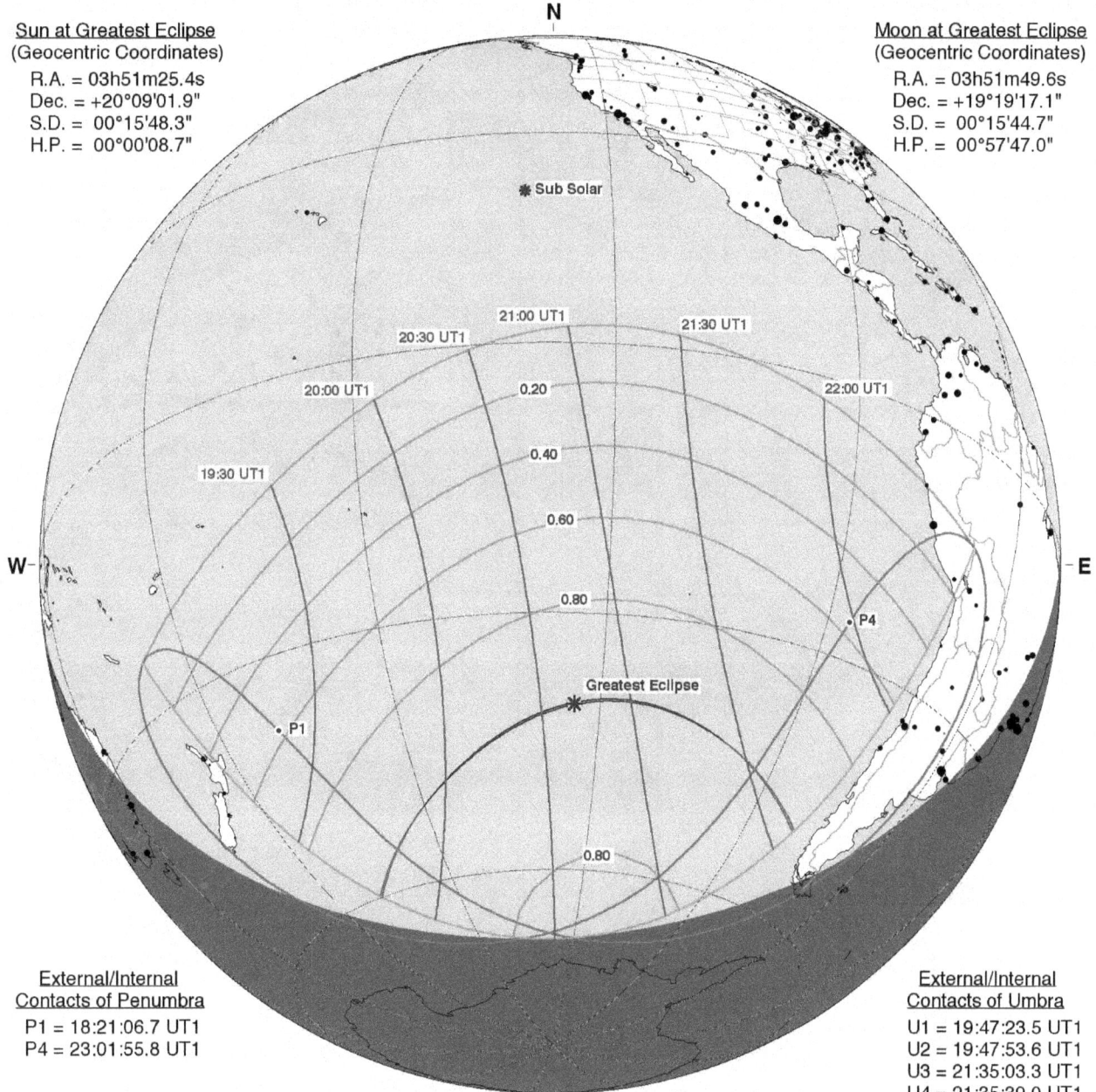

External/Internal
Contacts of Penumbra
P1 = 18:21:06.7 UT1
P4 = 23:01:55.8 UT1

External/Internal
Contacts of Umbra
U1 = 19:47:23.5 UT1
U2 = 19:47:53.6 UT1
U3 = 21:35:03.3 UT1
U4 = 21:35:39.0 UT1

ΔT =   84.3 s          Eph. = JPL DE405

Circumstances at Greatest Eclipse:  20:41:25.9 UT1
Lat. =  40°05.5'S          Sun Alt. =  29.4°
Long. = 123°45.1'W          Sun Azm. = 352.0°
Path Width = 26.7 km          Duration = 00m21.3s

Circumstances at Greatest Duration:  20:39:25.9 UT1
Lat. =  40°14.9'S          Sun Alt. =  29.3°
Long. = 124°36.1'W          Sun Azm. = 353.4°
Path Width = 26.6 km          Duration = 00m21.4s

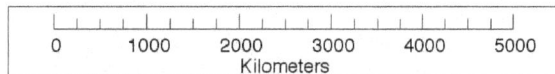

©2016  F. Espenak
www.EclipseWise.com

# Partial Solar Eclipse of 2050 Nov 14

Greatest Eclipse = 13:30:52.8 TD  (= 13:29:28.2 UT1)

| | |
|---|---|
| Eclipse Magnitude = 0.8874 | Saros Series = 153 |
| Gamma = 1.0447 | Saros Member = 11 of 70 |

**Sun at Greatest Eclipse**
(Geocentric Coordinates)
R.A. = 15h19m50.5s
Dec. = -18°21'19.2"
S.D. = 00°16'09.8"
H.P. = 00°00'08.9"

**Moon at Greatest Eclipse**
(Geocentric Coordinates)
R.A. = 15h20m29.5s
Dec. = -17°24'01.9"
S.D. = 00°15'10.6"
H.P. = 00°55'41.9"

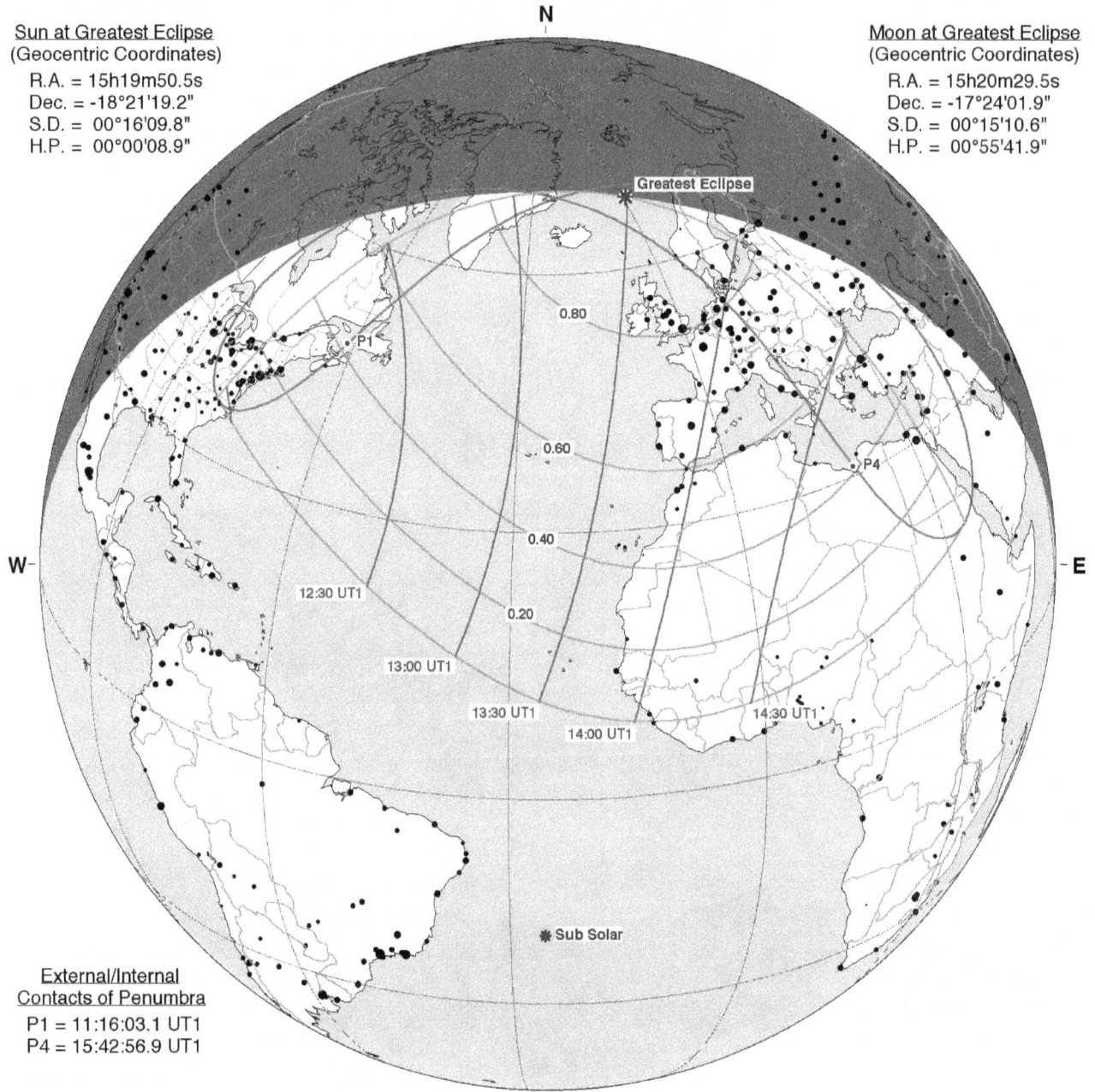

Greatest Eclipse

N

0.80

0.60

0.40

12:30 UT1

0.20

13:00 UT1

13:30 UT1

14:30 UT1

14:00 UT1

P1

P4

W

E

Sub Solar

**External/Internal
Contacts of Penumbra**
P1 = 11:16:03.1 UT1
P4 = 15:42:56.9 UT1

ΔT = 84.6 s

S

Eph. = JPL DE405

**Circumstances at Greatest Eclipse: 13:29:28.2 UT1**

| | |
|---|---|
| Lat. = 69°31.7'N | Sun Alt. = 0.0° |
| Long. = 000°59.0'E | Sun Azm. = 205.8° |

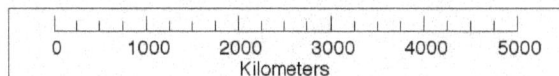

0    1000    2000    3000    4000    5000
Kilometers

©2016  F. Espenak
www.EclipseWise.com

# Partial Solar Eclipse of 2051 Apr 11

Greatest Eclipse = 02:10:38.6 TD   (= 02:09:13.8 UT1)

| Eclipse Magnitude = 0.9849 | Saros Series = 120 |
|---|---|
| Gamma = 1.0169 | Saros Member = 63 of 71 |

**N**

Sun at Greatest Eclipse
(Geocentric Coordinates)
R.A. = 01h18m13.3s
Dec. = +08°15'12.8"
S.D. = 00°15'57.8"
H.P. = 00°00'08.8"

Moon at Greatest Eclipse
(Geocentric Coordinates)
R.A. = 01h17m01.7s
Dec. = +09°14'52.8"
S.D. = 00°16'42.8"
H.P. = 01°01'20.2"

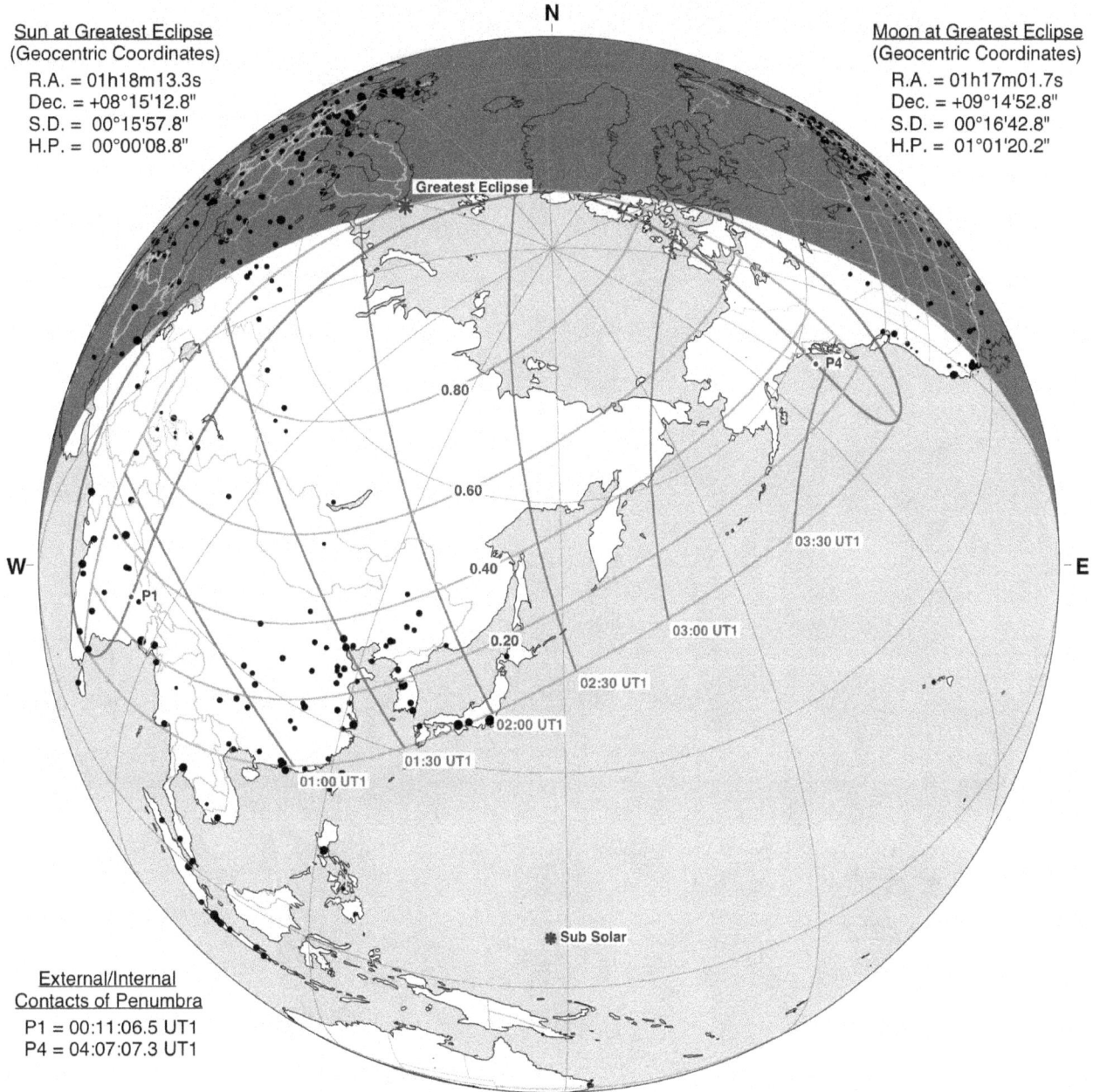

Greatest Eclipse

P4

0.80

0.60

03:30 UT1

**W**

0.40

**E**

03:00 UT1

0.20

P1

02:30 UT1

02:00 UT1

01:30 UT1

01:00 UT1

Sub Solar

External/Internal
Contacts of Penumbra
P1 = 00:11:06.5 UT1
P4 = 04:07:07.3 UT1

ΔT =   84.9 s        **S**        Eph. = JPL DE405

Circumstances at Greatest Eclipse: 02:09:13.8 UT1

| Lat. = 71°37.6'N | Sun Alt. =   0.0° |
|---|---|
| Long. = 032°08.7'E | Sun Azm. =  62.9° |

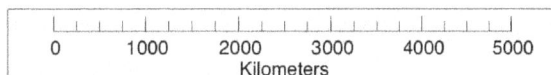

| 0 | 1000 | 2000 | 3000 | 4000 | 5000 |
|---|---|---|---|---|---|

Kilometers

©2016  F. Espenak
www.EclipseWise.com

# Partial Solar Eclipse of 2051 Oct 04

Greatest Eclipse = 21:02:14.5 TD  (= 21:00:49.4 UT1)

| | |
|---|---|
| Eclipse Magnitude = 0.6024 | Saros Series = 125 |
| Gamma = -1.2094 | Saros Member = 56 of 73 |

**Sun at Greatest Eclipse**
(Geocentric Coordinates)
R.A. = 12h42m39.3s
Dec. = -04°35'05.4"
S.D. = 00°15'59.2"
H.P. = 00°00'08.8"

**Moon at Greatest Eclipse**
(Geocentric Coordinates)
R.A. = 12h41m20.9s
Dec. = -05°37'21.2"
S.D. = 00°14'44.4"
H.P. = 00°54'05.8"

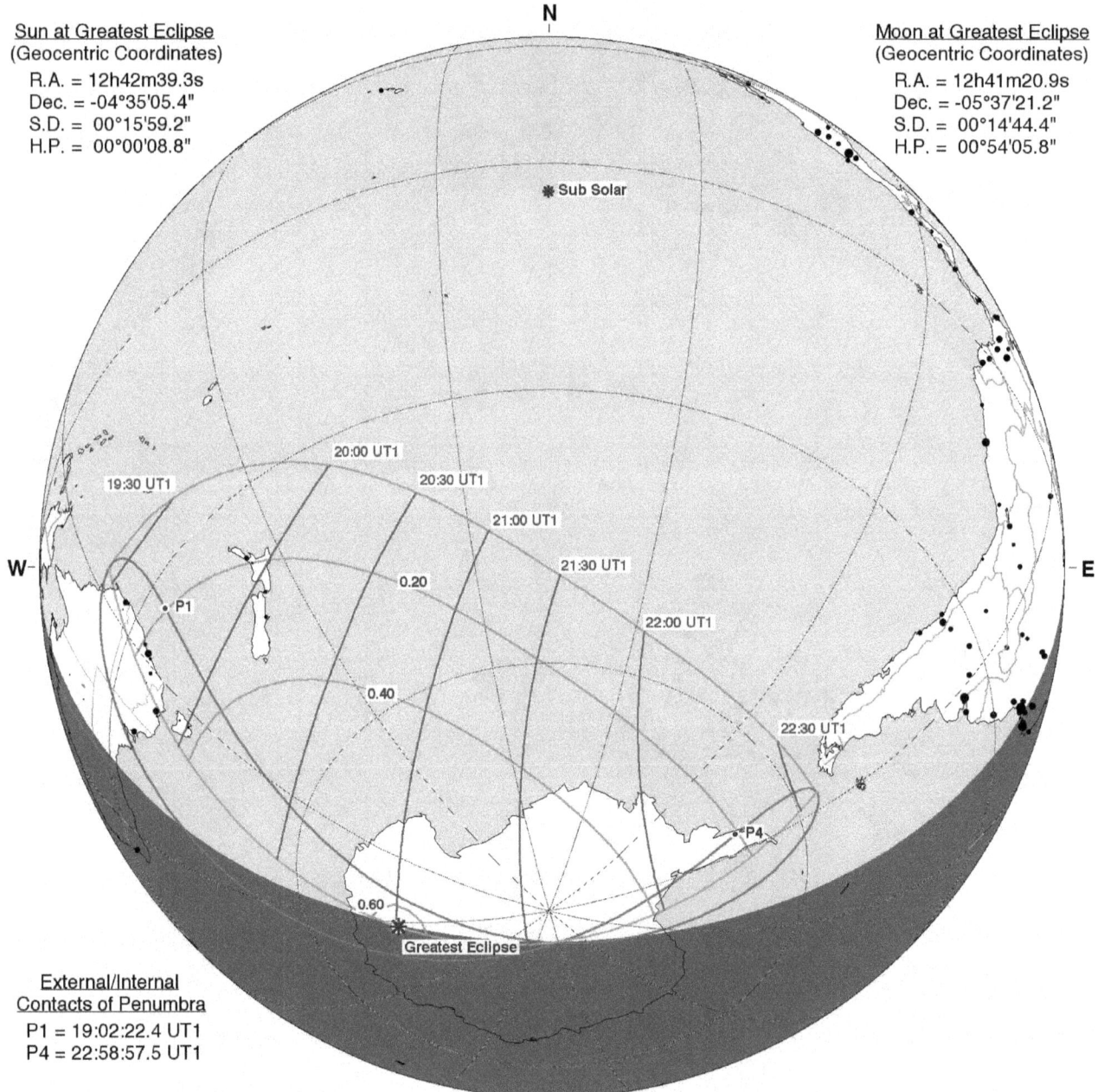

N

* Sub Solar

W

E

20:00 UT1
20:30 UT1
19:30 UT1
21:00 UT1
21:30 UT1
0.20
22:00 UT1
P1
0.40
22:30 UT1
P4
0.60
* Greatest Eclipse

**External/Internal Contacts of Penumbra**
P1 = 19:02:22.4 UT1
P4 = 22:58:57.5 UT1

ΔT = 85.1 s

S

Eph. = JPL DE405

Circumstances at Greatest Eclipse: 21:00:49.4 UT1

| | |
|---|---|
| Lat. = 72°02.7'S | Sun Alt. = 0.0° |
| Long. = 117°41.2'E | Sun Azm. = 105.0° |

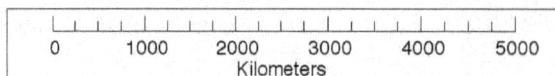

| 0 | 1000 | 2000 | 3000 | 4000 | 5000 |
|---|---|---|---|---|---|

Kilometers

©2016 F. Espenak
www.EclipseWise.com

# Total Solar Eclipse of 2052 Mar 30

Greatest Eclipse = 18:31:52.9 TD   (= 18:30:27.4 UT1)

Eclipse Magnitude = 1.0466          Saros Series = 130
Gamma = 0.3239                      Saros Member = 54 of 73

Sun at Greatest Eclipse
(Geocentric Coordinates)
R.A. = 00h39m33.8s
Dec. = +04°15'25.9"
S.D. = 00°16'00.7"
H.P. = 00°00'08.8"

Moon at Greatest Eclipse
(Geocentric Coordinates)
R.A. = 00h39m10.3s
Dec. = +04°34'05.5"
S.D. = 00°16'29.6"
H.P. = 01°00'31.8"

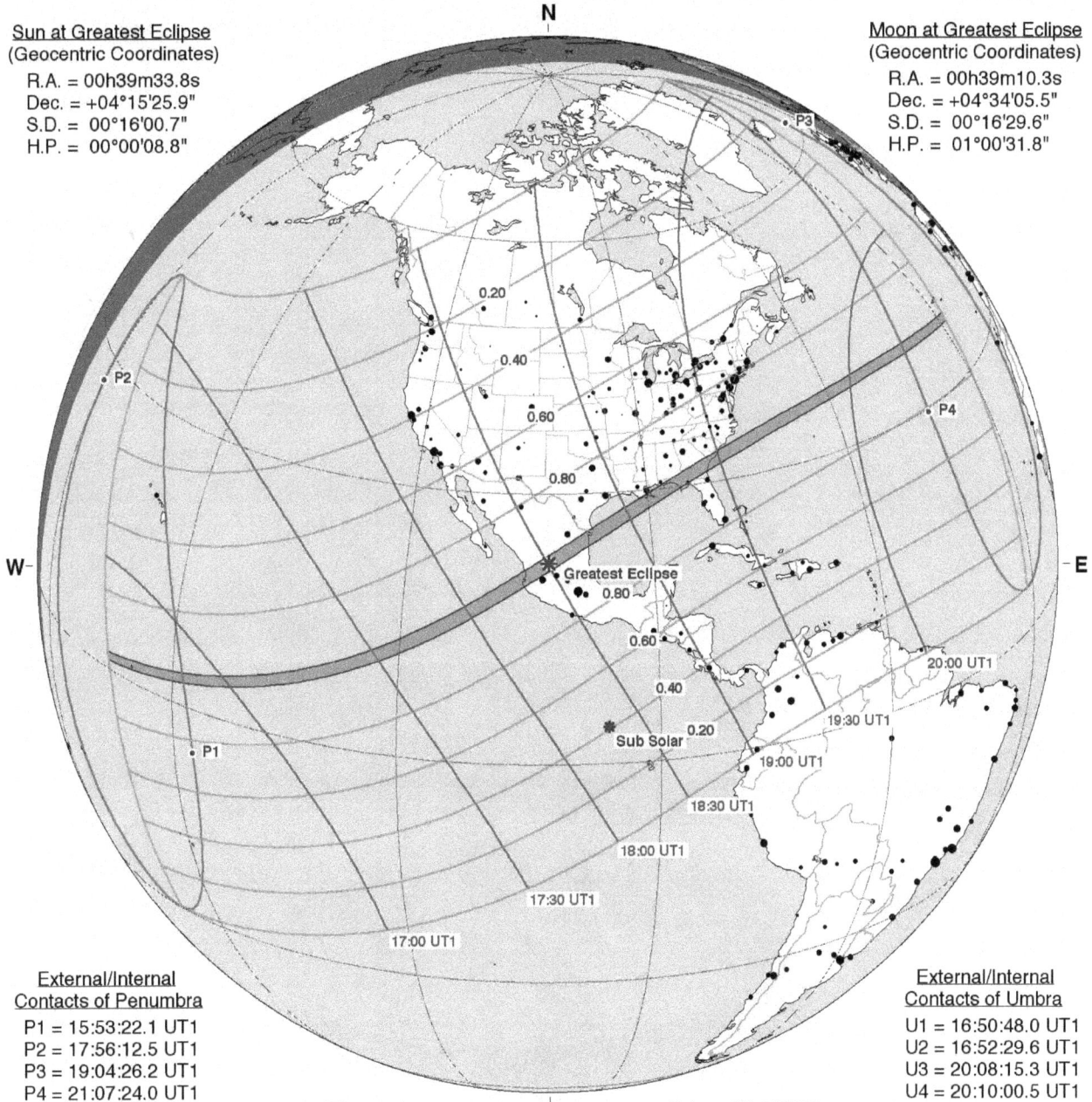

External/Internal
Contacts of Penumbra
P1 = 15:53:22.1 UT1
P2 = 17:56:12.5 UT1
P3 = 19:04:26.2 UT1
P4 = 21:07:24.0 UT1

External/Internal
Contacts of Umbra
U1 = 16:50:48.0 UT1
U2 = 16:52:29.6 UT1
U3 = 20:08:15.3 UT1
U4 = 20:10:00.5 UT1

ΔT =   85.4 s          Eph. = JPL DE405

Circumstances at Greatest Eclipse: 18:30:27.4 UT1
Lat. = 22°22.8'N          Sun Alt. =  71.0°
Long. = 102°34.9'W        Sun Azm. = 161.3°
Path Width = 163.7 km     Duration = 04m08.2s

Circumstances at Greatest Duration: 18:32:33.3 UT1
Lat. = 22°45.2'N          Sun Alt. =  70.9°
Long. = 101°55.6'W        Sun Azm. = 165.1°
Path Width = 163.5 km     Duration = 04m08.2s

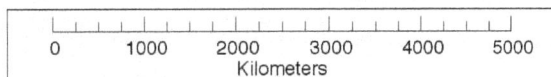

©2016  F. Espenak
www.EclipseWise.com

0   1000   2000   3000   4000   5000
Kilometers

169

# Annular Solar Eclipse of 2052 Sep 22

Greatest Eclipse = 23:39:09.7 TD (= 23:37:44.0 UT1)

| | |
|---|---|
| Eclipse Magnitude = 0.9734 | Saros Series = 135 |
| Gamma = -0.4480 | Saros Member = 41 of 71 |

**Sun at Greatest Eclipse**
(Geocentric Coordinates)
R.A. = 12h02m27.0s
Dec. = -00°15'55.5"
S.D. = 00°15'56.2"
H.P. = 00°00'08.8"

**Moon at Greatest Eclipse**
(Geocentric Coordinates)
R.A. = 12h01m56.4s
Dec. = -00°39'49.3"
S.D. = 00°15'17.9"
H.P. = 00°56'08.8"

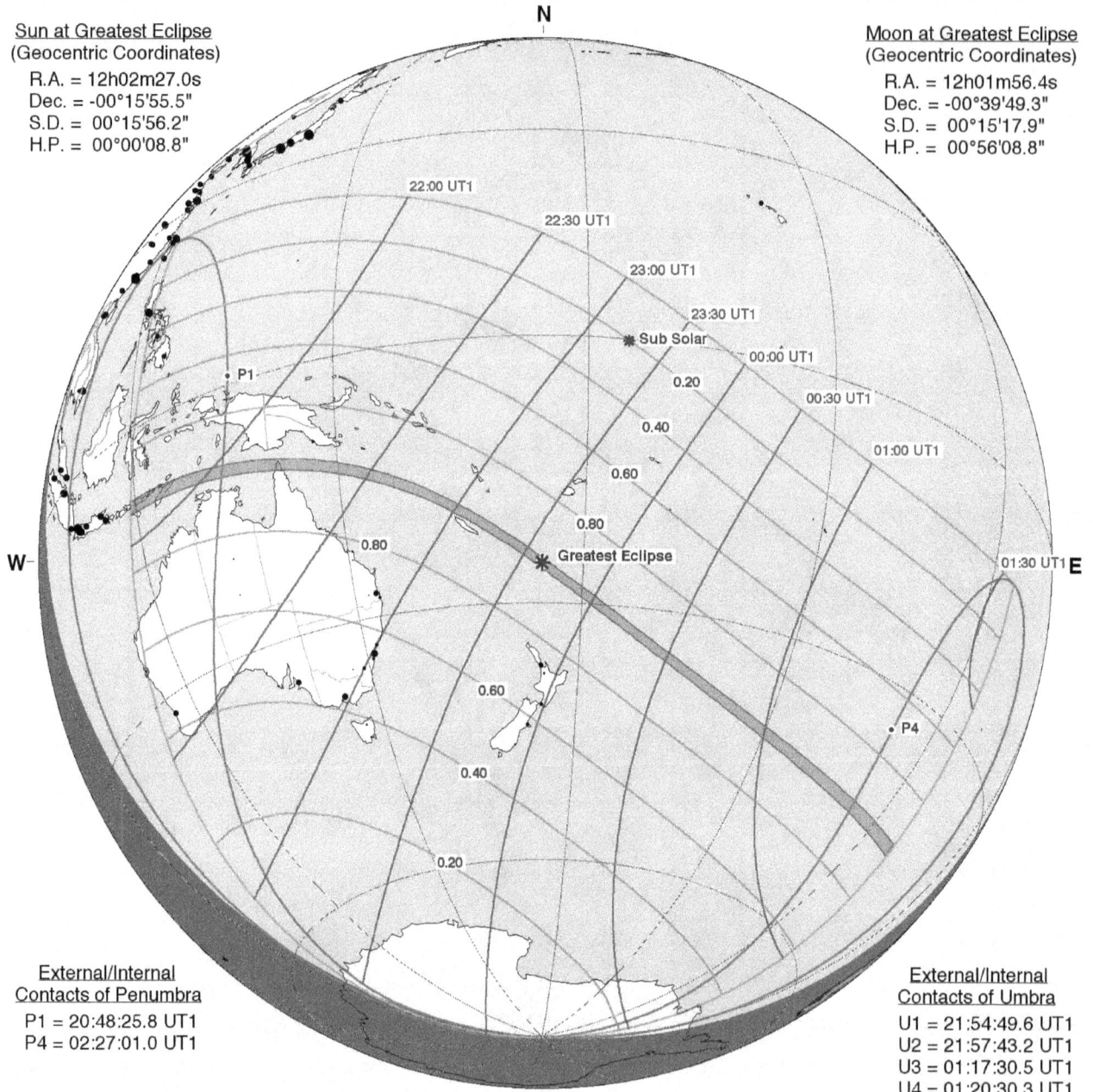

**External/Internal Contacts of Penumbra**
P1 = 20:48:25.8 UT1
P4 = 02:27:01.0 UT1

**External/Internal Contacts of Umbra**
U1 = 21:54:49.6 UT1
U2 = 21:57:43.2 UT1
U3 = 01:17:30.5 UT1
U4 = 01:20:30.3 UT1

ΔT = 85.7 s        Eph. = JPL DE405

**Circumstances at Greatest Eclipse: 23:37:44.0 UT1**

| | |
|---|---|
| Lat. = 25°41.3'S | Sun Alt. = 63.2° |
| Long. = 174°56.0'E | Sun Azm. = 19.6° |
| Path Width = 106.0 km | Duration = 02m50.9s |

**Circumstances at Greatest Duration: 00:04:03.2 UT1**

| | |
|---|---|
| Lat. = 30°17.6'S | Sun Alt. = 59.5° |
| Long. = 177°20.4'W | Sun Azm. = 348.9° |
| Path Width = 106.1 km | Duration = 02m51.5s |

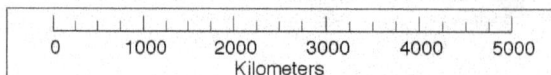

©2016 F. Espenak
www.EclipseWise.com

# Annular Solar Eclipse of  2053 Mar 20

Greatest Eclipse =  07:08:19.4 TD   (= 07:06:53.4 UT1)

Eclipse Magnitude =  0.9919          Saros Series =  140

Gamma = -0.4089          Saros Member =  31 of 71

Sun at Greatest Eclipse
(Geocentric Coordinates)
R.A. = 00h00m30.3s
Dec. = +00°03'17.2"
S.D. = 00°16'03.6"
H.P. = 00°00'08.8"

Moon at Greatest Eclipse
(Geocentric Coordinates)
R.A. = 00h00m59.0s
Dec. = -00°19'05.6"
S.D. = 00°15'41.9"
H.P. = 00°57'37.0"

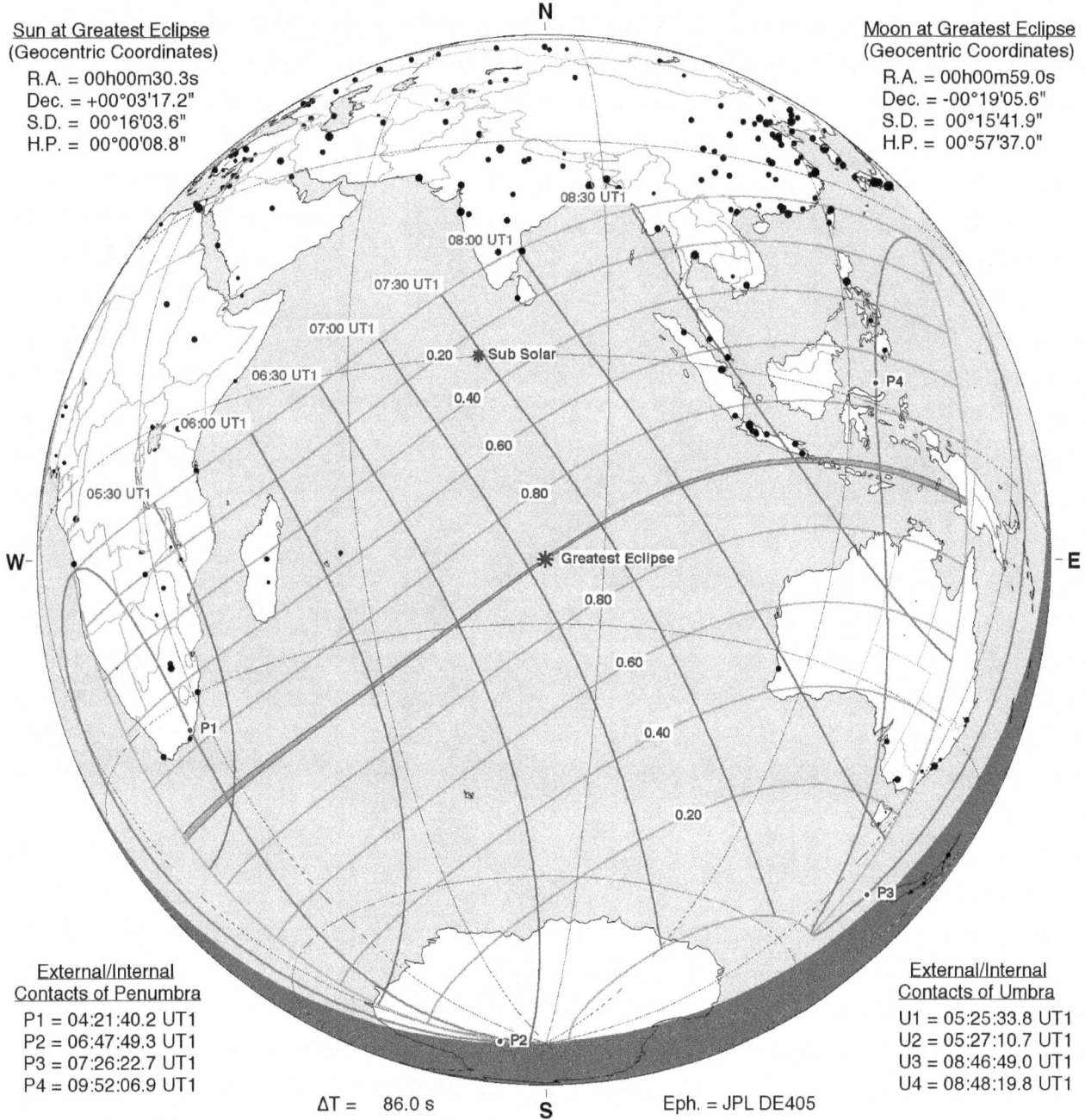

N

08:30 UT1
08:00 UT1
07:30 UT1
07:00 UT1
06:30 UT1
06:00 UT1
05:30 UT1

0.20  Sub Solar
0.40
0.60
0.80

Greatest Eclipse
0.80
0.60
0.40
0.20

P1
P2
P3
P4

W
E

S

External/Internal
Contacts of Penumbra
P1 = 04:21:40.2 UT1
P2 = 06:47:49.3 UT1
P3 = 07:26:22.7 UT1
P4 = 09:52:06.9 UT1

ΔT =    86.0 s

External/Internal
Contacts of Umbra
U1 = 05:25:33.8 UT1
U2 = 05:27:10.7 UT1
U3 = 08:46:49.0 UT1
U4 = 08:48:19.8 UT1

Eph. = JPL DE405

Circumstances at Greatest Eclipse:  07:06:53.4 UT1

| | |
|---|---|
| Lat. = 23°00.9'S | Sun Alt. =  65.7° |
| Long. = 082°54.1'E | Sun Azm. = 340.8° |
| Path Width =  31.1 km | Duration = 00m49.5s |

Circumstances at Greatest Duration:  05:26:22.2 UT1

| | |
|---|---|
| Lat. = 42°07.6'S | Sun Alt. =   0.0° |
| Long. = 010°16.4'E | Sun Azm. = 90.0° |
| Path Width =  93.4 km | Duration = 01m28.3s |

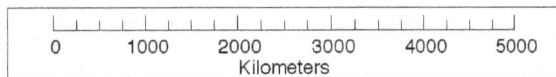

©2016  F. Espenak
www.EclipseWise.com

0    1000   2000   3000   4000   5000
Kilometers

# Total Solar Eclipse of 2053 Sep 12

Greatest Eclipse = 09:34:08.9 TD  (= 09:32:42.6 UT1)

| | |
|---|---|
| Eclipse Magnitude = 1.0328 | Saros Series = 145 |
| Gamma = 0.3140 | Saros Member = 24 of 77 |

**Sun at Greatest Eclipse**
(Geocentric Coordinates)
R.A. = 11h23m36.1s
Dec. = +03°55'14.2"
S.D. = 00°15'53.4"
H.P. = 00°00'08.7"

**Moon at Greatest Eclipse**
(Geocentric Coordinates)
R.A. = 11h23m58.5s
Dec. = +04°12'57.2"
S.D. = 00°16'09.4"
H.P. = 00°59'17.8"

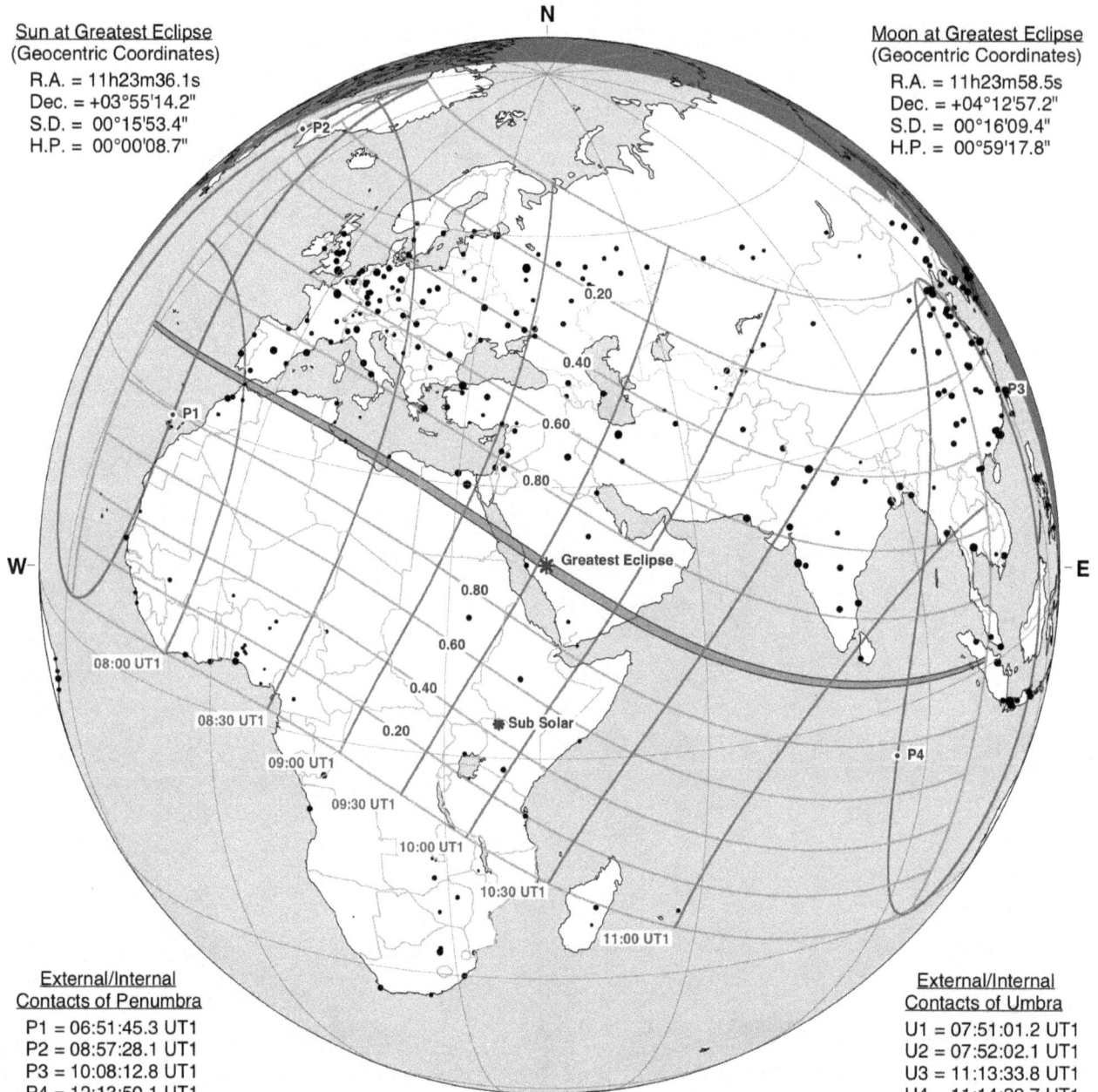

N

P2

0.20

0.40

0.60

0.80

P3

W — — E

Greatest Eclipse

0.80

0.60

P1

0.40

08:00 UT1

0.20

08:30 UT1

Sub Solar

09:00 UT1

P4

09:30 UT1

10:00 UT1

10:30 UT1

11:00 UT1

S

**External/Internal
Contacts of Penumbra**
P1 = 06:51:45.3 UT1
P2 = 08:57:28.1 UT1
P3 = 10:08:12.8 UT1
P4 = 12:13:50.1 UT1

**External/Internal
Contacts of Umbra**
U1 = 07:51:01.2 UT1
U2 = 07:52:02.1 UT1
U3 = 11:13:33.8 UT1
U4 = 11:14:29.7 UT1

ΔT = 86.3 s          Eph. = JPL DE405

**Circumstances at Greatest Eclipse: 09:32:42.6 UT1**

| | |
|---|---|
| Lat. = 21°28.1'N | Sun Alt. = 71.6° |
| Long. = 041°39.8'E | Sun Azm. = 198.6° |
| Path Width = 116.4 km | Duration = 03m04.1s |

**Circumstances at Greatest Duration: 09:29:56.2 UT1**

| | |
|---|---|
| Lat. = 21°56.9'N | Sun Alt. = 71.5° |
| Long. = 040°50.4'E | Sun Azm. = 193.6° |
| Path Width = 116.3 km | Duration = 03m04.2s |

0  1000  2000  3000  4000  5000
Kilometers

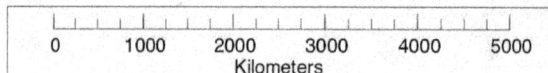

# Partial Solar Eclipse of 2054 Mar 09

Greatest Eclipse = 12:33:40.5 TD  (= 12:32:13.9 UT1)

| | |
|---|---|
| Eclipse Magnitude = 0.6678 | Saros Series = 150 |
| Gamma = -1.1711 | Saros Member = 19 of 71 |

**N**

Sun at Greatest Eclipse
(Geocentric Coordinates)
R.A. = 23h20m07.5s
Dec. = -04°17'25.4"
S.D. = 00°16'06.6"
H.P. = 00°00'08.9"

Moon at Greatest Eclipse
(Geocentric Coordinates)
R.A. = 23h21m24.6s
Dec. = -05°18'27.6"
S.D. = 00°14'55.7"
H.P. = 00°54'47.2"

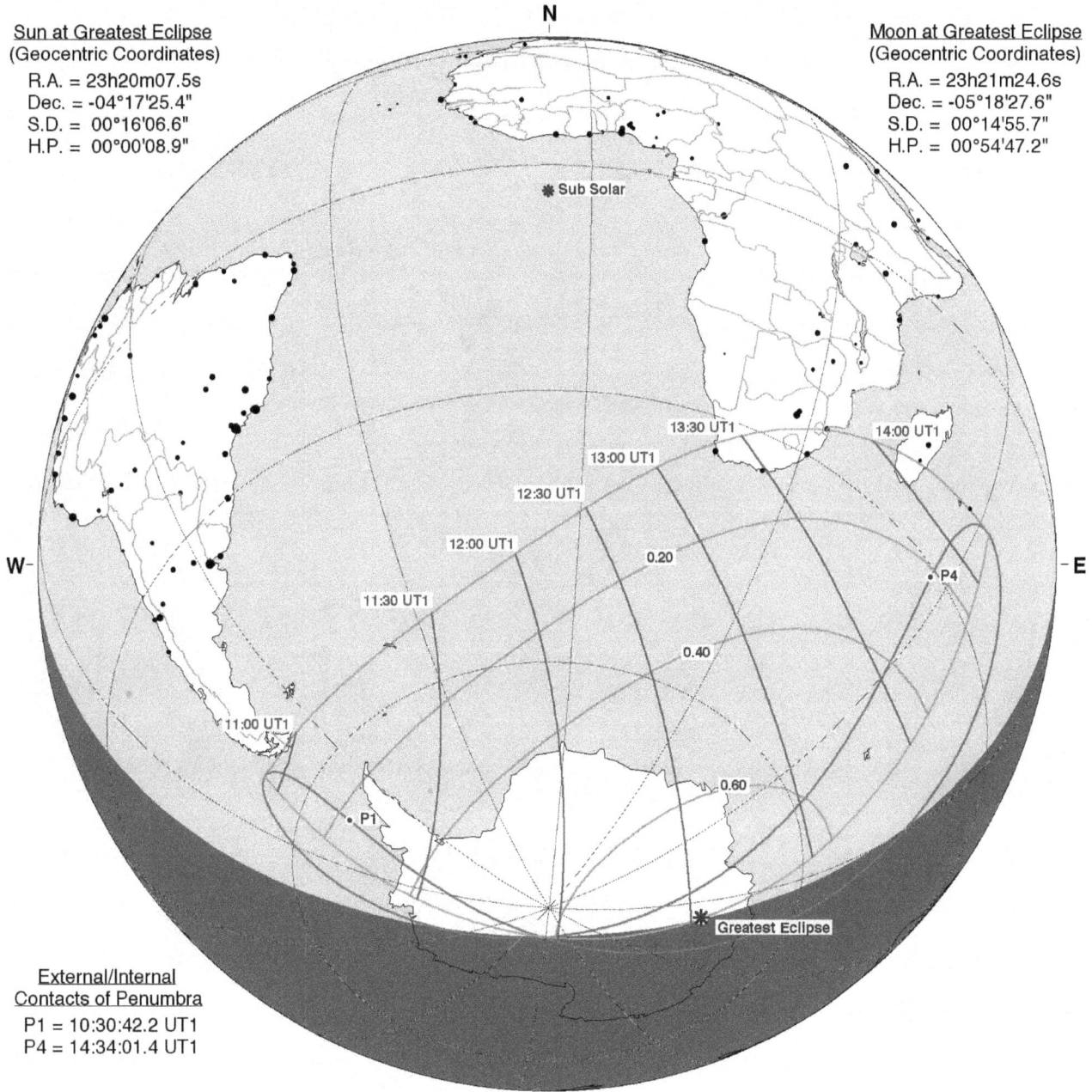

✳ Sub Solar

13:30 UT1
14:00 UT1
13:00 UT1
12:30 UT1
0.20
12:00 UT1
P4
**W** — — **E**
11:30 UT1
0.40
11:00 UT1
0.60
P1
✳ Greatest Eclipse

External/Internal
Contacts of Penumbra
P1 = 10:30:42.2 UT1
P4 = 14:34:01.4 UT1

ΔT = 86.6 s     **S**     Eph. = JPL DE405

Circumstances at Greatest Eclipse:  12:32:13.9 UT1

| | |
|---|---|
| Lat. = 72°02.6'S | Sun Alt. = 0.0° |
| Long. = 097°51.8'E | Sun Azm. = 256.0° |

| 0 | 1000 | 2000 | 3000 | 4000 | 5000 |
|---|---|---|---|---|---|

Kilometers

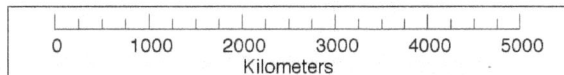

©2016 F. Espenak
www.EclipseWise.com

# Partial Solar Eclipse of 2054 Aug 03

Greatest Eclipse = 18:04:02.1 TD  (= 18:02:35.2 UT1)

Eclipse Magnitude = 0.0656          Saros Series = 117

Gamma = -1.4941          Saros Member = 71 of 71

Sun at Greatest Eclipse
(Geocentric Coordinates)
R.A. = 08h56m24.5s
Dec. = +17°17'09.2"
S.D. = 00°15'45.7"
H.P. = 00°00'08.7"

Moon at Greatest Eclipse
(Geocentric Coordinates)
R.A. = 08h55m14.2s
Dec. = +15°47'22.5"
S.D. = 00°16'41.7"
H.P. = 01°01'16.5"

N

Sub Solar

W

E

18:00 UT1

P1

P4

Greatest Eclipse

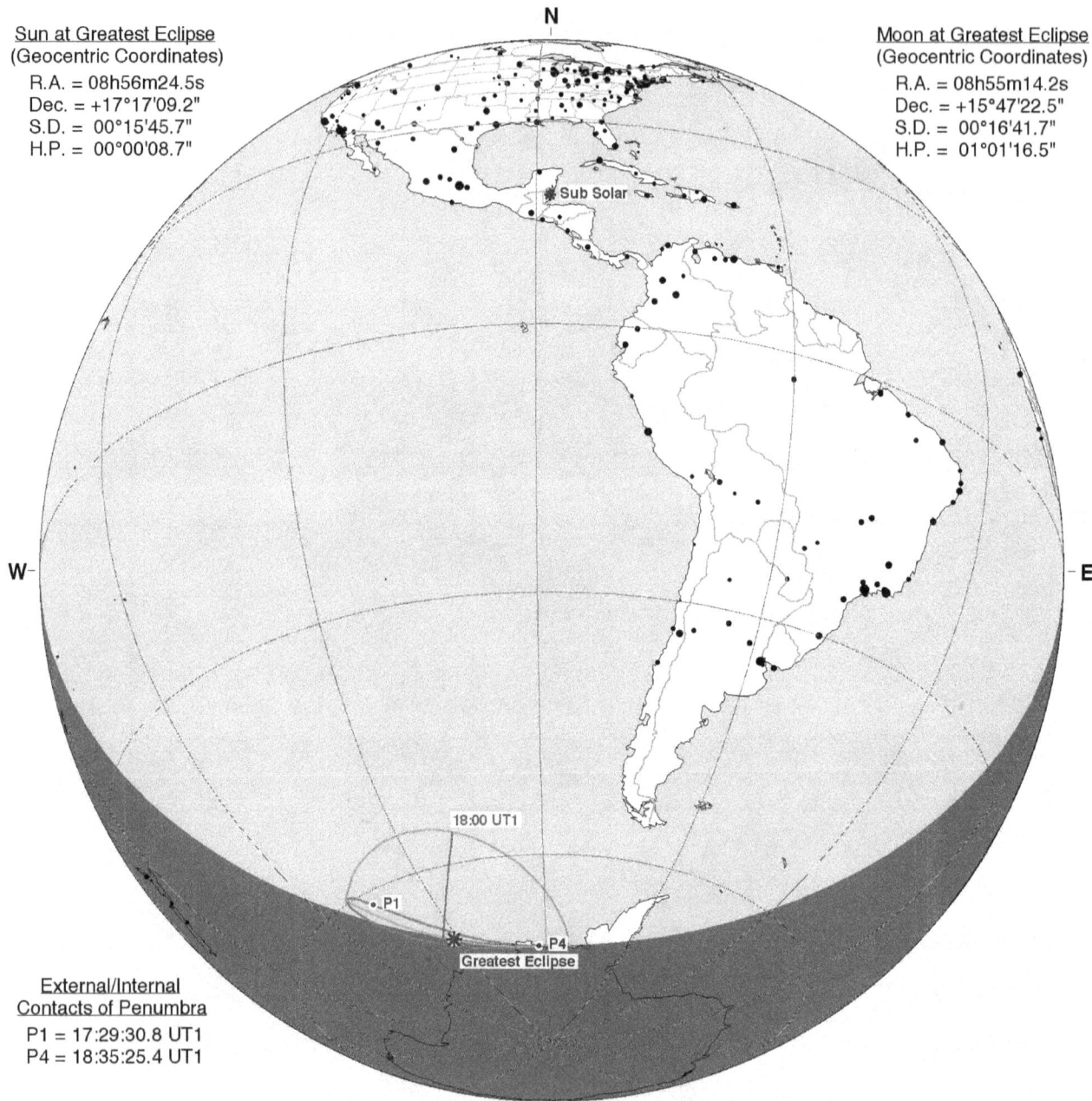

External/Internal
Contacts of Penumbra
P1 = 17:29:30.8 UT1
P4 = 18:35:25.4 UT1

ΔT =    86.9 s          S          Eph. = JPL DE405

Circumstances at Greatest Eclipse: 18:02:35.2 UT1
Lat. = 69°46.4'S          Sun Alt. =   0.0°
Long. = 121°24.7'W          Sun Azm. = 30.7°

0    1000    2000    3000    4000    5000
Kilometers

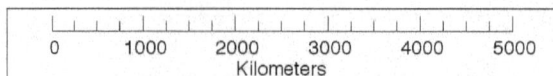

# Partial Solar Eclipse of 2054 Sep 02

Greatest Eclipse = 01:09:33.7 TD  (= 01:08:06.7 UT1)

Eclipse Magnitude = 0.9793

Gamma = 1.0215

Saros Series = 155

Saros Member = 8 of 71

**Sun at Greatest Eclipse**
(Geocentric Coordinates)
R.A. = 10h45m28.2s
Dec. = +07°52'58.6"
S.D. = 00°15'50.9"
H.P. = 00°00'08.7"

**Moon at Greatest Eclipse**
(Geocentric Coordinates)
R.A. = 10h46m40.4s
Dec. = +08°52'49.8"
S.D. = 00°16'42.0"
H.P. = 01°01'17.5"

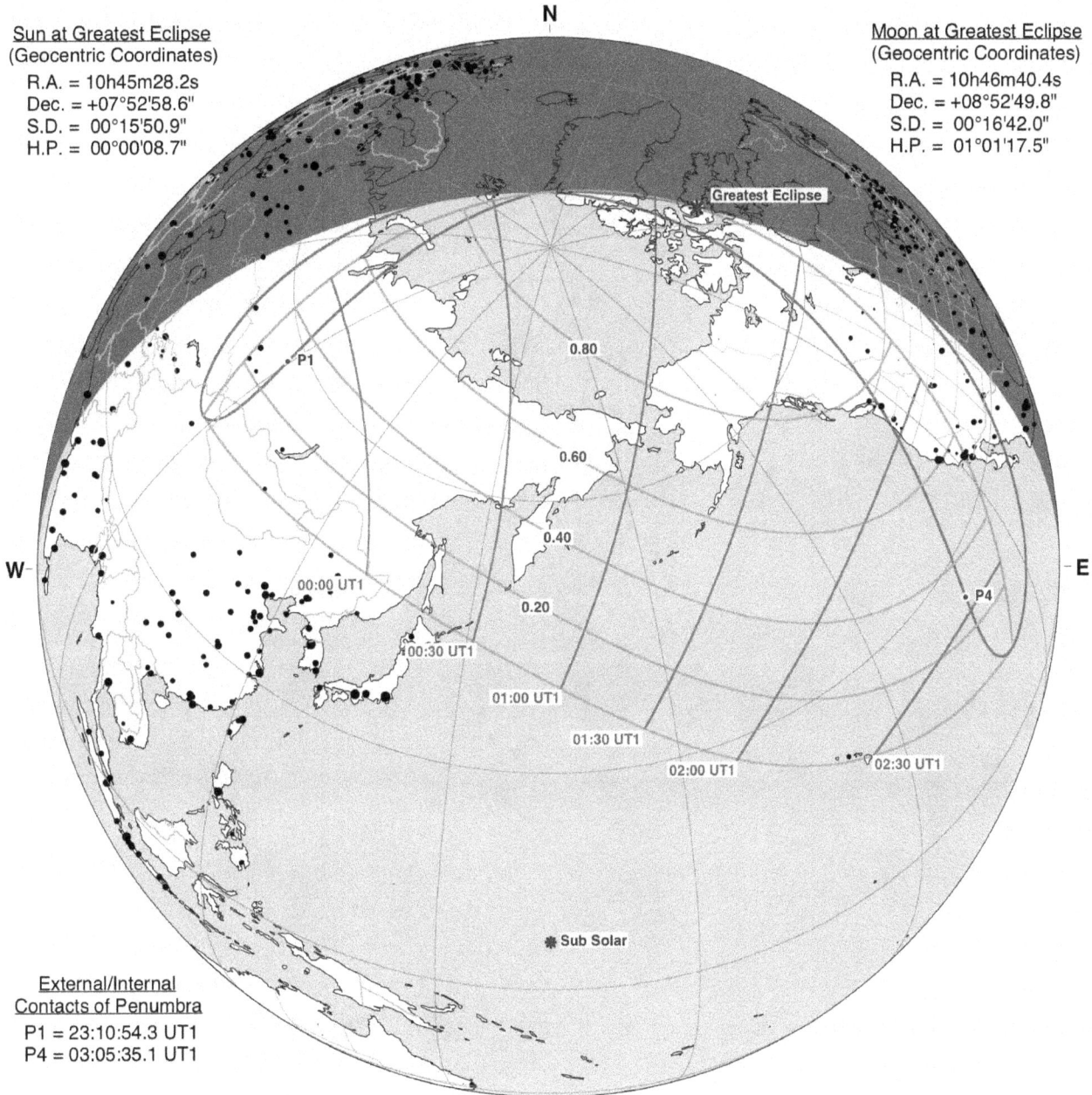

N

Greatest Eclipse

0.80

0.60

0.40

0.20

00:00 UT1

W

E

P1

P4

00:30 UT1

01:00 UT1

01:30 UT1

02:00 UT1

02:30 UT1

Sub Solar

**External/Internal
Contacts of Penumbra**
P1 = 23:10:54.3 UT1
P4 = 03:05:35.1 UT1

ΔT = 86.9 s

S

Eph. = JPL DE405

Circumstances at Greatest Eclipse: 01:08:06.7 UT1
Lat. = 71°40.9'N     Sun Alt. = 0.0°
Long. = 082°24.8'W     Sun Azm. = 295.9°

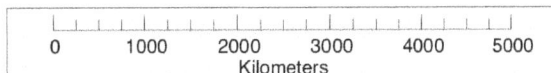

0 1000 2000 3000 4000 5000
Kilometers

# Partial Solar Eclipse of 2055 Jan 27

Greatest Eclipse = 17:54:05.3 TD (= 17:52:38.1 UT1)

Eclipse Magnitude = 0.6932      Saros Series = 122
Gamma = 1.1550      Saros Member = 60 of 70

**Sun at Greatest Eclipse**
(Geocentric Coordinates)
R.A. = 20h40m41.0s
Dec. = -18°19'18.9"
S.D. = 00°16'14.5"
H.P. = 00°00'08.9"

**Moon at Greatest Eclipse**
(Geocentric Coordinates)
R.A. = 20h39m58.6s
Dec. = -17°17'11.5"
S.D. = 00°14'53.3"
H.P. = 00°54'38.4"

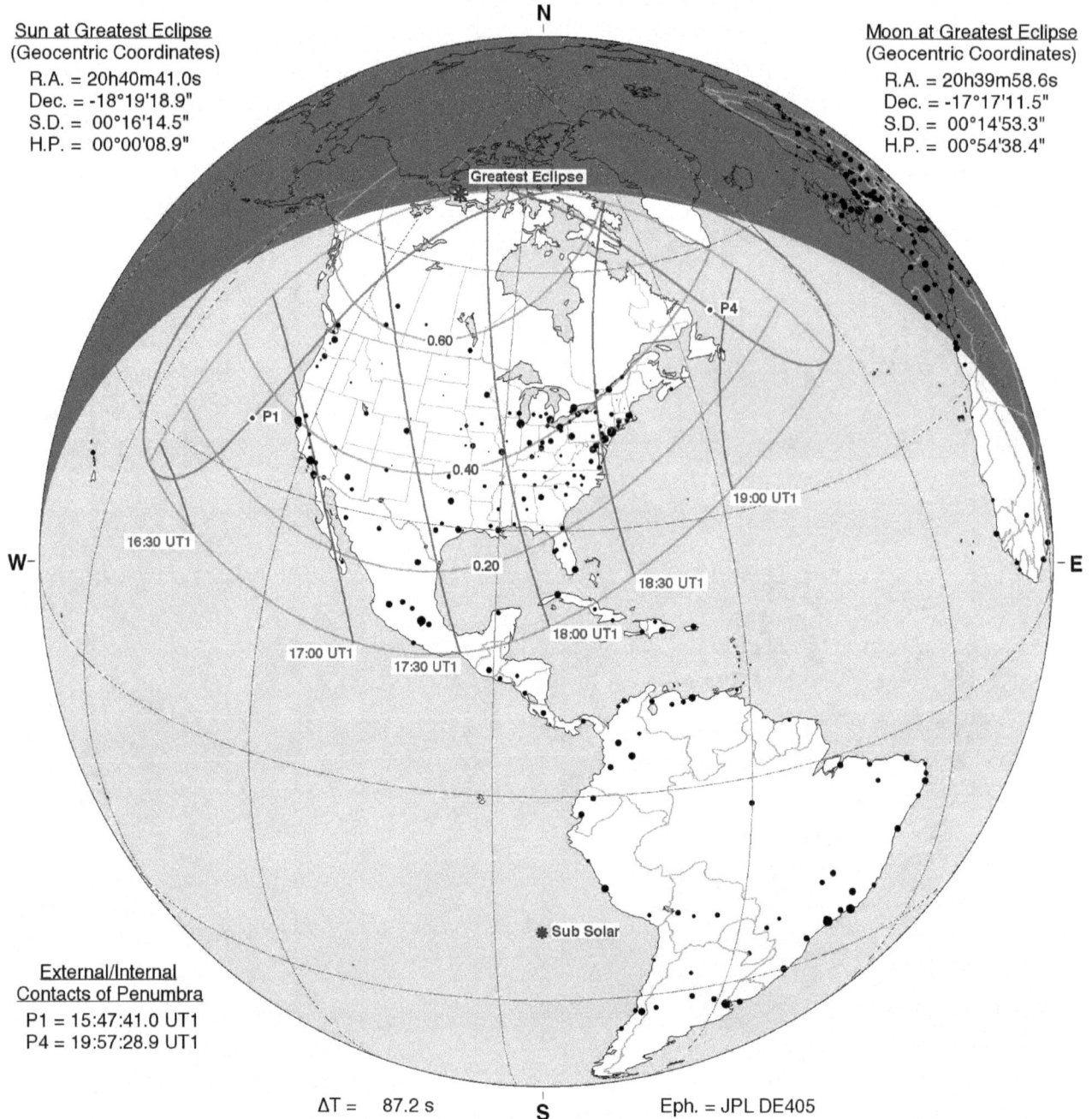

N

Greatest Eclipse

P4

0.60

P1

0.40

19:00 UT1

16:30 UT1

0.20

18:30 UT1

18:00 UT1

17:00 UT1

17:30 UT1

W

E

Sub Solar

**External/Internal**
**Contacts of Penumbra**
P1 = 15:47:41.0 UT1
P4 = 19:57:28.9 UT1

ΔT = 87.2 s

S

Eph. = JPL DE405

**Circumstances at Greatest Eclipse: 17:52:38.1 UT1**
Lat. = 69°32.9'N      Sun Alt. = 0.0°
Long. = 112°19.0'W      Sun Azm. = 154.1°

| 0 | 1000 | 2000 | 3000 | 4000 | 5000 |
Kilometers

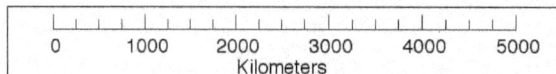

©2016 F. Espenak
www.EclipseWise.com

# Total Solar Eclipse of 2055 Jul 24

Greatest Eclipse = 09:57:50.3 TD   (= 09:56:22.7 UT1)

Eclipse Magnitude = 1.0359          Saros Series = 127
Gamma = -0.8012          Saros Member = 60 of 82

Sun at Greatest Eclipse
(Geocentric Coordinates)
R.A. = 08h15m04.2s
Dec. = +19°48'43.3"
S.D. = 00°15'44.6"
H.P. = 00°00'08.7"

Moon at Greatest Eclipse
(Geocentric Coordinates)
R.A. = 08h14m39.2s
Dec. = +19°01'42.7"
S.D. = 00°16'09.1"
H.P. = 00°59'16.7"

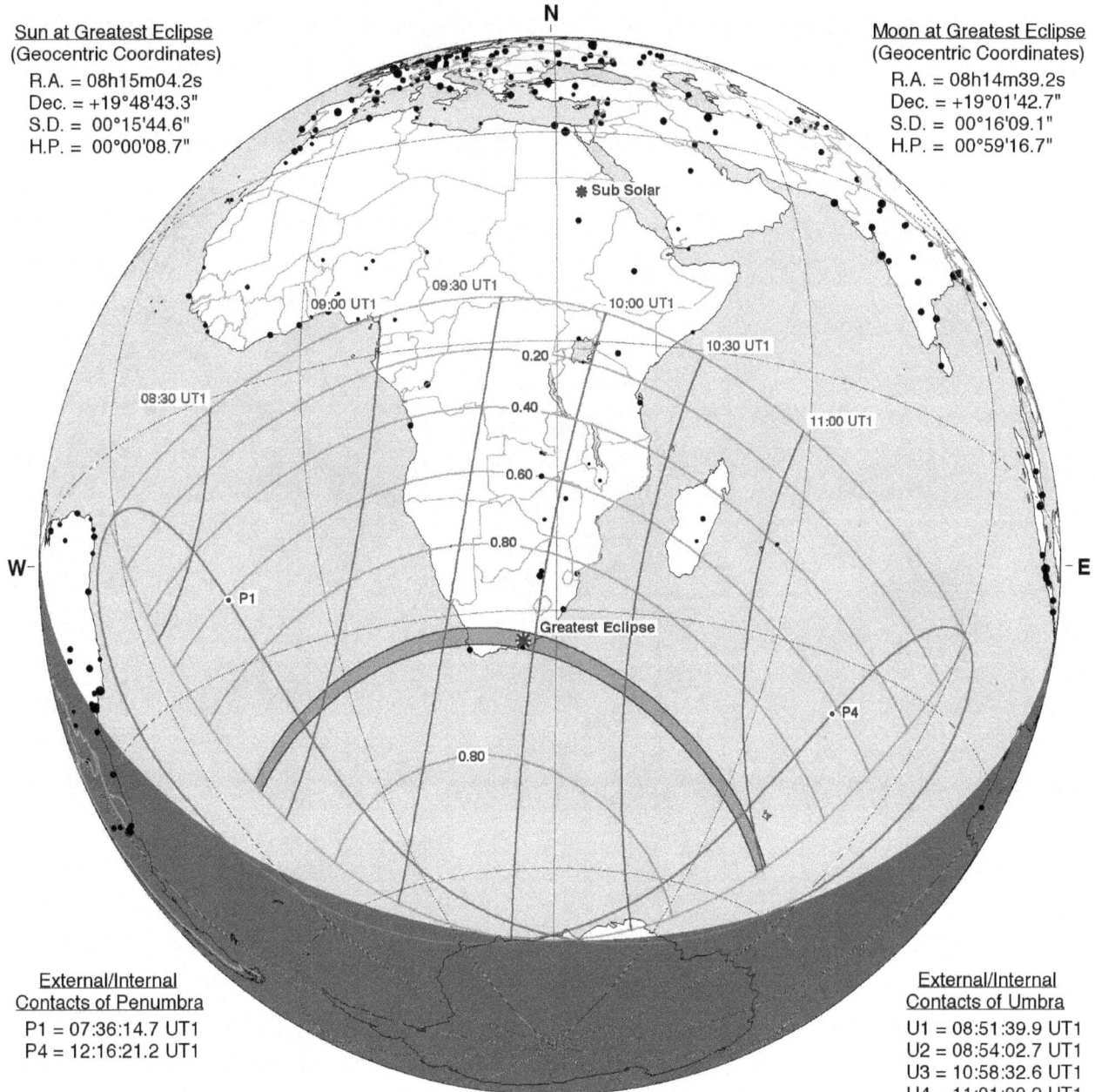

N

Sub Solar

09:30 UT1
09:00 UT1
10:00 UT1
10:30 UT1
08:30 UT1
0.20
0.40
11:00 UT1
0.60
0.80
W
E
P1
0.80
Greatest Eclipse
P4

External/Internal
Contacts of Penumbra
P1 = 07:36:14.7 UT1
P4 = 12:16:21.2 UT1

External/Internal
Contacts of Umbra
U1 = 08:51:39.9 UT1
U2 = 08:54:02.7 UT1
U3 = 10:58:32.6 UT1
U4 = 11:01:00.2 UT1

ΔT = 87.5 s     S     Eph. = JPL DE405

Circumstances at Greatest Eclipse: 09:56:22.7 UT1

| | |
|---|---|
| Lat. = 33°16.9'S | Sun Alt. = 36.5° |
| Long. = 025°43.3'E | Sun Azm. = 8.0° |
| Path Width = 201.8 km | Duration = 03m16.8s |

Circumstances at Greatest Duration: 09:56:38.1 UT1

| | |
|---|---|
| Lat. = 33°18.1'S | Sun Alt. = 36.5° |
| Long. = 025°49.2'E | Sun Azm. = 7.8° |
| Path Width = 201.7 km | Duration = 03m16.8s |

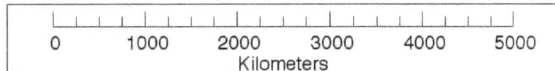

0   1000   2000   3000   4000   5000
Kilometers

©2016  F. Espenak
www.EclipseWise.com

# Annular Solar Eclipse of 2056 Jan 16

Greatest Eclipse = 22:16:45.2 TD  (= 22:15:17.3 UT1)

Eclipse Magnitude = 0.9760

Gamma = 0.4199

Saros Series = 132

Saros Member = 48 of 71

**Sun at Greatest Eclipse**
(Geocentric Coordinates)
R.A. = 19h54m06.4s
Dec. = -20°50'41.3"
S.D. = 00°16'15.5"
H.P. = 00°00'08.9"

**Moon at Greatest Eclipse**
(Geocentric Coordinates)
R.A. = 19h53m57.0s
Dec. = -20°26'45.0"
S.D. = 00°15'38.4"
H.P. = 00°57'23.8"

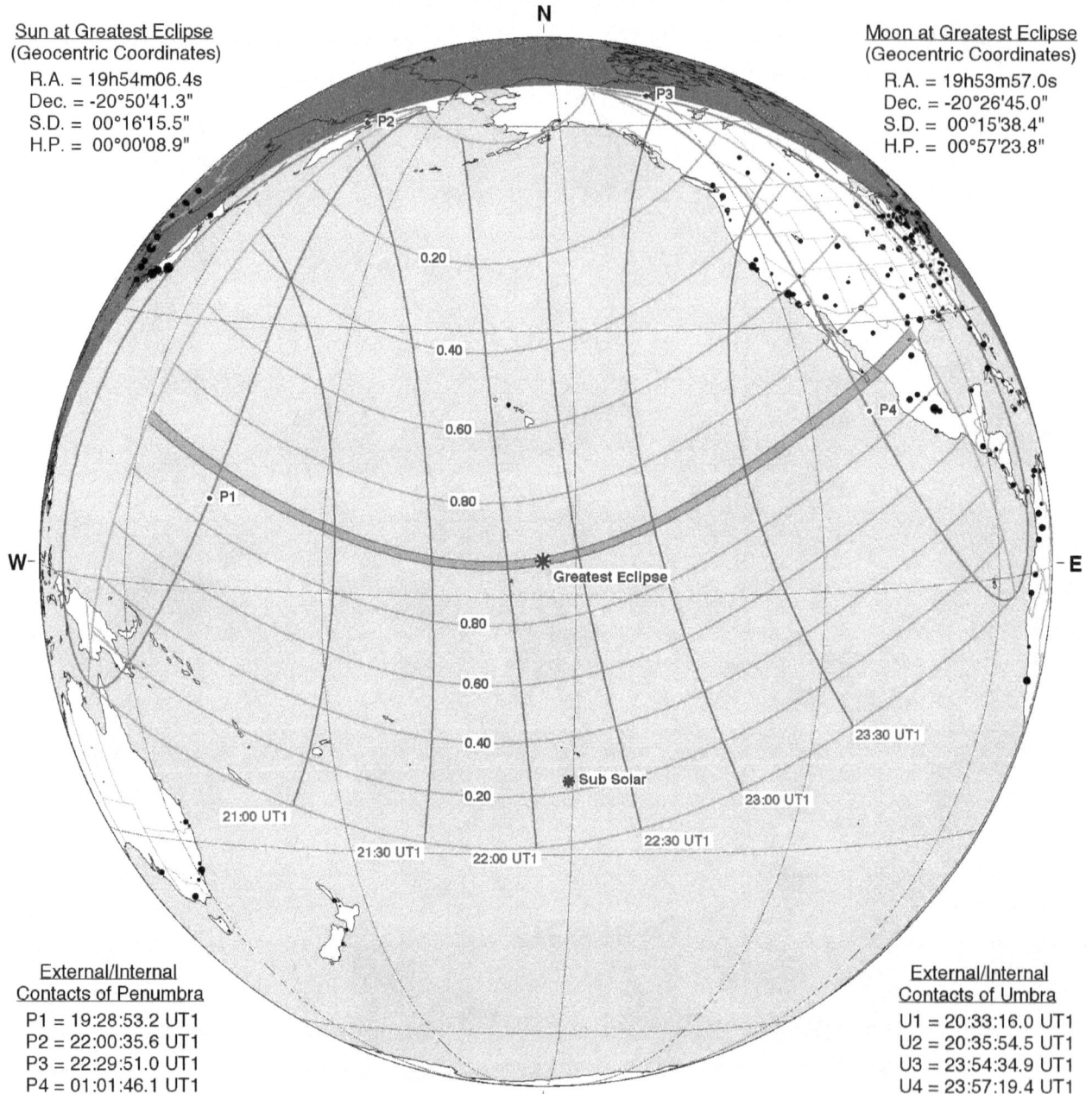

N

W

E

0.20
0.40
0.60
0.80
Greatest Eclipse
0.80
0.60
0.40
0.20
Sub Solar

P1
P2
P3
P4

23:30 UT1
23:00 UT1
22:30 UT1
22:00 UT1
21:30 UT1
21:00 UT1

S

**External/Internal
Contacts of Penumbra**
P1 = 19:28:53.2 UT1
P2 = 22:00:35.6 UT1
P3 = 22:29:51.0 UT1
P4 = 01:01:46.1 UT1

**External/Internal
Contacts of Umbra**
U1 = 20:33:16.0 UT1
U2 = 20:35:54.5 UT1
U3 = 23:54:34.9 UT1
U4 = 23:57:19.4 UT1

ΔT = 87.8 s

Eph. = JPL DE405

**Circumstances at Greatest Eclipse: 22:15:17.3 UT1**

| | |
|---|---|
| Lat. = 03°54.9'N | Sun Alt. = 65.1° |
| Long. = 153°36.2'W | Sun Azm. = 175.1° |
| Path Width = 94.5 km | Duration = 02m52.4s |

**Circumstances at Greatest Duration: 22:18:47.3 UT1**

| | |
|---|---|
| Lat. = 04°05.0'N | Sun Alt. = 65.1° |
| Long. = 152°40.5'W | Sun Azm. = 179.1° |
| Path Width = 94.5 km | Duration = 02m52.5s |

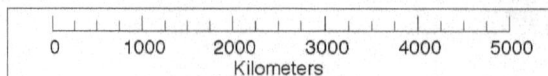

0 1000 2000 3000 4000 5000
Kilometers

©2016  F. Espenak
www.EclipseWise.com

# Annular Solar Eclipse of 2056 Jul 12

Greatest Eclipse = 20:21:59.4 TD  (= 20:20:31.2 UT1)

Eclipse Magnitude = 0.9878      Saros Series = 137

Gamma = -0.0426      Saros Member = 38 of 70

Sun at Greatest Eclipse
(Geocentric Coordinates)
R.A. = 07h31m42.7s
Dec. = +21°45'32.5"
S.D. = 00°15'44.0"
H.P. = 00°00'08.7"

Moon at Greatest Eclipse
(Geocentric Coordinates)
R.A. = 07h31m42.1s
Dec. = +21°43'09.5"
S.D. = 00°15'17.9"
H.P. = 00°56'08.9"

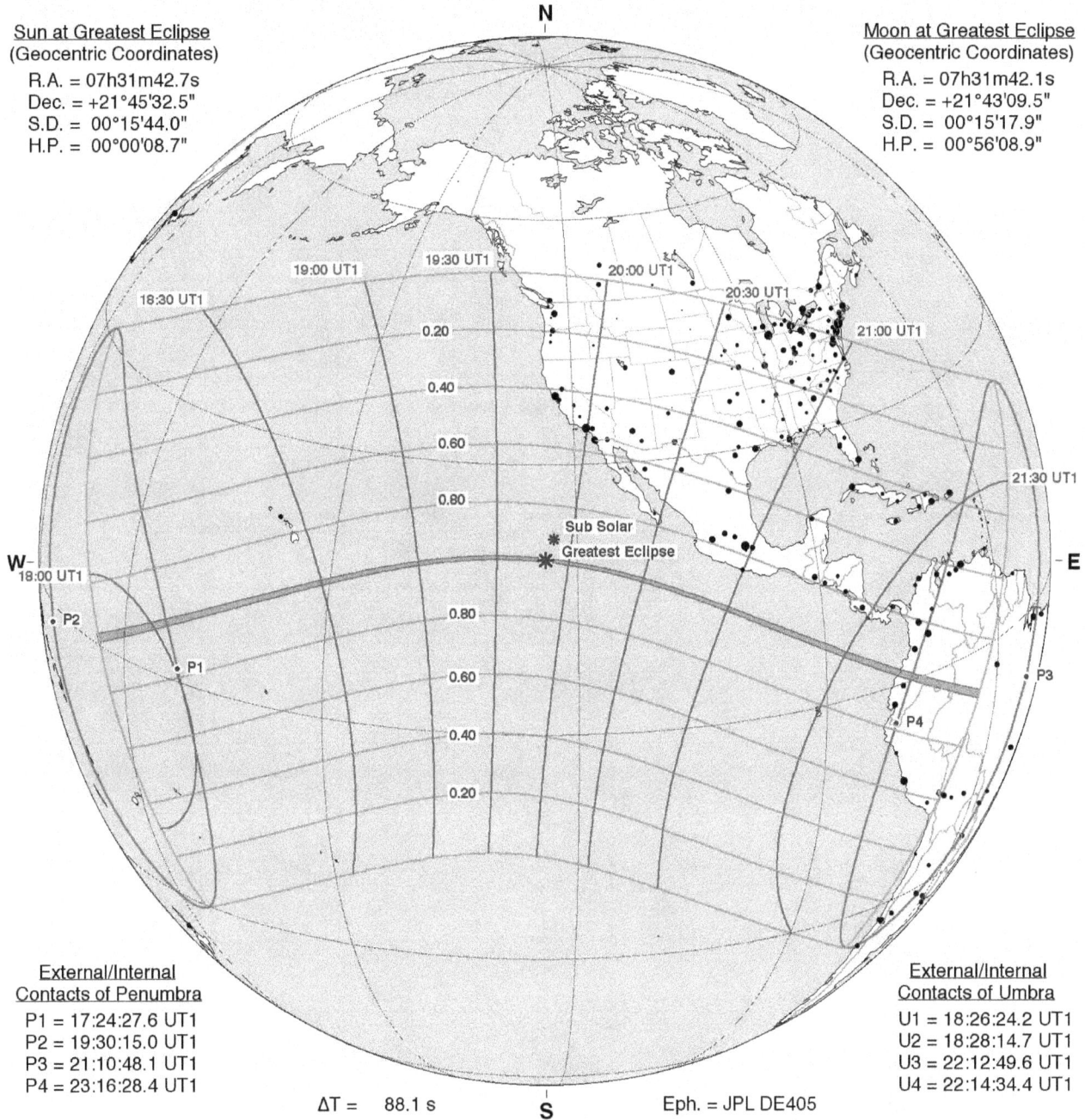

External/Internal
Contacts of Penumbra
P1 = 17:24:27.6 UT1
P2 = 19:30:15.0 UT1
P3 = 21:10:48.1 UT1
P4 = 23:16:28.4 UT1

External/Internal
Contacts of Umbra
U1 = 18:26:24.2 UT1
U2 = 18:28:14.7 UT1
U3 = 22:12:49.6 UT1
U4 = 22:14:34.4 UT1

ΔT = 88.1 s      Eph. = JPL DE405

Circumstances at Greatest Eclipse: 20:20:31.2 UT1

| | |
|---|---|
| Lat. = 19°26.4'N | Sun Alt. = 87.7° |
| Long. = 123°48.5'W | Sun Azm. = 3.3° |
| Path Width = 43.2 km | Duration = 01m25.9s |

Circumstances at Greatest Duration: 18:27:19.5 UT1

| | |
|---|---|
| Lat. = 00°41.1'N | Sun Alt. = 0.0° |
| Long. = 174°21.7'E | Sun Azm. = 68.2° |
| Path Width = 104.6 km | Duration = 01m49.4s |

©2016 F. Espenak
www.EclipseWise.com

# Total Solar Eclipse of 2057 Jan 05

Greatest Eclipse = 09:47:52.2 TD  (= 09:46:23.7 UT1)

| | |
|---|---|
| Eclipse Magnitude = 1.0287 | Saros Series = 142 |
| Gamma = -0.2837 | Saros Member = 25 of 72 |

**Sun at Greatest Eclipse**
(Geocentric Coordinates)
R.A. = 19h07m25.3s
Dec. = -22°31'37.8"
S.D. = 00°16'15.9"
H.P. = 00°00'08.9"

**Moon at Greatest Eclipse**
(Geocentric Coordinates)
R.A. = 19h07m26.6s
Dec. = -22°48'43.6"
S.D. = 00°16'27.8"
H.P. = 01°00'25.4"

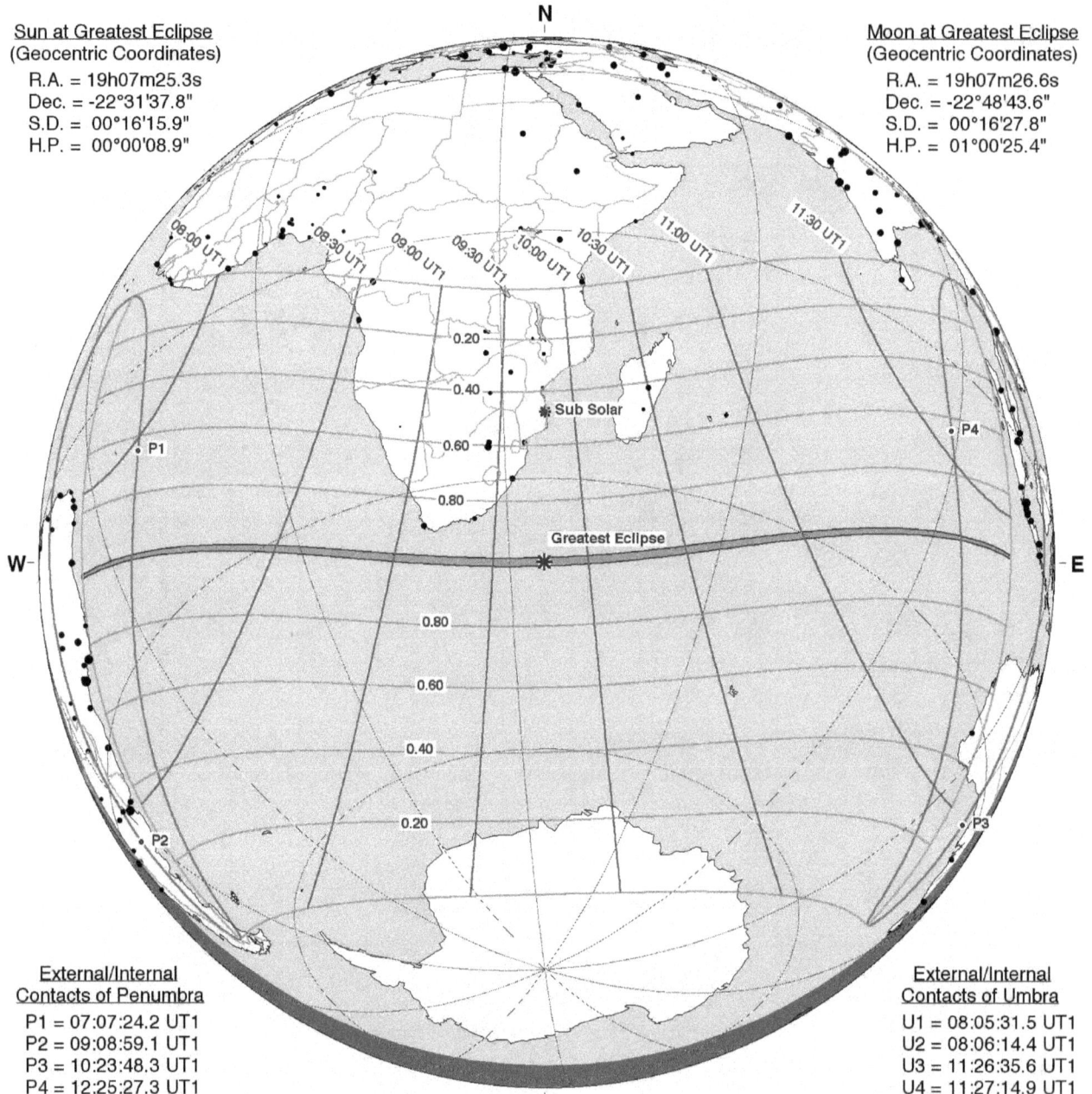

N

08:00 UT1  08:30 UT1  09:00 UT1  09:30 UT1  10:00 UT1  10:30 UT1  11:00 UT1  11:30 UT1

0.20
0.40
* Sub Solar
0.60
P1
0.80

W

Greatest Eclipse *

E

0.80
0.60
0.40
0.20

P2

P4

P3

S

**External/Internal Contacts of Penumbra**
P1 = 07:07:24.2 UT1
P2 = 09:08:59.1 UT1
P3 = 10:23:48.3 UT1
P4 = 12:25:27.3 UT1

**External/Internal Contacts of Umbra**
U1 = 08:05:31.5 UT1
U2 = 08:06:14.4 UT1
U3 = 11:26:35.6 UT1
U4 = 11:27:14.9 UT1

ΔT = 88.4 s          Eph. = JPL DE405

| Circumstances at Greatest Eclipse: 09:46:23.7 UT1 | | Circumstances at Greatest Duration: 09:44:47.7 UT1 | |
|---|---|---|---|
| Lat. = 39°13.0'S | Sun Alt. = 73.3° | Lat. = 39°13.8'S | Sun Alt. = 73.3° |
| Long. = 035°06.8'E | Sun Azm. = 358.9° | Long. = 034°23.1'E | Sun Azm. = 2.5° |
| Path Width = 101.6 km | Duration = 02m28.7s | Path Width = 101.6 km | Duration = 02m28.7s |

0    1000    2000    3000    4000    5000
Kilometers

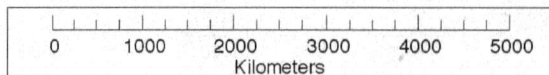

# Annular Solar Eclipse of 2057 Jul 01

Greatest Eclipse = 23:40:15.3 TD (= 23:38:46.6 UT1)

Eclipse Magnitude = 0.9464     Saros Series = 147

Gamma = 0.7455     Saros Member = 25 of 80

**N**

Sun at Greatest Eclipse
(Geocentric Coordinates)
R.A. = 06h46m13.5s
Dec. = +23°00'23.1"
S.D. = 00°15'43.9"
H.P. = 00°00'08.6"

Moon at Greatest Eclipse
(Geocentric Coordinates)
R.A. = 06h46m10.1s
Dec. = +23°40'36.4"
S.D. = 00°14'44.6"
H.P. = 00°54'06.5"

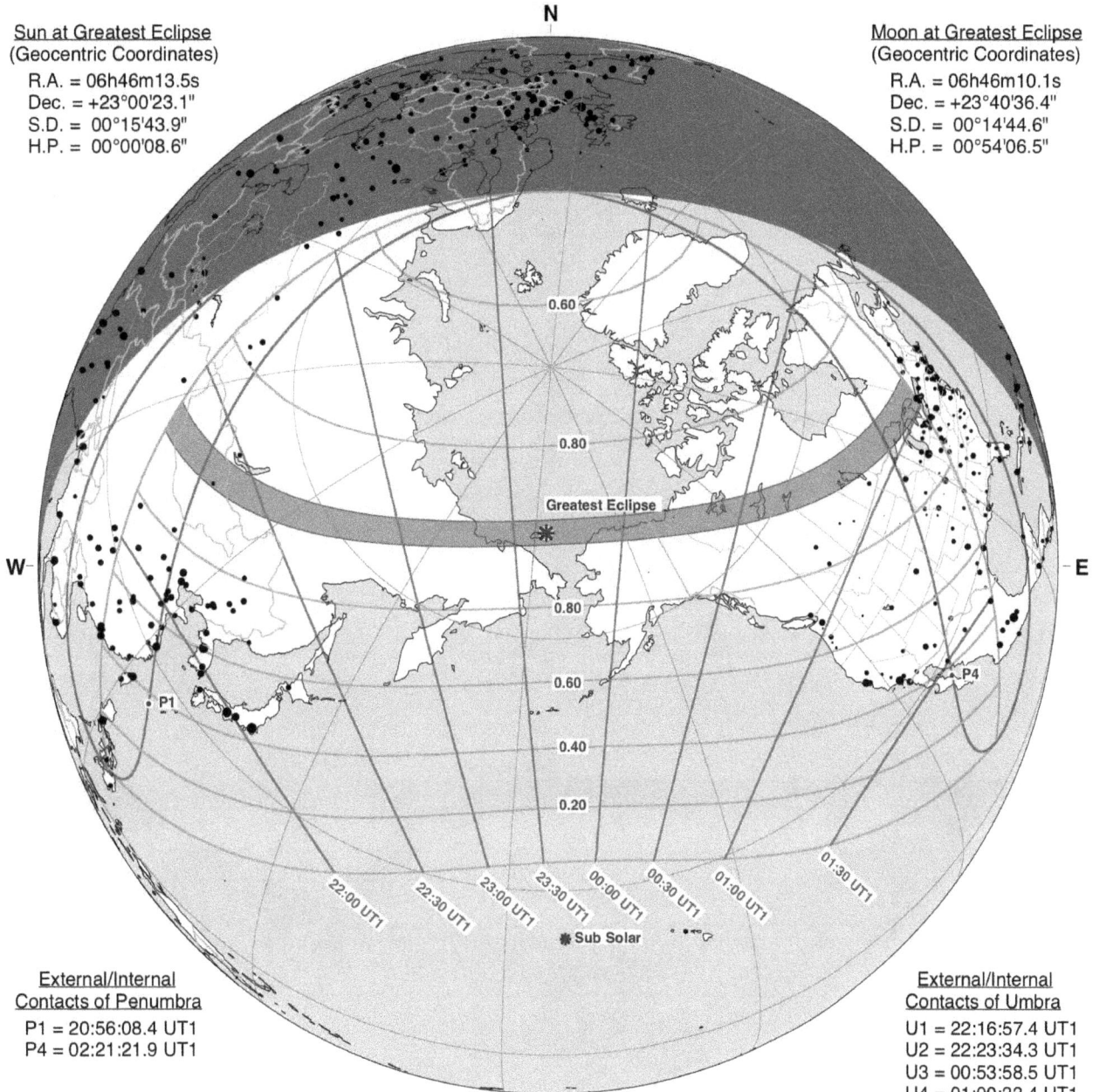

0.60

0.80

Greatest Eclipse

**W**     **E**

0.80

0.60

P1

P4

0.40

0.20

22:00 UT1   22:30 UT1   23:00 UT1   23:30 UT1   00:00 UT1   00:30 UT1   01:00 UT1   01:30 UT1

* Sub Solar

External/Internal
Contacts of Penumbra
P1 = 20:56:08.4 UT1
P4 = 02:21:21.9 UT1

External/Internal
Contacts of Umbra
U1 = 22:16:57.4 UT1
U2 = 22:23:34.3 UT1
U3 = 00:53:58.5 UT1
U4 = 01:00:33.4 UT1

ΔT = 88.8 s     **S**     Eph. = JPL DE405

| Circumstances at Greatest Eclipse: 23:38:46.6 UT1 | | Circumstances at Greatest Duration: 23:38:04.2 UT1 | |
|---|---|---|---|
| Lat. = 71°29.6'N | Sun Alt. = 41.5° | Lat. = 71°28.4'N | Sun Alt. = 41.5° |
| Long. = 176°17.0'W | Sun Azm. = 176.8° | Long. = 177°10.8'W | Sun Azm. = 175.4° |
| Path Width = 298.2 km | Duration = 04m22.5s | Path Width = 298.2 km | Duration = 04m22.5s |

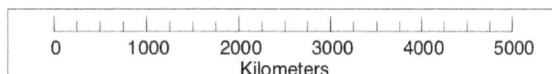

©2016 F. Espenak
www.EclipseWise.com

0   1000   2000   3000   4000   5000
Kilometers

# Total Solar Eclipse of 2057 Dec 26

Greatest Eclipse = 01:14:35.2 TD  (= 01:13:06.1 UT1)

| | |
|---|---|
| Eclipse Magnitude = 1.0348 | Saros Series = 152 |
| Gamma = -0.9405 | Saros Member = 15 of 70 |

Sun at Greatest Eclipse
(Geocentric Coordinates)
R.A. = 18h20m37.6s
Dec. = -23°20'50.0"
S.D. = 00°16'15.6"
H.P. = 00°00'08.9"

Moon at Greatest Eclipse
(Geocentric Coordinates)
R.A. = 18h20m22.1s
Dec. = -24°18'21.4"
S.D. = 00°16'44.3"
H.P. = 01°01'25.8"

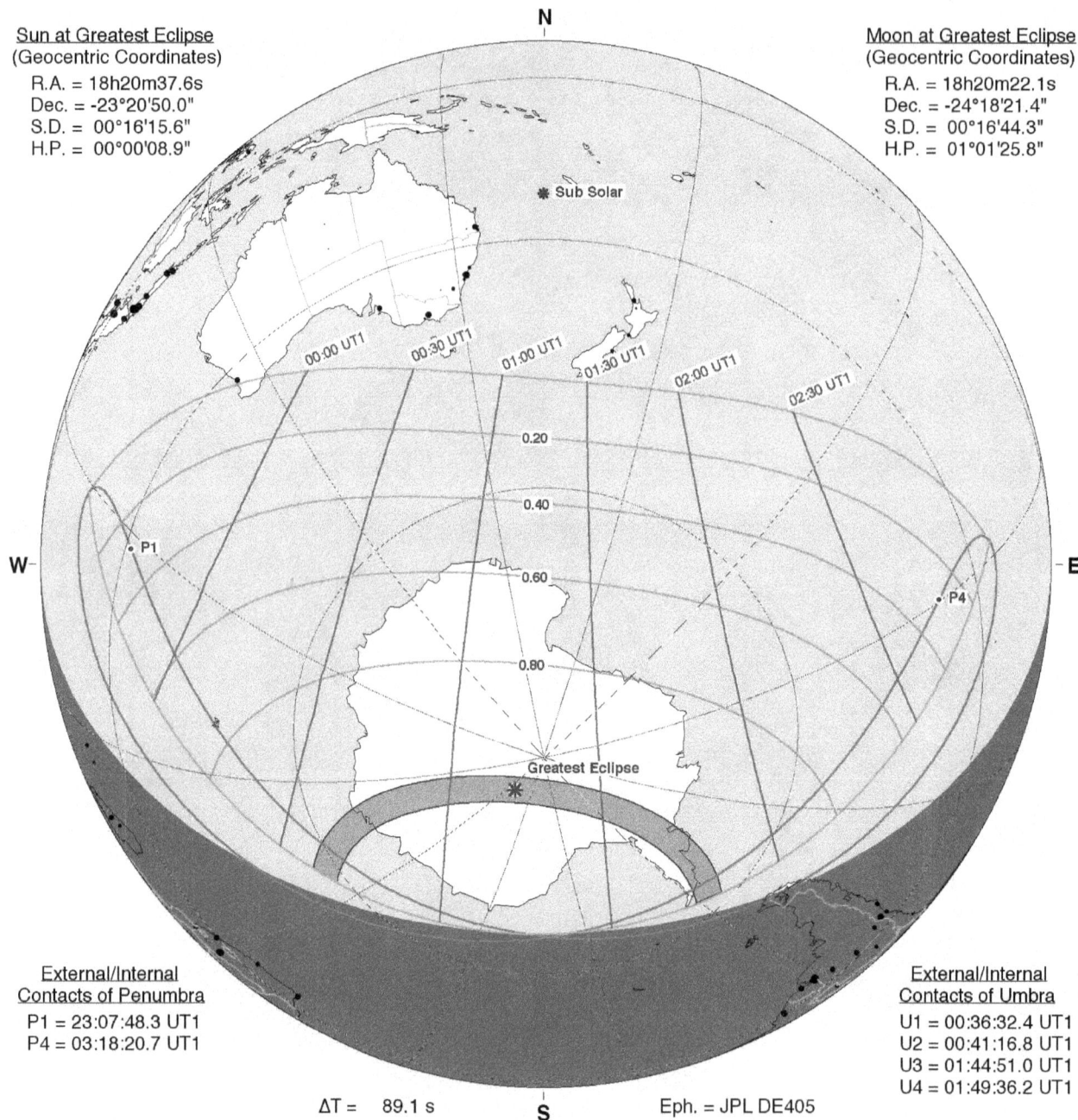

N

* Sub Solar

00:00 UT1  00:30 UT1  01:00 UT1  01:30 UT1  02:00 UT1  02:30 UT1

0.20

0.40

P1

0.60

P4

0.80

Greatest Eclipse
*

W — E

External/Internal
Contacts of Penumbra
P1 = 23:07:48.3 UT1
P4 = 03:18:20.7 UT1

External/Internal
Contacts of Umbra
U1 = 00:36:32.4 UT1
U2 = 00:41:16.8 UT1
U3 = 01:44:51.0 UT1
U4 = 01:49:36.2 UT1

ΔT = 89.1 s

S

Eph. = JPL DE405

Circumstances at Greatest Eclipse: 01:13:06.1 UT1

| | |
|---|---|
| Lat. = 84°51.1'S | Sun Alt. = 19.4° |
| Long. = 021°42.3'E | Sun Azm. = 141.4° |
| Path Width = 354.8 km | Duration = 01m49.9s |

Circumstances at Greatest Duration: 01:13:13.9 UT1

| | |
|---|---|
| Lat. = 84°53.7'S | Sun Alt. = 19.4° |
| Long. = 021°02.3'E | Sun Azm. = 142.0° |
| Path Width = 354.8 km | Duration = 01m49.9s |

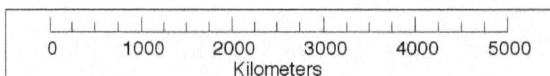

©2016  F. Espenak
www.EclipseWise.com

0   1000   2000   3000   4000   5000
Kilometers

# Partial Solar Eclipse of 2058 May 22

Greatest Eclipse = 10:39:25.5 TD   (= 10:37:56.2 UT1)

| | |
|---|---|
| Eclipse Magnitude = 0.4141 | Saros Series = 119 |
| Gamma = -1.3194 | Saros Member = 68 of 71 |

Sun at Greatest Eclipse
(Geocentric Coordinates)
R.A. = 03h58m00.8s
Dec. = +20°28'40.9"
S.D. = 00°15'48.1"
H.P. = 00°00'08.7"

Moon at Greatest Eclipse
(Geocentric Coordinates)
R.A. = 03h59m32.2s
Dec. = +19°18'44.2"
S.D. = 00°15'09.0"
H.P. = 00°55'36.1"

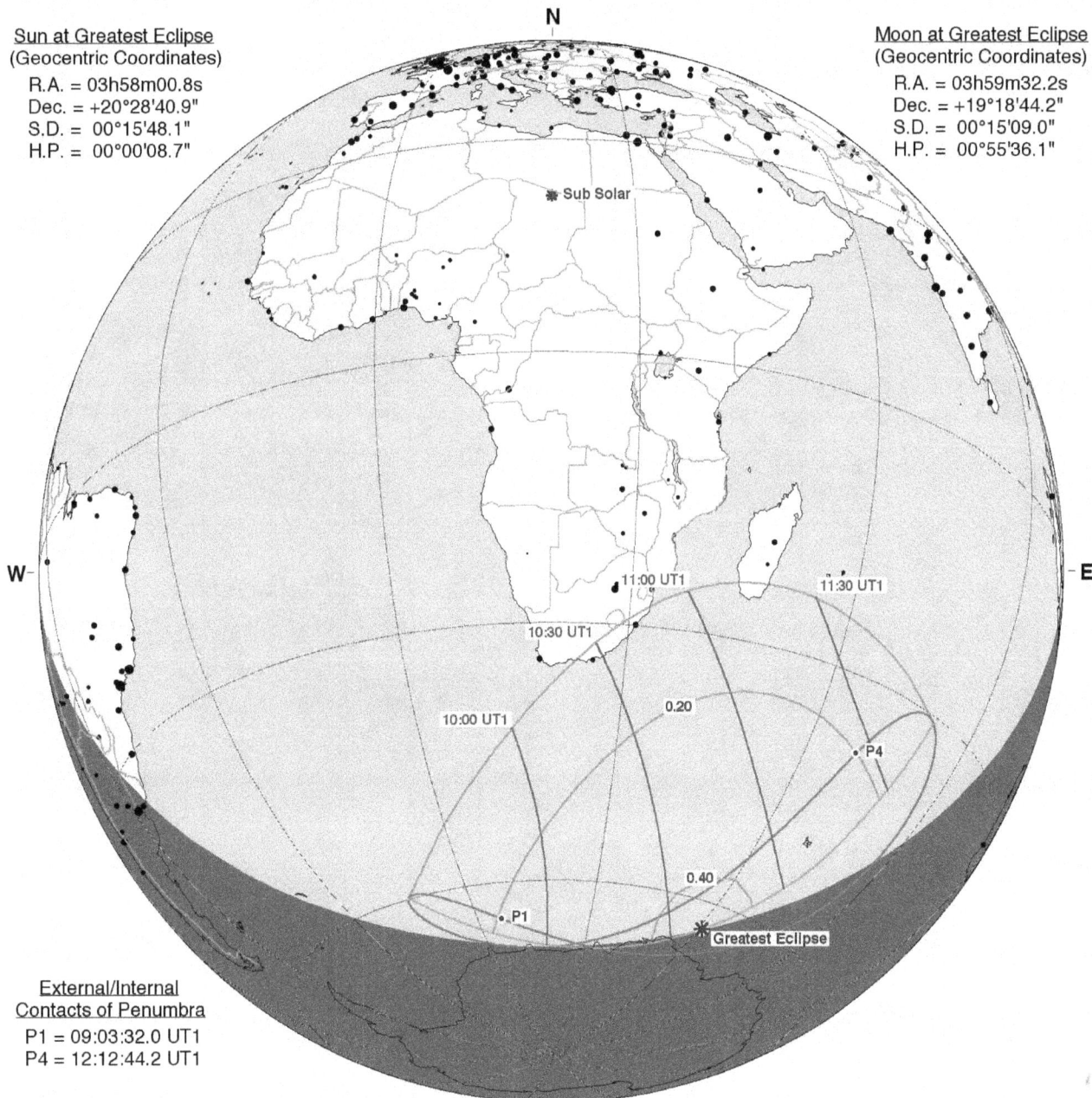

N

Sub Solar

11:00 UT1

11:30 UT1

W

E

10:30 UT1

10:00 UT1

0.20

P4

10:00 UT1

0.40

P1

Greatest Eclipse

External/Internal
Contacts of Penumbra
P1 = 09:03:32.0 UT1
P4 = 12:12:44.2 UT1

ΔT =   89.3 s

S

Eph. = JPL DE405

Circumstances at Greatest Eclipse: 10:37:56.2 UT1

| | |
|---|---|
| Lat. = 63°31.8'S | Sun Alt. =   0.0° |
| Long. = 061°03.2'E | Sun Azm. = 321.7° |

| 0 | 1000 | 2000 | 3000 | 4000 | 5000 |
|---|---|---|---|---|---|

Kilometers

©2016  F. Espenak
www.EclipseWise.com

# Partial Solar Eclipse of 2058 Jun 21

Greatest Eclipse = 00:19:34.6 TD  (= 00:18:05.3 UT1)

Eclipse Magnitude = 0.1261          Saros Series = 157
Gamma = 1.4869          Saros Member = 1 of 70

**Sun at Greatest Eclipse**
(Geocentric Coordinates)
R.A. = 05h59m41.6s
Dec. = +23°25'56.0"
S.D. = 00°15'44.3"
H.P. = 00°00'08.7"

**Moon at Greatest Eclipse**
(Geocentric Coordinates)
R.A. = 05h59m06.9s
Dec. = +24°46'21.8"
S.D. = 00°14'50.9"
H.P. = 00°54'29.6"

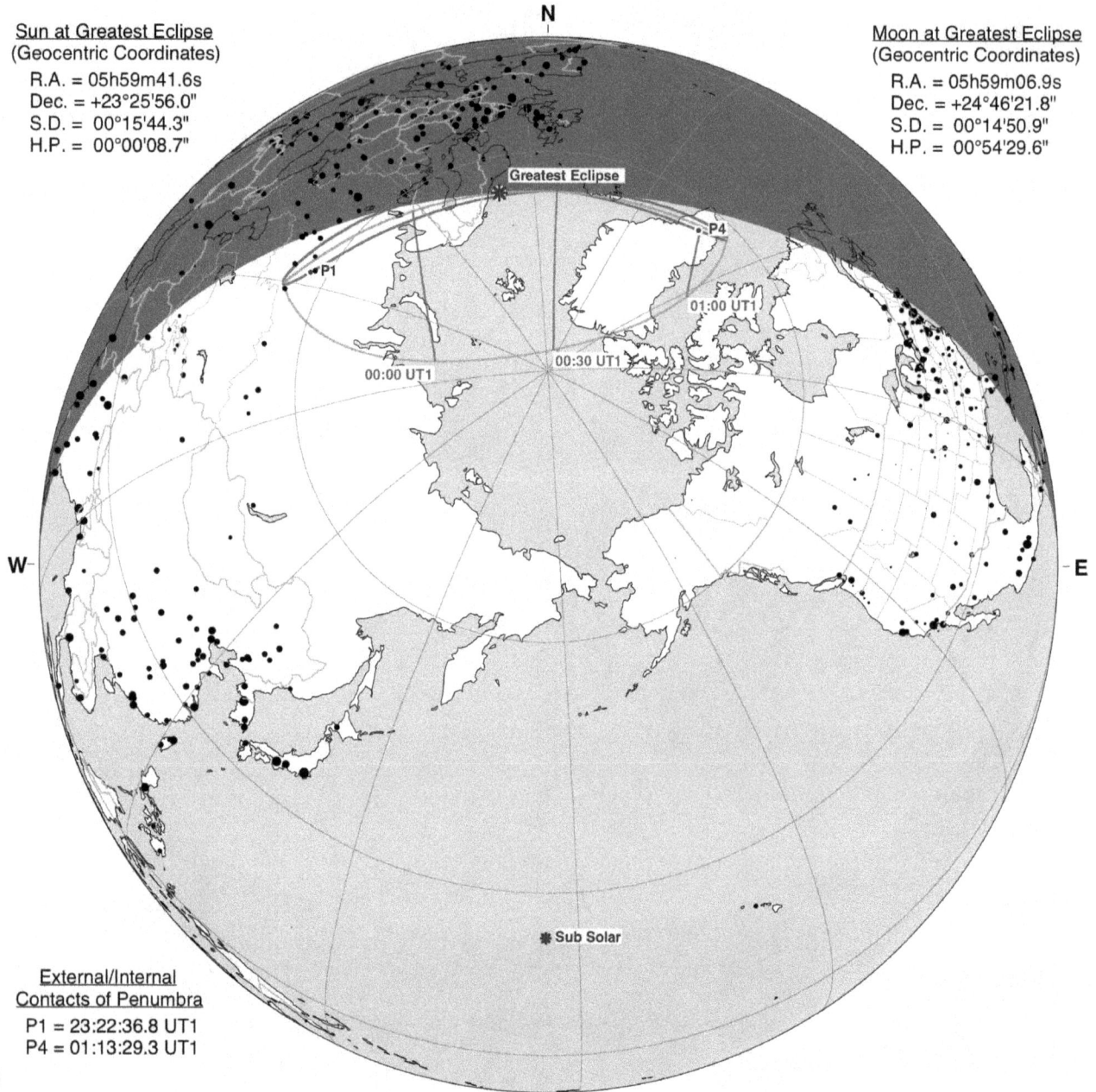

N

Greatest Eclipse

P1

P4

01:00 UT1

00:00 UT1     00:30 UT1

W

E

Sub Solar

**External/Internal**
**Contacts of Penumbra**
P1 = 23:22:36.8 UT1
P4 = 01:13:29.3 UT1

ΔT = 89.4 s          S          Eph. = JPL DE405

Circumstances at Greatest Eclipse: 00:18:05.3 UT1
Lat. = 65°57.0'N          Sun Alt. = 0.0°
Long. = 009°47.8'E          Sun Azm. = 12.7°

| 0 | 1000 | 2000 | 3000 | 4000 | 5000 |
|---|------|------|------|------|------|

Kilometers

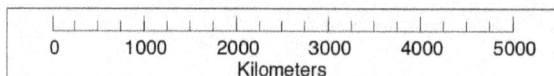

©2016 F. Espenak
www.EclipseWise.com

# Partial Solar Eclipse of 2058 Nov 16

Greatest Eclipse = 03:23:07.3 TD   (= 03:21:37.7 UT1)

Eclipse Magnitude = 0.7644          Saros Series = 124
Gamma = 1.1224            Saros Member = 57 of 73

**N**

Sun at Greatest Eclipse
(Geocentric Coordinates)
R.A. = 15h26m32.8s
Dec. = -18°46'09.8"
S.D. = 00°16'10.2"
H.P. = 00°00'08.9"

Moon at Greatest Eclipse
(Geocentric Coordinates)
R.A. = 15h28m05.5s
Dec. = -17°45'10.1"
S.D. = 00°15'47.0"
H.P. = 00°57'55.4"

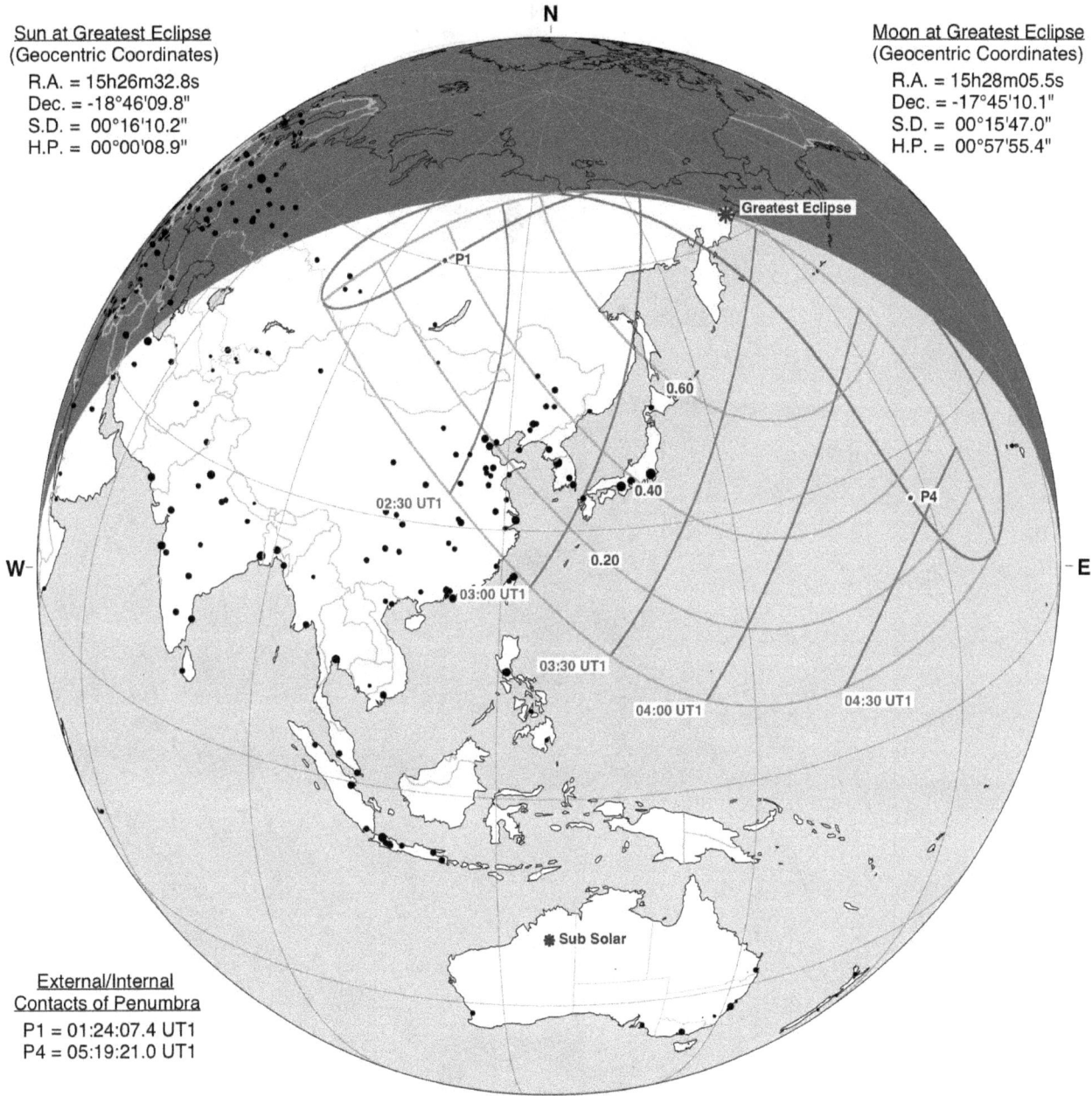

Greatest Eclipse

P1

0.60

02:30 UT1

0.40

0.20

P4

03:00 UT1

**W**                                                                **E**

03:30 UT1

04:00 UT1          04:30 UT1

External/Internal
Contacts of Penumbra
P1 = 01:24:07.4 UT1
P4 = 05:19:21.0 UT1

Sub Solar

ΔT =    89.7 s          **S**          Eph. = JPL DE405

Circumstances at Greatest Eclipse: 03:21:37.7 UT1
Lat. = 62°54.0'N          Sun Alt. =   0.0°
Long. = 174°05.5'E          Sun Azm. = 225.1°

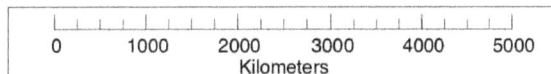

| 0 | 1000 | 2000 | 3000 | 4000 | 5000 |
|---|---|---|---|---|---|

Kilometers

©2016 F. Espenak
www.EclipseWise.com

# Total Solar Eclipse of 2059 May 11

Greatest Eclipse = 19:22:15.6 TD   (= 19:20:45.6 UT1)

Eclipse Magnitude = 1.0242          Saros Series = 129
Gamma = -0.5080                     Saros Member = 54 of 80

**Sun at Greatest Eclipse**
(Geocentric Coordinates)
R.A. = 03h14m47.9s
Dec. = +18°02'08.6"
S.D. = 00°15'50.2"
H.P. = 00°00'08.7"

**Moon at Greatest Eclipse**
(Geocentric Coordinates)
R.A. = 03h15m32.3s
Dec. = +17°34'20.5"
S.D. = 00°15'59.6"
H.P. = 00°58'41.8"

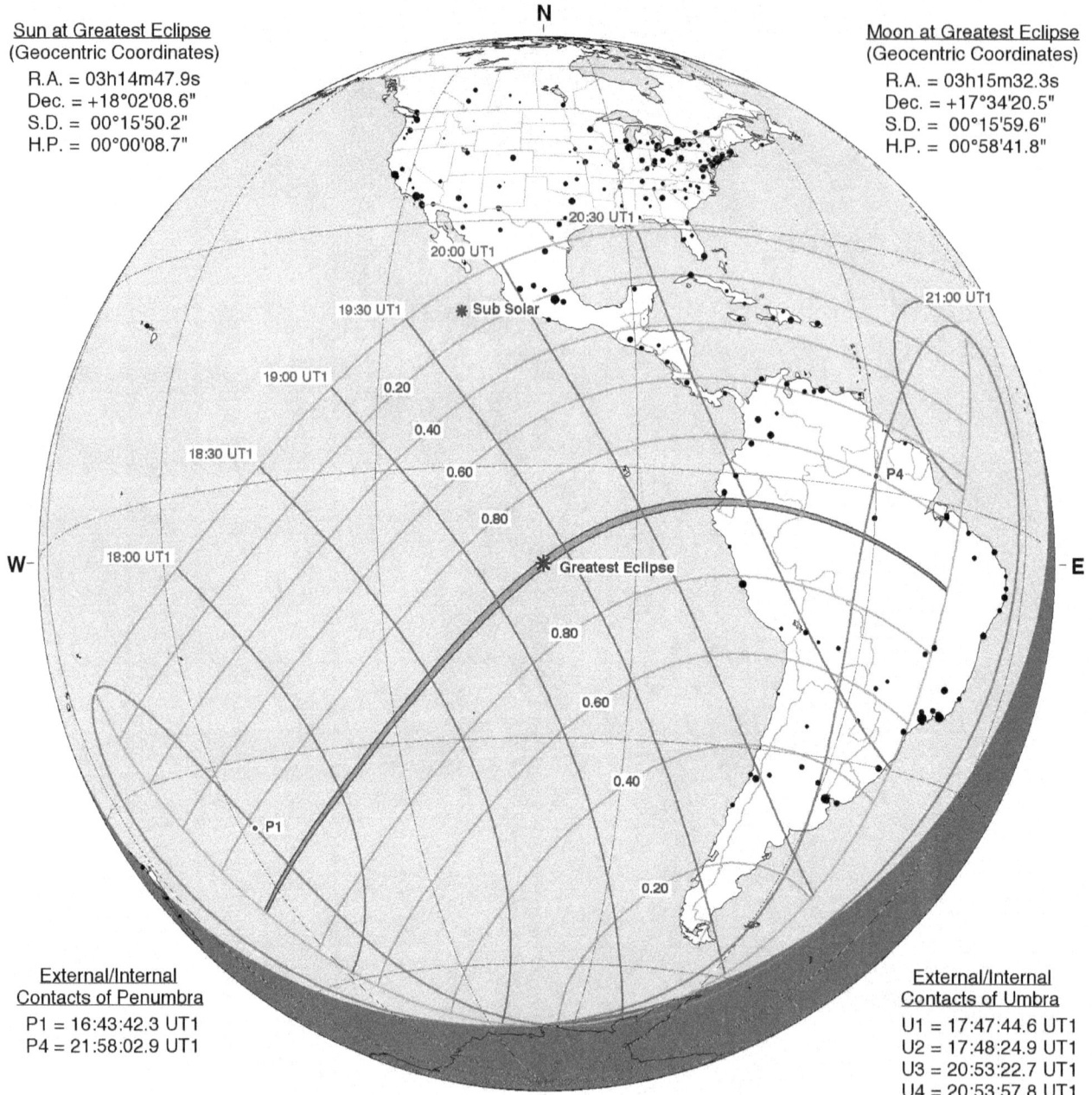

N

20:30 UT1
20:00 UT1
19:30 UT1   ✳ Sub Solar
19:00 UT1
0.20
18:30 UT1
0.40
0.60
21:00 UT1
0.80
W
✳ Greatest Eclipse
E
0.80
0.60
P4
0.40
P1
0.20

**External/Internal Contacts of Penumbra**
P1 = 16:43:42.3 UT1
P4 = 21:58:02.9 UT1

**External/Internal Contacts of Umbra**
U1 = 17:47:44.6 UT1
U2 = 17:48:24.9 UT1
U3 = 20:53:22.7 UT1
U4 = 20:53:57.8 UT1

ΔT = 90.0 s          S          Eph. = JPL DE405

**Circumstances at Greatest Eclipse: 19:20:45.6 UT1**
Lat. = 10°43.0'S          Sun Alt. = 59.4°
Long. = 100°29.3'W        Sun Azm. = 339.9°
Path Width = 94.6 km      Duration = 02m23.4s

**Circumstances at Greatest Duration: 19:22:42.0 UT1**
Lat. = 10°21.8'S          Sun Alt. = 59.4°
Long. = 099°59.3'W        Sun Azm. = 338.0°
Path Width = 94.8 km      Duration = 02m23.4s

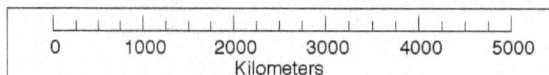

0      1000     2000     3000     4000     5000
Kilometers

# Annular Solar Eclipse of 2059 Nov 05

Greatest Eclipse = 09:18:14.6 TD  (= 09:16:44.3 UT1)

| | |
|---|---|
| Eclipse Magnitude = 0.9417 | Saros Series = 134 |
| Gamma = 0.4454 | Saros Member = 46 of 71 |

Sun at Greatest Eclipse
(Geocentric Coordinates)
R.A. = 14h42m02.6s
Dec. = -15°43'28.3"
S.D. = 00°16'07.6"
H.P. = 00°00'08.9"

Moon at Greatest Eclipse
(Geocentric Coordinates)
R.A. = 14h42m42.7s
Dec. = -15°21'02.7"
S.D. = 00°14'58.8"
H.P. = 00°54'58.7"

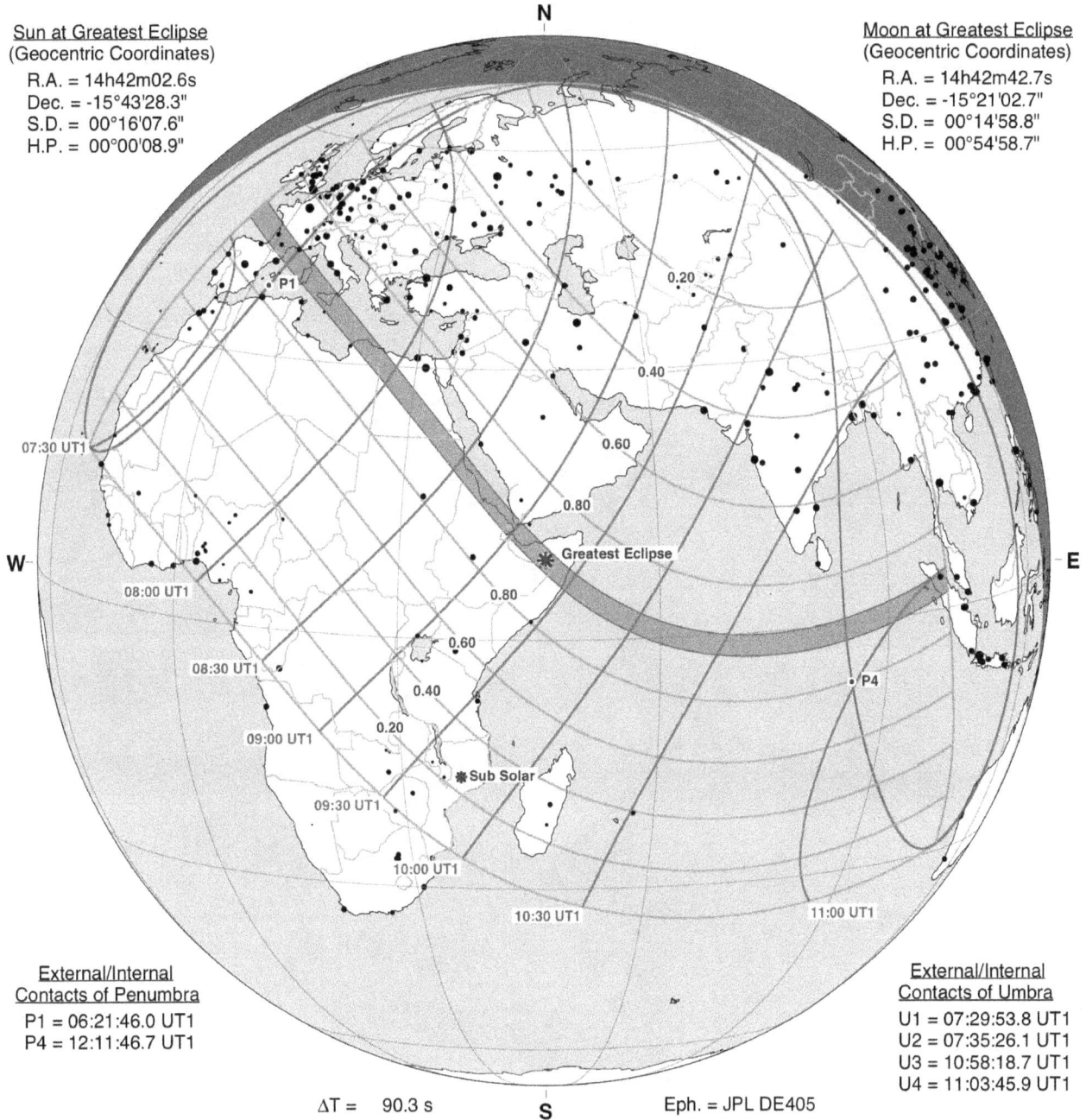

External/Internal
Contacts of Penumbra
P1 = 06:21:46.0 UT1
P4 = 12:11:46.7 UT1

External/Internal
Contacts of Umbra
U1 = 07:29:53.8 UT1
U2 = 07:35:26.1 UT1
U3 = 10:58:18.7 UT1
U4 = 11:03:45.9 UT1

ΔT = 90.3 s

Eph. = JPL DE405

Circumstances at Greatest Eclipse: 09:16:44.3 UT1

| | |
|---|---|
| Lat. = 08°44.6'N | Sun Alt. = 63.5° |
| Long. = 046°59.3'E | Sun Azm. = 202.7° |
| Path Width = 238.3 km | Duration = 06m59.6s |

Circumstances at Greatest Duration: 09:31:43.2 UT1

| | |
|---|---|
| Lat. = 06°02.0'N | Sun Alt. = 62.3° |
| Long. = 050°11.1'E | Sun Azm. = 217.9° |
| Path Width = 243.9 km | Duration = 07m01.5s |

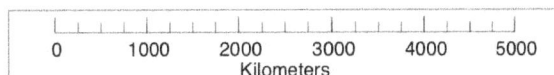

©2016 F. Espenak
www.EclipseWise.com

# Total Solar Eclipse of 2060 Apr 30

Greatest Eclipse = 10:09:59.8 TD  (= 10:08:29.2 UT1)

| | |
|---|---|
| Eclipse Magnitude = 1.0660 | Saros Series = 139 |
| Gamma = 0.2422 | Saros Member = 32 of 71 |

Sun at Greatest Eclipse
(Geocentric Coordinates)
R.A. = 02h33m38.4s
Dec. = +15°04'16.7"
S.D. = 00°15'52.6"
H.P. = 00°00'08.7"

Moon at Greatest Eclipse
(Geocentric Coordinates)
R.A. = 02h33m13.6s
Dec. = +15°17'46.8"
S.D. = 00°16'38.8"
H.P. = 01°01'05.8"

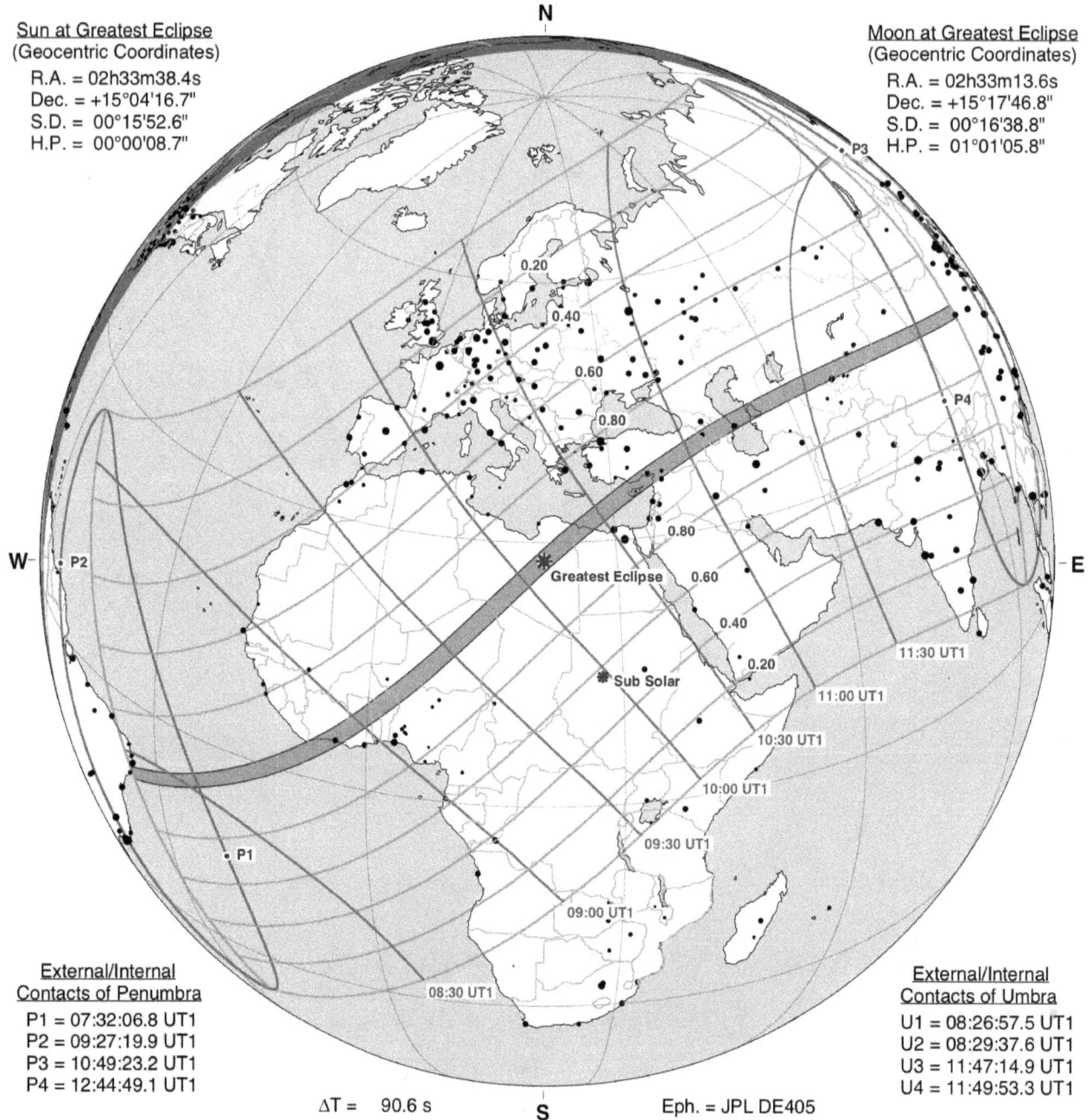

External/Internal
Contacts of Penumbra
P1 = 07:32:06.8 UT1
P2 = 09:27:19.9 UT1
P3 = 10:49:23.2 UT1
P4 = 12:44:49.1 UT1

External/Internal
Contacts of Umbra
U1 = 08:26:57.5 UT1
U2 = 08:29:37.6 UT1
U3 = 11:47:14.9 UT1
U4 = 11:49:53.3 UT1

ΔT = 90.6 s

Eph. = JPL DE405

Circumstances at Greatest Eclipse: 10:08:29.2 UT1

| | |
|---|---|
| Lat. = 27°57.4'N | Sun Alt. = 75.8° |
| Long. = 020°49.0'E | Sun Azm. = 154.1° |
| Path Width = 222.0 km | Duration = 05m15.2s |

Circumstances at Greatest Duration: 10:13:24.7 UT1

| | |
|---|---|
| Lat. = 29°09.3'N | Sun Alt. = 75.5° |
| Long. = 022°23.1'E | Sun Azm. = 166.1° |
| Path Width = 221.3 km | Duration = 05m15.5s |

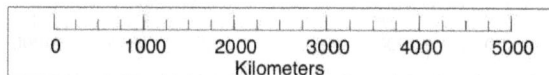

©2016  F. Espenak
www.EclipseWise.com

# Annular Solar Eclipse of 2060 Oct 24

Greatest Eclipse = 09:24:10.4 TD (= 09:22:39.4 UT1)

Eclipse Magnitude = 0.9277          Saros Series = 144

Gamma = -0.2625          Saros Member = 19 of 70

Sun at Greatest Eclipse
(Geocentric Coordinates)
R.A. = 13h58m17.5s
Dec. = -12°04'28.2"
S.D. = 00°16'04.8"
H.P. = 00°00'08.8"

Moon at Greatest Eclipse
(Geocentric Coordinates)
R.A. = 13h57m52.2s
Dec. = -12°17'09.7"
S.D. = 00°14'42.1"
H.P. = 00°53'57.3"

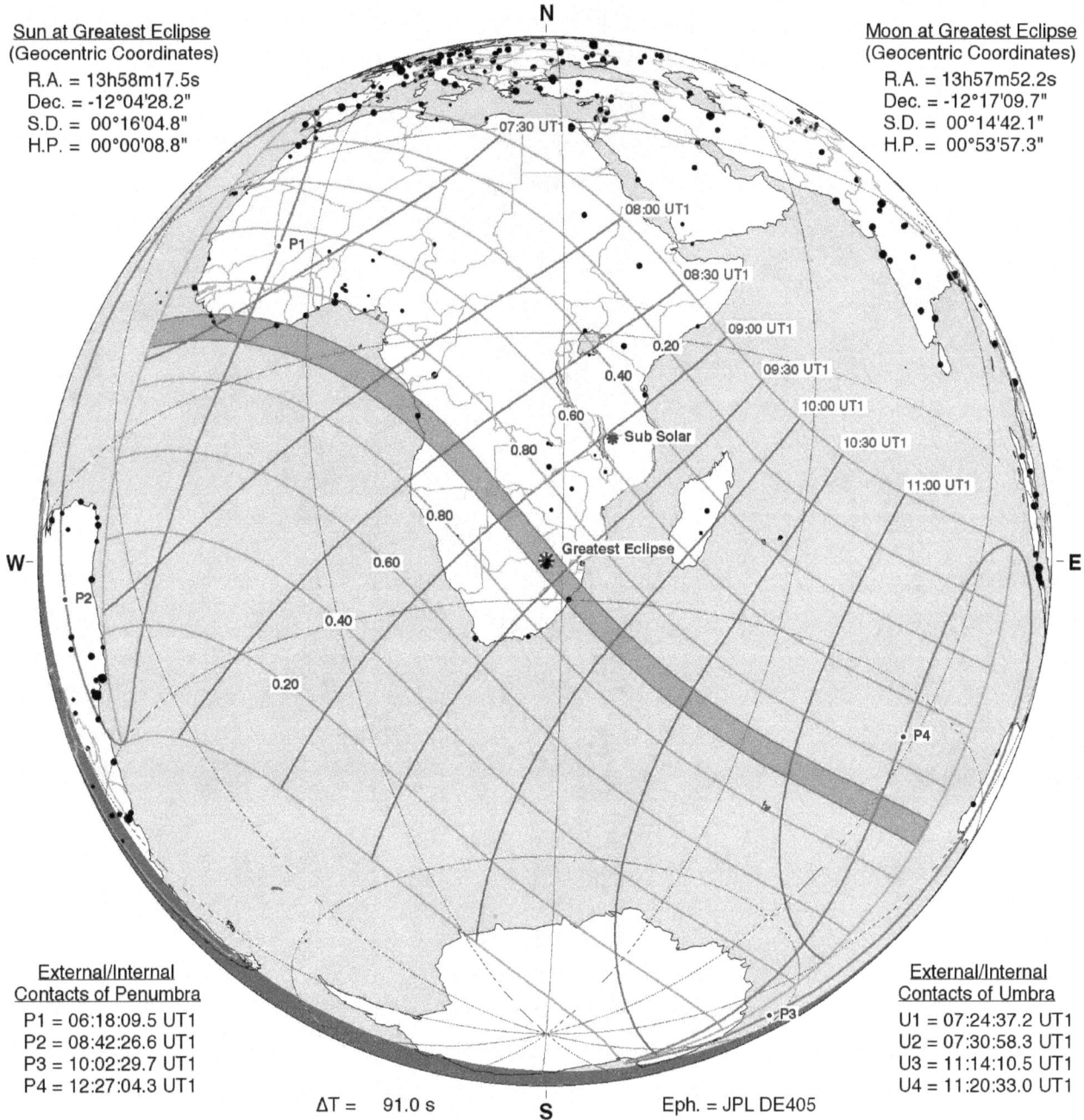

External/Internal
Contacts of Penumbra
P1 = 06:18:09.5 UT1
P2 = 08:42:26.6 UT1
P3 = 10:02:29.7 UT1
P4 = 12:27:04.3 UT1

External/Internal
Contacts of Umbra
U1 = 07:24:37.2 UT1
U2 = 07:30:58.3 UT1
U3 = 11:14:10.5 UT1
U4 = 11:20:33.0 UT1

ΔT = 91.0 s          Eph. = JPL DE405

Circumstances at Greatest Eclipse: 09:22:39.4 UT1

Lat. = 25°47.2'S          Sun Alt. = 74.6°
Long. = 028°01.1'E          Sun Azm. = 28.1°
Path Width = 280.5 km          Duration = 08m05.8s

Circumstances at Greatest Duration: 09:39:59.5 UT1

Lat. = 29°42.0'S          Sun Alt. = 72.4°
Long. = 032°03.8'E          Sun Azm. = 356.6°
Path Width = 278.5 km          Duration = 08m08.0s

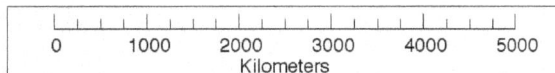

©2016 F. Espenak
www.EclipseWise.com

0   1000   2000   3000   4000   5000
Kilometers

# Total Solar Eclipse of 2061 Apr 20

Greatest Eclipse = 02:56:49.1 TD  (= 02:55:17.8 UT1)

| | |
|---|---|
| Eclipse Magnitude = 1.0476 | Saros Series = 149 |
| Gamma = 0.9578 | Saros Member = 23 of 71 |

**Sun at Greatest Eclipse**
(Geocentric Coordinates)
R.A. = 01h53m47.8s
Dec. = +11°39'59.8"
S.D. = 00°15'55.3"
H.P. = 00°00'08.8"

**Moon at Greatest Eclipse**
(Geocentric Coordinates)
R.A. = 01h52m03.2s
Dec. = +12°32'19.1"
S.D. = 00°16'36.4"
H.P. = 01°00'56.9"

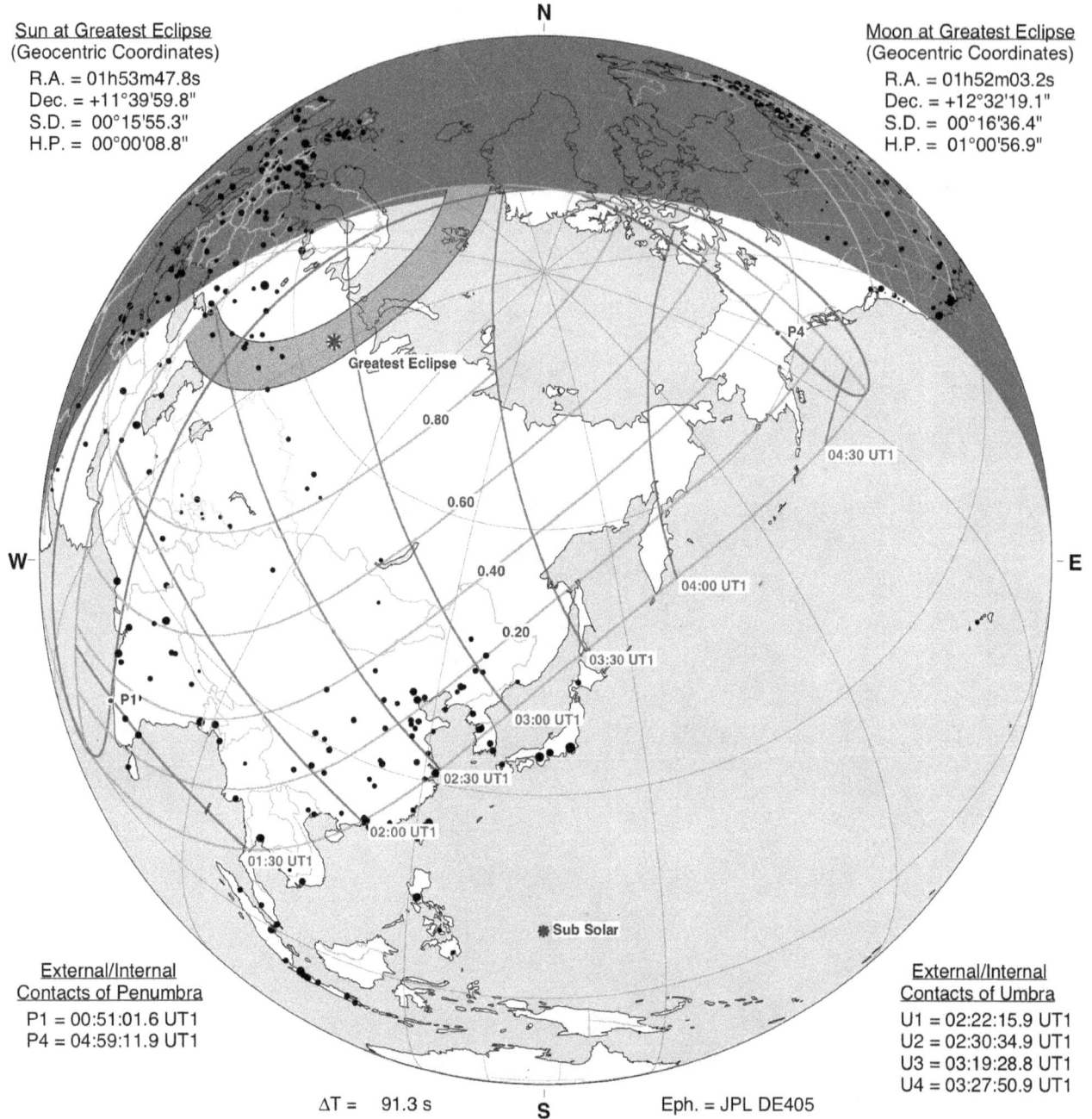

N

Greatest Eclipse

0.80

0.60

0.40

0.20

P4

04:30 UT1

04:00 UT1

03:30 UT1

03:00 UT1

02:30 UT1

02:00 UT1

01:30 UT1

P1

W

E

Sub Solar

**External/Internal
Contacts of Penumbra**
P1 = 00:51:01.6 UT1
P4 = 04:59:11.9 UT1

**External/Internal
Contacts of Umbra**
U1 = 02:22:15.9 UT1
U2 = 02:30:34.9 UT1
U3 = 03:19:28.8 UT1
U4 = 03:27:50.9 UT1

ΔT =  91.3 s    S    Eph. = JPL DE405

**Circumstances at Greatest Eclipse: 02:55:17.8 UT1**

| | |
|---|---|
| Lat. = 64°32.2'N | Sun Alt. = 16.2° |
| Long. = 059°03.3'E | Sun Azm. = 96.8° |
| Path Width = 558.9 km | Duration = 02m37.4s |

**Circumstances at Greatest Duration: 02:55:31.2 UT1**

| | |
|---|---|
| Lat. = 64°39.5'N | Sun Alt. = 16.2° |
| Long. = 059°01.8'E | Sun Azm. = 96.9° |
| Path Width = 558.2 km | Duration = 02m37.4s |

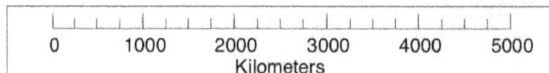

©2016  F. Espenak
www.EclipseWise.com

0   1000   2000   3000   4000   5000
Kilometers

# Annular Solar Eclipse of 2061 Oct 13

Greatest Eclipse = 10:32:09.7 TD   (= 10:30:38.0 UT1)

| | |
|---|---|
| Eclipse Magnitude = 0.9469 | Saros Series = 154 |
| Gamma = -0.9639 | Saros Member = 9 of 71 |

Sun at Greatest Eclipse
(Geocentric Coordinates)
R.A. = 13h16m11.1s
Dec. = -08°03'03.6"
S.D. = 00°16'01.7"
H.P. = 00°00'08.8"

Moon at Greatest Eclipse
(Geocentric Coordinates)
R.A. = 13h14m30.5s
Dec. = -08°50'16.1"
S.D. = 00°15'07.5"
H.P. = 00°55'30.4"

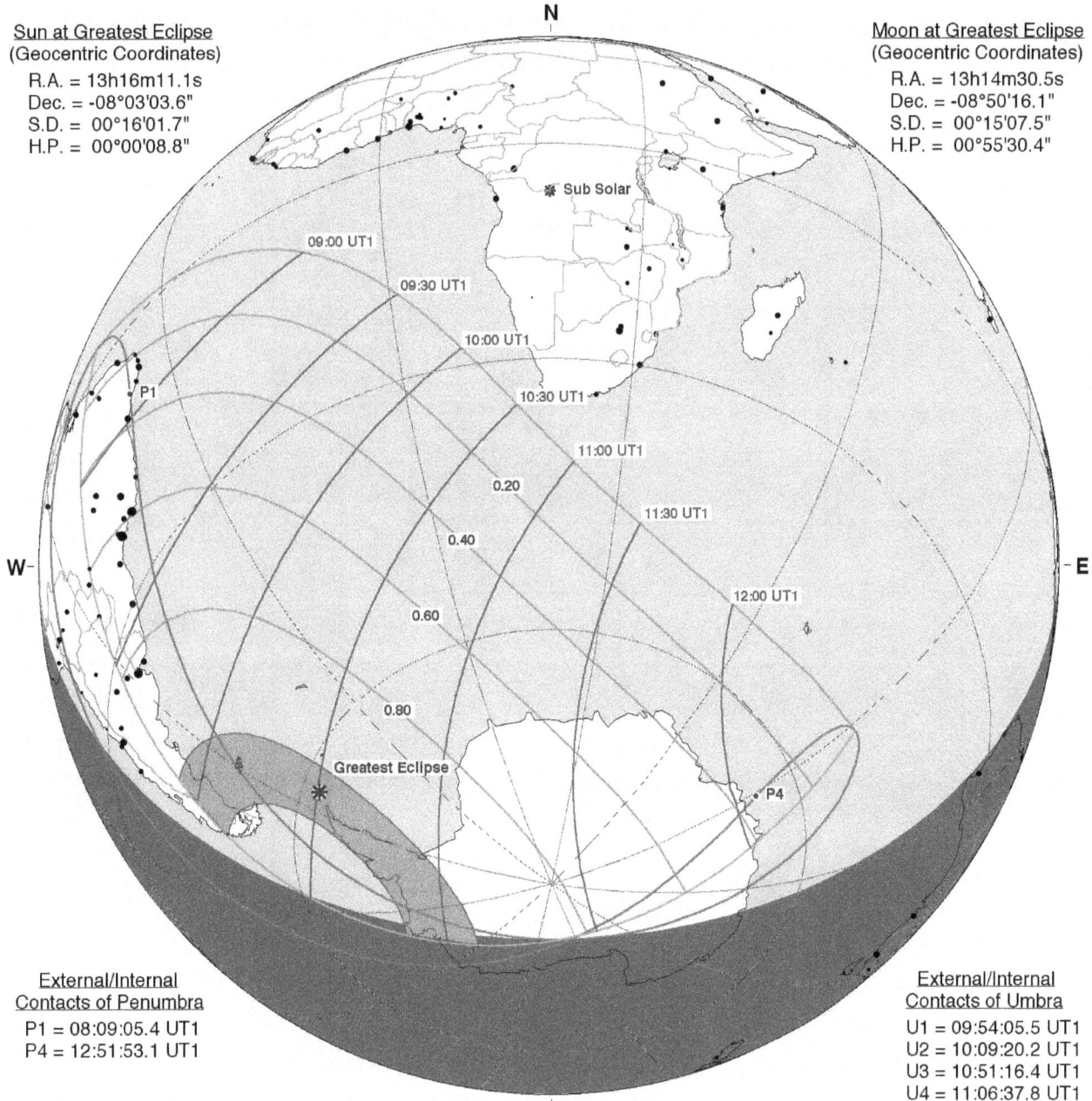

N

Sub Solar

09:00 UT1
09:30 UT1
10:00 UT1
10:30 UT1
11:00 UT1
11:30 UT1
12:00 UT1

0.20
0.40
0.60
0.80

P1

W

E

Greatest Eclipse

P4

External/Internal
Contacts of Penumbra
P1 = 08:09:05.4 UT1
P4 = 12:51:53.1 UT1

External/Internal
Contacts of Umbra
U1 = 09:54:05.5 UT1
U2 = 10:09:20.2 UT1
U3 = 10:51:16.4 UT1
U4 = 11:06:37.8 UT1

ΔT = 91.6 s     S     Eph. = JPL DE405

Circumstances at Greatest Eclipse: 10:30:38.0 UT1

| | |
|---|---|
| Lat. = 62°08.2'S | Sun Alt. = 14.9° |
| Long. = 054°29.2'W | Sun Azm. = 79.0° |
| Path Width = 742.6 km | Duration = 03m41.0s |

Circumstances at Greatest Duration: 10:33:54.1 UT1

| | |
|---|---|
| Lat. = 63°45.1'S | Sun Alt. = 14.7° |
| Long. = 054°50.1'W | Sun Azm. = 78.1° |
| Path Width = 728.7 km | Duration = 03m41.0s |

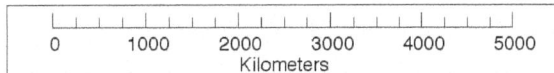

©2016 F. Espenak
www.EclipseWise.com

0    1000    2000    3000    4000    5000
Kilometers

# Partial Solar Eclipse of 2062 Mar 11

Greatest Eclipse = 04:26:16.2 TD   (= 04:24:44.3 UT1)

Eclipse Magnitude = 0.9331          Saros Series = 121
Gamma = -1.0238                     Saros Member = 63 of 71

**Sun at Greatest Eclipse**
(Geocentric Coordinates)
R.A. = 23h26m28.0s
Dec. = -03°36'57.3"
S.D. = 00°16'06.2"
H.P. = 00°00'08.9"

**Moon at Greatest Eclipse**
(Geocentric Coordinates)
R.A. = 23h28m20.0s
Dec. = -04°27'39.9"
S.D. = 00°15'26.8"
H.P. = 00°56'41.5"

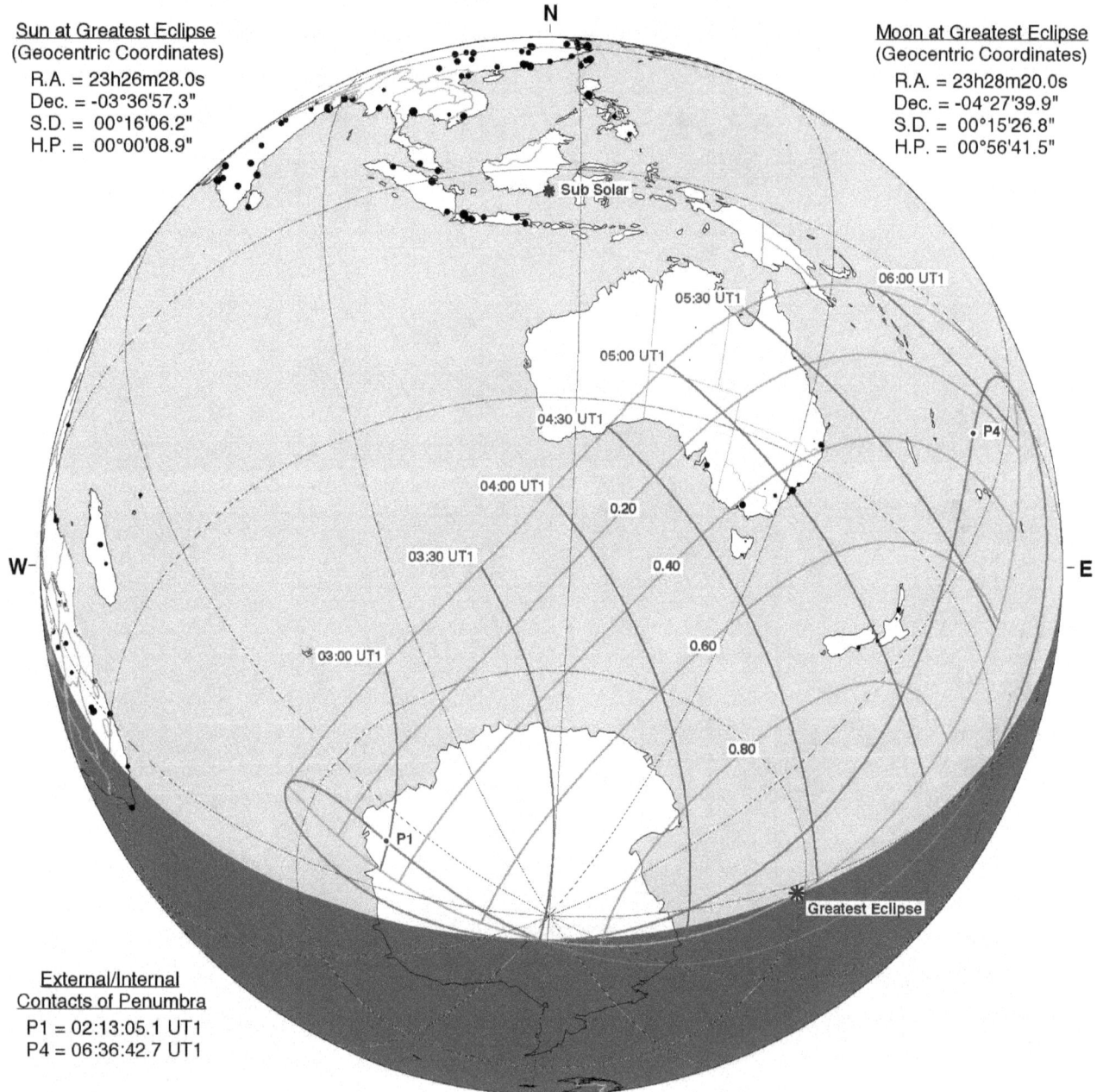

**External/Internal
Contacts of Penumbra**
P1 = 02:13:05.1 UT1
P4 = 06:36:42.7 UT1

ΔT = 91.9 s          Eph. = JPL DE405

**Circumstances at Greatest Eclipse: 04:24:44.3 UT1**
Lat. = 60°58.3'S          Sun Alt. = 0.0°
Long. = 147°14.6'W        Sun Azm. = 262.5°

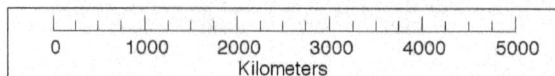

©2016 F. Espenak
www.EclipseWise.com

# Partial Solar Eclipse of 2062 Sep 03

Greatest Eclipse = 08:54:27.4 TD  (= 08:52:55.1 UT1)

Eclipse Magnitude = 0.9749          Saros Series = 126
Gamma = 1.0192              Saros Member = 50 of 72

Sun at Greatest Eclipse
(Geocentric Coordinates)
R.A. = 10h50m30.3s
Dec. = +07°22'28.5"
S.D. = 00°15'51.2"
H.P. = 00°00'08.7"

Moon at Greatest Eclipse
(Geocentric Coordinates)
R.A. = 10h52m25.5s
Dec. = +08°16'29.0"
S.D. = 00°16'22.2"
H.P. = 01°00'04.6"

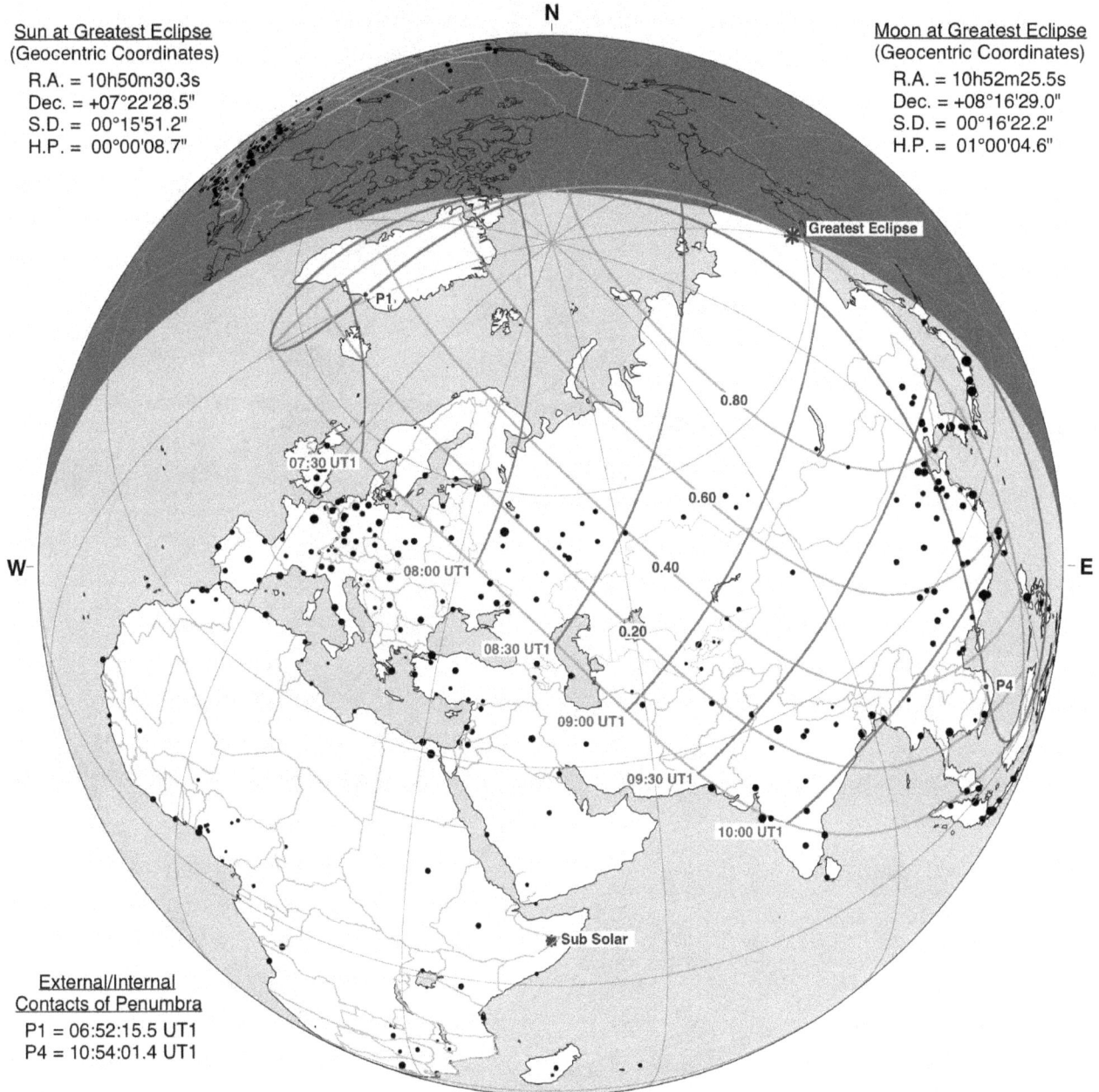

N

Greatest Eclipse

P1

0.80

07:30 UT1

0.60

08:00 UT1

0.40

08:30 UT1

0.20

09:00 UT1

P4

09:30 UT1

W

E

10:00 UT1

Sub Solar

External/Internal
Contacts of Penumbra
P1 = 06:52:15.5 UT1
P4 = 10:54:01.4 UT1

ΔT = 92.2 s        S        Eph. = JPL DE405

Circumstances at Greatest Eclipse: 08:52:55.1 UT1
Lat. = 61°18.3'N        Sun Alt. = 0.0°
Long. = 150°12.6'E        Sun Azm. = 285.5°

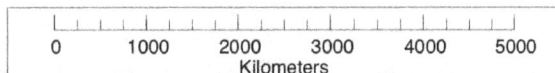

| | | | | | | | | | | |
|---|---|---|---|---|---|---|---|---|---|---|
| 0 | | 1000 | | 2000 | | 3000 | | 4000 | | 5000 |

Kilometers

©2016 F. Espenak
www.EclipseWise.com

# Annular Solar Eclipse of 2063 Feb 28

Greatest Eclipse = 07:43:30.0 TD  (= 07:41:57.4 UT1)

| | |
|---|---|
| Eclipse Magnitude = 0.9293 | Saros Series = 131 |
| Gamma = -0.3360 | Saros Member = 53 of 70 |

**Sun at Greatest Eclipse**
(Geocentric Coordinates)
R.A. = 22h45m11.8s
Dec. = -07°54'42.4"
S.D. = 00°16'08.9"
H.P. = 00°00'08.9"

**Moon at Greatest Eclipse**
(Geocentric Coordinates)
R.A. = 22h45m46.2s
Dec. = -08°10'47.1"
S.D. = 00°14'47.6"
H.P. = 00°54'17.7"

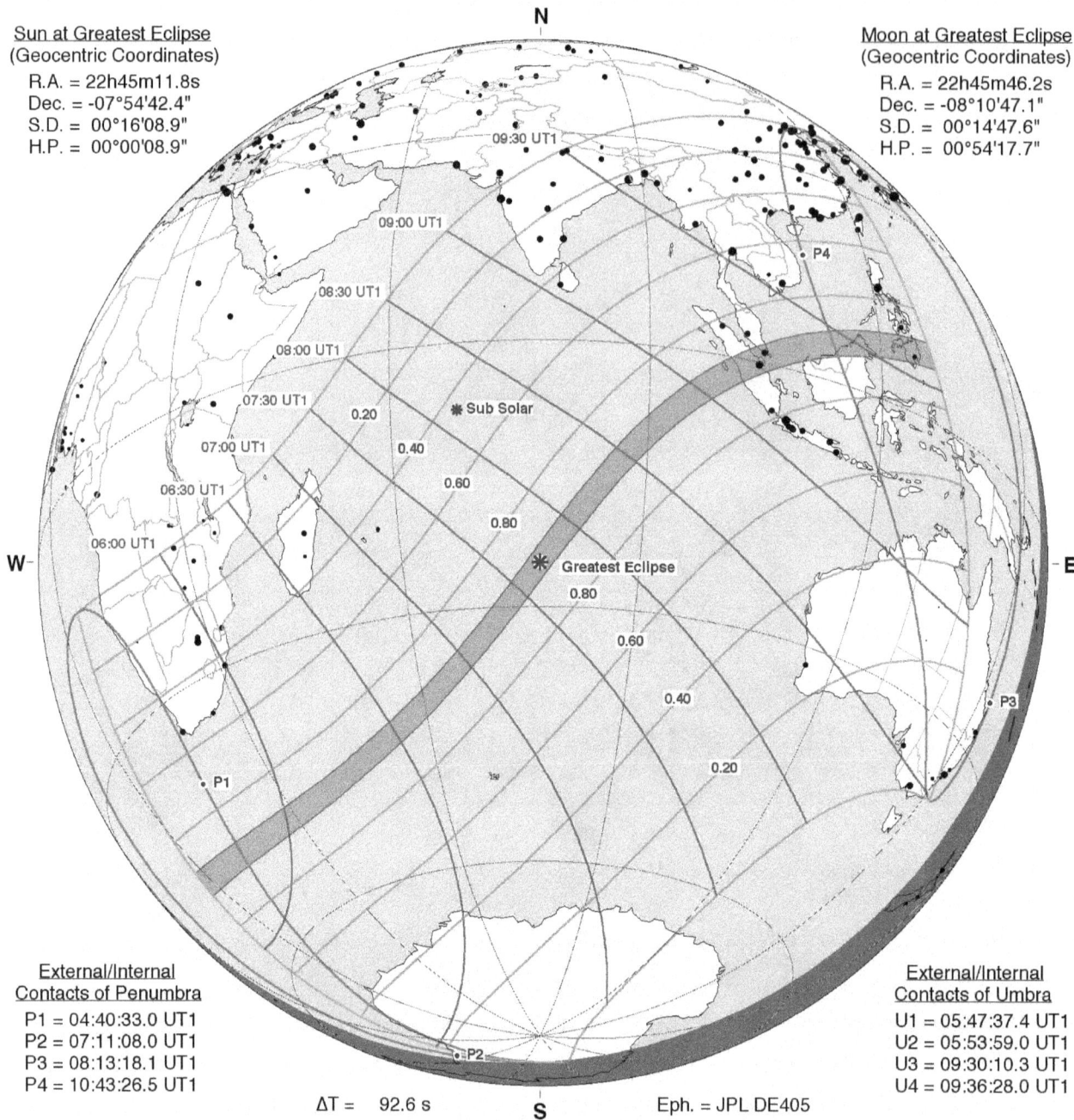

N

09:30 UT1
09:00 UT1
08:30 UT1
08:00 UT1
07:30 UT1
07:00 UT1
06:30 UT1
06:00 UT1

0.20
0.40
0.60
0.80

✳ Sub Solar

✳ Greatest Eclipse

0.80
0.60
0.40
0.20

P1
P2
P3
P4

W
E
S

**External/Internal Contacts of Penumbra**
P1 = 04:40:33.0 UT1
P2 = 07:11:08.0 UT1
P3 = 08:13:18.1 UT1
P4 = 10:43:26.5 UT1

**External/Internal Contacts of Umbra**
U1 = 05:47:37.4 UT1
U2 = 05:53:59.0 UT1
U3 = 09:30:10.3 UT1
U4 = 09:36:28.0 UT1

ΔT = 92.6 s          Eph. = JPL DE405

**Circumstances at Greatest Eclipse: 07:41:57.4 UT1**

| | |
|---|---|
| Lat. = 25°13.9'S | Sun Alt. = 70.2° |
| Long. = 077°37.8'E | Sun Azm. = 329.4° |
| Path Width = 279.6 km | Duration = 07m41.2s |

**Circumstances at Greatest Duration: 07:27:17.6 UT1**

| | |
|---|---|
| Lat. = 28°52.6'S | Sun Alt. = 68.8° |
| Long. = 074°24.5'E | Sun Azm. = 351.4° |
| Path Width = 276.8 km | Duration = 07m42.5s |

0  1000  2000  3000  4000  5000
Kilometers

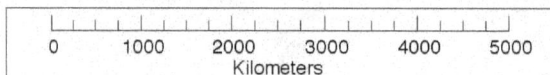

©2016  F. Espenak
www.EclipseWise.com

# Total Solar Eclipse of 2063 Aug 24

Greatest Eclipse = 01:22:10.6 TD  (= 01:20:37.6 UT1)

Eclipse Magnitude = 1.0750

Gamma = 0.2771

Saros Series = 136

Saros Member = 40 of 71

**Sun at Greatest Eclipse**
(Geocentric Coordinates)
R.A. = 10h12m03.7s
Dec. = +11°07'34.9"
S.D. = 00°15'48.9"
H.P. = 00°00'08.7"

**Moon at Greatest Eclipse**
(Geocentric Coordinates)
R.A. = 10h12m34.5s
Dec. = +11°22'46.8"
S.D. = 00°16'43.4"
H.P. = 01°01'22.6"

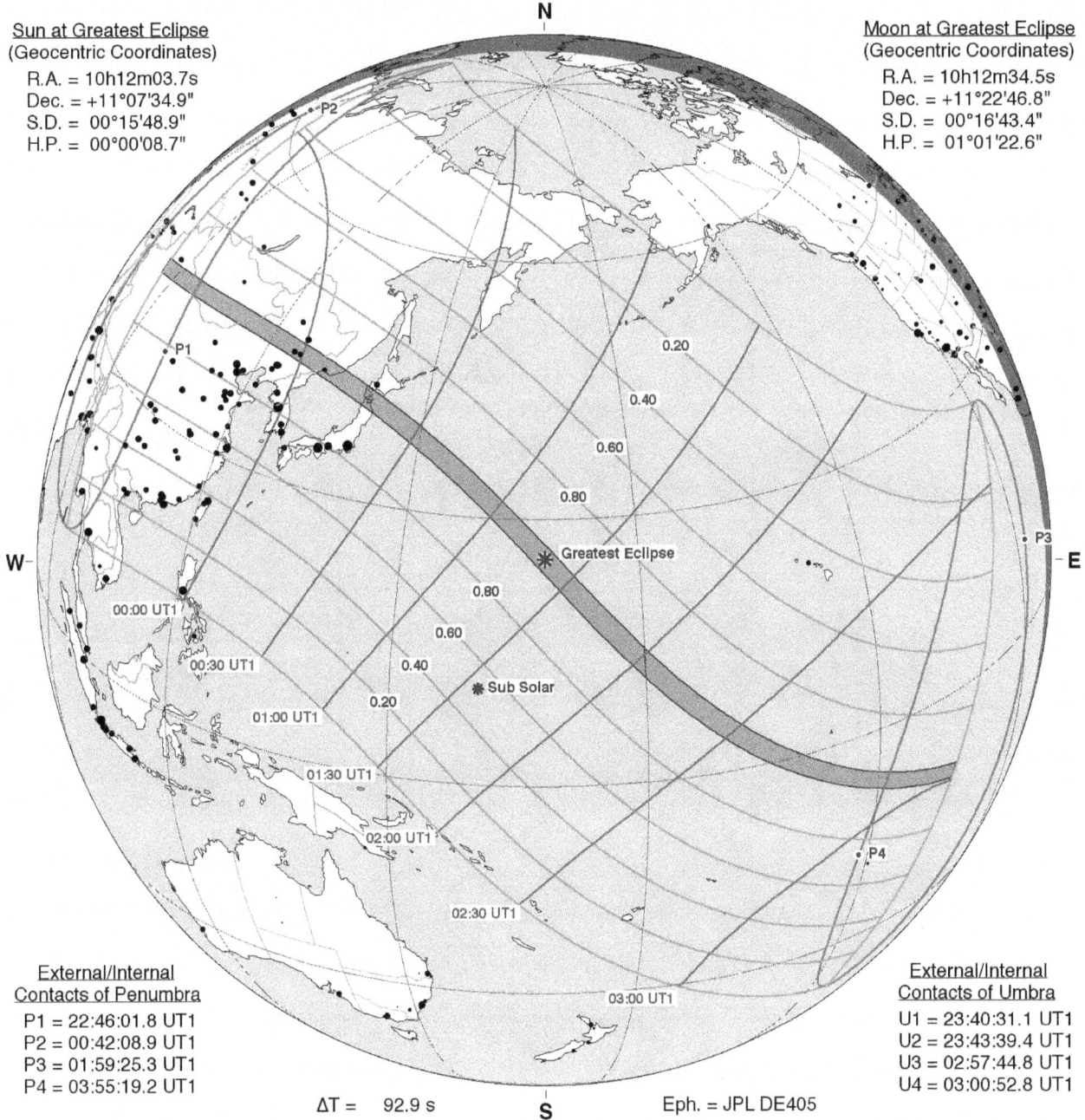

N

P2

P1

0.20
0.40
0.60
0.80

* Greatest Eclipse

P3

W

E

00:00 UT1
00:30 UT1
01:00 UT1
01:30 UT1
02:00 UT1
02:30 UT1
03:00 UT1

0.80
0.60
0.40
0.20

* Sub Solar

P4

**External/Internal
Contacts of Penumbra**
P1 = 22:46:01.8 UT1
P2 = 00:42:08.9 UT1
P3 = 01:59:25.3 UT1
P4 = 03:55:19.2 UT1

**External/Internal
Contacts of Umbra**
U1 = 23:40:31.1 UT1
U2 = 23:43:39.4 UT1
U3 = 02:57:44.8 UT1
U4 = 03:00:52.8 UT1

ΔT = 92.9 s

S

Eph. = JPL DE405

**Circumstances at Greatest Eclipse: 01:20:37.6 UT1**

| | |
|---|---|
| Lat. = 25°33.7'N | Sun Alt. = 73.8° |
| Long. = 168°19.2'E | Sun Azm. = 208.6° |
| Path Width = 252.2 km | Duration = 05m49.1s |

**Circumstances at Greatest Duration: 01:15:57.6 UT1**

| | |
|---|---|
| Lat. = 26°49.2'N | Sun Alt. = 73.5° |
| Long. = 166°56.6'E | Sun Azm. = 198.7° |
| Path Width = 251.3 km | Duration = 05m49.3s |

©2016  F. Espenak
www.EclipseWise.com

0  1000  2000  3000  4000  5000
Kilometers

# Annular Solar Eclipse of 2064 Feb 17

Greatest Eclipse = 07:00:23.3 TD  (= 06:58:50.1 UT1)

| | |
|---|---|
| Eclipse Magnitude = 0.9262 | Saros Series = 141 |
| Gamma = 0.3597 | Saros Member = 26 of 70 |

**N**

Sun at Greatest Eclipse
(Geocentric Coordinates)
R.A. = 22h02m13.8s
Dec. = -12°01'37.5"
S.D. = 00°16'11.3"
H.P. = 00°00'08.9"

Moon at Greatest Eclipse
(Geocentric Coordinates)
R.A. = 22h01m38.9s
Dec. = -11°44'08.3"
S.D. = 00°14'47.1"
H.P. = 00°54'15.6"

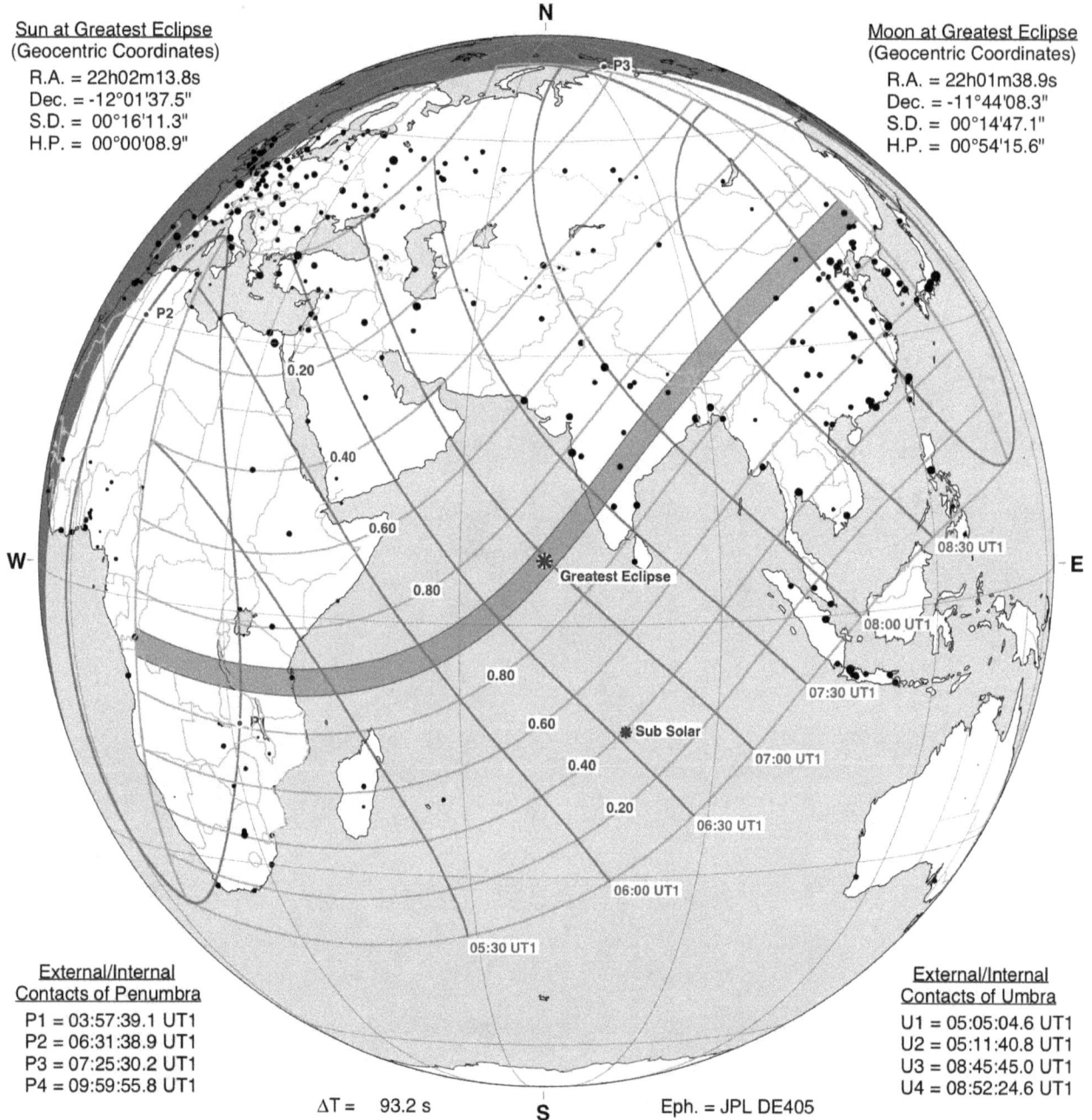

P3

P2

0.20

0.40

0.60

08:30 UT1

**W**

**E**

Greatest Eclipse

0.80

08:00 UT1

0.80

07:30 UT1

0.60

Sub Solar

0.40

07:00 UT1

P1

0.20

06:30 UT1

06:00 UT1

05:30 UT1

External/Internal
Contacts of Penumbra
P1 = 03:57:39.1 UT1
P2 = 06:31:38.9 UT1
P3 = 07:25:30.2 UT1
P4 = 09:59:55.8 UT1

External/Internal
Contacts of Umbra
U1 = 05:05:04.6 UT1
U2 = 05:11:40.8 UT1
U3 = 08:45:45.0 UT1
U4 = 08:52:24.6 UT1

ΔT = 93.2 s     **S**     Eph. = JPL DE405

| Circumstances at Greatest Eclipse: 06:58:50.1 UT1 | | Circumstances at Greatest Duration: 06:43:05.7 UT1 | |
|---|---|---|---|
| Lat. = 07°02.0'N | Sun Alt. = 68.9° | Lat. = 03°50.2'N | Sun Alt. = 67.4° |
| Long. = 069°37.3'E | Sun Azm. = 154.4° | Long. = 066°34.3'E | Sun Azm. = 134.9° |
| Path Width = 294.6 km | Duration = 08m56.4s | Path Width = 300.9 km | Duration = 08m58.7s |

0      1000    2000    3000    4000    5000
Kilometers

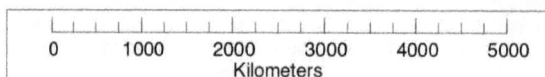

©2016 F. Espenak
www.EclipseWise.com

# Total Solar Eclipse of 2064 Aug 12

Greatest Eclipse = 17:46:06.3 TD  (= 17:44:32.7 UT1)

| | |
|---|---|
| Eclipse Magnitude = 1.0495 | Saros Series = 146 |
| Gamma = -0.4652 | Saros Member = 30 of 76 |

**Sun at Greatest Eclipse**
(Geocentric Coordinates)
R.A. = 09h32m49.7s
Dec. = +14°33'07.1"
S.D. = 00°15'47.0"
H.P. = 00°00'08.7"

**Moon at Greatest Eclipse**
(Geocentric Coordinates)
R.A. = 09h32m02.7s
Dec. = +14°07'45.3"
S.D. = 00°16'19.3"
H.P. = 00°59'54.2"

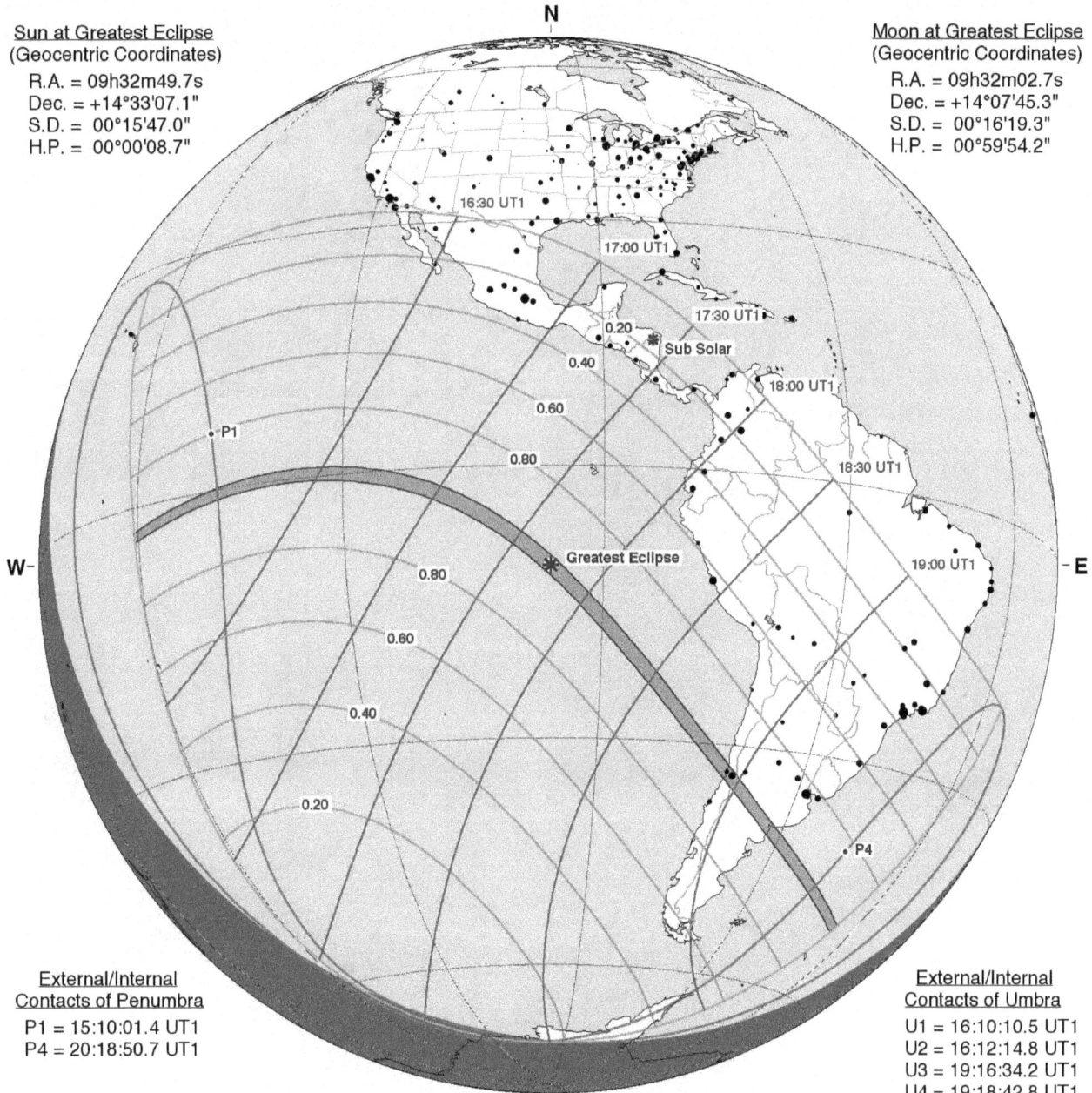

**External/Internal Contacts of Penumbra**
P1 = 15:10:01.4 UT1
P4 = 20:18:50.7 UT1

**External/Internal Contacts of Umbra**
U1 = 16:10:10.5 UT1
U2 = 16:12:14.8 UT1
U3 = 19:16:34.2 UT1
U4 = 19:18:42.8 UT1

ΔT =    93.6 s

Eph. = JPL DE405

**Circumstances at Greatest Eclipse:  17:44:32.7 UT1**

| | |
|---|---|
| Lat. =  10°55.8'S | Sun Alt. =  62.2° |
| Long. = 096°05.9'W | Sun Azm. =  23.8° |
| Path Width = 183.7 km | Duration = 04m27.7s |

**Circumstances at Greatest Duration:  17:40:44.1 UT1**

| | |
|---|---|
| Lat. =  10°06.0'S | Sun Alt. =  62.1° |
| Long. = 097°04.6'W | Sun Azm. =  28.0° |
| Path Width = 184.6 km | Duration = 04m27.8s |

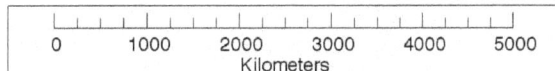

©2016  F. Espenak
www.EclipseWise.com

Kilometers
0    1000    2000    3000    4000    5000

# Partial Solar Eclipse of 2065 Aug 02

Greatest Eclipse = 05:34:16.6 TD  (= 05:32:42.3 UT1)

| | |
|---|---|
| Eclipse Magnitude = 0.4903 | Saros Series = 156 |
| Gamma = -1.2758 | Saros Member = 4 of 69 |

**Sun at Greatest Eclipse**
(Geocentric Coordinates)
R.A. = 08h51m52.4s
Dec. = +17°35'43.5"
S.D. = 00°15'45.5"
H.P. = 00°00'08.7"

**Moon at Greatest Eclipse**
(Geocentric Coordinates)
R.A. = 08h50m03.4s
Dec. = +16°28'16.4"
S.D. = 00°15'28.9"
H.P. = 00°56'49.3"

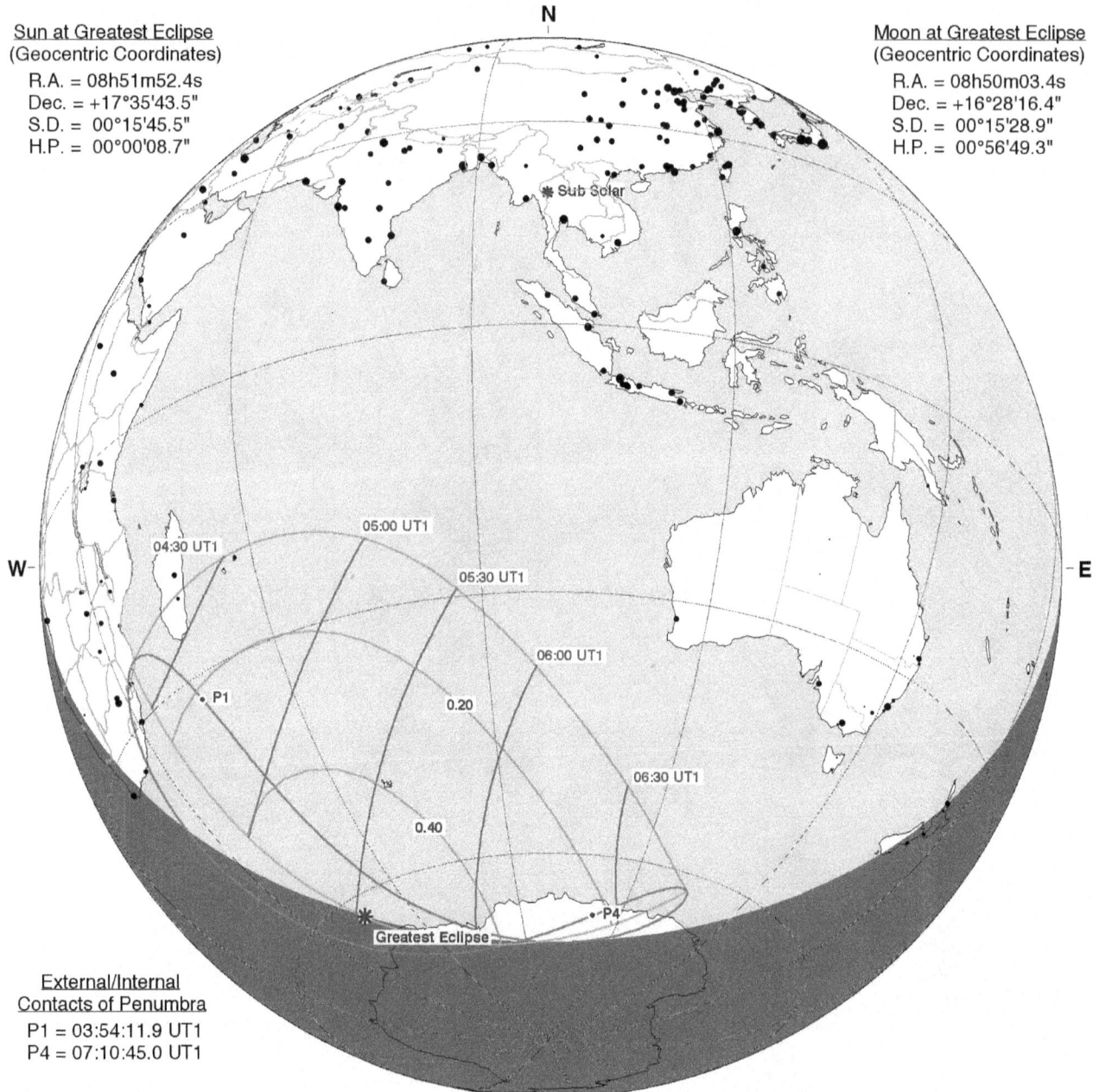

N

* Sub Solar

W

E

04:30 UT1
05:00 UT1
05:30 UT1
06:00 UT1
0.20
06:30 UT1
0.40

P1

P4

* Greatest Eclipse

**External/Internal
Contacts of Penumbra**
P1 = 03:54:11.9 UT1
P4 = 07:10:45.0 UT1

ΔT = 94.3 s

S

Eph. = JPL DE405

**Circumstances at Greatest Eclipse: 05:32:42.3 UT1**

| | |
|---|---|
| Lat. = 62°42.7'S | Sun Alt. = 0.0° |
| Long. = 046°24.5'E | Sun Azm. = 48.7° |

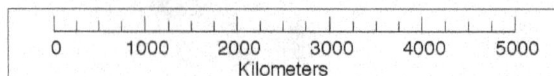

| | | | | | |
|---|---|---|---|---|---|
| 0 | 1000 | 2000 | 3000 | 4000 | 5000 |

Kilometers

©2016 F. Espenak
www.EclipseWise.com

# Partial Solar Eclipse of 2065 Dec 27

Greatest Eclipse = 08:39:55.7 TD  (= 08:38:21.2 UT1)

Eclipse Magnitude = 0.8769          Saros Series = 123
Gamma = -1.0688          Saros Member = 56 of 70

Sun at Greatest Eclipse
(Geocentric Coordinates)
R.A. = 18h26m44.9s
Dec. = -23°17'20.3"
S.D. = 00°16'15.7"
H.P. = 00°00'08.9"

Moon at Greatest Eclipse
(Geocentric Coordinates)
R.A. = 18h27m25.5s
Dec. = -24°21'42.8"
S.D. = 00°16'37.3"
H.P. = 01°01'00.2"

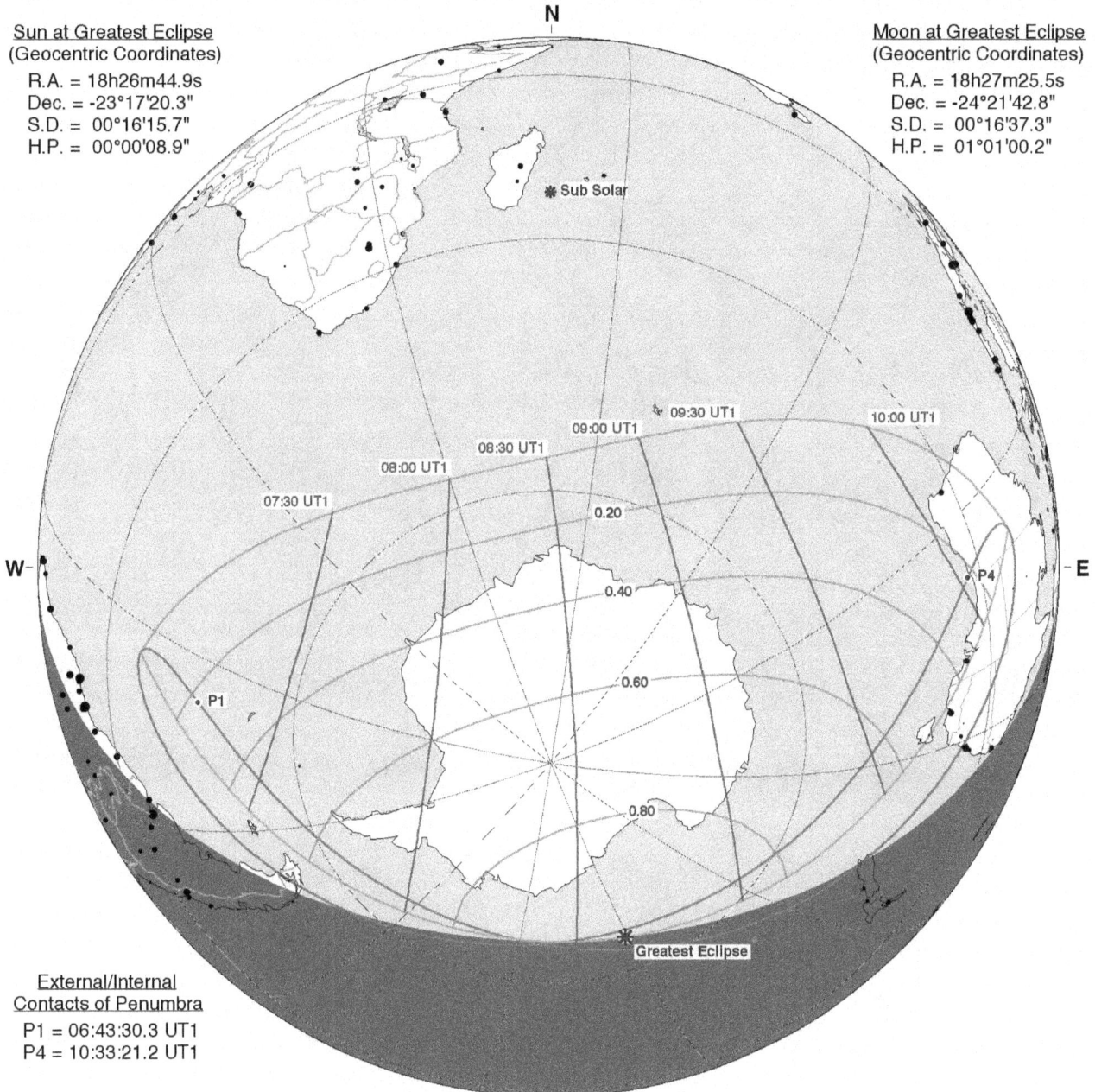

N

* Sub Solar

09:30 UT1
10:00 UT1
09:00 UT1
08:30 UT1
08:00 UT1
07:30 UT1
0.20
0.40
W
E
0.60
P4
P1
0.80

* Greatest Eclipse

External/Internal
Contacts of Penumbra
P1 = 06:43:30.3 UT1
P4 = 10:33:21.2 UT1

ΔT = 94.6 s          S          Eph. = JPL DE405

Circumstances at Greatest Eclipse: 08:38:21.2 UT1
Lat. = 65°23.8'S          Sun Alt. = 0.0°
Long. = 149°17.5'W          Sun Azm. = 198.3°

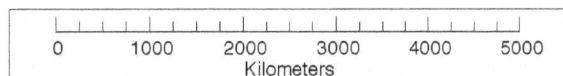

0   1000   2000   3000   4000   5000
Kilometers

©2016 F. Espenak
www.EclipseWise.com

# Partial Solar Eclipse of 2065 Feb 05

Greatest Eclipse = 09:52:25.5 TD  (= 09:50:51.6 UT1)

Eclipse Magnitude = 0.9123      Saros Series = 151

Gamma = 1.0336      Saros Member = 17 of 72

**Sun at Greatest Eclipse**
(Geocentric Coordinates)
R.A. = 21h18m22.7s
Dec. = -15°41'30.6"
S.D. = 00°16'13.3"
H.P. = 00°00'08.9"

**Moon at Greatest Eclipse**
(Geocentric Coordinates)
R.A. = 21h16m47.2s
Dec. = -14°47'52.5"
S.D. = 00°15'25.7"
H.P. = 00°56'37.5"

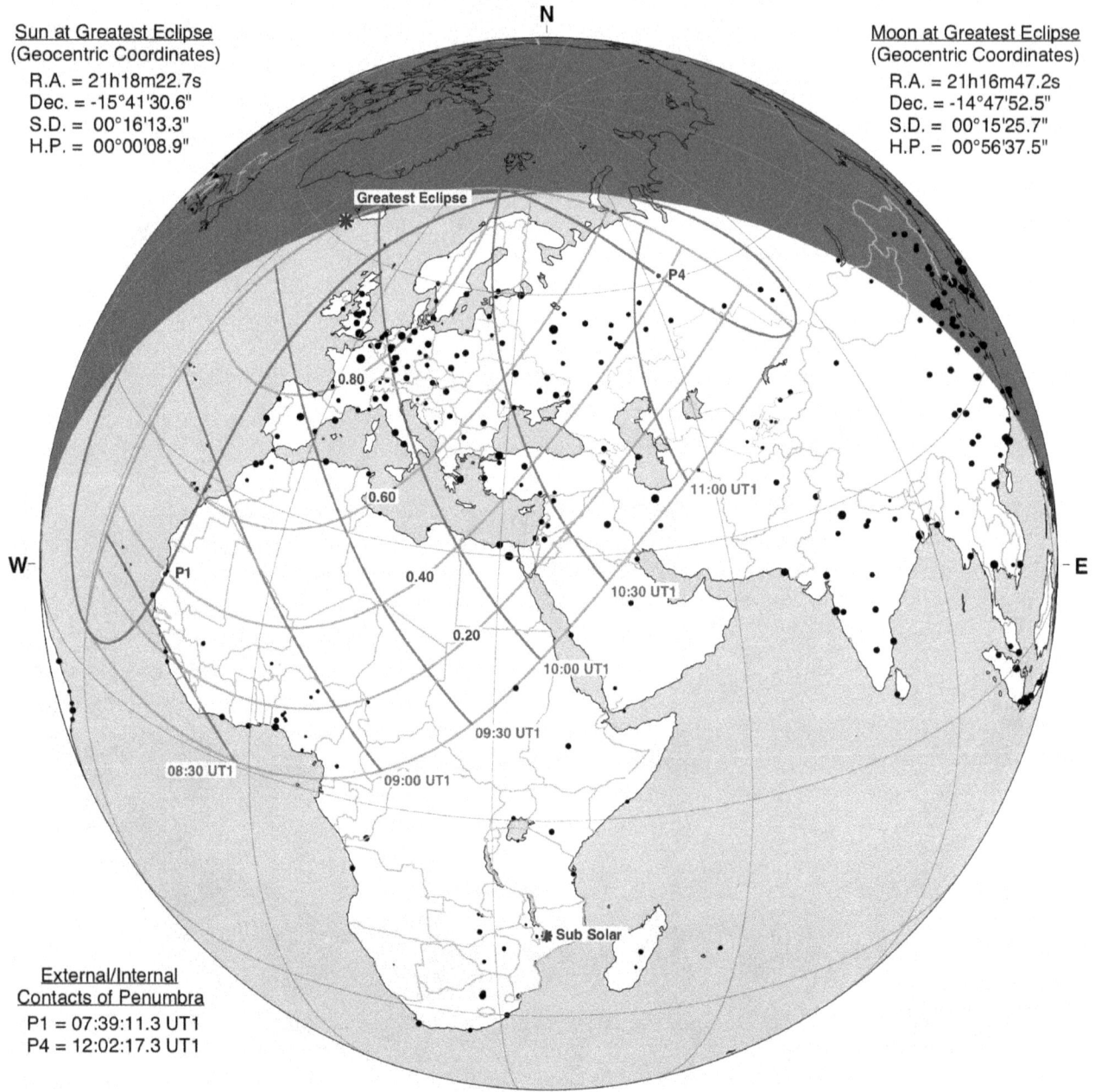

N

Greatest Eclipse

P4

0.80

0.60

11:00 UT1

P1

0.40

10:30 UT1

W

0.20

E

10:00 UT1

09:30 UT1

08:30 UT1

09:00 UT1

Sub Solar

**External/Internal
Contacts of Penumbra**
P1 = 07:39:11.3 UT1
P4 = 12:02:17.3 UT1

ΔT = 93.9 s     S     Eph. = JPL DE405

**Circumstances at Greatest Eclipse: 09:50:51.6 UT1**
Lat. = 62°09.7'N     Sun Alt. = 0.0°
Long. = 022°01.5'W     Sun Azm. = 125.4°

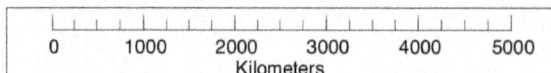

| | | | | | |
|---|---|---|---|---|---|
| 0 | 1000 | 2000 | 3000 | 4000 | 5000 |

Kilometers

©2016 F. Espenak
www.EclipseWise.com

# Partial Solar Eclipse of 2065 Jul 03

Greatest Eclipse = 17:33:52.5 TD   (= 17:32:18.3 UT1)

| | |
|---|---|
| Eclipse Magnitude = 0.1639 | Saros Series = 118 |
| Gamma = 1.4619 | Saros Member = 71 of 72 |

Sun at Greatest Eclipse
(Geocentric Coordinates)
R.A. = 06h53m43.9s
Dec. = +22°51'26.7"
S.D. = 00°15'43.9"
H.P. = 00°00'08.6"

Moon at Greatest Eclipse
(Geocentric Coordinates)
R.A. = 06h54m50.6s
Dec. = +24°10'43.8"
S.D. = 00°15'05.3"
H.P. = 00°55'22.6"

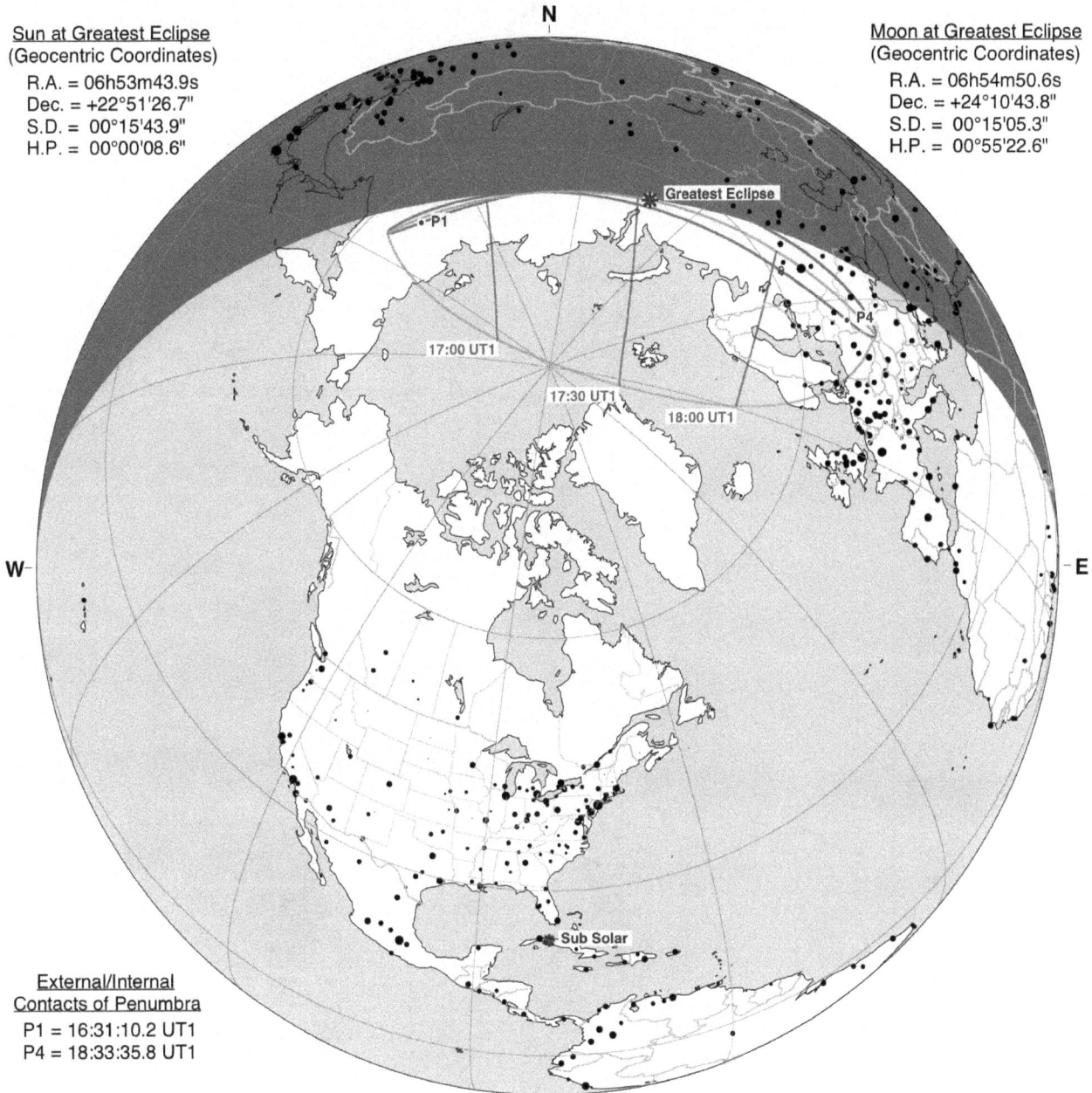

N

Greatest Eclipse

P1

17:00 UT1

17:30 UT1

18:00 UT1

P4

W

E

Sub Solar

External/Internal
Contacts of Penumbra
P1 = 16:31:10.2 UT1
P4 = 18:33:35.8 UT1

ΔT =   94.2 s          S          Eph. = JPL DE405

Circumstances at Greatest Eclipse: 17:32:18.3 UT1
Lat. = 64°49.3'N          Sun Alt. =   0.0°
Long. = 071°43.9'E          Sun Azm. = 335.9°

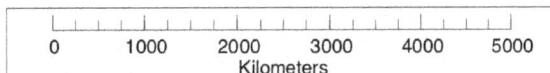

| | | | | | |
|---|---|---|---|---|---|
| 0 | 1000 | 2000 | 3000 | 4000 | 5000 |

Kilometers

©2016  F. Espenak
www.EclipseWise.com

# Annular Solar Eclipse of 2066 Jun 22

Greatest Eclipse = 19:25:47.7 TD  (= 19:24:12.8 UT1)

Eclipse Magnitude = 0.9435          Saros Series = 128
Gamma = 0.7330          Saros Member = 61 of 73

**Sun at Greatest Eclipse**
(Geocentric Coordinates)
R.A. = 06h07m28.7s
Dec. = +23°25'11.2"
S.D. = 00°15'44.2"
H.P. = 00°00'08.7"

**Moon at Greatest Eclipse**
(Geocentric Coordinates)
R.A. = 06h07m48.1s
Dec. = +24°04'22.4"
S.D. = 00°14'42.0"
H.P. = 00°53'57.0"

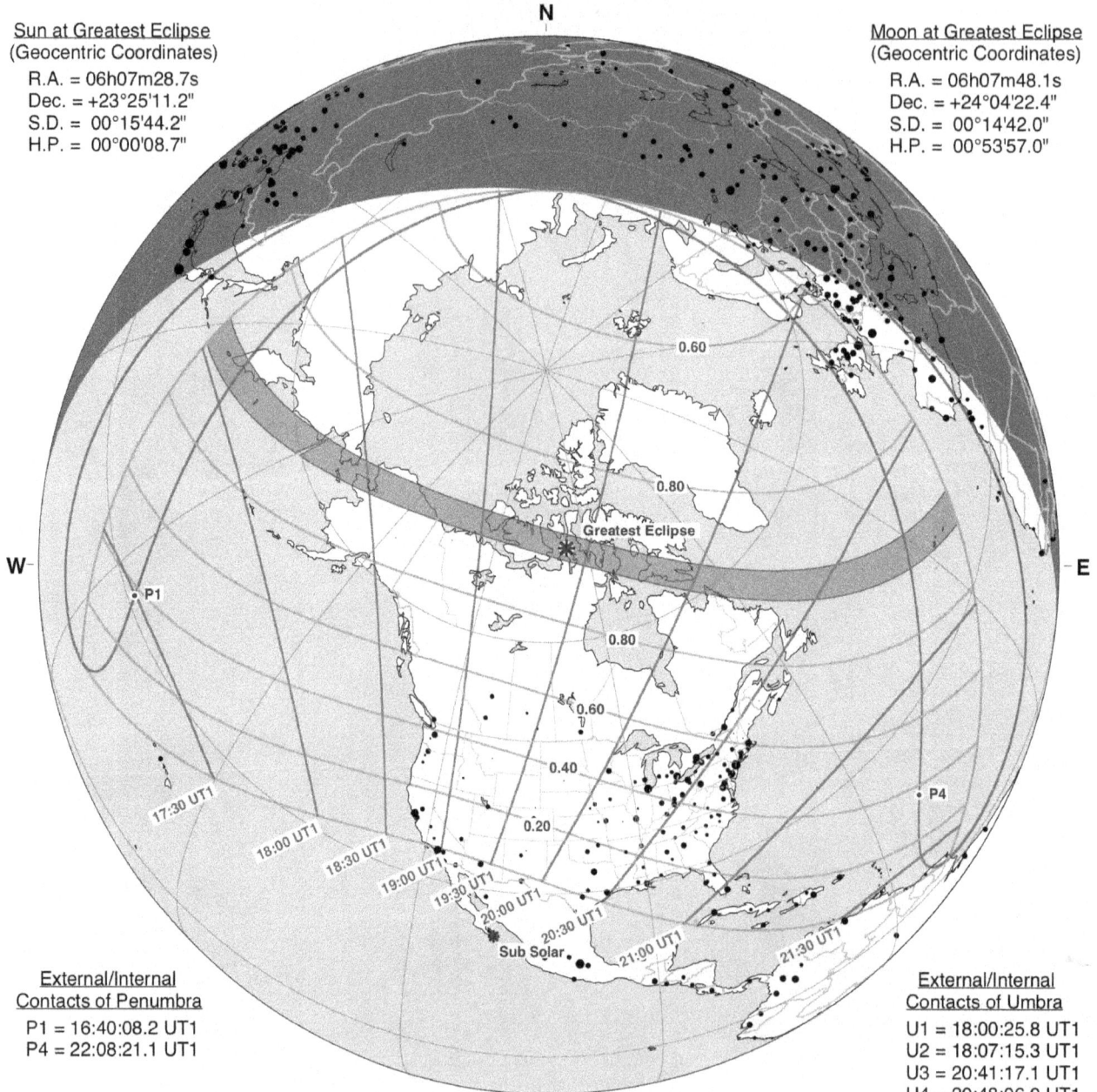

**External/Internal Contacts of Penumbra**
P1 = 16:40:08.2 UT1
P4 = 22:08:21.1 UT1

**External/Internal Contacts of Umbra**
U1 = 18:00:25.8 UT1
U2 = 18:07:15.3 UT1
U3 = 20:41:17.1 UT1
U4 = 20:48:06.0 UT1

ΔT = 94.9 s          Eph. = JPL DE405

**Circumstances at Greatest Eclipse: 19:24:12.8 UT1**
Lat. = 70°07.9'N          Sun Alt. = 42.6°
Long. = 096°29.9'W          Sun Azm. = 197.5°
Path Width = 308.5 km          Duration = 04m39.8s

**Circumstances at Greatest Duration: 19:21:23.9 UT1**
Lat. = 70°31.1'N          Sun Alt. = 42.5°
Long. = 099°39.3'W          Sun Azm. = 192.6°
Path Width = 308.3 km          Duration = 04m39.8s

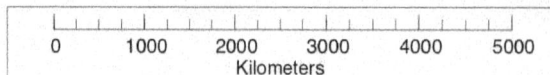

©2016 F. Espenak
www.EclipseWise.com

# Total Solar Eclipse of 2066 Dec 17

Greatest Eclipse = 00:23:39.9 TD  (= 00:22:04.7 UT1)

Eclipse Magnitude = 1.0416          Saros Series = 133
Gamma = -0.4043                     Saros Member = 48 of 72

Sun at Greatest Eclipse
(Geocentric Coordinates)
R.A. = 17h39m46.4s
Dec. = -23°20'56.0"
S.D. = 00°16'15.1"
H.P. = 00°00'08.9"

Moon at Greatest Eclipse
(Geocentric Coordinates)
R.A. = 17h39m53.3s
Dec. = -23°45'32.9"
S.D. = 00°16'39.9"
H.P. = 01°01'09.6"

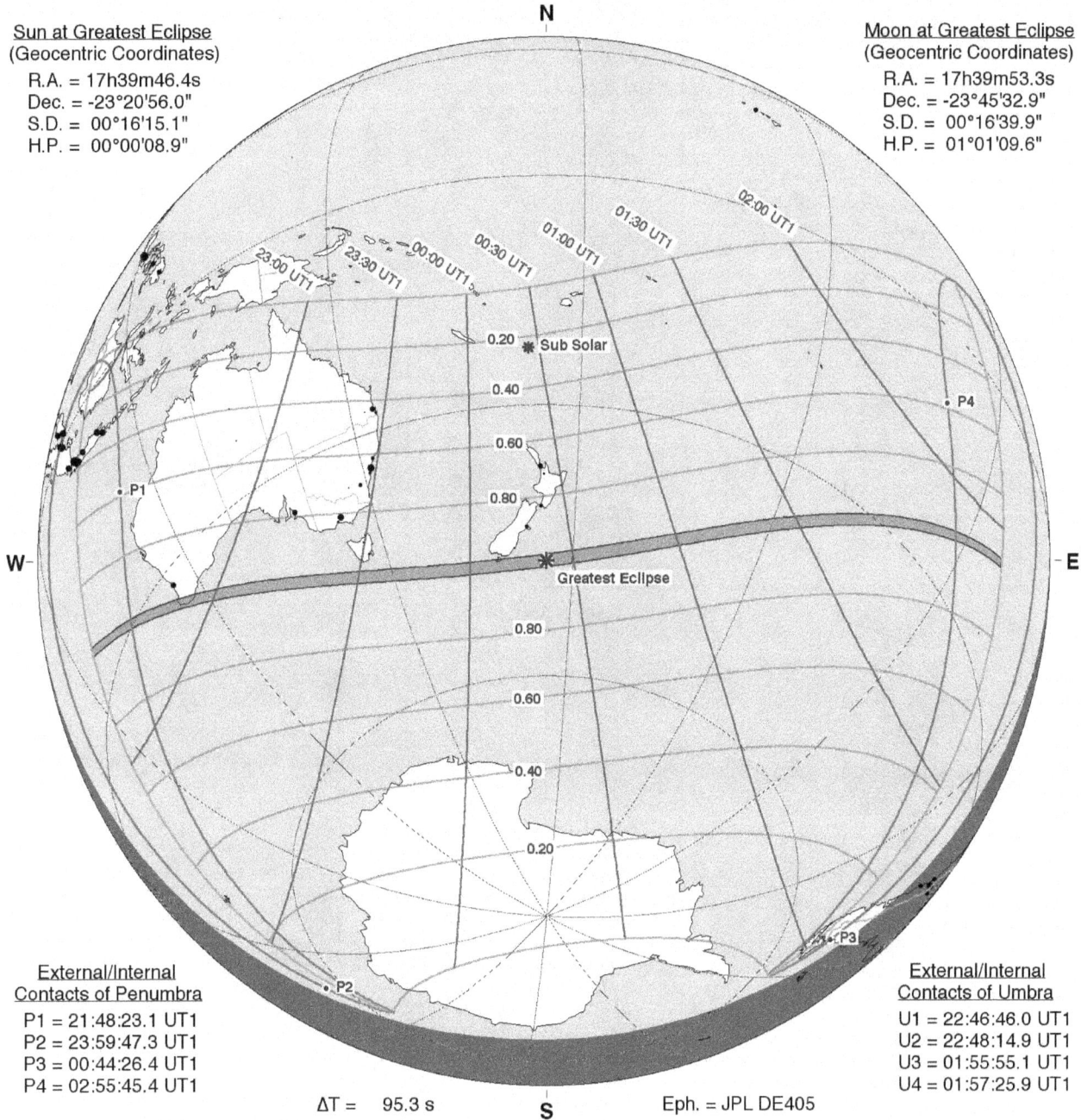

N

23:00 UT1  23:30 UT1  00:00 UT1  00:30 UT1  01:00 UT1  01:30 UT1  02:00 UT1

0.20  * Sub Solar
0.40
0.60
0.80

P4

W — — E

* Greatest Eclipse

0.80
0.60
0.40
0.20

P1

P3

P2

S

External/Internal
Contacts of Penumbra
P1 = 21:48:23.1 UT1
P2 = 23:59:47.3 UT1
P3 = 00:44:26.4 UT1
P4 = 02:55:45.4 UT1

ΔT = 95.3 s          Eph. = JPL DE405

External/Internal
Contacts of Umbra
U1 = 22:46:46.0 UT1
U2 = 22:48:14.9 UT1
U3 = 01:55:55.1 UT1
U4 = 01:57:25.9 UT1

Circumstances at Greatest Eclipse: 00:22:04.7 UT1

Lat. = 47°22.1'S       Sun Alt. = 65.9°
Long. = 175°38.4'E     Sun Azm. = 355.1°
Path Width = 152.2 km  Duration = 03m14.5s

Circumstances at Greatest Duration: 00:21:56.4 UT1

Lat. = 47°22.5'S       Sun Alt. = 65.9°
Long. = 175°33.7'E     Sun Azm. = 355.3°
Path Width = 152.2 km  Duration = 03m14.5s

©2016  F. Espenak
www.EclipseWise.com

Kilometers
0  1000  2000  3000  4000  5000

# Annular Solar Eclipse of 2067 Jun 11

Greatest Eclipse = 20:42:26.4 TD  (= 20:40:50.8 UT1)

| | |
|---|---|
| Eclipse Magnitude = 0.9670 | Saros Series = 138 |
| Gamma = -0.0387 | Saros Member = 34 of 70 |

**Sun at Greatest Eclipse**
(Geocentric Coordinates)
R.A. = 05h20m58.3s
Dec. = +23°07'36.6"
S.D. = 00°15'45.1"
H.P. = 00°00'08.7"

**Moon at Greatest Eclipse**
(Geocentric Coordinates)
R.A. = 05h20m58.0s
Dec. = +23°05'29.3"
S.D. = 00°15'00.0"
H.P. = 00°55'03.2"

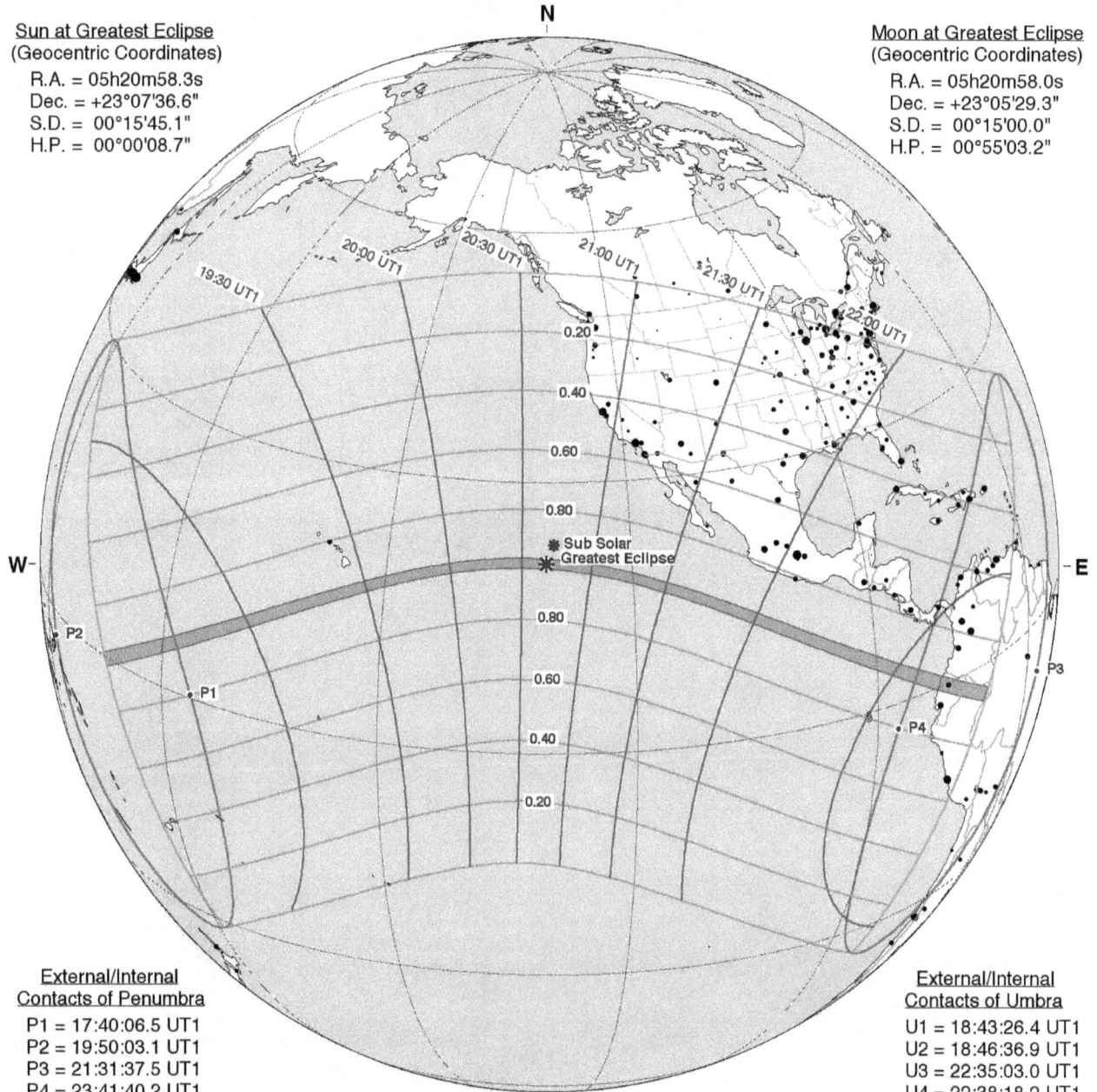

N

19:30 UT1
20:00 UT1
20:30 UT1
21:00 UT1
21:30 UT1
22:00 UT1

0.20
0.40
0.60
0.80

W

Sub Solar
Greatest Eclipse

0.80
0.60
0.40
0.20

E

P2
P1
P3
P4

**External/Internal Contacts of Penumbra**
P1 = 17:40:06.5 UT1
P2 = 19:50:03.1 UT1
P3 = 21:31:37.5 UT1
P4 = 23:41:40.2 UT1

**External/Internal Contacts of Umbra**
U1 = 18:43:26.4 UT1
U2 = 18:46:36.9 UT1
U3 = 22:35:03.0 UT1
U4 = 22:38:18.2 UT1

ΔT = 95.6 s

S

Eph. = JPL DE405

**Circumstances at Greatest Eclipse: 20:40:50.8 UT1**

| | |
|---|---|
| Lat. = 21°02.4'N | Sun Alt. = 87.9° |
| Long. = 130°19.2'W | Sun Azm. = 2.0° |
| Path Width = 118.9 km | Duration = 04m05.2s |

**Circumstances at Greatest Duration: 20:42:21.3 UT1**

| | |
|---|---|
| Lat. = 21°00.9'N | Sun Alt. = 87.8° |
| Long. = 129°53.9'W | Sun Azm. = 342.6° |
| Path Width = 119.0 km | Duration = 04m05.2s |

0   1000   2000   3000   4000   5000
Kilometers

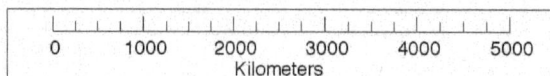

©2016  F. Espenak
www.EclipseWise.com

# Hybrid Solar Eclipse of 2067 Dec 06

Greatest Eclipse = 14:03:43.3 TD (= 14:02:07.3 UT1)

Eclipse Magnitude = 1.0011          Saros Series = 143

Gamma = 0.2845          Saros Member = 26 of 72

**Sun at Greatest Eclipse**
(Geocentric Coordinates)
R.A. = 16h52m45.8s
Dec. = -22°31'49.1"
S.D. = 00°16'13.8"
H.P. = 00°00'08.9"

**Moon at Greatest Eclipse**
(Geocentric Coordinates)
R.A. = 16h52m46.9s
Dec. = -22°15'09.9"
S.D. = 00°15'59.7"
H.P. = 00°58'42.2"

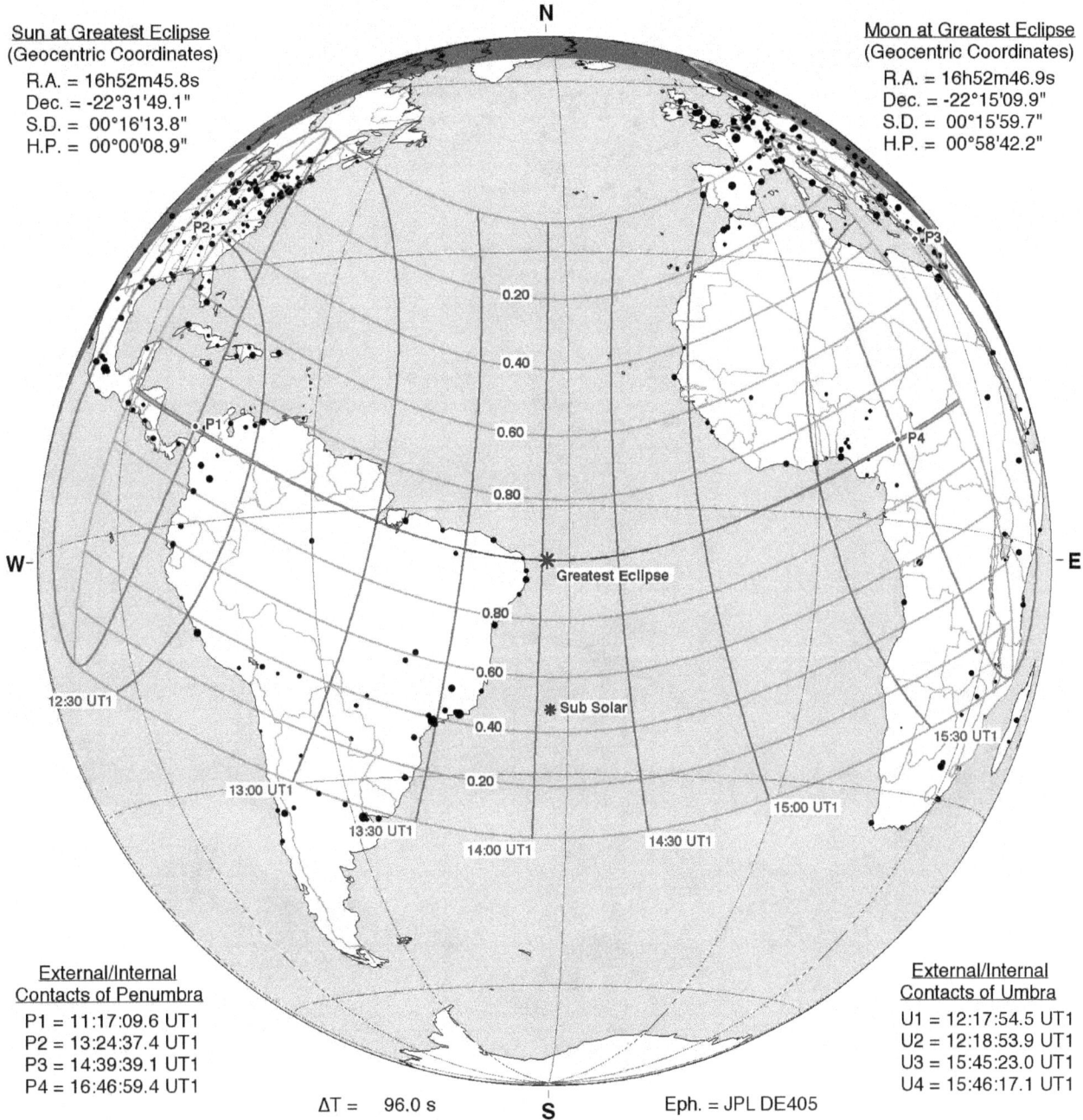

N

P2
P3
P1
P4

0.20
0.40
0.60
0.80

* Greatest Eclipse

W — — E

0.80
0.60

* Sub Solar

0.40

12:30 UT1
0.20

15:30 UT1

13:00 UT1          15:00 UT1

13:30 UT1          14:30 UT1

14:00 UT1          14:30 UT1

**External/Internal**
**Contacts of Penumbra**
P1 = 11:17:09.6 UT1
P2 = 13:24:37.4 UT1
P3 = 14:39:39.1 UT1
P4 = 16:46:59.4 UT1

**External/Internal**
**Contacts of Umbra**
U1 = 12:17:54.5 UT1
U2 = 12:18:53.9 UT1
U3 = 15:45:23.0 UT1
U4 = 15:46:17.1 UT1

ΔT = 96.0 s          S          Eph. = JPL DE405

**Circumstances at Greatest Eclipse: 14:02:07.3 UT1**

| | |
|---|---|
| Lat. = 06°02.8'S | Sun Alt. = 73.5° |
| Long. = 032°31.2'W | Sun Azm. = 180.9° |
| Path Width = 4.1 km | Duration = 00m07.5s |

**Circumstances at Greatest Duration: 12:18:24.2 UT1**

| | |
|---|---|
| Lat. = 16°11.9'N | Sun Alt. = 0.0° |
| Long. = 089°56.5'W | Sun Azm. = 113.5° |
| Path Width = 56.2 km | Duration = 00m59.0s |

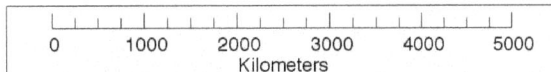

0     1000     2000     3000     4000     5000
Kilometers

©2016 F. Espenak
www.EclipseWise.com

# Total Solar Eclipse of 2068 May 31

Greatest Eclipse = 03:56:39.1 TD   (= 03:55:02.7 UT1)

| | |
|---|---|
| Eclipse Magnitude = 1.0110 | Saros Series = 148 |
| Gamma = -0.7970 | Saros Member = 24 of 75 |

**Sun at Greatest Eclipse**
(Geocentric Coordinates)
R.A. = 04h35m49.8s
Dec. = +22°01'13.9"
S.D. = 00°15'46.5"
H.P. = 00°00'08.7"

**Moon at Greatest Eclipse**
(Geocentric Coordinates)
R.A. = 04h35m58.7s
Dec. = +21°15'11.0"
S.D. = 00°15'47.8"
H.P. = 00°57'58.6"

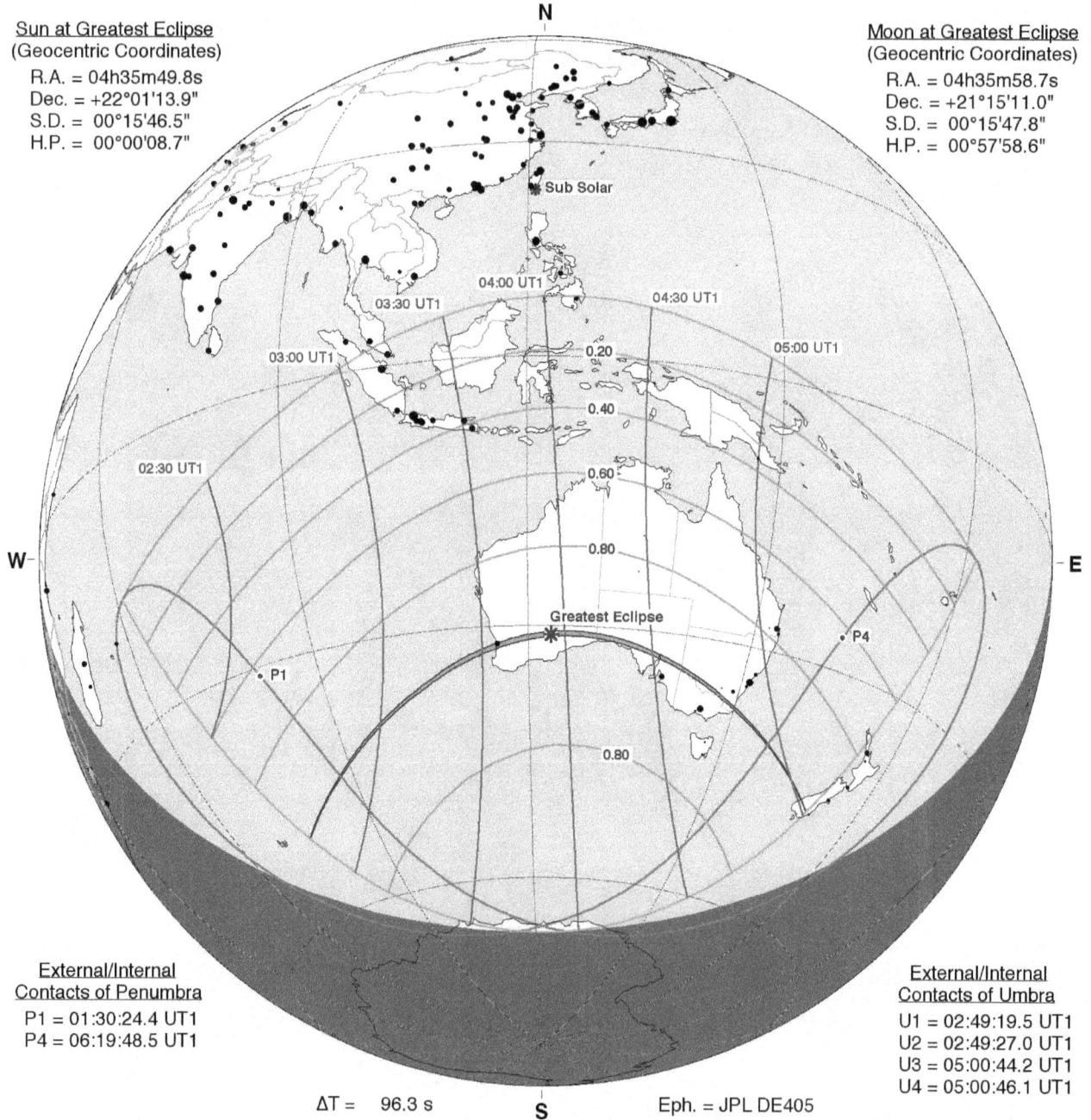

N

Sub Solar

04:00 UT1
03:30 UT1
04:30 UT1
03:00 UT1
05:00 UT1
0.20
0.40
02:30 UT1
0.60
0.80

W — — E

Greatest Eclipse

P1
P4
0.80

S

**External/Internal Contacts of Penumbra**
P1 = 01:30:24.4 UT1
P4 = 06:19:48.5 UT1

**External/Internal Contacts of Umbra**
U1 = 02:49:19.5 UT1
U2 = 02:49:27.0 UT1
U3 = 05:00:44.2 UT1
U4 = 05:00:46.1 UT1

ΔT =   96.3 s        Eph. = JPL DE405

**Circumstances at Greatest Eclipse: 03:55:02.7 UT1**

| | |
|---|---|
| Lat. = 31°00.8'S | Sun Alt. = 36.9° |
| Long. = 123°05.1'E | Sun Azm. = 357.2° |
| Path Width = 62.8 km | Duration = 01m05.8s |

**Circumstances at Greatest Duration: 03:53:13.6 UT1**

| | |
|---|---|
| Lat. = 31°03.9'S | Sun Alt. = 36.9° |
| Long. = 122°25.8'E | Sun Azm. = 358.5° |
| Path Width = 62.8 km | Duration = 01m05.9s |

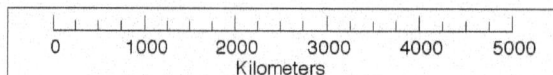

| | | | | | |
|---|---|---|---|---|---|
| 0 | 1000 | 2000 | 3000 | 4000 | 5000 |

Kilometers

©2016  F. Espenak
www.EclipseWise.com

# Partial Solar Eclipse of 2068 Nov 24

Greatest Eclipse = 21:32:29.6 TD  (= 21:30:52.9 UT1)

Eclipse Magnitude = 0.9109          Saros Series = 153
Gamma = 1.0299          Saros Member = 12 of 70

Sun at Greatest Eclipse
(Geocentric Coordinates)
R.A. = 16h05m39.1s
Dec. = -20°49'55.6"
S.D. = 00°16'12.0"
H.P. = 00°00'08.9"

Moon at Greatest Eclipse
(Geocentric Coordinates)
R.A. = 16h06m01.8s
Dec. = -19°53'06.5"
S.D. = 00°15'08.3"
H.P. = 00°55'33.5"

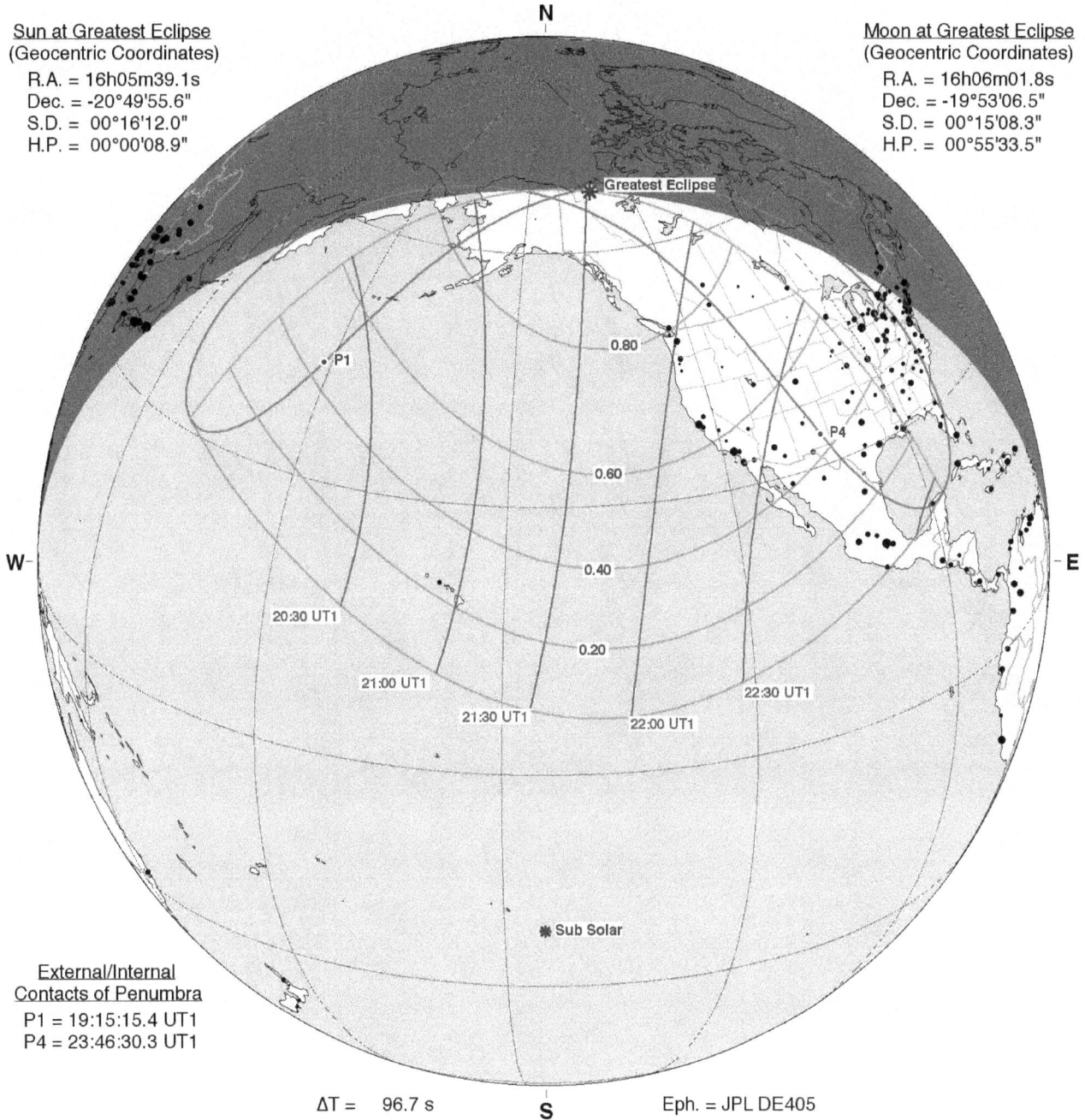

N

Greatest Eclipse

0.80

0.60

0.40

0.20

20:30 UT1

21:00 UT1

21:30 UT1

22:00 UT1

22:30 UT1

P1

P4

W

E

* Sub Solar

External/Internal
Contacts of Penumbra
P1 = 19:15:15.4 UT1
P4 = 23:46:30.3 UT1

ΔT = 96.7 s          S          Eph. = JPL DE405

Circumstances at Greatest Eclipse: 21:30:52.9 UT1
Lat. = 68°31.0'N          Sun Alt. =  0.0°
Long. = 131°13.5'W          Sun Azm. = 193.8°

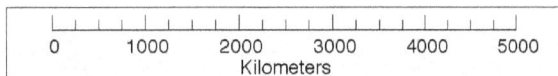

0   1000   2000   3000   4000   5000
Kilometers

©2016  F. Espenak
www.EclipseWise.com

# Partial Solar Eclipse of 2069 Apr 21

Greatest Eclipse = 10:11:08.9 TD (= 10:09:31.9 UT1)

| Eclipse Magnitude = 0.8992 | Saros Series = 120 |
|---|---|
| Gamma = 1.0624 | Saros Member = 64 of 71 |

**N**

Sun at Greatest Eclipse
(Geocentric Coordinates)
R.A. = 01h58m57.2s
Dec. = +12°07'52.1"
S.D. = 00°15'55.0"
H.P. = 00°00'08.8"

Moon at Greatest Eclipse
(Geocentric Coordinates)
R.A. = 01h57m49.5s
Dec. = +13°10'46.5"
S.D. = 00°16'43.2"
H.P. = 01°01'21.7"

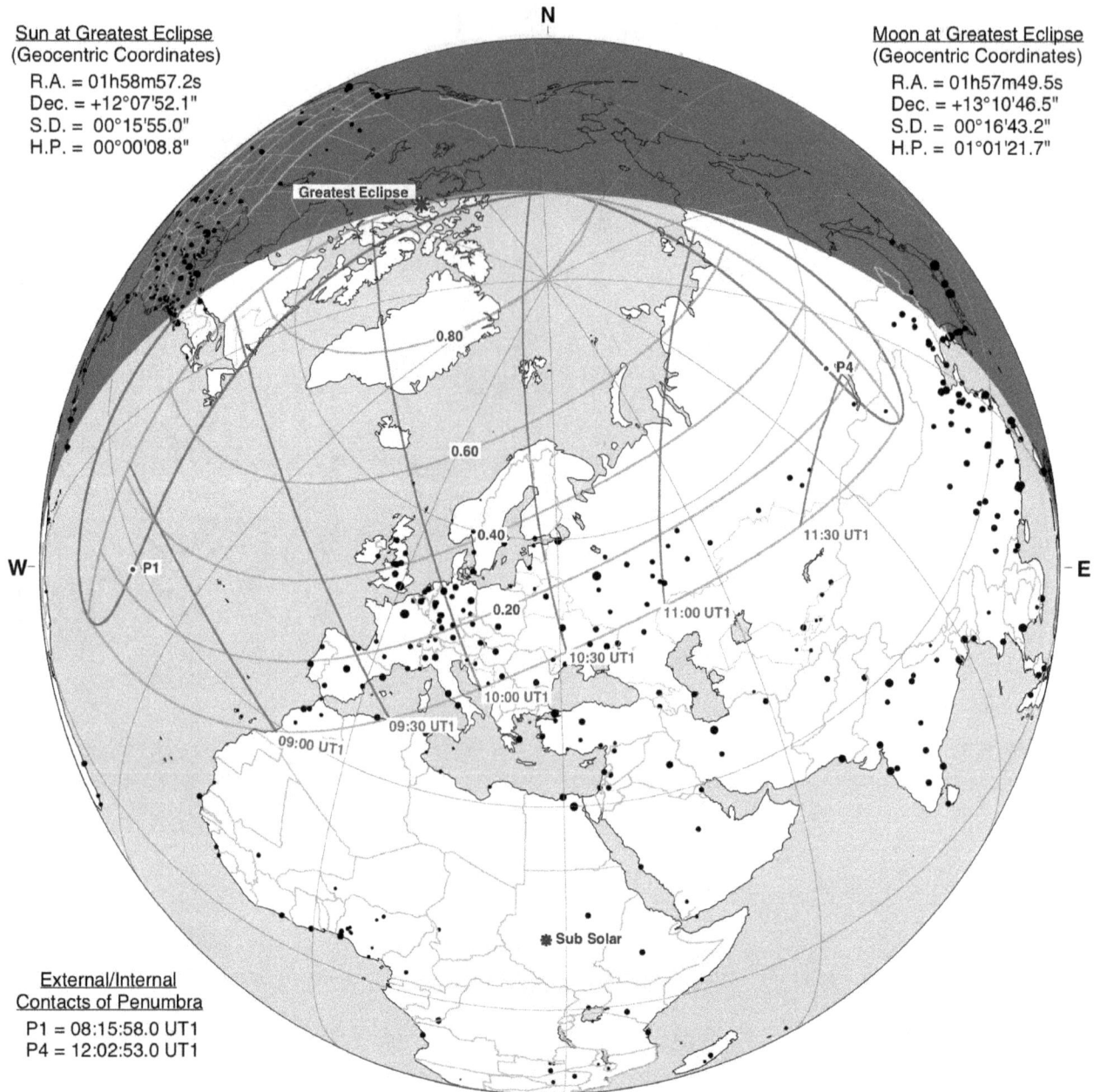

Greatest Eclipse

0.80

0.60

P4

0.40

**W** P1 **E**

0.20

11:30 UT1

11:00 UT1

10:30 UT1

10:00 UT1

09:30 UT1

09:00 UT1

Sub Solar

External/Internal
Contacts of Penumbra
P1 = 08:15:58.0 UT1
P4 = 12:02:53.0 UT1

ΔT = 97.0 s  **S**  Eph. = JPL DE405

Circumstances at Greatest Eclipse: 10:09:31.9 UT1

| Lat. = 71°02.7'N | Sun Alt. = 0.0° |
|---|---|
| Long. = 101°25.2'W | Sun Azm. = 49.7° |

0 1000 2000 3000 4000 5000
Kilometers

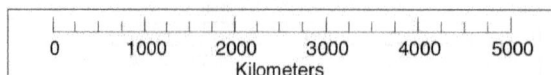

©2016 F. Espenak
www.EclipseWise.com

# Partial Solar Eclipse of 2069 May 20

Greatest Eclipse = 17:53:17.8 TD   (= 17:51:40.7 UT1)

Eclipse Magnitude = 0.0879          Saros Series = 158
Gamma = -1.4852                     Saros Member = 1 of 70

Sun at Greatest Eclipse
(Geocentric Coordinates)
R.A. = 03h52m35.6s
Dec. = +20°12'26.5"
S.D. = 00°15'48.3"
H.P. = 00°00'08.7"

Moon at Greatest Eclipse
(Geocentric Coordinates)
R.A. = 03h53m19.8s
Dec. = +18°43'03.9"
S.D. = 00°16'32.8"
H.P. = 01°00'43.6"

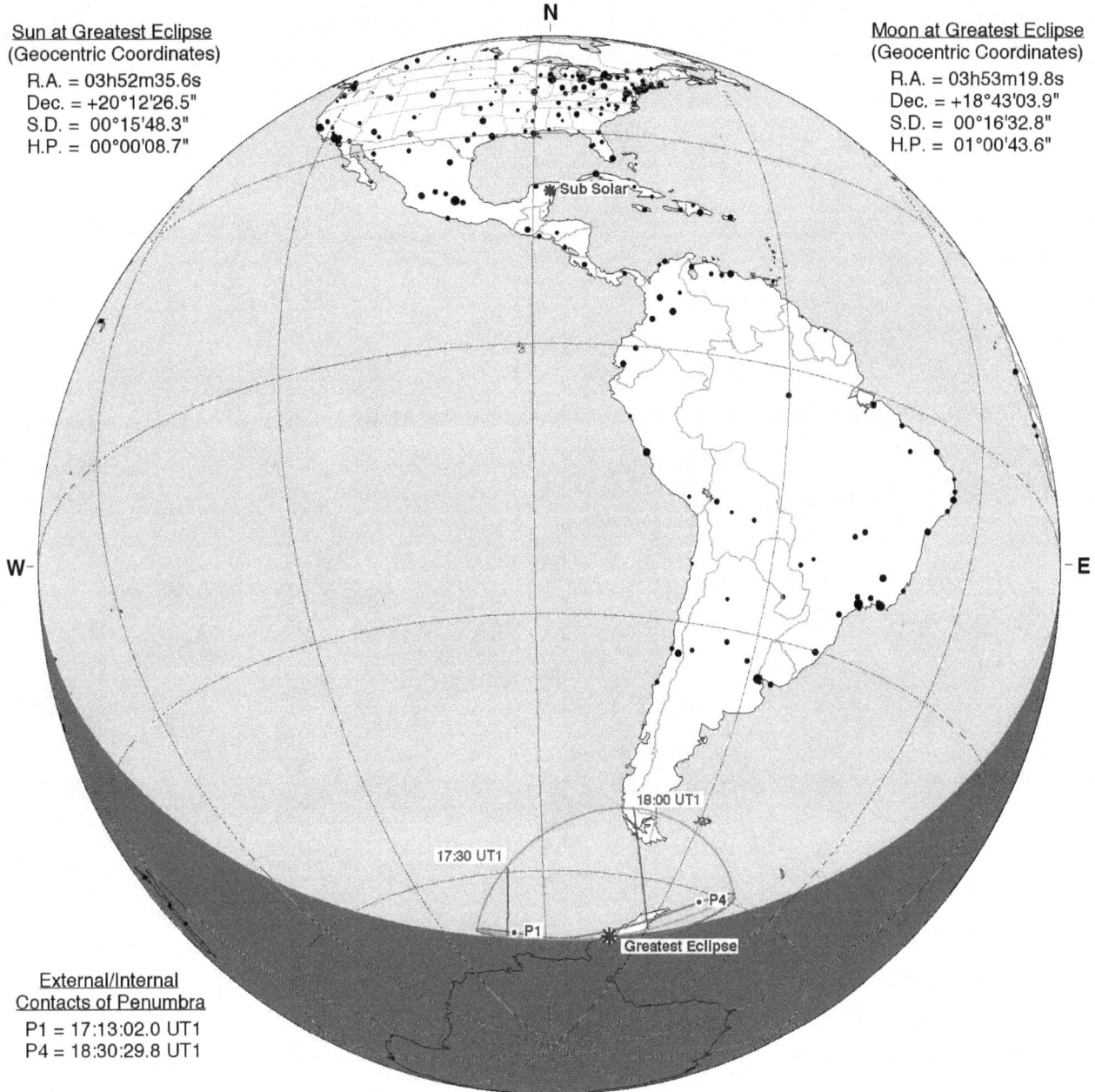

N

W

E

Sub Solar

18:00 UT1

17:30 UT1

P1

P4

Greatest Eclipse

External/Internal
Contacts of Penumbra
P1 = 17:13:02.0 UT1
P4 = 18:30:29.8 UT1

ΔT = 97.1 s          S          Eph. = JPL DE405

Circumstances at Greatest Eclipse: 17:51:40.7 UT1
Lat. = 68°45.6'S          Sun Alt. = 0.0°
Long. = 070°03.9'W        Sun Azm. = 342.5°

0   1000   2000   3000   4000   5000
Kilometers

©2016  F. Espenak
www.EclipseWise.com

# Partial Solar Eclipse of 2069 Oct 15

Greatest Eclipse = 04:19:56.3 TD   (= 04:18:19.0 UT1)

| | |
|---|---|
| Eclipse Magnitude = 0.5298 | Saros Series = 125 |
| Gamma = -1.2524 | Saros Member = 57 of 73 |

**Sun at Greatest Eclipse**
(Geocentric Coordinates)
R.A. = 13h22m54.2s
Dec. = -08°43'06.9"
S.D. = 00°16'02.2"
H.P. = 00°00'08.8"

**Moon at Greatest Eclipse**
(Geocentric Coordinates)
R.A. = 13h21m37.5s
Dec. = -09°48'03.1"
S.D. = 00°14'45.3"
H.P. = 00°54'09.1"

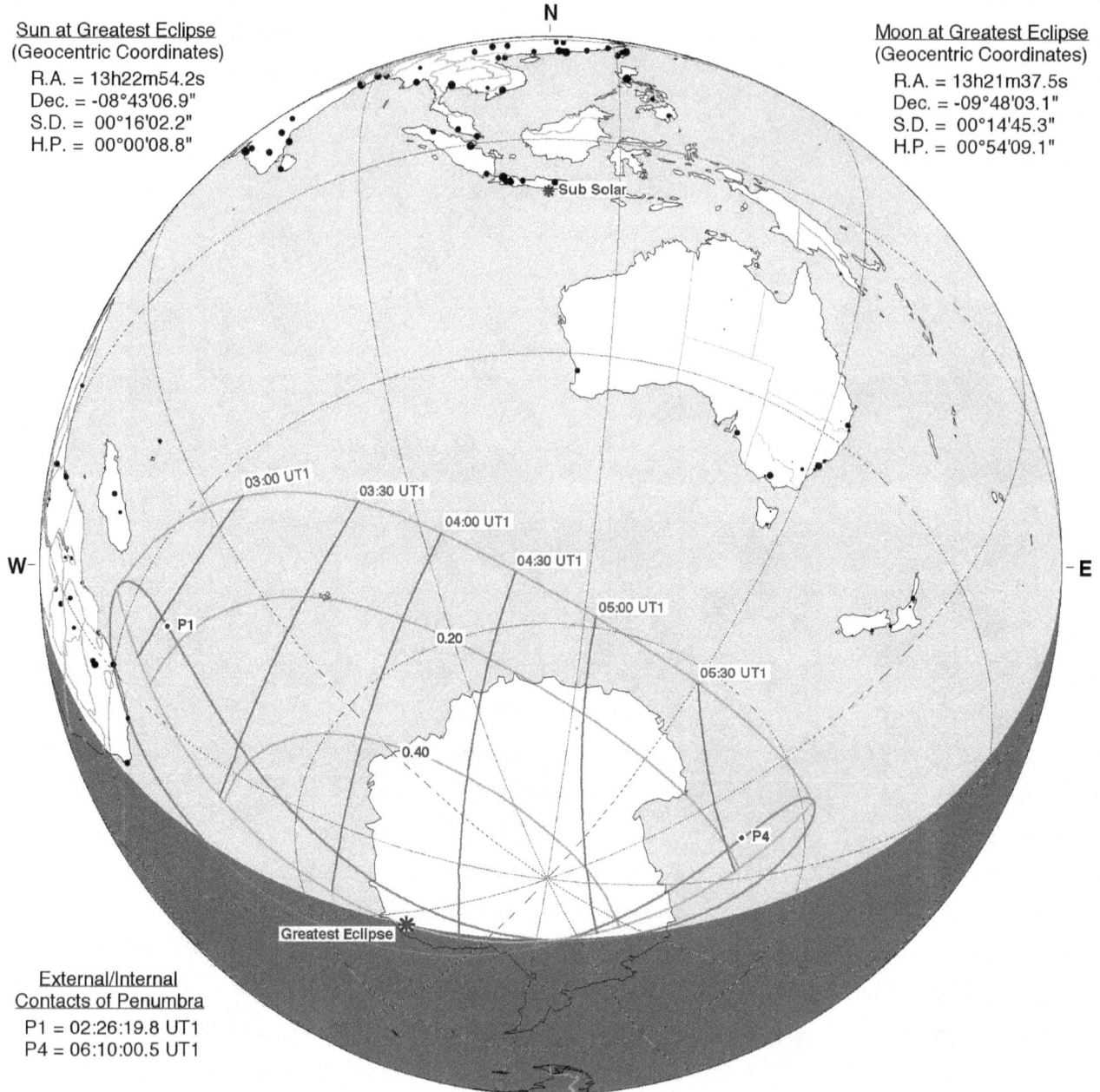

N

Sub Solar

03:00 UT1
03:30 UT1
04:00 UT1
04:30 UT1
05:00 UT1
0.20
05:30 UT1

W

E

P1
0.40
P4

Greatest Eclipse

**External/Internal**
**Contacts of Penumbra**
P1 = 02:26:19.8 UT1
P4 = 06:10:00.5 UT1

ΔT = 97.4 s

S

Eph. = JPL DE405

**Circumstances at Greatest Eclipse: 04:18:19.0 UT1**

| | |
|---|---|
| Lat. = 71°38.6'S | Sun Alt. = 0.0° |
| Long. = 005°36.9'W | Sun Azm. = 118.8° |

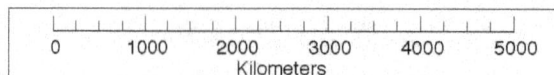

| | | | | | |
|---|---|---|---|---|---|
| 0 | 1000 | 2000 | 3000 | 4000 | 5000 |

Kilometers

©2016  F. Espenak
www.EclipseWise.com

# Total Solar Eclipse of 2070 Apr 11

Greatest Eclipse = 02:36:09.4 TD  (= 02:34:31.7 UT1)

Eclipse Magnitude = 1.0472
Gamma = 0.3652

Saros Series = 130
Saros Member = 55 of 73

**Sun at Greatest Eclipse**
(Geocentric Coordinates)
R.A. = 01h19m45.0s
Dec. = +08°24'18.3"
S.D. = 00°15'57.8"
H.P. = 00°00'08.8"

**Moon at Greatest Eclipse**
(Geocentric Coordinates)
R.A. = 01h19m20.0s
Dec. = +08°45'25.6"
S.D. = 00°16'27.4"
H.P. = 01°00'23.9"

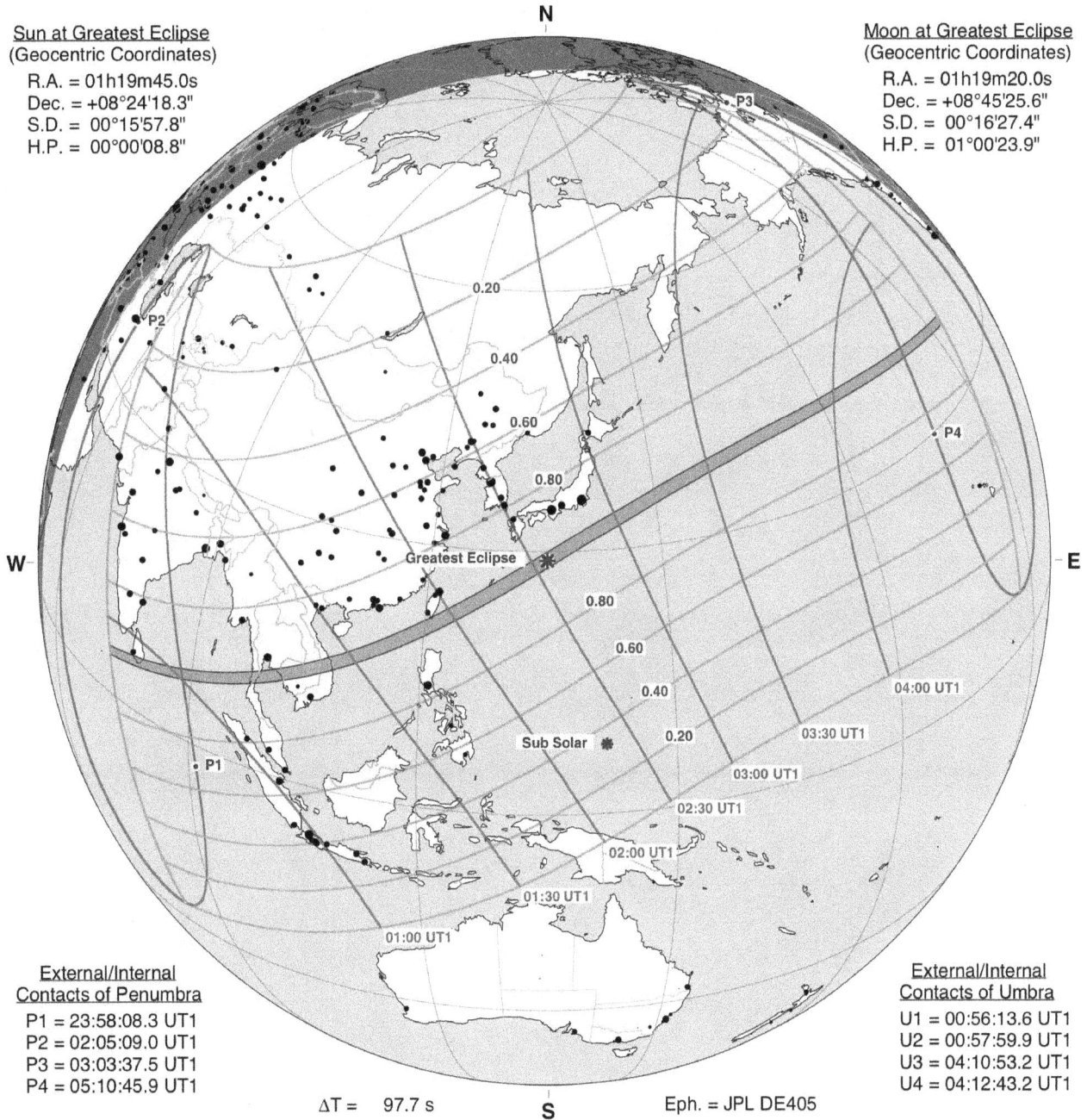

N

0.20
0.40
0.60
0.80

P3
P4
P2
P1

Greatest Eclipse

W — — E

0.80
0.60
0.40
0.20

Sub Solar

04:00 UT1
03:30 UT1
03:00 UT1
02:30 UT1
02:00 UT1
01:30 UT1
01:00 UT1

**External/Internal Contacts of Penumbra**
P1 = 23:58:08.3 UT1
P2 = 02:05:09.0 UT1
P3 = 03:03:37.5 UT1
P4 = 05:10:45.9 UT1

**External/Internal Contacts of Umbra**
U1 = 00:56:13.6 UT1
U2 = 00:57:59.9 UT1
U3 = 04:10:53.2 UT1
U4 = 04:12:43.2 UT1

ΔT = 97.7 s

S

Eph. = JPL DE405

**Circumstances at Greatest Eclipse: 02:34:31.7 UT1**
Lat. = 29°03.2'N
Long. = 134°54.7'E
Path Width = 168.2 km
Sun Alt. = 68.4°
Sun Azm. = 161.7°
Duration = 04m04.2s

**Circumstances at Greatest Duration: 02:37:25.2 UT1**
Lat. = 29°33.4'N
Long. = 135°54.6'E
Path Width = 167.9 km
Sun Alt. = 68.3°
Sun Azm. = 166.5°
Duration = 04m04.3s

0  1000  2000  3000  4000  5000
Kilometers

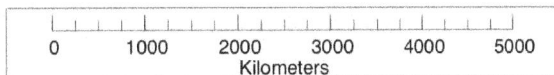

©2016  F. Espenak
www.EclipseWise.com

# Annular Solar Eclipse of 2070 Oct 04

Greatest Eclipse = 07:08:56.8 TD   (= 07:07:18.8 UT1)

| | |
|---|---|
| Eclipse Magnitude = 0.9731 | Saros Series = 135 |
| Gamma = -0.4950 | Saros Member = 42 of 71 |

**Sun at Greatest Eclipse**
(Geocentric Coordinates)
R.A. = 12h42m00.6s
Dec. = -04°30'57.6"
S.D. = 00°15'59.1"
H.P. = 00°00'08.8"

**Moon at Greatest Eclipse**
(Geocentric Coordinates)
R.A. = 12h41m27.3s
Dec. = -04°57'29.9"
S.D. = 00°15'20.7"
H.P. = 00°56'19.0"

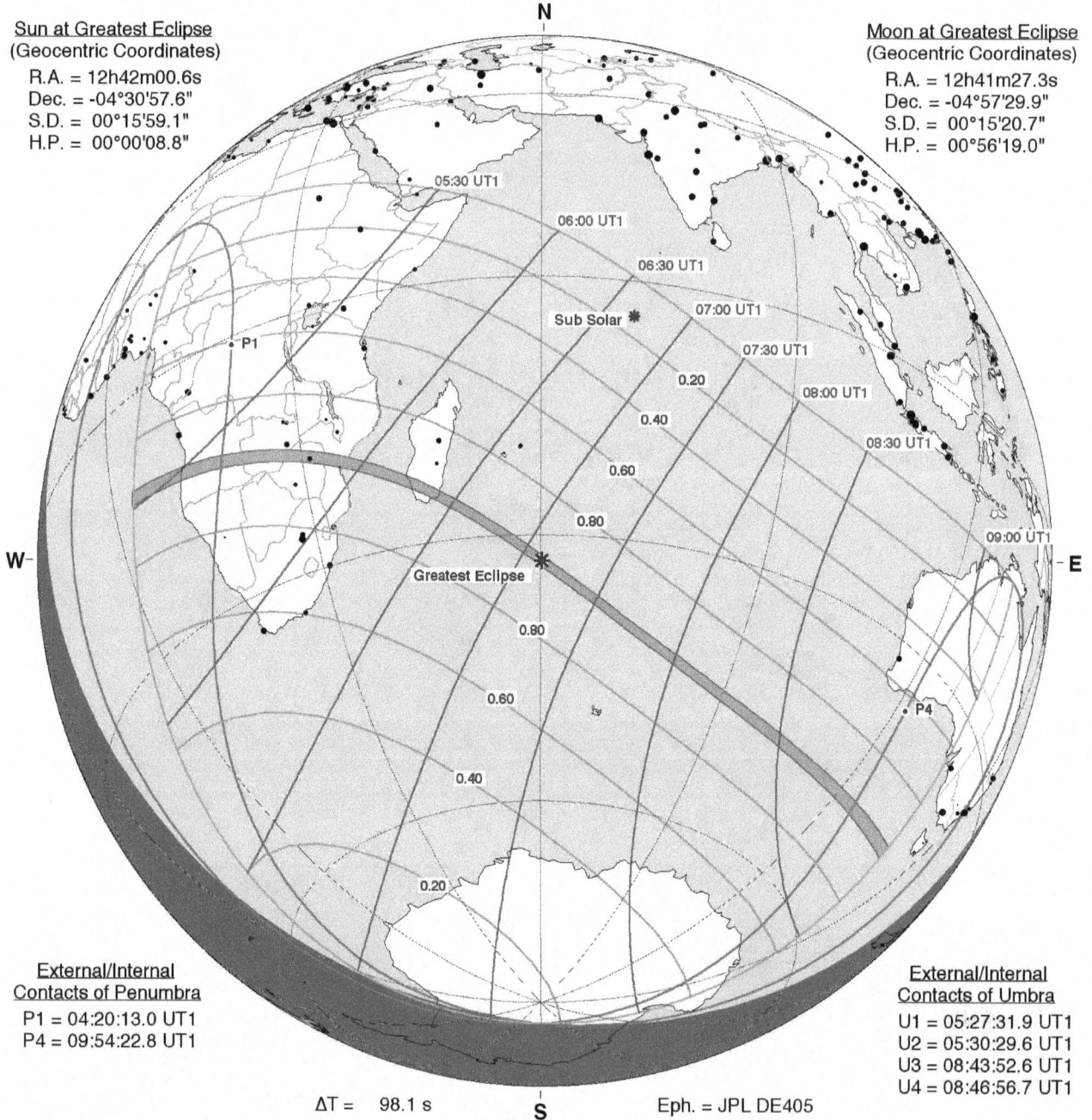

N

05:30 UT1
06:00 UT1
06:30 UT1
Sub Solar
07:00 UT1
07:30 UT1
0.20
08:00 UT1
0.40
08:30 UT1
0.60
09:00 UT1
0.80
P1
W
E
Greatest Eclipse
0.80
0.60
P4
0.40
0.20
S

**External/Internal
Contacts of Penumbra**
P1 = 04:20:13.0 UT1
P4 = 09:54:22.8 UT1

**External/Internal
Contacts of Umbra**
U1 = 05:27:31.9 UT1
U2 = 05:30:29.6 UT1
U3 = 08:43:52.6 UT1
U4 = 08:46:56.7 UT1

ΔT = 98.1 s        Eph. = JPL DE405

**Circumstances at Greatest Eclipse: 07:07:18.8 UT1**

| | |
|---|---|
| Lat. = 32°50.9'S | Sun Alt. = 60.1° |
| Long. = 060°14.5'E | Sun Azm. = 20.6° |
| Path Width = 110.2 km | Duration = 02m44.1s |

**Circumstances at Greatest Duration: 07:43:06.0 UT1**

| | |
|---|---|
| Lat. = 39°12.0'S | Sun Alt. = 53.8° |
| Long. = 072°52.3'E | Sun Azm. = 340.4° |
| Path Width = 112.1 km | Duration = 02m45.3s |

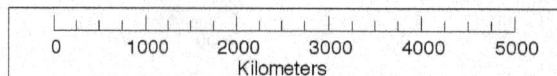

0    1000   2000   3000   4000   5000
Kilometers

©2016 F. Espenak
www.EclipseWise.com

# Annular Solar Eclipse of 2071 Mar 31

Greatest Eclipse = 15:01:06.4 TD  (= 14:59:27.9 UT1)

| | |
|---|---|
| Eclipse Magnitude = 0.9919 | Saros Series = 140 |
| Gamma = -0.3739 | Saros Member = 32 of 71 |

Sun at Greatest Eclipse
(Geocentric Coordinates)
R.A. = 00h40m27.6s
Dec. = +04°21'06.6"
S.D. = 00°16'00.8"
H.P. = 00°00'08.8"

Moon at Greatest Eclipse
(Geocentric Coordinates)
R.A. = 00h40m53.3s
Dec. = +04°00'40.0"
S.D. = 00°15'38.9"
H.P. = 00°57'25.9"

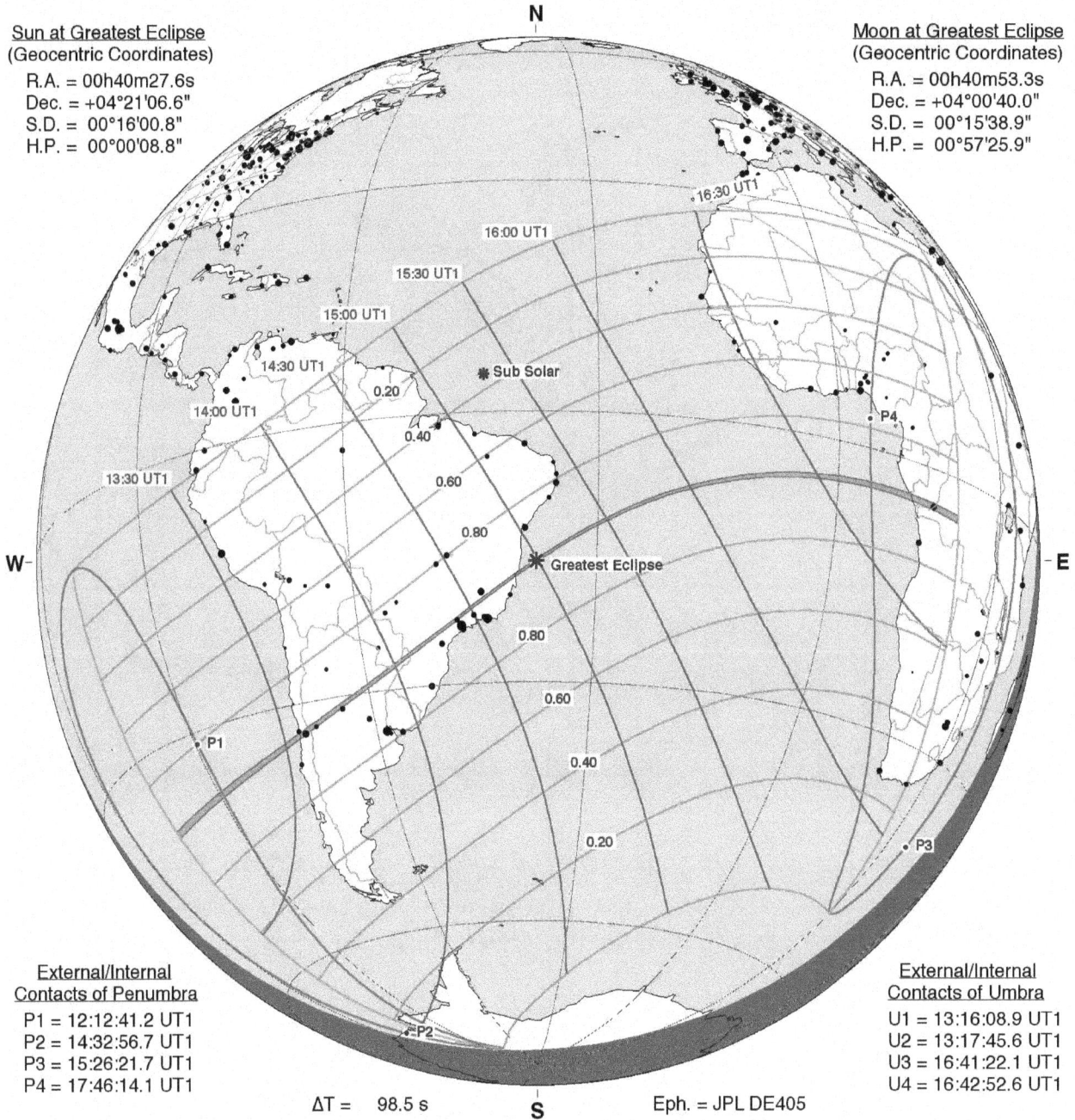

N

16:30 UT1
16:00 UT1
15:30 UT1
15:00 UT1
14:30 UT1
14:00 UT1
13:30 UT1

Sub Solar

0.20
0.40
0.60
0.80

Greatest Eclipse

0.80
0.60
0.40
0.20

P4
P1
P3
P2

W

E

External/Internal
Contacts of Penumbra
P1 = 12:12:41.2 UT1
P2 = 14:32:56.7 UT1
P3 = 15:26:21.7 UT1
P4 = 17:46:14.1 UT1

External/Internal
Contacts of Umbra
U1 = 13:16:08.9 UT1
U2 = 13:17:45.6 UT1
U3 = 16:41:22.1 UT1
U4 = 16:42:52.6 UT1

ΔT = 98.5 s

S

Eph. = JPL DE405

Circumstances at Greatest Eclipse: 14:59:27.9 UT1

| | |
|---|---|
| Lat. = 16°41.9'S | Sun Alt. = 67.9° |
| Long. = 037°09.6'W | Sun Azm. = 342.0° |
| Path Width = 30.7 km | Duration = 00m51.7s |

Circumstances at Greatest Duration: 13:16:57.2 UT1

| | |
|---|---|
| Lat. = 39°24.8'S | Sun Alt. = 0.0° |
| Long. = 104°39.1'W | Sun Azm. = 84.4° |
| Path Width = 91.6 km | Duration = 01m30.7s |

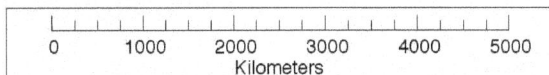

0   1000   2000   3000   4000   5000
Kilometers

©2016  F. Espenak
www.EclipseWise.com

# Total Solar Eclipse of 2071 Sep 23

Greatest Eclipse = 17:20:28.0 TD (= 17:18:49.2 UT1)

| | |
|---|---|
| Eclipse Magnitude = 1.0333 | Saros Series = 145 |
| Gamma = 0.2620 | Saros Member = 25 of 77 |

**Sun at Greatest Eclipse**
(Geocentric Coordinates)
R.A. = 12h02m56.7s
Dec. = -00°19'08.2"
S.D. = 00°15'56.2"
H.P. = 00°00'08.8"

**Moon at Greatest Eclipse**
(Geocentric Coordinates)
R.A. = 12h03m15.7s
Dec. = -00°04'20.0"
S.D. = 00°16'12.3"
H.P. = 00°59'28.3"

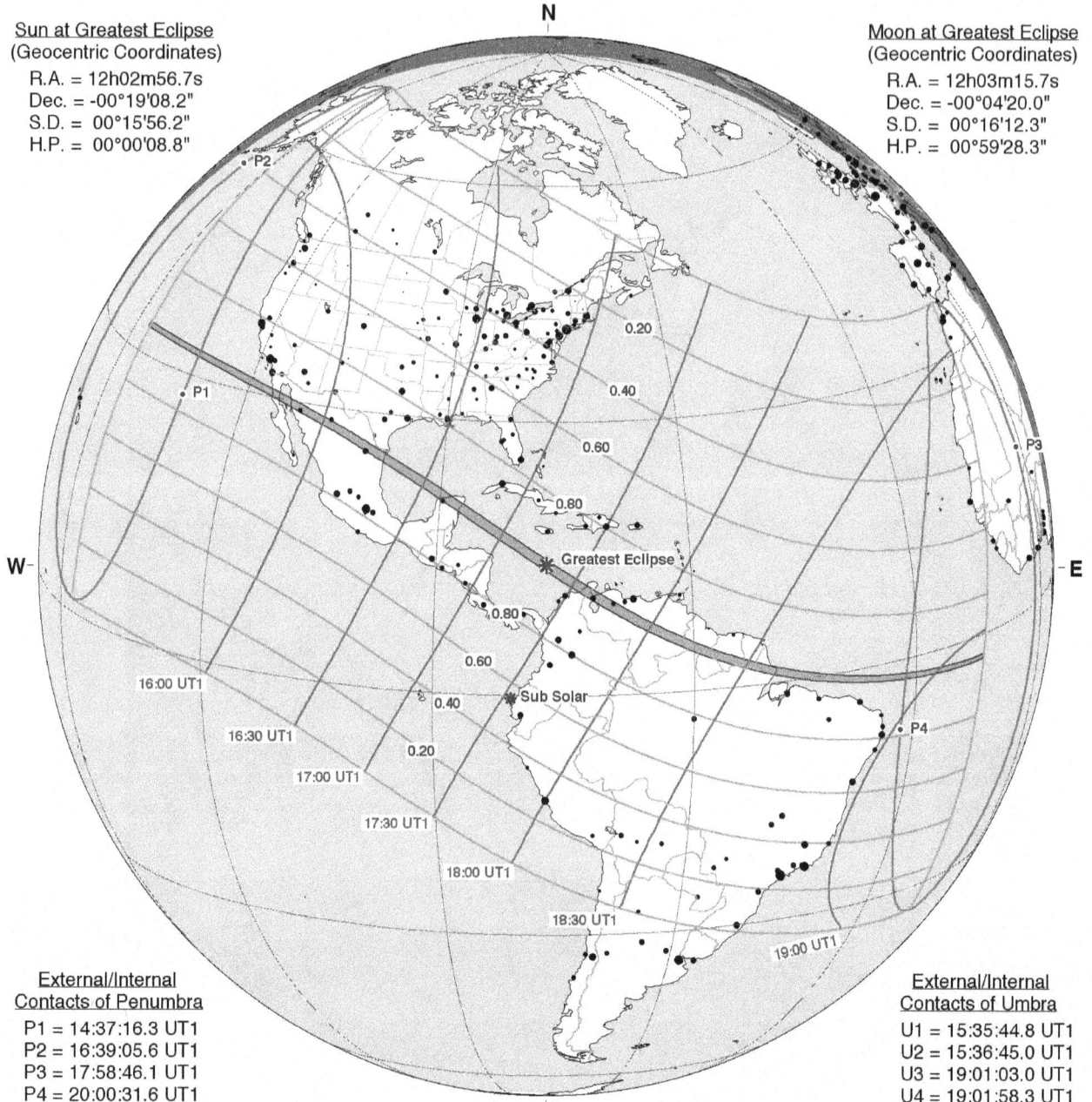

N

P2

P1

0.20
0.40
0.60
0.80
Greatest Eclipse

W

E

0.80
0.60
0.40
Sub Solar
0.20

16:00 UT1
16:30 UT1
17:00 UT1
17:30 UT1
18:00 UT1
18:30 UT1
19:00 UT1

P3

P4

**External/Internal
Contacts of Penumbra**
P1 = 14:37:16.3 UT1
P2 = 16:39:05.6 UT1
P3 = 17:58:46.1 UT1
P4 = 20:00:31.6 UT1

ΔT = 98.8 s

S

Eph. = JPL DE405

**External/Internal
Contacts of Umbra**
U1 = 15:35:44.8 UT1
U2 = 15:36:45.0 UT1
U3 = 19:01:03.0 UT1
U4 = 19:01:58.3 UT1

**Circumstances at Greatest Eclipse: 17:18:49.2 UT1**

| | |
|---|---|
| Lat. = 14°13.5'N | Sun Alt. = 74.7° |
| Long. = 076°54.7'W | Sun Azm. = 198.2° |
| Path Width = 116.1 km | Duration = 03m10.8s |

**Circumstances at Greatest Duration: 17:17:08.4 UT1**

| | |
|---|---|
| Lat. = 14°30.8'N | Sun Alt. = 74.7° |
| Long. = 077°22.7'W | Sun Azm. = 194.7° |
| Path Width = 116.0 km | Duration = 03m10.9s |

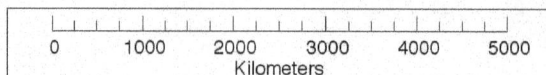

0 1000 2000 3000 4000 5000
Kilometers

©2016 F. Espenak
www.EclipseWise.com

# Partial Solar Eclipse of 2072 Mar 19

Greatest Eclipse = 20:10:31.1 TD  (= 20:08:51.9 UT1)

| | |
|---|---|
| Eclipse Magnitude = 0.7199 | Saros Series = 150 |
| Gamma = -1.1405 | Saros Member = 20 of 71 |

**N**

Sun at Greatest Eclipse
(Geocentric Coordinates)
R.A. = 00h00m16.3s
Dec. = +00°01'46.1"
S.D. = 00°16'03.7"
H.P. = 00°00'08.8"

Moon at Greatest Eclipse
(Geocentric Coordinates)
R.A. = 00h01m32.5s
Dec. = -00°57'26.5"
S.D. = 00°14'53.9"
H.P. = 00°54'40.7"

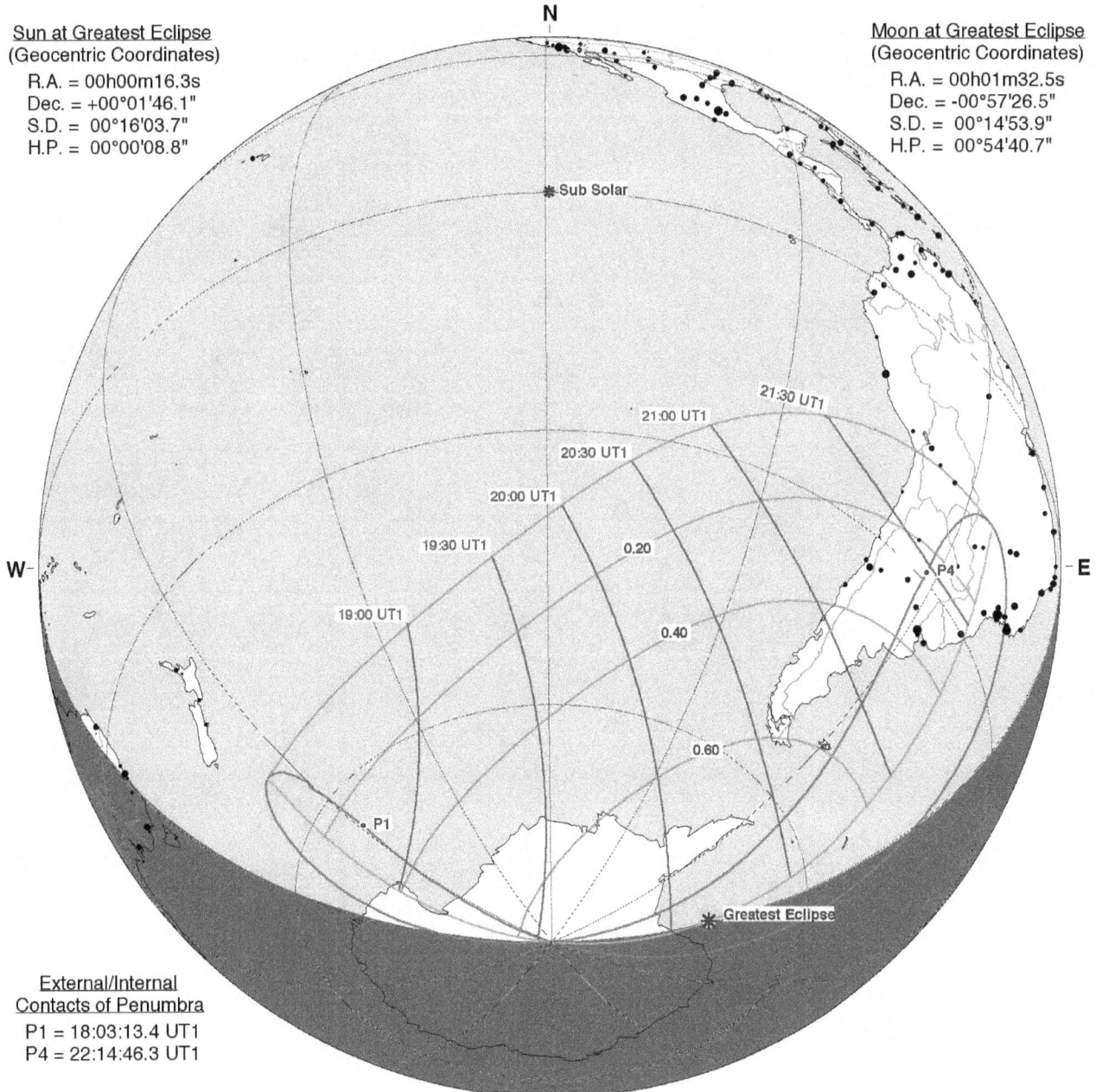

Sub Solar

21:30 UT1
21:00 UT1
20:30 UT1
20:00 UT1
19:30 UT1
0.20
19:00 UT1
0.40
P4
**W** — — **E**
0.60
P1
Greatest Eclipse

External/Internal
Contacts of Penumbra
P1 = 18:03:13.4 UT1
P4 = 22:14:46.3 UT1

ΔT = 99.2 s      **S**      Eph. = JPL DE405

Circumstances at Greatest Eclipse: 20:08:51.9 UT1

| | |
|---|---|
| Lat. = 72°11.0'S | Sun Alt. = 0.0° |
| Long. = 030°32.2'W | Sun Azm. = 270.1° |

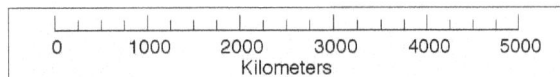

| 0 | 1000 | 2000 | 3000 | 4000 | 5000 |
|---|---|---|---|---|---|

Kilometers

©2016  F. Espenak
www.EclipseWise.com

# Total Solar Eclipse of 2072 Sep 12

Greatest Eclipse = 08:59:20.2 TD   (= 08:57:40.7 UT1)

| | |
|---|---|
| Eclipse Magnitude = 1.0558 | Saros Series = 155 |
| Gamma = 0.9655 | Saros Member = 9 of 71 |

**Sun at Greatest Eclipse**
(Geocentric Coordinates)
R.A. = 11h24m57.8s
Dec. = +03°46'31.0"
S.D. = 00°15'53.4"
H.P. = 00°00'08.7"

**Moon at Greatest Eclipse**
(Geocentric Coordinates)
R.A. = 11h26m09.4s
Dec. = +04°42'50.2"
S.D. = 00°16'42.8"
H.P. = 01°01'20.3"

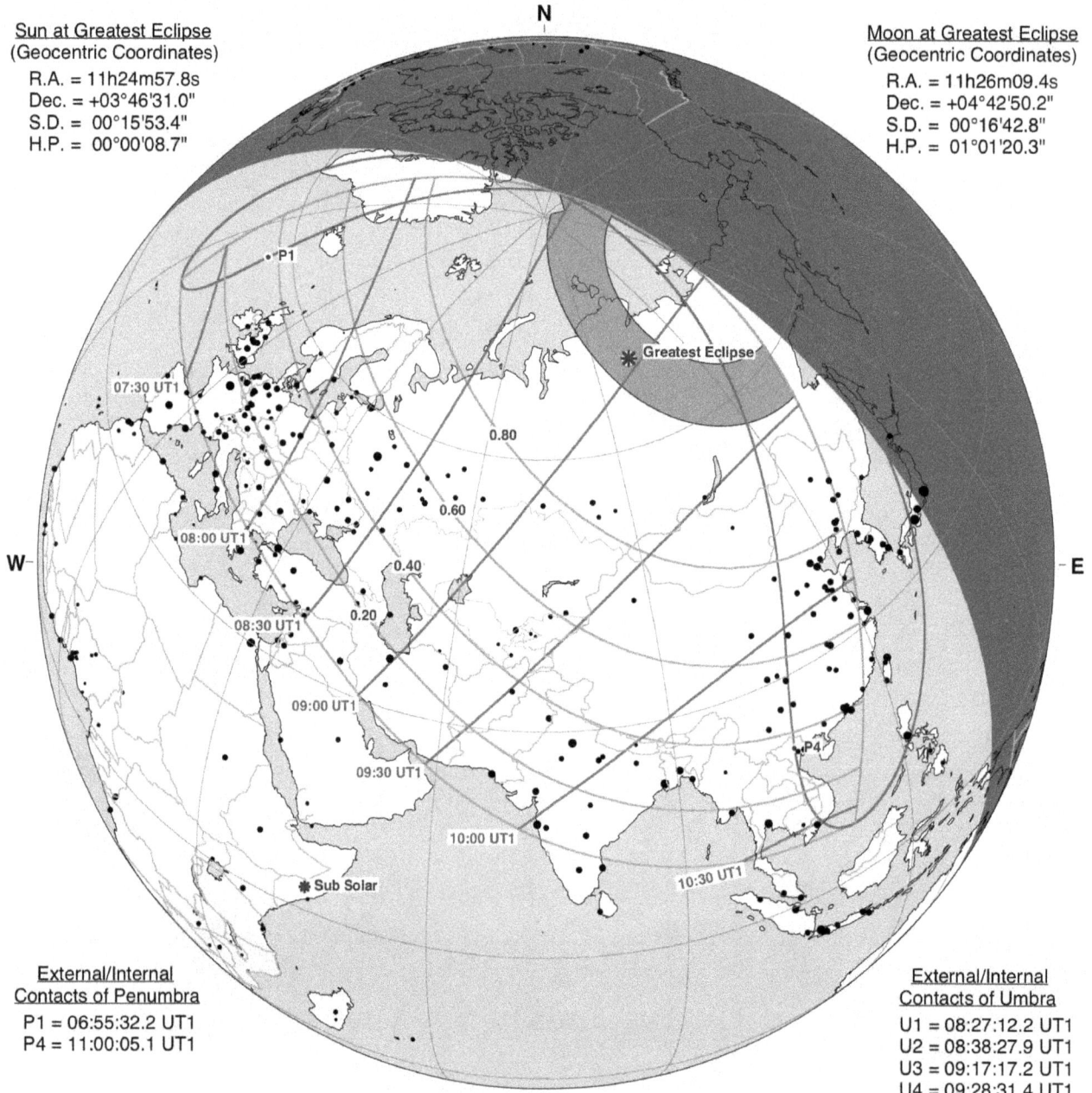

N

07:30 UT1
0.80
0.60
08:00 UT1
0.40
0.20
08:30 UT1
09:00 UT1
09:30 UT1
10:00 UT1
10:30 UT1

P1
Greatest Eclipse
Sub Solar
P4

W — — E

**External/Internal Contacts of Penumbra**
P1 = 06:55:32.2 UT1
P4 = 11:00:05.1 UT1

**External/Internal Contacts of Umbra**
U1 = 08:27:12.2 UT1
U2 = 08:38:27.9 UT1
U3 = 09:17:17.2 UT1
U4 = 09:28:31.4 UT1

ΔT = 99.6 s

S

Eph. = JPL DE405

**Circumstances at Greatest Eclipse: 08:57:40.7 UT1**

| | |
|---|---|
| Lat. = 69°46.3'N | Sun Alt. = 14.4° |
| Long. = 101°49.9'E | Sun Azm. = 240.0° |
| Path Width = 732.4 km | Duration = 03m12.9s |

**Circumstances at Greatest Duration: 08:57:45.4 UT1**

| | |
|---|---|
| Lat. = 69°44.0'N | Sun Alt. = 14.4° |
| Long. = 101°52.5'E | Sun Azm. = 240.1° |
| Path Width = 732.7 km | Duration = 03m12.9s |

0   1000   2000   3000   4000   5000
Kilometers

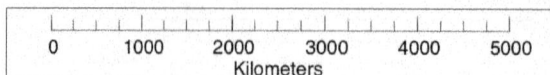

©2016 F. Espenak
www.EclipseWise.com

# Partial Solar Eclipse of 2073 Feb 07

Greatest Eclipse = 01:55:59.0 TD  (= 01:54:19.1 UT1)

| | |
|---|---|
| Eclipse Magnitude = 0.6768 | Saros Series = 122 |
| Gamma = 1.1651 | Saros Member = 61 of 70 |

Sun at Greatest Eclipse
(Geocentric Coordinates)
R.A. = 21h25m15.3s
Dec. = -15°09'16.6"
S.D. = 00°16'13.0"
H.P. = 00°00'08.9"

Moon at Greatest Eclipse
(Geocentric Coordinates)
R.A. = 21h24m18.6s
Dec. = -14°07'10.1"
S.D. = 00°14'54.9"
H.P. = 00°54'44.3"

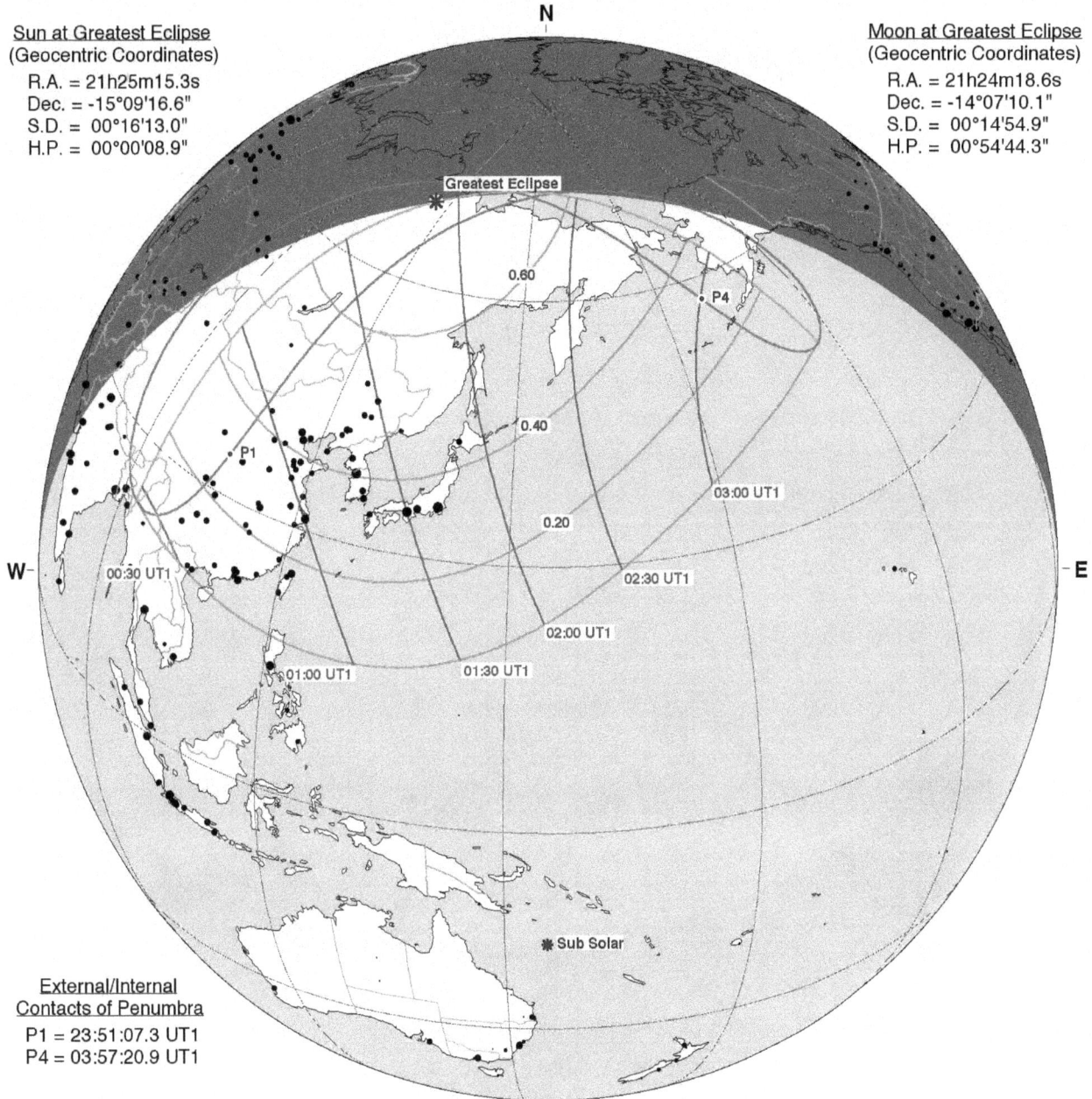

N

Greatest Eclipse

0.60

0.40

P4

03:00 UT1

0.20

02:30 UT1

P1

02:00 UT1

W

00:30 UT1

01:00 UT1    01:30 UT1

E

Sub Solar

External/Internal
Contacts of Penumbra
P1 = 23:51:07.3 UT1
P4 = 03:57:20.9 UT1

ΔT = 99.9 s          S          Eph. = JPL DE405

Circumstances at Greatest Eclipse: 01:54:19.1 UT1

| | |
|---|---|
| Lat. = 70°28.0'N | Sun Alt. = 0.0° |
| Long. = 114°44.4'E | Sun Azm. = 141.4° |

| | | | | | |
|---|---|---|---|---|---|
| 0 | 1000 | 2000 | 3000 | 4000 | 5000 |

Kilometers

217

# Total Solar Eclipse of 2073 Aug 03

Greatest Eclipse = 17:15:22.9 TD  (= 17:13:42.6 UT1)

| | |
|---|---|
| Eclipse Magnitude = 1.0294 | Saros Series = 127 |
| Gamma = -0.8763 | Saros Member = 61 of 82 |

**Sun at Greatest Eclipse**
(Geocentric Coordinates)
R.A. = 08h57m50.6s
Dec. = +17°11'06.4"
S.D. = 00°15'45.7"
H.P. = 00°00'08.7"

**Moon at Greatest Eclipse**
(Geocentric Coordinates)
R.A. = 08h57m11.2s
Dec. = +16°20'19.0"
S.D. = 00°16'06.2"
H.P. = 00°59'05.8"

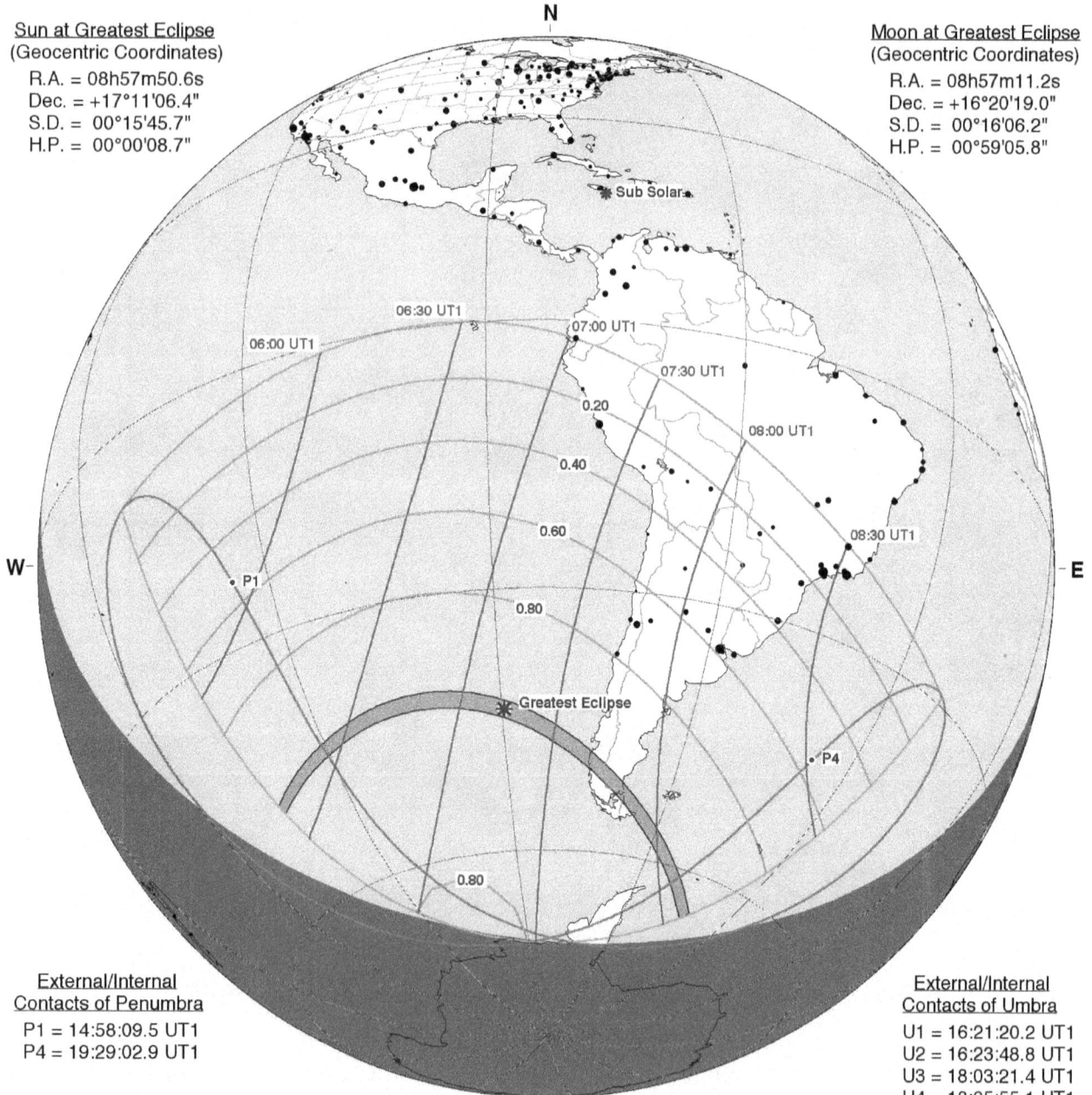

N

06:30 UT1
06:00 UT1
07:00 UT1
07:30 UT1
08:00 UT1
08:30 UT1

0.20
0.40
0.60
0.80

Sub Solar

P1
P4

0.80

Greatest Eclipse

W — 

— E

**External/Internal**
**Contacts of Penumbra**
P1 = 14:58:09.5 UT1
P4 = 19:29:02.9 UT1

**External/Internal**
**Contacts of Umbra**
U1 = 16:21:20.2 UT1
U2 = 16:23:48.8 UT1
U3 = 18:03:21.4 UT1
U4 = 18:05:55.1 UT1

ΔT = 100.3 s

S

Eph. = JPL DE405

**Circumstances at Greatest Eclipse: 17:13:42.6 UT1**

| | |
|---|---|
| Lat. = 43°14.5'S | Sun Alt. = 28.5° |
| Long. = 089°34.3'W | Sun Azm. = 13.8° |
| Path Width = 206.4 km | Duration = 02m29.2s |

**Circumstances at Greatest Duration: 17:14:07.2 UT1**

| | |
|---|---|
| Lat. = 43°17.7'S | Sun Alt. = 28.5° |
| Long. = 089°22.9'W | Sun Azm. = 13.5° |
| Path Width = 206.2 km | Duration = 02m29.2s |

0   1000   2000   3000   4000   5000
Kilometers

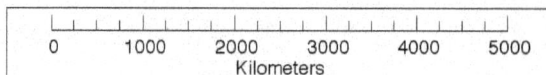

©2016  F. Espenak
www.EclipseWise.com

# Annular Solar Eclipse of 2074 Jan 27

Greatest Eclipse = 06:44:15.3 TD  (= 06:42:34.7 UT1)

Eclipse Magnitude = 0.9798

Gamma = 0.4251

Saros Series = 132

Saros Member = 49 of 71

**Sun at Greatest Eclipse**
(Geocentric Coordinates)
R.A. = 20h40m20.9s
Dec. = -18°20'28.4"
S.D. = 00°16'14.5"
H.P. = 00°00'08.9"

**Moon at Greatest Eclipse**
(Geocentric Coordinates)
R.A. = 20h40m04.5s
Dec. = -17°56'22.6"
S.D. = 00°15'41.1"
H.P. = 00°57'33.8"

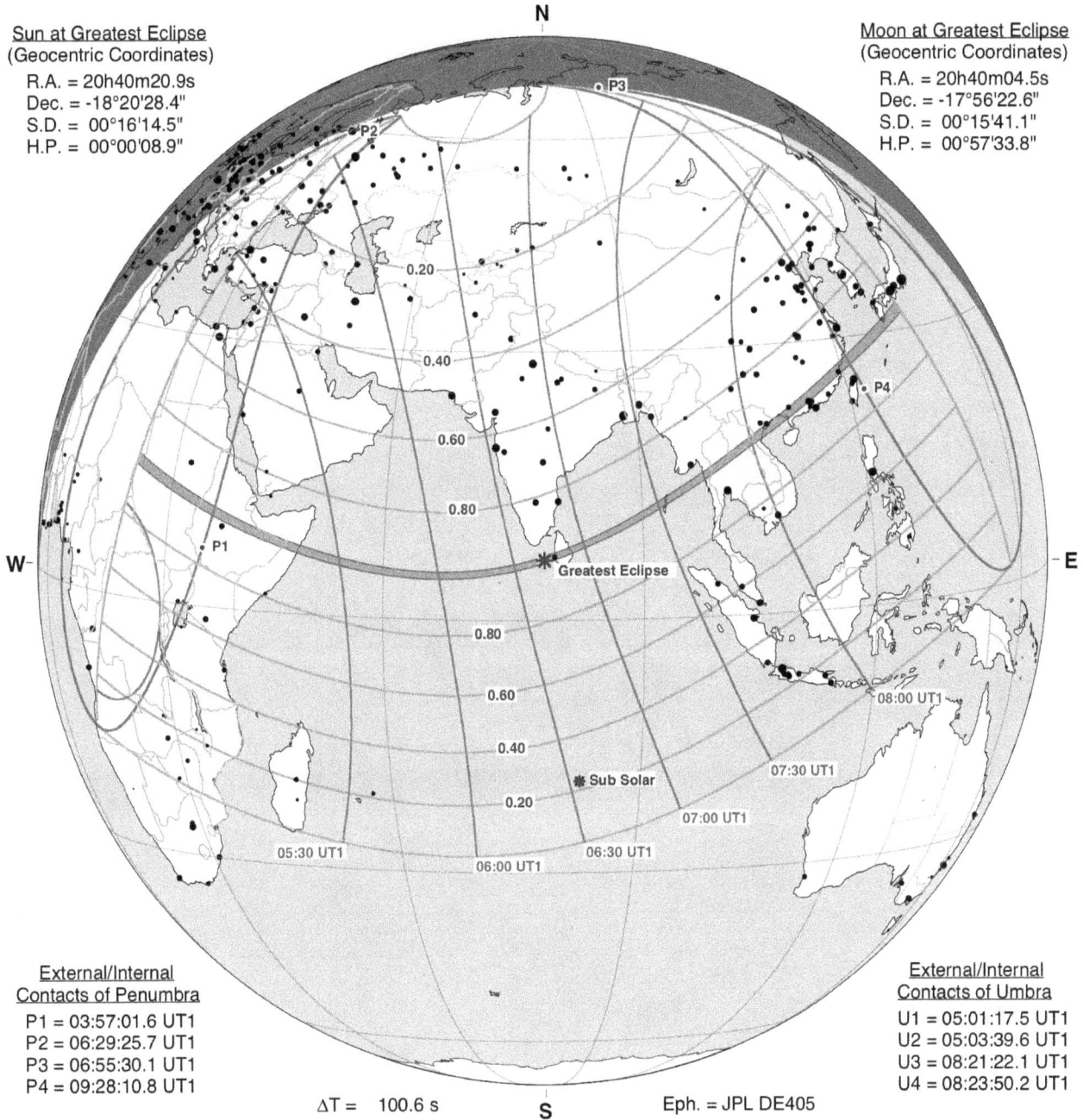

N

P3
P2
P4

0.20
0.40
0.60
0.80

W —

P1

Greatest Eclipse

— E

0.80
0.60
0.40
0.20

Sub Solar

08:00 UT1
07:30 UT1
07:00 UT1
06:30 UT1
06:00 UT1
05:30 UT1

**External/Internal
Contacts of Penumbra**
P1 = 03:57:01.6 UT1
P2 = 06:29:25.7 UT1
P3 = 06:55:30.1 UT1
P4 = 09:28:10.8 UT1

**External/Internal
Contacts of Umbra**
U1 = 05:01:17.5 UT1
U2 = 05:03:39.6 UT1
U3 = 08:21:22.1 UT1
U4 = 08:23:50.2 UT1

ΔT = 100.6 s

S

Eph. = JPL DE405

**Circumstances at Greatest Eclipse:  06:42:34.7 UT1**

Lat. = 06°33.8'N

Long. = 078°36.5'E

Path Width = 79.3 km

Sun Alt. = 64.8°

Sun Azm. = 171.3°

Duration = 02m21.1s

**Circumstances at Greatest Duration:  08:22:36.3 UT1**

Lat. = 32°32.7'N

Long. = 135°18.3'E

Path Width = 134.4 km

Sun Alt. = 0.0°

Sun Azm. = 248.1°

Duration = 02m21.6s

0  1000  2000  3000  4000  5000
Kilometers

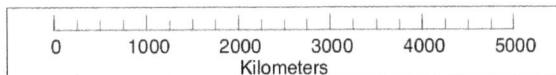

©2016  F. Espenak
www.EclipseWise.com

# Annular Solar Eclipse of 2074 Jul 24

Greatest Eclipse = 03:10:32.0 TD  (= 03:08:51.0 UT1)

Eclipse Magnitude = 0.9838          Saros Series = 137

Gamma = -0.1242          Saros Member = 39 of 70

**Sun at Greatest Eclipse**
(Geocentric Coordinates)
R.A. = 08h15m33.5s
Dec. = +19°47'02.4"
S.D. = 00°15'44.6"
H.P. = 00°00'08.7"

**Moon at Greatest Eclipse**
(Geocentric Coordinates)
R.A. = 08h15m29.9s
Dec. = +19°40'09.4"
S.D. = 00°15'15.0"
H.P. = 00°55'58.2"

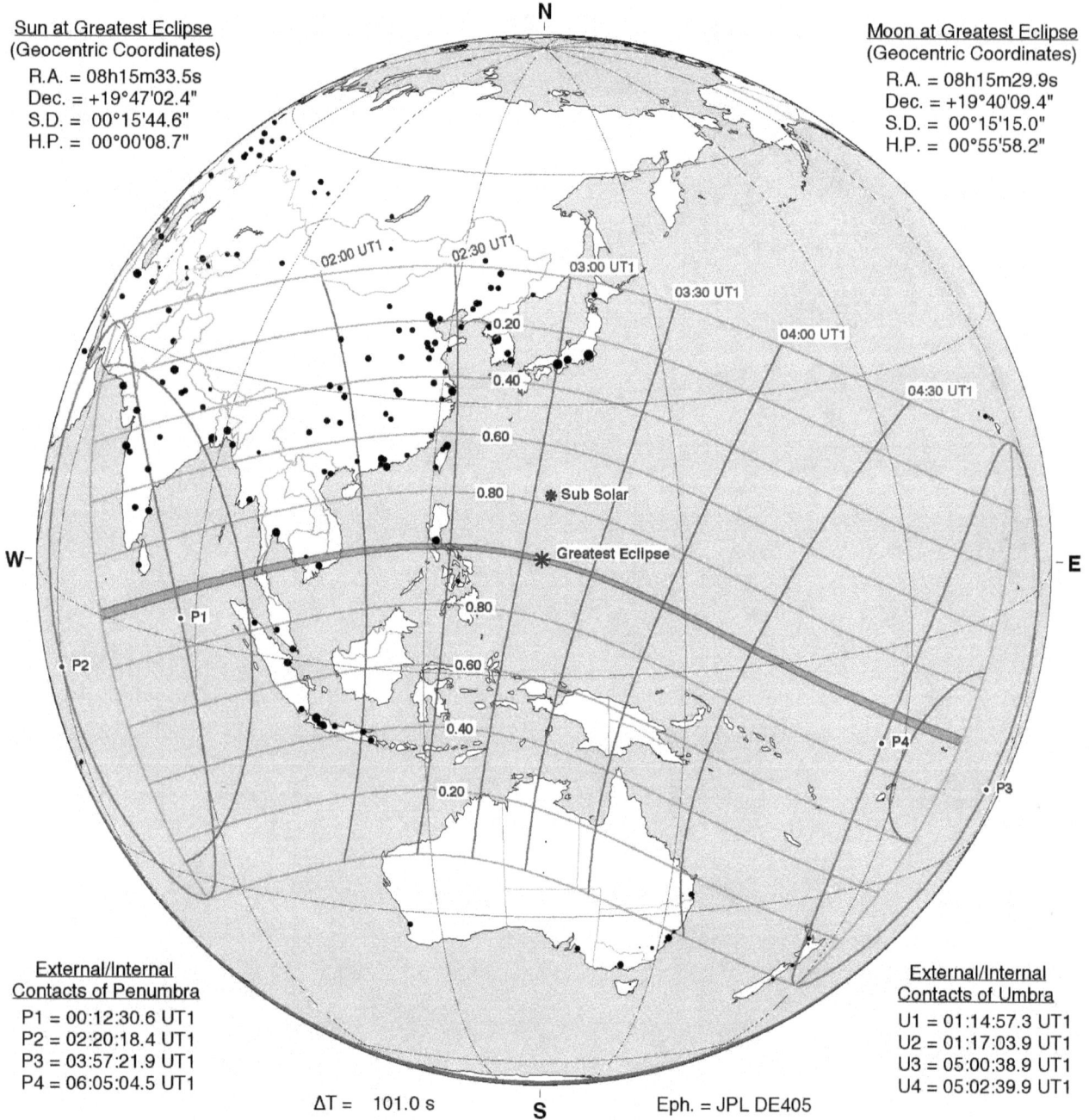

**External/Internal Contacts of Penumbra**
P1 = 00:12:30.6 UT1
P2 = 02:20:18.4 UT1
P3 = 03:57:21.9 UT1
P4 = 06:05:04.5 UT1

**External/Internal Contacts of Umbra**
U1 = 01:14:57.3 UT1
U2 = 01:17:03.9 UT1
U3 = 05:00:38.9 UT1
U4 = 05:02:39.9 UT1

ΔT = 101.0 s          Eph. = JPL DE405

**Circumstances at Greatest Eclipse: 03:08:51.0 UT1**

| | |
|---|---|
| Lat. = 12°46.9'N | Sun Alt. = 82.9° |
| Long. = 133°32.8'E | Sun Azm. = 6.9° |
| Path Width = 57.8 km | Duration = 01m57.4s |

**Circumstances at Greatest Duration: 01:16:00.6 UT1**

| | |
|---|---|
| Lat. = 00°03.8'S | Sun Alt. = 0.0° |
| Long. = 072°40.8'E | Sun Azm. = 70.2° |
| Path Width = 117.5 km | Duration = 02m06.4s |

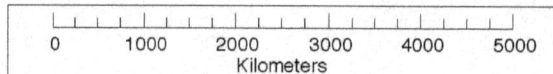

©2016  F. Espenak
www.EclipseWise.com

0  1000  2000  3000  4000  5000
Kilometers

# Total Solar Eclipse of 2075 Jan 16

Greatest Eclipse = 18:36:04.3 TD   (= 18:34:22.9 UT1)

| | |
|---|---|
| Eclipse Magnitude = 1.0311 | Saros Series = 142 |
| Gamma = -0.2799 | Saros Member = 26 of 72 |

Sun at Greatest Eclipse
(Geocentric Coordinates)
R.A. = 19h55m06.1s
Dec. = -20°47'51.3"
S.D. = 00°16'15.5"
H.P. = 00°00'08.9"

Moon at Greatest Eclipse
(Geocentric Coordinates)
R.A. = 19h55m12.8s
Dec. = -21°04'40.9"
S.D. = 00°16'29.7"
H.P. = 01°00'32.2"

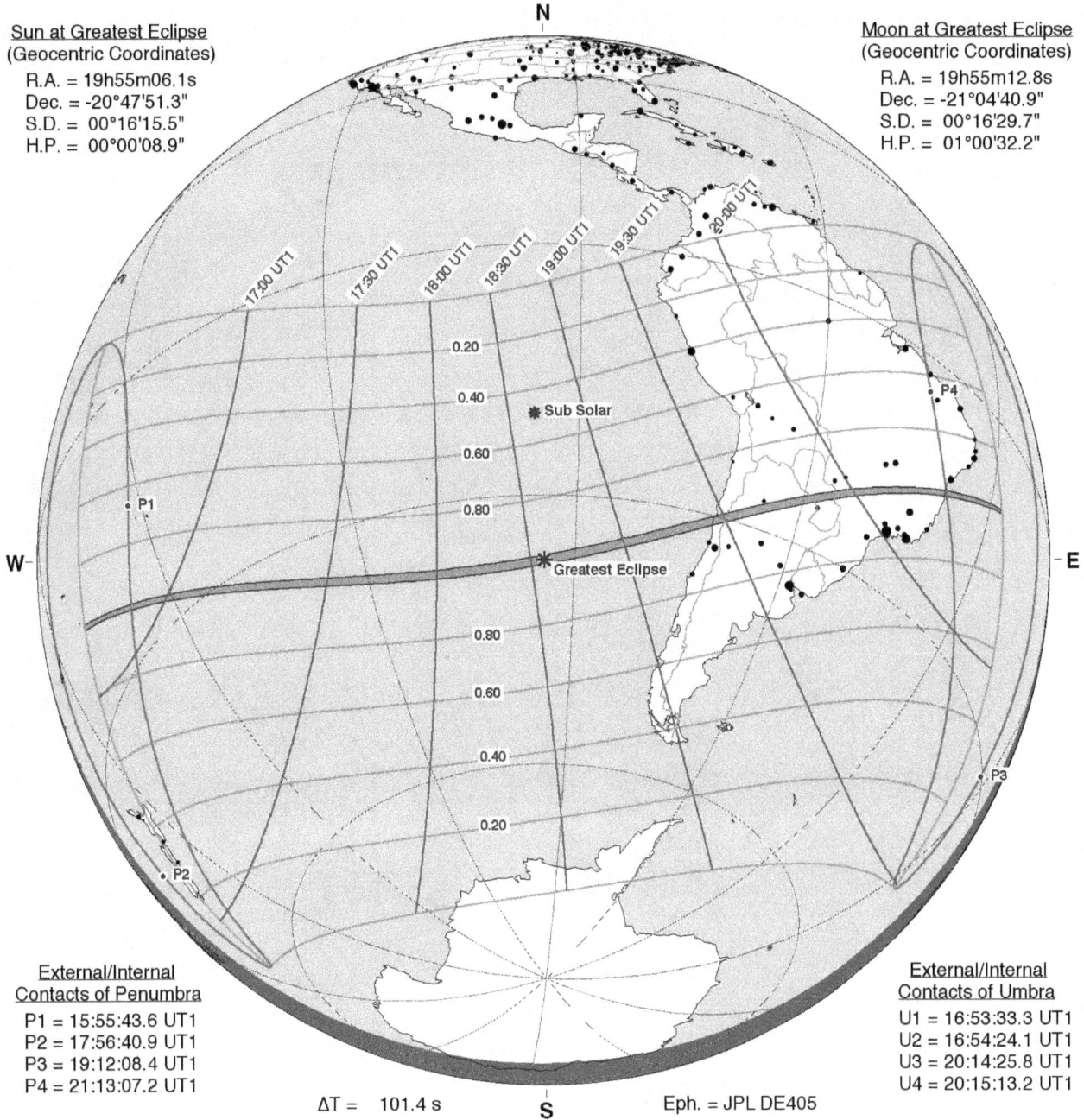

External/Internal
Contacts of Penumbra
P1 = 15:55:43.6 UT1
P2 = 17:56:40.9 UT1
P3 = 19:12:08.4 UT1
P4 = 21:13:07.2 UT1

External/Internal
Contacts of Umbra
U1 = 16:53:33.3 UT1
U2 = 16:54:24.1 UT1
U3 = 20:14:25.8 UT1
U4 = 20:15:13.2 UT1

ΔT = 101.4 s          Eph. = JPL DE405

| Circumstances at Greatest Eclipse: 18:34:22.9 UT1 | | Circumstances at Greatest Duration: 18:31:58.4 UT1 | |
|---|---|---|---|
| Lat. = 37°10.4'S | Sun Alt. = 73.5° | Lat. = 37°18.7'S | Sun Alt. = 73.5° |
| Long. = 094°16.9'W | Sun Azm. = 353.8° | Long. = 095°19.6'W | Sun Azm. = 359.3° |
| Path Width = 109.7 km | Duration = 02m42.0s | Path Width = 109.7 km | Duration = 02m42.1s |

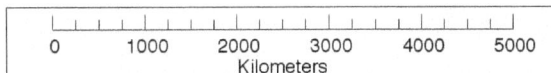

©2016 F. Espenak
www.EclipseWise.com

# Annular Solar Eclipse of 2075 Jul 13

Greatest Eclipse = 06:05:44.0 TD   (= 06:04:02.2 UT1)

| Eclipse Magnitude = 0.9467 | Saros Series = 147 |
|---|---|
| Gamma = 0.6583 | Saros Member = 26 of 80 |

**Sun at Greatest Eclipse**
(Geocentric Coordinates)
R.A. = 07h30m57.3s
Dec. = +21°47'03.5"
S.D. = 00°15'44.0"
H.P. = 00°00'08.7"

**Moon at Greatest Eclipse**
(Geocentric Coordinates)
R.A. = 07h31m05.6s
Dec. = +22°22'29.6"
S.D. = 00°14'43.7"
H.P. = 00°54'03.3"

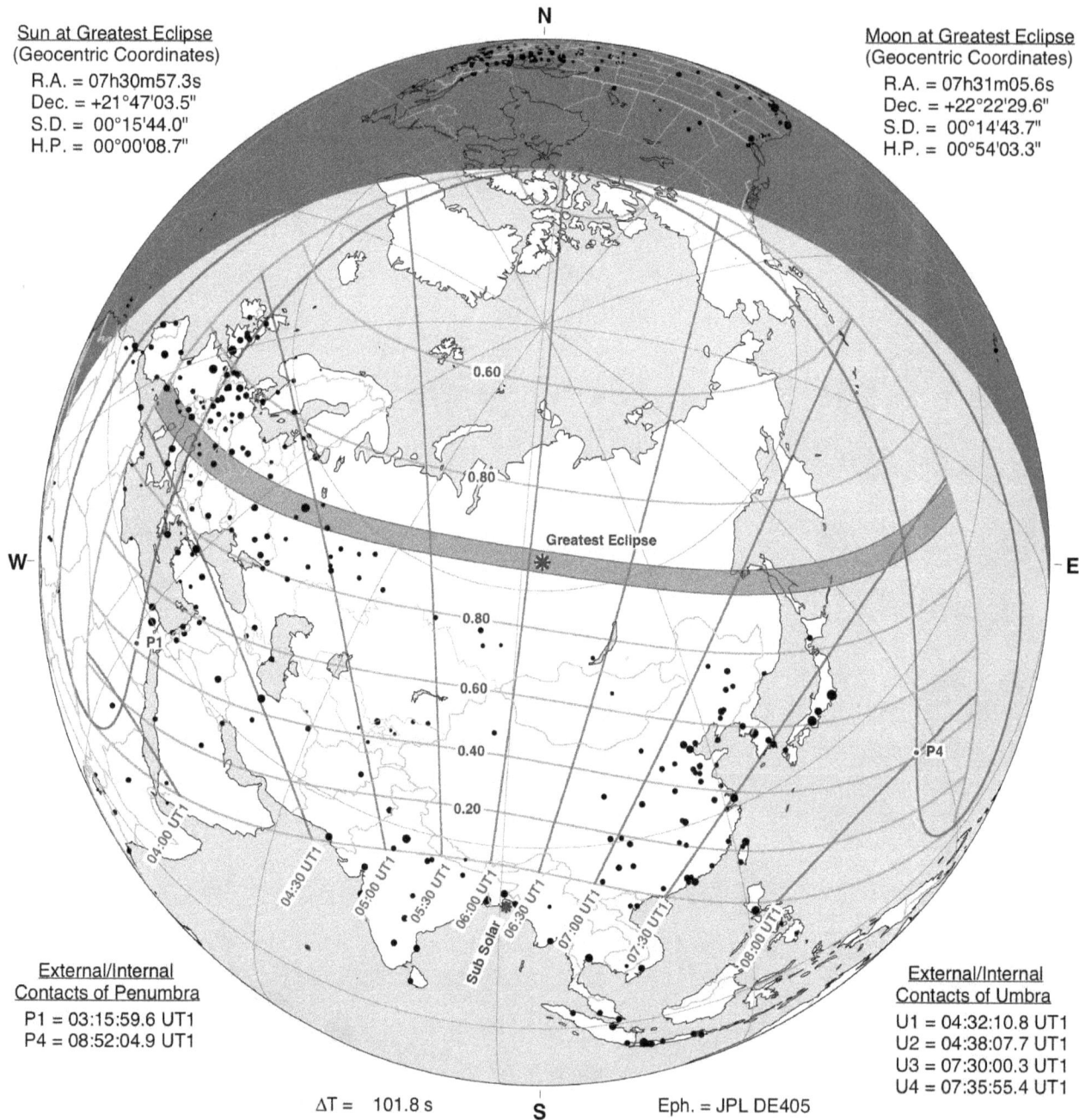

**External/Internal Contacts of Penumbra**
P1 = 03:15:59.6 UT1
P4 = 08:52:04.9 UT1

**External/Internal Contacts of Umbra**
U1 = 04:32:10.8 UT1
U2 = 04:38:07.7 UT1
U3 = 07:30:00.3 UT1
U4 = 07:35:55.4 UT1

ΔT = 101.8 s          Eph. = JPL DE405

**Circumstances at Greatest Eclipse: 06:04:02.2 UT1**

| | |
|---|---|
| Lat. = 63°07.6'N | Sun Alt. = 48.5° |
| Long. = 095°00.8'E | Sun Azm. = 186.4° |
| Path Width = 261.9 km | Duration = 04m44.6s |

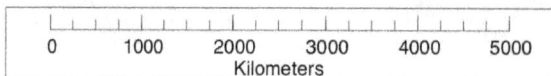

**Circumstances at Greatest Duration: 06:01:41.2 UT1**

| | |
|---|---|
| Lat. = 63°14.5'N | Sun Alt. = 48.5° |
| Long. = 093°06.0'E | Sun Azm. = 182.9° |
| Path Width = 261.8 km | Duration = 04m44.6s |

©2016 F. Espenak
www.EclipseWise.com

0   1000   2000   3000   4000   5000
Kilometers

# Total Solar Eclipse of 2076 Jan 06

Greatest Eclipse = 10:07:27.5 TD   (= 10:05:45.3 UT1)

| | |
|---|---|
| Eclipse Magnitude = 1.0342 | Saros Series = 152 |
| Gamma = -0.9373 | Saros Member = 16 of 70 |

**N**

Sun at Greatest Eclipse
(Geocentric Coordinates)
R.A. = 19h09m11.6s
Dec. = -22°28'36.7"
S.D. = 00°16'15.9"
H.P. = 00°00'08.9"

Moon at Greatest Eclipse
(Geocentric Coordinates)
R.A. = 19h09m16.9s
Dec. = -23°26'00.6"
S.D. = 00°16'43.8"
H.P. = 01°01'24.1"

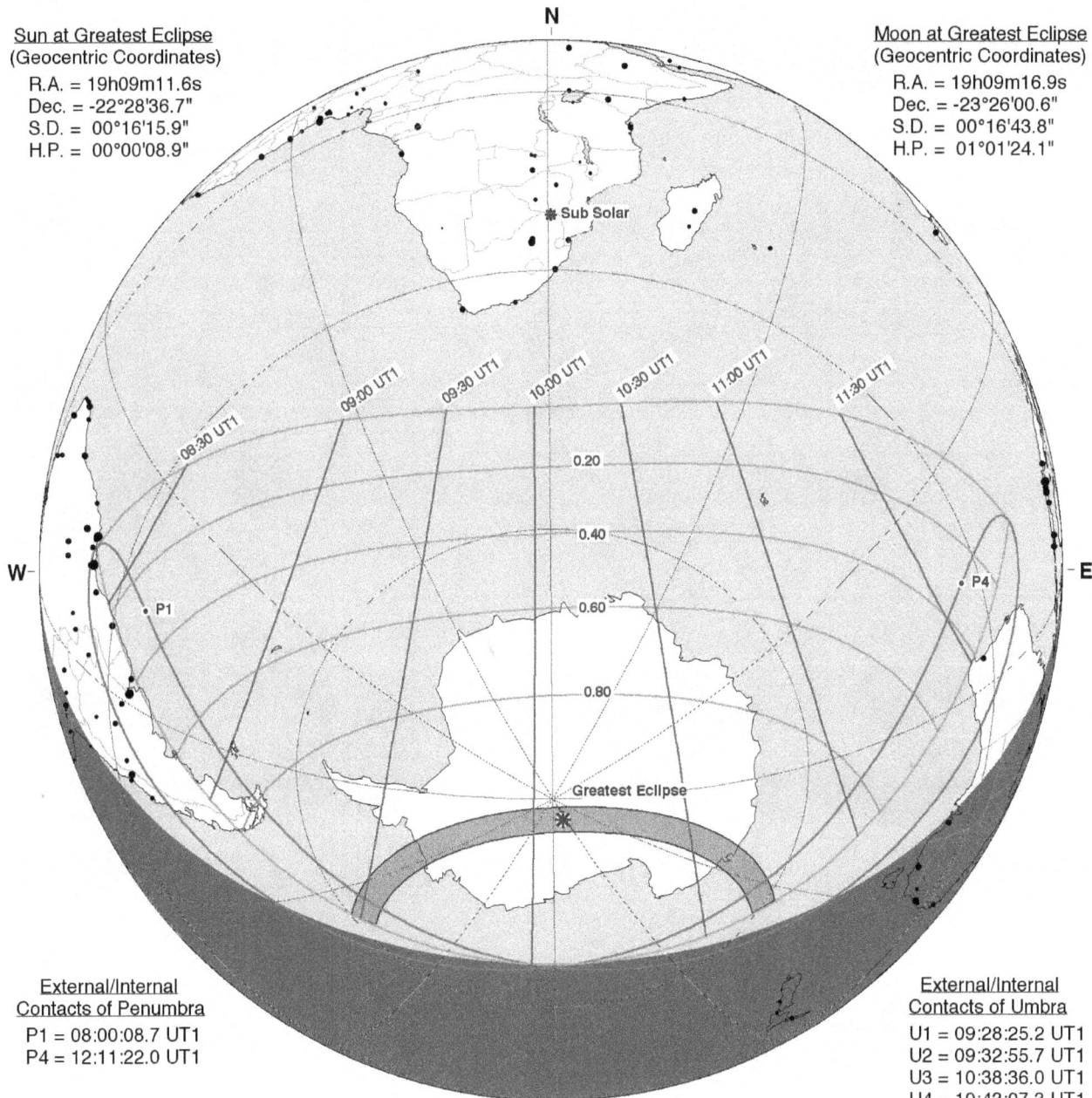

Sub Solar

08:30 UT1   09:00 UT1   09:30 UT1   10:00 UT1   10:30 UT1   11:00 UT1   11:30 UT1

0.20
0.40
0.60
0.80

**W**

**E**

P1

P4

Greatest Eclipse

**S**

External/Internal
Contacts of Penumbra
P1 = 08:00:08.7 UT1
P4 = 12:11:22.0 UT1

External/Internal
Contacts of Umbra
U1 = 09:28:25.2 UT1
U2 = 09:32:55.7 UT1
U3 = 10:38:36.0 UT1
U4 = 10:43:07.3 UT1

ΔT = 102.2 s          Eph. = JPL DE405

| Circumstances at Greatest Eclipse: 10:05:45.3 UT1 | | Circumstances at Greatest Duration: 10:05:54.8 UT1 | |
|---|---|---|---|
| Lat. = 87°10.6'S | Sun Alt. = 19.9° | Lat. = 87°08.4'S | Sun Alt. = 19.9° |
| Long. = 173°55.4'W | Sun Azm. = 203.5° | Long. = 175°33.7'W | Sun Azm. = 205.0° |
| Path Width = 340.4 km | Duration = 01m49.5s | Path Width = 340.4 km | Duration = 01m49.5s |

0   1000   2000   3000   4000   5000
Kilometers

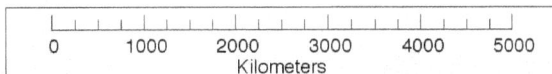

©2016  F. Espenak
www.EclipseWise.com

# Partial Solar Eclipse of 2076 Jun 01

Greatest Eclipse = 17:31:21.9 TD  (= 17:29:39.4 UT1)

Eclipse Magnitude = 0.2897          Saros Series = 119
Gamma = -1.3897          Saros Member = 69 of 71

**Sun at Greatest Eclipse**
(Geocentric Coordinates)
R.A. = 04h42m27.8s
Dec. = +22°14'01.6"
S.D. = 00°15'46.3"
H.P. = 00°00'08.7"

**Moon at Greatest Eclipse**
(Geocentric Coordinates)
R.A. = 04h43m42.6s
Dec. = +20°58'42.6"
S.D. = 00°15'11.7"
H.P. = 00°55'45.9"

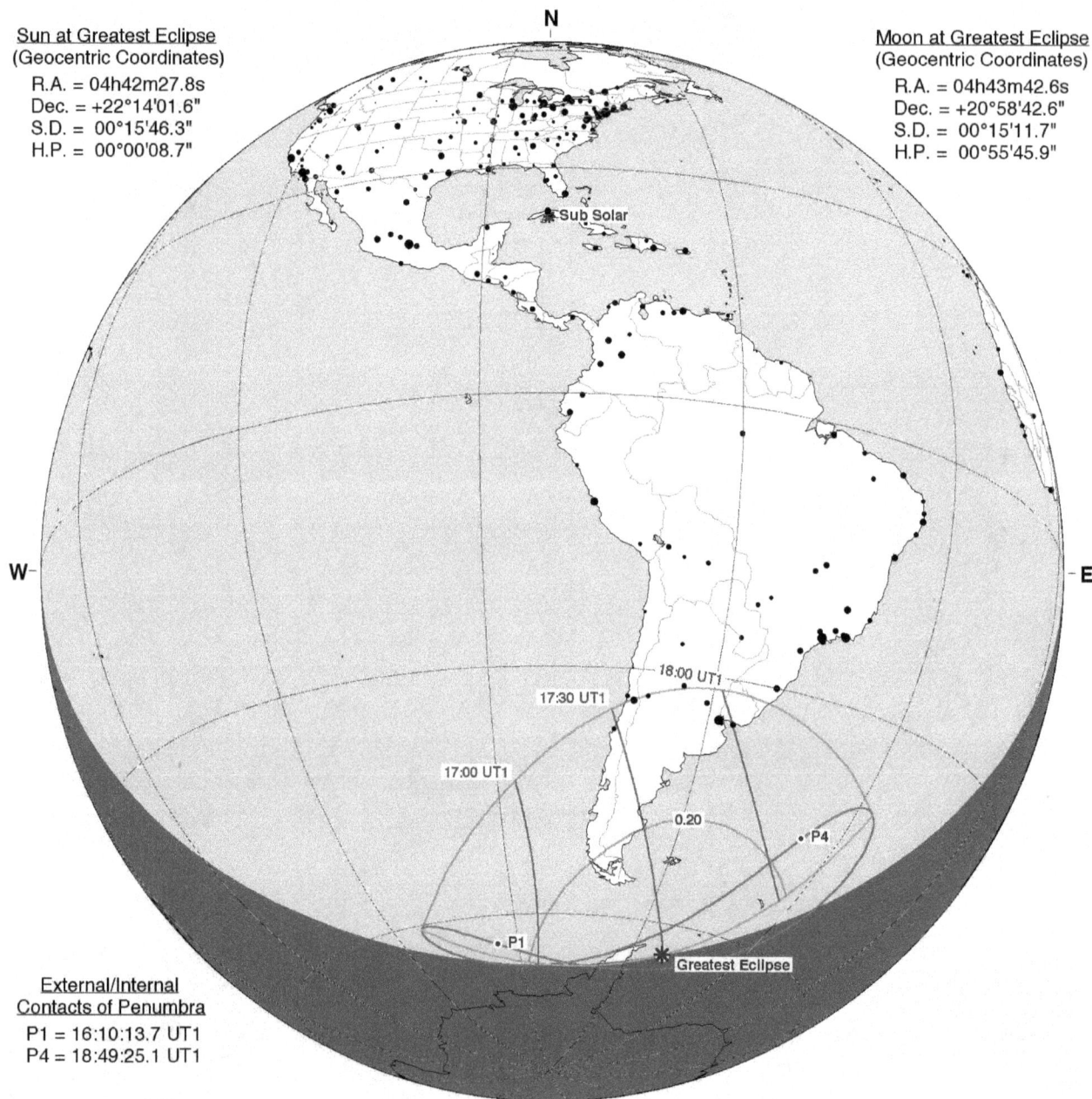

N

Sub Solar

W

E

18:00 UT1

17:30 UT1

17:00 UT1

0.20

P4

P1

Greatest Eclipse

**External/Internal**
**Contacts of Penumbra**
P1 = 16:10:13.7 UT1
P4 = 18:49:25.1 UT1

ΔT = 102.5 s          S          Eph. = JPL DE405

**Circumstances at Greatest Eclipse: 17:29:39.4 UT1**
Lat. = 64°23.0'S          Sun Alt. =  0.0°
Long. = 051°25.6'W          Sun Azm. = 331.1°

| 0 | 1000 | 2000 | 3000 | 4000 | 5000 |
Kilometers

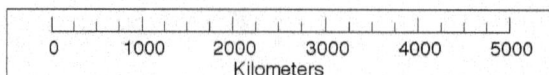

# Partial Solar Eclipse of 2076 Jul 01

Greatest Eclipse = 06:50:43.3 TD  (= 06:49:00.8 UT1)

| | |
|---|---|
| Eclipse Magnitude = 0.2746 | Saros Series = 157 |
| Gamma = 1.4005 | Saros Member = 2 of 70 |

Sun at Greatest Eclipse
(Geocentric Coordinates)
R.A. = 06h44m59.8s
Dec. = +23°01'35.5"
S.D. = 00°15'43.9"
H.P. = 00°00'08.6"

Moon at Greatest Eclipse
(Geocentric Coordinates)
R.A. = 06h44m53.2s
Dec. = +24°17'50.8"
S.D. = 00°14'52.7"
H.P. = 00°54'36.1"

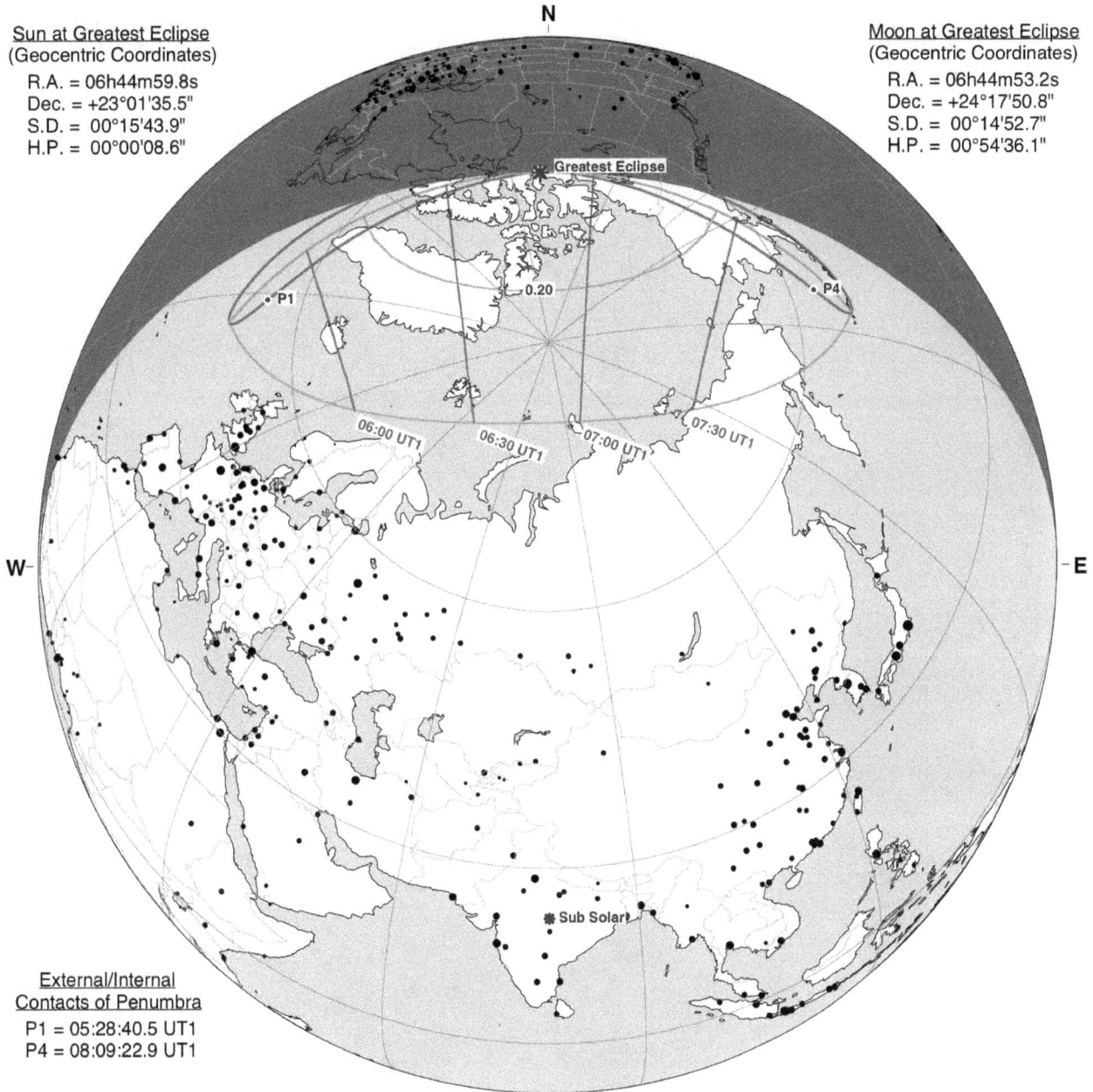

N

Greatest Eclipse

P1    0.20    P4

06:00 UT1    06:30 UT1    07:00 UT1    07:30 UT1

W —    — E

Sub Solar

External/Internal
Contacts of Penumbra
P1 = 05:28:40.5 UT1
P4 = 08:09:22.9 UT1

ΔT = 102.6 s    S    Eph. = JPL DE405

Circumstances at Greatest Eclipse: 06:49:00.8 UT1

| | |
|---|---|
| Lat. = 66°57.0'N | Sun Alt. = 0.0° |
| Long. = 098°18.8'W | Sun Azm. = 2.7° |

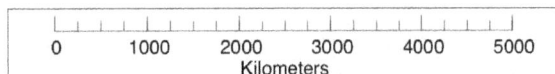

0    1000    2000    3000    4000    5000
Kilometers

©2016  F. Espenak
www.EclipseWise.com

# Partial Solar Eclipse of 2076 Nov 26

Greatest Eclipse = 11:43:00.9 TD  (= 11:41:18.0 UT1)

| | |
|---|---|
| Eclipse Magnitude = 0.7315 | Saros Series = 124 |
| Gamma = 1.1401 | Saros Member = 58 of 73 |

Sun at Greatest Eclipse
(Geocentric Coordinates)
R.A. = 16h12m39.7s
Dec. = -21°08'26.9"
S.D. = 00°16'12.3"
H.P. = 00°00'08.9"

Moon at Greatest Eclipse
(Geocentric Coordinates)
R.A. = 16h13m56.6s
Dec. = -20°05'16.9"
S.D. = 00°15'44.2"
H.P. = 00°57'45.3"

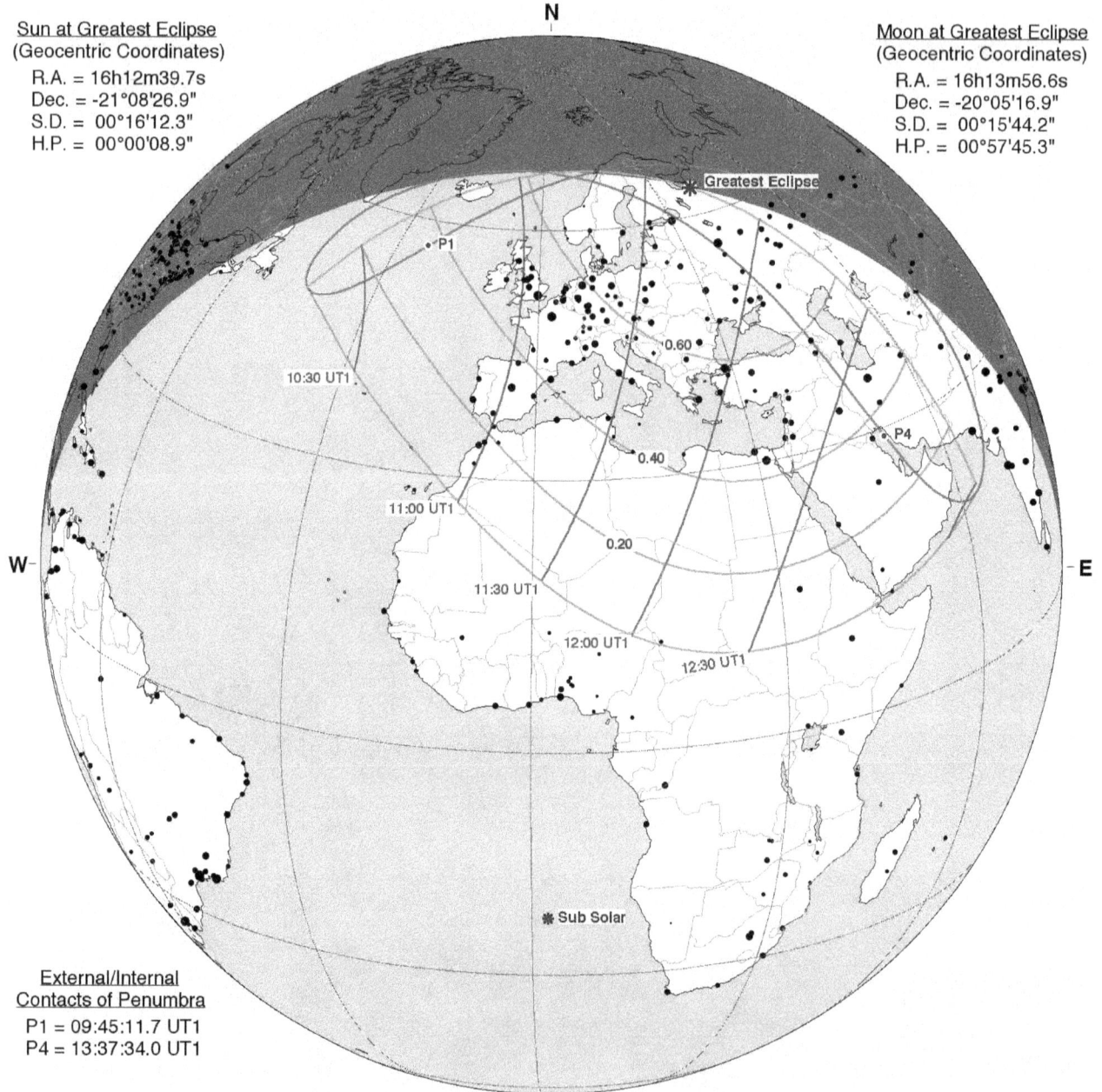

N

Greatest Eclipse

P1

0.60

10:30 UT1

0.40

P4

11:00 UT1

0.20

11:30 UT1

12:00 UT1

12:30 UT1

W

E

Sub Solar

External/Internal
Contacts of Penumbra
P1 = 09:45:11.7 UT1
P4 = 13:37:34.0 UT1

ΔT = 102.9 s

S

Eph. = JPL DE405

Circumstances at Greatest Eclipse: 11:41:18.0 UT1

| | |
|---|---|
| Lat. = 63°43.8'N | Sun Alt. = 0.0° |
| Long. = 039°55.2'E | Sun Azm. = 215.4° |

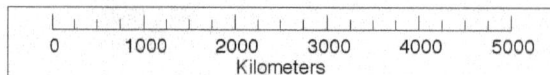

| 0 | 1000 | 2000 | 3000 | 4000 | 5000 |
|---|---|---|---|---|---|

Kilometers

©2016 F. Espenak
www.EclipseWise.com

# Total Solar Eclipse of 2077 May 22

Greatest Eclipse = 02:46:05.3 TD   (= 02:44:22.0 UT1)

| | |
|---|---|
| Eclipse Magnitude = 1.0290 | Saros Series = 129 |
| Gamma = -0.5725 | Saros Member = 55 of 80 |

Sun at Greatest Eclipse
(Geocentric Coordinates)
R.A. = 03h58m18.6s
Dec. = +20°29'25.4"
S.D. = 00°15'48.1"
H.P. = 00°00'08.7"

Moon at Greatest Eclipse
(Geocentric Coordinates)
R.A. = 03h59m01.0s
Dec. = +19°57'18.2"
S.D. = 00°16'02.6"
H.P. = 00°58'52.8"

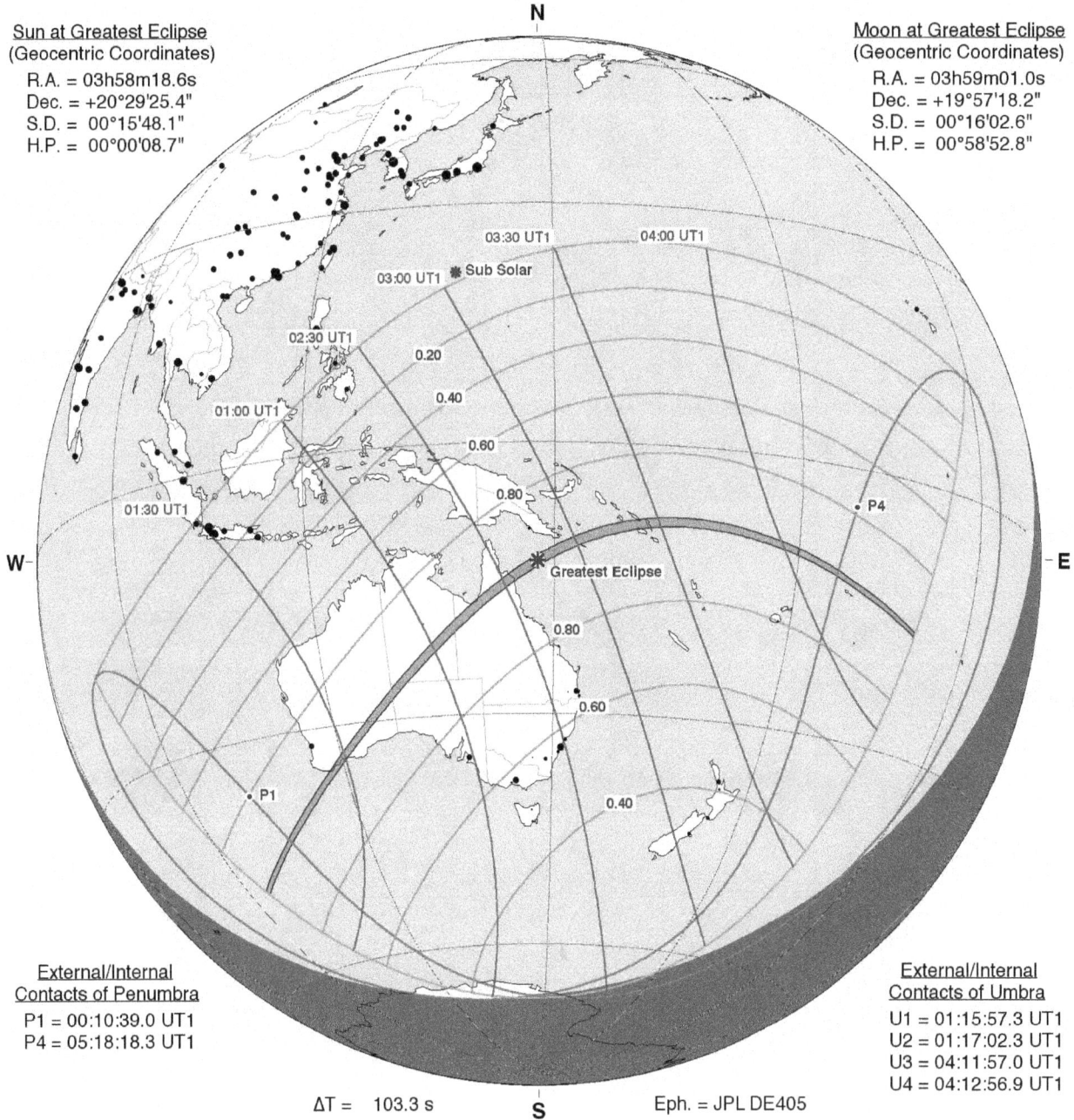

N

03:30 UT1

04:00 UT1

03:00 UT1   ✳ Sub Solar

02:30 UT1

0.20

0.40

01:00 UT1

0.60

0.80

01:30 UT1

P4

W —                                                                — E

✳
Greatest Eclipse

0.80

0.60

0.40

P1

S

External/Internal
Contacts of Penumbra
P1 = 00:10:39.0 UT1
P4 = 05:18:18.3 UT1

External/Internal
Contacts of Umbra
U1 = 01:15:57.3 UT1
U2 = 01:17:02.3 UT1
U3 = 04:11:57.0 UT1
U4 = 04:12:56.9 UT1

ΔT =   103.3 s          Eph. = JPL DE405

Circumstances at Greatest Eclipse: 02:44:22.0 UT1

| | |
|---|---|
| Lat. = 13°06.2'S | Sun Alt. = 55.0° |
| Long. = 148°07.3'E | Sun Azm. = 343.5° |
| Path Width = 118.9 km | Duration = 02m53.7s |

Circumstances at Greatest Duration: 02:46:17.2 UT1

| | |
|---|---|
| Lat. = 12°48.9'S | Sun Alt. = 55.0° |
| Long. = 148°38.9'E | Sun Azm. = 341.8° |
| Path Width = 119.2 km | Duration = 02m53.7s |

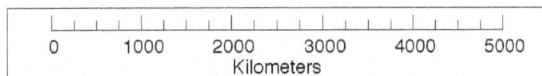

0   1000   2000   3000   4000   5000
Kilometers

©2016  F. Espenak
www.EclipseWise.com

# Annular Solar Eclipse of 2077 Nov 15

Greatest Eclipse = 17:07:56.2 TD   (= 17:06:12.6 UT1)

| | |
|---|---|
| Eclipse Magnitude = 0.9371 | Saros Series = 134 |
| Gamma = 0.4705 | Saros Member = 47 of 71 |

**Sun at Greatest Eclipse**
(Geocentric Coordinates)
R.A. = 15h26m24.3s
Dec. = -18°45'33.3"
S.D. = 00°16'10.1"
H.P. = 00°00'08.9"

**Moon at Greatest Eclipse**
(Geocentric Coordinates)
R.A. = 15h27m01.5s
Dec. = -18°21'22.3"
S.D. = 00°14'56.9"
H.P. = 00°54'51.8"

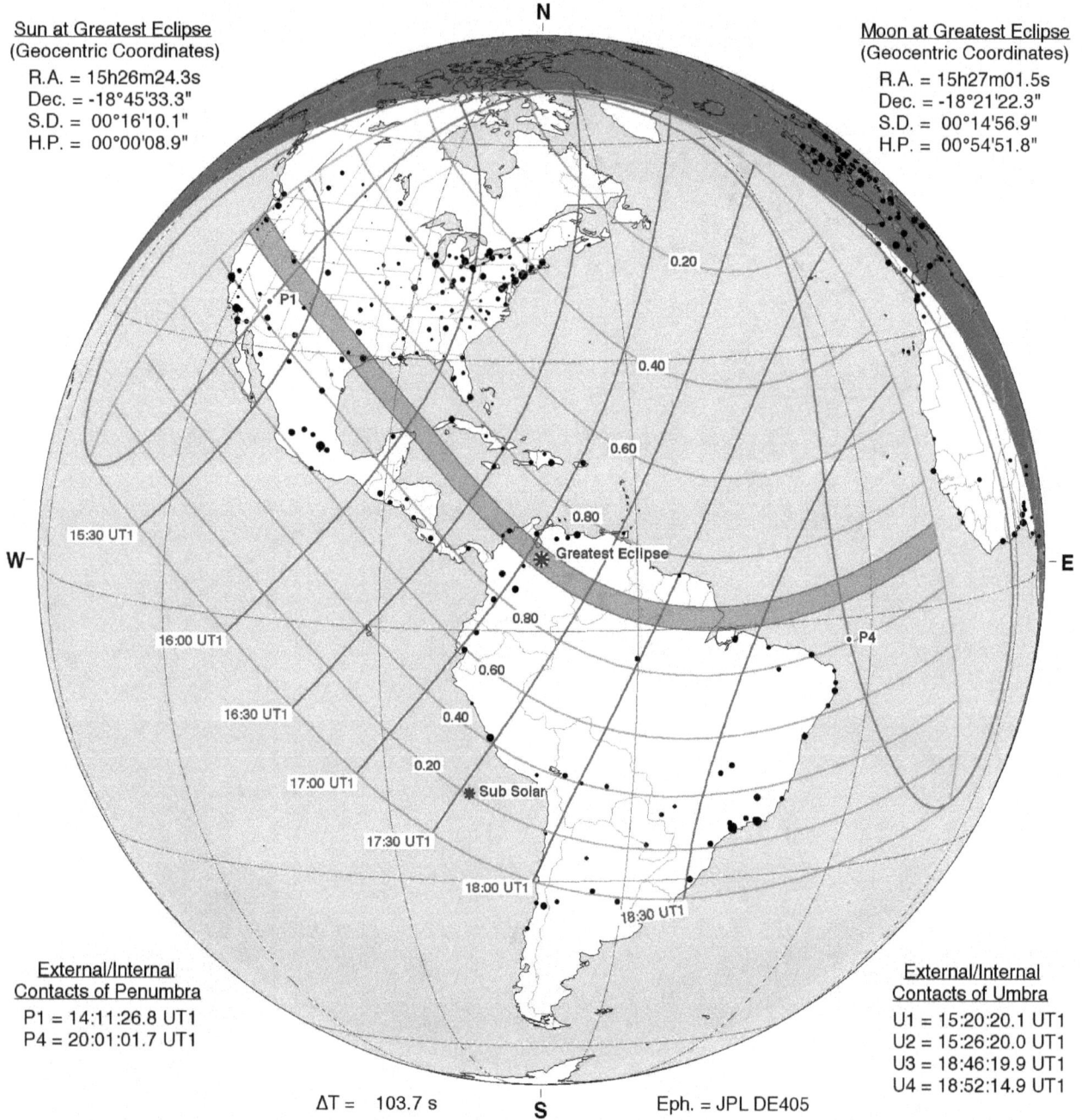

**External/Internal Contacts of Penumbra**
P1 = 14:11:26.8 UT1
P4 = 20:01:01.7 UT1

**External/Internal Contacts of Umbra**
U1 = 15:20:20.1 UT1
U2 = 15:26:20.0 UT1
U3 = 18:46:19.9 UT1
U4 = 18:52:14.9 UT1

ΔT = 103.7 s          Eph. = JPL DE405

**Circumstances at Greatest Eclipse: 17:06:12.6 UT1**

| | |
|---|---|
| Lat. = 07°47.5'N | Sun Alt. = 61.9° |
| Long. = 071°00.9'W | Sun Azm. = 199.1° |
| Path Width = 262.5 km | Duration = 07m53.9s |

**Circumstances at Greatest Duration: 17:19:19.0 UT1**

| | |
|---|---|
| Lat. = 05°48.5'N | Sun Alt. = 61.0° |
| Long. = 068°06.8'W | Sun Azm. = 211.6° |
| Path Width = 267.6 km | Duration = 07m55.9s |

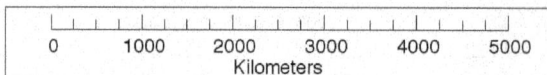

©2016 F. Espenak
www.EclipseWise.com

# Total Solar Eclipse of  2078 May 11

Greatest Eclipse =  17:56:54.8 TD   (= 17:55:10.7 UT1)

Eclipse Magnitude =  1.0701          Saros Series =  139
Gamma =  0.1838                      Saros Member =   33 of 71

Sun at Greatest Eclipse
(Geocentric Coordinates)
R.A. = 03h16m09.4s
Dec. = +18°07'17.6"
S.D. = 00°15'50.2"
H.P. = 00°00'08.7"

Moon at Greatest Eclipse
(Geocentric Coordinates)
R.A. = 03h15m52.6s
Dec. = +18°17'46.7"
S.D. = 00°16'39.9"
H.P. = 01°01'09.6"

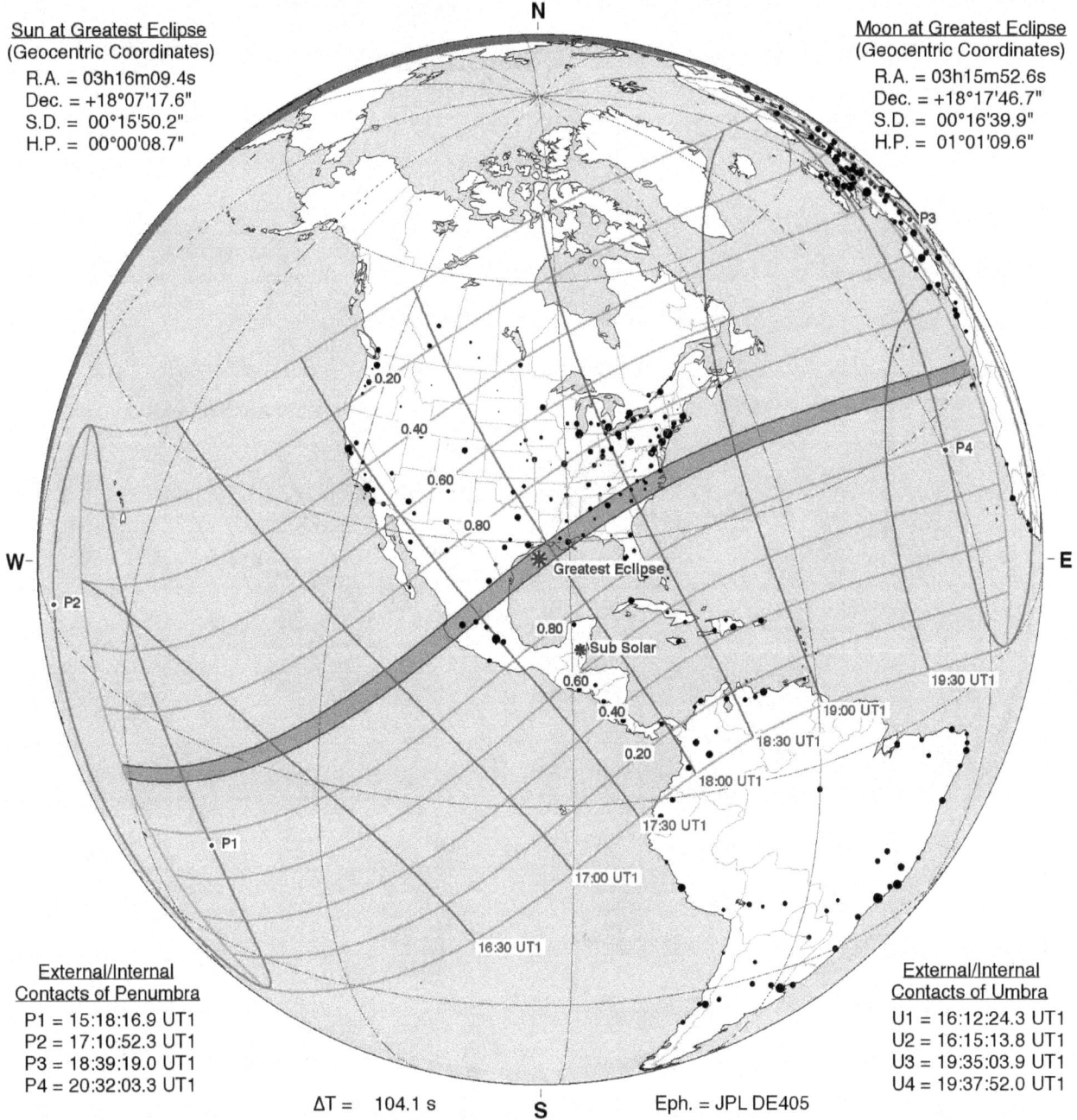

N

W

E

S

Greatest Eclipse

Sub Solar

0.20
0.40
0.60
0.80
0.80
0.60
0.40
0.20

P1
P2
P3
P4

19:30 UT1
19:00 UT1
18:30 UT1
18:00 UT1
17:30 UT1
17:00 UT1
16:30 UT1

External/Internal
Contacts of Penumbra
P1 = 15:18:16.9 UT1
P2 = 17:10:52.3 UT1
P3 = 18:39:19.0 UT1
P4 = 20:32:03.3 UT1

External/Internal
Contacts of Umbra
U1 = 16:12:24.3 UT1
U2 = 16:15:13.8 UT1
U3 = 19:35:03.9 UT1
U4 = 19:37:52.0 UT1

ΔT =   104.1 s          Eph. = JPL DE405

Circumstances at Greatest Eclipse:  17:55:10.7 UT1

| Lat. = 28°08.5'N | Sun Alt. =  79.3° |
| Long. = 093°55.1'W | Sun Azm. = 157.9° |
| Path Width = 232.2 km | Duration = 05m40.5s |

Circumstances at Greatest Duration:  18:00:33.5 UT1

| Lat. = 29°15.9'N | Sun Alt. =  78.8° |
| Long. = 092°06.2'W | Sun Azm. = 174.7° |
| Path Width = 231.6 km | Duration = 05m40.9s |

0  1000  2000  3000  4000  5000
Kilometers

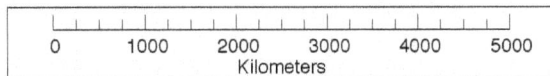

©2016  F. Espenak
www.EclipseWise.com

# Annular Solar Eclipse of 2078 Nov 04

Greatest Eclipse = 16:55:44.4 TD (= 16:53:59.9 UT1)

Eclipse Magnitude = 0.9255          Saros Series = 144
Gamma = -0.2285          Saros Member = 20 of 70

Sun at Greatest Eclipse
(Geocentric Coordinates)
R.A. = 14h40m53.9s
Dec. = -15°38'07.6"
S.D. = 00°16'07.5"
H.P. = 00°00'08.9"

Moon at Greatest Eclipse
(Geocentric Coordinates)
R.A. = 14h40m33.5s
Dec. = -15°49'24.5"
S.D. = 00°14'42.4"
H.P. = 00°53'58.5"

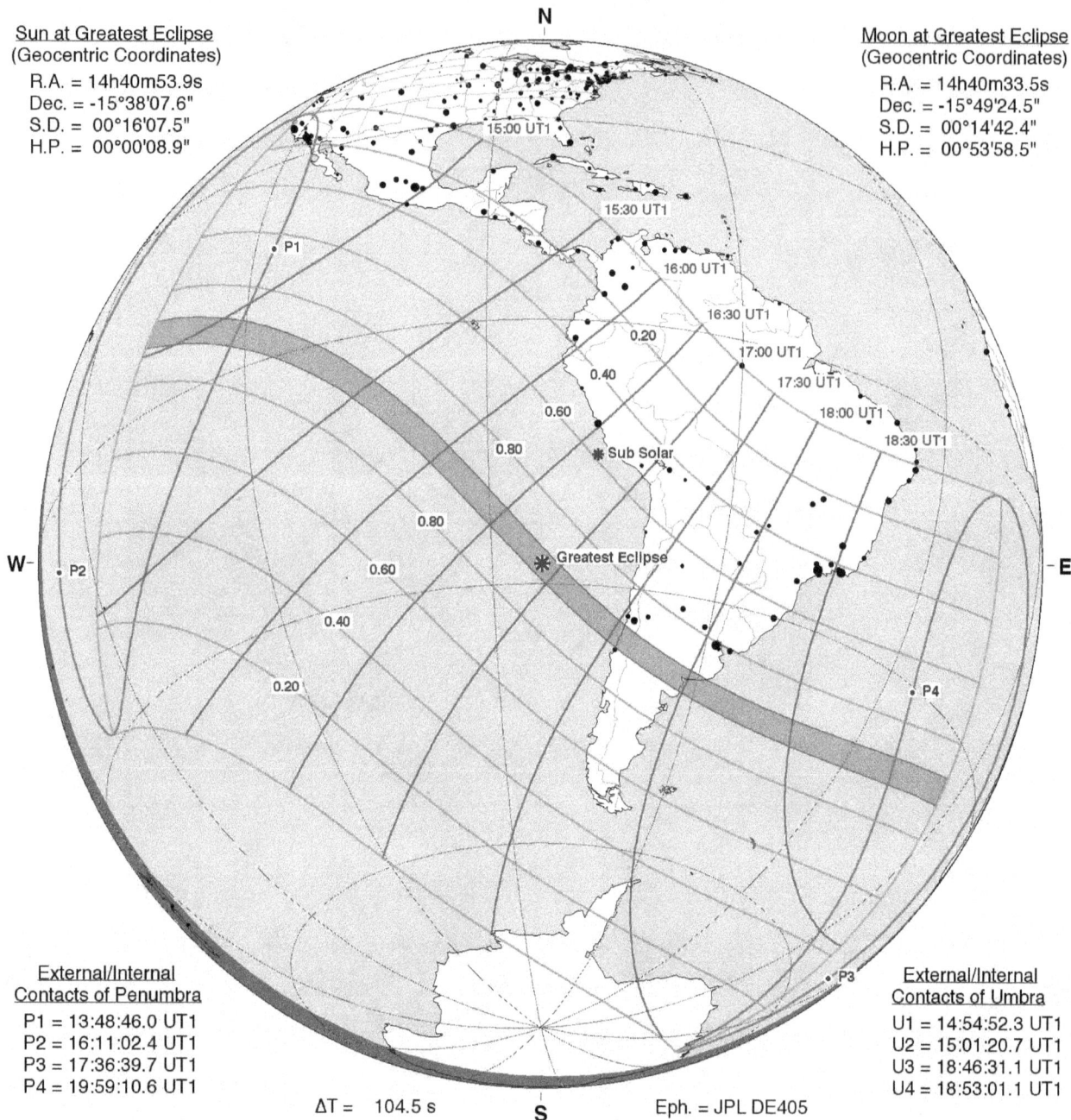

External/Internal
Contacts of Penumbra
P1 = 13:48:46.0 UT1
P2 = 16:11:02.4 UT1
P3 = 17:36:39.7 UT1
P4 = 19:59:10.6 UT1

External/Internal
Contacts of Umbra
U1 = 14:54:52.3 UT1
U2 = 15:01:20.7 UT1
U3 = 18:46:31.1 UT1
U4 = 18:53:01.1 UT1

ΔT = 104.5 s          Eph. = JPL DE405

Circumstances at Greatest Eclipse: 16:53:59.9 UT1
Lat. = 27°49.8'S          Sun Alt. = 76.6°
Long. = 083°31.2'W          Sun Azm. = 25.4°
Path Width = 287.5 km          Duration = 08m29.1s

Circumstances at Greatest Duration: 17:11:45.0 UT1
Lat. = 31°24.5'S          Sun Alt. = 74.0°
Long. = 078°59.0'W          Sun Azm. = 349.2°
Path Width = 286.0 km          Duration = 08m31.9s

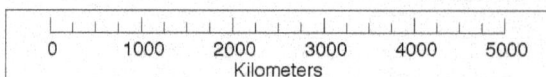

0          1000          2000          3000          4000          5000
Kilometers

# Total Solar Eclipse of 2079 May 01

Greatest Eclipse = 10:50:12.8 TD   (= 10:48:28.0 UT1)

| | |
|---|---|
| Eclipse Magnitude = 1.0512 | Saros Series = 149 |
| Gamma = 0.9081 | Saros Member = 24 of 71 |

**Sun at Greatest Eclipse**
(Geocentric Coordinates)
R.A. = 02h35m18.8s
Dec. = +15°12'06.8"
S.D. = 00°15'52.6"
H.P. = 00°00'08.7"

**Moon at Greatest Eclipse**
(Geocentric Coordinates)
R.A. = 02h33m47.0s
Dec. = +16°02'36.5"
S.D. = 00°16'34.7"
H.P. = 01°00'50.6"

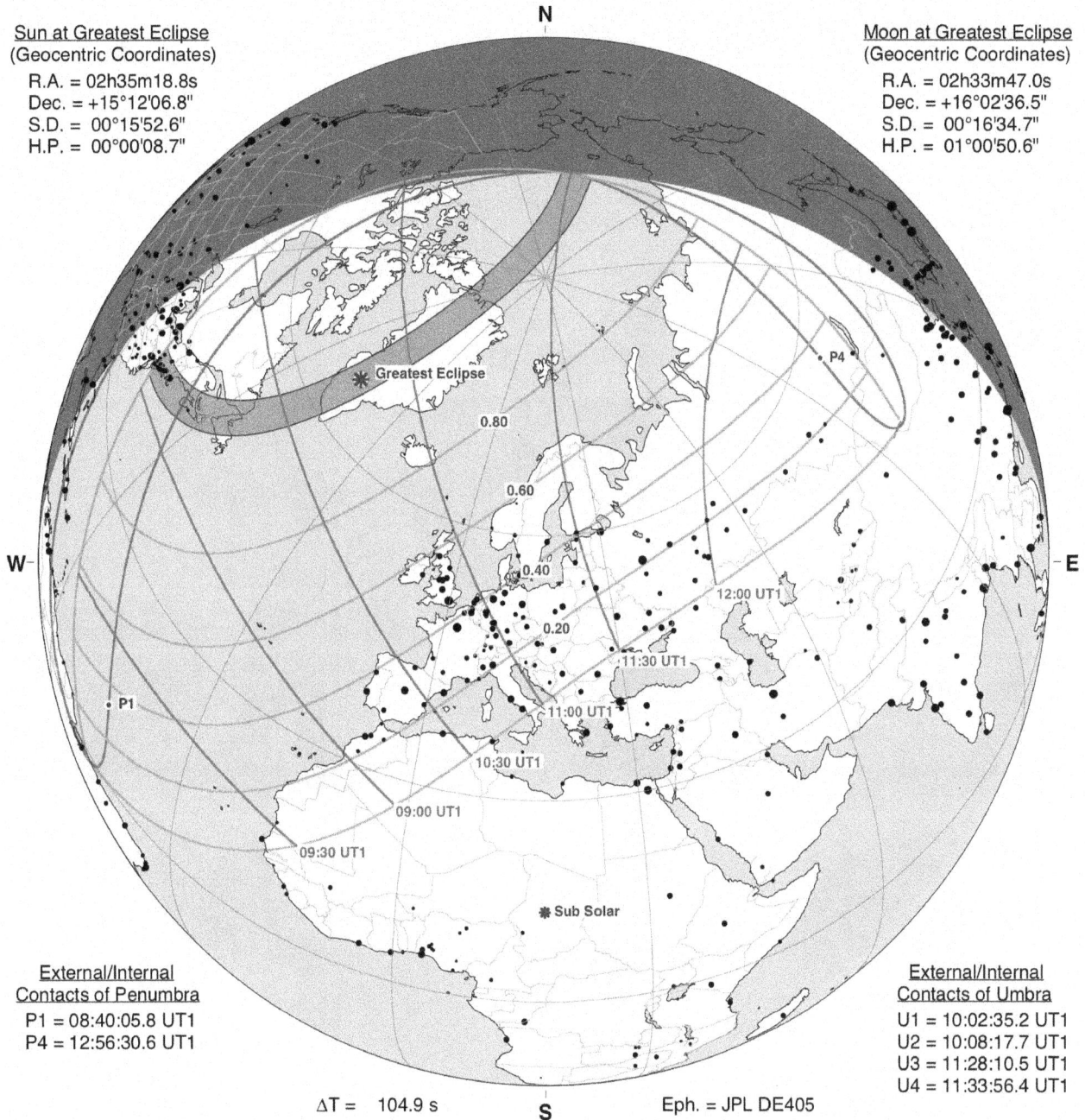

**External/Internal Contacts of Penumbra**
P1 = 08:40:05.8 UT1
P4 = 12:56:30.6 UT1

**External/Internal Contacts of Umbra**
U1 = 10:02:35.2 UT1
U2 = 10:08:17.7 UT1
U3 = 11:28:10.5 UT1
U4 = 11:33:56.4 UT1

ΔT = 104.9 s        Eph. = JPL DE405

**Circumstances at Greatest Eclipse: 10:48:28.0 UT1**

| | |
|---|---|
| Lat. = 66°10.0'N | Sun Alt. = 24.4° |
| Long. = 046°32.7'W | Sun Azm. = 108.2° |
| Path Width = 405.7 km | Duration = 02m54.9s |

**Circumstances at Greatest Duration: 10:49:13.2 UT1**

| | |
|---|---|
| Lat. = 66°33.8'N | Sun Alt. = 24.4° |
| Long. = 046°25.3'W | Sun Azm. = 108.7° |
| Path Width = 404.8 km | Duration = 02m54.9s |

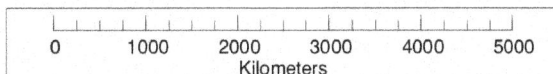

©2016 F. Espenak
www.EclipseWise.com

# Annular Solar Eclipse of 2079 Oct 24

Greatest Eclipse = 18:11:21.4 TD  (= 18:09:36.1 UT1)

| | |
|---|---|
| Eclipse Magnitude = 0.9484 | Saros Series = 154 |
| Gamma = -0.9243 | Saros Member = 10 of 71 |

**Sun at Greatest Eclipse**
(Geocentric Coordinates)
R.A. = 13h57m22.1s
Dec. = -11°59'23.6"
S.D. = 00°16'04.6"
H.P. = 00°00'08.8"

**Moon at Greatest Eclipse**
(Geocentric Coordinates)
R.A. = 13h55m50.0s
Dec. = -12°45'30.1"
S.D. = 00°15'09.9"
H.P. = 00°55'39.3"

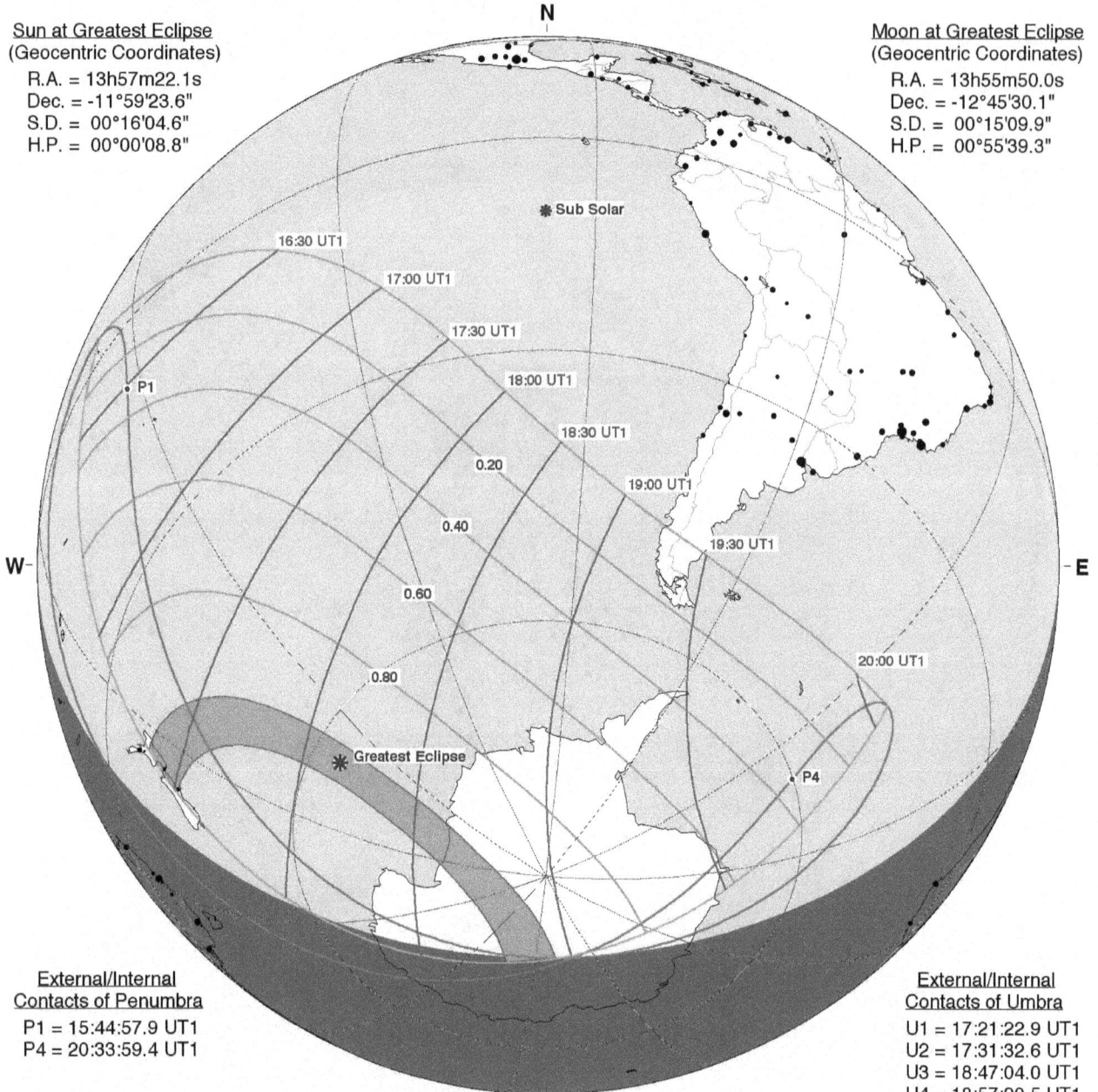

N

Sub Solar

16:30 UT1
17:00 UT1
17:30 UT1
18:00 UT1
18:30 UT1
19:00 UT1
19:30 UT1
20:00 UT1

P1

0.20
0.40
0.60
0.80

W

E

Greatest Eclipse

P4

**External/Internal Contacts of Penumbra**
P1 = 15:44:57.9 UT1
P4 = 20:33:59.4 UT1

**External/Internal Contacts of Umbra**
U1 = 17:21:22.9 UT1
U2 = 17:31:32.6 UT1
U3 = 18:47:04.0 UT1
U4 = 18:57:20.5 UT1

ΔT = 105.3 s     S     Eph. = JPL DE405

**Circumstances at Greatest Eclipse: 18:09:36.1 UT1**

| | |
|---|---|
| Lat. = 63°24.8'S | Sun Alt. = 22.0° |
| Long. = 160°46.3'W | Sun Azm. = 72.1° |
| Path Width = 494.6 km | Duration = 03m38.9s |

**Circumstances at Greatest Duration: 18:16:11.2 UT1**

| | |
|---|---|
| Lat. = 66°36.3'S | Sun Alt. = 21.7° |
| Long. = 160°30.0'W | Sun Azm. = 69.0° |
| Path Width = 484.5 km | Duration = 03m39.0s |

| 0 | 1000 | 2000 | 3000 | 4000 | 5000 |
|---|---|---|---|---|---|

Kilometers

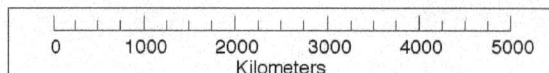

©2016 F. Espenak
www.EclipseWise.com

# Partial Solar Eclipse of 2080 Mar 21

Greatest Eclipse = 12:20:15.4 TD   (= 12:18:29.8 UT1)

| Eclipse Magnitude = 0.8734 | Saros Series = 121 |
|---|---|
| Gamma = -1.0578 | Saros Member = 64 of 71 |

Sun at Greatest Eclipse
(Geocentric Coordinates)
R.A. = 00h06m37.3s
Dec. = +00°43'02.5"
S.D. = 00°16'03.3"
H.P. = 00°00'08.8"

Moon at Greatest Eclipse
(Geocentric Coordinates)
R.A. = 00h08m33.2s
Dec. = -00°09'04.7"
S.D. = 00°15'24.0"
H.P. = 00°56'31.2"

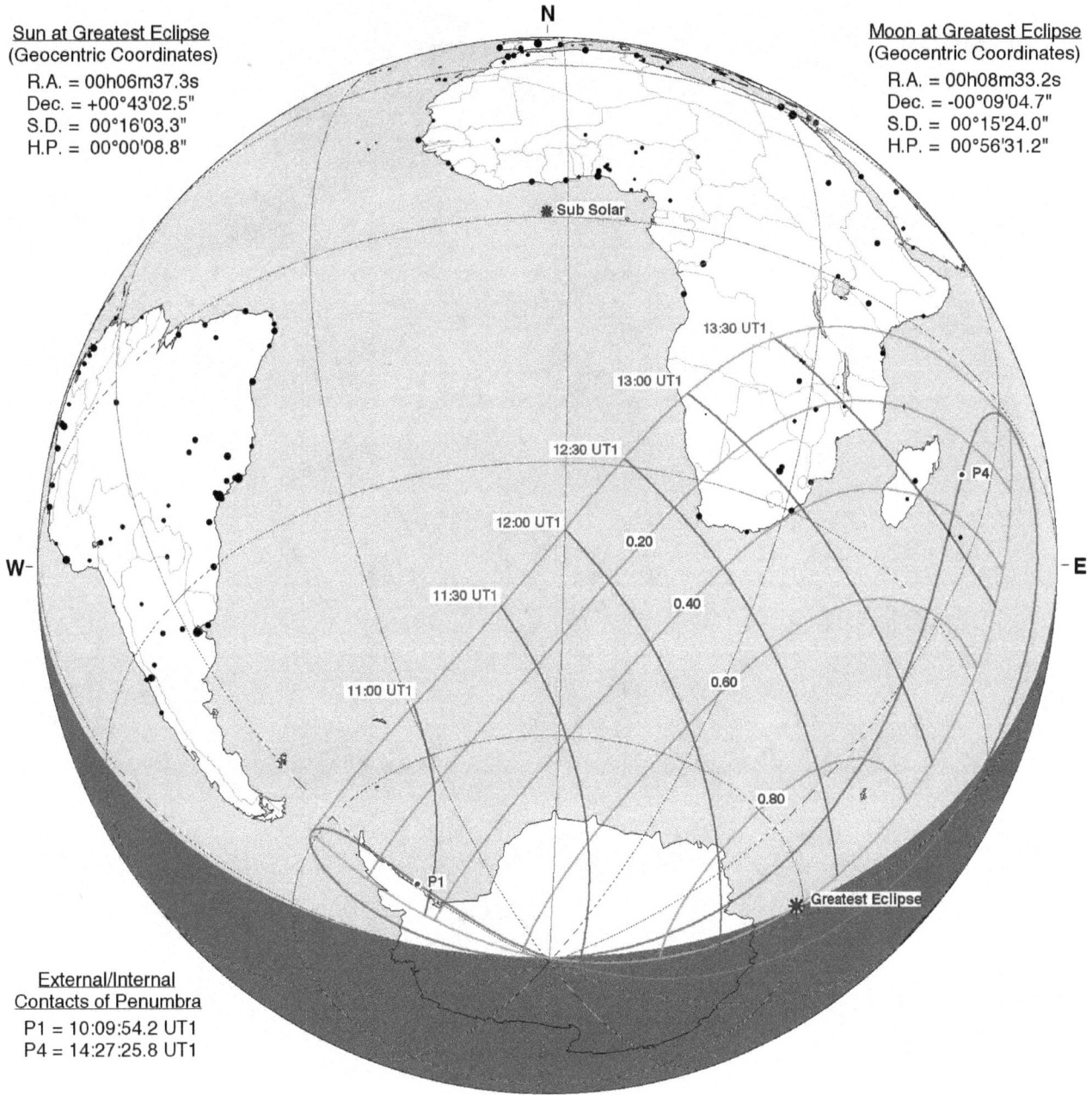

N

Sub Solar

13:30 UT1

13:00 UT1

12:30 UT1

12:00 UT1

0.20

11:30 UT1

0.40

11:00 UT1

0.60

0.80

P4

P1

Greatest Eclipse

W

E

External/Internal
Contacts of Penumbra
P1 = 10:09:54.2 UT1
P4 = 14:27:25.8 UT1

ΔT = 105.6 s       S       Eph. = JPL DE405

Circumstances at Greatest Eclipse: 12:18:29.8 UT1

| Lat. = 60°55.4'S | Sun Alt. = 0.0° |
|---|---|
| Long. = 085°42.9'E | Sun Azm. = 271.5° |

0   1000   2000   3000   4000   5000
Kilometers

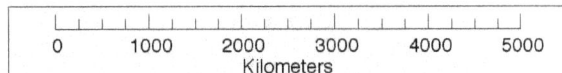

# Partial Solar Eclipse of 2080 Sep 13

Greatest Eclipse = 16:38:09.2 TD  (= 16:36:23.2 UT1)

| | |
|---|---|
| Eclipse Magnitude = 0.8743 | Saros Series = 126 |
| Gamma = 1.0724 | Saros Member = 51 of 72 |

Sun at Greatest Eclipse
(Geocentric Coordinates)
R.A. = 11h29m55.2s
Dec. = +03°14'46.9"
S.D. = 00°15'53.7"
H.P. = 00°00'08.7"

Moon at Greatest Eclipse
(Geocentric Coordinates)
R.A. = 11h31m59.3s
Dec. = +04°11'17.0"
S.D. = 00°16'24.7"
H.P. = 01°00'13.7"

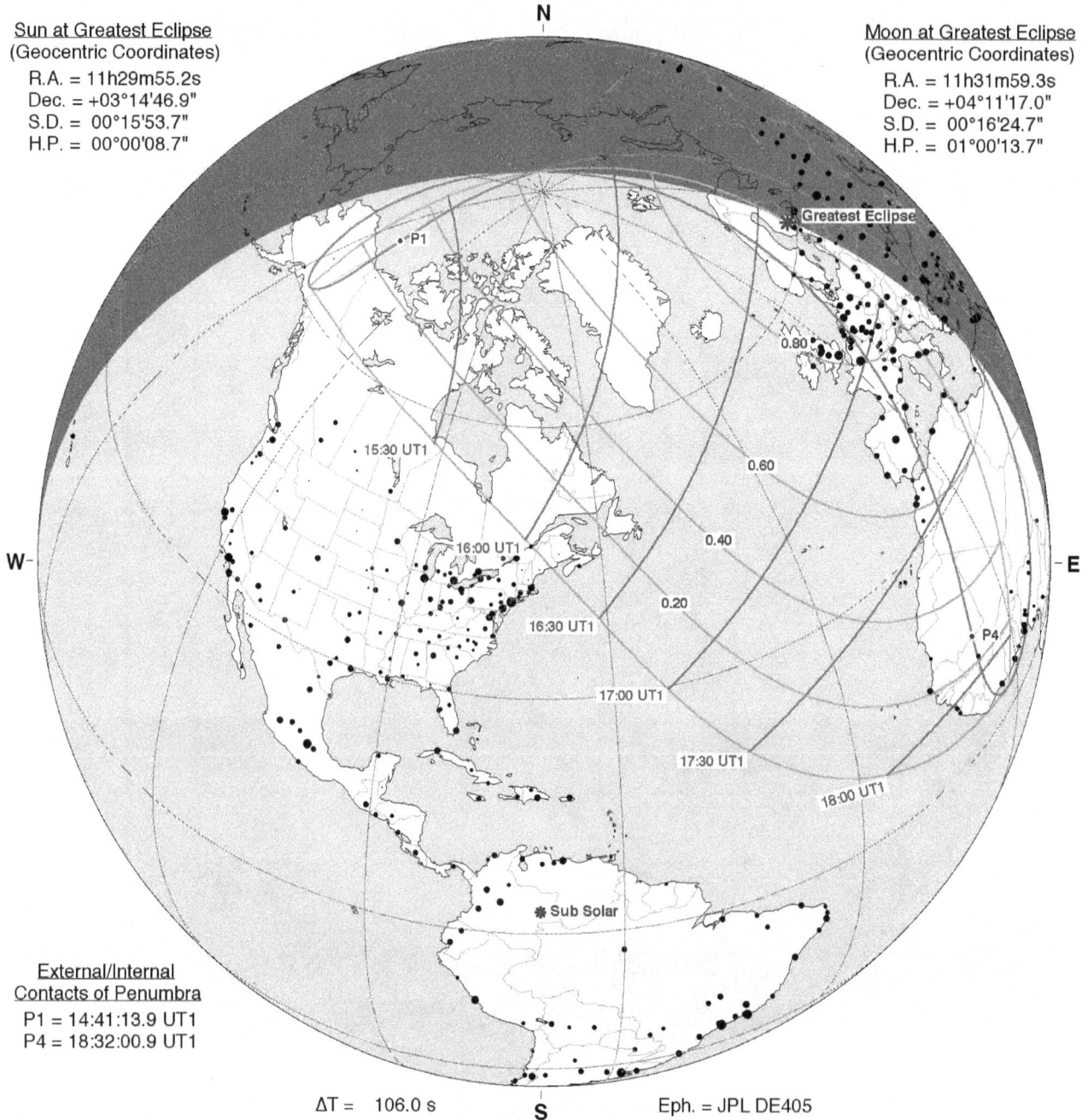

N

Greatest Eclipse

0.80
0.60
0.40
0.20

W

E

P1
15:30 UT1
16:00 UT1
16:30 UT1
17:00 UT1
17:30 UT1
18:00 UT1
P4

Sub Solar

External/Internal
Contacts of Penumbra
P1 = 14:41:13.9 UT1
P4 = 18:32:00.9 UT1

ΔT = 106.0 s

S

Eph. = JPL DE405

Circumstances at Greatest Eclipse: 16:36:23.2 UT1

| | |
|---|---|
| Lat. = 61°07.4'N | Sun Alt. = 0.0° |
| Long. = 025°36.9'E | Sun Azm. = 276.7° |

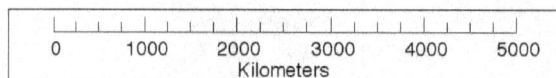

0  1000  2000  3000  4000  5000
Kilometers

©2016  F. Espenak
www.EclipseWise.com

# Annular Solar Eclipse of 2081 Mar 10

Greatest Eclipse = 15:23:30.7 TD   (= 15:21:44.3 UT1)

Eclipse Magnitude = 0.9304          Saros Series = 131
Gamma = -0.3653                     Saros Member = 54 of 70

**Sun at Greatest Eclipse**
(Geocentric Coordinates)
R.A. = 23h25m55.3s
Dec. = -03°40'25.8"
S.D. = 00°16'06.3"
H.P. = 00°00'08.9"

**Moon at Greatest Eclipse**
(Geocentric Coordinates)
R.A. = 23h26m33.6s
Dec. = -03°57'43.0"
S.D. = 00°14'46.5"
H.P. = 00°54'13.5"

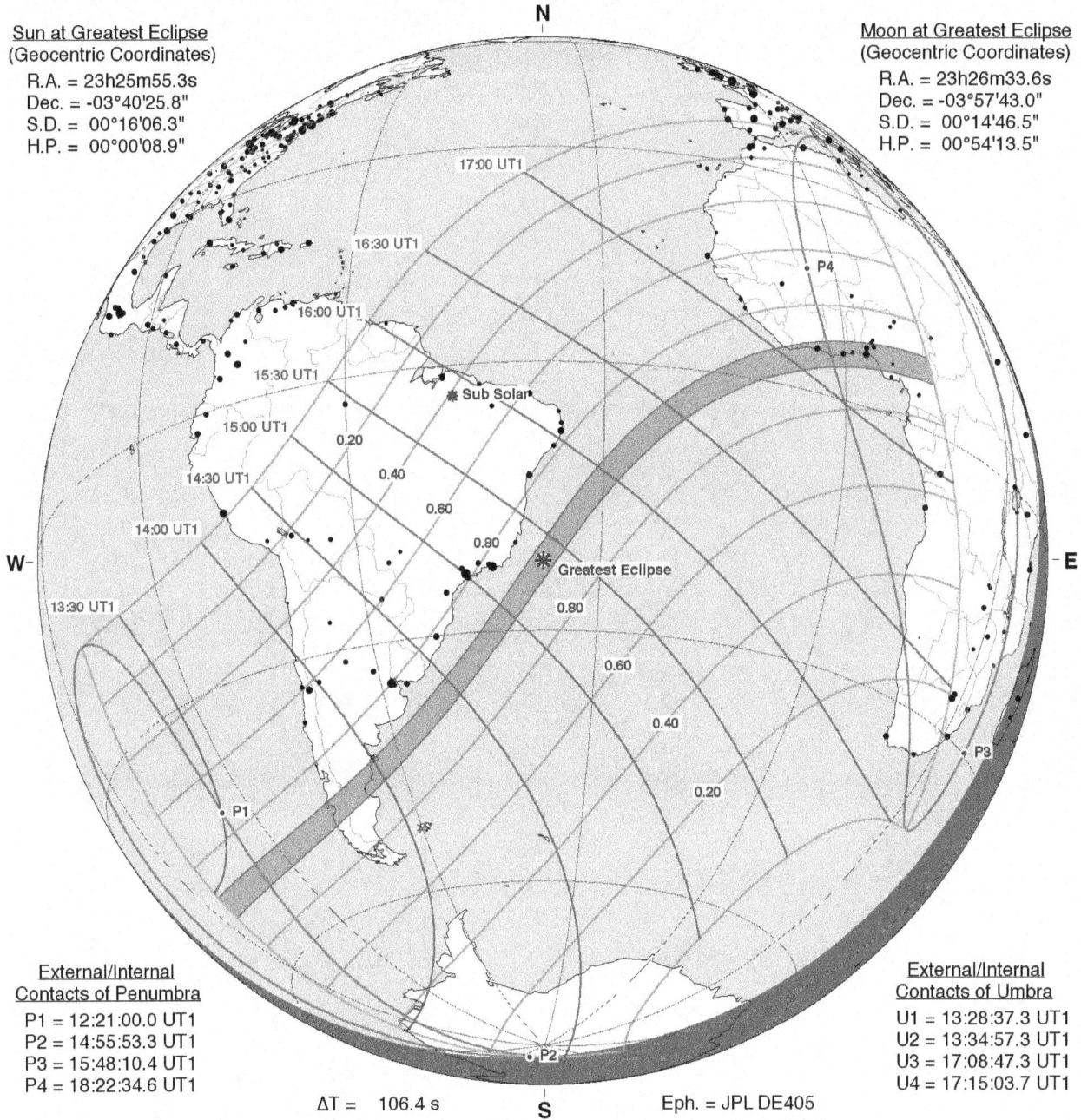

N

17:00 UT1
16:30 UT1
16:00 UT1
15:30 UT1
Sub Solar
15:00 UT1
0.20
14:30 UT1
0.40
0.60
14:00 UT1
0.80
W
Greatest Eclipse
0.80
13:30 UT1
0.60
0.40
0.20
P4
E
P3
P1
P2

**External/Internal
Contacts of Penumbra**
P1 = 12:21:00.0 UT1
P2 = 14:55:53.3 UT1
P3 = 15:48:10.4 UT1
P4 = 18:22:34.6 UT1

**External/Internal
Contacts of Umbra**
U1 = 13:28:37.3 UT1
U2 = 13:34:57.3 UT1
U3 = 17:08:47.3 UT1
U4 = 17:15:03.7 UT1

ΔT = 106.4 s

S

Eph. = JPL DE405

**Circumstances at Greatest Eclipse: 15:21:44.3 UT1**

| | |
|---|---|
| Lat. = 22°23.8'S | Sun Alt. = 68.4° |
| Long. = 036°55.8'W | Sun Azm. = 328.7° |
| Path Width = 277.4 km | Duration = 07m36.3s |

**Circumstances at Greatest Duration: 15:13:02.4 UT1**

| | |
|---|---|
| Lat. = 24°36.9'S | Sun Alt. = 68.0° |
| Long. = 038°40.2'W | Sun Azm. = 340.8° |
| Path Width = 275.1 km | Duration = 07m36.7s |

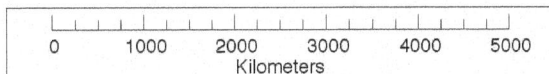

| | | | | | |
|---|---|---|---|---|---|
| 0 | 1000 | 2000 | 3000 | 4000 | 5000 |

Kilometers

# Total Solar Eclipse of 2081 Sep 03

Greatest Eclipse = 09:07:30.5 TD (= 09:05:43.7 UT1)

| | |
|---|---|
| Eclipse Magnitude = 1.0720 | Saros Series = 136 |
| Gamma = 0.3379 | Saros Member = 41 of 71 |

**Sun at Greatest Eclipse**
(Geocentric Coordinates)
R.A. = 10h52m00.4s
Dec. = +07°13'15.0"
S.D. = 00°15'51.2"
H.P. = 00°00'08.7"

**Moon at Greatest Eclipse**
(Geocentric Coordinates)
R.A. = 10h52m39.6s
Dec. = +07°31'30.8"
S.D. = 00°16'43.4"
H.P. = 01°01'22.4"

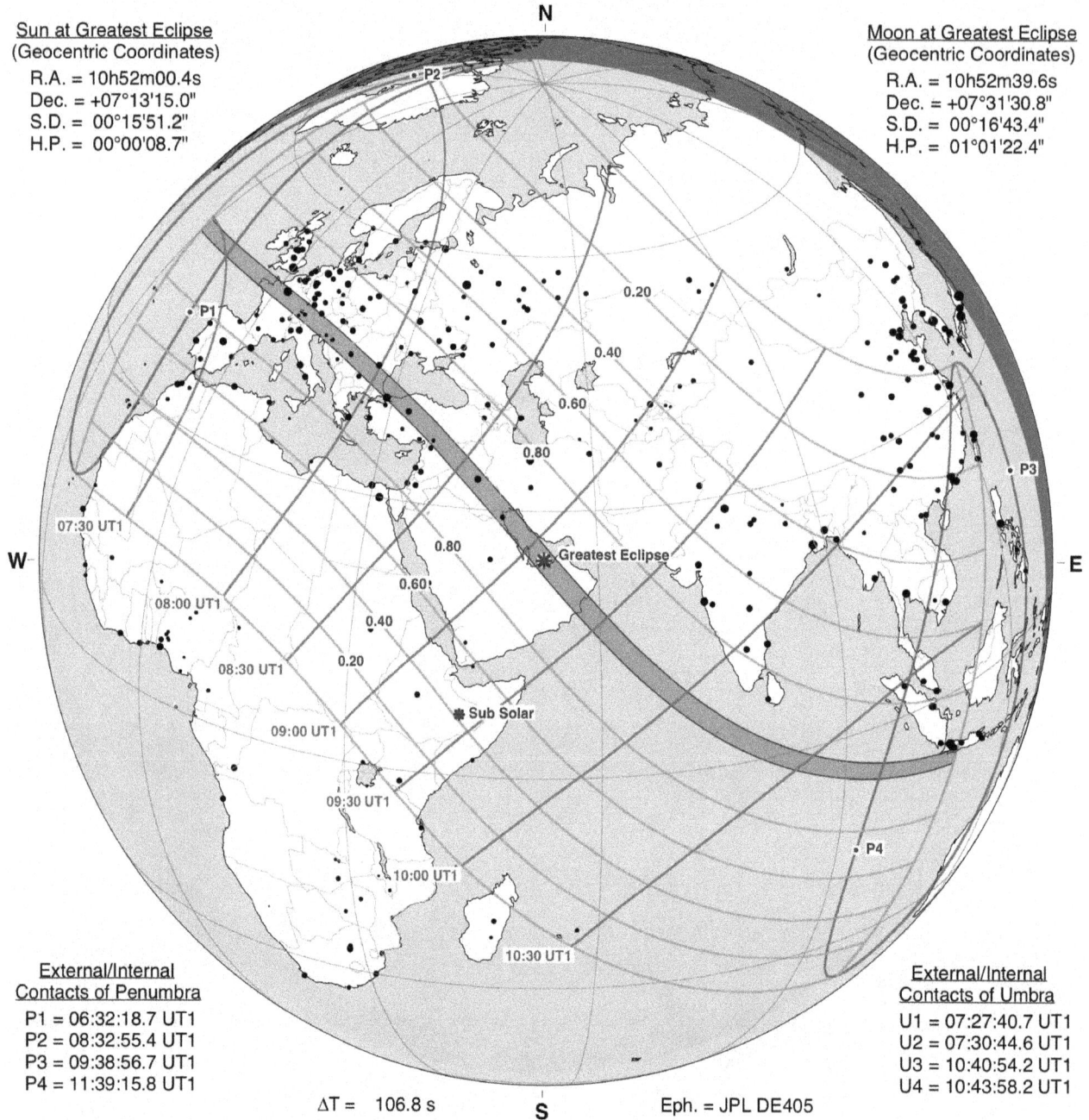

**External/Internal
Contacts of Penumbra**
P1 = 06:32:18.7 UT1
P2 = 08:32:55.4 UT1
P3 = 09:38:56.7 UT1
P4 = 11:39:15.8 UT1

ΔT = 106.8 s

**External/Internal
Contacts of Umbra**
U1 = 07:27:40.7 UT1
U2 = 07:30:44.6 UT1
U3 = 10:40:54.2 UT1
U4 = 10:43:58.2 UT1

Eph. = JPL DE405

**Circumstances at Greatest Eclipse: 09:05:43.7 UT1**
Lat. = 24°37.7'N    Sun Alt. = 70.1°
Long. = 053°25.2'E    Sun Azm. = 210.6°
Path Width = 247.3 km    Duration = 05m33.1s

**Circumstances at Greatest Duration: 09:02:43.3 UT1**
Lat. = 25°29.5'N    Sun Alt. = 70.0°
Long. = 052°34.9'E    Sun Azm. = 205.3°
Path Width = 246.6 km    Duration = 05m33.2s

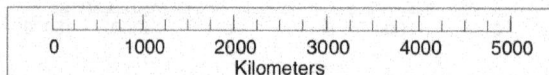

©2016 F. Espenak
www.EclipseWise.com

# Annular Solar Eclipse of 2082 Feb 27

Greatest Eclipse = 14:46:59.8 TD   (= 14:45:12.6 UT1)

Eclipse Magnitude = 0.9298          Saros Series = 141

Gamma = 0.3361          Saros Member =  27 of 70

Sun at Greatest Eclipse
(Geocentric Coordinates)
R.A. = 22h44m00.6s
Dec. = -08°01'49.1"
S.D. = 00°16'09.0"
H.P. = 00°00'08.9"

Moon at Greatest Eclipse
(Geocentric Coordinates)
R.A. = 22h43m26.2s
Dec. = -07°45'43.0"
S.D. = 00°14'48.3"
H.P. = 00°54'19.9"

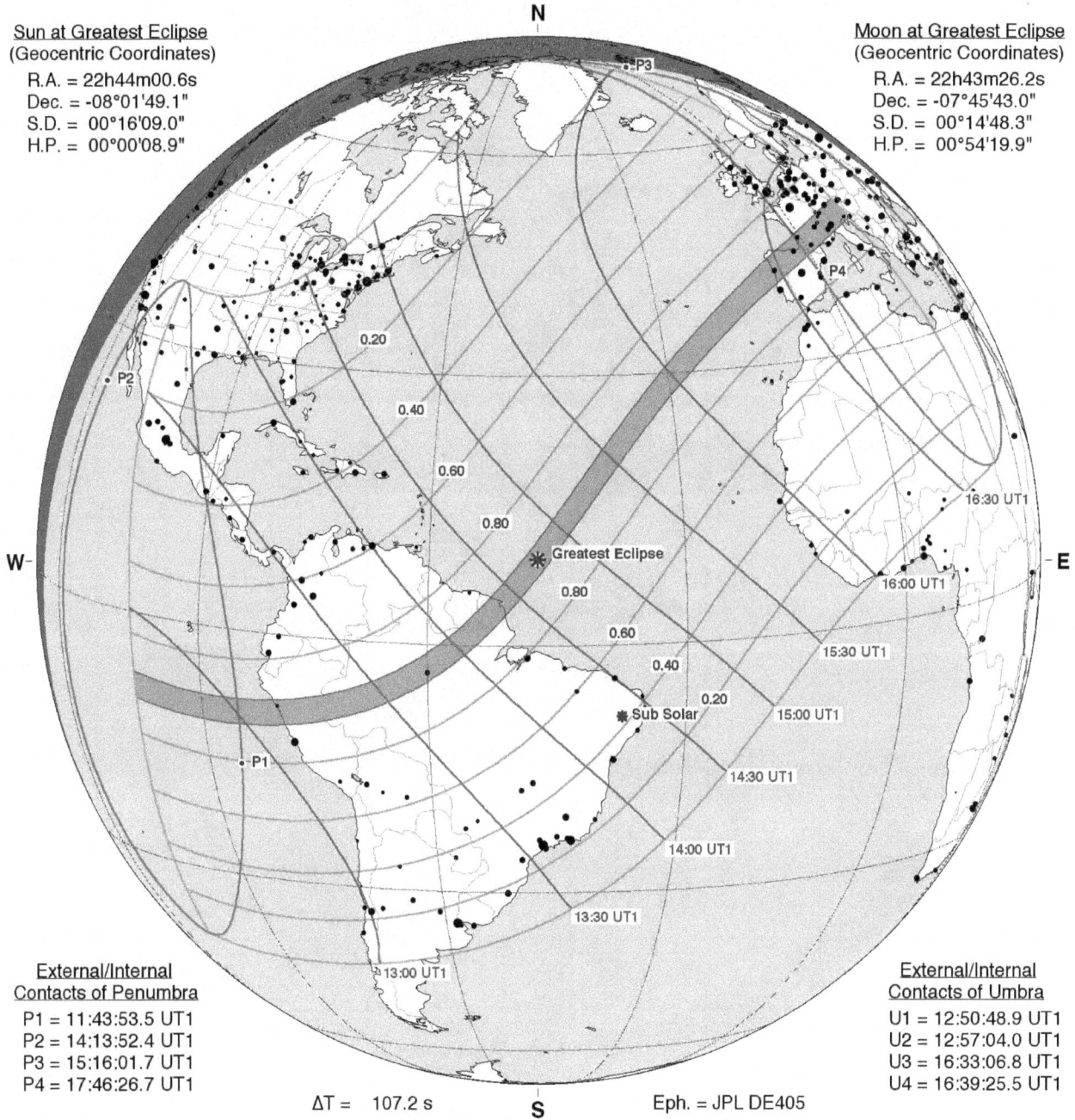

N

P3
P4
P2
0.20
0.40
0.60
0.80
Greatest Eclipse
0.80
16:30 UT1
16:00 UT1
0.60
15:30 UT1
0.40
0.20
Sub Solar
15:00 UT1
14:30 UT1
14:00 UT1
P1
13:30 UT1
13:00 UT1

W
E

S

External/Internal
Contacts of Penumbra
P1 = 11:43:53.5 UT1
P2 = 14:13:52.4 UT1
P3 = 15:16:01.7 UT1
P4 = 17:46:26.7 UT1

External/Internal
Contacts of Umbra
U1 = 12:50:48.9 UT1
U2 = 12:57:04.0 UT1
U3 = 16:33:06.8 UT1
U4 = 16:39:25.5 UT1

ΔT =  107.2 s          Eph. = JPL DE405

Circumstances at Greatest Eclipse:  14:45:12.6 UT1

| | |
|---|---|
| Lat. =  09°25.5'N | Sun Alt. =  70.3° |
| Long. = 047°20.5'W | Sun Azm. = 152.1° |
| Path Width = 276.9 km | Duration = 08m11.8s |

Circumstances at Greatest Duration:  14:32:21.4 UT1

| | |
|---|---|
| Lat. =  06°32.5'N | Sun Alt. =  69.3° |
| Long. = 049°44.1'W | Sun Azm. = 134.5° |
| Path Width = 281.5 km | Duration = 08m12.9s |

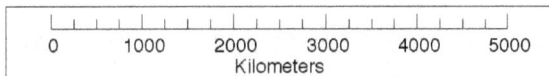

0   1000   2000   3000   4000   5000
Kilometers

# Total Solar Eclipse of 2082 Aug 24

Greatest Eclipse = 01:16:20.6 TD   (= 01:14:33.0 UT1)

| Eclipse Magnitude = 1.0452 | Saros Series = 146 |
|---|---|
| Gamma = -0.4004 | Saros Member = 31 of 76 |

Sun at Greatest Eclipse
(Geocentric Coordinates)
R.A. = 10h13m33.7s
Dec. = +10°59'07.1"
S.D. = 00°15'48.9"
H.P. = 00°00'08.7"

Moon at Greatest Eclipse
(Geocentric Coordinates)
R.A. = 10h12m50.5s
Dec. = +10°37'45.2"
S.D. = 00°16'16.8"
H.P. = 00°59'44.8"

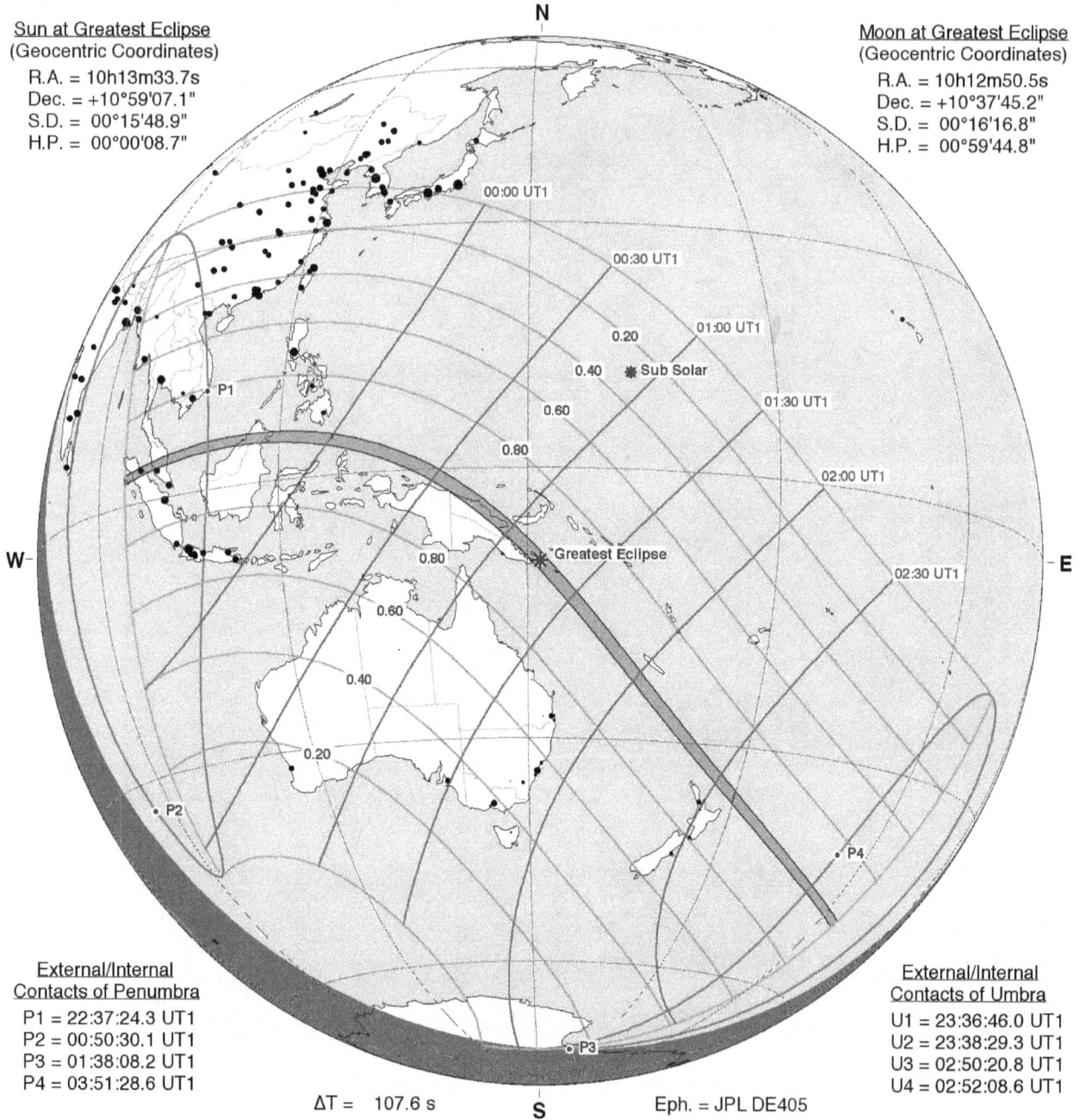

External/Internal
Contacts of Penumbra
P1 = 22:37:24.3 UT1
P2 = 00:50:30.1 UT1
P3 = 01:38:08.2 UT1
P4 = 03:51:28.6 UT1

ΔT = 107.6 s

Eph. = JPL DE405

External/Internal
Contacts of Umbra
U1 = 23:36:46.0 UT1
U2 = 23:38:29.3 UT1
U3 = 02:50:20.8 UT1
U4 = 02:52:08.6 UT1

Circumstances at Greatest Eclipse: 01:14:33.0 UT1

| Lat. = 10°16.7'S | Sun Alt. = 66.3° |
|---|---|
| Long. = 151°32.4'E | Sun Azm. = 26.3° |
| Path Width = 162.9 km | Duration = 04m01.0s |

Circumstances at Greatest Duration: 01:11:33.1 UT1

| Lat. = 09°33.4'S | Sun Alt. = 66.3° |
|---|---|
| Long. = 150°48.6'E | Sun Azm. = 30.3° |
| Path Width = 163.4 km | Duration = 04m01.1s |

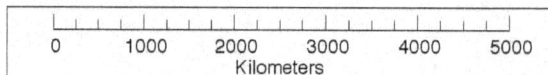

©2016  F. Espenak
www.EclipseWise.com

# Partial Solar Eclipse of 2083 Feb 16

Greatest Eclipse = 18:06:36.2 TD  (= 18:04:48.2 UT1)

| | |
|---|---|
| Eclipse Magnitude = 0.9433 | Saros Series = 151 |
| Gamma = 1.0170 | Saros Member = 18 of 72 |

**N**

Sun at Greatest Eclipse
(Geocentric Coordinates)
R.A. = 22h01m38.8s
Dec. = -12°04'40.8"
S.D. = 00°16'11.4"
H.P. = 00°00'08.9"

Moon at Greatest Eclipse
(Geocentric Coordinates)
R.A. = 21h59m56.3s
Dec. = -11°12'50.5"
S.D. = 00°15'28.4"
H.P. = 00°56'47.2"

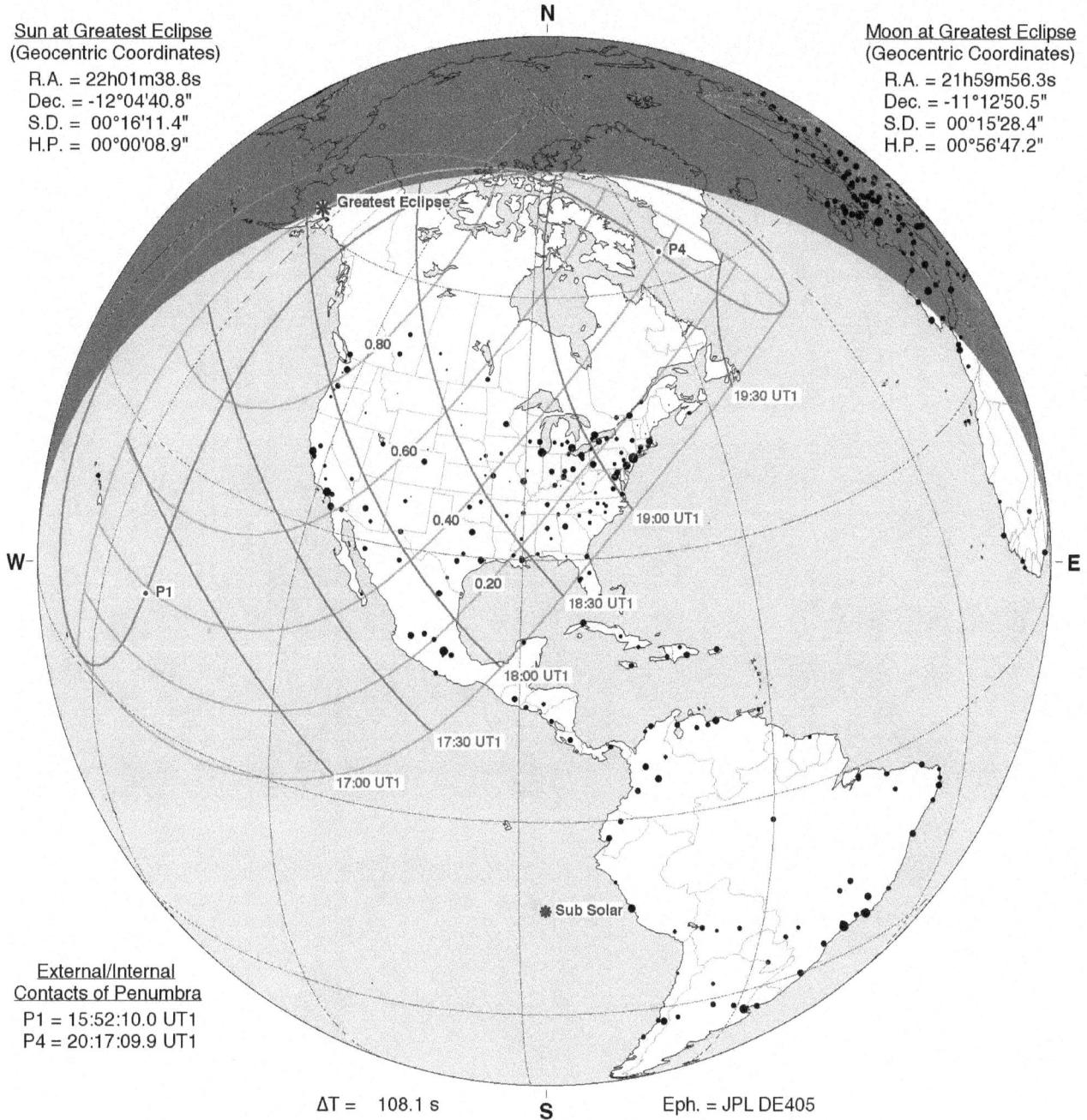

Greatest Eclipse

P4

0.80

0.60

19:30 UT1

0.40

19:00 UT1

**W**

0.20

18:30 UT1

P1

18:00 UT1

17:30 UT1

17:00 UT1

**E**

Sub Solar

External/Internal
Contacts of Penumbra
P1 = 15:52:10.0 UT1
P4 = 20:17:09.9 UT1

ΔT =   108.1 s          **S**          Eph. = JPL DE405

Circumstances at Greatest Eclipse: 18:04:48.2 UT1
Lat. = 61°36.4'N          Sun Alt. =   0.0°
Long. = 154°19.9'W          Sun Azm. = 116.1°

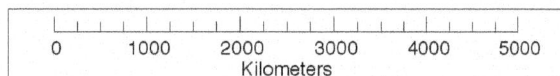

| 0 | 1000 | 2000 | 3000 | 4000 | 5000 |
|---|---|---|---|---|---|
Kilometers

©2016  F. Espenak
www.EclipseWise.com

# Partial Solar Eclipse of 2083 Jul 15

Greatest Eclipse = 00:14:22.9 TD (= 00:12:34.5 UT1)

| | |
|---|---|
| Eclipse Magnitude = 0.0169 | Saros Series = 118 |
| Gamma = 1.5464 | Saros Member = 72 of 72 |

**Sun at Greatest Eclipse**
(Geocentric Coordinates)
R.A. = 07h38m24.0s
Dec. = +21°30'20.2"
S.D. = 00°15'44.0"
H.P. = 00°00'08.7"

**Moon at Greatest Eclipse**
(Geocentric Coordinates)
R.A. = 07h40m00.2s
Dec. = +22°52'33.7"
S.D. = 00°15'02.8"
H.P. = 00°55'13.4"

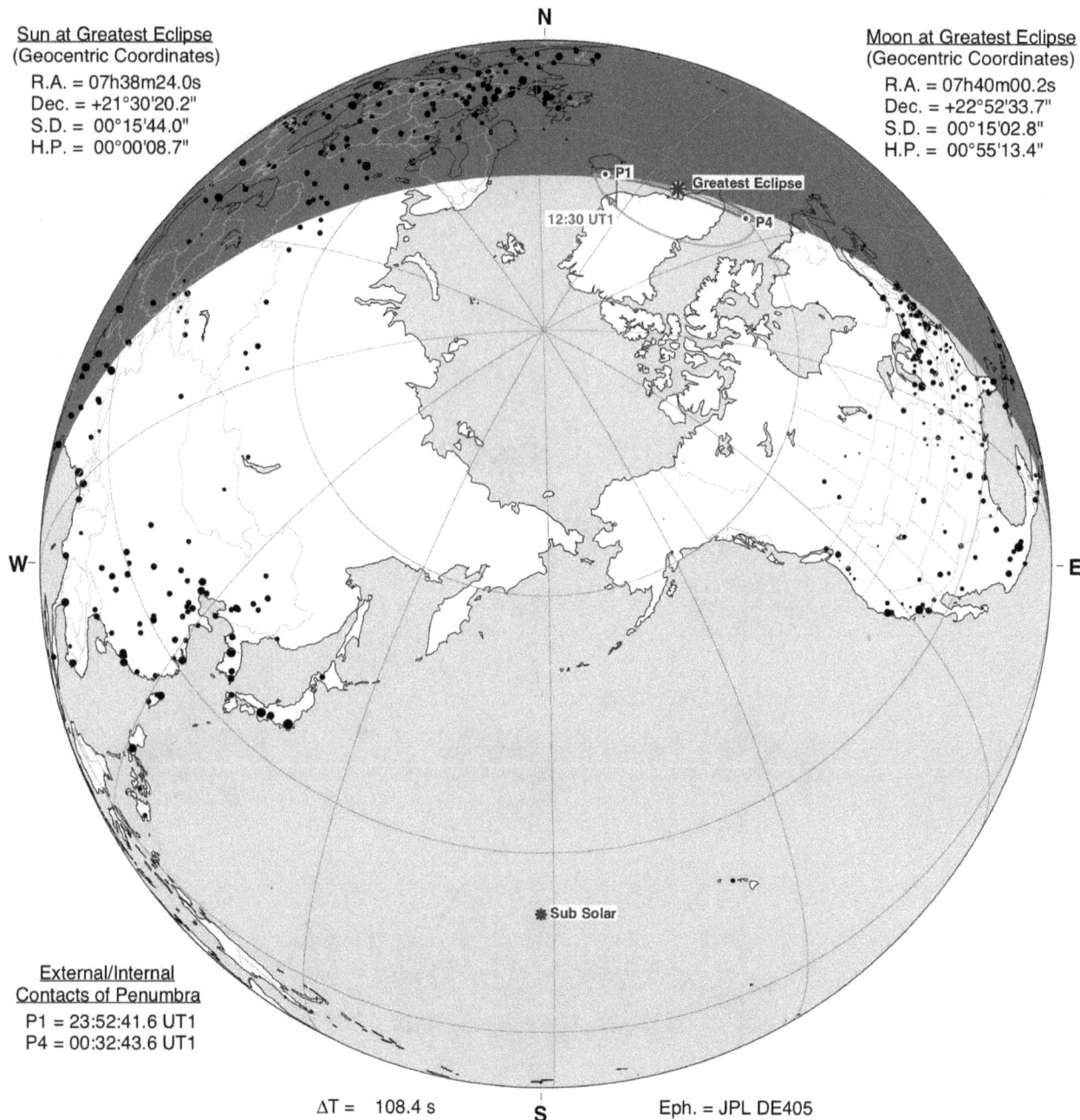

N

P1
Greatest Eclipse
12:30 UT1
P4

W

E

Sub Solar

**External/Internal**
**Contacts of Penumbra**
P1 = 23:52:41.6 UT1
P4 = 00:32:43.6 UT1

ΔT = 108.4 s

S

Eph. = JPL DE405

**Circumstances at Greatest Eclipse: 00:12:34.5 UT1**

| | |
|---|---|
| Lat. = 63°57.1'N | Sun Alt. = 0.0° |
| Long. = 037°57.0'W | Sun Azm. = 326.6° |

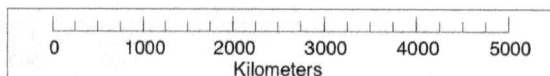

©2016 F. Espenak
www.EclipseWise.com

| | | | | | |
|---|---|---|---|---|---|
| 0 | 1000 | 2000 | 3000 | 4000 | 5000 |

Kilometers

# Partial Solar Eclipse of  2083 Aug 13

Greatest Eclipse =  12:34:41.2 TD   (= 12:32:52.7 UT1)

Eclipse Magnitude =  0.6146           Saros Series =  156
Gamma = -1.2064                       Saros Member =   5 of 69

Sun at Greatest Eclipse
(Geocentric Coordinates)
R.A. = 09h33m34.6s
Dec. = +14°29'20.4"
S.D. = 00°15'47.0"
H.P. = 00°00'08.7"

Moon at Greatest Eclipse
(Geocentric Coordinates)
R.A. = 09h31m40.1s
Dec. = +13°27'06.6"
S.D. = 00°15'25.9"
H.P. = 00°56'37.9"

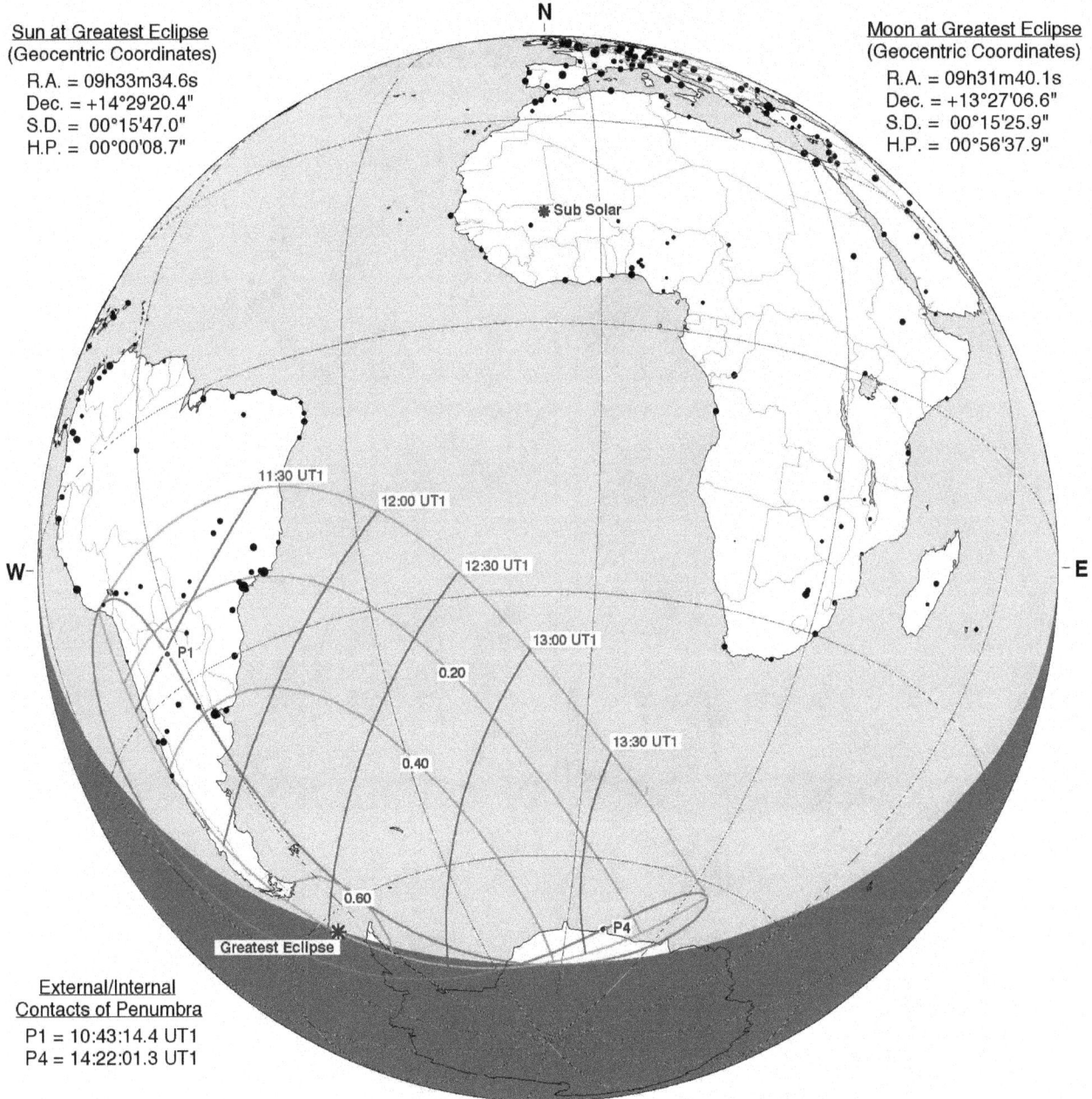

N

＊ Sub Solar

11:30 UT1
12:00 UT1
12:30 UT1
13:00 UT1
13:30 UT1

0.20
0.40
0.60

P1
P4

W

E

＊
Greatest Eclipse

External/Internal
Contacts of Penumbra
P1 = 10:43:14.4 UT1
P4 = 14:22:01.3 UT1

ΔT =    108.5 s          S          Eph. = JPL DE405

Circumstances at Greatest Eclipse:  12:32:52.7 UT1
Lat. = 62°05.6'S          Sun Alt. =    0.0°
Long. = 067°42.4'W        Sun Azm. =  57.7°

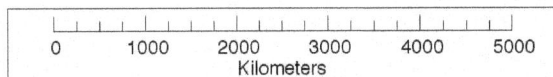

0    1000   2000   3000   4000   5000
Kilometers

©2016  F. Espenak
www.EclipseWise.com

241

# Partial Solar Eclipse of 2084 Jan 07

Greatest Eclipse = 17:30:23.5 TD (= 17:28:34.7 UT1)

Eclipse Magnitude = 0.8723          Saros Series = 123
Gamma = -1.0715          Saros Member = 57 of 70

Sun at Greatest Eclipse
(Geocentric Coordinates)
R.A. = 19h15m11.8s
Dec. = -22°18'17.9"
S.D. = 00°16'15.9"
H.P. = 00°00'08.9"

Moon at Greatest Eclipse
(Geocentric Coordinates)
R.A. = 19h16m15.2s
Dec. = -23°21'56.0"
S.D. = 00°16'38.6"
H.P. = 01°01'05.0"

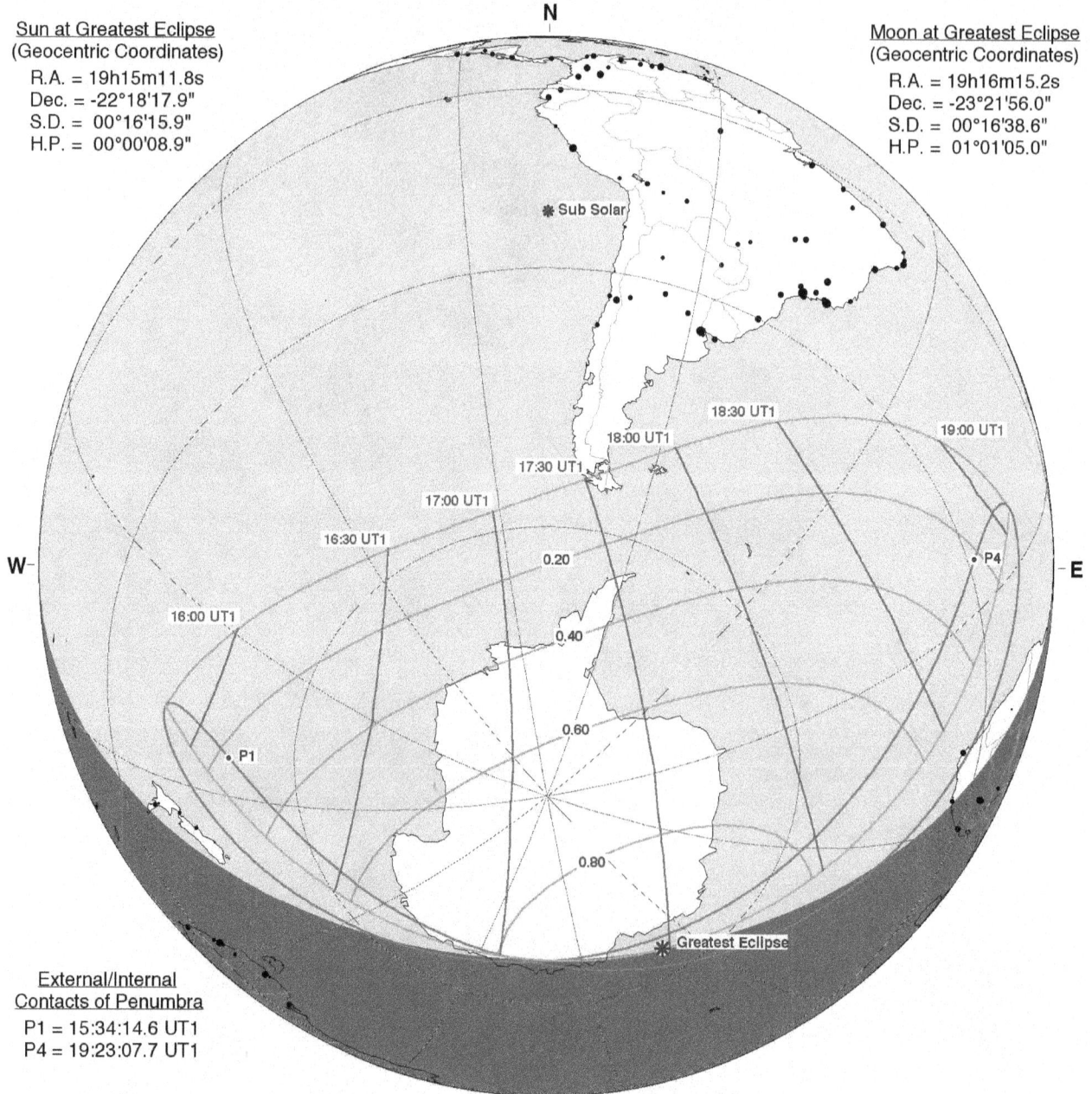

N

* Sub Solar

18:30 UT1          19:00 UT1
18:00 UT1
17:30 UT1
17:00 UT1
16:30 UT1
0.20
16:00 UT1          P4
0.40

W          E

0.60

P1

0.80

* Greatest Eclipse

External/Internal
Contacts of Penumbra
P1 = 15:34:14.6 UT1
P4 = 19:23:07.7 UT1

ΔT = 108.8 s          S          Eph. = JPL DE405

Circumstances at Greatest Eclipse: 17:28:34.7 UT1
Lat. = 64°24.6'S          Sun Alt. = 0.0°
Long. = 068°17.7'E          Sun Azm. = 208.5°

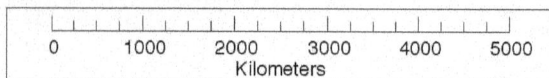

| | | | | | |
|---|---|---|---|---|---|
| 0 | 1000 | 2000 | 3000 | 4000 | 5000 |

Kilometers

©2016 F. Espenak
www.EclipseWise.com

# Annular Solar Eclipse of 2084 Jul 03

Greatest Eclipse = 01:50:25.9 TD  (= 01:48:36.6 UT1)

Eclipse Magnitude = 0.9421
Gamma = 0.8208

Saros Series = 128
Saros Member = 62 of 73

**Sun at Greatest Eclipse**
(Geocentric Coordinates)
R.A. = 06h52m43.5s
Dec. = +22°52'33.4"
S.D. = 00°15'43.9"
H.P. = 00°00'08.6"

**Moon at Greatest Eclipse**
(Geocentric Coordinates)
R.A. = 06h53m20.0s
Dec. = +23°35'54.8"
S.D. = 00°14'41.9"
H.P. = 00°53'56.6"

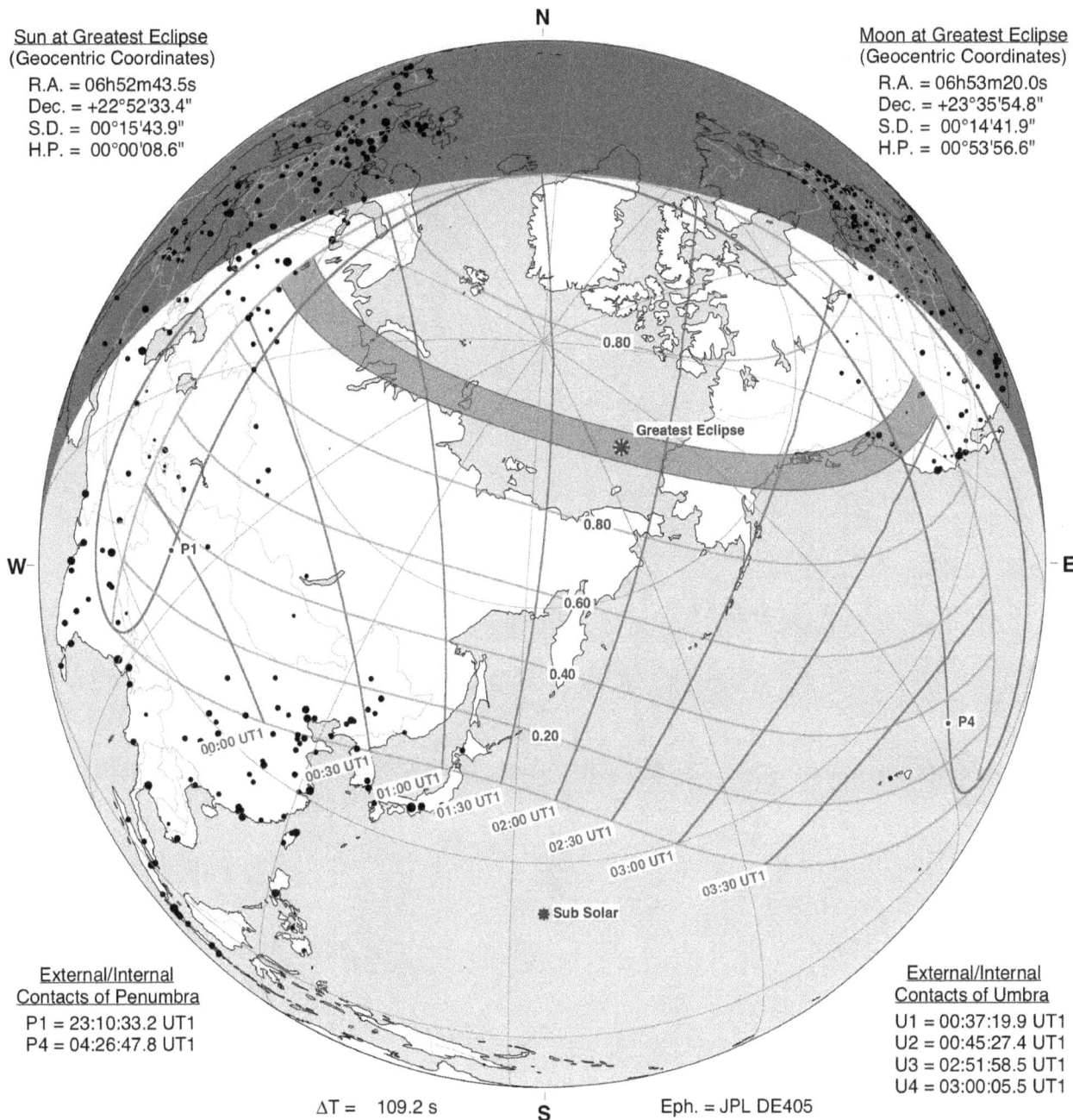

N

0.80

Greatest Eclipse

0.80

0.60

0.40

0.20

W

E

P1

P4

00:00 UT1
00:30 UT1
01:00 UT1
01:30 UT1
02:00 UT1
02:30 UT1
03:00 UT1
03:30 UT1

Sub Solar

**External/Internal Contacts of Penumbra**
P1 = 23:10:33.2 UT1
P4 = 04:26:47.8 UT1

**External/Internal Contacts of Umbra**
U1 = 00:37:19.9 UT1
U2 = 00:45:27.4 UT1
U3 = 02:51:58.5 UT1
U4 = 03:00:05.5 UT1

$\Delta T$ = 109.2 s

S

Eph. = JPL DE405

**Circumstances at Greatest Eclipse: 01:48:36.6 UT1**
Lat. = 75°00.6'N
Long. = 169°17.6'W
Path Width = 377.4 km
Sun Alt. = 34.5°
Sun Azm. = 222.0°
Duration = 04m25.1s

**Circumstances at Greatest Duration: 01:45:34.2 UT1**
Lat. = 75°58.3'N
Long. = 172°56.6'W
Path Width = 376.5 km
Sun Alt. = 34.5°
Sun Azm. = 216.7°
Duration = 04m25.2s

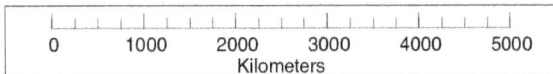

©2016 F. Espenak
www.EclipseWise.com

| | | | | | |
|---|---|---|---|---|---|
| 0 | 1000 | 2000 | 3000 | 4000 | 5000 |

Kilometers

# Total Solar Eclipse of 2084 Dec 27

Greatest Eclipse = 09:13:48.0 TD  (= 09:11:58.3 UT1)

| | |
|---|---|
| Eclipse Magnitude = 1.0396 | Saros Series = 133 |
| Gamma = -0.4094 | Saros Member = 49 of 72 |

Sun at Greatest Eclipse
(Geocentric Coordinates)
R.A. = 18h28m34.2s
Dec. = -23°15'58.0"
S.D. = 00°16'15.7"
H.P. = 00°00'08.9"

Moon at Greatest Eclipse
(Geocentric Coordinates)
R.A. = 18h28m50.3s
Dec. = -23°40'38.6"
S.D. = 00°16'38.7"
H.P. = 01°01'05.5"

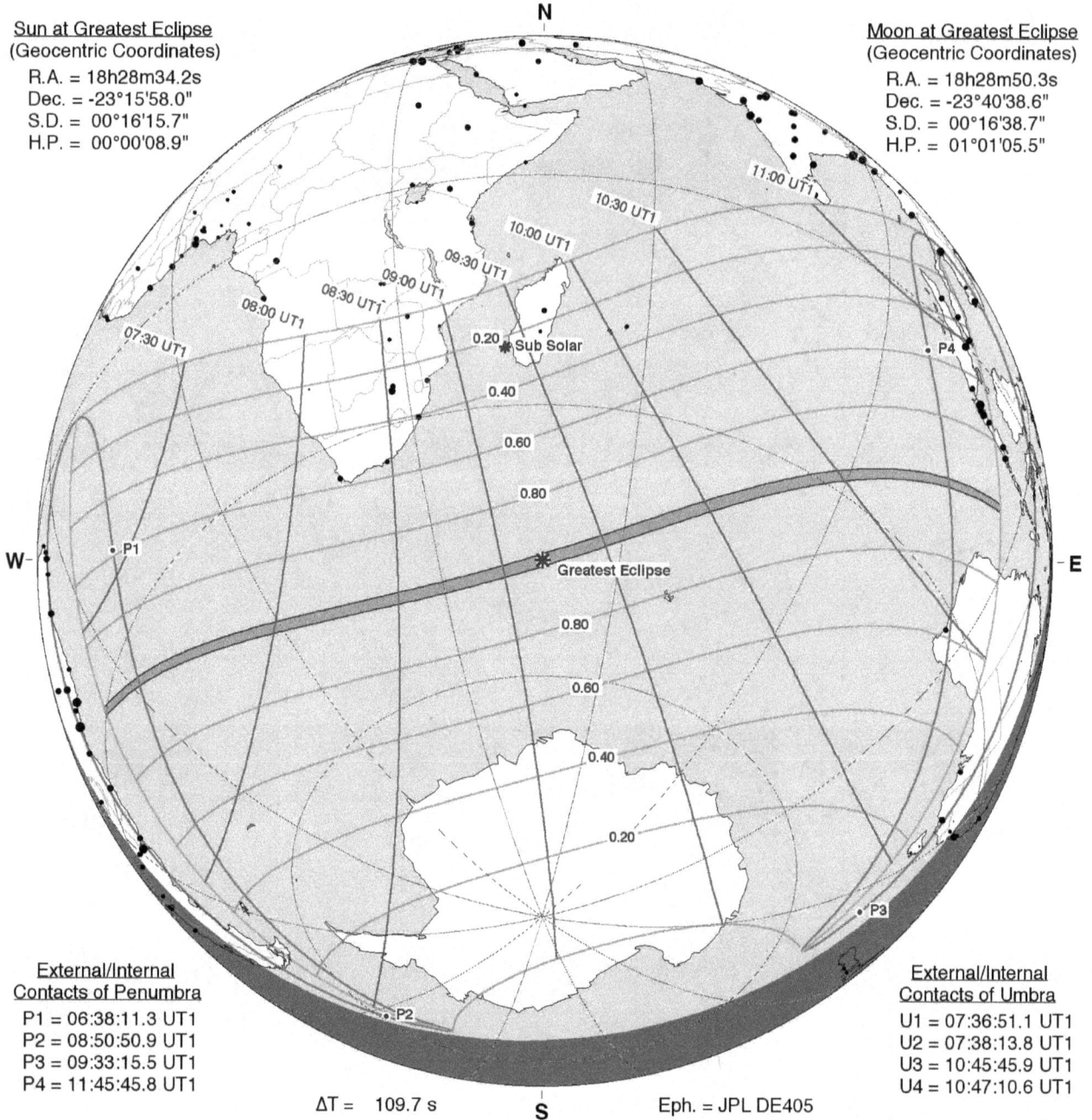

External/Internal
Contacts of Penumbra
P1 = 06:38:11.3 UT1
P2 = 08:50:50.9 UT1
P3 = 09:33:15.5 UT1
P4 = 11:45:45.8 UT1

External/Internal
Contacts of Umbra
U1 = 07:36:51.1 UT1
U2 = 07:38:13.8 UT1
U3 = 10:45:45.9 UT1
U4 = 10:47:10.6 UT1

ΔT = 109.7 s          Eph. = JPL DE405

Circumstances at Greatest Eclipse: 09:11:58.3 UT1

| | |
|---|---|
| Lat. = 47°19.4'S | Sun Alt. = 65.6° |
| Long. = 047°26.9'E | Sun Azm. = 348.6° |
| Path Width = 145.7 km | Duration = 03m04.4s |

Circumstances at Greatest Duration: 09:10:59.4 UT1

| | |
|---|---|
| Lat. = 47°25.8'S | Sun Alt. = 65.6° |
| Long. = 046°54.5'E | Sun Azm. = 350.3° |
| Path Width = 145.7 km | Duration = 03m04.4s |

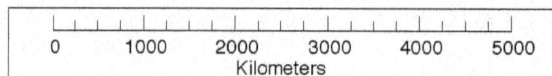

©2016  F. Espenak
www.EclipseWise.com

# Annular Solar Eclipse of 2085 Jun 22

Greatest Eclipse = 03:21:15.9 TD (= 03:19:25.8 UT1)

Eclipse Magnitude = 0.9704      Saros Series = 138
Gamma = 0.0453      Saros Member = 35 of 70

Sun at Greatest Eclipse
(Geocentric Coordinates)
R.A. = 06h06m22.6s
Dec. = +23°25'12.0"
S.D. = 00°15'44.3"
H.P. = 00°00'08.7"

Moon at Greatest Eclipse
(Geocentric Coordinates)
R.A. = 06h06m23.8s
Dec. = +23°27'40.5"
S.D. = 00°15'02.4"
H.P. = 00°55'11.8"

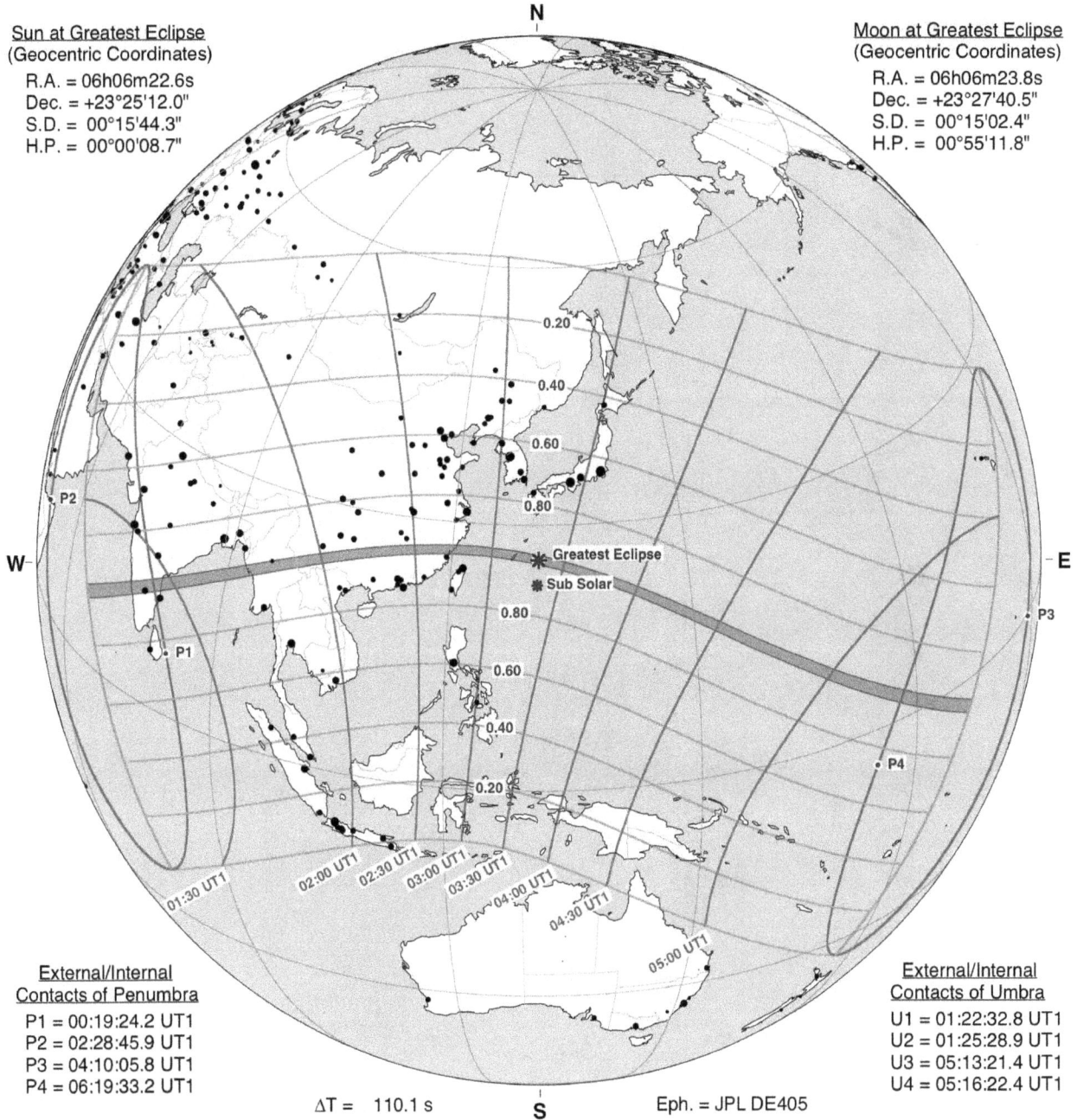

N

W

E

S

0.20
0.40
0.60
0.80
0.80
0.60
0.40
0.20

Greatest Eclipse
Sub Solar

P1
P2
P3
P4

01:30 UT1
02:00 UT1
02:30 UT1
03:00 UT1
03:30 UT1
04:00 UT1
04:30 UT1
05:00 UT1

External/Internal
Contacts of Penumbra
P1 = 00:19:24.2 UT1
P2 = 02:28:45.9 UT1
P3 = 04:10:05.8 UT1
P4 = 06:19:33.2 UT1

External/Internal
Contacts of Umbra
U1 = 01:22:32.8 UT1
U2 = 01:25:28.9 UT1
U3 = 05:13:21.4 UT1
U4 = 05:16:22.4 UT1

ΔT = 110.1 s      Eph. = JPL DE405

Circumstances at Greatest Eclipse: 03:19:25.8 UT1
Lat. = 26°09.1'N    Sun Alt. = 87.3°
Long. = 131°01.3'E    Sun Azm. = 186.1°
Path Width = 106.4 km    Duration = 03m29s

Circumstances at Greatest Duration: 03:15:53.1 UT1
Lat. = 26°20.3'N    Sun Alt. = 86.7°
Long. = 129°57.6'E    Sun Azm. = 152.8°
Path Width = 106.3 km    Duration = 03m29.0s

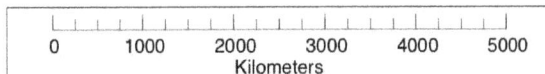

©2016 F. Espenak
www.EclipseWise.com

0 1000 2000 3000 4000 5000
Kilometers

# Annular Solar Eclipse of 2085 Dec 16

Greatest Eclipse = 22:37:47.8 TD  (= 22:35:57.3 UT1)

| | |
|---|---|
| Eclipse Magnitude = 0.9971 | Saros Series = 143 |
| Gamma = 0.2786 | Saros Member = 27 of 72 |

Sun at Greatest Eclipse
(Geocentric Coordinates)
R.A. = 17h41m09.8s
Dec. = -23°21'25.3"
S.D. = 00°16'15.0"
H.P. = 00°00'08.9"

Moon at Greatest Eclipse
(Geocentric Coordinates)
R.A. = 17h41m05.1s
Dec. = -23°05'11.3"
S.D. = 00°15'57.1"
H.P. = 00°58'32.5"

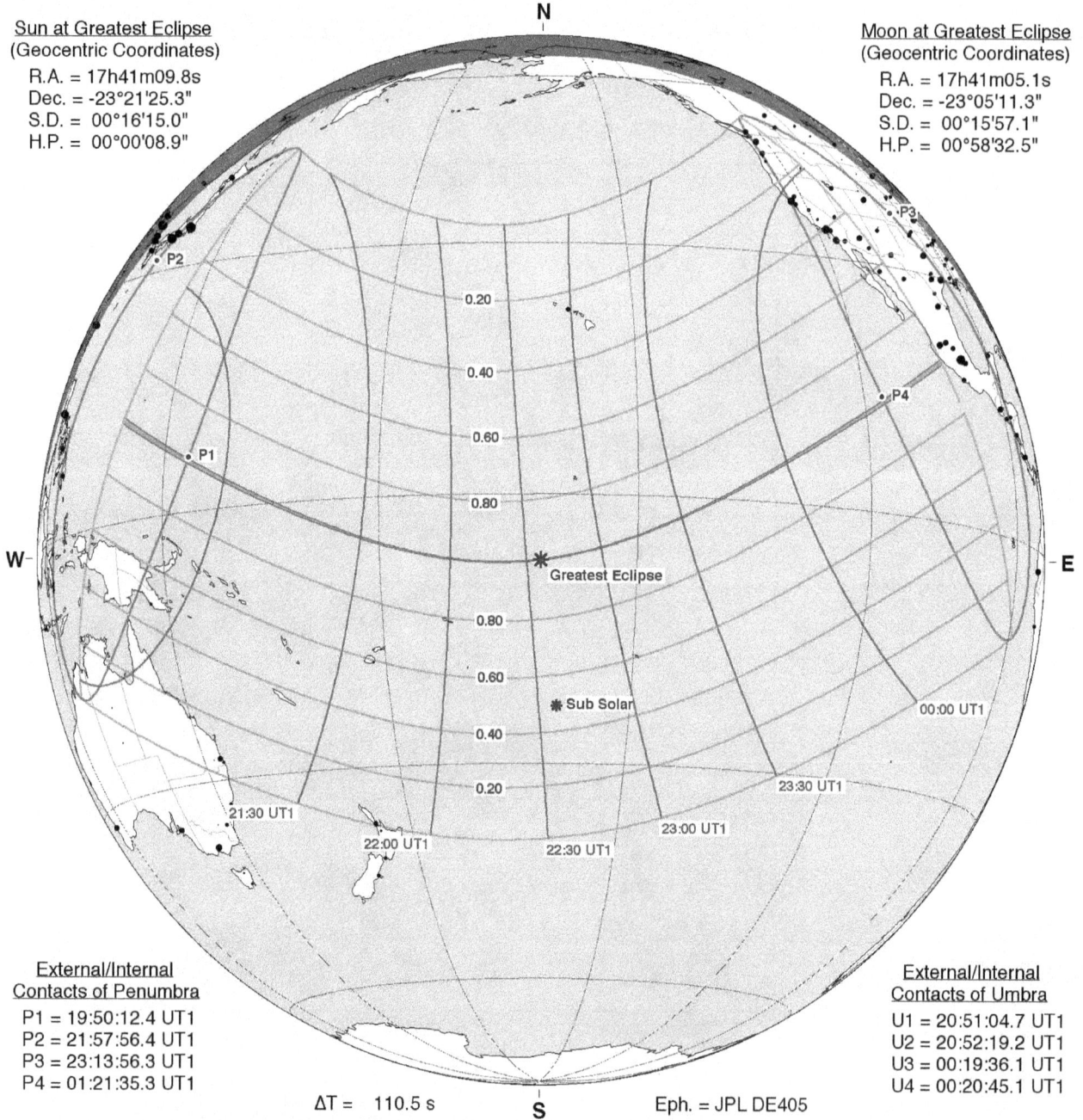

External/Internal
Contacts of Penumbra
P1 = 19:50:12.4 UT1
P2 = 21:57:56.4 UT1
P3 = 23:13:56.3 UT1
P4 = 01:21:35.3 UT1

ΔT = 110.5 s

Eph. = JPL DE405

External/Internal
Contacts of Umbra
U1 = 20:51:04.7 UT1
U2 = 20:52:19.2 UT1
U3 = 00:19:36.1 UT1
U4 = 00:20:45.1 UT1

Circumstances at Greatest Eclipse: 22:35:57.3 UT1

| | |
|---|---|
| Lat. = 07°15.5'S | Sun Alt. = 73.9° |
| Long. = 161°03.2'W | Sun Azm. = 176.5° |
| Path Width = 10.4 km | Duration = 00m19.2s |

Circumstances at Greatest Duration: 20:51:41.9 UT1

| | |
|---|---|
| Lat. = 11°25.9'N | Sun Alt. = 0.0° |
| Long. = 141°05.0'E | Sun Azm. = 113.9° |
| Path Width = 70.3 km | Duration = 01m14.1s |

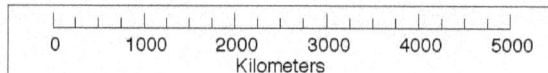

©2016 F. Espenak
www.EclipseWise.com

0  1000  2000  3000  4000  5000
Kilometers

# Total Solar Eclipse of 2086 Jun 11

Greatest Eclipse = 11:07:13.9 TD   (= 11:05:22.9 UT1)

Eclipse Magnitude = 1.0174          Saros Series = 148
Gamma = -0.7215                     Saros Member = 25 of 75

**Sun at Greatest Eclipse**
(Geocentric Coordinates)
R.A. = 05h20m59.8s
Dec. = +23°07'28.2"
S.D. = 00°15'45.2"
H.P. = 00°00'08.7"

**Moon at Greatest Eclipse**
(Geocentric Coordinates)
R.A. = 05h20m54.1s
Dec. = +22°25'37.5"
S.D. = 00°15'51.0"
H.P. = 00°58'10.3"

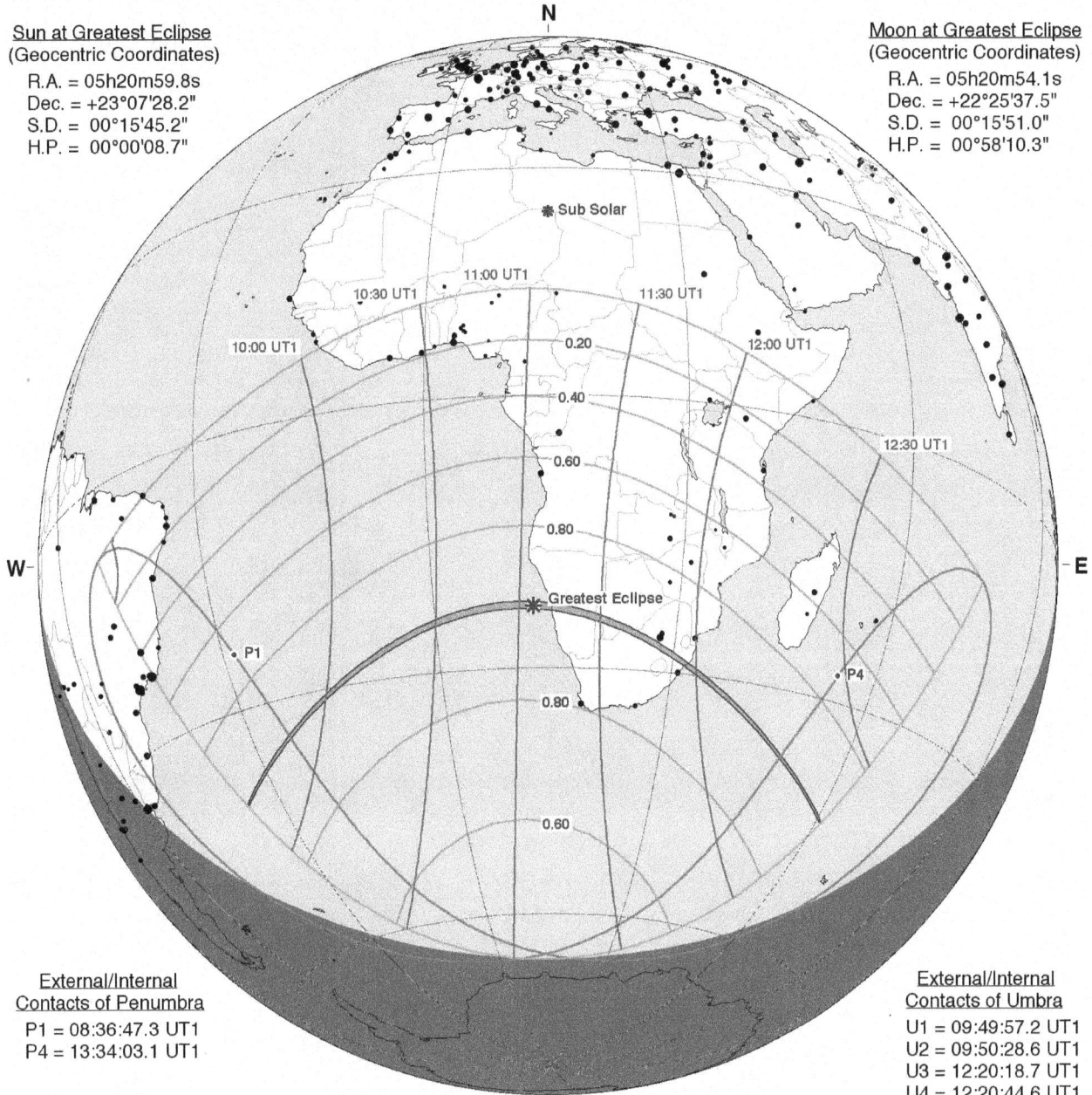

N

Sub Solar

11:00 UT1
10:30 UT1
11:30 UT1
10:00 UT1
12:00 UT1
0.20
0.40
12:30 UT1
0.60
0.80

W — E

Greatest Eclipse

0.80

P1
P4

0.60

**External/Internal Contacts of Penumbra**
P1 = 08:36:47.3 UT1
P4 = 13:34:03.1 UT1

**External/Internal Contacts of Umbra**
U1 = 09:49:57.2 UT1
U2 = 09:50:28.6 UT1
U3 = 12:20:18.7 UT1
U4 = 12:20:44.6 UT1

ΔT = 110.9 s          S          Eph. = JPL DE405

**Circumstances at Greatest Eclipse: 11:05:22.9 UT1**

| | |
|---|---|
| Lat. = 23°12.1'S | Sun Alt. = 43.7° |
| Long. = 012°13.5'E | Sun Azm. = 1.8° |
| Path Width = 85.9 km | Duration = 01m48.1s |

**Circumstances at Greatest Duration: 11:03:17.5 UT1**

| | |
|---|---|
| Lat. = 23°10.4'S | Sun Alt. = 43.6° |
| Long. = 011°33.2'E | Sun Azm. = 3.3° |
| Path Width = 86.0 km | Duration = 01m48.1s |

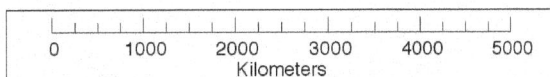

©2016 F. Espenak
www.EclipseWise.com

0    1000   2000   3000   4000   5000
Kilometers

# Partial Solar Eclipse of 2086 Dec 06

Greatest Eclipse = 05:38:55.4 TD (= 05:37:04.0 UT1)

Eclipse Magnitude = 0.9271      Saros Series = 153

Gamma = 1.0194      Saros Member = 13 of 70

<u>Sun at Greatest Eclipse</u>
(Geocentric Coordinates)
R.A. = 16h52m56.6s
Dec. = -22°31'57.0"
S.D. = 00°16'13.7"
H.P. = 00°00'08.9"

<u>Moon at Greatest Eclipse</u>
(Geocentric Coordinates)
R.A. = 16h53m00.7s
Dec. = -21°35'36.6"
S.D. = 00°15'06.1"
H.P. = 00°55'25.4"

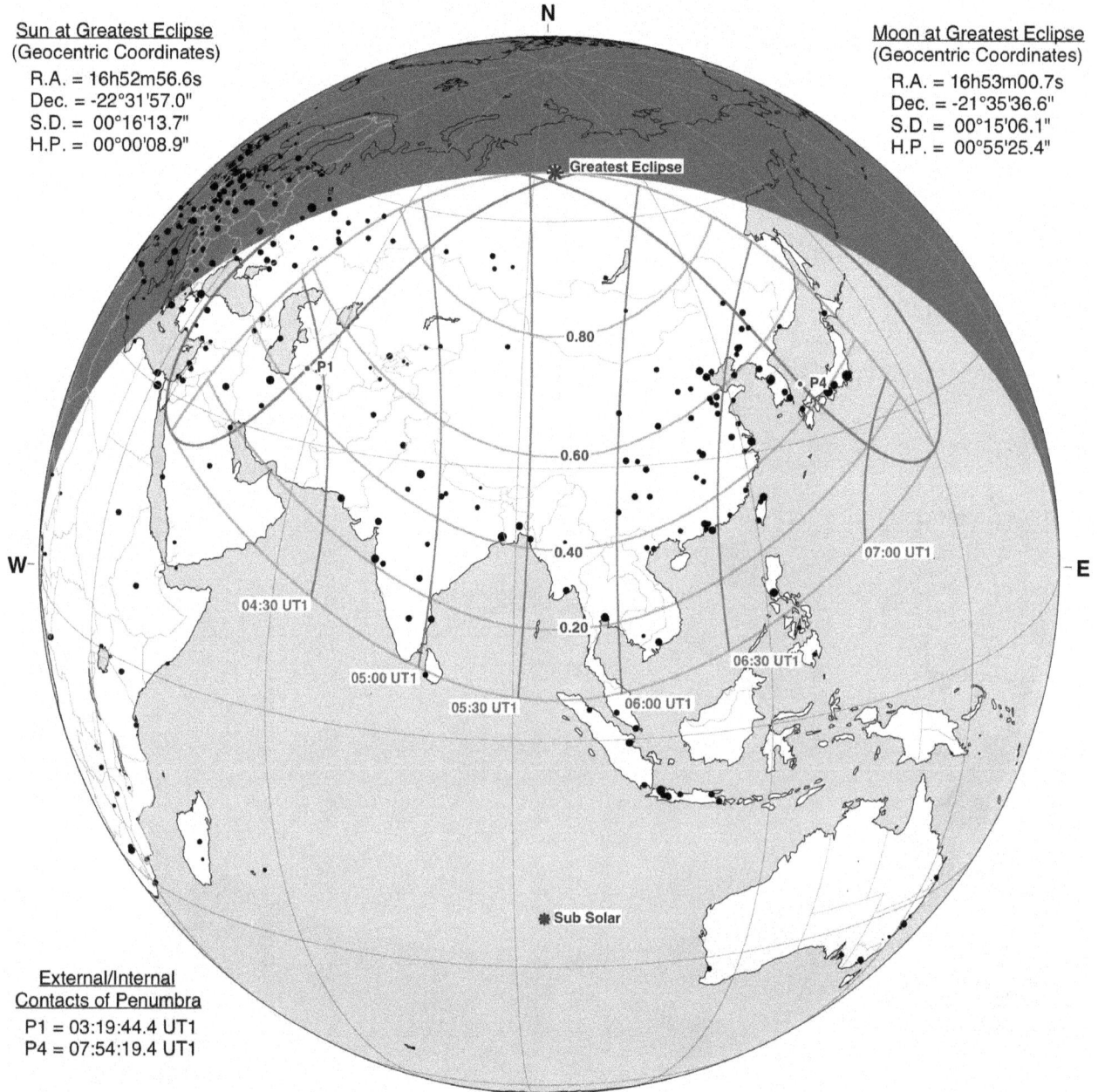

N

Greatest Eclipse

0.80

0.60

P1

P4

0.40

W   E

07:00 UT1

04:30 UT1

0.20

05:00 UT1

06:30 UT1

05:30 UT1

06:00 UT1

Sub Solar

<u>External/Internal</u>
<u>Contacts of Penumbra</u>
P1 = 03:19:44.4 UT1
P4 = 07:54:19.4 UT1

ΔT = 111.4 s    S    Eph. = JPL DE405

<u>Circumstances at Greatest Eclipse: 05:37:04.0 UT1</u>
Lat. = 67°26.7'N      Sun Alt. = 0.0°
Long. = 095°59.5'E      Sun Azm. = 182.3°

0   1000   2000   3000   4000   5000
Kilometers

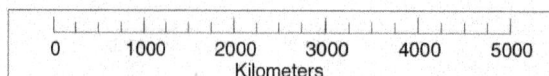

©2016 F. Espenak
www.EclipseWise.com

# Partial Solar Eclipse of 2087 May 02

Greatest Eclipse = 18:04:42.0 TD  (= 18:02:50.3 UT1)

Eclipse Magnitude = 0.8011          Saros Series = 120
Gamma = 1.1139          Saros Member = 65 of 71

**Sun at Greatest Eclipse**
(Geocentric Coordinates)
R.A. = 02h40m34.0s
Dec. = +15°36'24.7"
S.D. = 00°15'52.3"
H.P. = 00°00'08.7"

**Moon at Greatest Eclipse**
(Geocentric Coordinates)
R.A. = 02h39m33.9s
Dec. = +16°43'04.4"
S.D. = 00°16'43.4"
H.P. = 01°01'22.6"

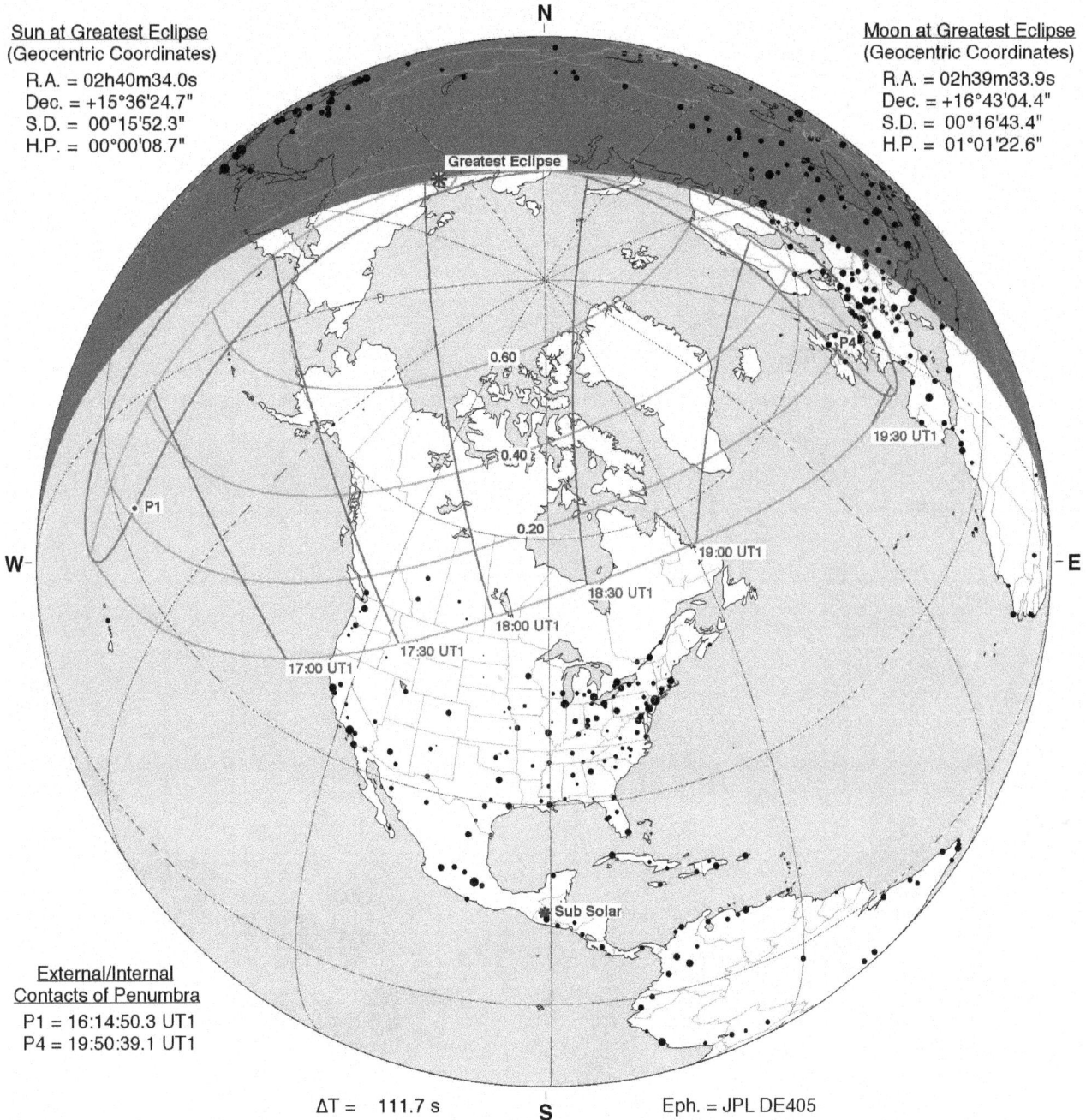

N

Greatest Eclipse

0.60
0.40
0.20

P1

P4

19:30 UT1
19:00 UT1
18:30 UT1
18:00 UT1
17:30 UT1
17:00 UT1

W

E

Sub Solar

**External/Internal
Contacts of Penumbra**
P1 = 16:14:50.3 UT1
P4 = 19:50:39.1 UT1

ΔT =   111.7 s          S          Eph. = JPL DE405

Circumstances at Greatest Eclipse:  18:02:50.3 UT1
Lat. =  70°17.6'N          Sun Alt. =   0.0°
Long. = 127°20.2'E          Sun Azm. =  37.1°

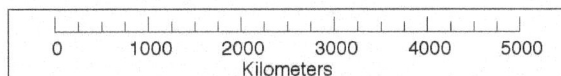

0   1000   2000   3000   4000   5000
Kilometers

©2016  F. Espenak
www.EclipseWise.com

# Partial Solar Eclipse of 2087 Jun 01

Greatest Eclipse = 01:27:14.3 TD   (= 01:25:22.5 UT1)

| | |
|---|---|
| Eclipse Magnitude = 0.2146 | Saros Series = 158 |
| Gamma = -1.4186 | Saros Member = 2 of 70 |

**Sun at Greatest Eclipse**
(Geocentric Coordinates)
R.A. = 04h37m04.0s
Dec. = +22°03'32.2"
S.D. = 00°15'46.5"
H.P. = 00°00'08.7"

**Moon at Greatest Eclipse**
(Geocentric Coordinates)
R.A. = 04h37m20.9s
Dec. = +20°37'32.1"
S.D. = 00°16'34.5"
H.P. = 01°00'49.8"

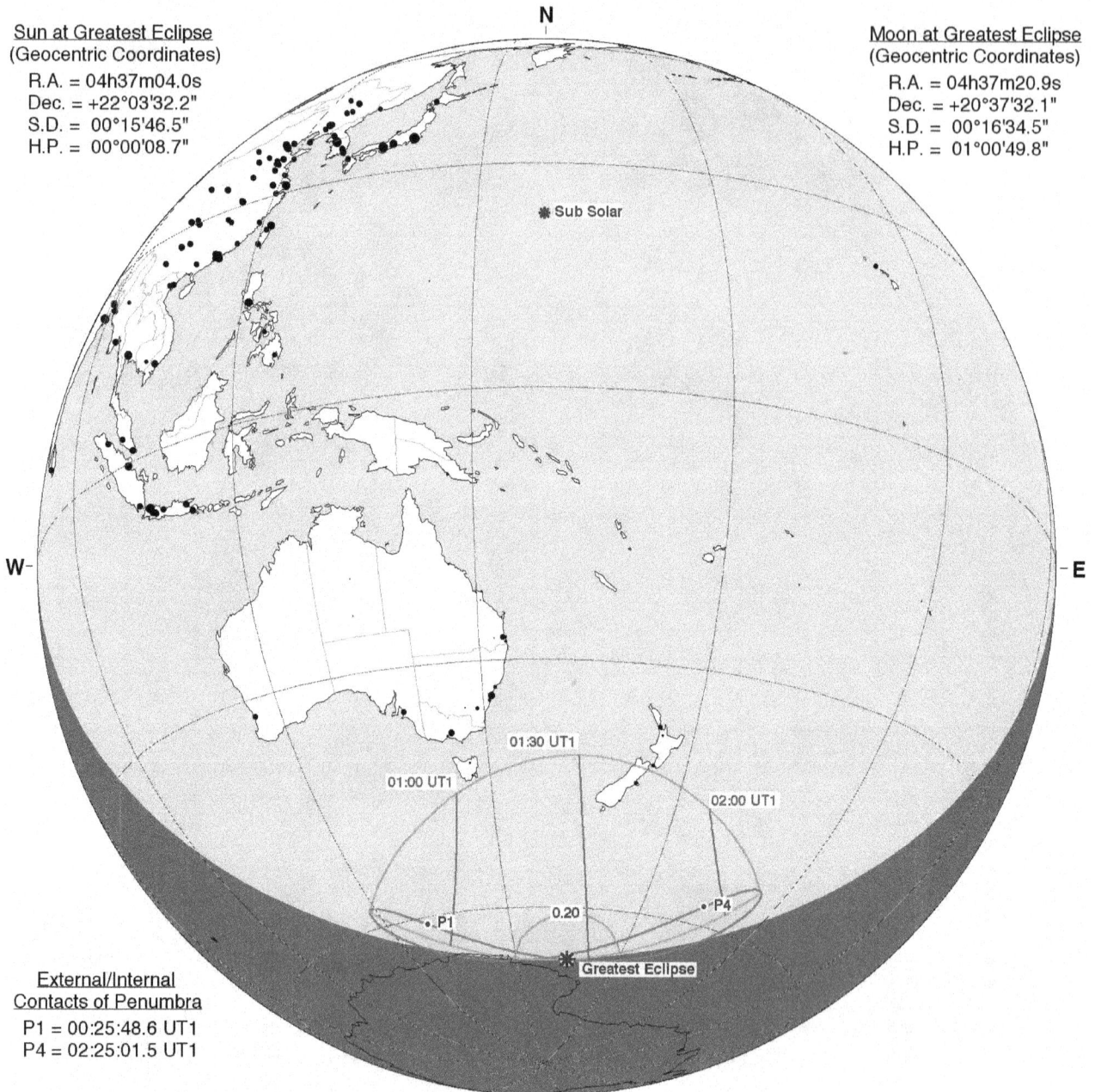

N

❋ Sub Solar

W — — E

01:30 UT1

01:00 UT1

02:00 UT1

0.20

• P1

• P4

❋ Greatest Eclipse

**External/Internal**
**Contacts of Penumbra**
P1 = 00:25:48.6 UT1
P4 = 02:25:01.5 UT1

ΔT = 111.8 s          S          Eph. = JPL DE405

Circumstances at Greatest Eclipse: 01:25:22.5 UT1
Lat. = 67°47.3'S          Sun Alt. = 0.0°
Long. = 165°07.4'E          Sun Azm. = 353.5°

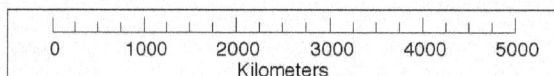

| 0 | 1000 | 2000 | 3000 | 4000 | 5000 |
|---|------|------|------|------|------|

Kilometers

# Partial Solar Eclipse of 2087 Oct 26

Greatest Eclipse = 11:46:56.7 TD   (= 11:45:04.6 UT1)

Eclipse Magnitude = 0.4696          Saros Series = 125
Gamma = -1.2882                      Saros Member = 58 of 73

Sun at Greatest Eclipse
(Geocentric Coordinates)
R.A. = 14h04m17.3s
Dec. = -12°36'18.7"
S.D. = 00°16'05.1"
H.P. = 00°00'08.8"

Moon at Greatest Eclipse
(Geocentric Coordinates)
R.A. = 14h03m06.4s
Dec. = -13°43'47.4"
S.D. = 00°14'46.3"
H.P. = 00°54'12.8"

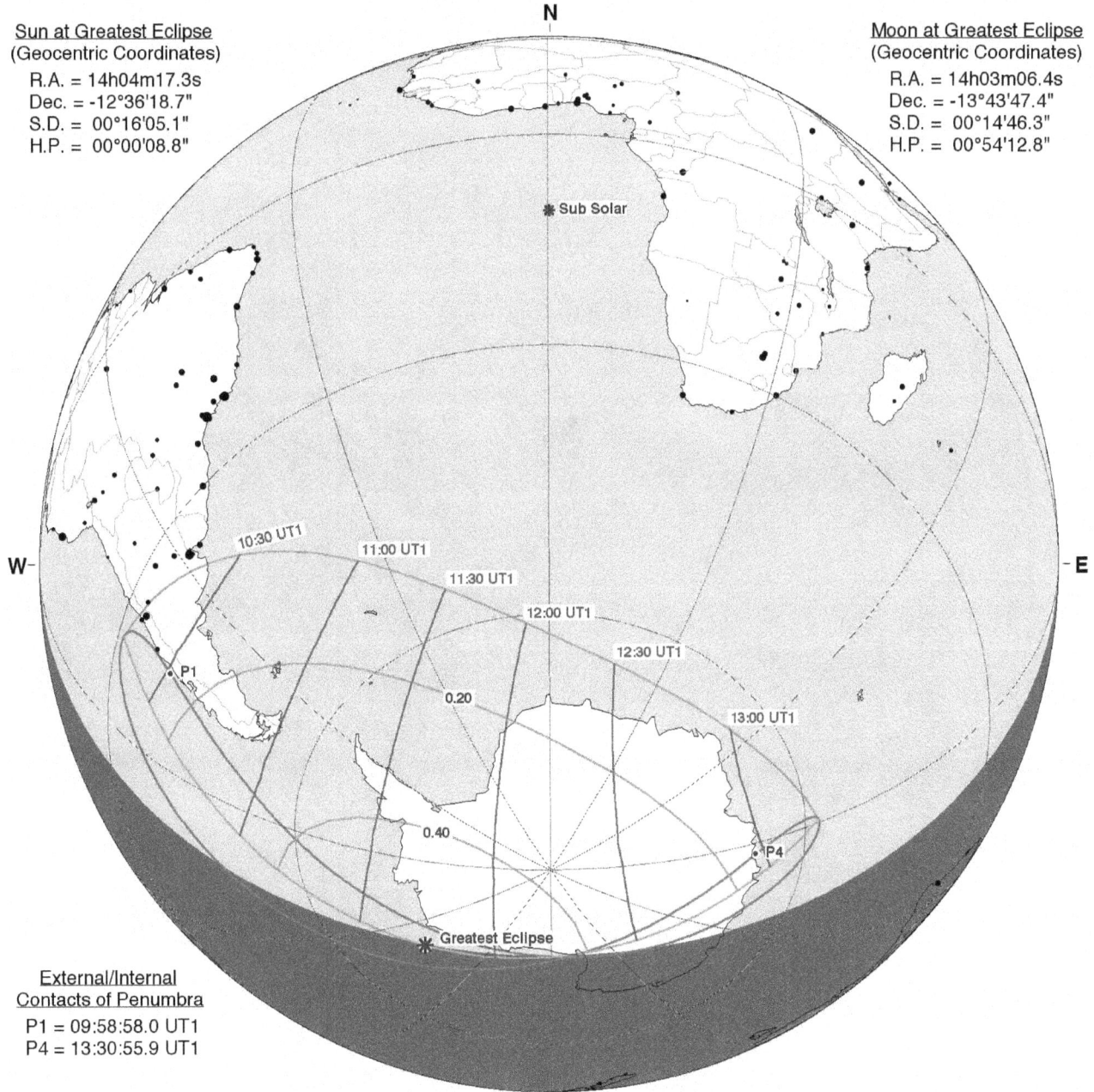

N

Sub Solar

10:30 UT1
11:00 UT1
11:30 UT1
12:00 UT1
12:30 UT1
13:00 UT1

W

E

P1

0.20

0.40

P4

Greatest Eclipse

External/Internal
Contacts of Penumbra
P1 = 09:58:58.0 UT1
P4 = 13:30:55.9 UT1

ΔT = 112.1 s       S       Eph. = JPL DE405

Circumstances at Greatest Eclipse: 11:45:04.6 UT1
Lat. = 71°01.3'S          Sun Alt. =  0.0°
Long. = 130°47.4'W        Sun Azm. = 132.1°

0    1000   2000   3000   4000   5000
Kilometers

©2016  F. Espenak
www.EclipseWise.com

# Total Solar Eclipse of 2088 Apr 21

Greatest Eclipse = 10:31:49.5 TD   (= 10:29:56.9 UT1)

| | |
|---|---|
| Eclipse Magnitude = 1.0474 | Saros Series = 130 |
| Gamma = 0.4135 | Saros Member = 56 of 73 |

**Sun at Greatest Eclipse**
(Geocentric Coordinates)
R.A. = 02h00m30.3s
Dec. = +12°16'07.2"
S.D. = 00°15'54.9"
H.P. = 00°00'08.8"

**Moon at Greatest Eclipse**
(Geocentric Coordinates)
R.A. = 02h00m04.9s
Dec. = +12°40'11.3"
S.D. = 00°16'25.1"
H.P. = 01°00'15.3"

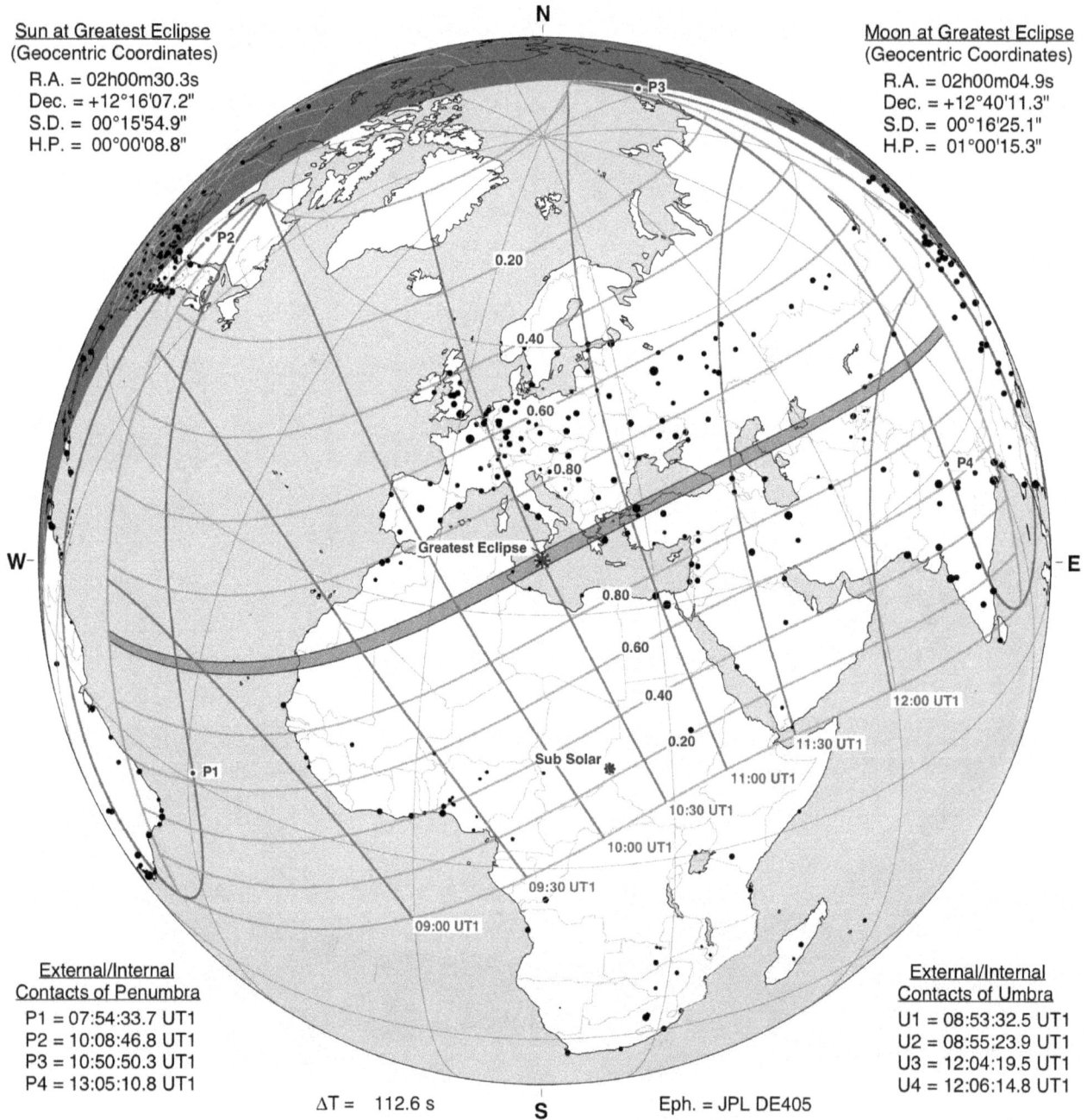

**External/Internal Contacts of Penumbra**
P1 = 07:54:33.7 UT1
P2 = 10:08:46.8 UT1
P3 = 10:50:50.3 UT1
P4 = 13:05:10.8 UT1

**External/Internal Contacts of Umbra**
U1 = 08:53:32.5 UT1
U2 = 08:55:23.9 UT1
U3 = 12:04:19.5 UT1
U4 = 12:06:14.8 UT1

ΔT = 112.6 s          Eph. = JPL DE405

**Circumstances at Greatest Eclipse: 10:29:56.9 UT1**

| | |
|---|---|
| Lat. = 35°59.8'N | Sun Alt. = 65.4° |
| Long. = 014°50.1'E | Sun Azm. = 162.6° |
| Path Width = 173.1 km | Duration = 03m57.8s |

**Circumstances at Greatest Duration: 10:33:08.9 UT1**

| | |
|---|---|
| Lat. = 36°31.2'N | Sun Alt. = 65.3° |
| Long. = 016°06.1'E | Sun Azm. = 167.6° |
| Path Width = 172.8 km | Duration = 03m58.0s |

0      1000      2000      3000      4000      5000
Kilometers

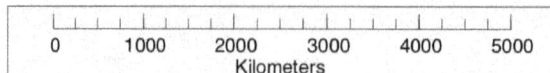

©2016  F. Espenak
www.EclipseWise.com

# Annular Solar Eclipse of 2088 Oct 14

Greatest Eclipse = 14:48:05.1 TD   (= 14:46:12.1 UT1)

| | |
|---|---|
| Eclipse Magnitude = 0.9727 | Saros Series = 135 |
| Gamma = -0.5349 | Saros Member = 43 of 71 |

**Sun at Greatest Eclipse**
(Geocentric Coordinates)
R.A. = 13h22m16.6s
Dec. = -08°39'19.3"
S.D. = 00°16'02.1"
H.P. = 00°00'08.8"

**Moon at Greatest Eclipse**
(Geocentric Coordinates)
R.A. = 13h21m42.6s
Dec. = -09°08'15.8"
S.D. = 00°15'23.5"
H.P. = 00°56'29.4"

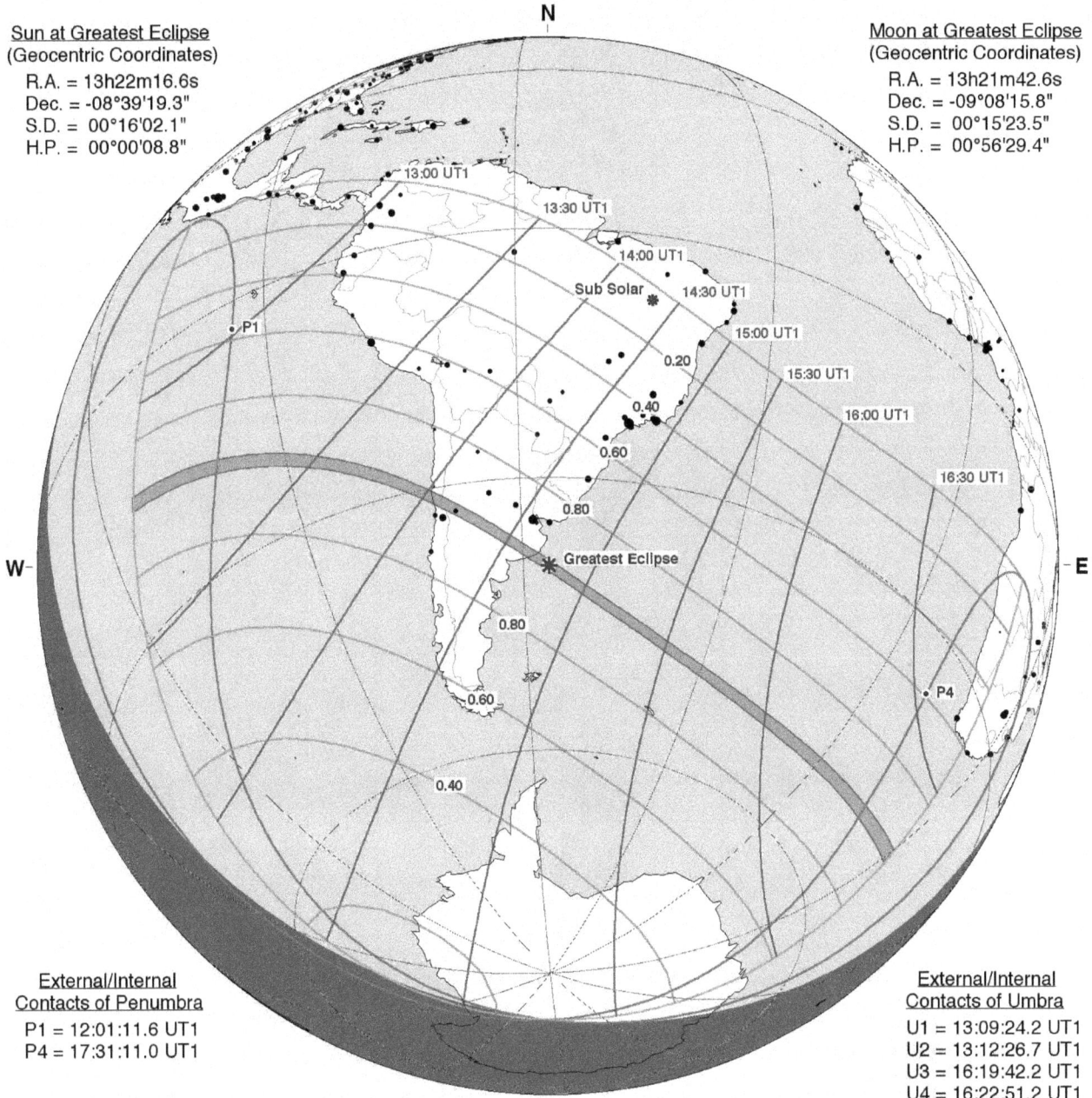

N

13:00 UT1
13:30 UT1
14:00 UT1
Sub Solar
14:30 UT1
15:00 UT1
0.20
15:30 UT1
0.40
16:00 UT1
0.60
16:30 UT1
0.80
Greatest Eclipse

P1

W

E

0.80

0.60

P4

0.40

**External/Internal**
**Contacts of Penumbra**
P1 = 12:01:11.6 UT1
P4 = 17:31:11.0 UT1

**External/Internal**
**Contacts of Umbra**
U1 = 13:09:24.2 UT1
U2 = 13:12:26.7 UT1
U3 = 16:19:42.2 UT1
U4 = 16:22:51.2 UT1

ΔT =   113.0 s

S

Eph. = JPL DE405

**Circumstances at Greatest Eclipse: 14:46:12.1 UT1**

| | |
|---|---|
| Lat. = 39°39.3'S | Sun Alt. = 57.4° |
| Long. = 056°15.9'W | Sun Azm. = 20.8° |
| Path Width = 115.2 km | Duration = 02m38.2s |

**Circumstances at Greatest Duration: 15:26:14.1 UT1**

| | |
|---|---|
| Lat. = 46°24.9'S | Sun Alt. = 49.8° |
| Long. = 039°00.3'W | Sun Azm. = 334.8° |
| Path Width = 119.0 km | Duration = 02m39.8s |

0    1000   2000   3000   4000   5000
Kilometers

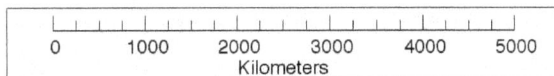

©2016  F. Espenak
www.EclipseWise.com

# Annular Solar Eclipse of 2089 Apr 10

Greatest Eclipse = 22:44:41.5 TD   (= 22:42:48.1 UT1)

Eclipse Magnitude = 0.9919          Saros Series = 140
Gamma = -0.3319                     Saros Member = 33 of 71

Sun at Greatest Eclipse
(Geocentric Coordinates)
R.A. = 01h20m36.9s
Dec. = +08°29'24.5"
S.D. = 00°15'57.8"
H.P. = 00°00'08.8"

Moon at Greatest Eclipse
(Geocentric Coordinates)
R.A. = 01h20m58.3s
Dec. = +08°11'12.9"
S.D. = 00°15'35.9"
H.P. = 00°57'14.7"

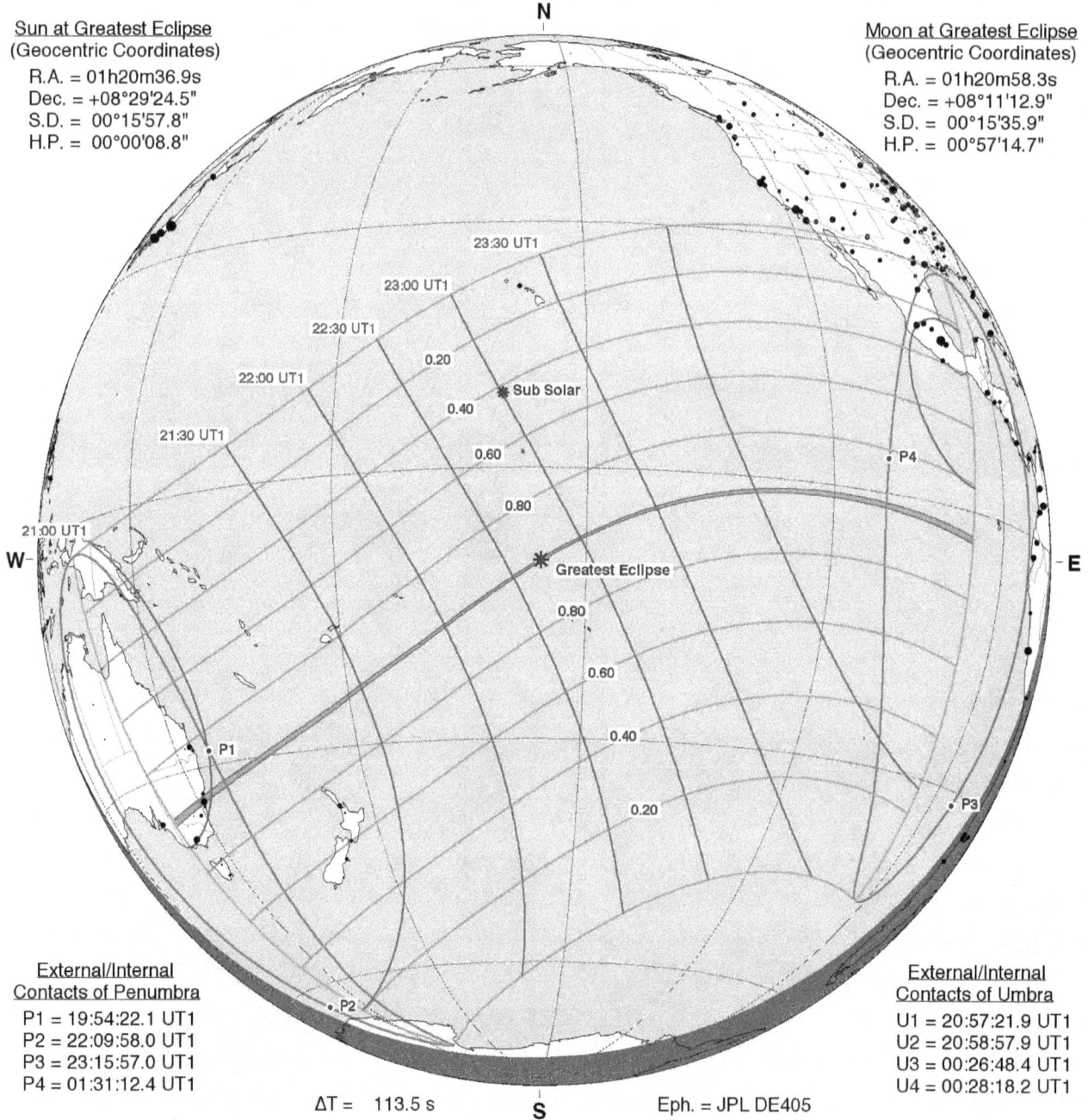

External/Internal
Contacts of Penumbra
P1 = 19:54:22.1 UT1
P2 = 22:09:58.0 UT1
P3 = 23:15:57.0 UT1
P4 = 01:31:12.4 UT1

External/Internal
Contacts of Umbra
U1 = 20:57:21.9 UT1
U2 = 20:58:57.9 UT1
U3 = 00:26:48.4 UT1
U4 = 00:28:18.2 UT1

ΔT = 113.5 s          Eph. = JPL DE405

Circumstances at Greatest Eclipse: 22:42:48.1 UT1
Lat. = 10°11.8'S        Sun Alt. = 70.6°
Long. = 155°02.6'W      Sun Azm. = 343.7°
Path Width = 30.0 km    Duration = 00m53.2s

Circumstances at Greatest Duration: 20:58:09.9 UT1
Lat. = 35°22.6'S        Sun Alt. = 0.0°
Long. = 141°46.8'E      Sun Azm. = 79.6°
Path Width = 89.8 km    Duration = 01m32.4s

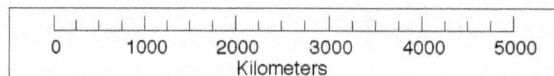

©2016 F. Espenak
www.EclipseWise.com

# Total Solar Eclipse of 2089 Oct 04

Greatest Eclipse = 01:15:23.2 TD  (= 01:13:29.3 UT1)

| | |
|---|---|
| Eclipse Magnitude = 1.0333 | Saros Series = 145 |
| Gamma = 0.2167 | Saros Member = 26 of 77 |

Sun at Greatest Eclipse
(Geocentric Coordinates)
R.A. = 12h42m34.2s
Dec. = -04°34'29.0"
S.D. = 00°15'59.1"
H.P. = 00°00'08.8"

Moon at Greatest Eclipse
(Geocentric Coordinates)
R.A. = 12h42m49.6s
Dec. = -04°22'10.5"
S.D. = 00°16'15.0"
H.P. = 00°59'38.5"

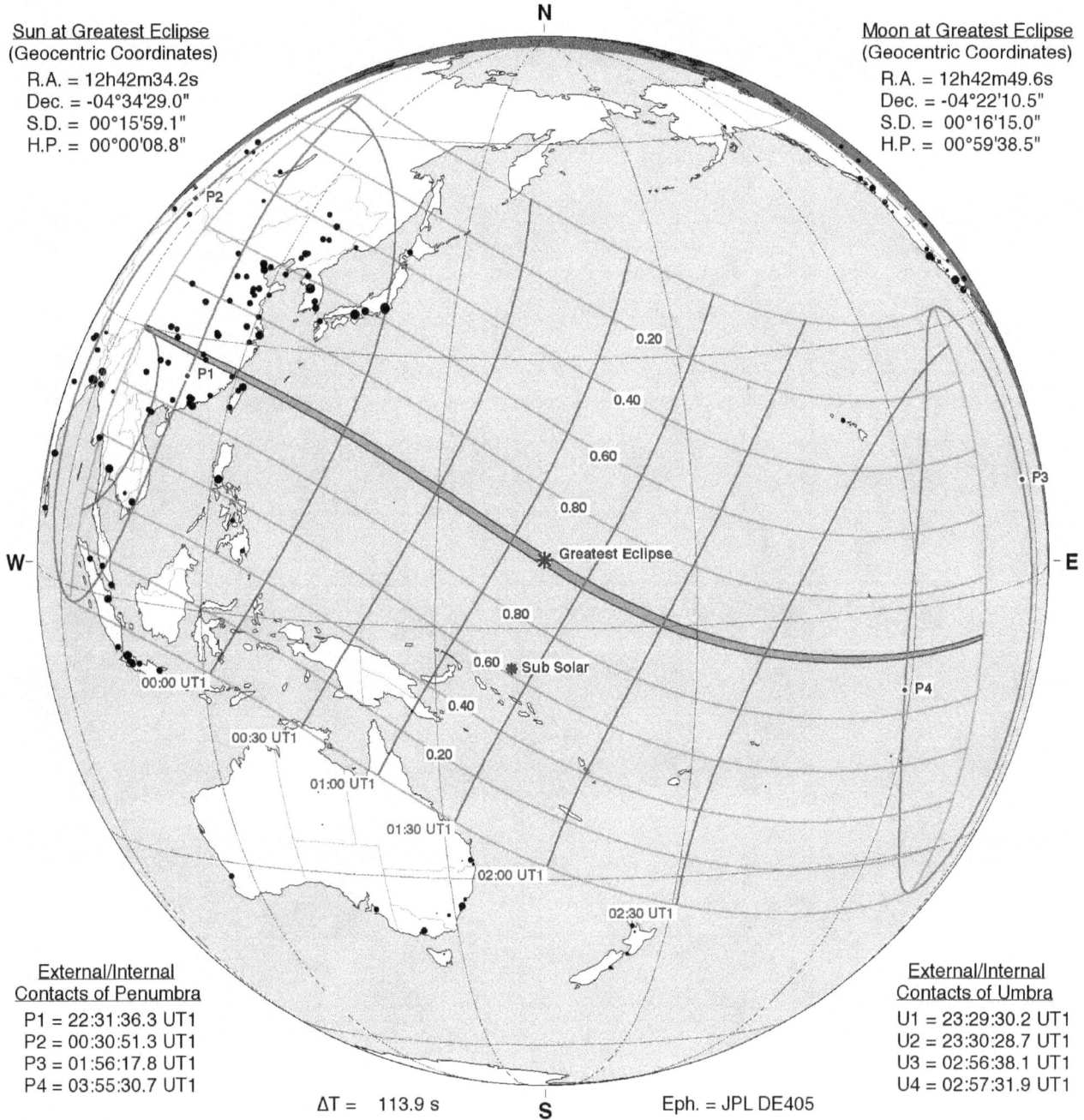

External/Internal
Contacts of Penumbra
P1 = 22:31:36.3 UT1
P2 = 00:30:51.3 UT1
P3 = 01:56:17.8 UT1
P4 = 03:55:30.7 UT1

External/Internal
Contacts of Umbra
U1 = 23:29:30.2 UT1
U2 = 23:30:28.7 UT1
U3 = 02:56:38.1 UT1
U4 = 02:57:31.9 UT1

ΔT =   113.9 s

Eph. = JPL DE405

Circumstances at Greatest Eclipse: 01:13:29.3 UT1

| | |
|---|---|
| Lat. = 07°25.5'N | Sun Alt. =  77.4° |
| Long. = 162°31.1'E | Sun Azm. = 197.3° |
| Path Width = 114.9 km | Duration = 03m14.2s |

Circumstances at Greatest Duration: 01:13:02.0 UT1

| | |
|---|---|
| Lat. = 07°29.9'N | Sun Alt. =  77.4° |
| Long. = 162°23.7'E | Sun Azm. = 196.2° |
| Path Width = 114.9 km | Duration = 03m14.2s |

©2016  F. Espenak
www.EclipseWise.com

# Partial Solar Eclipse of 2090 Mar 31

Greatest Eclipse = 03:38:07.9 TD   (= 03:36:13.5 UT1)

Eclipse Magnitude = 0.7843          Saros Series = 150
Gamma = -1.1028                     Saros Member = 21 of 71

**Sun at Greatest Eclipse**
(Geocentric Coordinates)
R.A. = 00h40m11.0s
Dec. = +04°19'18.8"
S.D. = 00°16'00.8"
H.P. = 00°00'08.8"

**Moon at Greatest Eclipse**
(Geocentric Coordinates)
R.A. = 00h41m23.0s
Dec. = +03°22'02.4"
S.D. = 00°14'52.2"
H.P. = 00°54'34.6"

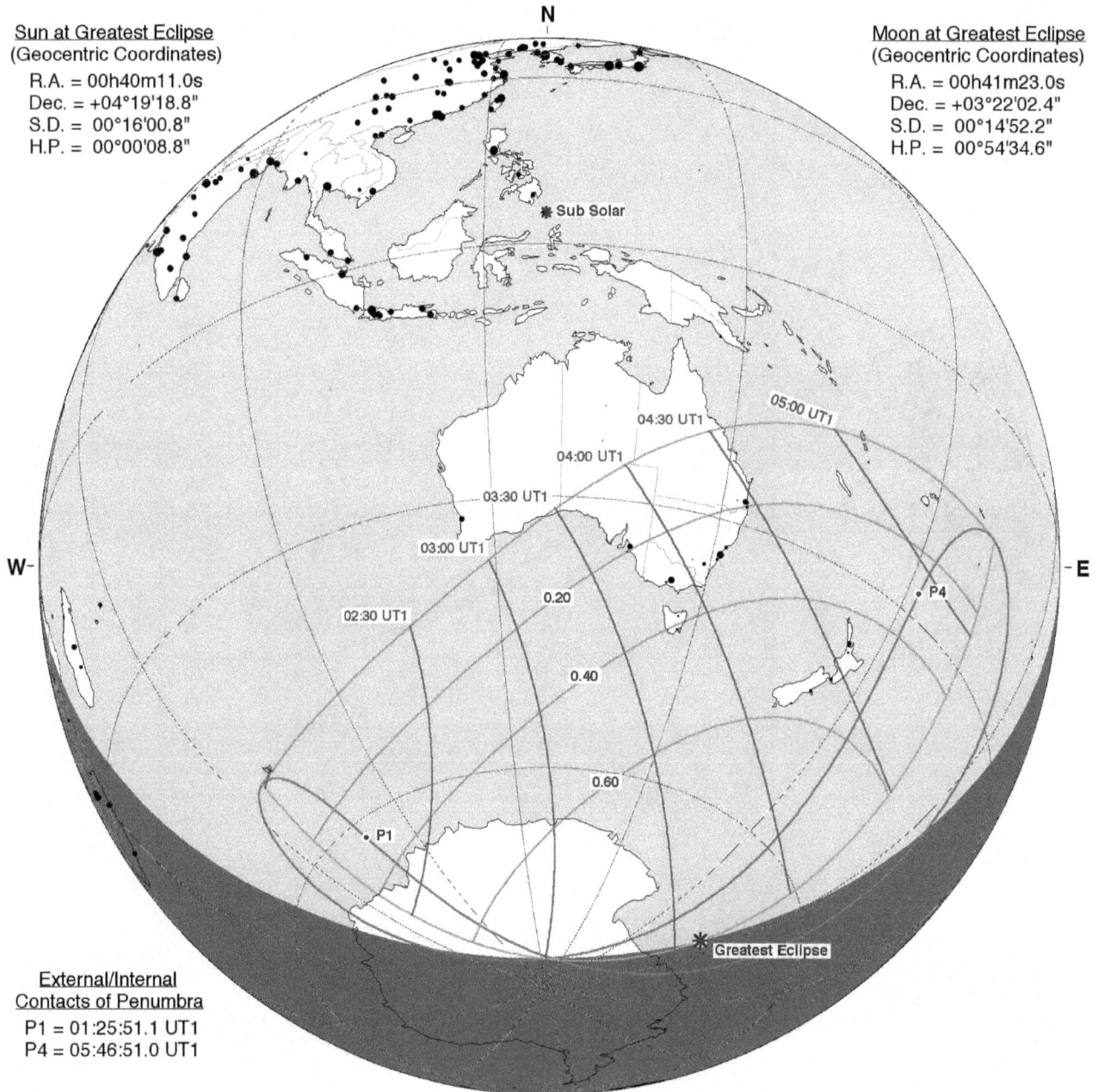

**External/Internal Contacts of Penumbra**
P1 = 01:25:51.1 UT1
P4 = 05:46:51.0 UT1

ΔT = 114.3 s          Eph. = JPL DE405

**Circumstances at Greatest Eclipse: 03:36:13.5 UT1**
Lat. = 72°03.9'S          Sun Alt. = 0.0°
Long. = 156°36.2'W        Sun Azm. = 284.2°

| | | | | | |
|---|---|---|---|---|---|
| 0 | 1000 | 2000 | 3000 | 4000 | 5000 |

Kilometers

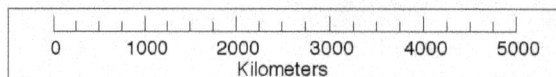

©2016  F. Espenak
www.EclipseWise.com

# Total Solar Eclipse of 2090 Sep 23

Greatest Eclipse = 16:56:36.3 TD  (= 16:54:41.5 UT1)

Eclipse Magnitude = 1.0562          Saros Series = 155
Gamma = 0.9157          Saros Member = 10 of 71

Sun at Greatest Eclipse
(Geocentric Coordinates)
R.A. = 12h04m19.6s
Dec. = -00°28'06.5"
S.D. = 00°15'56.2"
H.P. = 00°00'08.8"

Moon at Greatest Eclipse
(Geocentric Coordinates)
R.A. = 12h05m28.3s
Dec. = +00°25'15.5"
S.D. = 00°16'43.4"
H.P. = 01°01'22.6"

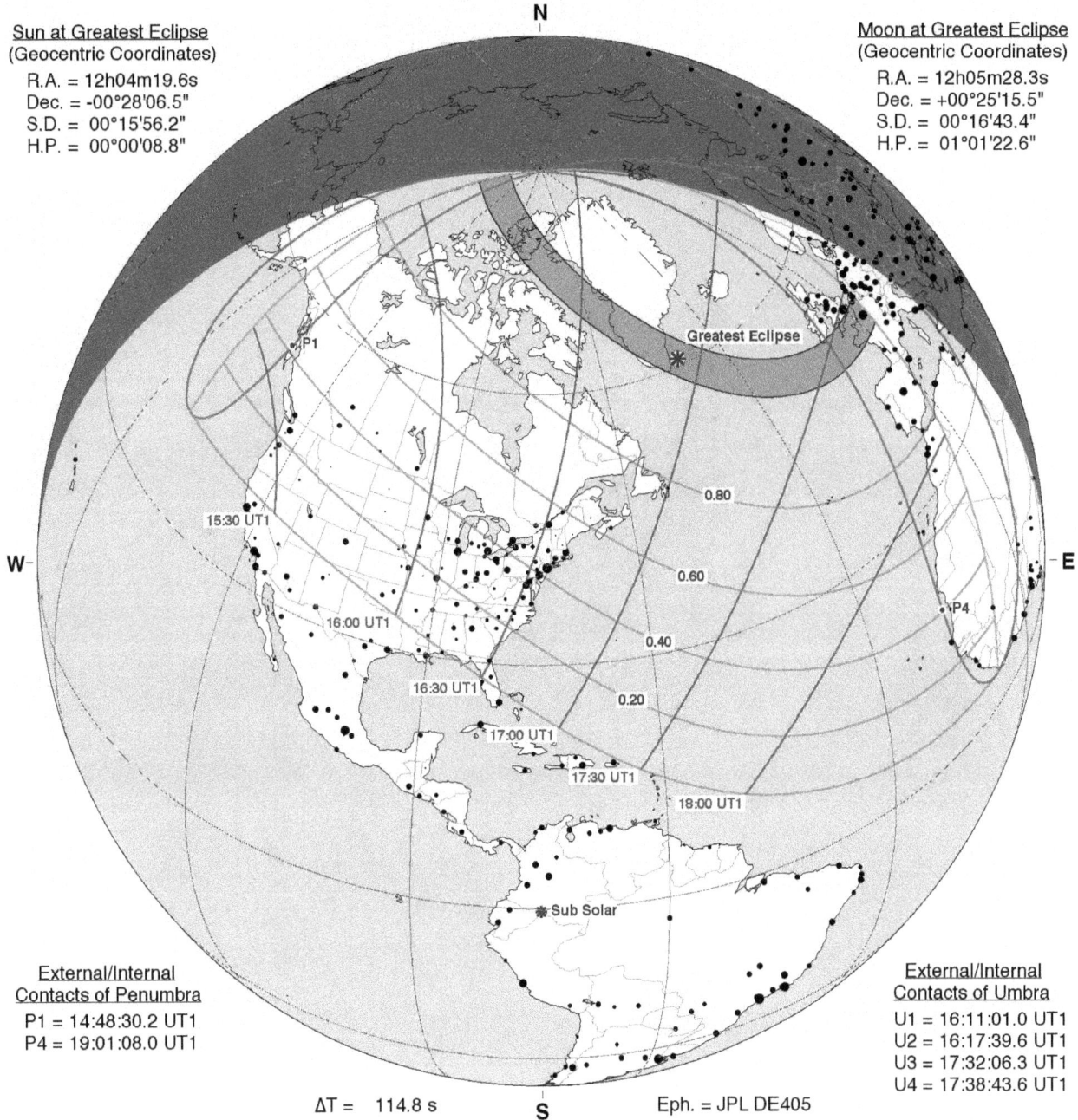

N

Greatest Eclipse

0.80
0.60
0.40
0.20

P1

15:30 UT1
16:00 UT1
16:30 UT1
17:00 UT1
17:30 UT1
18:00 UT1

P4

W

E

Sub Solar

External/Internal
Contacts of Penumbra
P1 = 14:48:30.2 UT1
P4 = 19:01:08.0 UT1

External/Internal
Contacts of Umbra
U1 = 16:11:01.0 UT1
U2 = 16:17:39.6 UT1
U3 = 17:32:06.3 UT1
U4 = 17:38:43.6 UT1

ΔT = 114.8 s          S          Eph. = JPL DE405

Circumstances at Greatest Eclipse: 16:54:41.5 UT1

Lat. = 60°41.2'N          Sun Alt. = 23.2°
Long. = 040°45.3'W          Sun Azm. = 218.5°
Path Width = 462.9 km          Duration = 03m35.9s

Circumstances at Greatest Duration: 16:54:48.5 UT1

Lat. = 60°38.7'N          Sun Alt. = 23.2°
Long. = 040°40.8'W          Sun Azm. = 218.6°
Path Width = 463.1 km          Duration = 03m35.9s

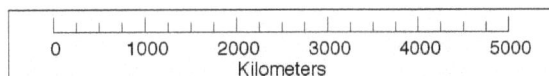

0          1000          2000          3000          4000          5000
Kilometers

# Partial Solar Eclipse of 2091 Feb 18

Greatest Eclipse = 09:54:39.8 TD (= 09:52:44.6 UT1)

| | |
|---|---|
| Eclipse Magnitude = 0.6558 | Saros Series = 122 |
| Gamma = 1.1779 | Saros Member = 62 of 70 |

**Sun at Greatest Eclipse**
(Geocentric Coordinates)
R.A. = 22h08m17.5s
Dec. = -11°28'13.5"
S.D. = 00°16'11.1"
H.P. = 00°00'08.9"

**Moon at Greatest Eclipse**
(Geocentric Coordinates)
R.A. = 22h07m09.8s
Dec. = -10°25'58.4"
S.D. = 00°14'56.6"
H.P. = 00°54'50.7"

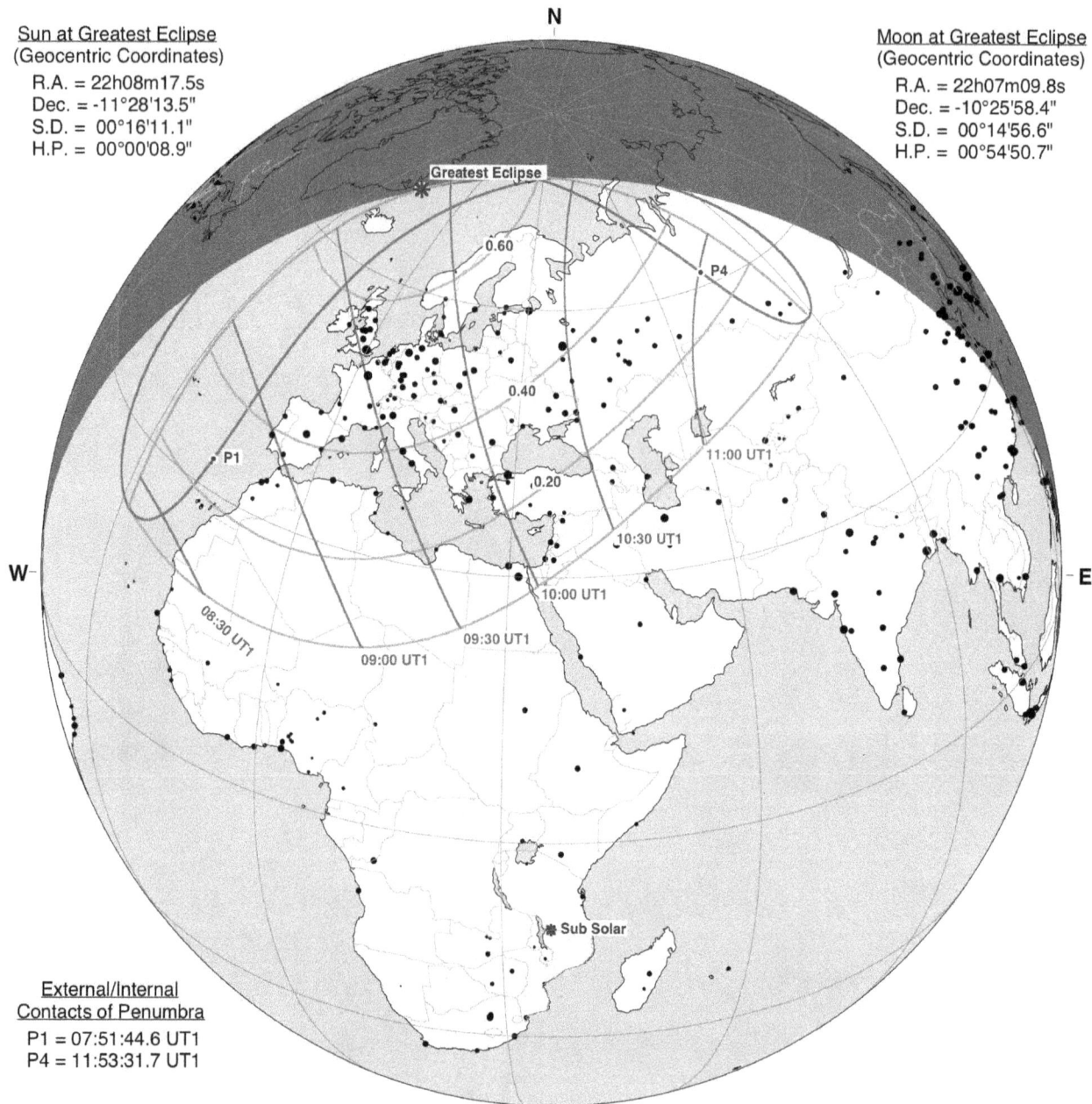

**External/Internal**
**Contacts of Penumbra**
P1 = 07:51:44.6 UT1
P4 = 11:53:31.7 UT1

ΔT = 115.1 s          Eph. = JPL DE405

**Circumstances at Greatest Eclipse: 09:52:44.6 UT1**
Lat. = 71°13.5'N          Sun Alt. = 0.0°
Long. = 018°02.1'W          Sun Azm. = 128.2°

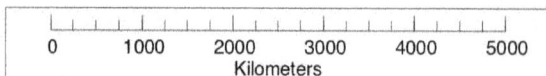

©2016 F. Espenak
www.EclipseWise.com

# Total Solar Eclipse of  2091 Aug 15

Greatest Eclipse = 00:34:42.9 TD   (= 00:32:47.3 UT1)

Eclipse Magnitude = 1.0216          Saros Series = 127
Gamma = -0.9490          Saros Member = 62 of 82

Sun at Greatest Eclipse
(Geocentric Coordinates)
R.A. = 09h39m24.9s
Dec. = +14°00'16.0"
S.D. = 00°15'47.2"
H.P. = 00°00'08.7"

Moon at Greatest Eclipse
(Geocentric Coordinates)
R.A. = 09h38m32.0s
Dec. = +13°05'59.9"
S.D. = 00°16'03.2"
H.P. = 00°58'54.9"

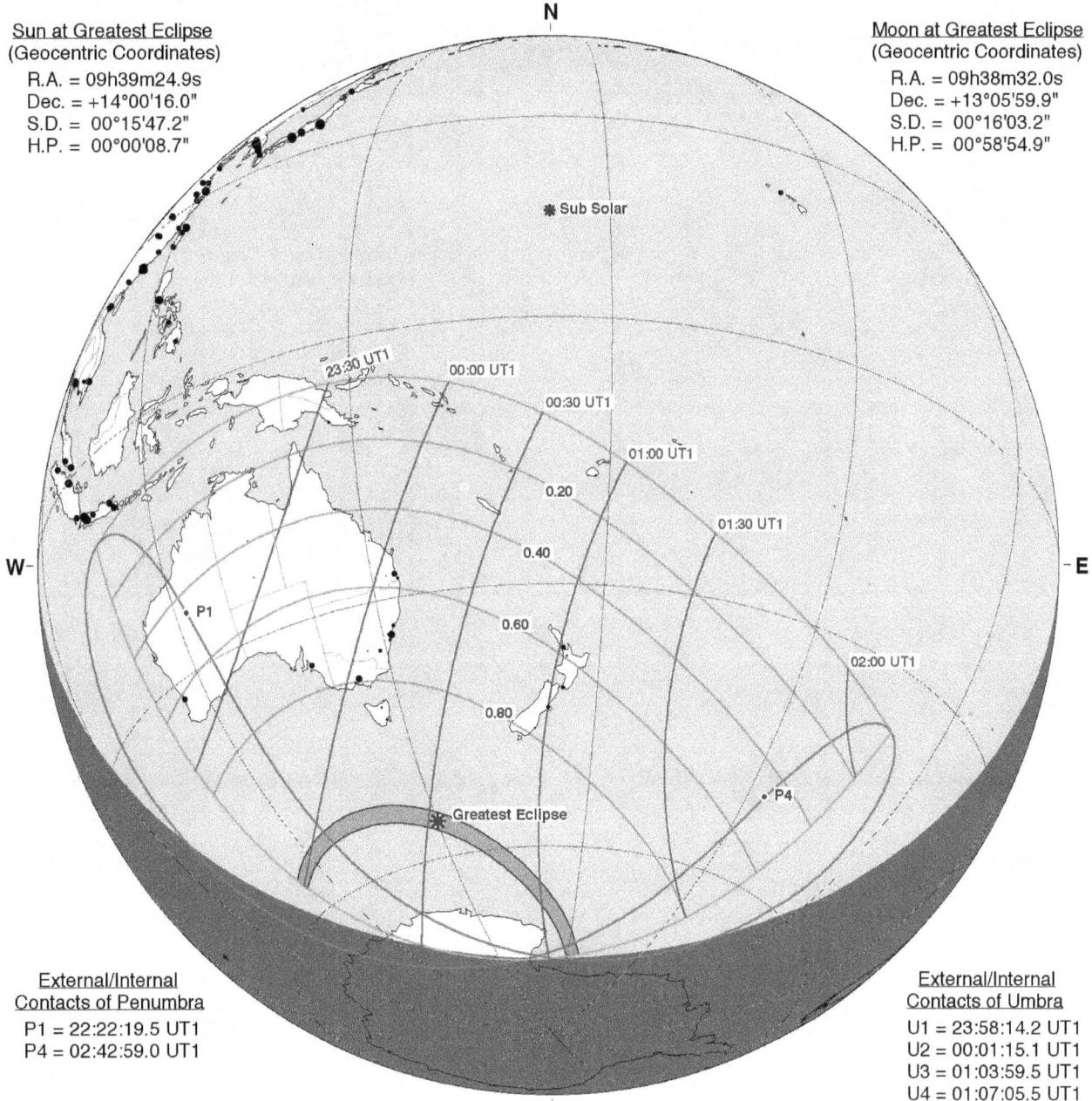

N

✳ Sub Solar

23:30 UT1
00:00 UT1
00:30 UT1
01:00 UT1
0.20
01:30 UT1
0.40
W — — E
0.60
02:00 UT1
0.80
P1
P4
✳ Greatest Eclipse

External/Internal
Contacts of Penumbra
P1 = 22:22:19.5 UT1
P4 = 02:42:59.0 UT1

External/Internal
Contacts of Umbra
U1 = 23:58:14.2 UT1
U2 = 00:01:15.1 UT1
U3 = 01:03:59.5 UT1
U4 = 01:07:05.5 UT1

ΔT =   115.6 s          S          Eph. = JPL DE405

Circumstances at Greatest Eclipse:  00:32:47.3 UT1

| | |
|---|---|
| Lat. = 55°34.9'S | Sun Alt. =  17.8° |
| Long. = 150°12.1'E | Sun Azm. = 23.2° |
| Path Width = 236.5 km | Duration = 01m38.1s |

Circumstances at Greatest Duration:  00:33:18.2 UT1

| | |
|---|---|
| Lat. = 55°41.5'S | Sun Alt. =  17.8° |
| Long. = 150°31.0'E | Sun Azm. = 22.8° |
| Path Width = 236.0 km | Duration = 01m38.1s |

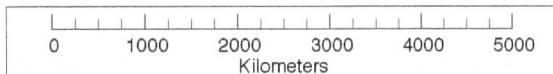

©2016  F. Espenak
www.EclipseWise.com

```
|   |   |   |   |   |   |
0  1000  2000  3000  4000  5000
           Kilometers
```

# Annular Solar Eclipse of 2092 Feb 07

Greatest Eclipse = 15:10:20.2 TD  (= 15:08:24.1 UT1)

Eclipse Magnitude = 0.9840          Saros Series = 132
Gamma = 0.4322          Saros Member = 50 of 71

Sun at Greatest Eclipse
(Geocentric Coordinates)
R.A. = 21h25m01.6s
Dec. = -15°10'15.3"
S.D. = 00°16'13.1"
H.P. = 00°00'08.9"

Moon at Greatest Eclipse
(Geocentric Coordinates)
R.A. = 21h24m39.5s
Dec. = -14°45'56.9"
S.D. = 00°15'43.8"
H.P. = 00°57'43.8"

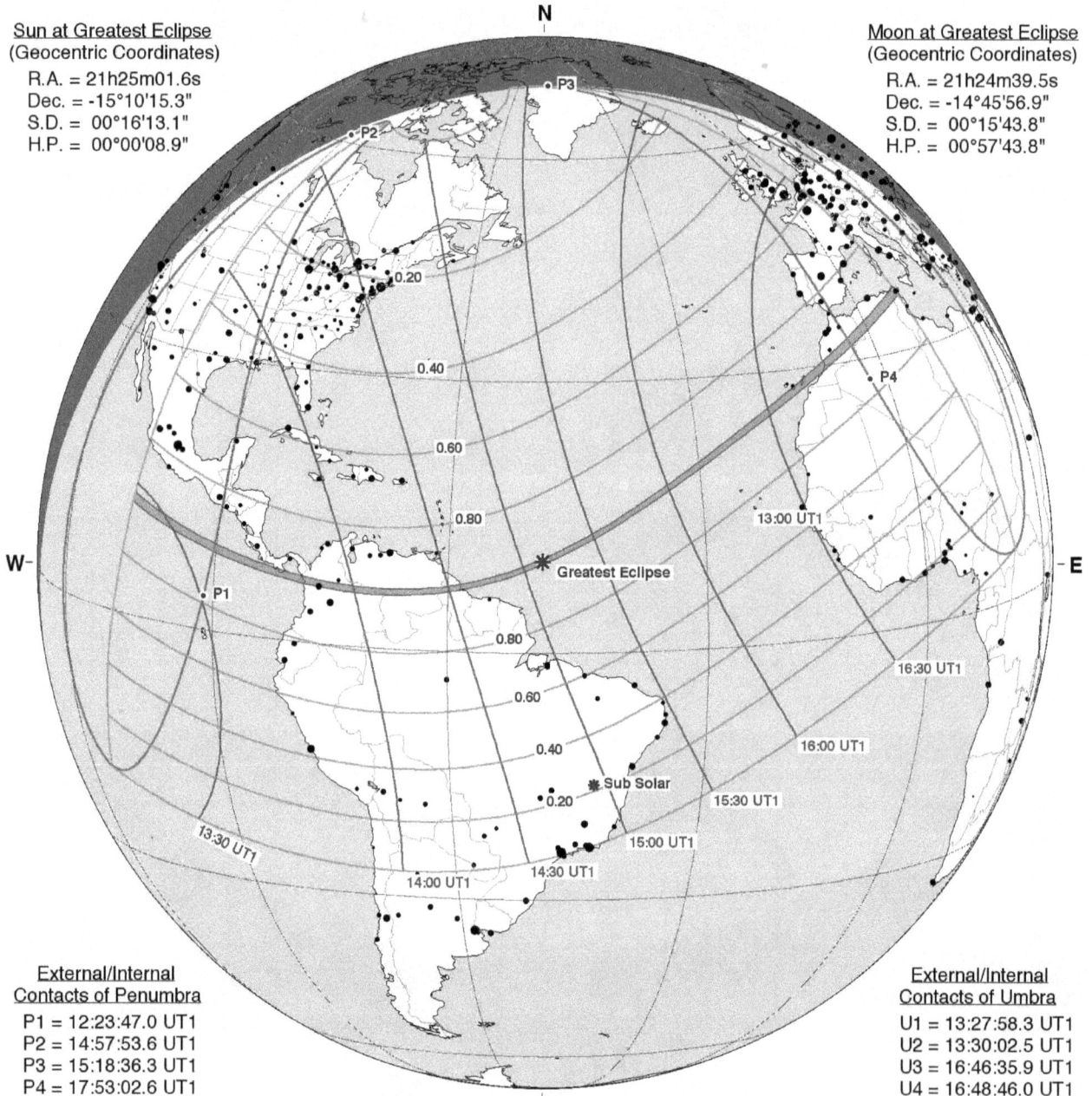

External/Internal
Contacts of Penumbra
P1 = 12:23:47.0 UT1
P2 = 14:57:53.6 UT1
P3 = 15:18:36.3 UT1
P4 = 17:53:02.6 UT1

External/Internal
Contacts of Umbra
U1 = 13:27:58.3 UT1
U2 = 13:30:02.5 UT1
U3 = 16:46:35.9 UT1
U4 = 16:48:46.0 UT1

ΔT = 116.0 s          Eph. = JPL DE405

Circumstances at Greatest Eclipse: 15:08:24.1 UT1

| | |
|---|---|
| Lat. = 09°56.2'N | Sun Alt. = 64.3° |
| Long. = 049°00.6'W | Sun Azm. = 167.9° |
| Path Width = 62.5 km | Duration = 01m47.7s |

Circumstances at Greatest Duration: 13:29:00.3 UT1

| | |
|---|---|
| Lat. = 12°50.4'N | Sun Alt. = 0.0° |
| Long. = 105°12.6'W | Sun Azm. = 105.6° |
| Path Width = 113.9 km | Duration = 01m57.5s |

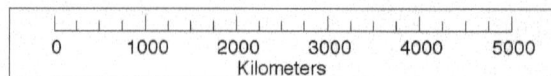

©2016  F. Espenak
www.EclipseWise.com

# Annular Solar Eclipse of 2092 Aug 03

Greatest Eclipse = 09:59:32.8 TD  (= 09:57:36.3 UT1)

| Eclipse Magnitude = 0.9794 | Saros Series = 137 |
|---|---|
| Gamma = -0.2044 | Saros Member = 40 of 70 |

**N**

Sun at Greatest Eclipse
(Geocentric Coordinates)
R.A. = 08h58m14.3s
Dec. = +17°09'21.7"
S.D. = 00°15'45.7"
H.P. = 00°00'08.7"

Moon at Greatest Eclipse
(Geocentric Coordinates)
R.A. = 08h58m05.6s
Dec. = +16°58'10.4"
S.D. = 00°15'12.2"
H.P. = 00°55'47.9"

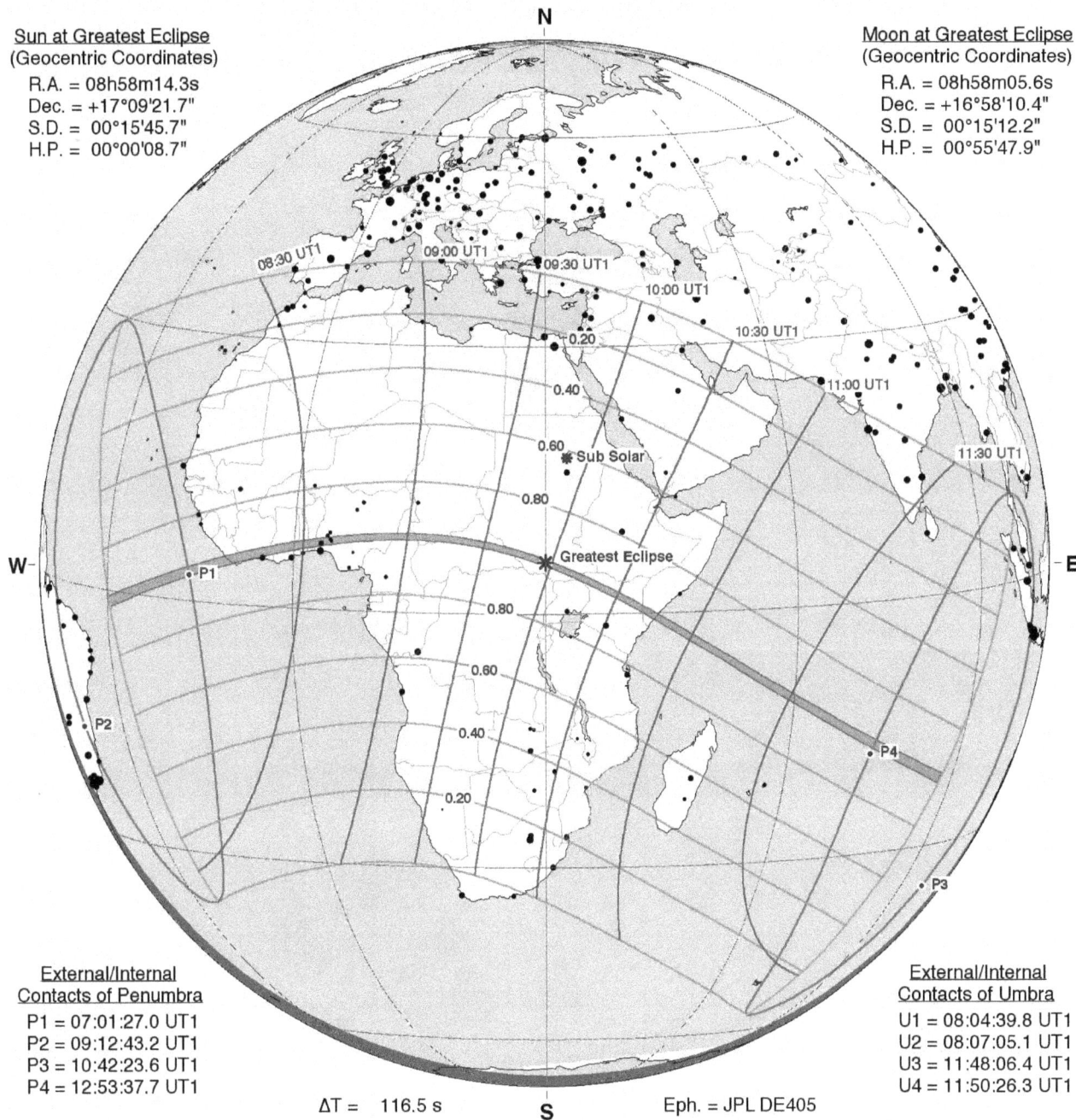

08:30 UT1  09:00 UT1  09:30 UT1  10:00 UT1  10:30 UT1  11:00 UT1  11:30 UT1

0.20  0.40  0.60 Sub Solar  0.80

**W**  P1  Greatest Eclipse  **E**

0.80  0.60  0.40  0.20

P2  P4  P3

External/Internal
Contacts of Penumbra
P1 = 07:01:27.0 UT1
P2 = 09:12:43.2 UT1
P3 = 10:42:23.6 UT1
P4 = 12:53:37.7 UT1

External/Internal
Contacts of Umbra
U1 = 08:04:39.8 UT1
U2 = 08:07:05.1 UT1
U3 = 11:48:06.4 UT1
U4 = 11:50:26.3 UT1

ΔT = 116.5 s

**S**

Eph. = JPL DE405

| Circumstances at Greatest Eclipse: 09:57:36.3 UT1 | | Circumstances at Greatest Duration: 09:16:14.1 UT1 | |
|---|---|---|---|
| Lat. = 05°35.1'N | Sun Alt. = 78.2° | Lat. = 08°03.3'N | Sun Alt. = 65.4° |
| Long. = 030°01.2'E | Sun Azm. = 10.1° | Long. = 019°05.1'E | Sun Azm. = 66.0° |
| Path Width = 74.7 km | Duration = 02m31.4s | Path Width = 80.1 km | Duration = 02m33.9s |

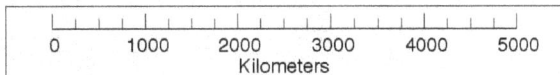

| 0 | 1000 | 2000 | 3000 | 4000 | 5000 |
|---|---|---|---|---|---|

Kilometers

©2016  F. Espenak
www.EclipseWise.com

261

# Total Solar Eclipse of 2093 Jan 27

Greatest Eclipse = 03:22:16.1 TD   (= 03:20:19.1 UT1)

| | |
|---|---|
| Eclipse Magnitude = 1.0340 | Saros Series = 142 |
| Gamma = -0.2737 | Saros Member = 27 of 72 |

**N**

Sun at Greatest Eclipse
(Geocentric Coordinates)
R.A. = 20h41m22.6s
Dec. = -18°16'28.3"
S.D. = 00°16'14.6"
H.P. = 00°00'08.9"

Moon at Greatest Eclipse
(Geocentric Coordinates)
R.A. = 20h41m33.8s
Dec. = -18°32'48.9"
S.D. = 00°16'31.4"
H.P. = 01°00'38.7"

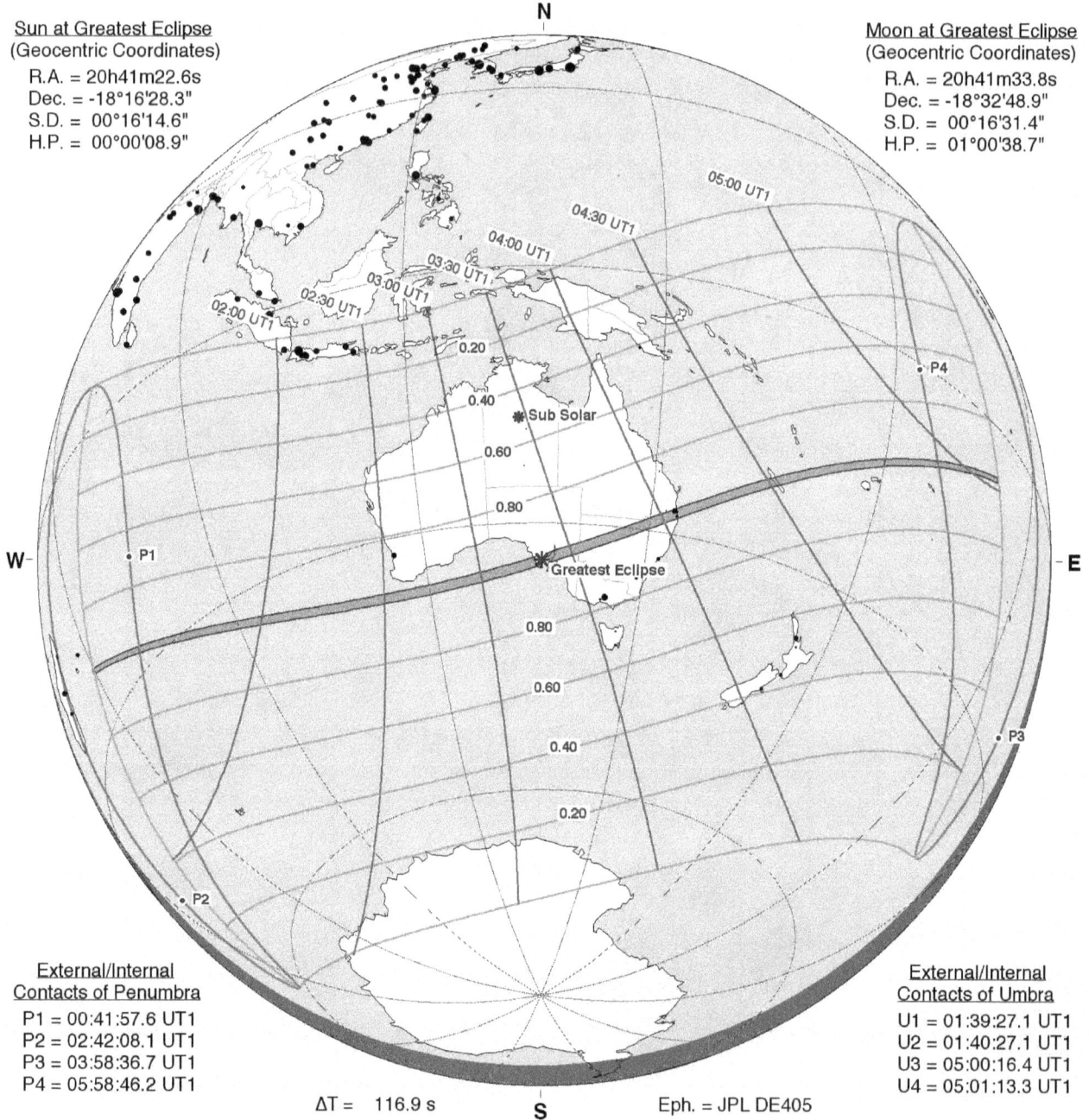

**W**

**E**

05:00 UT1
04:30 UT1
04:00 UT1
03:30 UT1
03:00 UT1
02:30 UT1
02:00 UT1

0.20
0.40
0.60
0.80
Sub Solar
Greatest Eclipse
0.80
0.60
0.40
0.20

P1
P2
P3
P4

External/Internal
Contacts of Penumbra
P1 = 00:41:57.6 UT1
P2 = 02:42:08.1 UT1
P3 = 03:58:36.7 UT1
P4 = 05:58:46.2 UT1

External/Internal
Contacts of Umbra
U1 = 01:39:27.1 UT1
U2 = 01:40:27.1 UT1
U3 = 05:00:16.4 UT1
U4 = 05:01:13.3 UT1

ΔT = 116.9 s

**S**

Eph. = JPL DE405

| Circumstances at Greatest Eclipse: 03:20:19.1 UT1 | | Circumstances at Greatest Duration: 03:17:17.4 UT1 | |
|---|---|---|---|
| Lat. = 34°07.3'S | Sun Alt. = 73.9° | Lat. = 34°25.2'S | Sun Alt. = 73.8° |
| Long. = 136°07.0'E | Sun Azm. = 349.5° | Long. = 134°53.8'E | Sun Azm. = 356.4° |
| Path Width = 119.3 km | Duration = 02m58.0s | Path Width = 119.2 km | Duration = 02m58.1s |

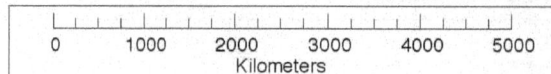

0    1000    2000    3000    4000    5000
Kilometers

©2016  F. Espenak
www.EclipseWise.com

# Annular Solar Eclipse of 2093 Jul 23

Greatest Eclipse = 12:32:03.8 TD   (= 12:30:06.4 UT1)

| Eclipse Magnitude = 0.9463 | Saros Series = 147 |
|---|---|
| Gamma = 0.5717 | Saros Member = 27 of 80 |

**N**

Sun at Greatest Eclipse
(Geocentric Coordinates)
R.A. = 08h14m45.4s
Dec. = +19°49'29.6"
S.D. = 00°15'44.6"
H.P. = 00°00'08.7"

Moon at Greatest Eclipse
(Geocentric Coordinates)
R.A. = 08h15m01.4s
Dec. = +20°20'03.3"
S.D. = 00°14'43.0"
H.P. = 00°54'00.8"

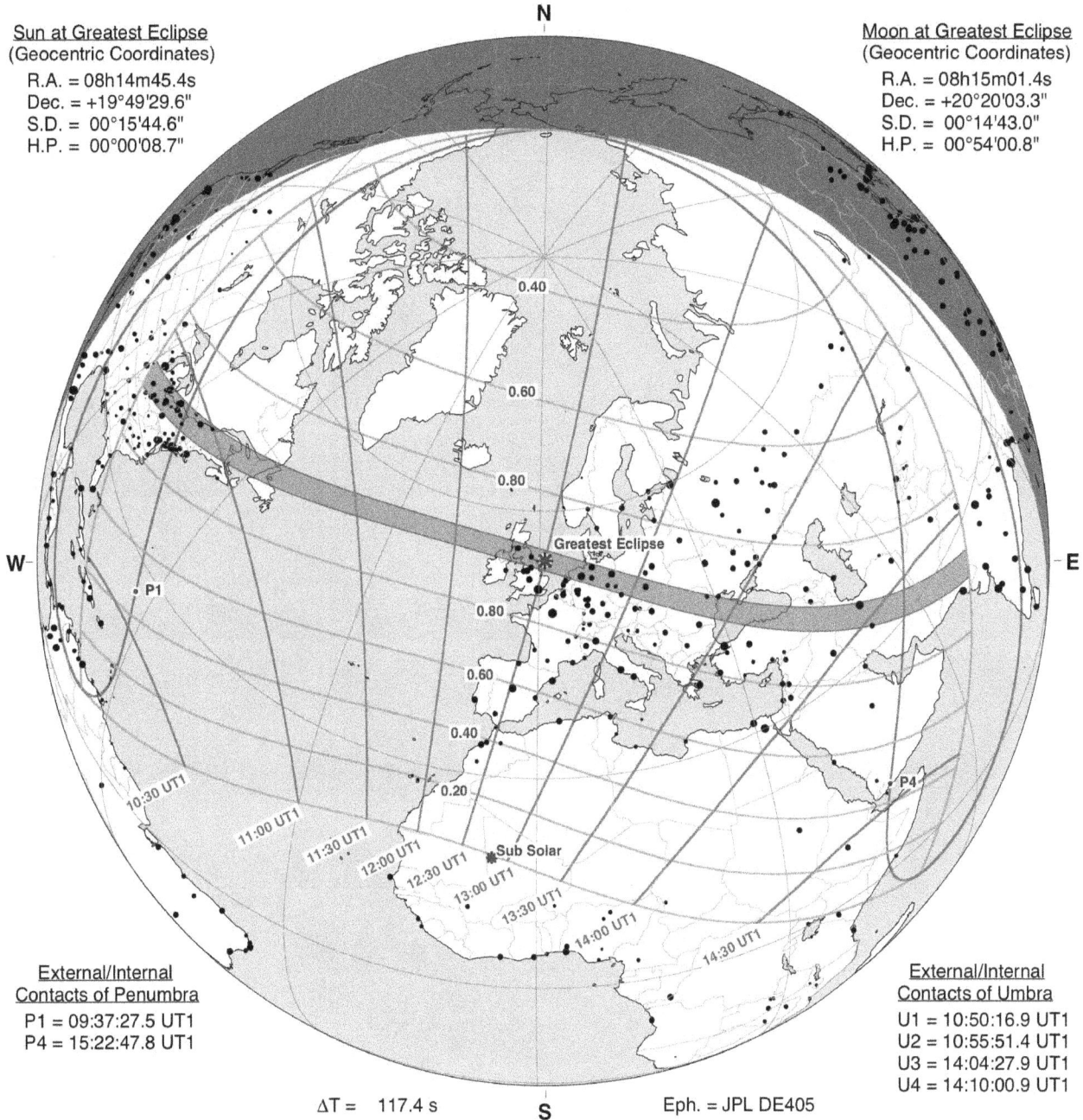

0.40
0.60
0.80

**W**                                                                    **E**

Greatest Eclipse

0.80
0.60
0.40
0.20

P1

P4

10:30 UT1
11:00 UT1
11:30 UT1
12:00 UT1
12:30 UT1
13:00 UT1
13:30 UT1
14:00 UT1
14:30 UT1

Sub Solar

External/Internal
Contacts of Penumbra
P1 = 09:37:27.5 UT1
P4 = 15:22:47.8 UT1

External/Internal
Contacts of Umbra
U1 = 10:50:16.9 UT1
U2 = 10:55:51.4 UT1
U3 = 14:04:27.9 UT1
U4 = 14:10:00.9 UT1

ΔT = 117.4 s          **S**          Eph. = JPL DE405

| Circumstances at Greatest Eclipse: 12:30:06.4 UT1 | | Circumstances at Greatest Duration: 12:25:45.4 UT1 | |
|---|---|---|---|
| Lat. = 54°33.7'N | Sun Alt. = 54.9° | Lat. = 54°57.1'N | Sun Alt. = 54.8° |
| Long. = 001°02.6'E | Sun Azm. = 191.3° | Long. = 001°26.7'W | Sun Azm. = 185.4° |
| Path Width = 241.3 km | Duration = 05m11.3s | Path Width = 241.0 km | Duration = 05m11.4s |

0       1000      2000      3000      4000      5000
Kilometers

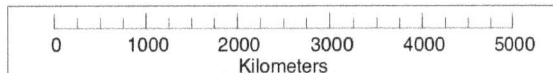

©2016 F. Espenak
www.EclipseWise.com

# Total Solar Eclipse of 2094 Jan 16

Greatest Eclipse = 18:59:03.4 TD  (= 18:57:05.6 UT1)

| | |
|---|---|
| Eclipse Magnitude = 1.0342 | Saros Series = 152 |
| Gamma = -0.9333 | Saros Member = 17 of 70 |

**N**

**Sun at Greatest Eclipse**
(Geocentric Coordinates)
R.A. = 19h56m48.4s
Dec. = -20°43'02.8"
S.D. = 00°16'15.5"
H.P. = 00°00'08.9"

**Moon at Greatest Eclipse**
(Geocentric Coordinates)
R.A. = 19h57m12.4s
Dec. = -21°39'54.3"
S.D. = 00°16'43.2"
H.P. = 01°01'21.9"

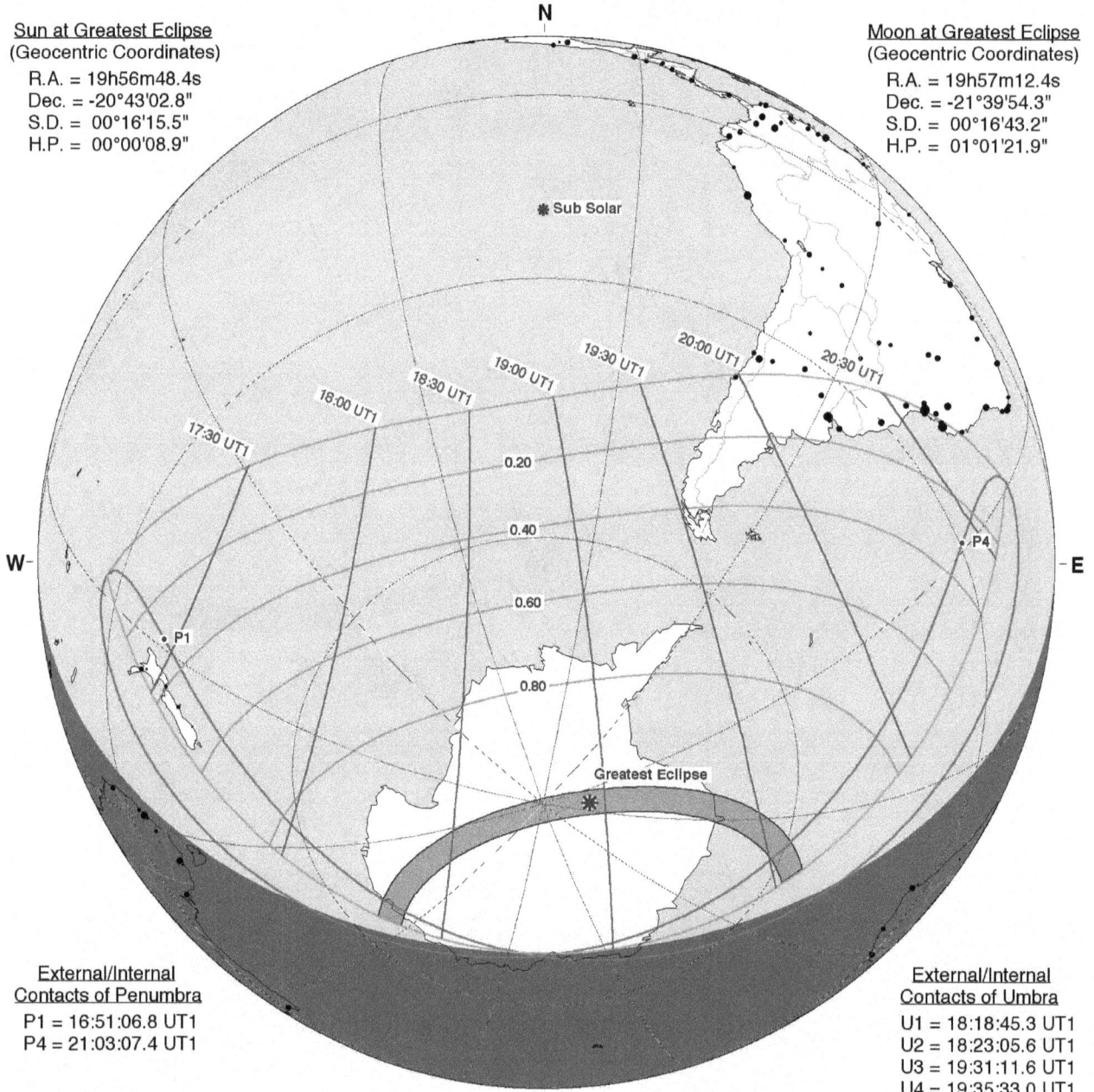

Sub Solar

20:00 UT1
20:30 UT1
19:30 UT1
19:00 UT1
18:30 UT1
18:00 UT1
17:30 UT1

0.20
0.40
0.60
0.80

P4

**W**

**E**

P1

Greatest Eclipse

**External/Internal Contacts of Penumbra**
P1 = 16:51:06.8 UT1
P4 = 21:03:07.4 UT1

**External/Internal Contacts of Umbra**
U1 = 18:18:45.3 UT1
U2 = 18:23:05.6 UT1
U3 = 19:31:11.6 UT1
U4 = 19:35:33.0 UT1

ΔT = 117.8 s

**S**

Eph. = JPL DE405

**Circumstances at Greatest Eclipse: 18:57:05.6 UT1**

| | |
|---|---|
| Lat. = 84°46.4'S | Sun Alt. = 20.5° |
| Long. = 010°51.9'W | Sun Azm. = 267.1° |
| Path Width = 329.4 km | Duration = 01m51.5s |

**Circumstances at Greatest Duration: 18:57:16.6 UT1**

| | |
|---|---|
| Lat. = 84°40.3'S | Sun Alt. = 20.5° |
| Long. = 010°57.2'W | Sun Azm. = 267.1° |
| Path Width = 329.5 km | Duration = 01m51.5s |

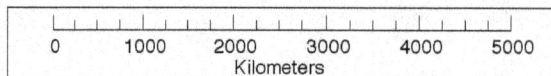

0    1000    2000    3000    4000    5000
Kilometers

©2016  F. Espenak
www.EclipseWise.com

# Partial Solar Eclipse of 2094 Jun 13

Greatest Eclipse = 00:22:11.3 TD   (= 00:20:13.1 UT1)

Eclipse Magnitude = 0.1618          Saros Series = 119
Gamma = -1.4613                     Saros Member = 70 of 71

Sun at Greatest Eclipse
(Geocentric Coordinates)
R.A. = 05h27m41.4s
Dec. = +23°13'07.5"
S.D. = 00°15'45.0"
H.P. = 00°00'08.7"

Moon at Greatest Eclipse
(Geocentric Coordinates)
R.A. = 05h28m34.2s
Dec. = +21°52'30.5"
S.D. = 00°15'14.5"
H.P. = 00°55'56.1"

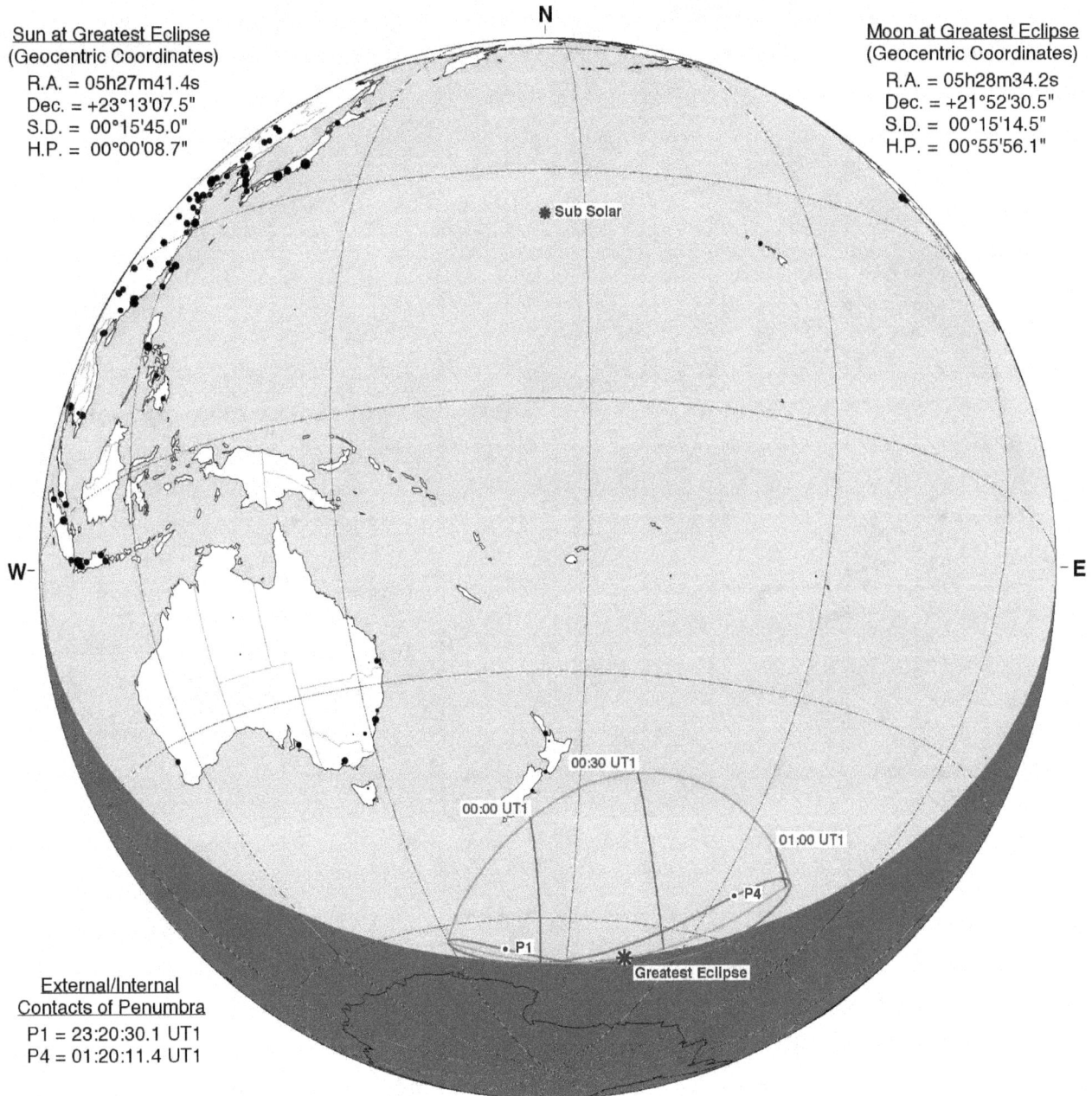

N

* Sub Solar

W

E

00:30 UT1
00:00 UT1
01:00 UT1
• P4
• P1
* Greatest Eclipse

External/Internal
Contacts of Penumbra
P1 = 23:20:30.1 UT1
P4 = 01:20:11.4 UT1

ΔT = 118.2 s        S        Eph. = JPL DE405

Circumstances at Greatest Eclipse: 00:20:13.1 UT1
Lat. = 65°18.3'S        Sun Alt. =   0.0°
Long. = 163°56.0'W      Sun Azm. = 340.7°

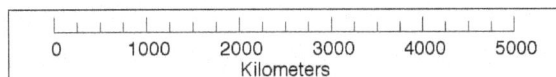

| 0 | 1000 | 2000 | 3000 | 4000 | 5000 |
Kilometers

©2016 F. Espenak
www.EclipseWise.com

# Partial Solar Eclipse of 2094 Jul 12

Greatest Eclipse = 13:24:34.9 TD   (= 13:22:36.6 UT1)

Eclipse Magnitude = 0.4225            Saros Series = 157
Gamma = 1.3149            Saros Member =   3 of 70

Sun at Greatest Eclipse
(Geocentric Coordinates)
R.A. = 07h29m49.1s
Dec. = +21°49'23.2"
S.D. = 00°15'43.9"
H.P. = 00°00'08.7"

Moon at Greatest Eclipse
(Geocentric Coordinates)
R.A. = 07h30m06.1s
Dec. = +23°01'02.4"
S.D. = 00°14'54.5"
H.P. = 00°54'43.0"

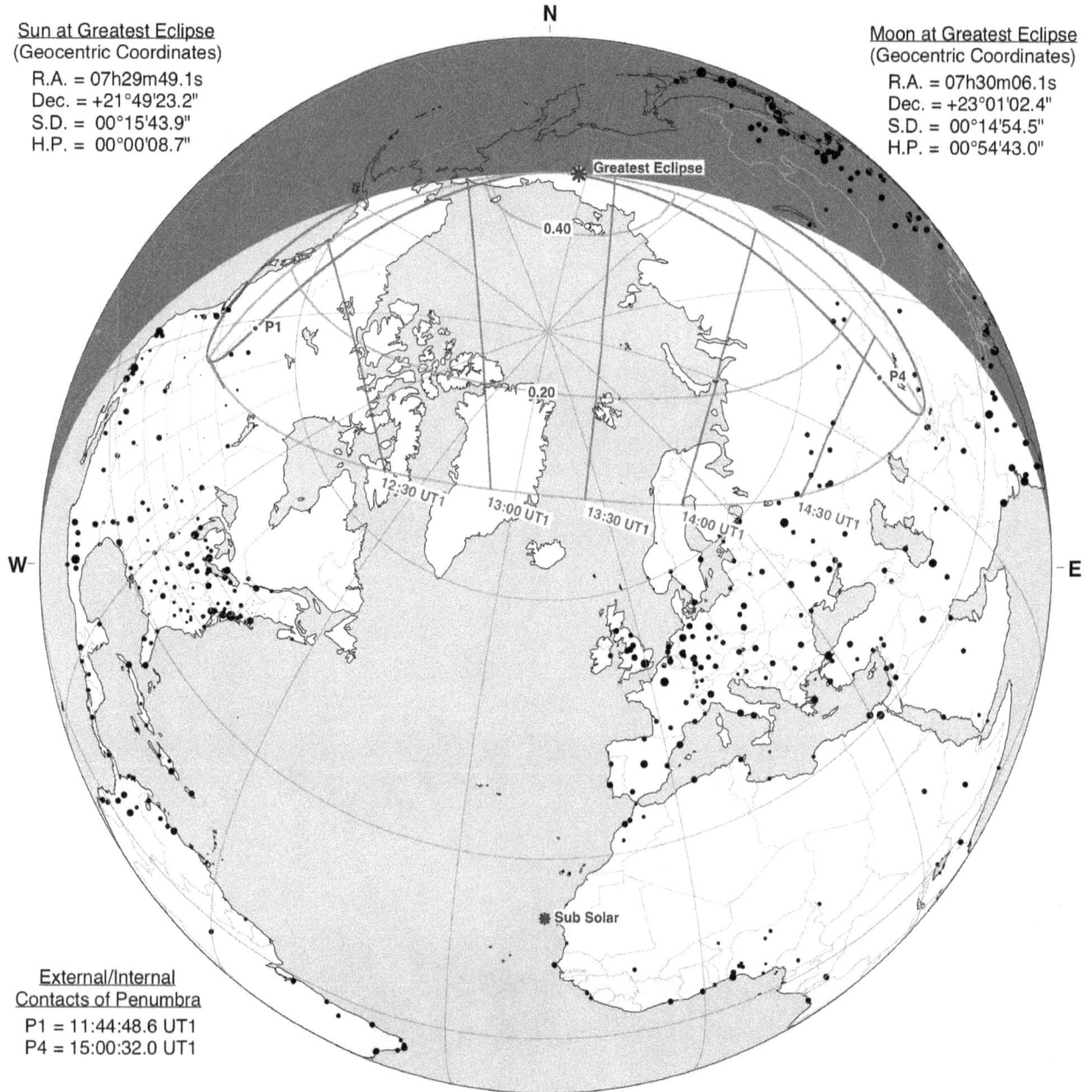

N

Greatest Eclipse

0.40

P1

P4

0.20

12:30 UT1    13:00 UT1    13:30 UT1    14:00 UT1    14:30 UT1

W                                                                E

Sub Solar

External/Internal
Contacts of Penumbra
P1 = 11:44:48.6 UT1
P4 = 15:00:32.0 UT1

ΔT =   118.3 s            S            Eph. = JPL DE405

Circumstances at Greatest Eclipse: 13:22:36.6 UT1
Lat. = 67°58.1'N            Sun Alt. =   0.0°
Long. = 152°27.1'E            Sun Azm. = 352.2°

| 0 | 1000 | 2000 | 3000 | 4000 | 5000 |
Kilometers

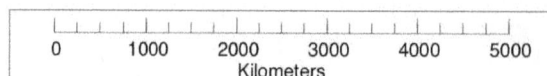

©2016  F. Espenak
www.EclipseWise.com

# Partial Solar Eclipse of  2094 Dec 07

Greatest Eclipse =  20:05:55.6 TD   (= 20:03:56.9 UT1)

| Eclipse Magnitude =  0.7046 | Saros Series =  124 |
|---|---|
| Gamma =  1.1547 | Saros Member =   59 of 73 |

**N**

Sun at Greatest Eclipse
(Geocentric Coordinates)
R.A. = 17h00m09.4s
Dec. = -22°42'52.2"
S.D. = 00°16'13.9"
H.P. = 00°00'08.9"

Moon at Greatest Eclipse
(Geocentric Coordinates)
R.A. = 17h01m06.4s
Dec. = -21°37'52.2"
S.D. = 00°15'41.5"
H.P. = 00°57'35.2"

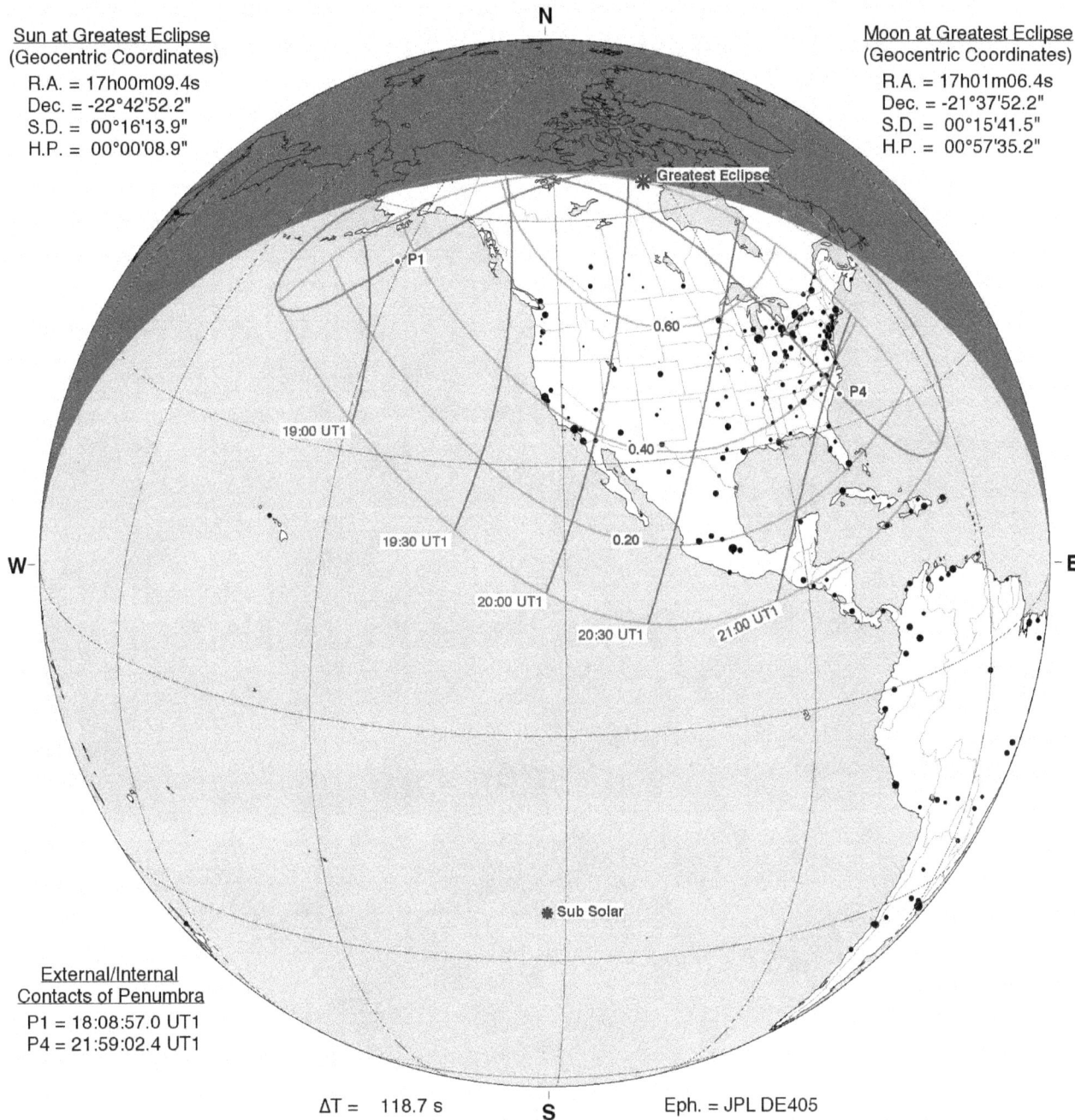

Greatest Eclipse

P1

0.60

P4

19:00 UT1

19:30 UT1

0.40

**W** — — **E**

20:00 UT1

0.20

20:30 UT1

21:00 UT1

Sub Solar

External/Internal
Contacts of Penumbra
P1 = 18:08:57.0 UT1
P4 = 21:59:02.4 UT1

ΔT =   118.7 s      **S**      Eph. = JPL DE405

Circumstances at Greatest Eclipse:  20:03:56.9 UT1

| Lat. = 64°40.1'N | Sun Alt. =   0.0° |
|---|---|
| Long. = 095°17.3'W | Sun Azm. = 205.5° |

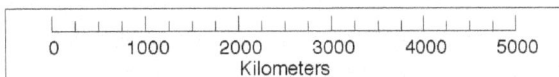

©2016  F. Espenak
www.EclipseWise.com

0    1000    2000    3000    4000    5000
Kilometers

# Total Solar Eclipse of 2095 Jun 02

Greatest Eclipse = 10:07:39.9 TD  (= 10:05:40.7 UT1)

| | |
|---|---|
| Eclipse Magnitude = 1.0332 | Saros Series = 129 |
| Gamma = -0.6396 | Saros Member = 56 of 80 |

**Sun at Greatest Eclipse**
(Geocentric Coordinates)
R.A. = 04h42m53.4s
Dec. = +22°14'41.8"
S.D. = 00°15'46.4"
H.P. = 00°00'08.7"

**Moon at Greatest Eclipse**
(Geocentric Coordinates)
R.A. = 04h43m30.2s
Dec. = +21°37'59.7"
S.D. = 00°16'05.6"
H.P. = 00°59'03.8"

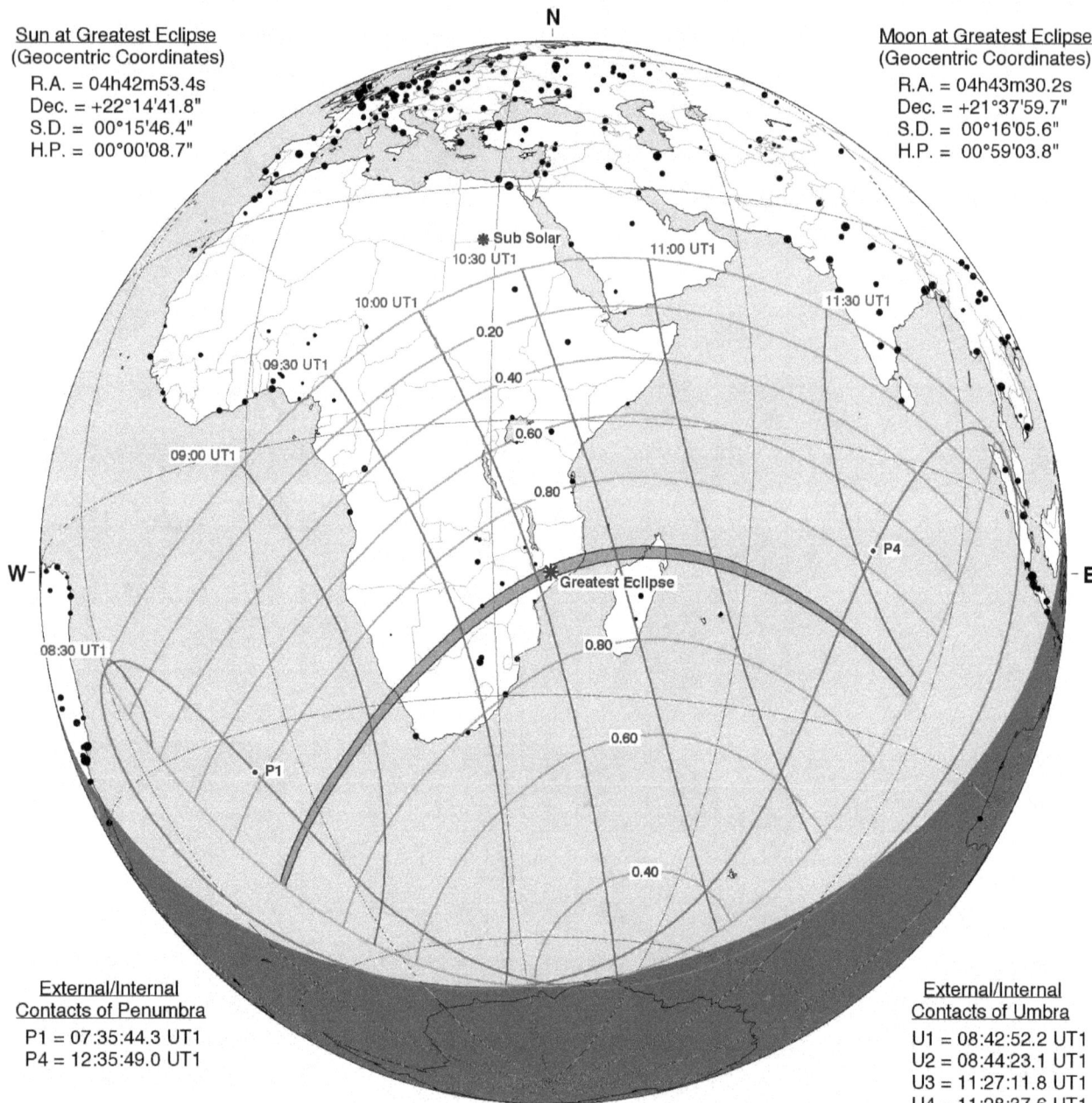

N

Sub Solar
10:30 UT1
11:00 UT1
10:00 UT1
0.20
09:30 UT1
0.40
0.60
09:00 UT1
0.80
11:30 UT1

W — — E

Greatest Eclipse
0.80
08:30 UT1
0.60
P4
P1
0.40

**External/Internal Contacts of Penumbra**
P1 = 07:35:44.3 UT1
P4 = 12:35:49.0 UT1

**External/Internal Contacts of Umbra**
U1 = 08:42:52.2 UT1
U2 = 08:44:23.1 UT1
U3 = 11:27:11.8 UT1
U4 = 11:28:37.6 UT1

ΔT = 119.2 s

S

Eph. = JPL DE405

**Circumstances at Greatest Eclipse: 10:05:40.7 UT1**

| | |
|---|---|
| Lat. = 16°42.3'S | Sun Alt. = 50.1° |
| Long. = 036°51.2'E | Sun Azm. = 347.4° |
| Path Width = 144.8 km | Duration = 03m18.4s |

**Circumstances at Greatest Duration: 10:06:58.5 UT1**

| | |
|---|---|
| Lat. = 16°33.3'S | Sun Alt. = 50.1° |
| Long. = 037°14.1'E | Sun Azm. = 346.3° |
| Path Width = 145.0 km | Duration = 03m18.5s |

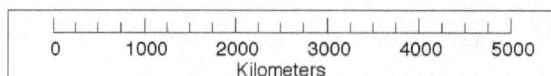

0    1000    2000    3000    4000    5000
Kilometers

©2016 F. Espenak
www.EclipseWise.com

# Annular Solar Eclipse of  2095 Nov 27

Greatest Eclipse =  01:02:57.4 TD   (= 01:00:57.8 UT1)

Eclipse Magnitude =  0.9330          Saros Series =  134
Gamma =  0.4903          Saros Member =  48 of 71

**N**

Sun at Greatest Eclipse
(Geocentric Coordinates)
R.A. = 16h12m24.6s
Dec. = -21°07'41.4"
S.D. = 00°16'12.2"
H.P. = 00°00'08.9"

Moon at Greatest Eclipse
(Geocentric Coordinates)
R.A. = 16h12m56.4s
Dec. = -20°41'58.0"
S.D. = 00°14'55.2"
H.P. = 00°54'45.3"

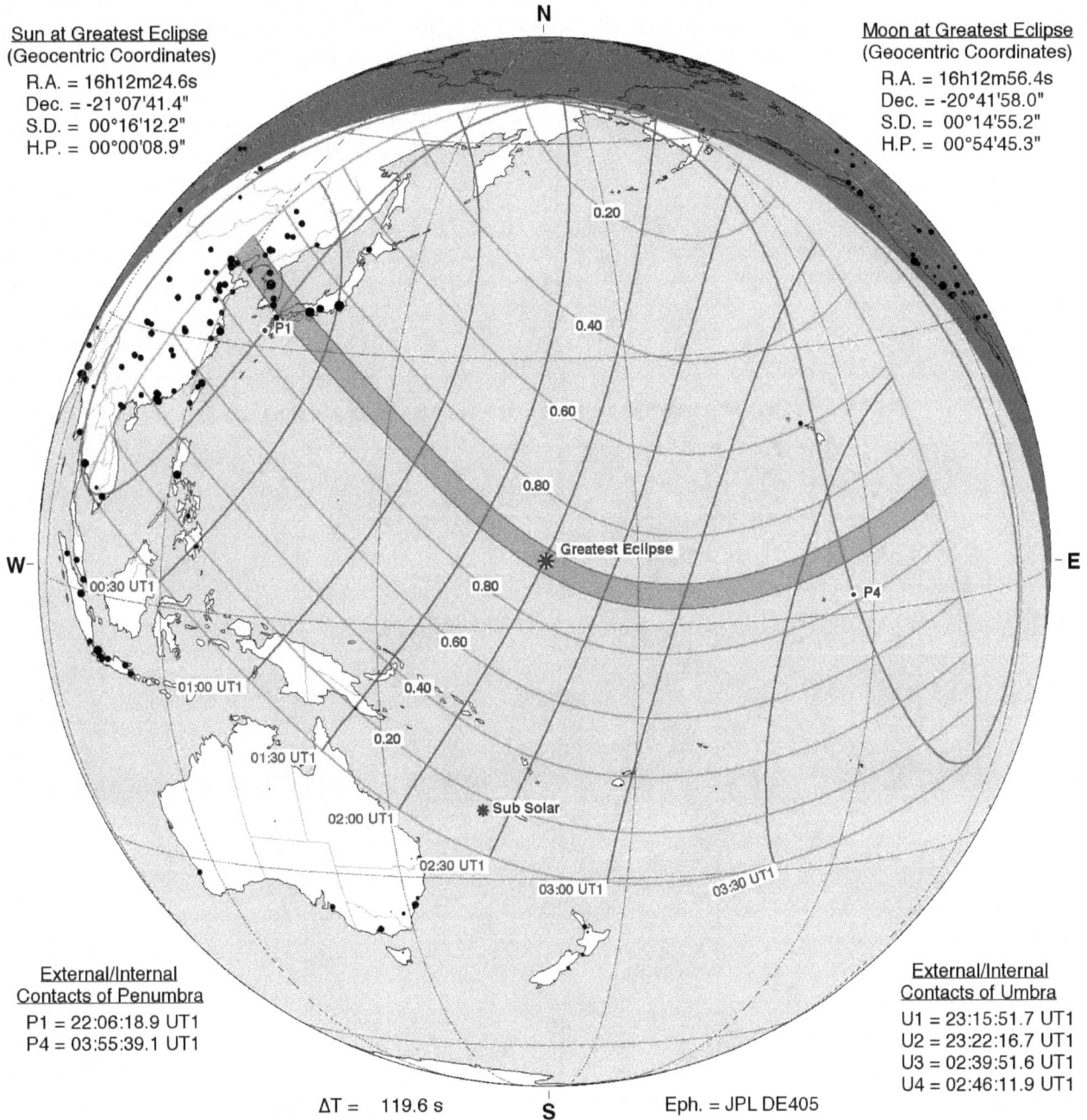

0.20
0.40
P1
0.60
0.80
Greatest Eclipse

**W** —                                                                        — **E**

00:30 UT1
P4
0.80
01:00 UT1
0.60
0.40
01:30 UT1
0.20
02:00 UT1
Sub Solar
02:30 UT1
03:00 UT1    03:30 UT1

External/Internal
Contacts of Penumbra
P1 = 22:06:18.9 UT1
P4 = 03:55:39.1 UT1

External/Internal
Contacts of Umbra
U1 = 23:15:51.7 UT1
U2 = 23:22:16.7 UT1
U3 = 02:39:51.6 UT1
U4 = 02:46:11.9 UT1

ΔT =   119.6 s          **S**          Eph. = JPL DE405

| Circumstances at Greatest Eclipse: 01:00:57.8 UT1 | | Circumstances at Greatest Duration: 01:11:24.9 UT1 | |
| --- | --- | --- | --- |
| Lat. = 07°14.4'N | Sun Alt. =  60.6° | Lat. = 05°59.6'N | Sun Alt. =  60.1° |
| Long. = 169°30.1'E | Sun Azm. = 195.1° | Long. = 171°53.4'E | Sun Azm. = 204.6° |
| Path Width = 284.8 km | Duration = 08m46.7s | Path Width = 288.6 km | Duration = 08m48.4s |

0     1000     2000     3000     4000     5000
Kilometers

*©2016  F. Espenak*
*www.EclipseWise.com*

# Total Solar Eclipse of 2096 May 22

Greatest Eclipse = 01:37:14.1 TD  (= 01:35:14.1 UT1)

| | |
|---|---|
| Eclipse Magnitude = 1.0737 | Saros Series = 139 |
| Gamma = 0.1196 | Saros Member = 34 of 71 |

Sun at Greatest Eclipse
(Geocentric Coordinates)
R.A. = 03h59m45.5s
Dec. = +20°33'28.2"
S.D. = 00°15'48.1"
H.P. = 00°00'08.7"

Moon at Greatest Eclipse
(Geocentric Coordinates)
R.A. = 03h59m36.3s
Dec. = +20°40'26.9"
S.D. = 00°16'40.8"
H.P. = 01°01'13.0"

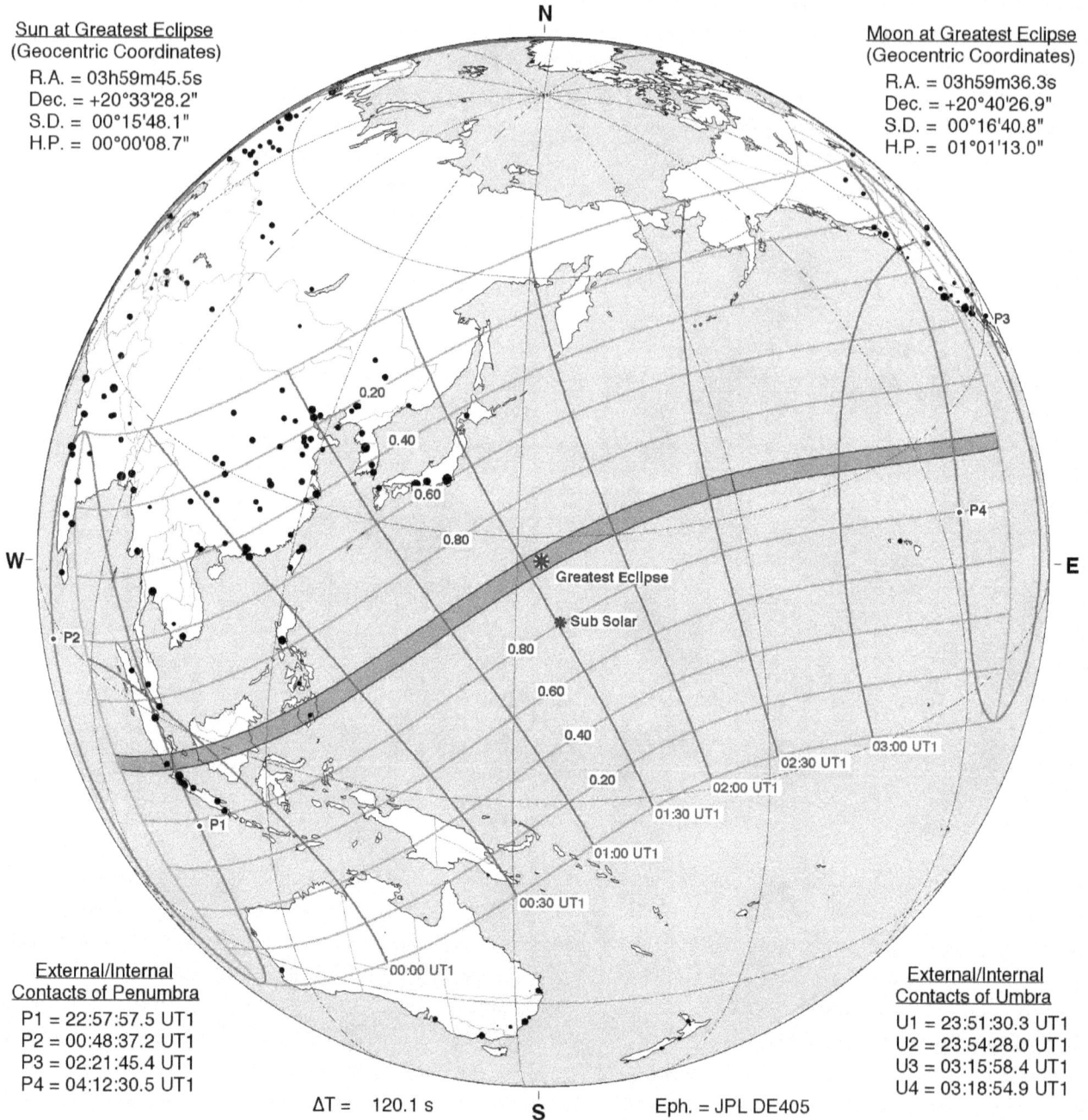

N

W

E

S

0.20
0.40
0.60
0.80
Greatest Eclipse
Sub Solar
0.80
0.60
0.40
0.20

P1
P2
P3
P4

03:00 UT1
02:30 UT1
02:00 UT1
01:30 UT1
01:00 UT1
00:30 UT1
00:00 UT1

External/Internal
Contacts of Penumbra
P1 = 22:57:57.5 UT1
P2 = 00:48:37.2 UT1
P3 = 02:21:45.4 UT1
P4 = 04:12:30.5 UT1

External/Internal
Contacts of Umbra
U1 = 23:51:30.3 UT1
U2 = 23:54:28.0 UT1
U3 = 03:15:58.4 UT1
U4 = 03:18:54.9 UT1

ΔT = 120.1 s          Eph. = JPL DE405

Circumstances at Greatest Eclipse: 01:35:14.1 UT1

| | |
|---|---|
| Lat. = 27°16.0'N | Sun Alt. = 83.0° |
| Long. = 153°07.8'E | Sun Azm. = 162.3° |
| Path Width = 240.9 km | Duration = 06m06.5s |

Circumstances at Greatest Duration: 01:40:24.6 UT1

| | |
|---|---|
| Lat. = 28°08.3'N | Sun Alt. = 82.4° |
| Long. = 154°56.2'E | Sun Azm. = 185.9° |
| Path Width = 240.6 km | Duration = 06m06.9s |

0   1000   2000   3000   4000   5000
Kilometers

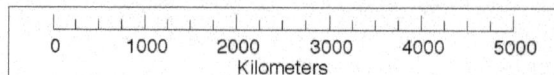

# Annular Solar Eclipse of 2096 Nov 15

Greatest Eclipse = 00:36:14.8 TD   (= 00:34:14.3 UT1)

Eclipse Magnitude = 0.9237          Saros Series = 144
Gamma = -0.2018          Saros Member = 21 of 70

Sun at Greatest Eclipse
(Geocentric Coordinates)
R.A. = 15h25m10.4s
Dec. = -18°40'58.6"
S.D. = 00°16'10.0"
H.P. = 00°00'08.9"

Moon at Greatest Eclipse
(Geocentric Coordinates)
R.A. = 15h24m54.6s
Dec. = -18°51'10.6"
S.D. = 00°14'42.9"
H.P. = 00°54'00.1"

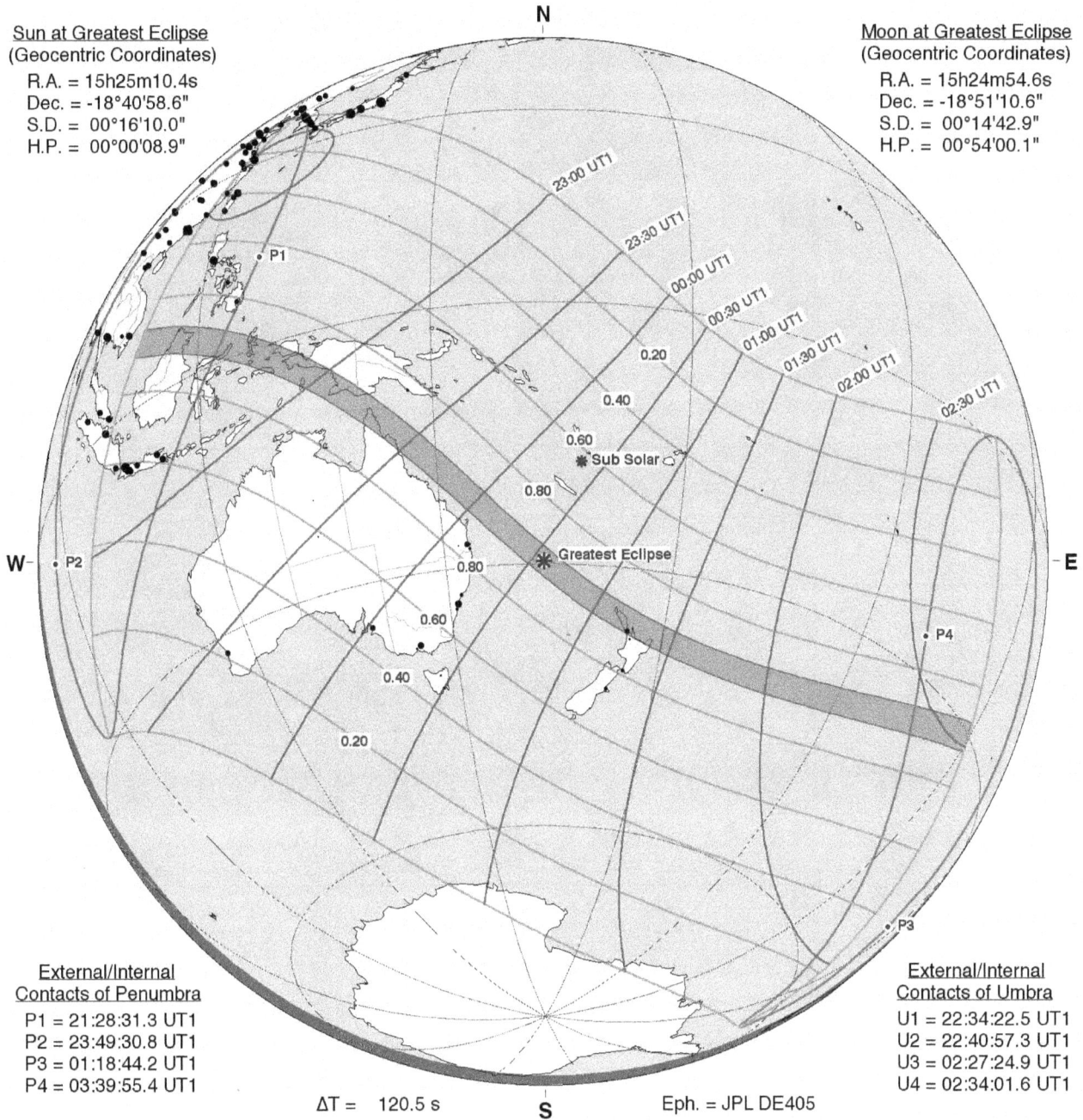

N

W

E

S

P1

P2

P3

P4

Sub Solar

Greatest Eclipse

23:00 UT1
23:30 UT1
00:00 UT1
00:30 UT1
01:00 UT1
01:30 UT1
02:00 UT1
02:30 UT1

0.20
0.40
0.60
0.80
0.80
0.60
0.40
0.20

External/Internal
Contacts of Penumbra
P1 = 21:28:31.3 UT1
P2 = 23:49:30.8 UT1
P3 = 01:18:44.2 UT1
P4 = 03:39:55.4 UT1

External/Internal
Contacts of Umbra
U1 = 22:34:22.5 UT1
U2 = 22:40:57.3 UT1
U3 = 02:27:24.9 UT1
U4 = 02:34:01.6 UT1

ΔT = 120.5 s          Eph. = JPL DE405

Circumstances at Greatest Eclipse: 00:34:14.3 UT1
Lat. = 29°43.4'S          Sun Alt. = 78.2°
Long. = 162°59.4'E          Sun Azm. = 21.8°
Path Width = 293.6 km          Duration = 08m52.5s

Circumstances at Greatest Duration: 00:50:32.4 UT1
Lat. = 32°31.4'S          Sun Alt. = 75.7°
Long. = 167°30.9'E          Sun Azm. = 344.5°
Path Width = 292.7 km          Duration = 08m55.4s

©2016 F. Espenak
www.EclipseWise.com

0   1000   2000   3000   4000   5000
Kilometers

# Total Solar Eclipse of 2097 May 11

Greatest Eclipse = 18:34:31.4 TD   (= 18:32:30.4 UT1)

| | |
|---|---|
| Eclipse Magnitude = 1.0538 | Saros Series = 149 |
| Gamma = 0.8516 | Saros Member = 25 of 71 |

**Sun at Greatest Eclipse**
(Geocentric Coordinates)
R.A. = 03h17m49.7s
Dec. = +18°13'35.1"
S.D. = 00°15'50.2"
H.P. = 00°00'08.7"

**Moon at Greatest Eclipse**
(Geocentric Coordinates)
R.A. = 03h16m33.2s
Dec. = +19°01'53.1"
S.D. = 00°16'32.8"
H.P. = 01°00'43.7"

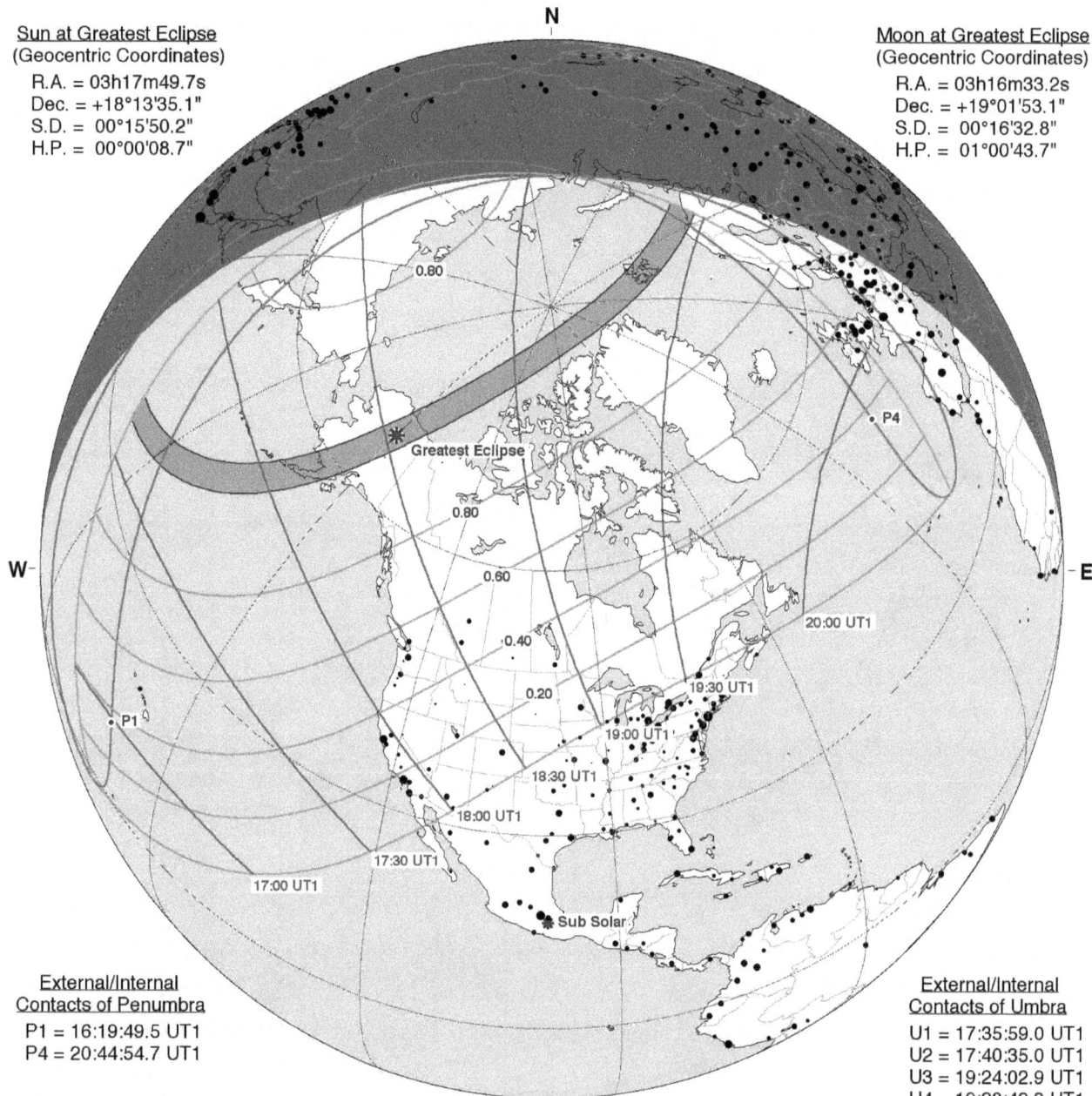

**External/Internal Contacts of Penumbra**
P1 = 16:19:49.5 UT1
P4 = 20:44:54.7 UT1

**External/Internal Contacts of Umbra**
U1 = 17:35:59.0 UT1
U2 = 17:40:35.0 UT1
U3 = 19:24:02.9 UT1
U4 = 19:28:42.3 UT1

ΔT = 121.0 s          Eph. = JPL DE405

**Circumstances at Greatest Eclipse: 18:32:30.4 UT1**

| | |
|---|---|
| Lat. = 67°24.7'N | Sun Alt. = 31.3° |
| Long. = 149°50.0'W | Sun Azm. = 120.5° |
| Path Width = 339.5 km | Duration = 03m10.0s |

**Circumstances at Greatest Duration: 18:33:49.5 UT1**

| | |
|---|---|
| Lat. = 68°02.4'N | Sun Alt. = 31.3° |
| Long. = 149°12.2'W | Sun Azm. = 121.8° |
| Path Width = 338.6 km | Duration = 03m10.0s |

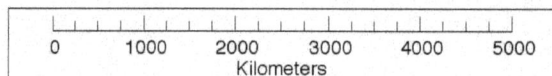

©2016  F. Espenak
www.EclipseWise.com

# Annular Solar Eclipse of 2097 Nov 04

Greatest Eclipse = 02:01:25.2 TD  (= 01:59:23.7 UT1)

Eclipse Magnitude = 0.9494      Saros Series = 154
Gamma = -0.8926      Saros Member = 11 of 71

**Sun at Greatest Eclipse**
(Geocentric Coordinates)
R.A. = 14h40m01.3s
Dec. = -15°33'59.2"
S.D. = 00°16'07.3"
H.P. = 00°00'08.9"

**Moon at Greatest Eclipse**
(Geocentric Coordinates)
R.A. = 14h38m39.0s
Dec. = -16°19'33.5"
S.D. = 00°15'12.3"
H.P. = 00°55'48.3"

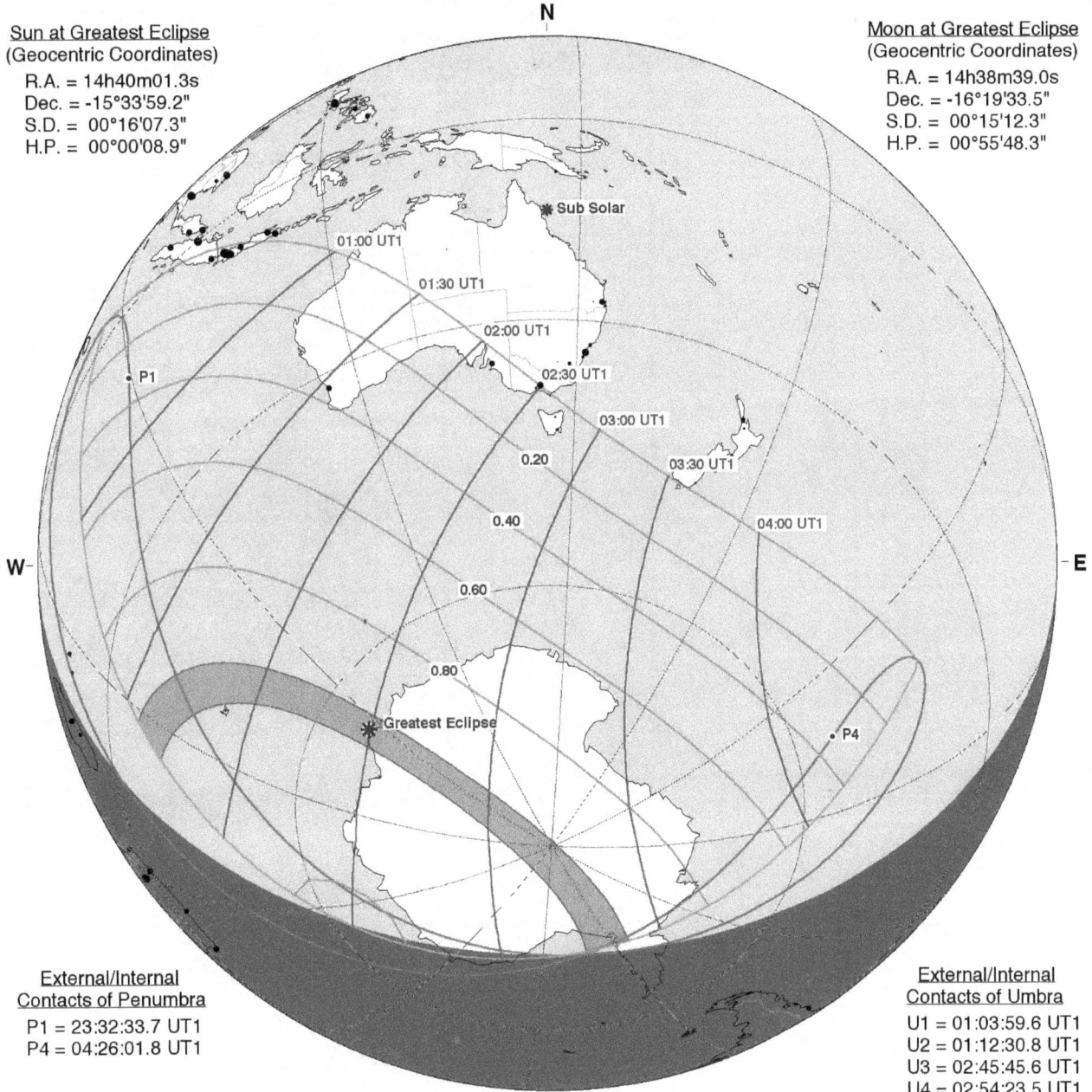

N

Sub Solar

01:00 UT1
01:30 UT1
02:00 UT1
02:30 UT1
03:00 UT1
03:30 UT1
04:00 UT1

P1

0.20
0.40
0.60
0.80

Greatest Eclipse

P4

W — — E

**External/Internal Contacts of Penumbra**
P1 = 23:32:33.7 UT1
P4 = 04:26:01.8 UT1

**External/Internal Contacts of Umbra**
U1 = 01:03:59.6 UT1
U2 = 01:12:30.8 UT1
U3 = 02:45:45.6 UT1
U4 = 02:54:23.5 UT1

ΔT = 121.5 s      S      Eph. = JPL DE405

**Circumstances at Greatest Eclipse: 01:59:23.7 UT1**

| | |
|---|---|
| Lat. = 65°46.2'S | Sun Alt. = 26.4° |
| Long. = 086°28.0'E | Sun Azm. = 68.0° |
| Path Width = 410.7 km | Duration = 03m35.6s |

**Circumstances at Greatest Duration: 02:08:26.3 UT1**

| | |
|---|---|
| Lat. = 70°04.1'S | Sun Alt. = 25.9° |
| Long. = 088°06.1'E | Sun Azm. = 62.2° |
| Path Width = 403.5 km | Duration = 03m35.8s |

©2016 F. Espenak
www.EclipseWise.com

0      1000     2000     3000     4000     5000
Kilometers

# Partial Solar Eclipse of 2098 Apr 01

Greatest Eclipse = 20:02:30.8 TD (= 20:00:29.0 UT1)

| Eclipse Magnitude = 0.7984 | Saros Series = 121 |
|---|---|
| Gamma = -1.1005 | Saros Member = 65 of 71 |

Sun at Greatest Eclipse
(Geocentric Coordinates)
R.A. = 00h46m32.1s
Dec. = +04°59'38.4"
S.D. = 00°16'00.4"
H.P. = 00°00'08.8"

Moon at Greatest Eclipse
(Geocentric Coordinates)
R.A. = 00h48m30.7s
Dec. = +04°05'18.4"
S.D. = 00°15'21.2"
H.P. = 00°56'20.9"

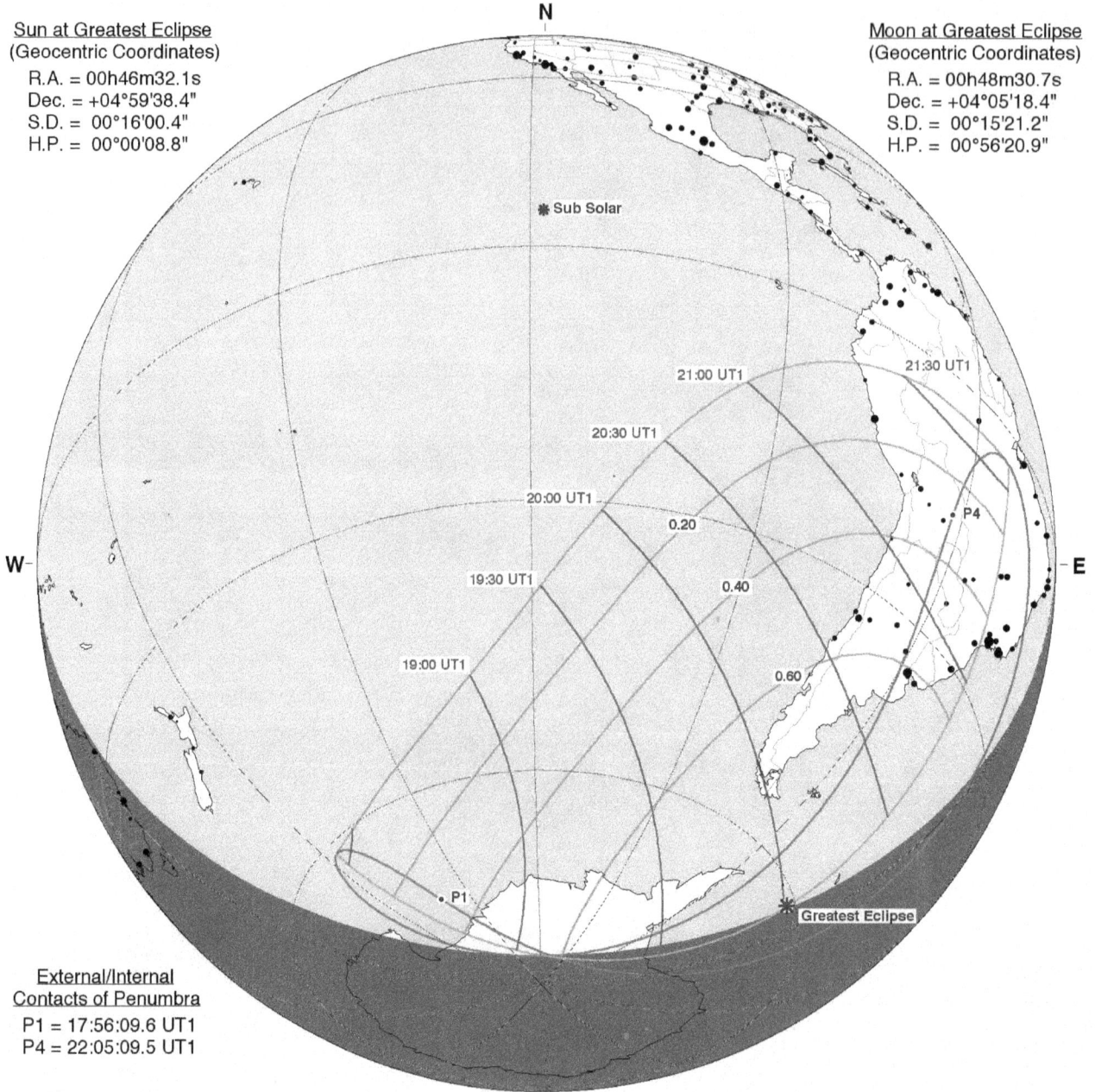

External/Internal
Contacts of Penumbra
P1 = 17:56:09.6 UT1
P4 = 22:05:09.5 UT1

ΔT = 121.9 s      Eph. = JPL DE405

Circumstances at Greatest Eclipse: 20:00:29.0 UT1

| Lat. = 61°02.5'S | Sun Alt. = 0.0° |
|---|---|
| Long. = 038°24.1'W | Sun Azm. = 280.4° |

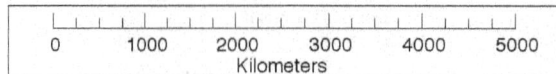

| 0 | 1000 | 2000 | 3000 | 4000 | 5000 |
|---|---|---|---|---|---|

Kilometers

# Partial Solar Eclipse of 2098 Sep 25

Greatest Eclipse = 00:31:16.2 TD  (= 00:29:13.8 UT1)

| Eclipse Magnitude = 0.7871 | Saros Series = 126 |
|---|---|
| Gamma = 1.1184 | Saros Member = 52 of 72 |

Sun at Greatest Eclipse
(Geocentric Coordinates)
R.A. = 12h09m17.5s
Dec. = -01°00'22.0"
S.D. = 00°15'56.5"
H.P. = 00°00'08.8"

Moon at Greatest Eclipse
(Geocentric Coordinates)
R.A. = 12h11m27.7s
Dec. = -00°01'23.2"
S.D. = 00°16'27.0"
H.P. = 01°00'22.3"

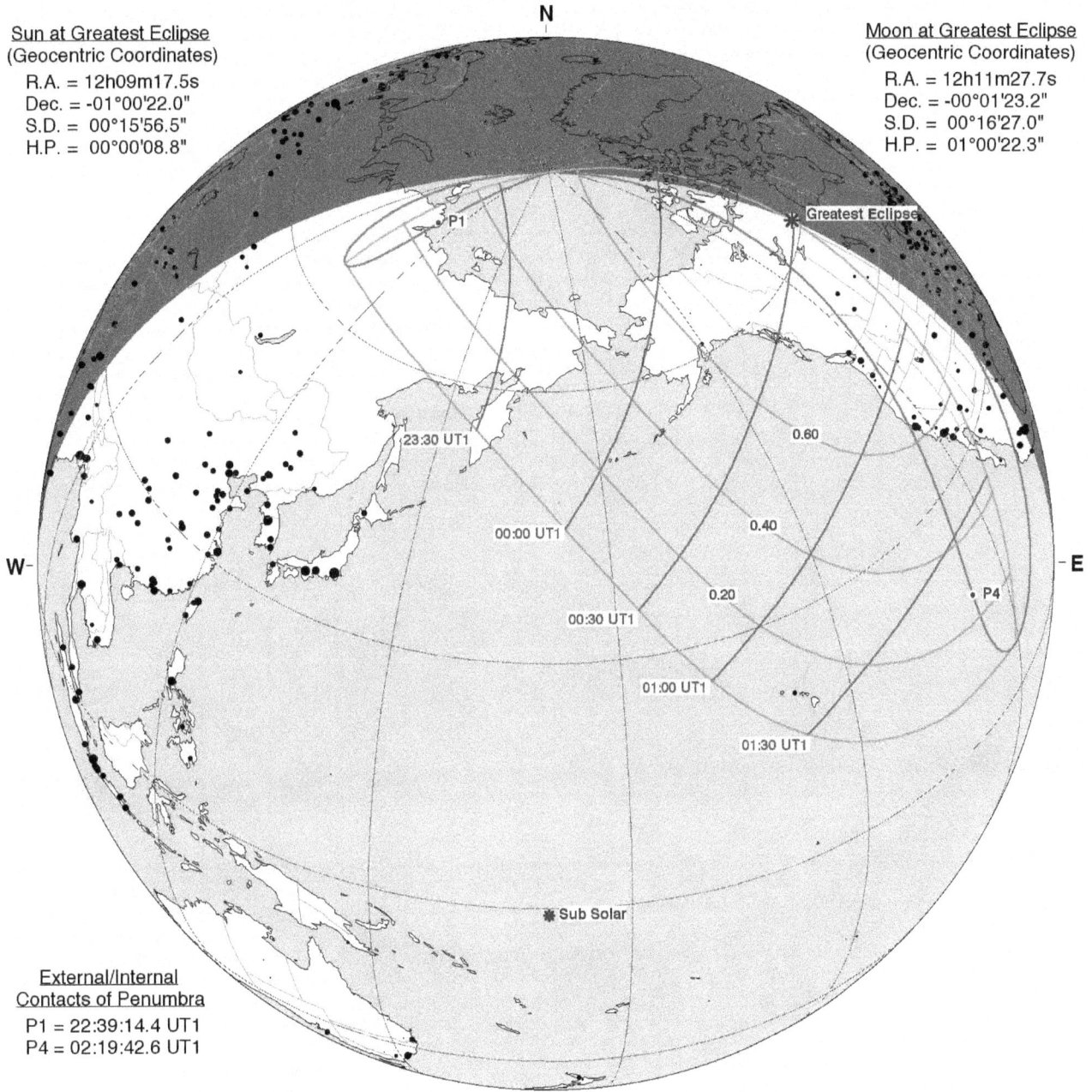

N

Greatest Eclipse

P1

23:30 UT1

0.60

00:00 UT1

0.40

W

0.20

00:30 UT1

E

P4

01:00 UT1

01:30 UT1

Sub Solar

External/Internal
Contacts of Penumbra
P1 = 22:39:14.4 UT1
P4 = 02:19:42.6 UT1

ΔT = 122.3 s

S

Eph. = JPL DE405

Circumstances at Greatest Eclipse: 00:29:13.8 UT1

| Lat. = 61°05.6'N | Sun Alt. =  0.0° |
|---|---|
| Long. = 101°17.4'W | Sun Azm. = 267.9° |

0  1000  2000  3000  4000  5000
Kilometers

©2016  F. Espenak
www.EclipseWise.com

# Partial Solar Eclipse of 2098 Oct 24

Greatest Eclipse = 10:36:10.8 TD   (= 10:34:08.4 UT1)

| | |
|---|---|
| Eclipse Magnitude = 0.0057 | Saros Series = 164 |
| Gamma = -1.5407 | Saros Member = 1 of 80 |

**Sun at Greatest Eclipse**
(Geocentric Coordinates)
R.A. = 13h57m42.1s
Dec. = -12°01'06.6"
S.D. = 00°16'04.5"
H.P. = 00°00'08.8"

**Moon at Greatest Eclipse**
(Geocentric Coordinates)
R.A. = 13h55m00.2s
Dec. = -13°22'41.3"
S.D. = 00°16'04.0"
H.P. = 00°58'57.8"

N

* Sub Solar

W

E

10:30 UT1

P1

Greatest Eclipse

P4

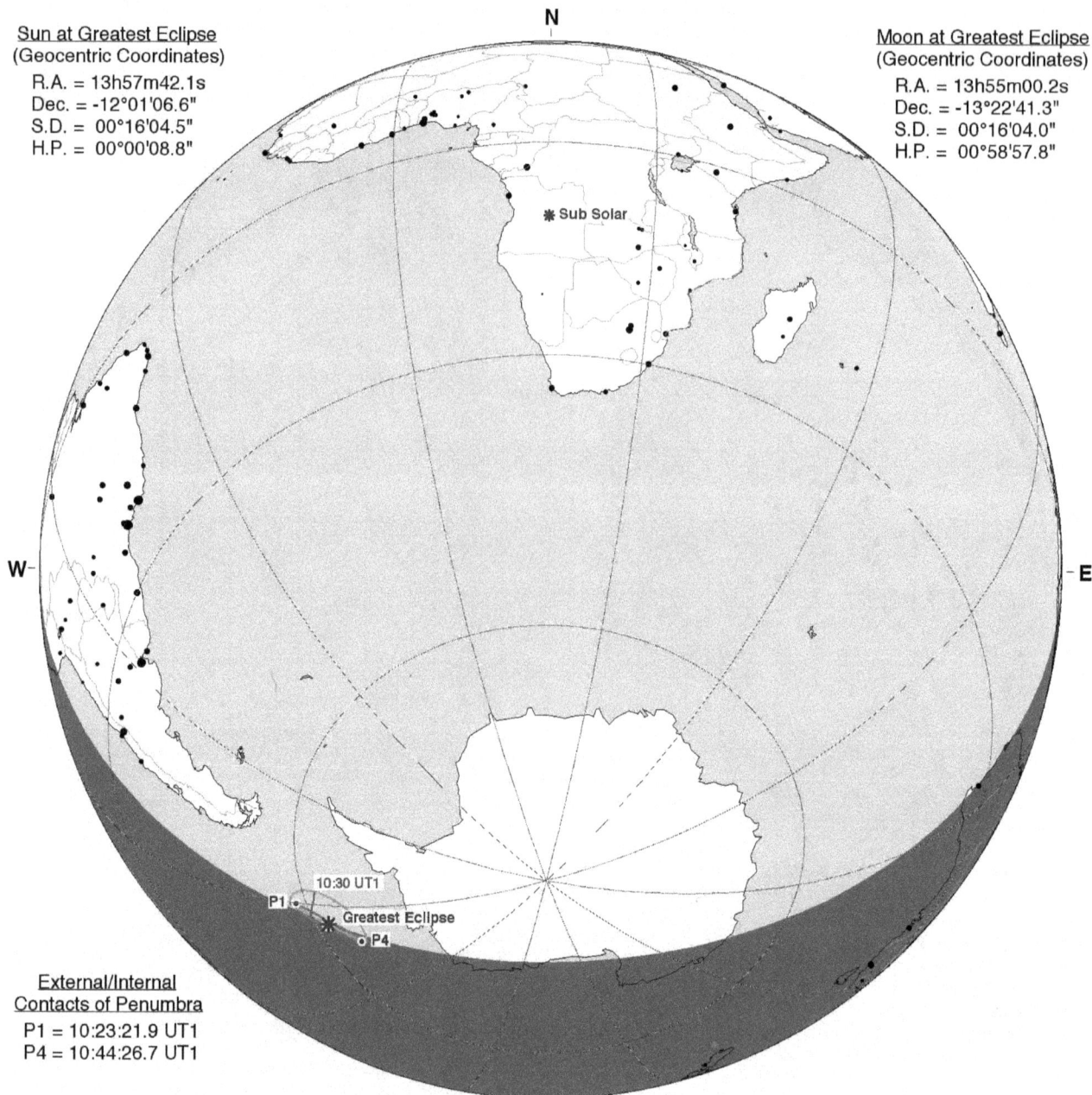

**External/Internal**
**Contacts of Penumbra**
P1 = 10:23:21.9 UT1
P4 = 10:44:26.7 UT1

ΔT = 122.4 s     S     Eph. = JPL DE405

**Circumstances at Greatest Eclipse: 10:34:08.4 UT1**

| | |
|---|---|
| Lat. = 61°45.7'S | Sun Alt. = 0.0° |
| Long. = 095°47.2'W | Sun Azm. = 116.1° |

| | | | | | |
|---|---|---|---|---|---|
| 0 | 1000 | 2000 | 3000 | 4000 | 5000 |

Kilometers

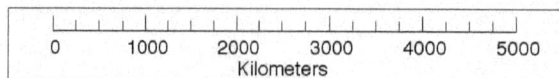

©2016  F. Espenak
www.EclipseWise.com

# Annular Solar Eclipse of 2099 Mar 21

Greatest Eclipse = 22:54:32.0 TD (= 22:52:29.2 UT1)

Eclipse Magnitude = 0.9318          Saros Series = 131
Gamma = -0.4016          Saros Member = 55 of 70

Sun at Greatest Eclipse
(Geocentric Coordinates)
R.A. = 00h06m00.9s
Dec. = +00°39'05.5"
S.D. = 00°16'03.5"
H.P. = 00°00'08.8"

Moon at Greatest Eclipse
(Geocentric Coordinates)
R.A. = 00h06m43.3s
Dec. = +00°20'09.1"
S.D. = 00°14'45.5"
H.P. = 00°54'09.7"

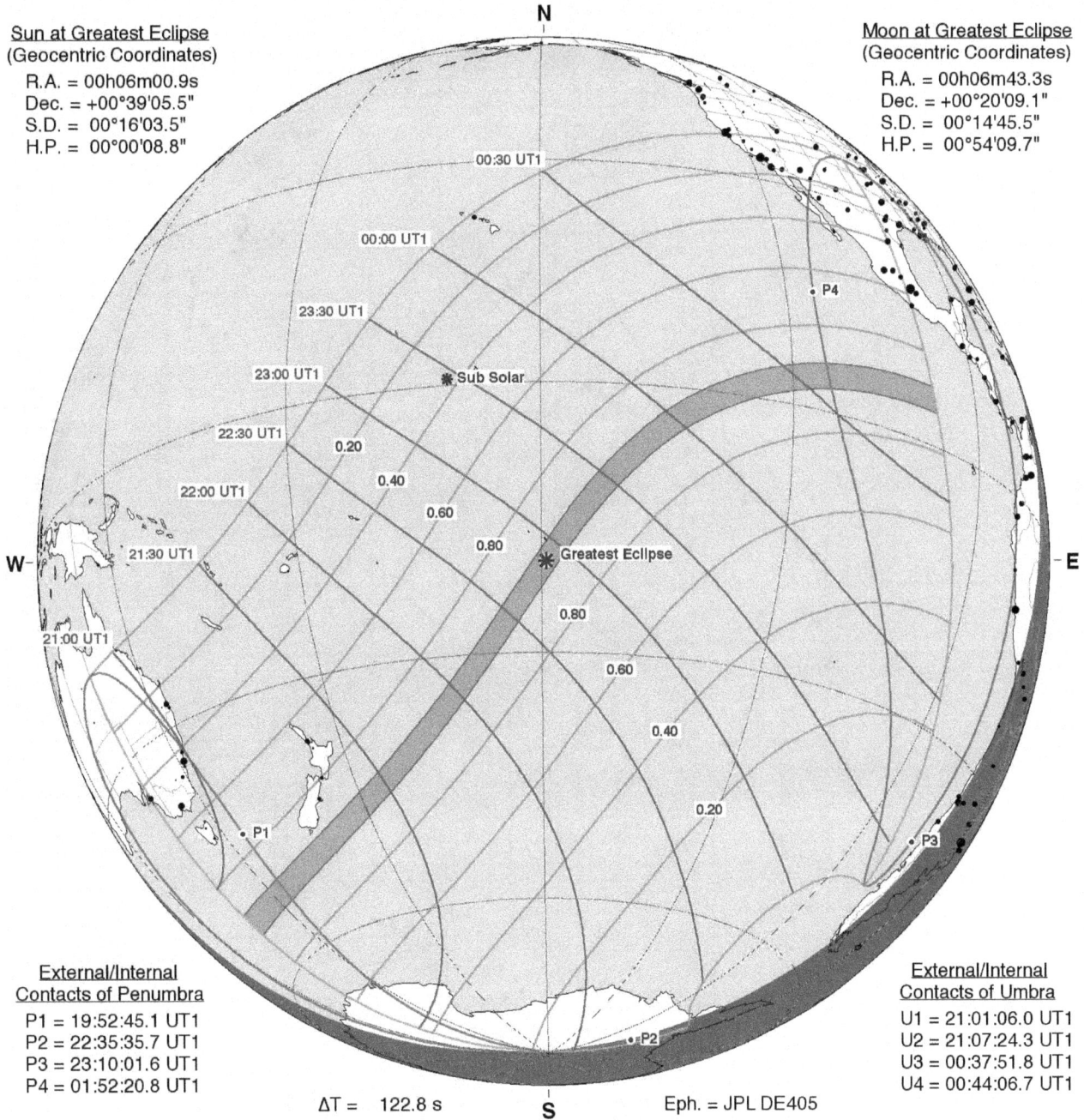

External/Internal
Contacts of Penumbra
P1 = 19:52:45.1 UT1
P2 = 22:35:35.7 UT1
P3 = 23:10:01.6 UT1
P4 = 01:52:20.8 UT1

External/Internal
Contacts of Umbra
U1 = 21:01:06.0 UT1
U2 = 21:07:24.3 UT1
U3 = 00:37:51.8 UT1
U4 = 00:44:06.7 UT1

ΔT = 122.8 s          Eph. = JPL DE405

Circumstances at Greatest Eclipse: 22:52:29.2 UT1

| | |
|---|---|
| Lat. = 20°01.2'S | Sun Alt. = 66.2° |
| Long. = 149°22.4'W | Sun Azm. = 328.9° |
| Path Width = 275.0 km | Duration = 07m32.3s |

Circumstances at Greatest Duration: 22:52:12.1 UT1

| | |
|---|---|
| Lat. = 20°05.6'S | Sun Alt. = 66.2° |
| Long. = 149°25.7'W | Sun Azm. = 329.3° |
| Path Width = 274.9 km | Duration = 07m32.3s |

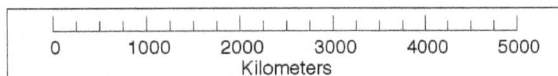

©2016  F. Espenak
www.EclipseWise.com

# Total Solar Eclipse of 2099 Sep 14

Greatest Eclipse = 16:57:53.0 TD  (= 16:55:49.7 UT1)

| | |
|---|---|
| Eclipse Magnitude = 1.0684 | Saros Series = 136 |
| Gamma = 0.3942 | Saros Member = 42 of 71 |

**Sun at Greatest Eclipse**
(Geocentric Coordinates)
R.A. = 11h31m25.7s
Dec. = +03°05'04.1"
S.D. = 00°15'53.8"
H.P. = 00°00'08.7"

**Moon at Greatest Eclipse**
(Geocentric Coordinates)
R.A. = 11h32m12.4s
Dec. = +03°26'11.8"
S.D. = 00°16'43.1"
H.P. = 01°01'21.6"

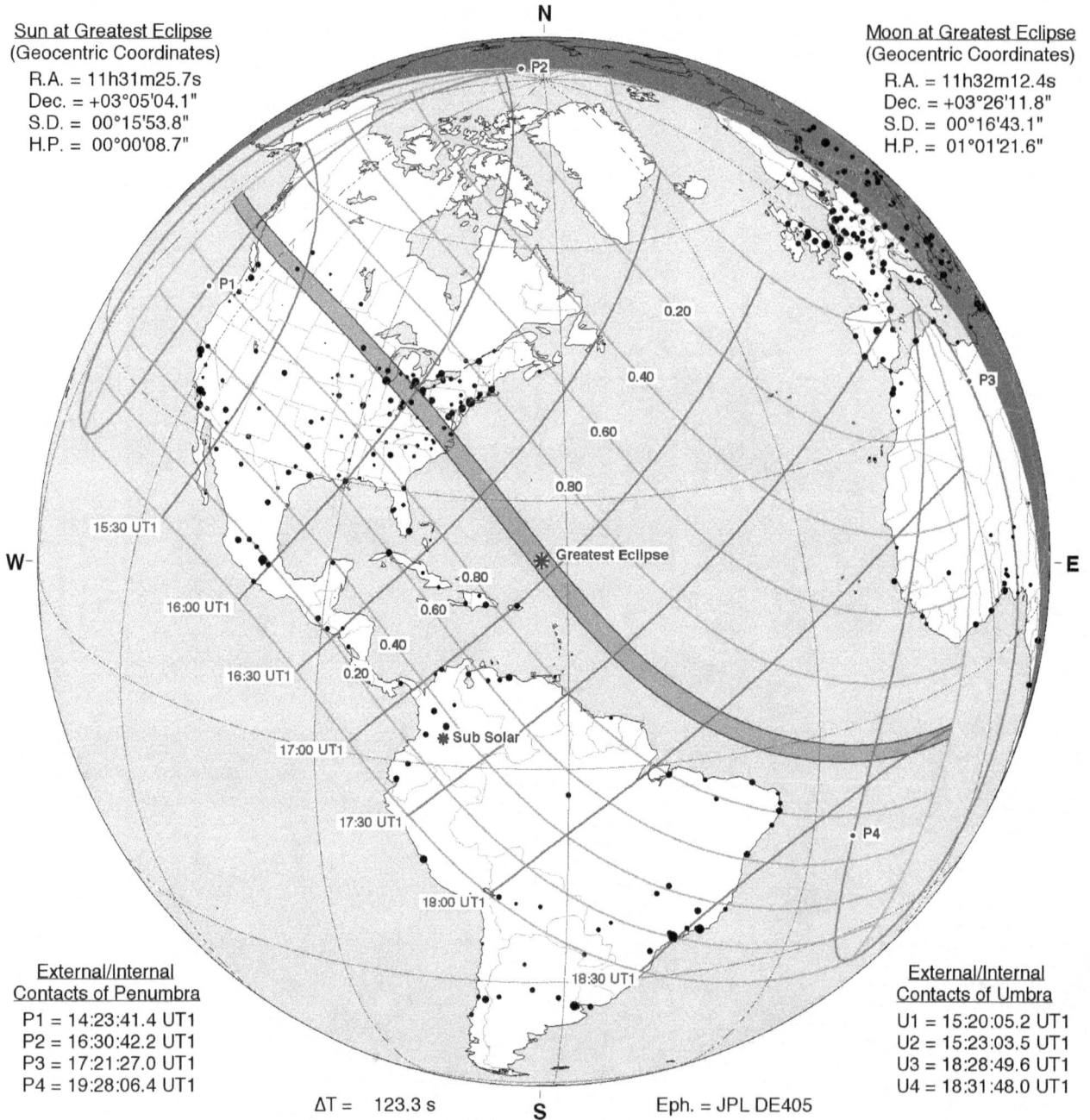

N

P2

0.20

0.40

0.60

0.80

Greatest Eclipse

15:30 UT1

W

E

16:00 UT1

0.80

0.60

0.40

16:30 UT1

0.20

17:00 UT1

Sub Solar

17:30 UT1

P4

18:00 UT1

18:30 UT1

P1

P3

S

**External/Internal**
**Contacts of Penumbra**
P1 = 14:23:41.4 UT1
P2 = 16:30:42.2 UT1
P3 = 17:21:27.0 UT1
P4 = 19:28:06.4 UT1

**External/Internal**
**Contacts of Umbra**
U1 = 15:20:05.2 UT1
U2 = 15:23:03.5 UT1
U3 = 18:28:49.6 UT1
U4 = 18:31:48.0 UT1

ΔT =  123.3 s

Eph. = JPL DE405

**Circumstances at Greatest Eclipse: 16:55:49.7 UT1**

| | |
|---|---|
| Lat. = 23°21.8'N | Sun Alt. =  66.7° |
| Long. = 063°08.2'W | Sun Azm. = 211.5° |
| Path Width = 241.0 km | Duration = 05m17.8s |

**Circumstances at Greatest Duration: 16:54:46.0 UT1**

| | |
|---|---|
| Lat. = 23°40.5'N | Sun Alt. =  66.6° |
| Long. = 063°25.2'W | Sun Azm. = 209.9° |
| Path Width = 240.7 km | Duration = 05m17.8s |

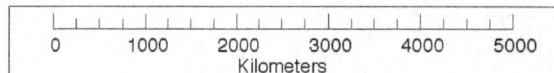

©2016  F. Espenak
www.EclipseWise.com

0      1000    2000    3000    4000    5000
Kilometers

# Annular Solar Eclipse of 2100 Mar 10

Greatest Eclipse = 22:28:11.0 TD   (= 22:26:07.3 UT1)

Eclipse Magnitude = 0.9338          Saros Series = 141
Gamma = 0.3077              Saros Member = 28 of 70

Sun at Greatest Eclipse
(Geocentric Coordinates)
R.A. = 23h24m46.6s
Dec. = -03°47'43.4"
S.D. = 00°16'06.4"
H.P. = 00°00'08.9"

Moon at Greatest Eclipse
(Geocentric Coordinates)
R.A. = 23h24m14.3s
Dec. = -03°33'06.4"
S.D. = 00°14'49.6"
H.P. = 00°54'24.7"

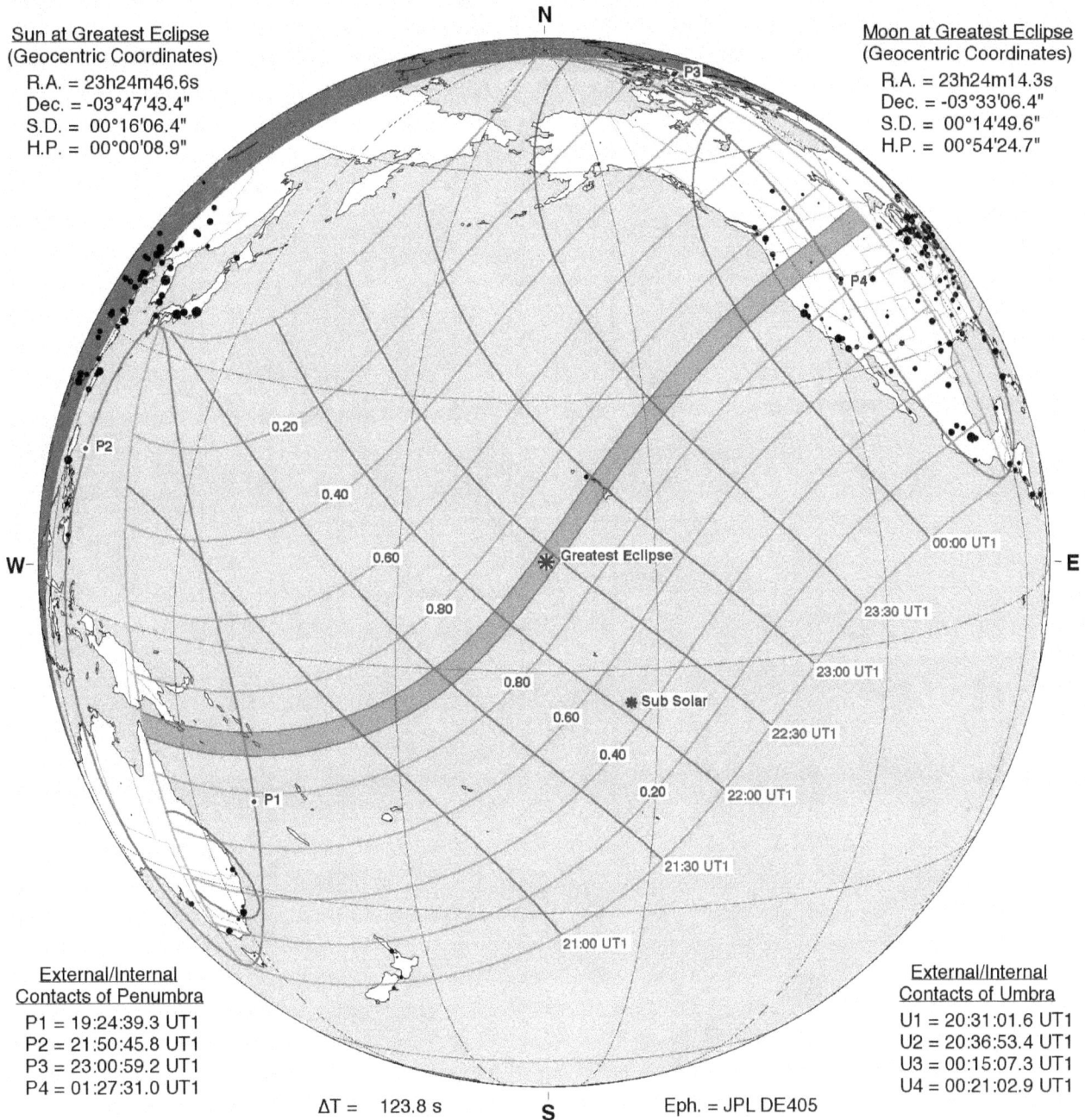

N

P3

P4

P2

0.20

0.40

0.60

0.80

W

Greatest Eclipse

00:00 UT1

E

23:30 UT1

23:00 UT1

0.80

Sub Solar

0.60

22:30 UT1

0.40

22:00 UT1

0.20

P1

21:30 UT1

21:00 UT1

External/Internal
Contacts of Penumbra
P1 = 19:24:39.3 UT1
P2 = 21:50:45.8 UT1
P3 = 23:00:59.2 UT1
P4 = 01:27:31.0 UT1

External/Internal
Contacts of Umbra
U1 = 20:31:01.6 UT1
U2 = 20:36:53.4 UT1
U3 = 00:15:07.3 UT1
U4 = 00:21:02.9 UT1

ΔT = 123.8 s

S

Eph. = JPL DE405

Circumstances at Greatest Eclipse: 22:26:07.3 UT1

| | |
|---|---|
| Lat. = 11°57.3'N | Sun Alt. = 72.0° |
| Long. = 162°45.8'W | Sun Azm. = 150.6° |
| Path Width = 257.5 km | Duration = 07m29.2s |

Circumstances at Greatest Duration: 22:19:52.0 UT1

| | |
|---|---|
| Lat. = 10°27.3'N | Sun Alt. = 71.8° |
| Long. = 163°54.1'W | Sun Azm. = 140.8° |
| Path Width = 259.2 km | Duration = 07m29.4s |

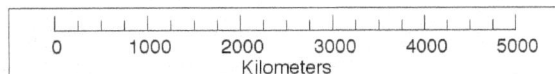

©2016  F. Espenak
www.EclipseWise.com

0    1000    2000    3000    4000    5000
Kilometers

# Total Solar Eclipse of  2100 Sep 04

Greatest Eclipse =  08:49:20.3 TD   (= 08:47:16.0 UT1)

| | |
|---|---|
| Eclipse Magnitude =  1.0402 | Saros Series =  146 |
| Gamma = -0.3384 | Saros Member =  32 of 76 |

**Sun at Greatest Eclipse**
(Geocentric Coordinates)
R.A. = 10h53m24.7s
Dec. = +07°04'34.9"
S.D. = 00°15'51.3"
H.P. = 00°00'08.7"

**Moon at Greatest Eclipse**
(Geocentric Coordinates)
R.A. = 10h52m46.7s
Dec. = +06°46'49.6"
S.D. = 00°16'14.1"
H.P. = 00°59'35.2"

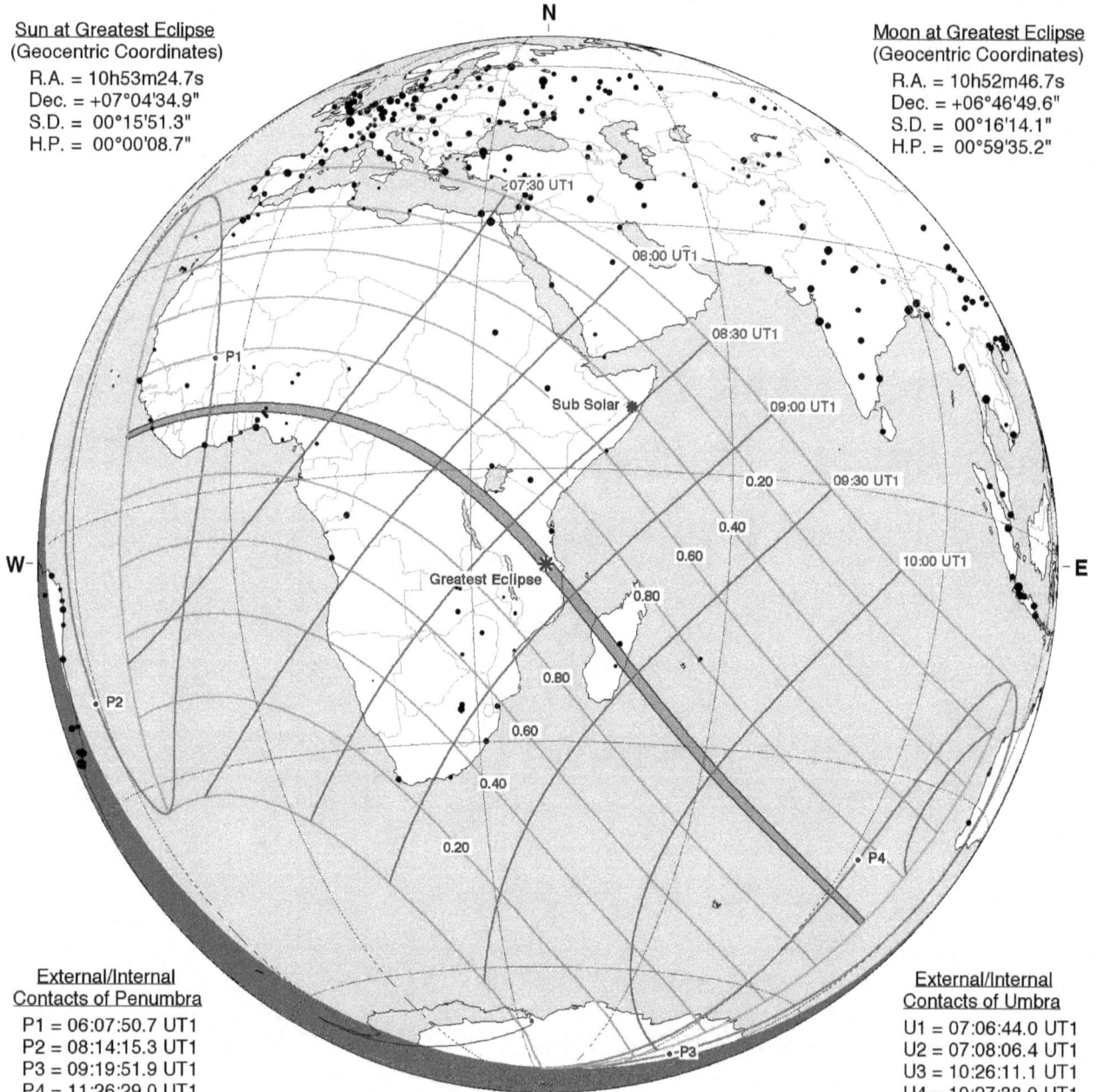

N

07:30 UT1
08:00 UT1
08:30 UT1
Sub Solar
09:00 UT1
0.20    09:30 UT1
0.40
0.60    10:00 UT1
0.80
Greatest Eclipse
0.80
0.60
0.40
0.20

P1
P2
P3
P4

W
E
S

**External/Internal
Contacts of Penumbra**
P1 = 06:07:50.7 UT1
P2 = 08:14:15.3 UT1
P3 = 09:19:51.9 UT1
P4 = 11:26:29.0 UT1

**External/Internal
Contacts of Umbra**
U1 = 07:06:44.0 UT1
U2 = 07:08:06.4 UT1
U3 = 10:26:11.1 UT1
U4 = 10:27:38.0 UT1

ΔT =   124.3 s          Eph. = JPL DE405

**Circumstances at Greatest Eclipse:  08:47:16.0 UT1**
Lat. =  10°28.1'S       Sun Alt. =  70.2°
Long. = 038°39.7'E      Sun Azm. =  28.2°
Path Width = 142.2 km   Duration = 03m32.5s

**Circumstances at Greatest Duration:  08:45:50.6 UT1**
Lat. =  10°06.1'S       Sun Alt. =  70.1°
Long. = 038°19.7'E      Sun Azm. =  30.5°
Path Width = 142.3 km   Duration = 03m32.5s

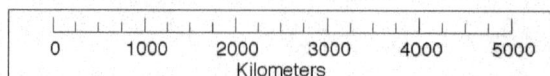

0   1000   2000   3000   4000   5000
Kilometers

©2016  F. Espenak
www.EclipseWise.com

# Appendix D

## Central Solar Eclipse Maps: 2001 to 2100

# Key to Central Solar Eclipse Maps

## Key to Continental Solar Eclipse Maps (Appendix D)

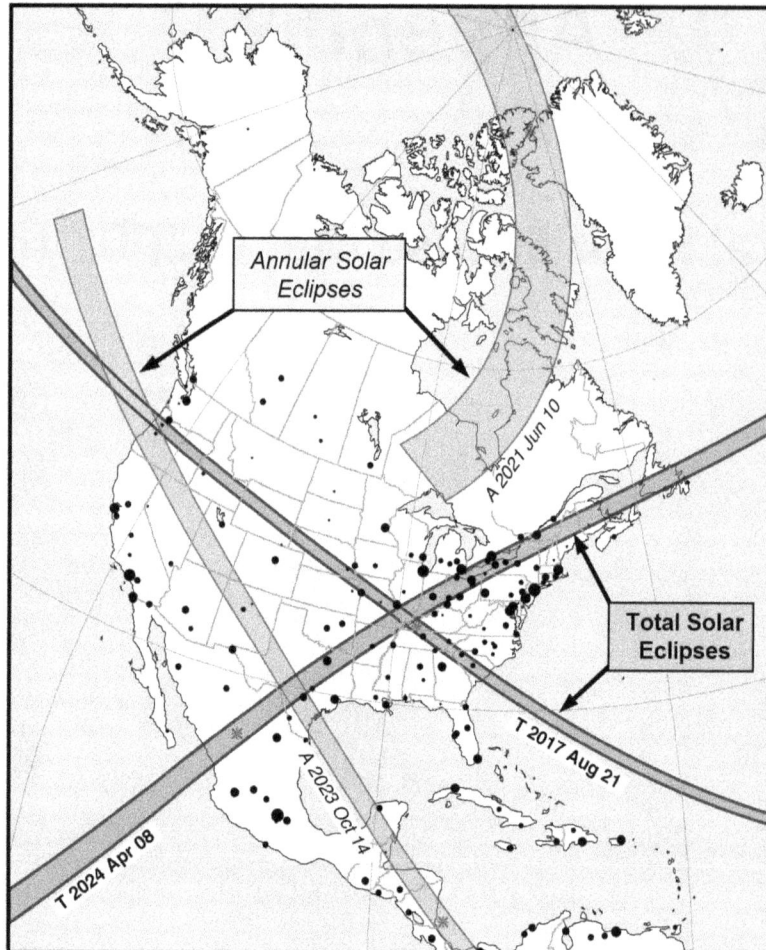

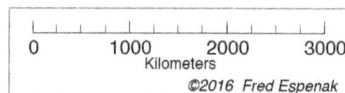

**Total Eclipses –**     the path is darkly shaded (blue in Color & Deluxe Editions);
the date appears in bold and is preceded by the letter **T**

**Annular Eclipses –**     the path is lightly shaded (orange in Color & Deluxe Editions);
the date appears in italics and is preceded by the letter **A**

**Hybrid Eclipses –**     the path is darkly shaded (or blue) along the total sections and
lightly shaded (or orange) along the annular sections;
the date appears in bold-italics and is preceded by the letter **H**

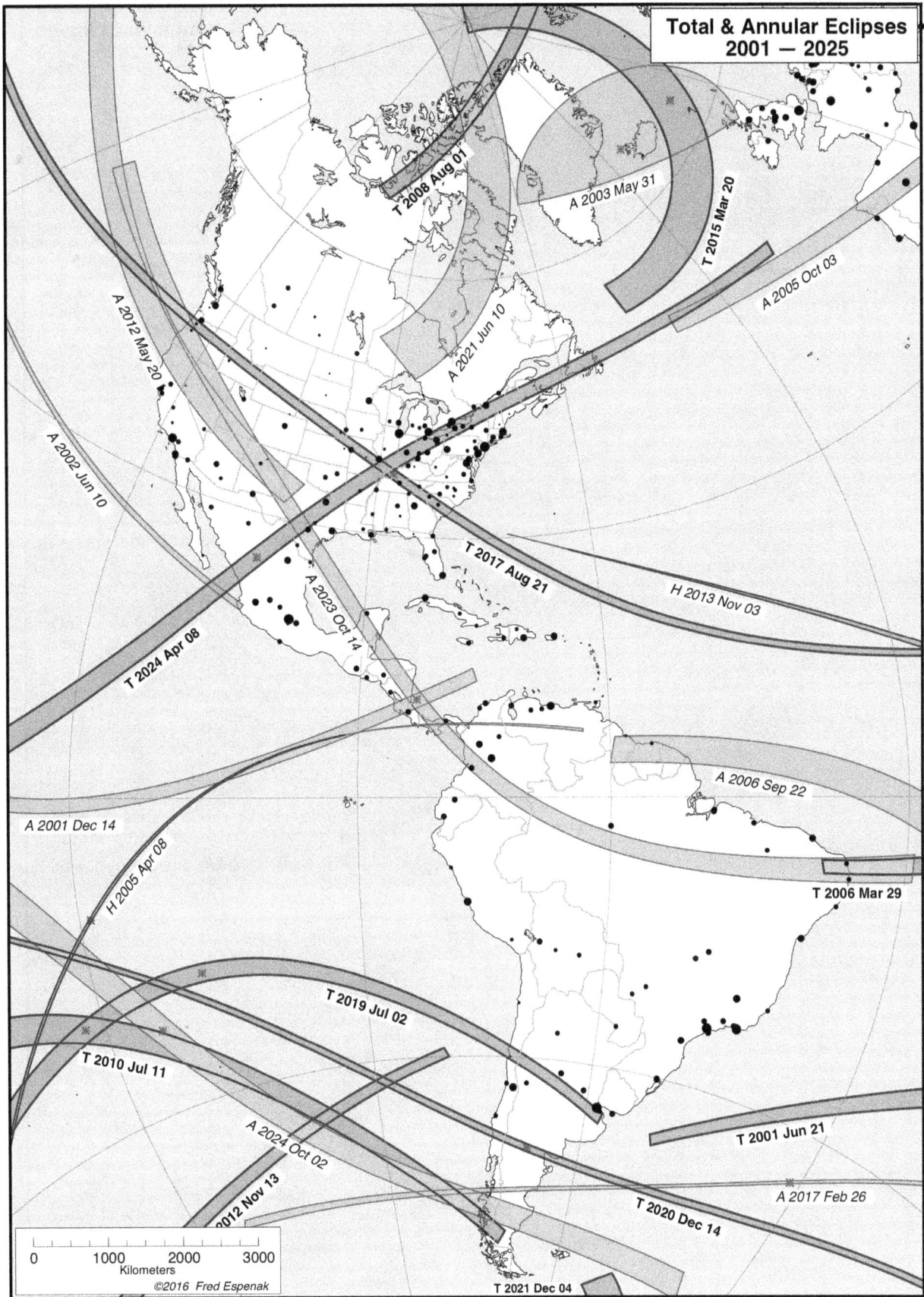

Total & Annular Eclipses
2001 — 2025

T 2008 Aug 01

A 2003 May 31

T 2015 Mar 20

A 2005 Oct 03

A 2012 May 20

A 2021 Jun 10

A 2002 Jun 10

T 2017 Aug 21

H 2013 Nov 03

A 2023 Oct 14

T 2024 Apr 08

A 2001 Dec 14

H 2005 Apr 08

A 2006 Sep 22

T 2006 Mar 29

T 2019 Jul 02

T 2010 Jul 11

A 2024 Oct 02

2012 Nov 13

T 2001 Jun 21

A 2017 Feb 26

T 2020 Dec 14

T 2021 Dec 04

0    1000    2000    3000
Kilometers

©2016  Fred Espenak

**Total & Annular Eclipses
2001 — 2025**

A 2021 Jun 10
A 2003 May 31
T 2015 Mar 20
T 2008 Aug 01
T 2024 Apr 08
T 2009 Jul 22
A 2019 Dec 26
A 2020 Jun 21
T 2006 Mar 29
A 2005 Oct 03
H 2013 Nov 03
A 2010 Jan 15
T 2001 Jun 21
A 2016 Sep 01
A 2006 Sep 22
A 2017 Feb 26
T 2020 Dec 14
T 2002 Dec 04
A 2009 Jan 26

0    1000    2000    3000
Kilometers
©2016  Fred Espenak

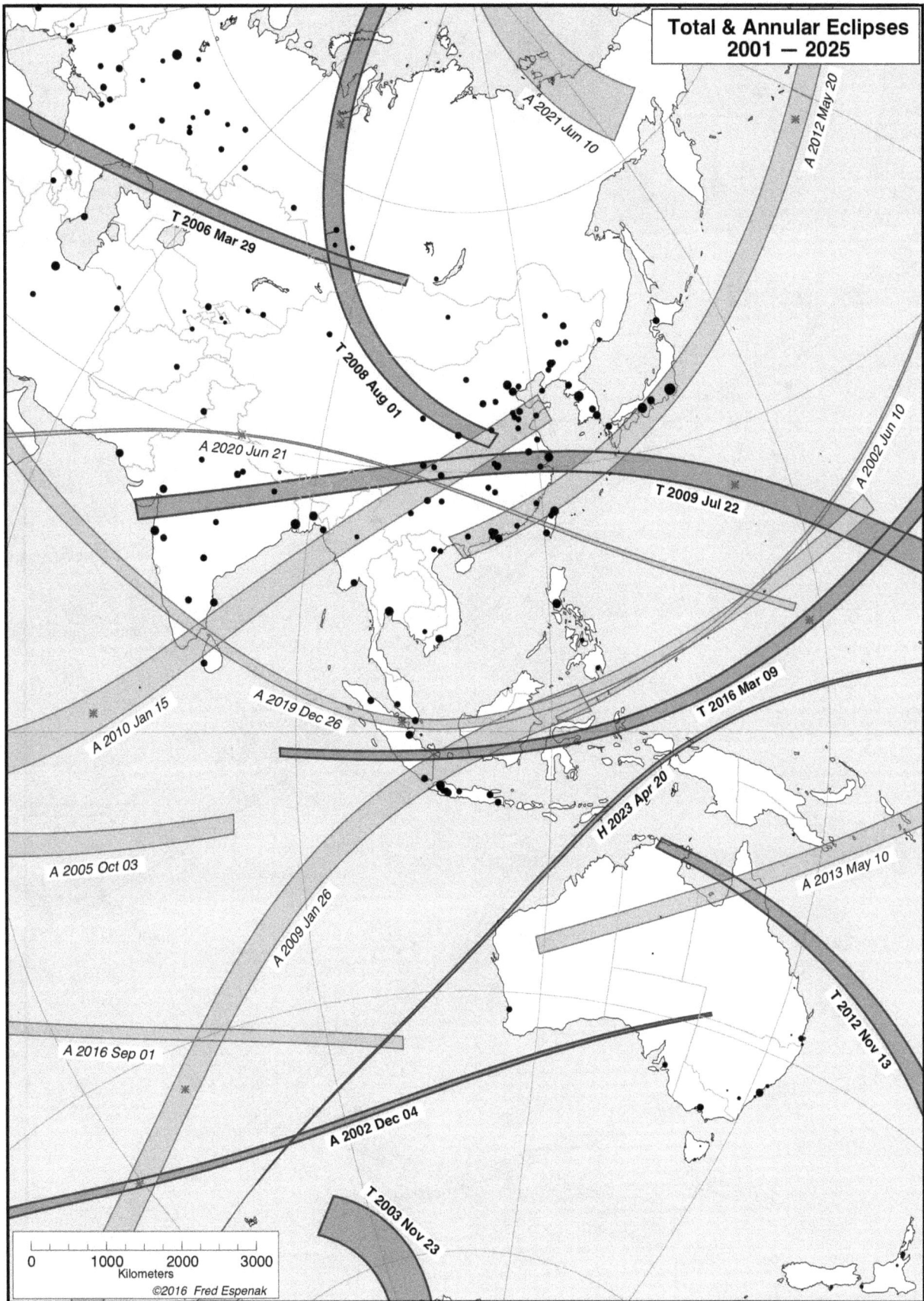

**Total & Annular Eclipses 2001 — 2025**

A 2021 Jun 10
A 2012 May 20
T 2006 Mar 29
T 2008 Aug 01
A 2020 Jun 21
T 2009 Jul 22
A 2002 Jun 10
A 2010 Jan 15
A 2019 Dec 26
T 2016 Mar 09
H 2023 Apr 20
A 2005 Oct 03
A 2009 Jan 26
A 2013 May 10
A 2016 Sep 01
T 2012 Nov 13
A 2002 Dec 04
T 2003 Nov 23

0    1000    2000    3000
Kilometers
©2016 Fred Espenak

Total & Annular Eclipses
2026 — 2050

T 2033 Mar 30

A 2039 Jun 21

T 2042 Apr 20

T 2026 Aug 12

A 2048 Jun 11

T 2044 Aug 23

A 2046 Feb 05

T 2027 Aug 02

A 2038 Jul 02

A 2049 May 31

T 2045 Aug 12

A 2045 Feb 16

A 2038 Jan 05

H 2031 Nov 14

T 2034 Mar 20

A 2028 Jan 26

A 2035 Mar 09

A 2034 Sep 12

T 2046 Aug 02

T 2038 Dec 26

A 2027 Feb 06

H 2050 May 20

A 2027 Feb 06

T 2048 Dec 05

T 2041 Apr 30

0    1000    2000    3000
Kilometers

©2016  Fred Espenak

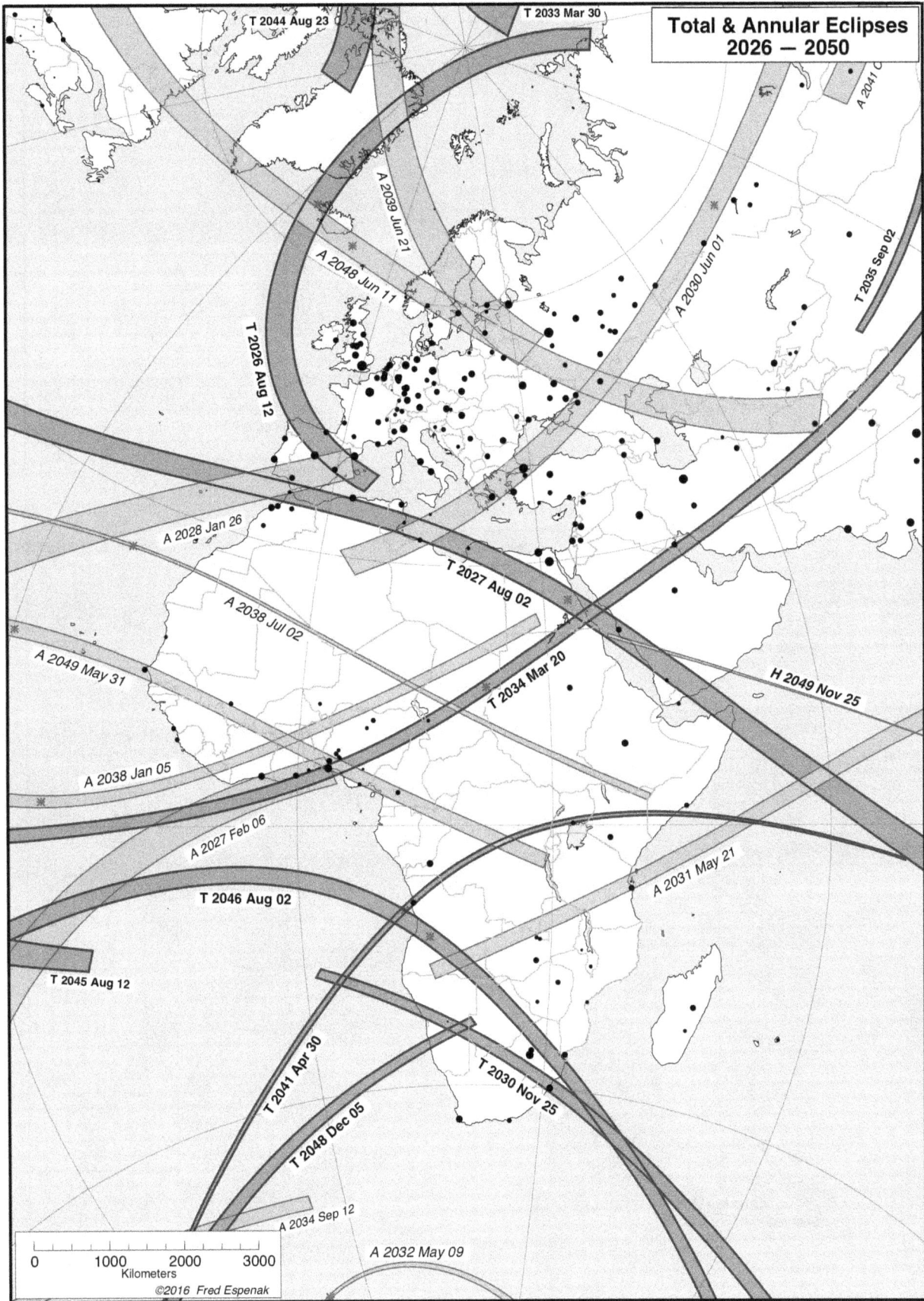

**Total & Annular Eclipses 2026 — 2050**

T 2044 Aug 23

T 2033 Mar 30

A 2041 C

A 2039 Jun 21

A 2030 Jun 01

T 2035 Sep 02

A 2048 Jun 11

T 2026 Aug 12

A 2028 Jan 26

T 2027 Aug 02

A 2038 Jul 02

A 2034 Mar 20

H 2049 Nov 25

A 2049 May 31

A 2038 Jan 05

A 2027 Feb 06

A 2031 May 21

T 2046 Aug 02

T 2045 Aug 12

T 2041 Apr 30

T 2030 Nov 25

T 2048 Dec 05

A 2034 Sep 12

A 2032 May 09

0    1000    2000    3000
Kilometers

©2016  Fred Espenak

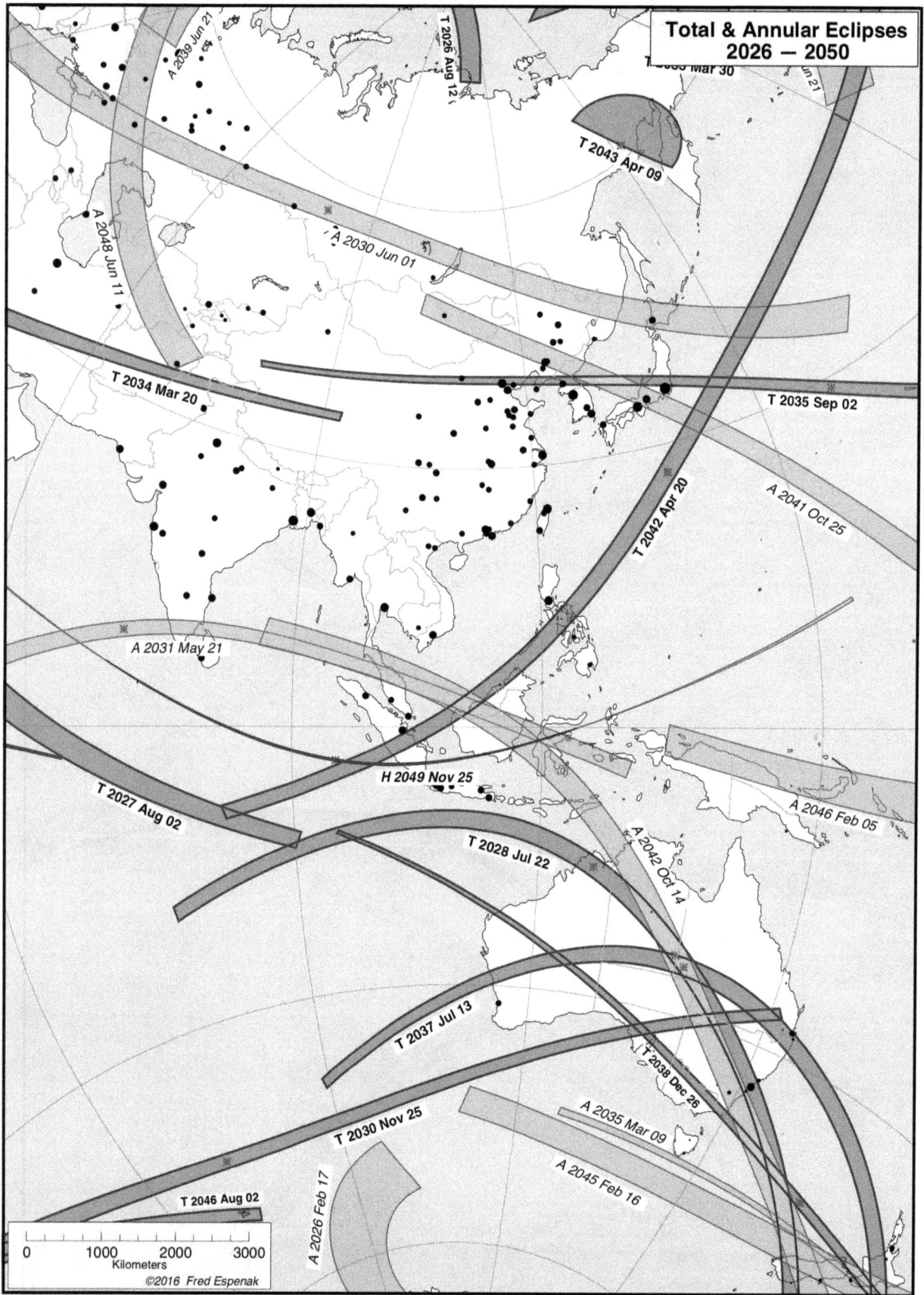

**Total & Annular Eclipses 2026 — 2050**

A 2039 Jun 21

T 2026 Aug 12

T 2043 Apr 09

A 2030 Jun 01

A 2048 Jun 11

T 2034 Mar 20

T 2035 Sep 02

A 2041 Oct 25

T 2042 Apr 20

A 2031 May 21

A 2046 Feb 05

T 2027 Aug 02

H 2049 Nov 25

T 2028 Jul 22

A 2042 Oct 14

T 2037 Jul 13

T 2038 Dec 26

T 2030 Nov 25

A 2035 Mar 09

A 2026 Feb 17

A 2045 Feb 16

T 2046 Aug 02

0    1000    2000    3000
Kilometers

©2016  Fred Espenak

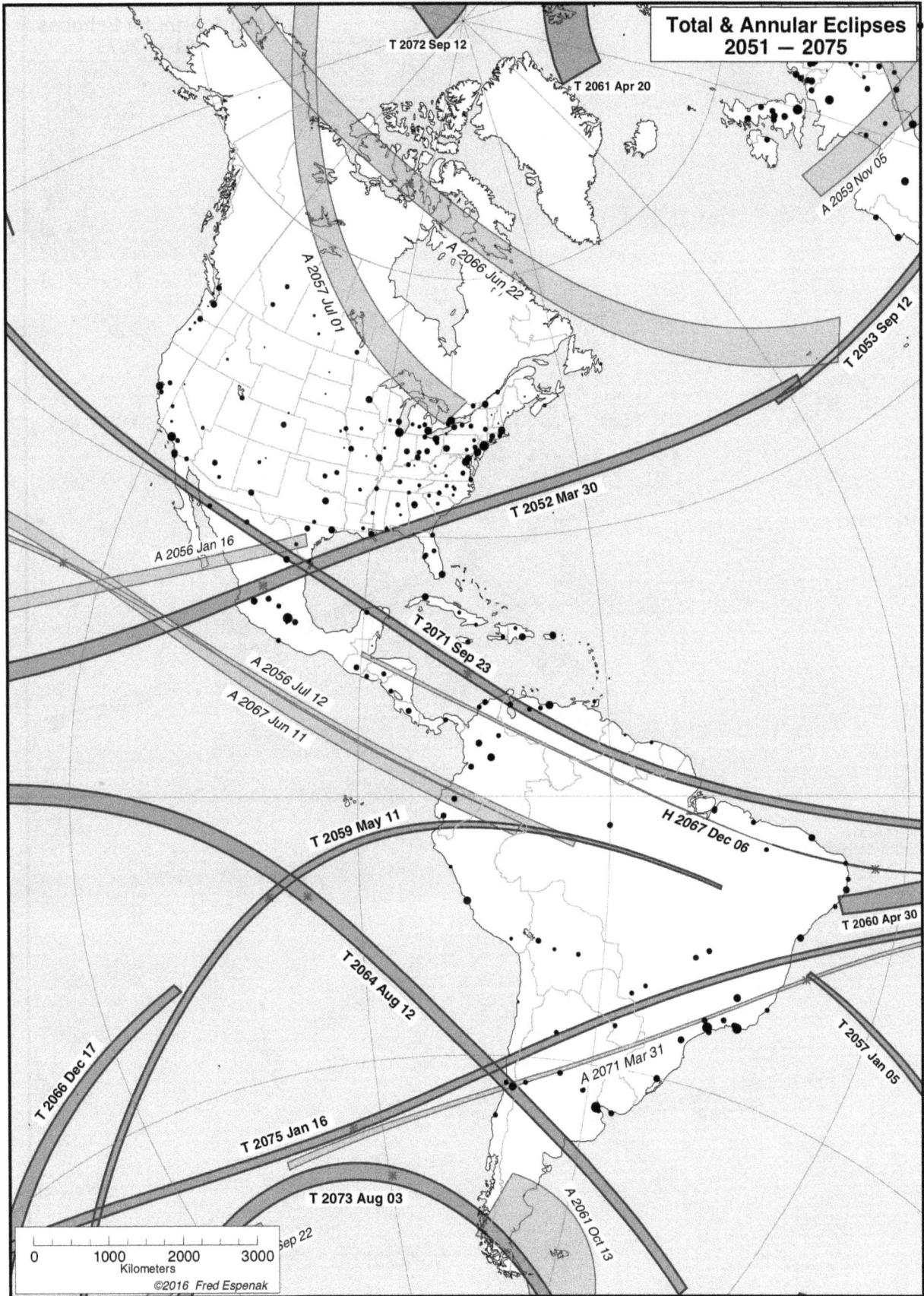

**Total & Annular Eclipses**
**2051 — 2075**

T 2072 Sep 12

T 2061 Apr 20

A 2059 Nov 05

A 2057 Jul 01

A 2066 Jun 22

T 2053 Sep 12

T 2052 Mar 30

A 2056 Jan 16

T 2071 Sep 23

A 2056 Jul 12

A 2067 Jun 11

H 2067 Dec 06

T 2059 May 11

T 2060 Apr 30

T 2064 Aug 12

T 2057 Jan 05

T 2066 Dec 17

A 2071 Mar 31

T 2075 Jan 16

T 2073 Aug 03

A 2061 Oct 13

Sep 22

| 0 | 1000 | 2000 | 3000 |
Kilometers

©2016  Fred Espenak

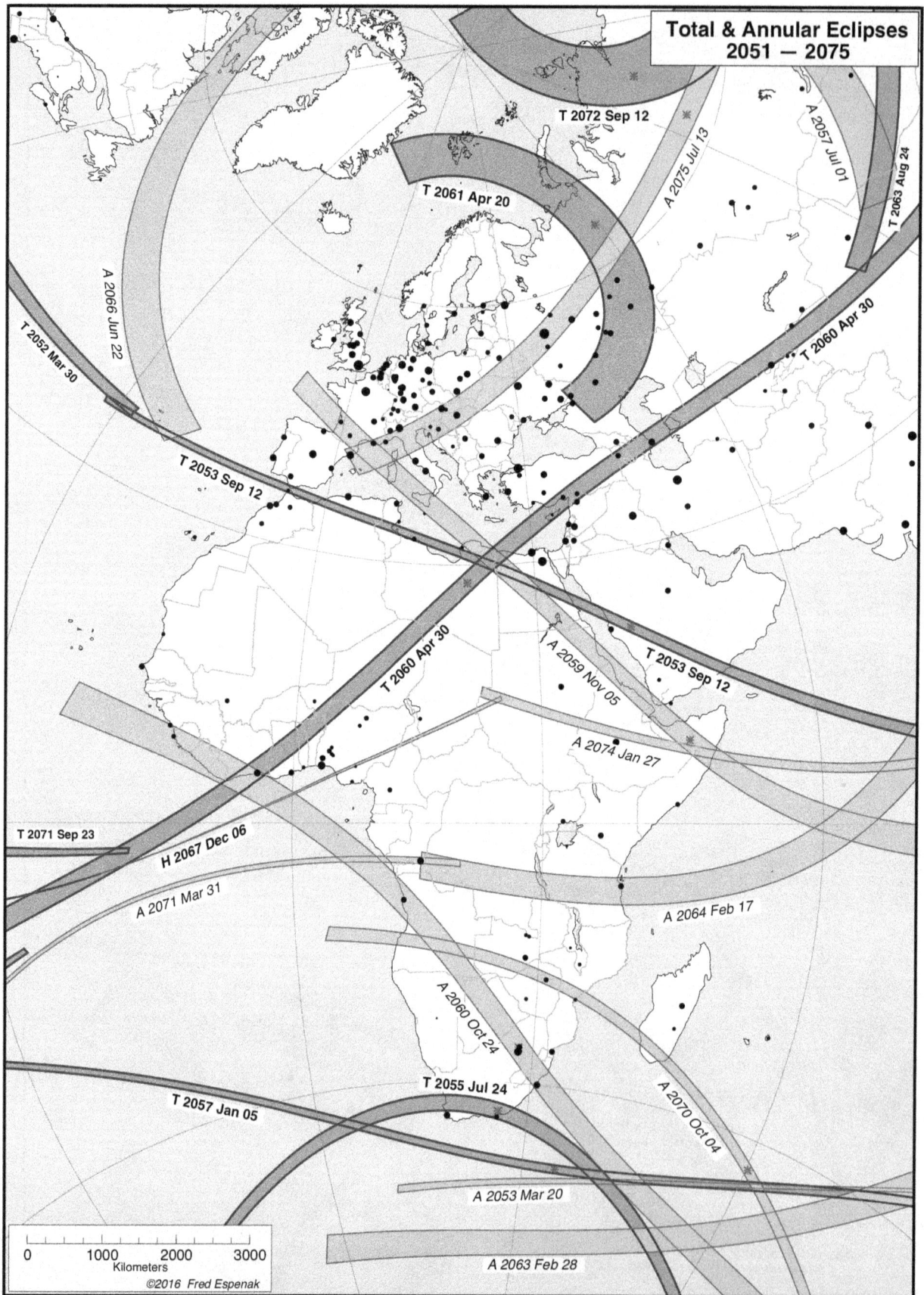

**Total & Annular Eclipses
2051 — 2075**

T 2072 Sep 12

A 2057 Jul 01

A 2075 Jul 13

T 2063 Aug 24

T 2061 Apr 20

A 2066 Jun 22

T 2052 Mar 30

T 2060 Apr 30

T 2053 Sep 12

T 2060 Apr 30

A 2059 Nov 05

T 2053 Sep 12

A 2074 Jan 27

T 2071 Sep 23

H 2067 Dec 06

A 2071 Mar 31

A 2064 Feb 17

A 2060 Oct 24

A 2070 Oct 04

T 2055 Jul 24

T 2057 Jan 05

A 2053 Mar 20

0    1000    2000    3000
Kilometers

©2016  Fred Espenak

A 2063 Feb 28

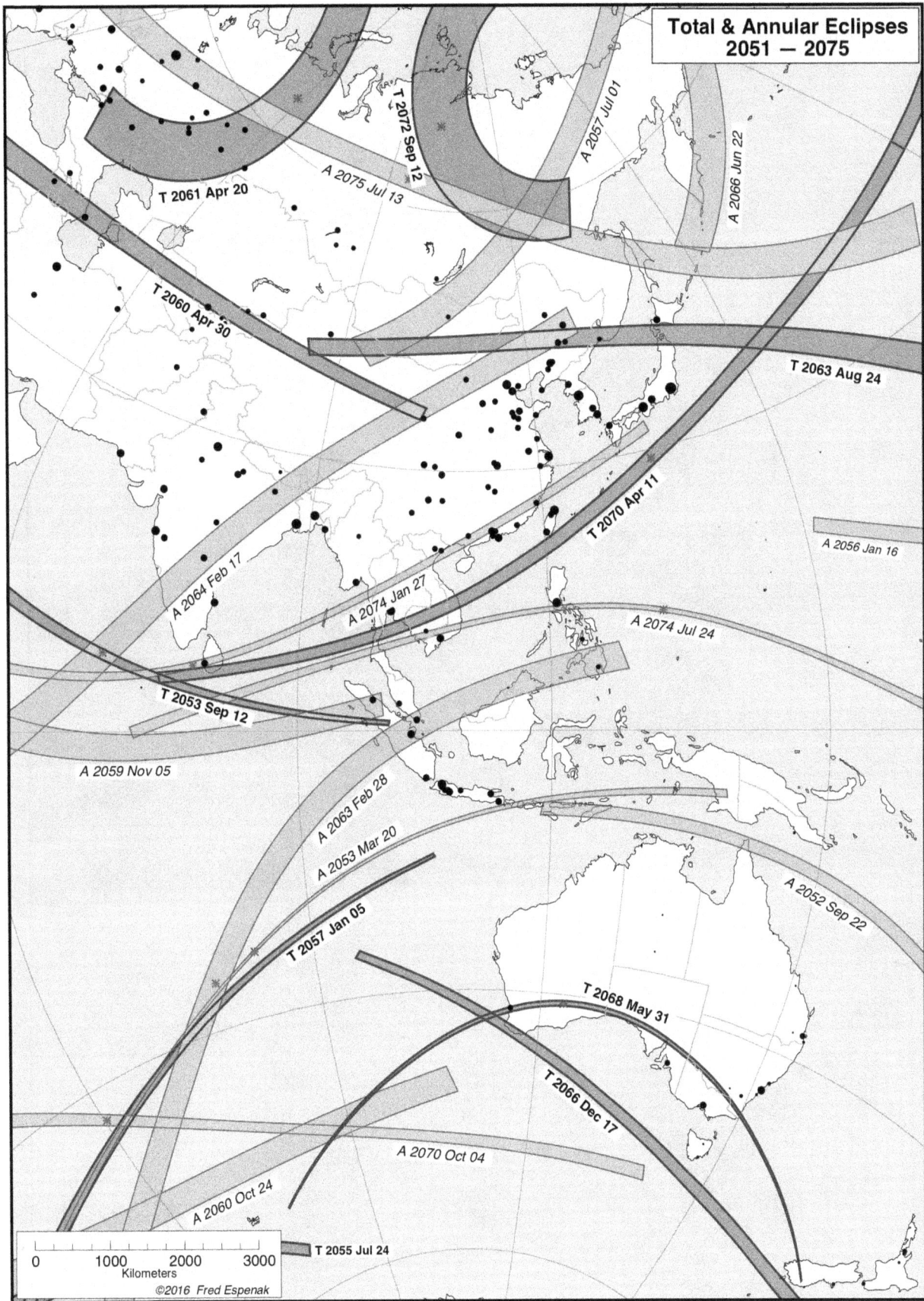

**Total & Annular Eclipses 2051 — 2075**

T 2072 Sep 12

A 2057 Jul 01

A 2066 Jun 22

A 2075 Jul 13

T 2061 Apr 20

T 2060 Apr 30

T 2063 Aug 24

T 2070 Apr 11

A 2056 Jan 16

A 2064 Feb 17

A 2074 Jan 27

A 2074 Jul 24

T 2053 Sep 12

A 2059 Nov 05

A 2063 Feb 28

A 2053 Mar 20

A 2052 Sep 22

T 2057 Jan 05

T 2068 May 31

T 2066 Dec 17

A 2070 Oct 04

A 2060 Oct 24

T 2055 Jul 24

0 1000 2000 3000
Kilometers

©2016 Fred Espenak

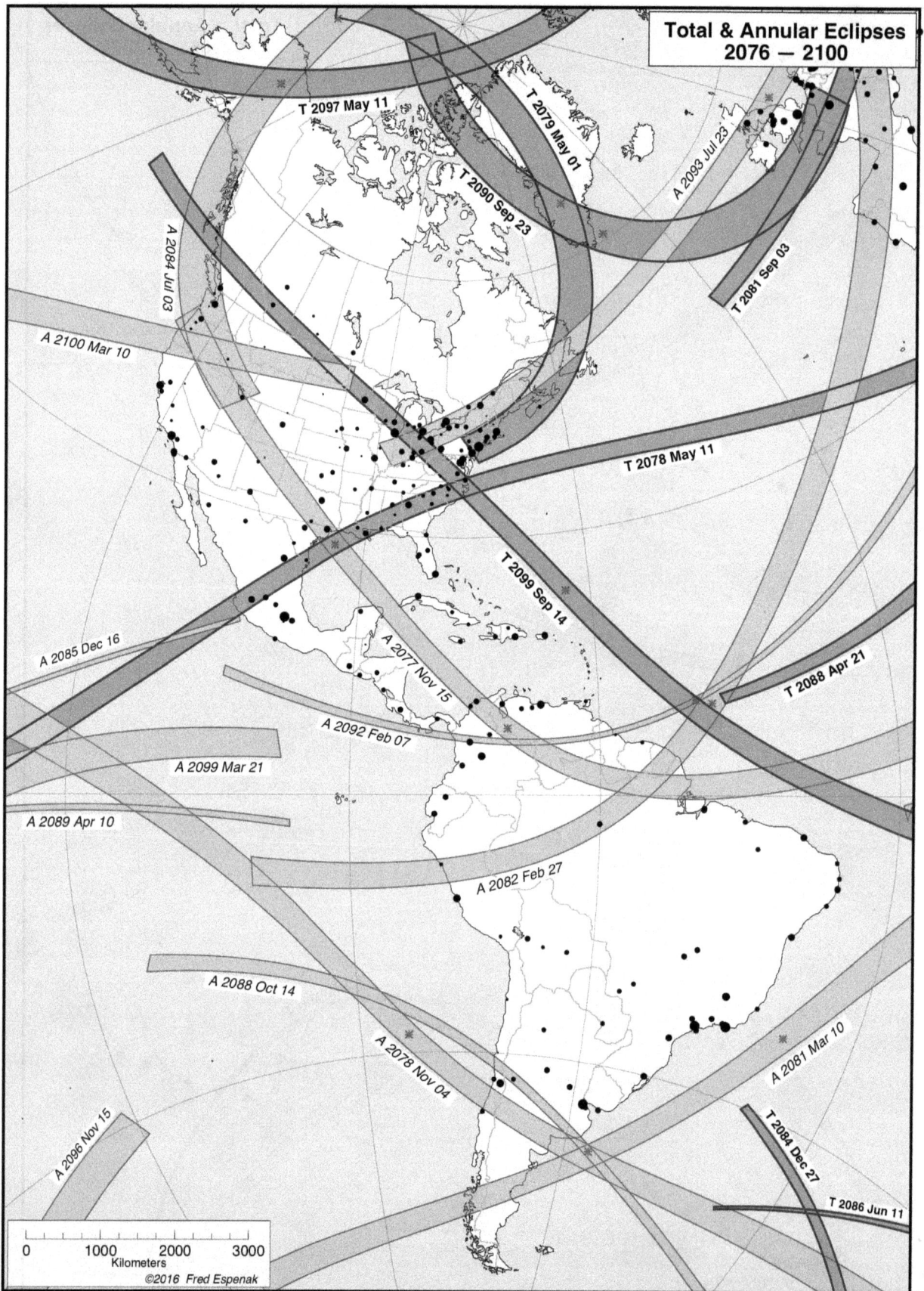

**Total & Annular Eclipses 2076 — 2100**

T 2097 May 11
T 2079 May 01
A 2093 Jul 23
T 2090 Sep 23
T 2081 Sep 03
A 2084 Jul 03
A 2100 Mar 10
T 2078 May 11
T 2099 Sep 14
A 2085 Dec 16
A 2077 Nov 15
T 2088 Apr 21
A 2092 Feb 07
A 2099 Mar 21
A 2089 Apr 10
A 2082 Feb 27
A 2088 Oct 14
A 2081 Mar 10
A 2078 Nov 04
A 2096 Nov 15
T 2084 Dec 27
T 2086 Jun 11

0    1000    2000    3000
Kilometers

©2016  Fred Espenak

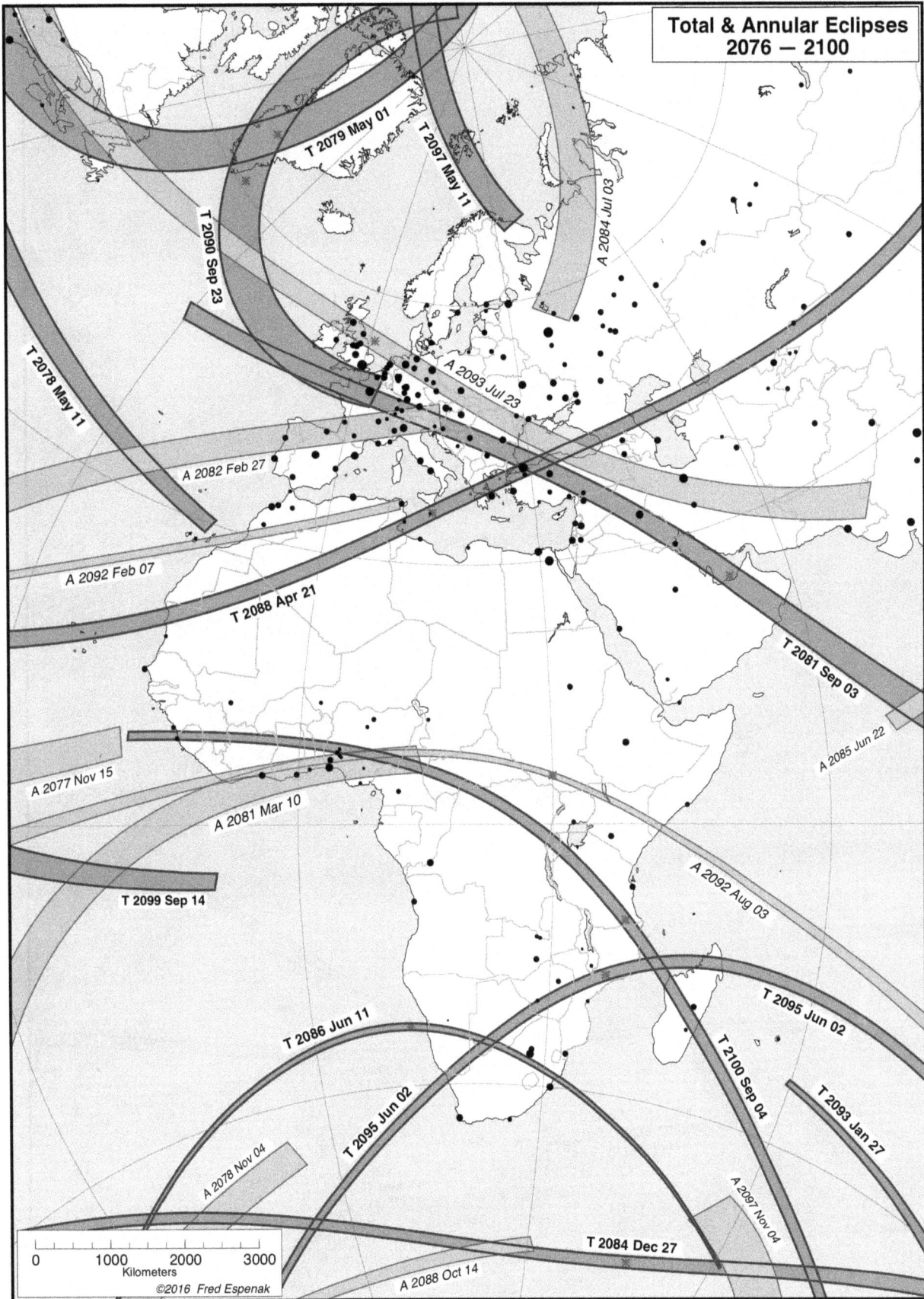

Total & Annular Eclipses
2076 — 2100

T 2079 May 01
T 2097 May 11
A 2084 Jul 03
T 2090 Sep 23
T 2078 May 11
A 2093 Jul 23
A 2082 Feb 27
A 2092 Feb 07
T 2088 Apr 21
T 2081 Sep 03
A 2085 Jun 22
A 2077 Nov 15
A 2081 Mar 10
T 2099 Sep 14
A 2092 Aug 03
T 2086 Jun 11
T 2095 Jun 02
T 2100 Sep 04
T 2093 Jan 27
A 2078 Nov 04
T 2095 Jun 02
A 2097 Nov 04
T 2084 Dec 27
A 2088 Oct 14

0   1000   2000   3000
Kilometers
©2016  Fred Espenak

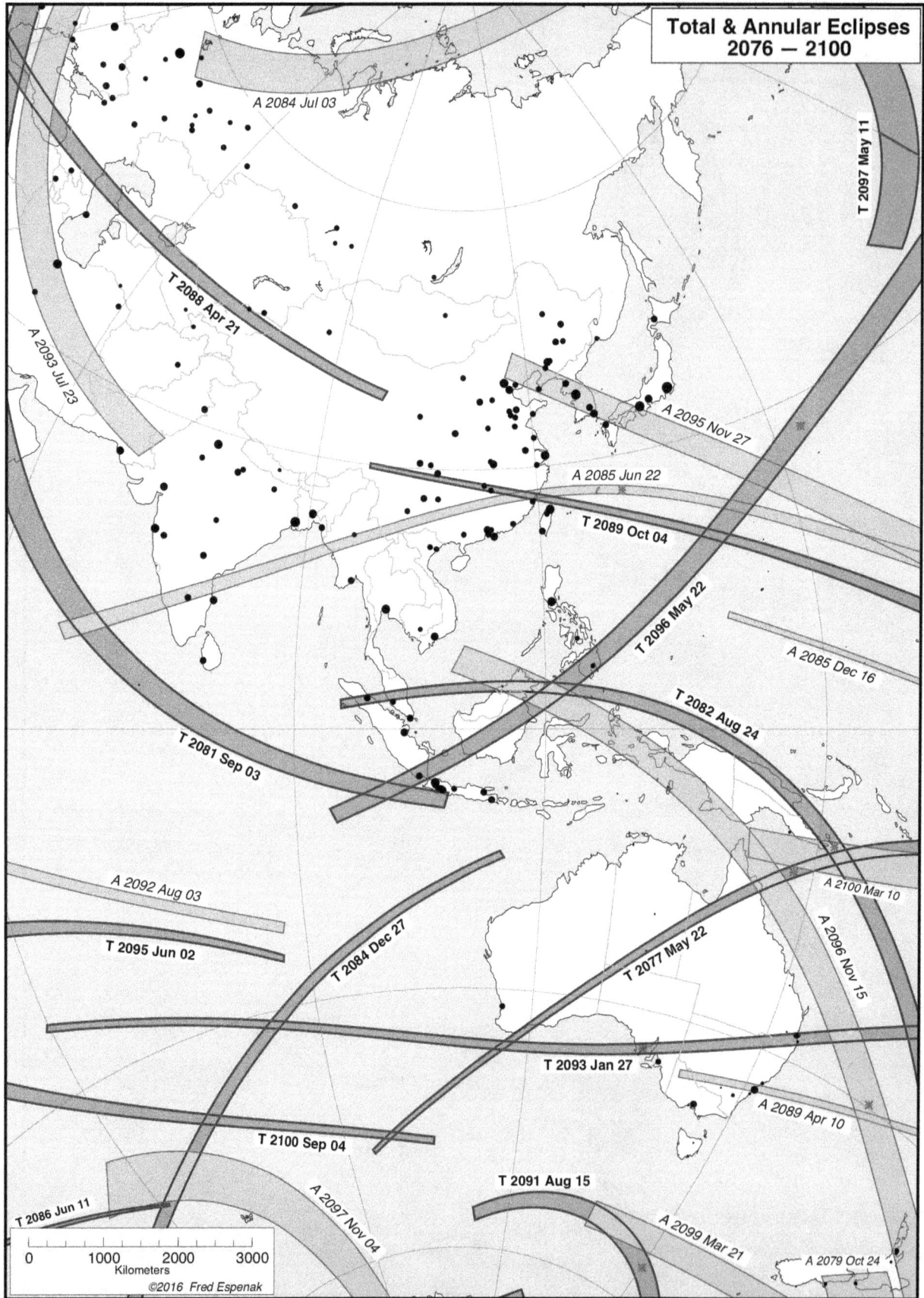

**Total & Annular Eclipses 2076 — 2100**

A 2084 Jul 03

T 2097 May 11

T 2088 Apr 21

A 2093 Jul 23

A 2095 Nov 27

A 2085 Jun 22

T 2089 Oct 04

T 2096 May 22

A 2085 Dec 16

T 2082 Aug 24

T 2081 Sep 03

A 2092 Aug 03

A 2100 Mar 10

T 2095 Jun 02

T 2084 Dec 27

T 2077 May 22

A 2096 Nov 15

T 2093 Jan 27

A 2089 Apr 10

T 2100 Sep 04

T 2091 Aug 15

T 2086 Jun 11

A 2097 Nov 04

A 2099 Mar 21

A 2079 Oct 24

0    1000    2000    3000
Kilometers
©2016 Fred Espenak

# Astropixels Publications

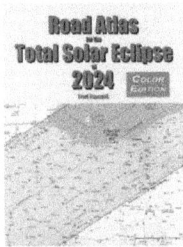

**Road Atlas for the Total Solar Eclipse of 2024** (Fred Espenak) contains a comprehensive series of 26 high-resolution maps of the path of totality across the USA, Mexico and Canada. The large scale (1 inch = 22 miles) shows both major and minor roads, towns and cities, rivers, lakes, parks, national forests, wilderness areas and mountain ranges. The duration of totality is plotted in 30-second steps, making it easy to estimate the length of the total eclipse from any location in the eclipse path.

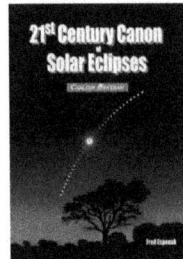

**21st Century Canon of Solar Eclipses** (Fred Espenak) contains maps and data for all 224 solar eclipses occurring during the 100-year period from 2001 through 2100. Appendix A is comprehensive catalog listing the essential characteristics of each eclipse. Appendix B contains maps depicting the geographic regions of visibility of each eclipse with 12 maps per page. Appendix C has detailed full-page maps of every eclipse from 2017 through 2066. Appendix D plots the track of every central eclipse (total, annular and hybrid) on large-scale maps with countries borders and major cities.

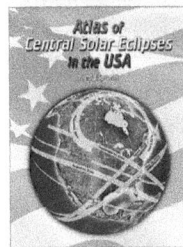

**Atlas of Central Solar Eclipses in the USA** (Fred Espenak) contains of a series of 499 global maps showing the track of every total and annular solar eclipse across the USA from 1001 through 3000. It is accompanied by a catalog that lists the major characteristics of each eclipse. A set of 20 detailed maps, each covering a 50-year period and centered on the lower 48 states, shows the path of every total and annular eclipse. The maps include state boundaries and major cities. These maps also cover southern Canada and northern Mexico.

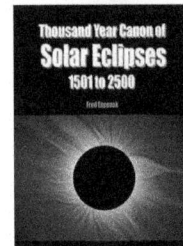

**Thousand Year Canon of Solar Eclipses 1501 to 2500** (Fred Espenak) contains maps and data for each of the 2,389 solar eclipses occurring over the ten-century period centered on the present era. A comprehensive catalog lists the essential characteristics of each eclipse while a series of global maps show the exact geographic extent of each eclipse.

**Thousand Year Canon of Lunar Eclipses 1501 to 2500** (Fred Espenak) contains diagrams, maps and data for each of the 2,424 lunar eclipses occurring over the ten-century period centered on the present era. A comprehensive catalog lists the essential characteristics of each eclipse while a series of diagrams and maps illustrate the Moon-shadow geometry and geographic regions of visibility of each eclipse.

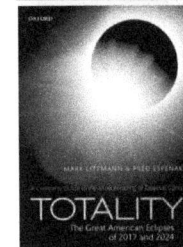

**Totality: The Great American Eclipses of 2017 and 2024** (Mark Littmann & Fred Espenak) is the ultimate guide to the most stunning of celestial sights, total eclipses of the Sun  The book provides information, photographs, and illustrations to help the public understand and safely enjoy all aspects of these eclipses including how to observe a total eclipse of the Sun, how to photograph and video record an eclipse, why solar eclipses happen, and more. Several chapters focus exclusively on the total eclipses of 2017 and 2024 though the USA.

For complete details and ordering information on the above *Astropixels Publications*, visit:

## *astropixels.com/pubs/*

www.ingramcontent.com/pod-product-compliance
Lightning Source LLC
Chambersburg PA
CBHW080231270326
41926CB00020B/4205